Solar Energy Desalination Technology

Solar Energy Desalination Technology

Hongfei Zheng
Beijing Institute of Technology, Haidian, Beijing

ELSEVIER
elsevier.com

Elsevier
Radarweg 29, PO Box 211, 1000 AE Amsterdam, Netherlands
The Boulevard, Langford Lane, Kidlington, Oxford OX5 1GB, United Kingdom
50 Hampshire Street, 5th Floor, Cambridge, MA 02139, United States

Library of Congress Cataloging-in-Publication Data
A catalog record for this book is available from the Library of Congress

British Library Cataloguing-in-Publication Data
A catalogue record for this book is available from the British Library

ISBN: 978-0-12-805411-6

Publisher: John Fedor
Acquisition Editor: Anita Koch
Editorial Project Manager: Sarah Jane Watson
Production Project Manager: Paul Prasad Chandramohan
Designer: Mathew Limbert

Typeset by TNQ Books and Journals

Contents

Preface

Many places in the world are suffering the freshwater shortage. Desalination is one of the important means to solve the shortage of fresh water. However, huge fossil fuels will be consumed to carry out desalination of sea water using conventional energy. At the same time, a huge amount of greenhouse gas will be emitted to atmospheric environment. Therefore, it is of social significance and economic value to desalinate with solar energy.

With the increasing maturity of solar energy utilization technology, not just solar collector but also solar photovoltaic power generation has reached the practical stage In particular, the technology of medium and low temperature solar heat collection has become a popular technology for solar energy utilization in many countries, and the collection temperature can easily reach up to 150 °C, which is very suitable for desalination. In this regard, the practical application has shown great social and economic benefits, and also has great development potential.

In order to promote the application of new energy and renewable energy in the field of seawater desalination, I wrote this book *Solar Desalination Technology* on the basis of many years of research in this field. The book provides new thoughts, concepts and feasible solutions in the desalination field and the thermal and economic efficiency according to the current technology. The book discusses about the classification of solar desalination technology—passive type, positive type, humidification-dehumidification type, absorption or adsorption type, multiple process compound type and combined with modern desalination industrial technology, etc. This book also introduces the problems faced by various kinds of solar desalination technology, and points out the future development direction. Researchers in the fields of solar energy application technology, desalination technology and engineering and technical personnel can use this book as a reference for their research. Also this can be treated as a reference textbook for the senior undergraduate and graduate students.

When writing this book the author does his best to make this book rigorous and combine with practice. But due to the limited knowledge of the author, omissions and fallacies in this book are unavoidable. Author earnestly requests peers and readers to find and correct these faults.

A large number of the previous literatures and ideas in the book are used. Most of the literatures are listed in the reference section of every chapter. Here, the author expresses his deepest gratitude for these contributions and if any of the references are overlooked, the author expresses his sincerest apology.

The author would like to thank many researchers and colleagues who helped and assisted him during the development phases of this book. They have proposed many useful suggestions. The author would also like to thank the National Natural Science Foundation of China (No. 51576017) and the National High Technology Research and Development Program "863" of China (No. 2013AA102407-2) for their support.

Hongfei Zheng

10 Jan. 2017

CHAPTER 1

General Problems in Seawater Desalination

Chapter Outline

Solar Energy Desalination Technology. http://dx.doi.org/10.1016/B978-0-12-805411-6.00001-4

Water is available in abundant quantities in nature. However, there is a shortage of potable water in many countries in the world. Desalination seems to be the most suitable solution. However, traditional desalination technology needs a large amount of traditional energy. Even if we can pay for the energy, the pollution due to burning the fossil fuels requires us to act with caution. Fortunately, water shortage is an area that often has rich solar energy resources. For example, the Arab Gulf region and Latin America are suffering a severe shortage of fresh water but have rich solar energy resources. Recently, solar desalination is gradually emerging as a successful renewable energy source for producing fresh water.

In fact, China is also a fresh water shortage country. China's per capita water resource is less than 2400 m^3, which is roughly a quarter of the world's average. With 40% of the population and more than 60% of the gross national product (GNP), the coastal areas constitute only 13% of the country's territory. Due to the high-density population, the per capita water resource of the coastal cities is less than 500 m^3, and even below 200 m^3 in some areas. Water shortage is hindering the economic development of these regions. Therefore making the most of the coastal advantages and taking the road of desalination is an important way to solve the water shortage problem for the coastal cities, and applying the desalination technology to develop a supplemental water supply has been commonly used all over the world.

1.1 Fresh Water Resource Shortage and Its Solutions

Fresh water is one of the basic substances on which the human society relies for survival and development. Sixty percent of the human body is liquid, which is primarily water. Water is so vital to human health that the loss of 10% of the body's water may seriously impair the physiological function, and 20% water loss will lead to instant death. Water is the source of life. It is indispensable to the social economy as well, for example, crops will die and industrial production will stop without water. So water is also the origin of all civilization. The surface area of the Earth is approximately 510 million square kilometers, 70.8% of which is the ocean. The average depth of the ocean is approximately 3800 m, so the total amount of water on Earth would be nearly 1.4 billion km^3 [1–3]. Seen from the per capita possession of water, the Earth is rich in water resources, and there seems to be no danger of water shortage. However, the high salinity of seawater, which cannot be directly used for drinking or irrigation, forms more than 97% of the total water quantity, and hardly 3% of fresh water is left, with extremely uneven distribution. Three-fourths of the fresh water is frozen in the glacier at the poles and the alpine zones, and the rest is

Table 1.1: Storage quantities of the different types of water resources [4].

Water storage place	Total volume (km^3)	Percentage of total water	Percentage of fresh water
Atmosphere	12,900	0.001	0.01
Glacier	24,064,000	1.72	68.7
Surface water	300,000	0.021	0.006
Rivers	2,120	0.0002	0.006
Lakes	176,400	0.013	0.26
Moors	11,470	0.0008	0.03
Underground soil	16,500	0.0012	0.05
Poles	10,530,000	0.75	30.1
Lithosphere	23,400,000	1.68	—
Ocean	138,000,000	95.18	—
Total quantity	1,396,513,390		

found more (around 37 times) in the groundwater than the surface water. The fresh water in rivers, lakes, and groundwater available for direct human use is less than 0.36% [1–3]. For storage quantities of the different types of water resources and their percentages by volume of the total water supply on Earth, see Table 1.1.

The distribution and water quality condition of water resources on the Earth is given in Table 1.2, while the quality requirements of water used for industry, agriculture, and daily life is shown in Table 1.3.

As shown in the earlier two tables, though seawater is abundant, it cannot be directly used because of its high salinity. The fresh water on Earth available for human use is extremely maldistributed. There are many desert and semidesert areas on land, which have very low rainfall and serious water scarcity. The high-salinity water in these areas cannot be drunk directly or used for industrial and agricultural production. More

Table 1.2: The distribution and water quality condition of water resources on the earth [2].

Distribution		Quantity of water		Quality of water (salinity/ppm[a])
		Volume (10^9 m^3)	Percentage (%)	
Vapor in the atmosphere		12,900	0.001	
Surface water	Rivers, Lakes,	230,000	0.017	100–500[b]
	Glacier	29,120,000	2.157	
	Ocean	1,318,720,000	97.2	28,000–35,000
Groundwater		8,616,600	0.625	300–10,000
Total quantity		1,356,699,500	100.00	

Some of inland lakes' salinity reaches more than 40,000 ppm.
[a]1 ppm = 1/1000,000 = 1 mg/L.
[b]Some low rainfall region, salinity of surface water reaches 1000–5000 ppm.

Table 1.3: The quality requirements of water used for industry, agriculture, and daily life [2].

Applications	Permission impurity content (ppm)
Drinking	<500
Industry	1–3000[a]
Agriculture	~3000

[a]Water used for medicine, electronic, and some science experiments require higher quality.

seriously, with the world's population increase and the rapid development of modern industry, on one hand, demand for fresh water continues to pyramid; on the other hand, the contradiction between water supply and demand intensifies with the destruction of fresh water resources caused by industrial wastewater pollution. It is estimated that the fresh water demand per person per year is about 1500–1800 m^3 by average (including industrial and agricultural water), which is extremely difficult to meet in the water-deficient areas. At present, many countries in the world are facing not only the shortage of fresh water for industrial and agricultural use, but also the difficulty in finding potable water in many areas. Conflicts and even wars have been caused by the scramble for fresh water resources. Every year in the world, 3.575×10^6 people die of diseases caused by unclean water [5]. Therefore the supply of fresh water has become the fundamental issue that must be considered.

According to statistics, China's total amount of water resources ranks sixth in the world, but the per capita water resource is only one-fourth of the world's average. Seven relevant organizations, such as the Commission on Sustainable Development, have collected the statistics for 153 countries and regions. China's per capita water resource ranks 121st in the world, which has been listed as one of the 13 severe water-poor countries by the United Nations. Some islands, saline coastal regions, and inland brackish water areas along the coastline of China are all poor in fresh water. People in these areas have been suffering from many kinds of diseases caused by the long-term ingestion of unhealthy water, which directly affects their health and the local economic development. So China also faces the serious challenge to settle the water supply crisis.

At present, China's urban water shortage amount is 6 billion m^3. Especially for the coastal areas with more than 40% population and 60% GNP, the water shortage amount takes up one-third of the country's total. Water resource deficiency has become a bottleneck for the economic and social development of these areas. As predicted by experts, the annual increasing rate of water consumption in coastal cities was 4.5–6.5% between 1998 and 2007, while between 2010 and 2015, the rate from 1998 would double, and the water shortage situation could worsen. Take the new engine for China's economy. Tianjin Binhai New Area, as an example: the per capita water resource of Tianjin is 153 m^3/year, accounting for 6.9% of the figure per capita nationwide, which has already seriously restricted the areas' development [6].

To increase the supply of fresh water, besides the conventional measures like nearby or interbasin water diversion, an effective way is to desalinate seawater or brackish water nearby, especially for those remote areas with widely distributed water. Seawater is the most abundant water resource on Earth, and brackish water is also sufficient in some areas like the Qinghai-Tibetan Plateau of China. In many coastal and distant inland regions, seawater or brackish water desalination is much more economically efficient than distant water transportation.

There are many ways of desalination for seawater or brackish water. However, the routine methods like distillation, the ion-exchange method, the dialysis method, the reverse osmosis membrane, and the freezing process require a large consumption of fuel or electrical power. It was reported that, as of 1990, the production capacity of all desalination units installed worldwide was 13×10^6 m^3/day [7–8]. However, in 2008, the capacity of installed seawater desalination systems around the world was already 87×10^6 m^3/day, and the capacity of the running systems was 63×10^6 m^3/day according to the report of the International Desalination Association [9,10]. Obviously this number increased astonishingly, and it basically doubled (or more) every 10 years. However, the rapid increase of desalinated water will produce a series of problems, with the most outstanding being the energy consumption. The production of 13×10^6 m^3 desalinated water per day is estimated to consume 130 million tons of fuel per year [1]. Given the big technology progress at the current stage, producing 63×10^6 m^3 of fresh water per day still requires the consumption of 3.78×10^8 kWh of electricity and the emission of 400,000 tons of CO_2. Even if such fuel costs are economically affordable, the greenhouse effect and air pollution still need to be concerned about, so desalination with solar energy will certainly draw much attention. The Chinese situation is similar. Many rural areas, isolated islands, and some other regions in China are still lacking electricity. Under such energy shortage conditions, solar desalination becomes one of the most important approaches to overcome fresh water deficiency. Therefore solar desalination has extensive application prospects.

From an economic point of view, desalination with solar energy is getting more and more competitive in the market. In the United States, fresh water produced by solar desalination costs only 11 cents per kilogram (one-quarter of the normal pure water price), while in China, the price is also only about 0.1 Yuan per kilogram, which is in accord with China's current level of consumption.

1.2 Composition and Properties of Seawater

Seawater is a kind of liquid with complex components, containing over 80 known chemical elements mostly existing in the form of ions. During the processes of concentration and crystallization of seawater, the ions precipitate as salt, among which the contents of 11 components exceed 1 ppm (see Table 1.4). These components are regarded as the major elements of seawater, accounting for 99.58% of the total salinity.

Besides, there are many kinds of dissolved gases in seawater (see Table 1.5), with the most abundant being CO_2, N_2, and O_2.

Influencing factors directly related to the operation of desalination include suspended solids, dissolved gases, and pH value, which have significant effects on seawater pretreatment and corrosion and scaling in the production process.

Physicochemical properties of seawater are important parameters to be considered during the design, calculation, and operation processes of the desalination unit. For the general chemical properties of seawater, see Table 1.6, and the physical properties can be found in Table 1.7.

Table 1.4: The major elements of seawater [2].

Elements	Cl	Na	Mg	S	Ca	K	Br	Sr	B	Si	F
Existence form	Cl^-	Na^+	Mg^{2+}	$(SO_4)^{2-}$	Ca^{2+}	K^+	Br^-	Sr^{2+}	$B(OH)_3$	$Si(OH)_4$	F^-
Content (ppm)	19,000	10,500	1350	885	400	380	65	8.0	4.6	3.0	1.3

Table 1.5: Dissolved gases in seawater [2].

Gases	CO_2	N_2	O_2	Ar
Content (ppm)	102.5	12.82	8.05	0.479

Table 1.6: The general chemical properties of seawater [2].

pH value	Chlorinity (‰)	Salinity (‰)	Total salt content (‰)
7.5–8.4	~19.38	~34.85	~35.07

Note: (1) Chlorinity refers to grams of Ag consumed to subside all halogen in 0.3285233 kg seawater under a standard state;
(2) Salinity refers to the total grams of oxide and chloride of 1 kg seawater, after all carbonate and organics oxidated, all the haloid converted to chloride in those seawater, ‰;
(3) Total salt content is the sum of concentration of dissolved salts in seawater, ‰.

Table 1.7: The general physical properties of seawater (25°C, 1 atm under standard state) [2].

Density (kg/m^3)	Specific heat (kJ/kg°C)	Heat of vaporization (kJ/kg)	Ice point (°C)	Boiling point (°C)	Vapor pressure (Pa)	Osmotic pressure (Pa)	Dynamic viscosity (kg/m s)
1023.4	3.90	2436.3	−1.91	100.56	$0.9812p_0$	$-0.034T$	0.96×10^{-3}

Note: p_0, Vapor pressure of pure water at the same temperature; T, temperature of seawater (K).

1.3 Summary of General Seawater Desalination Methods

Desalination can be done either by extracting fresh water or removing salt from seawater, which are also basically the two categories in the actual desalination process. The current desalination methods applied in practice primarily belong to the former, and only electrodialysis and the ion-exchange method fall into the latter. The main methods used at present are listed in Table 1.8 [2].

1.3.1 Distillation

Distillation is a process of separating fresh water from seawater by heating evaporation and condensation. It can also be divided into four principal modes according to the energy, process, and equipment that are used.

Table 1.8: General desalination methods.

Classification	Methods
Extracting fresh water	**1.** Distillation **a.** Vertical tube distillation **b.** Multistage flash **c.** Vapor compression distillation **d.** Solar distillation **e.** Humidification–dehumidification desalination **f.** Absorption desalination **2.** Freezing(crystallizer based) **a.** Vapor compression freezing **b.** Vapor absorption freezing **c.** Direct refrigerant freezing **3.** Hydrate desalination (crystallizer based) **4.** Solvent extraction desalination **5.** Reverse osmosis (membrane based)
Separate salt from seawater	**1.** Electrodialysis (membrane based) **2.** Ion exchanging

1.3.1.1 Vertical Tube Distillation

As the process, shown in Fig. 1.1, makes multiple evaporators in a series and gets the seawater feed into the first series evaporator with indirect steam heating, the vapor generated there is called the "second steam"; the second steam is then led into the next evaporator for heating, and the seawater concentrate will also evaporate accordingly. The number of the series evaporator is named "series" (also known as "number of effect"), and the last series is connected with the pressure relief system to ensure that the boiling point of desalinated seawater will decrease from the first to the last series, so that the second steam from the former series can heat the desalinated seawater in the next series. The condensate after the second series is the produced fresh water.

The advantage of vertical tube distillation is the high efficiency of heat utilization, and the drawback is the large investment in equipment caused by the damage of boiler scale.

1.3.1.2 Vapor Compression Distillation

The principle is illustrated in Fig. 1.2. The preheated seawater is gradually vaporized with the increase of temperature, and the generated second steam condenses back in the jacket of the evaporator after compression, which provides the energy in the process. A characteristic of this method is using no external heat source, except the auxiliary heat needed as initial drive. This kind of system is suitable for the areas with abundant electricity.

The advantage of vapor compression distillation is the high efficiency of heat utilization, while the disadvantage is also the serious damage from boiler scale.

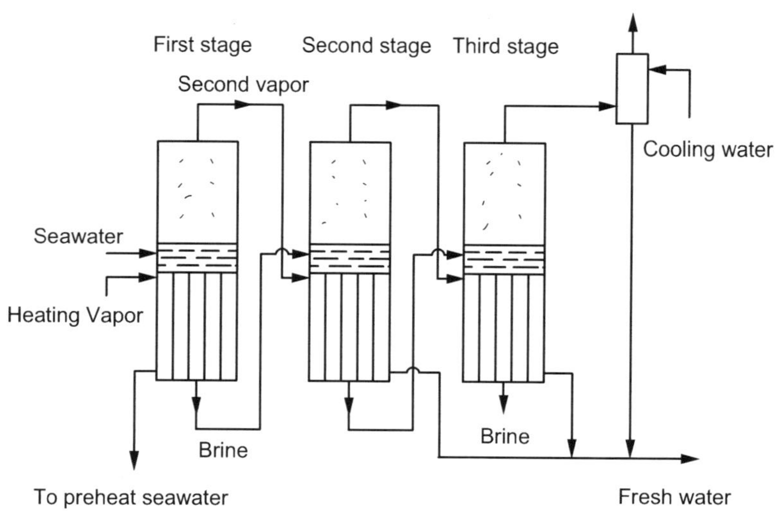

Figure 1.1
Vertical tube distillation process.

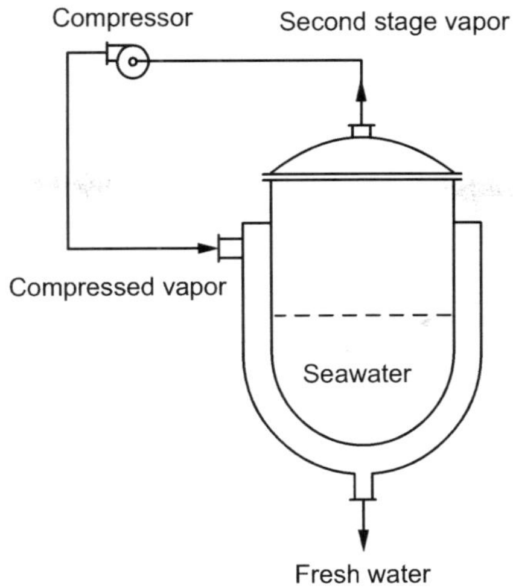

Figure 1.2
Vapor compression distillation schematic.

1.3.1.3 Multistage Flash Distillation

One of the various flow schemes for this method is introduced in Fig. 1.3. After preheating, the seawater first goes to a low-pressure device and evaporates immediately, then keeps evaporating in the next stage with a lower pressure. The generated vapor condenses to fresh water outside the preheating tube, while the gradually preheated seawater will be heated in the main heater by the vapor from the heat source. The special features of this method are the separation of heating and evaporating, simple equipment structure, and low cost, which is suitable for mass production. The large circulation volume of salt water and the high operating costs are the possible shortcomings.

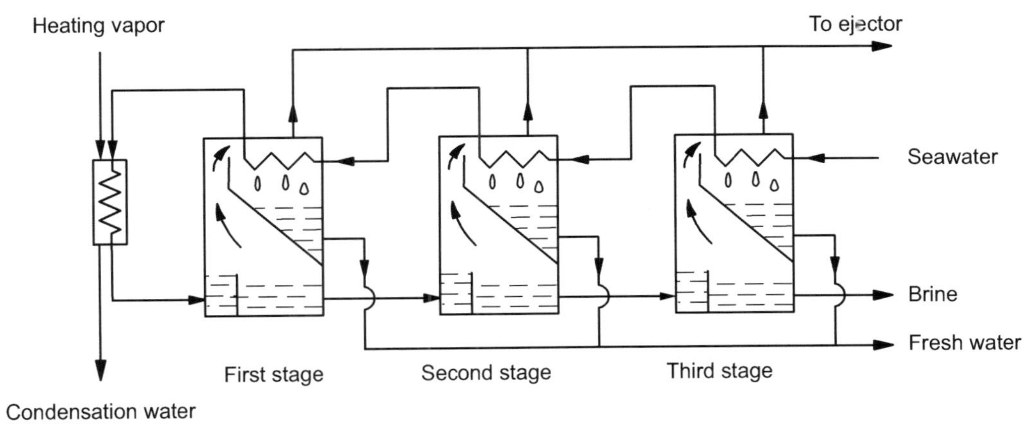

Figure 1.3
Multistage flash distillation process.

1.3.1.4 Solar Distillation

The principle is to do the desalination by solar radiant heat. The simplest solar distiller is illustrated in Fig. 1.4. To increase the heat absorption efficiency, the seawater batholith needs to be painted in black, so that the vapor can be condensed into freshwater on glass panels or plastic films with high solar transmittance. As the oldest desalination method, solar distillation has its advantages of low operating cost and simple device, but the disadvantages are that device space occupancy is large, yield per unit is low, and the thermal property can be influenced by different areas and climate conditions. At present, freshwater production per day by a simple device is only 3–4 kg/(m^2 d).

With no limitation to the initial concentration and high fresh water purity, solar distillation is well suited for the water-deficient areas with high temperature and long sunshine time. Fresh water can also be obtained by a similar device in areas where the heat source or available waste heat is sufficient.

1.3.2 Freezing Method

When the seawater is frozen, only pure water precipitates in the form of ice crystals, while salt precipitates at different times, so the fresh water can be gained in ice melting. The "freezing" is the desalination method based on this principle. It can be divided into three ways based on different refrigerants and relevant processing treatments.

1.3.2.1 Vacuum Freeze-Vapor Compression Method

As shown in Fig. 1.5, after it is precooled to around zero degrees, spray the seawater to a high vacuum freezer; part of the water absorbs heat by vaporization, and the remaining seawater produces ice crystals by freezing (water itself can be served as refrigerant). The solid–liquid mixtures of crystal and salt water melt into fresh water after removing the salt attached on the surface and enwrapped inside the ice crystal by separation and

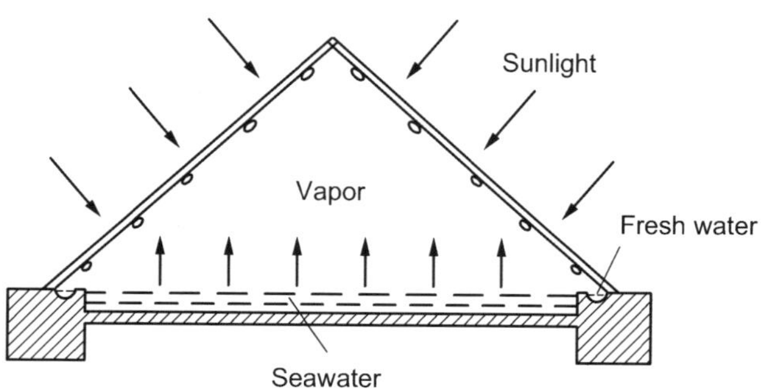

Figure 1.4
Solar distillation diagram.

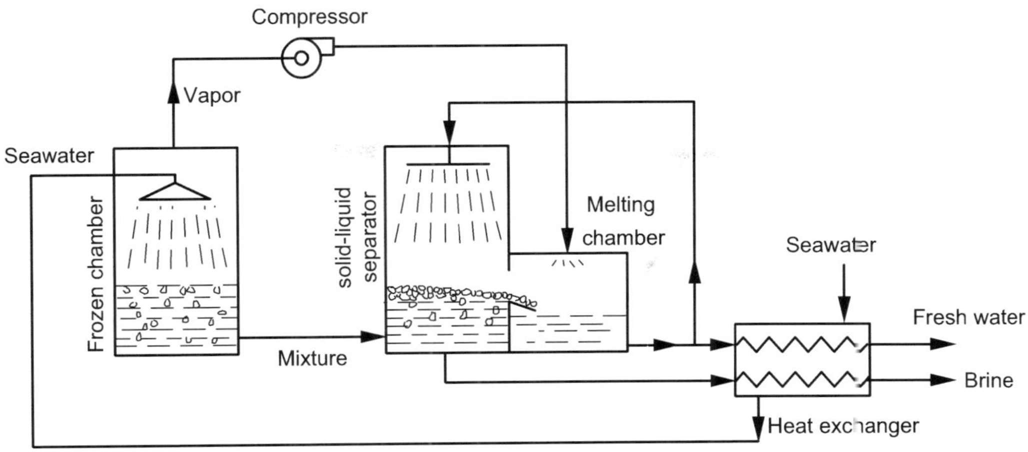

Figure 1.5
Vacuum freeze-vapor compression desalination schematic.

washing. Then condensation occurs when the compressed vapor goes into the melting device. One part of the fresh water generated by ice melting and vapor condensing will be released as a product, and the other part will be returned to the scrubbing separator for washing the mixtures.

1.3.2.2 Vacuum Freeze-Vapor Absorption Method

As illustrated in Fig. 1.6, this method is to continuously evaporate and freeze the seawater by absorbing the vapor generated in a freezer. The diluted absorbent will be regenerated by concentration for recycling use. Except for the substitution of an absorption system for compressor, others are the same as the vacuum freeze-vapor compression method.

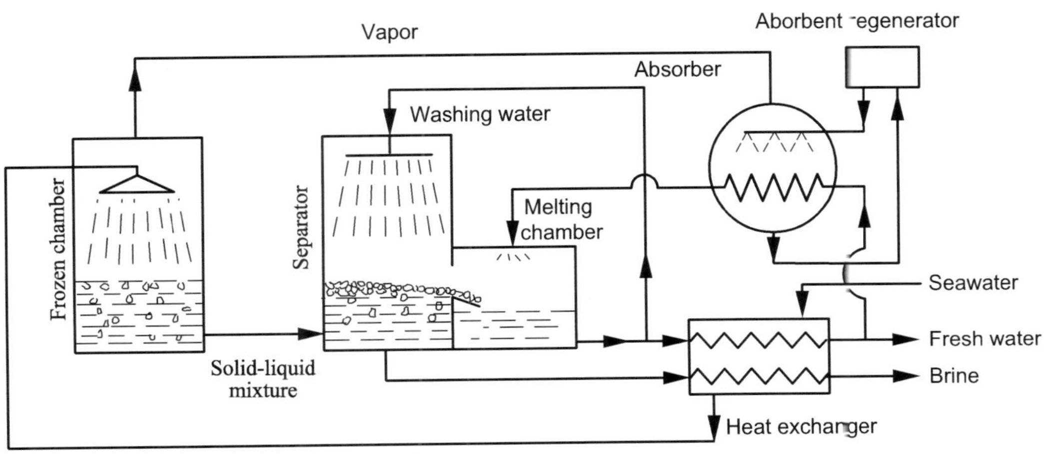

Figure 1.6
Vacuum freeze-vapor absorption desalination schematic.

1.3.2.3 Direct Refrigerant Freezing Process

As explained in Fig. 1.7, take the n-butane, insoluble in water and close to the freezing point of seawater, as the refrigerant and put it in the freezer after seawater mixing and preheating. In the freezer with temperature around $-3°C$ and slightly lower pressure than atmospheric pressure, the n-butane will absorb heat and evaporate, which makes the seawater freeze. Compress the n-butane vapor to above 1 atm and let it directly contact the ice, then the n-butane will liquefy and the ice will freeze. In the immiscible system, water and n-butane separate by difference densities, which releases the water as a product and keeps the n-butane for recycling use.

This method's merits are low required energy with small crystal latent heat, low operating temperature, and little fouling and corrosion. The disadvantages lie in the high operation technique standard to convey and separate the ice crystal suspension, the difficulty in washing the crystal, and the consumption of partial product fresh water.

1.3.3 Electrodialysis

As shown in Fig. 1.8, arrange the anion-exchange and cation-exchange membranes in alternate, and separate them into multiple intervals filled with seawater. Under external DC electric field, the anions and cations desalinate the seawater in some of the intervals and condense the adjacent intervals at the same time by permeation through the membranes. Electrodialysis is characterized by removing salt ions from the seawater. When applied to the desalination for salt water with salinity below 5000 ppm, this method is more economical than others for its reasonable energy transformation. However, it cannot dispose impurities carrying no charge, and its economic performance significantly

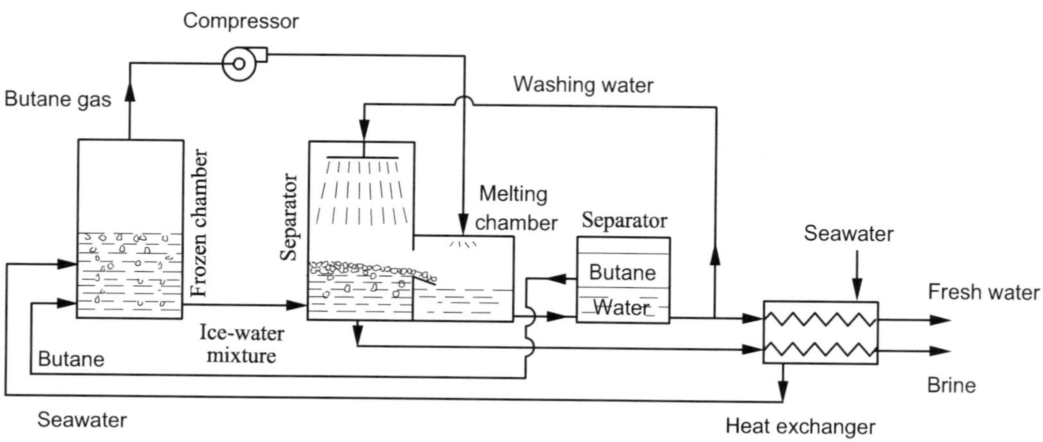

Figure 1.7
Direct refrigerant freezing process.

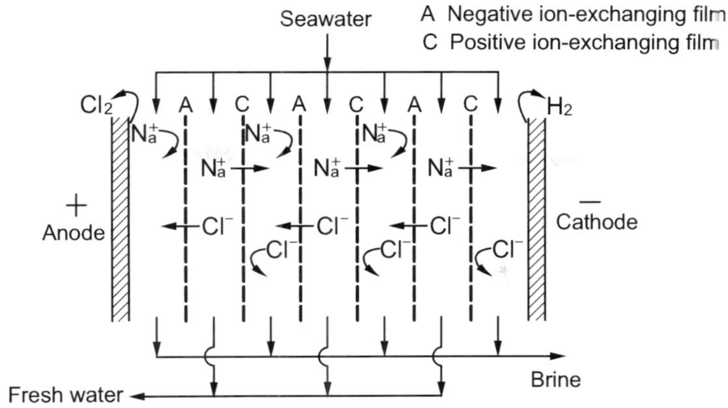

Figure 1.8
Electrodialysis desalination schematic.

decreases with the increase of salinity. The current high-temperature electrodialysis used for desalination is low in energy consumption.

The ion-exchange membrane is crucial to electrodialysis. A good ion-exchange membrane is supposed to have excellent permselectivity (i.e., anion membrane allows only anions, and cation can only permeate through cation membrane), outstanding electrochemical properties, sufficient mechanical strength, and chemical stability. The membrane used for high-temperature electrodialysis should also be fire-resistant.

1.3.4 Reverse Osmosis

When separated from the seawater by a semipermeable membrane, the fresh water will automatically permeate to the seawater side; if this side is applied with a pressure larger than the seawater osmolarity (approximately 25 atm, 1 atm = 101,325 Pa), then the pure water will go into the fresh water by reverse osmosis (RO), which is the principle of reverse osmosis desalination, as illustrated in Fig. 1.9. This method is noted for no change of phase, so energy consumption is relatively low, and the separation can also be achieved

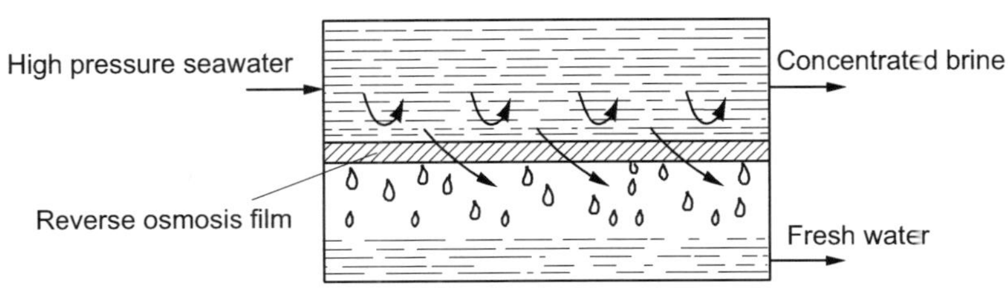

Figure 1.9
Reverse osmosis desalination.

for organic and neutral impurities. But due to the limitation of high-pressure operation and semipermeable membrane property, large output per desalination unit and high source water concentration are not suitable.

1.3.5 Solvent Extraction

Some organic solvent, for example, triethylamine, is able to dissolve a certain amount of water under low temperature and separate the dissolved water when the temperature increases. The desalination method based on this feature is called "Solvent Extraction." Fig. 1.10 shows how this method works: after reverse contacting with solvents in the extraction tower, the seawater is transported back to the separator through regenerator and heat exchanger, then separates out the water. The solvent is recyclable. Advantages of this method include optimum operation temperature close to the room temperature, low energy consumption, and little equipment corrosion. The advantage is that the solvent's solubility in water can affect the quantity of fresh water, which shows good selectivity only for the low-salinity salt water. The key to this method lies in the ideal extraction solvent.

1.3.6 Hydrate Method

Some gaseous compounds like propane, though immiscible with water, can form a multimolecular hydrated crystal with water under low temperature (e.g., $C_3H_8 \cdot 17H_2O$). The method using this kind of character is named "Hydrate."

What Fig. 1.11 shows is the principle of "Hydrate." Mixed in the crystal slot and with conditions suitable for the formation of hydrate, the precooling seawater and liquid propane form a saline solid–liquid mixture containing 10–15% hydrate crystal. After separating, washing, and melting, the fresh water can be obtained. The propane can be recycled. The character of this method is similar to the method of "Freezing." Its

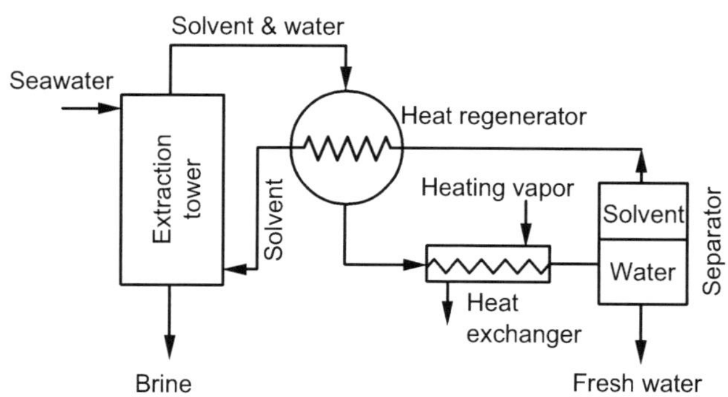

Figure 1.10
Solvent extraction desalination schematic.

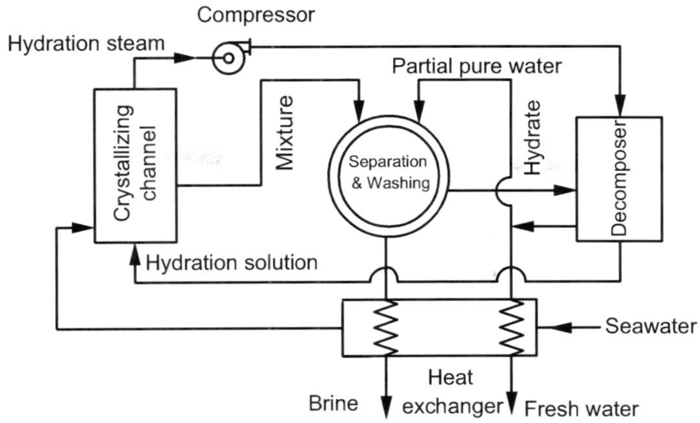

Figure 1.11
Hydrate desalination schematic.

disadvantages are the difficulty in separation and washing because of the small particles in crystal, the poor fresh water quality, and the consumption of fresh water during the washing process.

1.3.7 Ion-Exchange Method

Ion exchange is the method to obtain fresh water by exchanging the active group of ion-exchange resin and the anions and cations in seawater. As it is shown in Fig. 1.12, the cation is replaced with the H^+ in resin by using cation-exchange resin; the anion is then substituted by the OH^-. Water is formed after the combination of H^+ and OH^-, while salt is therefore removed.

The higher the saline ion concentration is, the more the resin regeneration cost will be. So this method is normally not applicable to seawater and salt water with high concentration but is widely used in low-salinity and high-purity water processing. There is an inorganic ion-exchange resin called "Siler Zeolite." After it gets mixed with seawater and forms precipitate with the main ions like Na, Mg, Ca, Cl, and sulfate, fresh water can be separated.

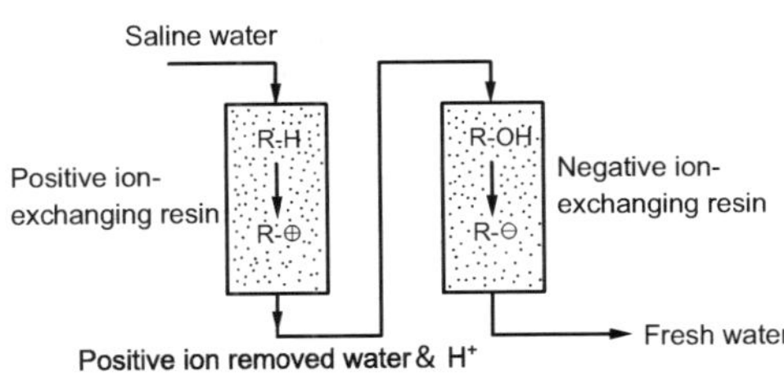

Figure 1.12
Ion-exchange desalination schematic.

Due to the high cost, this method is only for producing a small amount of water for emergency use.

1.3.8 Absorption Method

The cooling problem of the system terminal exists in all conventional desalination systems; even for the vapor with pretty low temperature, the emission may cause a huge waste of energy. How to the recycle the energy of low-temperature vapor has always been a major challenge. The absorption-type desalination system can partly solve this problem.

Some metallic salt solution or porous material solids exhibit relatively intensive absorption to vapor or say that these materials have low partial pressure of vapor on the surface, which enables them to absorb vapor; when the temperature rises to a certain degree, they can also give out water, obtaining the fresh water. Fig. 1.13 shows how an absorption-type desalination system works.

The desalination system shown in Fig. 1.13 consists of four main parts: absorber, generator, condenser, and evaporator. The operation principle of the system is as follows: the absorber takes in low-temperature vapor from the evaporator (which can also be obtained from the last stage of the cooling system in the low-temperature, multiple-effect desalination system) to intensify the last stage of the cooling process. And the diluted solution in the absorber will be sent to the generator and heated up to boiling, then the released hot vapor will be transported to the condenser for cooling, and then the fresh water can be produced. Certainly, to use the energy of the hot vapor repeatedly, send the hot vapor to the first stage of the low-temperature, multiple-effect desalination system and

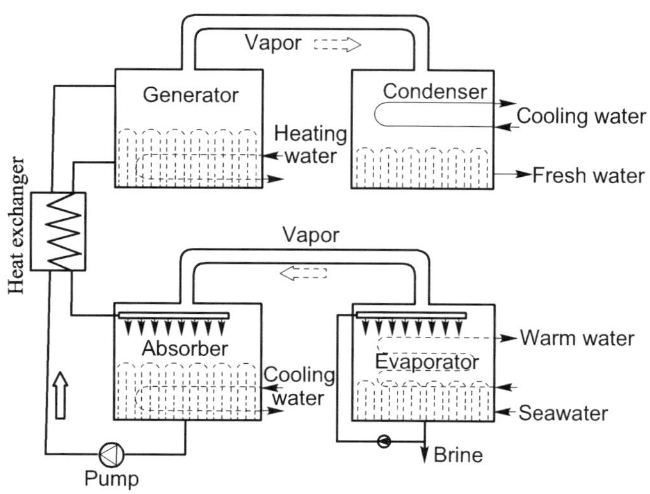

Figure 1.13
Absorption-type desalination system.

connect the last stage to the condenser after multiple-utilization, which can largely improve the system efficiency. So the condenser and evaporator of the absorption-type desalination system can be regarded as a low-temperature, multiple-effect desalination system.

The principle of the solid absorption-type desalination system is similar and is just the moist absorption system changed from liquid to solid.

1.3.9 Air Humidification and Dehumidification Method

Generally the natural air contains vapor, and the amount of vapor varies with the temperature. The air humidification and dehumidification desalination system operates based on this property of air. Its basic principle is indicated in Fig. 1.14. After heated externally, the hot seawater first enters into the system from the top of the humidifier, then flows downward through the porous material or by spraying and transfers heat and mass by convection with the upflow dry air and eventually increases the temperature and humidity of the dry air. The warm and humid air will then be fed in the dehumidifier and heat transferred with the condenser coil pipe or other surface air cooler. Afterward the air temperature will fall below the dew point, and fresh water can be obtained by condensation, thus realizing the desalination process.

The major advantages of this kind of system: (1) it separates the heating, evaporating, and condensing processes and operates in different components, so that each process can have better efficiency by taking pertinent measures; (2) it is convenient to recycle part of the latent heat of the condensation of vapor, by which the overall efficiency can be achieved; (3) the system's reaction to the heating temperature is not strong, so it's well adaptable to

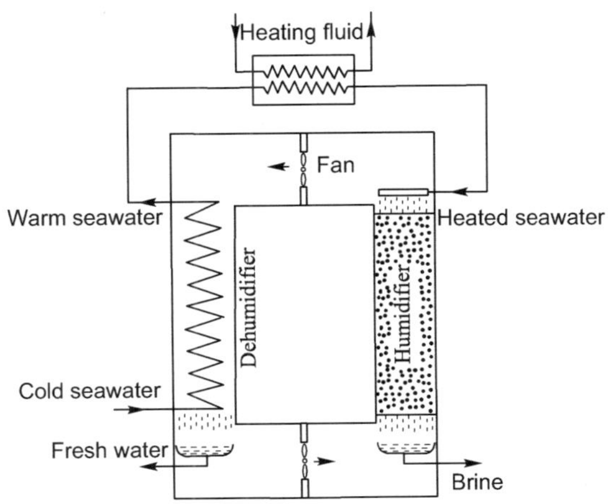

Figure 1.14
Humidification–dehumidification desalination system.

the solar energy with violent temperature change and other waste heat; (4) the system operates under low temperature, and the destructive factors such as scaling and corrosion are much less serious than the high-temperature system, so the system cost can be reduced by using low-cost construction materials; and (5) the system operates under atmospheric pressure, is easy to control, and there is no need to use strong materials.

1.3.10 Membrane Distillation

Membrane distillation (MD) is a separation process in which only vapors are allowed to pass through a porous hydrophobic membrane. The separation is achieved because of the vapor pressure differences between the membrane surfaces. The membrane consists of so many pores in which the pores cannot be wetted by the feed solution. In generation, one side of the membrane directly contacts with hot saline water to form the hot side. Another side of the membrane directly or indirectly comes in contact with cold, fresh water to form the cold side. The vapor passing through the membrane condenses on the cold side to produce the distillate. A typical membrane distillation unit is shown in Fig. 1.15.

There are mainly four types of membrane distillation processes, namely, direct contact membrane distillation, air gap membrane distillation, sweeping gas distillation, and vacuum membrane distillation. In all these processes the hot feed solution is in direct contact with the membrane surface, shown in Fig. 1.16.

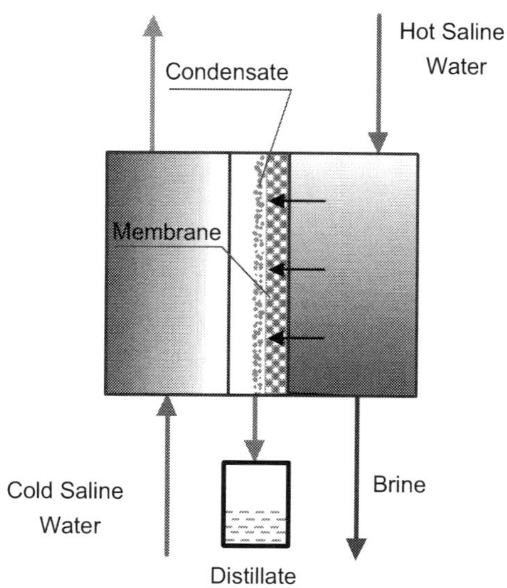

Figure 1.15
Schematic diagram of the membrane distillation process.

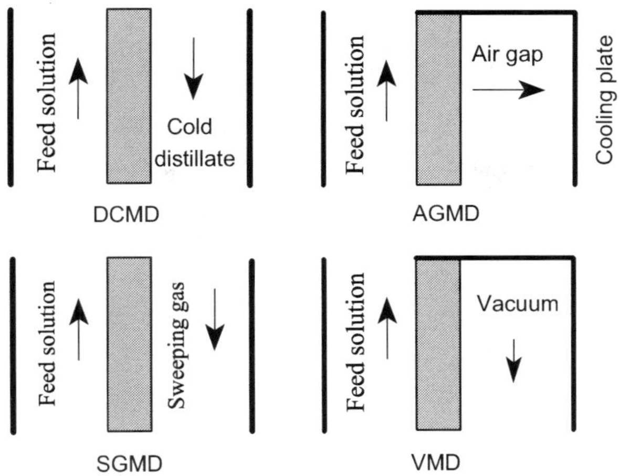

Figure 1.16
Four types of membrane distillation processes.

The membranes used for the membrane distillation process must be highly hydrophobic and highly porous and must have low thermal conductivity. The permeate flux increases with the increase in pore size and the reduction in membrane thickness. Generally, the pore size is in the range of 0.1–0.4 μm, and the thickness of the membrane is in the range of 30–60 μm. The operating parameters, like feed water temperature, flow rate, thickness of air gap, thickness of membrane, thermal conductivity of membrane, porosity, tortuosity, and long-term operations, have effects on distillate yield. Membrane distillation has more advantages compared to RO:

1. pipe lines are thinner due to the absence of high pressure;
2. reduced leakage compared to RO;
3. high saline feed can be treated;
4. operation is simpler compared to RO;
5. feed water pretreatment is not necessary; and
6. system efficiency is independent of feed water salinity.

1.3.11 Forward Osmosis

Forward osmosis (FO) has a similar working process with reverse osmosis, but the moving direction of the water molecules is just the opposite. In forward osmosis, the water molecules from saline water move through the semipermeable membrane toward the draw solution, which is maintained at a higher concentration than the feed solution. FO mainly uses osmotic pressure gradient and not the hydraulic pressure gradient.

The operation principle of FO and RO is shown in Fig. 1.17. Water and the saline water with low chemical potential are separated by a semipermeable membrane. Water

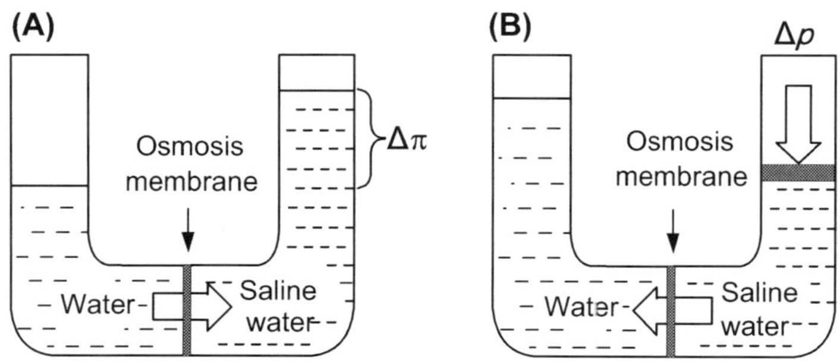

Figure 1.17
Operation principle of (A) forward osmosis and (B) reverse osmosis.

molecules spontaneously move from water to saline water through the semipermeable membrane, as shown in Fig. 1.17A, which is the forward osmosis process. The liquid level of the saline water will be elevated until the liquid level pressure difference between the two sides is the same as the osmotic pressure difference ($\Delta p = \Delta\pi$). If the water is replaced by the draw solution with lower chemical potential, water molecules spontaneously will move from saline water to draw solution through the semipermeable membrane. It is still the forward osmosis process. Oppositely, the RO process will happen when outer pressure ($\Delta p > \Delta\pi$) is brought to saline water, such as Fig. 1.17B, in which the water molecules move from saline water to the water side through the semipermeable membrane. The feature of forward osmosis is the process driving force just coming from chemical potential difference but no outer force.

Forward osmosis is being widely studied for desalination. However, it was concluded that the conventional RO membrane was not suitable for FO because of higher internal concentration polarization caused by a higher concentration of draw solution. The performance of the FO unit can be further improved by improving the membrane to withstand higher internal concentration polarization. The schematic representation of the FO system is shown in Fig. 1.18.

Also, solar energy can be used in the FO system. The commercial solar FO system has been developed by Trevi systems and uses a novel draw solution (RTS-2300). The osmotic driving force of draw solution was over 5000 psi, and as a result, the membrane of less surface area is sufficient for treating feed waters. The schematic representation of their system is shown in Fig. 1.19. The energy consumption of the forward osmosis system is lower than the reverse osmosis system of the same capacity.

Forward osmosis has more advantages compared to RO:

1. The operation is under low or no hydraulic pressure so that the energy consumption is relatively low.

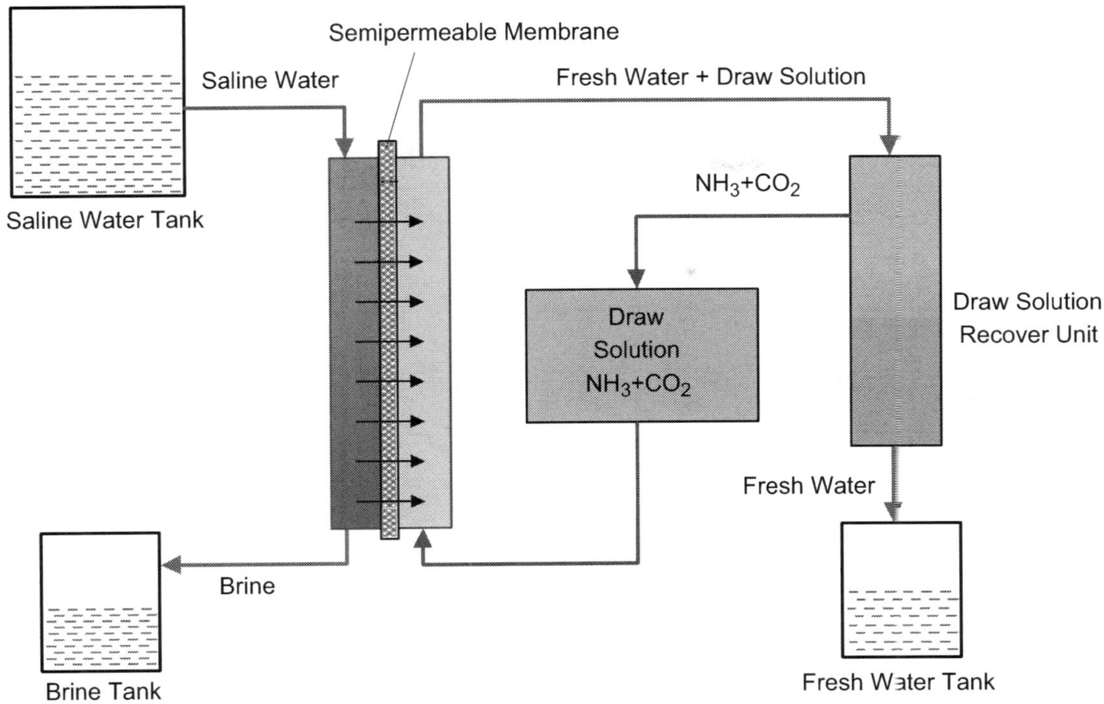

Figure 1.18
Forward osmosis unit using NH_3-CO_2 as draw solution [28].

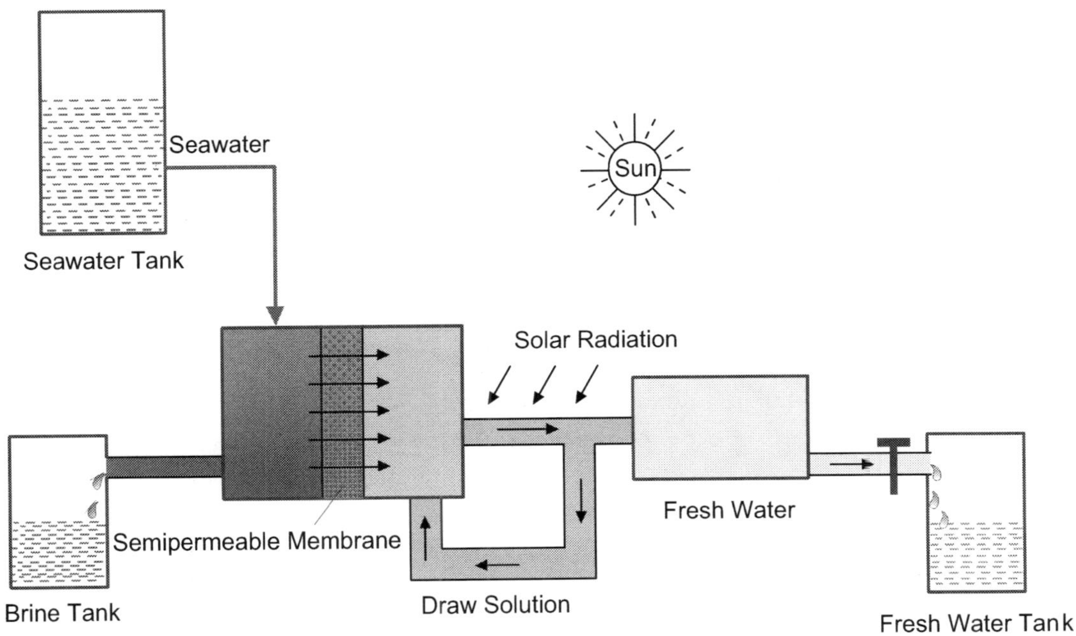

Figure 1.19
Forward osmosis unit driven by solar energy [29].

2. High salt rejection: the water recovery rate reaches 75% but only 35–50% in traditional reverse osmosis processes.
3. Less membrane fouling because the forward osmosis membrane is hydrophilic, so it can effectively reduce membrane pollution.
4. Equipment is simple.
5. Low operating temperature, which is suitable to use the solar energy or waste heat.

1.4 The Historical Evolution of the Solar Energy Seawater Desalination

Human beings have a long history in studying seawater desalination by solar energy. In one of his meteorology books, ancient Greek philosopher and physicist Aristotle (BC 384–322) described the water cycle process for the first time: the condensed water is "sweet," he wrote. People surmise that what Aristotle described is the desalination process of brackish water or seawater [12]. However, the earliest recorded solar desalination method was invented by an Arabian alchemist in the Middle Ages (about AD 1453–1476), which first appeared in the data book of the famous French solar engineer Mouchot [13]. Since the high-reflectivity mirror was still not available in the medieval period, the Arabian alchemist focused the sunlight on a polished Damascus mirror and projected the high-powered sunlight to a glass bottle filled with original seawater; the fresh water was gained from the vapor produced. Mouchot also mentioned in his book the experimental research he did in person, including alcohol evaporation by solar energy and the heating experiment on the line-focus metallic-reflective concentrating collector [12,13].

During the Renaissance period (AD 1533–1615), the famous scientist Della Porta described a kind of solar still used for brackish water desalination in his monograph Magiae Naturalis, volume 19. The structure of this unit is similar to the experimental device of Nebbia et al., as shown in Fig. 1.20 [13]. He also wrote, in volume 20 of his book, about how to get fresh water from the air, which is the embryo of today's humidification–dehumidification method.

In 1862, Lavoisier focused the sunlight in a flask by using large lenses installed on a support to do solar distillation [12]. In 1869, Mouchot described the whole process of solar distillation done by a silvered glass reflector [13].

No influential solar desalination projects have appeared over a long time. After the Industrial Revolution (1870), Swedish engineer Carlos Wilson designed and built the world's first large-scale solar brackish water desalination plant in the Northern Chile region Las Salinas in 1872 (picture of the plant is shown in Fig. 1.21). The plant distilled the high-salinity wastewater (140,000 ppm) from mineral smelting to get fresh water. This earliest desalination plant covered an area of 7896 m^2, consisting of many basin-type stills with the width of 1.14 m and length of 61 m, and had an effective collector area of

Figure 1.20
Della Porta's solar desalination unit.

Figure 1.21
The world's first large-scale solar brackish water desalination plant.

4450 m^2 and daily water production of 22.7 m^3(4.9 L/m^2 day). It kept operating for almost 40 years until the mineral was mined out [13].

Solar distillation has been ignored for a long time due to its low output and high initial cost. After World War I, it provoked people's attention again. By then, many new types of solar stills were invented, for example, roof type, tilted wick type, tilted basin type, aeration type, and so on, which solved problems, to a great extent, for sea rescue and domestic water supply.

In 1928, Pasteur first reported the experiment of solar distillation done by spherical reflector [13]. He focused the sun's rays in a copper ebullator by a spherical reflector. The generated vapor was then led into a conventional water-cooling condenser and produced distilled water in there.

Only few scholars studied the solar distillation technology before the end of World War II (before 1945). For a time in that period, there was a stirring of interest in the construction of solar desalination plants, but it was soon repressed by the economic turndown. Shortly after World War II, with the support of a massive influx of funding, Maris Telkes from MIT led his team to study the solar distillation technology in many ways [11]. At the same time, the US Department of Defense noticed that the solar desalination technology could help solve the fresh water supply problem for the army's offshore works, and they also financed the research on this technology. After that, the solar desalination technology has attracted increasing attention from governments and scholars around the world.

As of 1974, there were 25 large-scale solar stills in the world, among which a dozen of stills had a daily water output of more than 1 ton/day, with the largest one covering an area of 8640 m^2 and producing 28 tons/day. After the 1980s, the reports about building large-scale solar desalination devices were few [14].

Since the 1980s, there has also been much research and application of solar desalination in China. As early as the 1970s, China had already constructed one solar still with an evaporation area of 385 m^2 and daily water output of 1 ton and another still with an evaporation area of 50 m^2 and daily output of 0.2 ton on the Hainan Wu Qi Island and the Xisha Zhongjian Island separately. In the 1980s, Guangzhou Energy Research Institute finished the experimental study on the saturated air solar still and built a large desalination device with a solar lighting area of hundreds of square meters on the Shengsi Island in 1982 [14].

Recently, from the point of energy saving and device efficiency improvement, there is a tendency to combine the modern desalination technology with the use of solar energy, which is helpful to use the advanced experience and technology of the current desalination industry to improve the overall efficiency of solar desalination.

Early in the year of 1984, Abu Dhabi built a multieffect evaporation solar desalination system driven by a flat-plate collector. It was made up of 18 vertical multieffect stills, and each effect had a preheating device with a water generation ratio of 12.4 [15] to improve efficiency. But the low efficiency of the solar heat collector affected the extensive engineering application of the multieffect evaporation solar desalination system, which has become a hot research topic. Jimmy Leblanc [16] designed and installed a multieffect evaporation solar desalination device, which was heated by a salinity-gradient solar collector, with temperature of the first effect kept at 68°C. The daily water production was approximately 2300 L/day. This system was able to purify the high-salinity seawater discharged from the solar collector and produce salt, by which the economic efficiency of this device could be improved.

In this aspect, there are still a number of application examples of solar desalination worldwide [17]. For instance, in Mexico, the desalination system with the collector area of 680 m^2, desalination capacity of 12 m^3/day, and operation temperature of the solar collector of 35–110°C has been installed; in France, the desalination engineering project has a collector area of 670 m^2, and its fresh water output is 40 m^3/day; in Japan, the Oaisyo company has developed a highly efficient "seawater desalination and salt making system," which produces fresh water by using solar energy and the strong capillarity of porous ceramic; in Kuwait, the 10-ton multistage flash solar system has been built. All these systems are the result of the close integration of the conventional desalination industry and solar collector technology, and they have brought good economic and social benefits.

Remarkable advances have been made in solar distillation, with various new types of solar distillation systems emerging in an endless stream, among which many can be economically comparable to the conventional desalination device. The solar distillation system is sure to bring more, better, and more economical quality fresh water in the near future.

1.5 The Classification and Development of Solar Desalination Technology

1.5.1 The Classification of Solar Energy Seawater Devices

Throughout the history of solar desalination study, there has been mainly two ways of desalination using solar energy. One is to drive the seawater phase change process by the thermal energy produced by solar energy, which is called thermal desalination; the other is to drive the electrodialysis process by solar power generation, which is called membrane desalination. From the point of thermodynamics, the former is driven by low heat and the latter is by high heat.

With the development of solar technology, a lot of devices have been derived from their combination with the desalination technique. It's almost certain that the current solar technology can be integrated into any kind of conventional desalination system to form a new device. Fig. 1.22 provides some concepts and thinking of the integration of solar technology and the desalination technique.

Because of the easy combination of solar system and desalination technology, the deadlock narrowing desalination technology to solar distillation has been largely broken

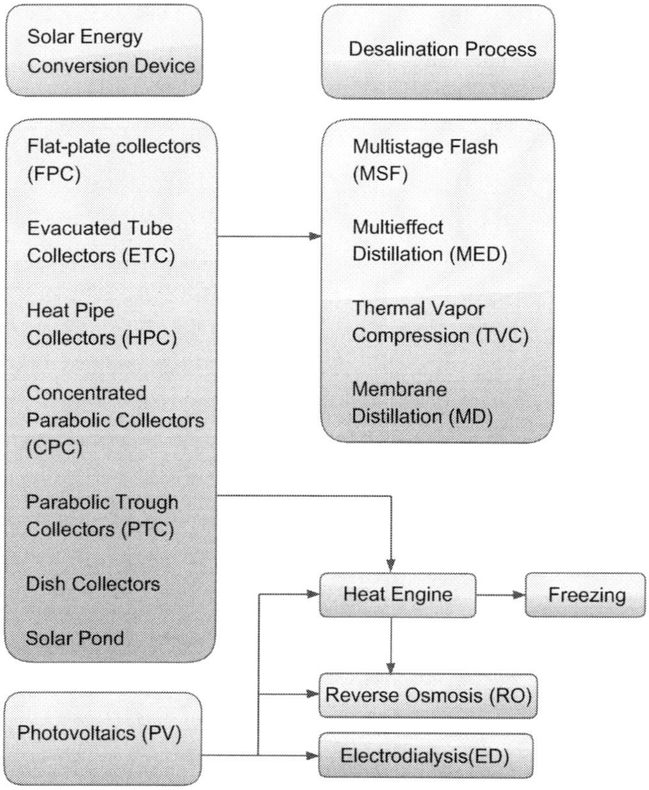

Figure 1.22
Desalination technologies coupled with solar energy technologies [17].

and the diversity of energy consumption and structure has been achieved, by which the solar desalination technology is gradually growing up. To summarize the ways of using solar energy, the current solar desalination system can easily be categorized as in Fig. 1.23.

Fig. 1.23 is a rough classification, and conventional methods can still be classified into thermal desalination and membrane desalination. The solar technology is also studied around these two classifications. Therefore the following discussion of desalination technology will focus on both methods and emphasize thermal desalination, which is now the mainstream. The "Thermal Desalination-Solar Energy" is to heat seawater with the thermal energy converted from solar energy through the solar collector, then evaporate the seawater and obtain fresh water by condensation. Typical thermal desalination methods are Multieffect Desalination-Solar Energy, Multieffect Humidification-Solar Energy, and Multistage Flash-Solar Energy (MSF-SE), etc. Other methods possible to be combined with solar energy include Thermal Vapor Compression-Solar Energy and Electrodialysis Desalination-Solar Energy.

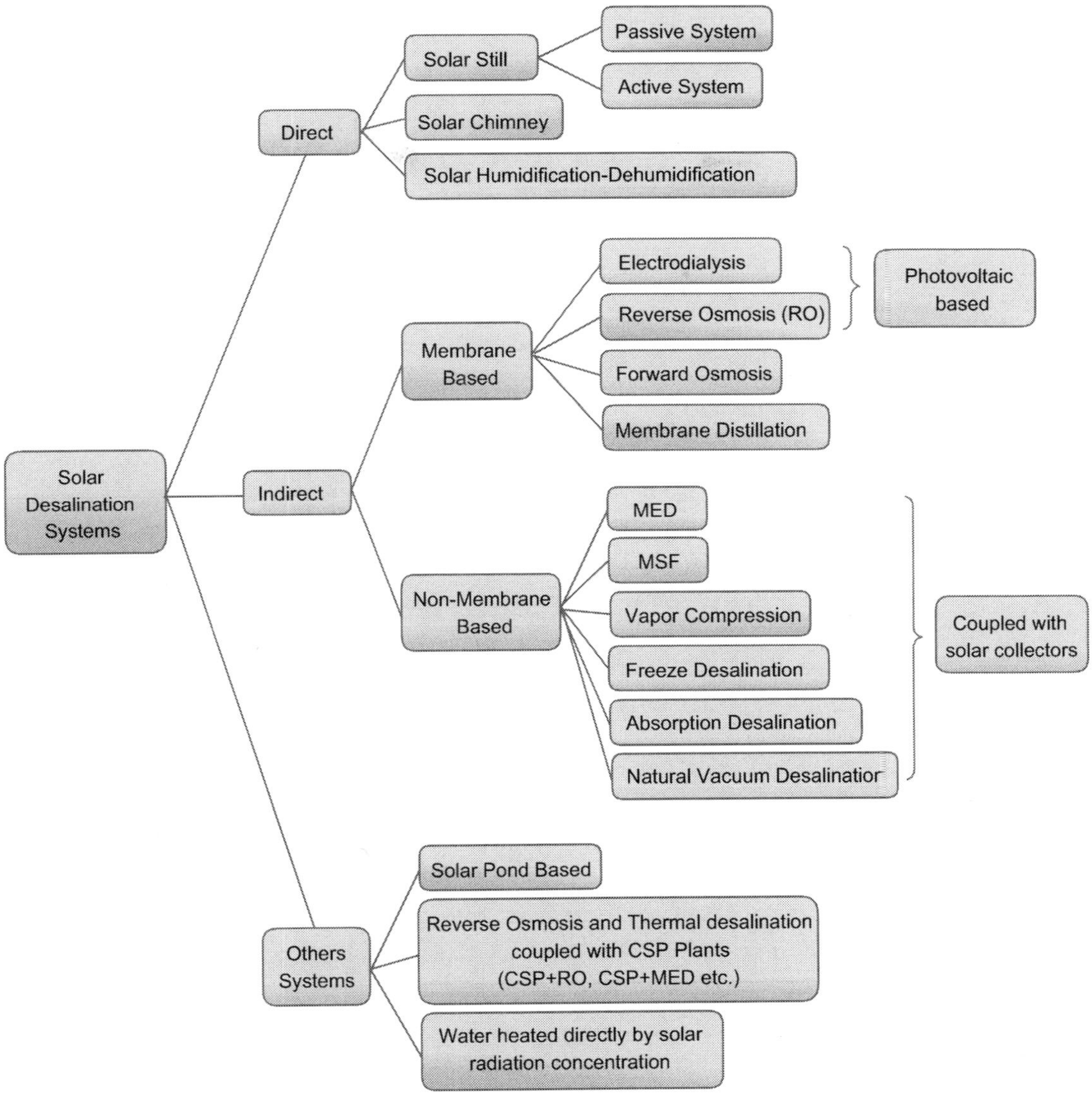

Figure 1.23
Classification of solar desalination [17].

The "Membrane Desalination-Solar Energy" first converts solar energy into electrical energy by the solar photovoltaic cell, then uses the high-pressure pump driven by electricity to compress seawater to pass through the specially made membrane (with the mechanism of ion reverse osmosis under a certain pressure) and get fresh water at the other side of the membrane. Such a method is also known as Photovoltaic-Reverse Osmosis (PV-RO). Besides the two methods mentioned earlier, there are still other methods, such as Natural Vacuum Desalination and Solar Thermal-Reverse Osmosis. More details will be introduced in the following section.

1.5.2 Thermal Method Solar Energy Seawater Desalination Technology

Thermal Desalination-Solar Energy (TD-SE) is the solar desalination technology with the earliest practical application. Regarding the system driven by the thermal energy produced by solar energy, it can be divided into direct and indirect methods according to the ways of using energy. As the name implies, the direct-method system is to distill with solar energy directly in the collector, while the indirect-method system separates the solar collector from the distiller. However, since 1996, quite a few scholars have made further study on the system mixing the direct and indirect methods. For example, Soliman (1976) first put forward the idea of adding heater in the basin-type solar still to improve the output per unit area [18]. Rai and Tiwari (1982) developed laboratory study on the device combining the flat-type solar collector and the basin-type still, which has provided more impetus for such work [19]. In view of the condition that the combination of solar system and modern industry device was widely applied, and all kinds of mixture devices had become the mainstream of the solar still system, Tiwari (1992) came up with a widely accepted concept to divide the solar distillation system into the passive type and the active type, depending on whether the accessories of water pump, fan, and solar collector would be used [20]. Thus the current solar still mainly includes two types: the active system and the passive system.

The passive TD-SE device does not rely on the additional accessories, but rather the thermal energy from the sunlight to passively complete the desalination process; the active TD-SE device has added some auxiliary elements to improve the internal heat and mass transfer process, and its water productivity may be several times as much as that of the passive TD-SE device by recycling the latent heat from condensation.

1.5.2.1 Passive Solar Distillation System

The most typical example of the passive solar distillation system is the basin-type solar still. The history of this type dates back to the 16th century, and it's still widely used because of its simple structure and convenient drawing materials. Current studies on the basin-type still mainly focus on the drawing of new materials, improvements of all kinds of thermal performance, and its co-use with other solar heat collectors. The ideal efficiency of the basin-solar still is around 35%, and the daily water output is approximately 3–4 kg/m^2 depending on the solar radiation. Distilled water production can be increased by 29% after adding the black naphthylamine with a concentration of 172.5 ppm. To reduce the heat capacity and increase the evaporation area of seawater, some researchers have done experiments by adding charcoal, sponge in seawater, or letting the seawater evaporate through the submerged core, and the results demonstrate the

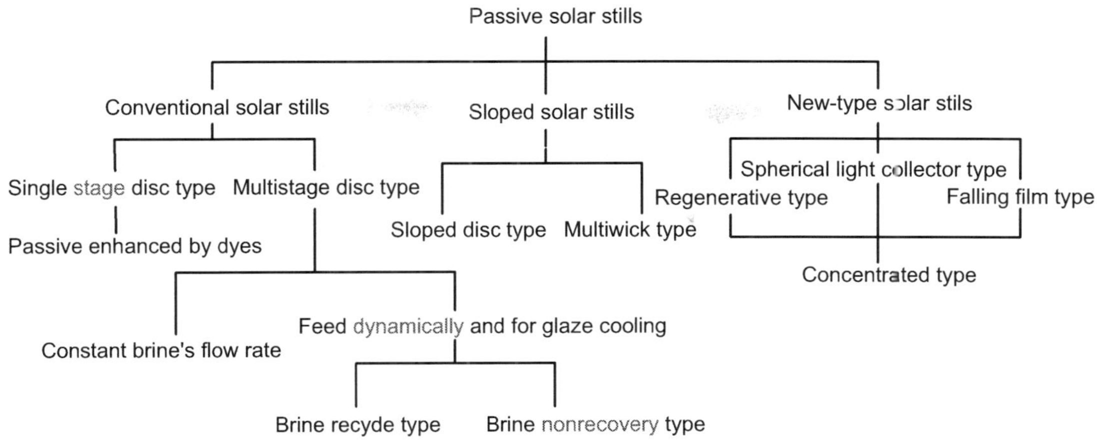

Figure 1.24
Classification of passive solar stills [19,20].

improvement of water production. The passive solar distillation system can also be subdivided as shown in Fig. 1.24.

1.5.2.2 Active Solar Distillation System

Serious shortcomings of the passive solar distillation system are low operating temperature and water production, inconvenient night operation, and utilization of surplus heat. Therefore the idea of active solar still was originated by Soliman (1976) [18]. Dozens of the active still design schemes and studies have been done. The initiative solar still can be roughly divided into some categories (see Fig. 1.25).

In the system of initiative solar distillation, because of its accessory equipment, the operating temperature has significantly increased, or the internal heat and mass transfer process has been improved. Moreover, recovering the latent heat of vaporization in the condensation process can be achieved in most initiative solar distillation systems, which produce one time or more water than the conventional solar still. This is the underlying reason for the extensive attention given to the active solar still.

All the early TD-SE methods are passive type. The passive TD-SE devices have still been widely used by now for their advantages of simple design, easy operation, convenient material drawing, low operating cost, and stable water productivity, etc. However, the heat and mass transfer inside the passive TD-SE device is mainly in the form of natural convection, which lacks efficiency. Although heat and mass transfer have been enhanced in some devices, influenced by big thermal inertia of the internal seawater and low utilization of the condensation latent heat of vapor, the device operation temperature is difficult to be increased, which leads to the low water output per unit lighting area, so the occupied areas of such devices are usually big.

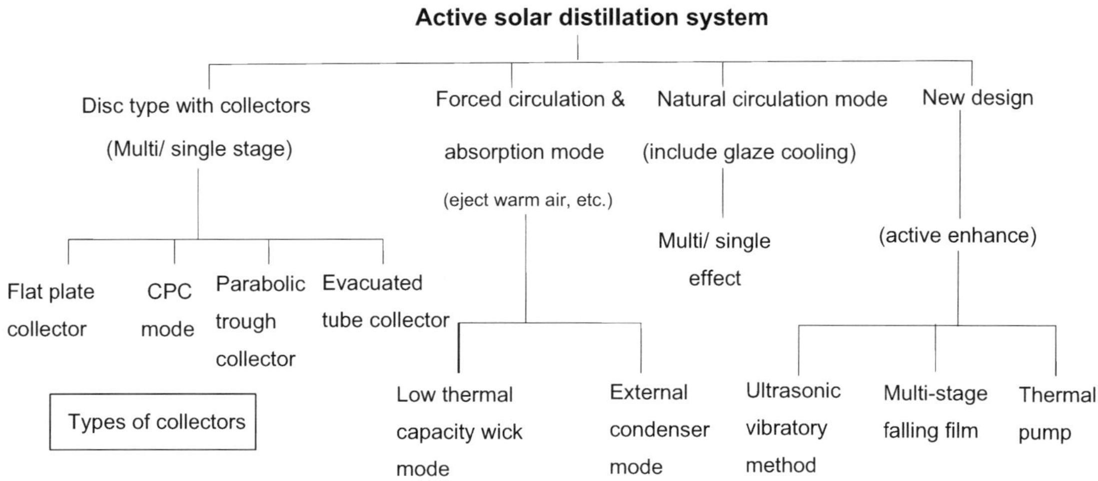

Figure 1.25
Classification of active solar distillation [12,19].

1.5.3 Membrane Method Solar Energy Seawater Desalination Technology

The Reverse Osmosis-Solar Energy (RO-SE) Method is one of the most important technologies at present, and it has received increasing attention because of its low energy consumption for water yield per unit. In the RO system, seawater is fed in by high pressure, only the fresh water can pass through the membrane, and salt will be exported. The RO technology has gained great development, with its energy recovery rate more than 50%, salt exclusion rate over 90%, and energy consumption rate of water yield per unit below 5 km/m^3 for some desalination plants where the RO technology is used on a mass scale. However, the RO technology also has its disadvantages; for example, it requires more strict seawater pretreatment, it's more susceptible to environmental pollution (which increases the maintenance cost), it has a higher cost due to the short membrane service life, and the large amounts of wastewater it produces may have an impact on the environment. The current solar reverse osmosis desalination technology is to power the system mostly by solar thermal or photovoltaic power. In 2003, Greece built an RO desalination device driven by solar photovoltaic power and wind power; due to the lack of an energy recovery device, the price of the produced fresh water is up to \$23–27/m^3, which cannot compete with the conventional RO technology yet [21].

There is another membrane desalination method (MD-SE: Membrane Desalination-Solar Energy), which is to drive the membrane filtration device by the electric power from the photovoltaic cell to realize seawater desalination.

When using RO-SE and MD-SE, their solar power generation parts and the desalination parts are basically separate [21], so they cannot be regarded as the absolute solar desalination methods, and the electric energy that drives the desalination device can also be obtained by power grid transmission. However, with the development of solar photovoltaic technology and the reduction of photovoltaic generation cost, the cost to produce fresh water by such methods will be decreasing year by year, and these methods will become highly competitive when the future photovoltaic generation cost is close to the conventional electricity price. Such methods have the following advantages over the thermal desalination methods.

Better independency: in the thermal system, the heat collection part and desalination part cannot be separated, so only in the coastal areas or islands with good solar radiation energy and no shelter can the system operate efficiently; water production will be poor in areas with low irradiance and shelters. The membrane system can choose to install the photovoltaic cell at places with better irradiance and the reverse osmosis device at places convenient for drawing water, then connect these two with the power grid, so that the water production efficiency will not be reduced by the shading from plants or high mountains.

Stable water production: in wet weather or at night, electric power can be obtained from an interconnected electric power system, so that the water production will not be influenced.

No waste of excess energy: in the condition of electricity excess, surplus electric energy can be stored in cells or exported by an interconnected electric power system to avoid energy waste. The storage and transmission of thermal energy are relatively difficult.

The MD-SE method also has its disadvantages.

Complicated system, limited water processing capacity: generally MD-SE requires some control system and moving parts, so such a system is relatively complicated and needs to be manned. The fresh water quality will not be satisfactory when the salinity of raw water is much too high, and the membrane components need to be replaced regularly. The quality of the water generated from the thermal method is better than that of the membrane method.

High cost: since the current price of the photovoltaic cell is still expensive, the water production cost of MD-SE is much higher than that of TD-SE. The application of such a method has been severely limited by the high cost, and the method can only stay in the development stage for a period of time in the future.

Unsuitable for miniaturization, highly restricted by regions: in some remote areas or islands where the miniaturized solar desalination system is needed for fresh water, the MD-SE method is uneconomical and even infeasible, while the TD-SE method has a clearer advantage.

To summarize, the MD-SE method is a kind of progressive solar desalination technology. Before promoting such a method, there are still many problems to be solved, such as how to increase the solar panel efficiency, reduce the production cost of the photovoltaic cell, and develop membrane desalination devices matching the photovoltaic power generation.

1.5.4 Combination of Thermal and Membrane Method

In the process of photovoltaic generation, the residual heat is usually discharged directly into the atmosphere, which is not only a waste of energy, but also the release of heat may have a negative effect on the environment. To solve this problem, in 2004, Van Helden [22,23] put forward a desalination method that can make the most use of the residual heat produced from the photovoltaic generation process. Afterward, many studies have been made on this subject. At present, there are two classifications, depending on whether there is concentration of the incident light during the photovoltaic generation process, Concentrating Photovoltaic and Thermal (CPV/T) and PV-Thermal (PV/T). The basic principle is to use the thermal and electric energy generated from the photovoltaic panels at the same time to drive the corresponding device to carry out desalination. In the previous photovoltaic generation process, cooling water is commonly used to dissipate the heat of the solar panel, then the overheated cooling water will be led into the radiator through a circulation pipeline to release the waste heat into the air. This way of heat dissipation results in energy waste as well as environmental damage. If the thermal-electric parallel method is used, the heat in the cooling water and the electric energy from the photovoltaic cell can drive the equipment at the same time to produce fresh water, and the total output will increase. The system flow is shown in Fig. 1.26.

It is to be noted that using the cooling water in the PV/T system for desalination cannot achieve an ideal effect due to its relatively low temperature, generally between 40–60°C;

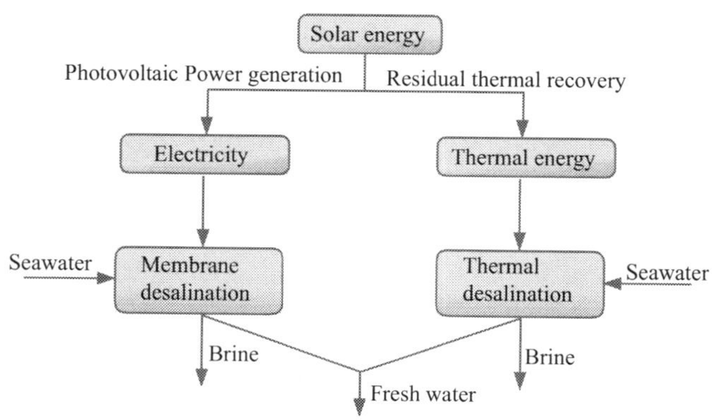

Figure 1.26
Desalination process with hybrid photovoltaic/thermal system [22].

by comparison, the cooling water temperature in the CPV/T system is much higher, and the residual heat used for desalination can bring satisfactory fresh water output.

The aforementioned desalination method is to convert solar energy into electric energy by using the photovoltaic cell, then drive the PV-RO system with the electric energy. The solar energy in such a method is roughly the equivalent of being converted twice, and eventually doing work in the form of mechanical energy to produce fresh water. Efficiency of the solar energy will become pretty low, because of the two times of energy conversion. To improve the solar efficiency, some scholars are dedicated to the study of the RO method that only requires one time of energy conversion. The thermal RO method stems from the solar water pump invented by Solomon de Cauxin in 1615, which is to heat the air by solar energy at one end of the pipe, then the heated air will do work by expansion to force the water out of the pump at the other end. In 1978, T. Pytlinski held forth the thermal RO method according to the principle of the solar water pump [24]. As of now, a lot of scholars have done research on this method. In 1981, T.E. Bowman first designed and did some experimental research on the solar thermal RO device functioning normally. In the conditions of collector area of 79.5 m^2 and the whole-day heat-collecting time of 8.5 h, 29.2 m^3 fresh water has been obtained [24]. To increase the pressure during the RO process, in the follow-up studies, D. Manolakos first used two sets of independent circulation loops in 2005. First, use the vacuum tube collector to heat water in the circulation loop, then deliver the heat through heat exchanger to low boiling point refrigerant HFC-134a in the second circulation loop; since the phase transition temperature of HFC-134a is much lower than water, it can convert to steam after gaining heat to drive the device to produce water [24]. The energy consumption of water production by the thermal RO method has been reduced to 2.1 kWh/m^3 after successive improvements. In summary, the thermal RO method is a new desalination method based on the principle of the solar water pump, and it shares a similar internal structure with other desalination methods, so there are a lot of experience references for the study of the thermal RO method. For the moment, RO is the most mature energy-saving technology in the world, therefore the thermal RO method has particular advantages over other desalination methods.

There is another simple structure of the thermal RO method, which takes advantage of the property that the conventional RO desalination system is sensitive to the seawater temperature. The solar collecting system is added to warm up the seawater, so that water-production efficiency of the RO system can be increased. As for the conventional RO desalination system, the water productivity will increase by 15% when the fed-in seawater temperature rises from 15°C to 35°C and increase by 22% when the seawater temperature reaches 40°C. But the temperature of the seawater can only increase to 43°C at the highest, or its impact on the membrane can be destructive [25]. In many places, the seawater temperature is only around 20°C, even in summer, and it is lower in winter.

Therefore to improve efficiency of the RO system, it is necessary to increase the inlet temperature when the solar collecting system can help a lot. Certainly, the heat energy of the concentrated salt water released out of the system should be recycled by the inlet seawater.

In a word, the solar desalination method is different from the conventional methods, because the solar energy density is low and unstable. So while spreading the solar desalination technology, it is also necessary to correlate with the properties of solar energy and develop a desalination system that can operate efficiently under low energy density. Meanwhile, to settle the instability of solar energy, usually the system should be equipped with an energy storage device.

1.5.5 Solar System Combined With Traditional Desalination Device

The wide application of the medium-temperature solar collector, for example, the vacuum tube type, parabolic trough type, and the medium-temperature solar cell, has enabled the solar still to operate at a higher temperature (≥75°C). This also makes it possible for solar energy as an energy source to integrate with other conventional desalination systems, which is becoming an active subject in the study of desalination [1]. With the heating temperature increase of the solar collector, solar energy can practically be integrated with all conventional seawater desalination systems (excluding the desalination system powered by electric energy). The solar systems with phased achievements and broad prospects mainly include multistage flash (Figs. 1.27 and 1.28), multiboiling distillation, and vapor compression, etc. For example, Kuwait has already built a 12-stage flash desalination system powered by a 220 m^2 parabolic trough solar collector and a 7000 L heat storage tank, with the production of 10 tons of fresh water [26] per day. This device can keep operating continuously in the nighttime and under the condition of limited solar radiation, with the daily water production per lighting area even 10 times larger than the one of conventional solar still. Thus there exists a great potential in the integration of the solar system and the conventional desalination device. Theoretical calculation shows that the

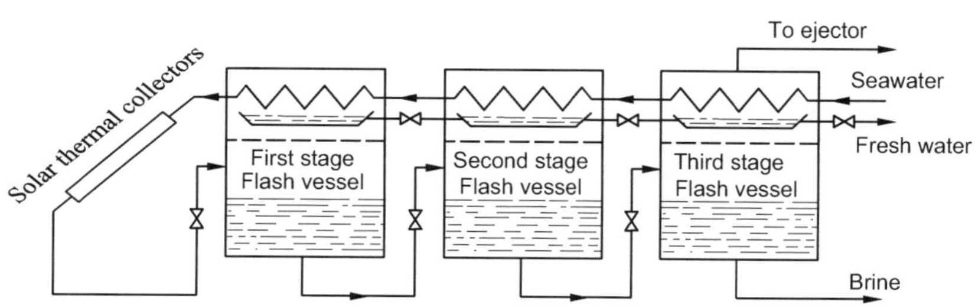

Figure 1.27
Multistage flash-solar energy system.

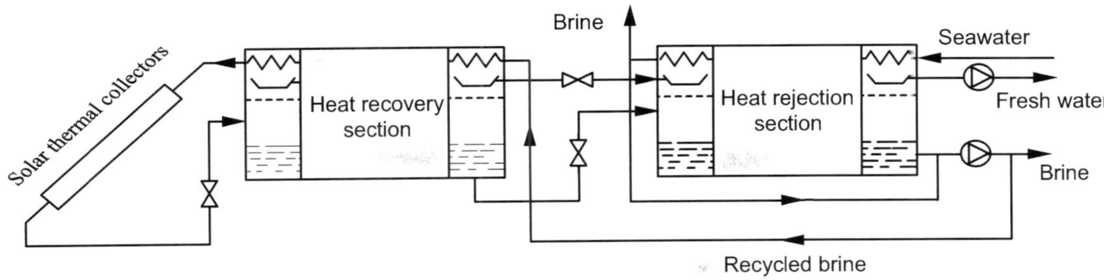

Figure 1.28
Multistage flash-solar energy system coupled with regenerator.

multiboiling distillation has more advantages over the multiflash system. Under the given condition of the same coefficient of performance, the multiboiling distillation requires lower numbers of stages, energy consumption, and external work consumption. The solar vapor compression system also has broad prospects, especially in the areas with relatively inexpensive electric power. It's worth noting that the technology is still immature and many problems are yet to be further studied in these areas, where researchers can take the spotlight.

Reports show that, among all kinds of multiboiling distillation systems, the multipacking tube distillation system is most suitable for using solar energy as the heat source [1]. This device has many advantages, the most important of which is that it can keep steady operation at any point with 0–100% input steam quantity and automatically adjust its operation status according to the quantity. What is more, the required heating temperature of 70–100°C is easy to reach for the parabolic trough-type or vacuum tube-type solar collector. Therefore further study on this is quite necessary.

Nevertheless, there are still problems pressing for solutions in the integrated systems introduced earlier, for instance, unfavorable to miniaturization, high equipment investment cost, existing corrosion, and scaling, etc. There is a lot of work to do to solve these problems. Figs. 1.27–1.30 illustrate some typical integrated systems of solar energy and conventional desalination devices [1].

1.5.6 Solar Energy Power-Desalting Cogeneration Technology

Solar concentrating technology has been developed greatly, which continuously increases the heat temperature of the solar collector, and many solar power plants have been set up worldwide. On one hand, fresh water is needed for cooling and producing steam in the solar power plants; on the other hand, the high-temperature waste heat generated from the plants can be used for desalination. Therefore the energy power-desalting cogeneration technology has been paid much attention by researchers recently. The most important

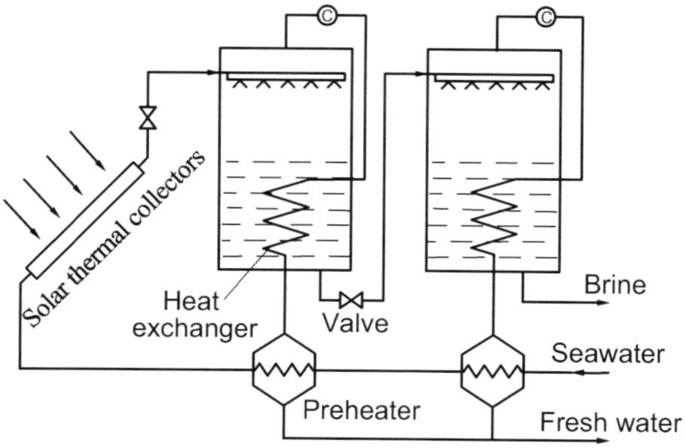

Figure 1.29
Solar vapor compression distillation process.

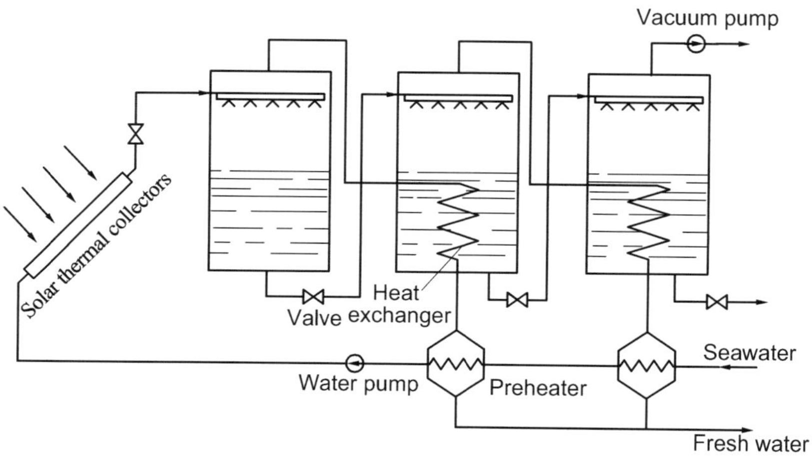

Figure 1.30
Solar vapor compression distillation process.

reason for this is that water shortage usually exists in the areas with abundant solar energy resources. Similar to conventional power stations, solar power plants also require cooling at the cold end of the steam turbine cycle. The barren desert areas are rich in solar energy but poor in water resources, so the water supply for solar desalination is always the focus of attention. Though the two ways of water cooling and air cooling are normally used in solar power plants, the cost of air cooling is higher and it may decline the power generation. The water consumption of a current large-scale solar power plant is generally between 2.27 and 3.02 m^3/MWh, while the number for the tower-type plant is about 2.27 m^3/MWh, and the trough-type is approximately 3.02 m^3/MWh. Although water consumption by using the air-cooling technology may largely be reduced to about 0.299 m^3/MWh, the investment cost increase and power generation decrease should be considered. As calculated, if the investment cost goes up by 7–9%, the power generation will fall by 5%.

Energy power-desalting cogeneration refers to the united production and supply system of solar seawater or brackish water desalination and solar power generation. There are two objectives of the system, power generation and fresh water production, so it is also called a "double objective" power plant. This system was originated in the 1970s and primarily applied in the Shevchenko industrial area of the former Soviet Union, the Texas chemical area of the United States, the aluminum industrial base in the Virgin Islands, the Torres petrochemical base in Italy, the Rotterdam Port in the Netherlands, and the Gulf states. Compared to the "single objective" power plant, the energy power-desalting cogeneration plant has better economic efficiency; the early investment drops sharply by the "double objective" sharing parts of the equipment. Corresponding energy consumption and labor cost declines by being calculated for cost allocation of the whole system.

Water produced from the power-desalting cogeneration system is primarily used as boiler feed water in a power plant or other process water. Some fresh water from good systems can even be turned into bottled drinking water or residential water, which provides experience and examples to integrate seawater desalination into the urban pipe network system and points out a new way to explore a sustainable fresh water solution for coastal cities.

To further improve the economic efficiency of the power-desalting cogeneration system, some people put forward to separate and purify the chemical materials such as sodium, magnesium, potassium, bromine, and bittern from the system by-product, i.e., the waste concentrated salt water, and to get the third or fourth product. For example, the energy consumption for bromine extraction from the salt water of one cycle of concentration is 70% lower than from seawater; the extraction of potassium sulfate for agricultural use from the concentrated salt water can save 50% of equipment investment than directly from seawater; and the remaining salt water after separation and purification can supply salt fields, by which a comprehensive utilization chain can be formed to increase economic benefit, lower the desalination costs, and help protect the environment by reducing the amount of waste salt water discharged into the sea.

At present, the annual average efficiency of the trough-type solar thermal power station with conventional water-cooling technology is around 14%, and a typical wet-cooling power station consumes 2–3 t/MWh of fresh water used for turbine condensation, mirror washing, and heat engineering system discharging, which is very difficult for the dry areas. However, the air-cooling method certainly will reduce the photoelectric conversion efficiency of solar power plants. The combined operation of solar thermal power generation and seawater desalination is one of the effective ways to increase the photothermal efficiency of the solar power generation system to more than 50%, reduce the generation cost, and provide self-consuming water of the power plant, as well as produce fresh water and enlarge the location range of the solar thermal plant, which is an inevitable trend in future scale development of the solar thermal power generation.

From a technical point of view, it is completely feasible to combine the two objectives. The steam generation system based on the Rankine cycle is used in all existing commercialized solar energy thermoelectric technologies, and the turbine outlet pressure of the system is about 18 kPa. A good integration with the low-temperature multieffect system can be achieved by appropriately increasing the pressure and temperature of exhaust gas and is superior to the power-desalting cogeneration system based on the reverse osmosis technology.

1.5.7 The Main Solar Desalination Systems Operating in the World

Governments and scientists all around the world have made a tremendous effort to push forward the development of solar desalination technology. Since the 1970s, thousands of solar seawater desalination experimental systems have been built, which has greatly contributed to the increase of water production and the deduction of energy consumption of unit output. Until now, whether it is in the water-deficient Arab countries or the developed countries, a lot of universities have research organizations specialized in solar desalination and water processing, and many important international conferences on seawater desalination are held every year.

The constant development and cost reduction enables the solar energy technology to be combined with more conventional desalination methods, which has lowered the water production cost and increased the scale year after year. All this progress has encouraged scientists to devote more to the solar desalination area. Tables 1.9–1.13 have listed the

Table 1.9: Solar-multistage flash desalination plants.

Location	Year	Energy source	Feed water type	Energy source details	Capacity (m^3/day)
El Paso, USA	1987	Solar pond	Seawater	3000 m^2 with 3.75 m depth	2.35–7.2
Gaza, Palestine	1999	FPC-PV	Brackish water	5.1 m^2 FPC, PV with battery storage	0.2
Berken, Germany		Solar collectors	Seawater		10
Lempedusa island, Italy		Solar collectors			0.3
Bari, Italy		Solar collectors	Seawater		5
Island of Cape Verde	1999	Solar pond			5
Suez, Egypt	2005	FPC		2.39 m^2 FPC	0.009

FPC, flat plat collector; *PV*, Photovoltaic.

Table 1.10: Solar-multieffect desalination plants.

Location	Year	Energy source	Feed water type	Energy source details	Capacity (m^3/day)
University of Ancona, Italy	1997	Solar pond	Seawater	625 m^2 with 3.5 m depth	30
Near Dead Sea, Israel		Solar pond			3000
Plataforma Solar de Almeria, Spain	2004	CPC	Seawater	500 m^2, gas boiler back-up with 30% continuous operation	72

CPC, compound parabolic concentrator.

main solar desalination systems around the world built since 1996 [17], from which the development prospects and trends of solar desalination are preliminarily revealed.

1.6 Facing Problems and Development Direction of Solar Desalination

1.6.1 Facing Problems of Conventional Solar Distillation Technology

Advantages of a conventional basin-type solar still include simple structure, convenient material drawing, low operating cost, and practically unmanned management. However, the disadvantages of low operating temperature, low production of water, and high equipment investment per unit of water production have limited its spreading and application. Therefore scientists around the world have been working on the theoretical analysis and experimental research to improve its performance for a long time, and many effective measures have been put forward, for example, to select cover board with more superior light transmission and condensation surface with better wettability, to increase seawater's absorption of sunlight, to reduce the emissivity of the cover board, to define the best thickness of the seawater awaiting distillation in the water container, and to lower system's heat loss. All these measures have played a role in the performance improvement of the conventional solar still. Yet there are no tremendous advances in the thermal efficiency, and the water production is only increased by 5%–10% [27].

To overcome the problem of low operating temperature, scientists have designed lots of schemes to equip the conventional solar still with other solar collectors, such as the solar still with flat-plate solar collector, CPC, or parabolic trough, which have greatly improved the water production per lighting area. However, another problem occurs, which is the increase of equipment investment per unit production, which is obviously unsatisfactory.

Comprehensive analysis on the reasons for low per unit yield indicates three serious defects in the conventional solar still. First, instead of reuse, the condensation latent heat of vapor dissipates through the cover board. Second, the heat exchange mode of natural convection seriously limits the improvement of the still's thermal performance. To

Table 1.11: Solar-reverse osmosis desalination plants.

Location	Year	Energy source	Feed water type	Energy source details	Capacity (m^3/day)
Gillen Bore, Australia	1993	PV	Brackish water	0.52 kWp PV	1.2
Lempedusa island, Italy		PV	Seawater	100 kWp PV	3 + 2 m^3/h
St. Lucie Inlet State Park, Florida, USA	1995	PV	Seawater	2.7 kWp PV, diesel generator	0.64
Lipari Island, Italy		PV	Seawater	63 kWp PV	2 m^3/h
Sadous Riyadh, Saudi Arabia	1995	PV	Brackish water	10.1 kWp PV, 264.24 kWh battery	10 + 5 (with still)
Heelat Ar Rakah, Oman	1995	PV	Brackish water	3.25 kWp PV, 9.6 kWh battery	5–7.5
Mudroch University, Australia	1997	PV	Brackish water	0.12 kWp PV	0.4
Canary Island, Spain	1998	PV	Seawater	4.8 kWp PV, 59.52 kWh battery storage	0.8–4.2
Lisbon, Portugal	2000	PV	Brackish water	0.15 kWp PV	0.08
Haifa, Israel	2000	PV, wind	Brackish water	3.5 kWp PV, 0.6 kWp wind, 36 kWh battery	3
Ceara, Brazil	2000	PV	Brackish water	1.1 kWp PV, 9.6 kWh battery	6
White Cliffs, Australia	2002	PV	Brackish water	0.34 kWp PV	0.5
Keratea, Greece	2003	PV-wind	Seawater	3.96 kWp PV, 0.9 kWp wind, 44.4 kWh battery	3.12
Massawa, Eritrea	2003	PV	Seawater	2.4 kWp PV with single-axis tracking	3
Baja California Sur, Mexico	2003	PV	Seawater		19
Canary Island, Spain	2004	PV	Seawater	5.6 kWp PV with tracking, 41 kWh battery	10
Agricultural University of Athens, Greece	2004	PV-wind	Seawater	0.846 kWp PV, 1 kWp wind, 7.56 kWh battery	2.2
Canary Island, Spain	2005	PV-wind	Seawater	0.6 kWp PV, 0.89 kWp wind, 21 kWh battery	1
Northwest of Sicily, Italy	2005	PV	Seawater	125 kWp PV, 160 kVA diesel generator, 1236 kWh battery	36
Agricultural University of Athens, Greece	2005	ETC-heat engine	Seawater	162 m^2 ETC, 100 kW heat engine, R-134a as working fluid	1.8
Cooper Pedy, Australia	2005	PV	Brackish water	3.2 kWp PV	0.764
Rajasthan, India	2006	PV	Brackish water	2.5 kWp PV	3.6

Table 1.11: Solar-reverse osmosis desalination plants.—cont'd

Location	Year	Energy source	Feed water type	Energy source details	Capacity (m^3/day)
Solarflow, Australia		PV	Brackish water	0.12 kWp PV	0.4
Agricultural University of Athens, Greece	2006	PV	Seawater	0.846 kWp PV, 7.56 kWh battery	0.8 (w/o battery), 0.9 (with battery)
Marett Island, Italy		PV	Seawater		5
San Nicola, Italy		PV	Seawater		12
Ras Ejder, Libya	2006	PV-wind-diesel	Seawater	50 kWp PV, 275 kWp wind, grid back-up	300
Ksar Ghilene, Tunisia	2006	PV	Brackish water	10.5 kWp PV, 72 kWh battery	50
Irbid, Jordan	2007	PV	Brackish water	0.136 kWp PV, 0.744 kWh battery	0.192
Fethiye area, Turkey	2007	PV	Brackish water	6 kWp PV, 4.8 kWh battery	2
Tangarfa, Agadir, Azla, and Marrakech, Morocco	2008	PV	Brackish water	4 kWp PV	8
Alexandria University, Egypt	2009	PV-wind-diesel	Brackish water	7.6 kWp PV, 5 kW wind, 5 kVA diesel generator	30

ETC, evacuated tube collector; PV, Photovoltaic.

radically change its internal heat and mass transfer mechanism to further improve the thermal performance, the utilization of the new technology of enhancing heat and mass transfer in the modern industry is necessary. Third, thermal capacity of the seawater awaiting distillation in the conventional solar still is too large to increase the operating temperature, which has weakened the driving force of the evaporation.

To make full use of the condensation latent heat of vapor, a number of scholars have designed various new solar stills, with a few notable characteristics. One is the design, with multiple evaporating and condensation surfaces. Condensation latent heat from the last stage transfers to the next evaporating surface and repeats, until the last stage of condensation surface. The other is the reuse of condensation latent heat by preheating the seawater in the evaporation chamber. To improve the heat and mass transfer process inside the still, many researchers have developed a variety of forced circulation solar distillation systems, in which not only passive enhancement of the condensation and evaporating surface structure takes place, but also the change of the original heat and mass transfer of natural convection forced by external energy have been performed. This helps significantly increase the heat and mass transfer coefficient of the system

Table 1.12: Solar-membrane distillation desalination plants.

Location	Year	Energy source	Feed water type	Energy source details	Capacity (m^3/day)
Hzag, Tunisia		Solar collectors			0.1–0.35
University of New South Wales, Australia	1991	FPC	Seawater	3 m^2 FPC	0.05
Island of Ibzia, Spain	1993	ETC	Seawater	51 m^2 ETC, 10 m^3 hot water storage	2
Tokyo, Japan	1994	Solar collectors and PV	Seawater		0.96
El Paso, USA	1999	Solar pond	Seawater	3000 m^2 solar pond with 3.75 m depth	0.4
Canary Island, Spain	2003	FPC and PV	Seawater	6 m^2 FPC, 0.08–0.096 kWp PV	0.08
Alexandria, Egypt	2005	FPC	Brackish water	5.73 m^2 FPC	0.064
Kelaa De Sraghna, Morocco	2005	FPC		5.73 m^2 FPC	
Porto Santo Island, Portugal		Direct heating			300
Gran Canary, Spain	2005	FPC-PV	Seawater	90 m^2 FPC, 4 m^3 hot water tank, 1.92 kWp PV, no battery	0.15
Irbid, Jordan	2005	FPC-PV	Brackish water	5.73 m^2 FPC, 0.106 kWp PV	0.1
Aqaba, Jordan	2006	FPC-PV	Seawater	72 m^2 FPC, 1.44 kWp PV, 3 m^3 water storage, battery storage	0.44
Tenerife, Spain	2007	FPC-PV			0.12

ETC, evacuated tube collector; *FPC*, flat plat collector; PV, Photovoltaic.

Table 1.13: Solar-electrodialysis desalination plants.

Location	Year	Energy source	Feed water type	Energy source details	Capacity (m^3/day)
Fukue City, Japan	1990	PV	Brackish water	65 kWp PV, 1.2 Ah battery storage	200
New Mexico, Mexico	1996	PV	Brackish water	2.3 kWp PV, 600 Ah battery	18
Isa Town, Bahrain	2002	PV	Brackish water	0.132 kWp PV	1.14
University of Alicante, Spain	2006	PV	Brackish water	0.272 kWp PV	1.32

PV, Photovoltaic.

and accomplish the purpose to improve the thermal performance of the distillation system. To reduce the thermal capacity of the seawater awaiting evaporation in the system and rapidly increase the temperature of evaporating surface, the most novel design is to utilize the new technology of falling film evaporation and condensation in the modern desalination industry, so that the solar energy's direct effect on the falling film seawater can make it quickly raise the temperature and evaporate, which solves the problem of low operating temperature.

Overcoming the three defects of the conventional solar still mentioned earlier will certainly improve its thermal performance. Moreover, if the three schemes can be combined together, it will undoubtedly cause a revolution of solar still. Such a kind of still not only utilizes the vapor's condensation latent several times, but also enhances its inside heat and mass transfer process, which relatively increases the operating temperature to guarantee a high production of water. And this is exactly what has been expected.

1.6.2 Development Direction of Solar Desalination

As already known, the development of solar technology is diversified. Solar energy as a kind of basic energy can convert to other energy in multiple forms, and any form of energy can be used for seawater desalination, so it is obvious that the solar technology can be integrated with various desalination technologies. The recently developed new solar desalination technologies are very refreshing. For example, the solar energy does work by organic Rankine cycle and powers the reverse osmosis system; the direct light in the tower-concentrating system stimulates evaporation; and the solar thermoacoustic system provides cold energy or electric power to the desalination system. With the combination of solar energy and seawater desalination technology becoming more and more closed and all types of new devices are constantly emerging, the high-production and low-cost solar desalination system will be invented for sure, when the time of large-scale production of fresh water by using solar energy is coming. However, during the process of developing the desalination technology, the following problems have to be noted.

First of all, the development of solar desalination technology must be combined with future demands. According to the different demands from large- and medium-sized cities, towns, villages, and remote areas not covered by the power networks and island areas, solar desalination will be developed in two directions of large scale and miniaturization to adapt to different market situations. As for the miniaturized desalination system, how to simplify the system structure and operation, use local materials, reduce system investment cost, and shorten the system construction period are problems to be improved and solved. For the solar desalination technology suitable for miniaturization, further studies on heat transfer intensification and energy recycling need to be made to increase the system efficiency and reduce the system cost. Regarding the medium- or large-sized solar desalination system, the heat supply system with low cost and high efficiency needs to be developed, especially the solar collecting system, such as solar vacuum tube collector, CPC, trough-type collector, and medium-temperature solar pond, etc.; problems like corrosion and scaling of the heat supply system, microbe development, and heat exchange efficiency have to be solved to increase heat supply efficiency and prolong the service life; the operation and structure parameters need to be optimized gradually; and the device and system must be modularized and simplified to improve system efficiency, so that the system can adapt to different requirements from customers.

Secondly, the medium- and large-sized solar desalination systems need to be combined with the current highly efficient desalination technology to reduce the equipment investment and increase the system stability. The conventional seawater desalination industry has been developed for several decades with mature technologies and solutions, and many technologies are easy to be combined with the solar system. For example, based on the current desalination technologies of MED and MSF and the fluctuating characteristics of solar energy, a new solar thermal desalination technology with variable condition, self-adaption, and sliding temperature operation can be developed to improve the system adaptability. Moreover, to intensify the heat exchange efficiency and heat source stability, and to ensure the operating stability and high efficiency of the system, further study on the coupling technology of the solar heat source and the seawater desalination system is needed.

Thirdly, research on the energy storage system and multienergy technology is necessary to ensure the stability of the heat supply system. Besides, an economic and reliable energy storage system is to be developed, the system operating time is to be prolonged to realize day and night continuous operation, utilization efficiency in the life circle of the system is to be improved, and water production cost is yet to be reduced.

The combination of solar desalination and the solar thermal power system is an important development direction because 60% of the desalination cost is spent on energy consumption. With the rapid development of solar thermal power technology, the key to the future application of solar desalination technology in large cities also lies in the development of a large-scale system coupling the solar energy thermoelectric system and the reverse osmosis plus multieffect distillation system, by which the overall heat utilization efficiency and economic benefit of the solar thermal power system can be improved. Also, electricity generation cost of the photovoltaic power system and fresh water production cost can be reduced, so this solution is the best way for the development of a large-scale solar desalination system.

References

[1] Kalogirou S. Survey of solar desalination systems and system selection. Energy 1996 22(1):69–81.

[2] Junhe W. Seawater desalination (in Chinese). Beijing: Science Press; 1978.

[3] Xinshi G. Solar energy engineering (in Chinese). Beijing: Academic Journal Press; 1988.

[4] EL-Dessouky HT, Ettouney HM. Fundamentals of salt water desalination. New York: Elsevier Science, B V; 2002.

[5] Sampathkumar K, V Arjunan T, Pitchandi P, Senthilkumar P. Active solar distillation—a detailed review. Renew Sustain Energy Rev 2010;14:1503–26.

[6] Qi L. Seawater desalination—a important strategy of water source sustainable development of the coastal cities. Sci Manag Chin 2010;28(1):58–60.

[7] Marinos D, Assimacopoulos D, Provatas F. In: Proceedings of the new technologies for the use of renewable energy sources in water desalination. Sec V. Athens; 1991. p. 48–56.

[8] Morris RM, Hanbury WT. In: Proceedings of the new technologies for the use of renewable energy sources in water desalination. Sec I. Athens; 1991. p. 30–50.

[9] IDA. Desalination in 2008: global market snapshot. 21st global water intelligence/international desalination association worldwide desalting plant inventory. 2008. p. 21.

[10] Sideman S. In: Kaksc S, editor. Some design aspects of thin film evaporator-condensers in water desalination, Heat Exchangers. New York: Hemisphere Publishing Corporation; 1980 p. 681–703.

[11] Fuyi H. The principle and application (Translated by Xie Yitang. In Chinese). Beijing: Electric Power Industry Press; 1982. p. 257–8.

[12] Belessiotis V, Delyannis E. The history of renewable energies for water desalination. Desalination 2000;128:147–59.

[13] Malic MAS, Tiwari GN, Kumar A, Sodha AS. Solar distillation. Oxford: Pergamon Press; 1982. p. 8–17.

[14] Jingtang L. Reviewing to the solar technology development in twentieth century (in Chinese). 2012-7-20. http://www.docin.com/p-6309566.html.

[15] El-Nashar AM, Samad M. The solar desalination plant in Abu Dhabi: 13 years of performance and operation history. Renew Energy 1998;14:263–74.

[16] Leblance J, Andrews J, Akbarzadeh A. Low-temperature solar-thermal multi-effect evaporation desalination systems. Int J Energy Res 2010;34:393–403.

[17] Ali MT, Fath HES, Armstrong PR. A comprehensive techno-economical review of indirect solar desalination. Renew Sustain Energy Rev 2011;15:4187–99.

[18] Soliman HS. Solar still coupled with a solar water heater. Mosul: Mosul University; 1976. p. 34.

[19] Rai SN, Tiwari GN. Single basin solar still coupled with flat plate collector. Energy Convers Manag 1982;23:145.

[20] Tiwari GN. Recent advances in solar distillation, in solar energy and energy conversation. New Delhi: Wiley Easten; 1992.

[21] Tzen E, Theofilloyianakos D, Kologios Z. Autonomous reverse osmosis units driven by RE sources experiences and lessons learned. Desalination 2008;221(1–3):29–36.

[22] Van Helden WGJ, van Zolingen RJC, Zondag HA. PV, thermal systems: PV panels supplying renewable electricity and heat. Prog. Photovolt Res. Appl. 2004;12:415–26.

[23] Hrayshat ES. Brackish water desalination by a stand alone reverse osmosis desalination unit powered by photovoltaic solar energy. Renew Energy 2008;33:1784–90.

[24] Nafey AS, Sharaf MA, García-Rodríguez L. Thermo-economic analysis of a combined solar organic Rankine cycle-reverse osmosis desalination process with different energy recovery configurations. Desalination 2010;261:138–47.

[25] Xianqiu Z, Mingxia Du, Yong Z. An experimental research on reverse osmosis desalination system assisted by solar energy. J Sol Energy (in Chinese) 2010;31(11):1419–22.

[26] Al-Hallaj S, Parekh S, Farid MM, Selman JR. Solar desalination with humidification–dehumidification cycle: review of economics. Desalination 2006;195(1–3):169–86.
[27] Tiwari GN, Dhiman NK. Performance study of a high temperature distillation system. Energy Convers Magt 1991;32(3):282–91.
[28] http://www.greentechmedia.com/articales/read/gust-post-desalination-realization.
[29] http://www.trevisystem.com/energy-comparison/.

CHAPTER 2

Solar Energy Utilization and Its Collection Devices

Chapter Outline

Solar Energy Desalination Technology. http://dx.doi.org/10.1016/B978-0-12-805411-6.00002-6

2.1 Solar Radiation Energy

2.1.1 The Sun and Its Energy

The sun is a sphere of intensely hot gaseous matter with a diameter of 1.39×10^9 m; it is on average 1.5×10^{11} m from the earth. As seen from the earth, the sun rotates on its axis about once every 4 weeks. However, it does not rotate as a solid body. The equator takes about 27 days and the polar regions take about 30 days for each rotation.

The sun has an effective blackbody temperature of 5777K. Here, the effective blackbody temperature is that of a blackbody radiating the same amount of energy as does the sun. The temperature in the central interior regions is variously estimated at 8×10^6 to 40×10^6K and the density is estimated to be about 100 times that of water. The energy produced in the interior of the solar sphere at a temperature of many millions of degrees must be transferred to the surface and then be radiated into space.

As we know, except for frequency, there is no essential difference between solar radiation and common radiation on earth, such as thermal radiation and electromagnetic waves. Sunlight contains enormous energy, and our main concern is how to use it. Before interacting with materials, the energy of sunlight, which is mainly photon energy, exists in the form of radiation energy and luminous energy. After interacting with materials, most sunlight energy is converted into thermal energy; the rest is transformed into electric, magnetic, and other forms of energy. Moreover, different spectrums of sunlight have different percentages of conversion. Thus, some people put forward the concept of spectrum splitting solar energy.

Sun illumines the universe. Its light is so glaring that people do not dare to look straight into it. For humans, sunlight is the brightest natural light. Observations and computations have shown us that the total light of the sun on the earth is about 2.5×10^{27} cd/m^2. After taking the 100-km-thick atmosphere surrounding the earth into consideration, 20% of sunlight is attenuated. Therefore, the light outside the atmosphere of the earth is about 3×10^{27} cd/m^2.

The energy density of sunlight is diluted when it is propagated in space. The further distance is from the sun, the thinner the energy density is going to be. As shown in Fig. 2.1, if we assume that the radius of the sun is r; the distance between the observation point and the sun surface is L, and the total radiation energy power of sun surface is E_0, we can get the energy flow density (Eq. (2.1)):

$$e_0 = \frac{E_0}{\frac{4}{3}\pi r^2}\left(\mathrm{W/m^2}\right) \tag{2.1}$$

By dividing the total energy the sun emits by its surface area, we can get the energy flow density of the sun's surface as 6.15×10^7 W/m^2, which is 45,000 times larger than the earth's density.

The energy flow density at the point that is L (m) from the sun can be calculated as (Eq. (2.2)):

$$e = \frac{E_0}{\frac{4}{3}\pi (r+L)^2}\left(\mathrm{W/m^2}\right) \tag{2.2}$$

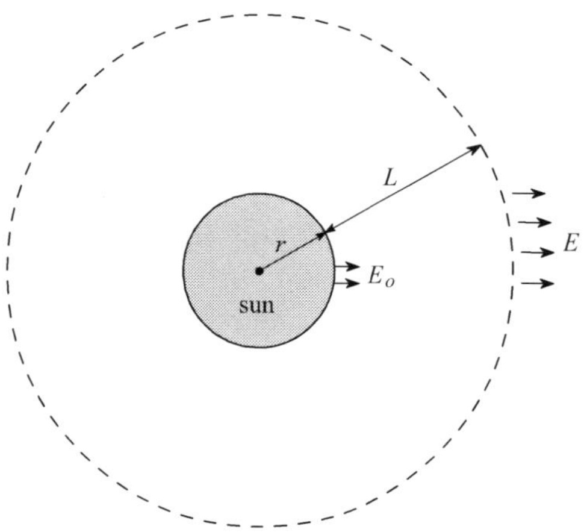

Figure 2.1
The process of solar radiation into space.

Solar radiation propagates in space through a process by which thermal energy flows from a 5700-K isothermal object to a 2.7-K cooler object. If we assume that the cosmic background and surface of the sun are both "blackbodies" and consider that there is no radiative heat transfer from the cosmic background to the sun, the law of energy transfer can be expressed as (Eq. (2.3)):

$$E_0 = \frac{4}{3}\pi r^2 \cdot \sigma \cdot T_s^4 (W) \tag{2.3}$$

This is the Stefan–Boltzmann law, where σ is the Stefan–Boltzmann constant and $\sigma = 5.67 \times 10^{-8} = \text{W/m}^2\ \text{K}^4$.

2.1.2 Solar Constant

Fig. 2.2 schematically shows the geometry of the sun–earth relationship. The eccentricity of the earth's orbit is such that the distance between the sun and the earth varies by 1.7%. At a distance of 1 astronomical unit, 1.495×10^{11} m, the mean earth–sun distance, the sun subtends an angle of $32'$. The radiation emitted by the sun and its spatial relationship to the earth result in the nearly fixed intensity of solar radiation outside the earth's atmosphere. The solar constant G_{sc} is the energy from the sun per unit time received on a unit area of surface perpendicular to the direction of propagation of the radiation at mean earth–sun distance outside the atmosphere.

Before rockets and spacecraft, estimates of the solar constant had to be made from ground-based measurements of solar radiation after it had been transmitted through the atmosphere and thus in part absorbed and scattered by components of the atmosphere. Extrapolations from terrestrial measurements made from high mountains were based on estimates of atmospheric transmission in various portions of the solar spectrum. Pioneering studies were by C. G. Abbot and colleagues at the Smithsonian Institution. These studies

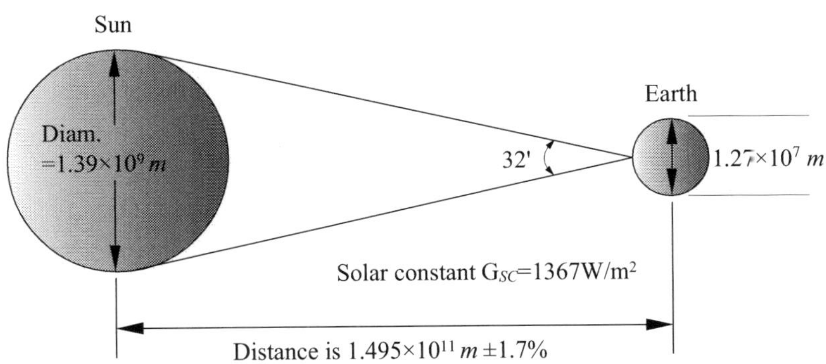

Figure 2.2
Sun–earth relationship.

and later measurements from rockets were summarized by Johnson (1954) [1]. Johnson revised Abbot's value of the solar constant of 1322 W/m^2 upward to 1395 W/m^2.

The availability of very−high altitude aircraft, balloons, and spacecraft has permitted direct measurements of solar radiation outside most or the entire earth's atmosphere. These measurements were made with a variety of the solar constant G_{sc} of 1353 W/m^2 with an estimated error of ±1.5%. This standard value was accepted by NASA (1971) and by the American Society of Testing and Materials [2].

The World Radiation Center (WRC) adopted a value of 1367 W/m^2, with an uncertainly on the order of 1% [3]. The value of G_{sc} of 1367 W/m^2 is used in this book.

2.1.3 Variation in Extraterrestrial Radiation

Two sources of variation in extraterrestrial radiation must be considered. The first is the variation in the radiation emitted by the sun. There are conflicting reports in the literature on periodic variations in intrinsic solar radiation. It has been suggested that there are small variations (less than ±1.5%) with different periodicities and variations related to sunspot activities. Willson et al. (1981) reported variances of up to 0.2% correlated with the development of sunspots [4]. For engineering purposes, in view of uncertainties and variability in atmospheric transmissions, the energy emitted by the sun can be considered to be fixed.

Variations in the earth−sun distance, however, lead to variations in extraterrestrial radiation flux in the range of ±3.3%. The dependence of extraterrestrial radiation on the time of year is shown in Fig. 2.3. A simple equation with adequate accuracy for most

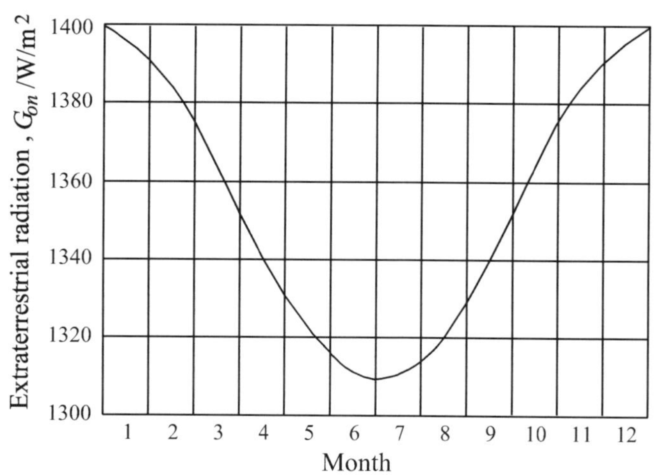

Figure 2.3
Variation in extraterrestrial solar radiation with time of year.

engineering calculations is given in Eq. (2.3a). Spencer (1971), as cited by Iqbal (1983), provided a more accurate equation (±0.01%) in the form of Eq. (2.3b) [5,6]:

$$G_{on} = G_{sc}\left(1 + 0.033\cos\frac{360n}{365}\right) \tag{2.3a}$$

$$G_{on} = G_{sc}(1.000110 + 0.034221\cos B + 0.001280\sin B + 0.000719\cos 2B + 0.000077\sin 2B) \tag{2.3b}$$

where G_{on} is the extraterrestrial radiation incident on the plane normal to the radiation on the nth day of the year, and B is given by (Eq. (2.4)):

$$B = (n-1)\frac{360}{365} \tag{2.4}$$

2.1.4 Attenuation of Solar Radiation in Atmosphere

As solar radiation penetrates the atmosphere and clouds, it is depleted by absorption, scattering, and reflecting. Although not all of it is lost, solar radiation that gets to the ground is around 30% less than solar radiation outside the atmosphere even on cloudless sunny days. Moreover, this percentage of decay may increase to above 90% on cloudy days. The magnitude of the solar constant might not seem large; however, the sun is 150 million kilometers away from the earth, and thus only 1 in 2.2 billion energies released can reach the earth. The radiant power of the sun is 3.75×10^{26} W. The energy released by the sun per second equals the energy released per second by burning 1.32×10^{16} tons of standard coal.

Based on its characteristics, the solar radiation that reaches the surface of the earth can be divided into direct and diffuse solar radiation. Direct solar radiation means that its direction does not change when it arrives at the earth's surface. To determine the size of direct solar radiation, people define the energy received by the unit area of the plane perpendicular to the direction of the solar radiation to be direct solar irradiance.

Besides the direct part of solar radiation, there is a diffuse part existing in the solar radiation that arrives at the surface of the earth. Diffuse solar radiation refers to the solar radiation that is received by the earth's surface, the direction of which has been changed owing to atmospheric or cloud reflection and scattering. It comes from the hemispheric sky in all directions; therefore diffuse solar radiation is also known as diffuse sky radiation. Normally diffuse radiation is much weaker than direct radiation. During the hot season of the midlatitude area, the diffuse solar radiation received by a horizontal plane at noon is only about 25% of direct radiation.

After penetrating the atmosphere, sunlight goes through many processes to arrive at the earth ground, and then finally it will be used and shared by living things and human beings. Its distributive processes in the atmosphere and on earth's surface are shown in Fig. 2.4. In these processes, solar radiation absorbed by atmosphere is about 4.0×10^{13} kW, making up 23% of the total solar radiation on earth. The solar radiation reflected by atmospheric molecules and particles is about 5.2×10^{13} kW, accounting for 30%. Around 8.1×10^{13} kW solar radiation survives a trip through the atmosphere and arrives at the surface of the earth, which accounts for 47% of the solar energy that reaches the earth, whereas the land surface of the earth only gets 1.7×10^{13} kW solar radiation, which accounts about 10% of solar radiation reaching the earth. In fact, only about 10% of solar radiation reaching the earth can be actually used by humans; therefore we should consider how to use it.

Because the earth revolves around the sun on an elliptical orbit, and the earth's axis is not perpendicular to its plane of motion trajectory, most areas on the earth have changes in season with the sun shining on the surface of the earth. Besides, the atmosphere that covers the earth makes weather change; for example, sunny, cloudy, rainy, or snowy. These

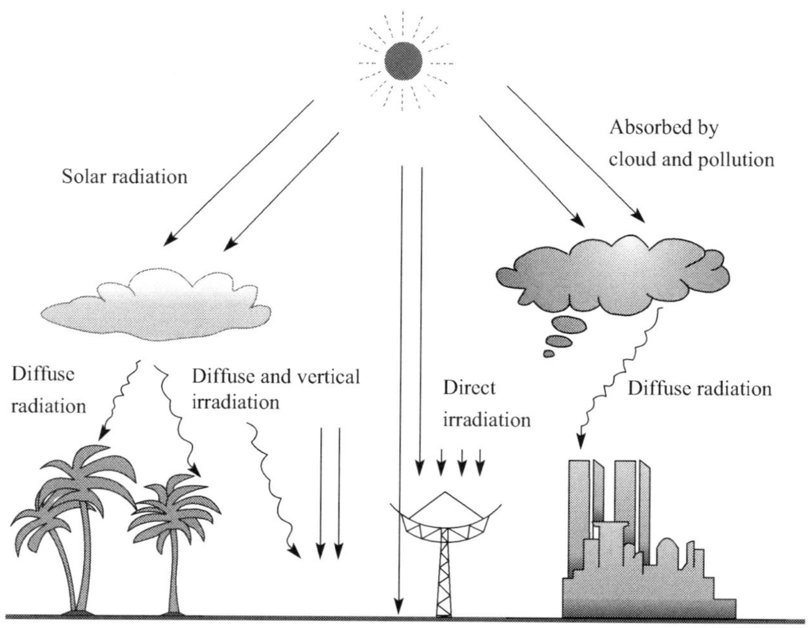

Figure 2.4
Distribution process of solar energy near the earth's surface.

changes in weather cause unfavorable factors for the use of solar energy on the earth's surface:

1. Solar radiation intensity is not very high. The average energy flow density of the sun is around 800 W/m^2, and this can increase to only around 1000 W/m^2 in good weather. Thus, the low solar radiation intensity makes it costly to collect and use solar energy on the earth's surface.
2. Although the total energy that the sun irradiates on earth is tremendous, the energy is unevenly distributed geographically. For instances, some areas almost have no rain and extremely hot weather whereas other areas have hardly any sunshine all year round. Thus unequal sunshine distribution is an unfavorable factor for collecting solar energy on the earth's surface.
3. Differences among seasons and weather, and changes between day and night keep solar thermal collecting devices from working continuously and satisfying customers' needs.
4. For now, solar energy collection technology can perform at relative higher efficiency only within the mid to low temperature range and has unsatisfactory efficiency within the high temperature range. Thus for solar thermal use, many relevant devices such as solar water heaters operate at a low temperature range.

Theoretically, because the temperature of the sun's surface reaches 5700K, solar collectors installed on the earth's surface can gather solar energy and its temperature can near 5700K. However, because of various factors involving heat loss, the limit is far from being reached. So far, the highest temperature that humans have collected from solar thermal energy is around 4000K. Even so, this temperature also can satisfy most human needs. Thus, it is both possible and meaningful to realize the use of solar thermal energy with high efficiency and high temperatures.

2.1.5 Spectral Distribution of Extraterrestrial Radiation

The three basic methods of heat transfer are conduction, convection, and radiation. Among them, radiation is a special type of heat transfer way because it does not require a contactor medium. It can even transfer heat in a vacuum. That the sunlight propagates and spreads in the universe is the phenomenon of radiation heat transfer. Hence sunlight is not essentially different from the thermal radiation or electromagnetic waves that we often see. The differences among them are just the frequencies and ranges of wavelengths. Thus we can study problems about sunlight propagation and the interaction of materials with sophisticated heat radiation, electromagnetic radiation, and optics theories.

The bandwidths of spectrums that make up the sunlight are large. The spectrum of sunlight consists of ultraviolet, visible, and infrared regions; the range of frequencies and the relevant percentage of each region is shown in Table 2.1.

Table 2.1: Range of frequencies and percentage of sunlight.

	Ultraviolet rays	Visible light	Infrared
Wavelength range (μm)	0–0.38	0.38–0.78	0.78–∞
Fraction within range (%)	6.4	48.0	45.6
Energy within range (W/m^2)	87	656	623

Visible light, which shows up as white, is the portion of the sunlight that is visible to human eyes. Its spectrum spans the relatively narrow range from 0.4 to 0.76 μm. The region whose spectrum is below 0.4 μm is called ultraviolet, and the region whose spectrum is above 0.76 μm is called infrared. About 95% of the energy of solar radiation is contained within the range of 0.3–2.5 μm. Energy that has a wavelength below 0.3 μm is only 1.2% of total energy. Energy that has a wavelength above 2.5 μm is not more than 4% of total energy. When parallel solar radiation penetrates the atmosphere, some is absorbed or distracted, and only radiation with a relative long-wavelength (ultraviolet light, visible light, infrared light, and radio waves) is able to survive. X-rays and other kinds of radiation with shorter wavelengths rapidly decay in the ionosphere because of the strong absorbent effect caused by nitrogen, oxygen, and other molecules; most ultraviolet rays are absorbed by ozone. Far-infrared radiation, whose wavelength is above 3 μm, is already extremely low before it goes through the atmosphere. Its amount can be neglected because it is absorbed by water vapor when it penetrates the atmosphere. For these reasons, ultraviolet rays only make up only about 8.03% of the solar radiation that successfully hits the ground, whereas visible rays and infrared rays make up 46.43% and 45.54%, respectively, of total solar radiation. Therefore the range of solar radiation that can be used on earth is mainly concentrated in the visible and infrared regions.

Normally, people calculate the solar radiation spectrum by arranging sunlight according to the frequency of each kind of light or wavelength. Fig. 2.5 is the most common solar radiation spectrum curve.

Table 2.2 provides the same information on the WRC spectrum in numerical form. The average energy, $G_{sc,\lambda}$ (in W/m^2 μm), over small bandwidths centered at wavelength λ is given in the second column. The fraction, $f_{0\sim\lambda}$, of total energy in the spectrum that is between wavelengths 0 and λ is given in the third column. $G_{sc,\lambda}$ is the average solar irradiance over the interval from the middle of the preceding wavelength interval to the middle of the following wavelength interval. For example, at 0.600 μm, 1748.8 W/m^2 μm is the average value between 0.595 and 0.610 μm.

2.2 Catalog of Solar Collectors and Their Fundamental Technologies

According to methods used to collect solar energy, the solar concentrator may be divided into two types: solar concentration and nonconcentration, as shown in Fig. 2.6.

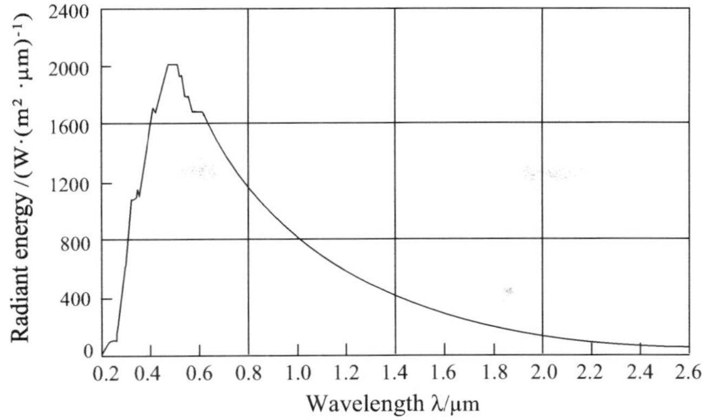

Figure 2.5
Solar radiation spectrum (NASA, 1971).

Table 2.2: Exoatmospheric solar irradiance (solar constant = 1353 W/m^2).

λ (μm)	$G_{sc,\lambda}$ (W/m^2 μm)	$f_{0\sim\lambda}$ (-)	λ (μm)	$G_{sc,\lambda}$ (W/m^2 μm)	$f_{0\sim\lambda}$ (-)	λ (μm)	$G_{sc,\lambda}$ (W/m^2 μm)	$f_{0\sim\lambda}$ (-)
0.24	63.0	0.001 4	0.47	2033	0.181 7	1.0	748	0.694 9
0.25	70.9	0.001 9	0.48	2074	0.196 8	1.2	485	0.784 0
0.26	130	0.002 7	0.49	1950	0.211 5	1.4	337	0.843 3
0.27	232	0.004 1	0.50	1942	0.226 0	1.6	245	0.886 1
0.28	222	0.005 6	0.51	1882	0.240 1	1.8	159	0.915 9
0.29	482	0.008 1	0.52	1833	0.253 8	2.0	103	0.934 9
0.30	514	0.012 1	0.53	1842	0.267 4	2.2	79	0.948 3
0.31	689	0.016 6	0.54	1783	0.280 8	2.4	62	0.958 6
0.32	830	0.022 2	0.55	1725	0.293 8	2.6	48	0.966 7
0.33	1059	0.029 3	0.56	1695	0.306 5	2.8	39	0.973 1
0.34	1074	0.037 2	0.57	1712	0.319 1	3.0	31	0.978 3
0.35	1093	0.045 2	0.58	1715	0.331 8	3.2	22.6	0.982 2
0.36	1068	0.053 2	0.59	1700	0.344 4	3.4	16.6	0.985 0
0.37	1181	0.061 5	0.60	1666	0.356 8	3.6	13.5	0.987 2
0.38	1120	0.070 0	0.62	1602	0.381 0	3.8	11.1	0.989 1
0.39	1098	0.078 2	0.64	1544	0.404 2	4.0	9.5	0.990 6
0.40	1429	0.087 3	0.66	1486	0.426 6	4.5	5.9	0.993 4
0.41	1751	0.099 2	0.68	1427	0.448 1	5.0	3.8	0.995 1
0.42	1747	0.112 2	0.70	1369	0.468 8	6.0	1.8	0.997 2
0.43	1639	0.124 7	0.72	1314	0.488 6	7.0	1.0	0.998 2
0.44	1810	0.137 3	0.75	1235	0.516 9	8.0	0.59	0.998 8
0.45	2006	0.151 4	0.80	1109	0.560 2	10.0	0.24	0.999 4
0.46	2066	0.166 5	0.90	891	0.633 7	50.0	3.9×10^{-4}	1.000 0

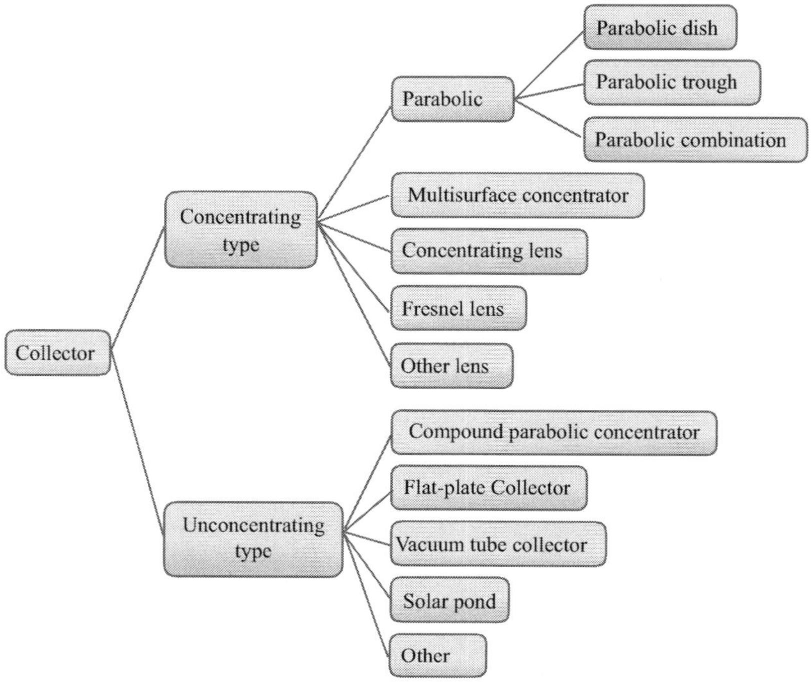

Figure 2.6
Classification map of solar collectors.

The main characteristic of solar concentration collectors is their high thermal efficiency, which can lead to a relatively high collection temperature; the highest collection temperature is reported to be 3000−4000°C [7]. The disadvantage of this type of collector is that only direct solar radiation can be collected. Therefore a solar-tracking system is required. The solar concentration collector is composed of the following types: (1) Parabolic concentrator: this type of concentrator uses paraboloids and its varieties within the three-dimensional space. It includes a rotating parabolic concentrator with a focus point (Fig. 2.7); a parabolic trough concentrator with a focus line (Fig. 2.8); and compound parabolic dishes (Fig. 2.9). Among these 3 types, the rotating paraboloid with a focus point is the most widely used. (2) Condensing lens: the condensing lens uses the refraction of transparent material to concentrate sunlight. In fact, sunlight will not concentrate one focus point because dispersion will happen when the multicolor of sunlight passes through the lens. The condensing lens is normally made of glass, which is not suitable for large-scale products. Therefore it may be applied only to collect solar energy with a small light transmission area. (3) Fresnel lens: in fact, this is a flat lens, as shown in Fig. 2.10. The disadvantage of an abnormally heavy lens is overcome. To strengthen this advantage, the Fresnel lens is normally made of light transparent plastic whose light transmittance is low. The shape precision for Fresnel lens with a large caliber

Figure 2.7
Parabolic dish solar concentrator.

Figure 2.8
Parabolic trough concentrator.

is hard to be guaranteed. (4) Others include the folding surface concentrator, conical reflector, spherical reflector, bar reflector, bucket trough reflector, and other solar concentration systems with combined devices.

Figure 2.9
Parabolic dish combination solar concentrator.

The rotating parabolic concentrator is the earliest reflective concentrator used to concentrate sunlight, and a dual-axis sun tracking is required. Ideally it can focus parallel light perfectly, although in fact, owing to the curvature errors of the paraboloid and nonparallel sunlight, a focal spot would be formed after the reflection. The receiver is placed on the focus point of the concentrator. The concentration ratio of this type of concentrator ranges from 100 to 1500, and the temperature ranges from 500 to 3000°C.

The parabolic trough concentrator is currently one of the most mature and widely applied solar concentration technologies, with the best commercial future. It can be conveniently assembled in series or parallel to achieve the requested thermal collector area. Only single-axis sun tracking is required. The concentration ratio is 15–70 and the temperature

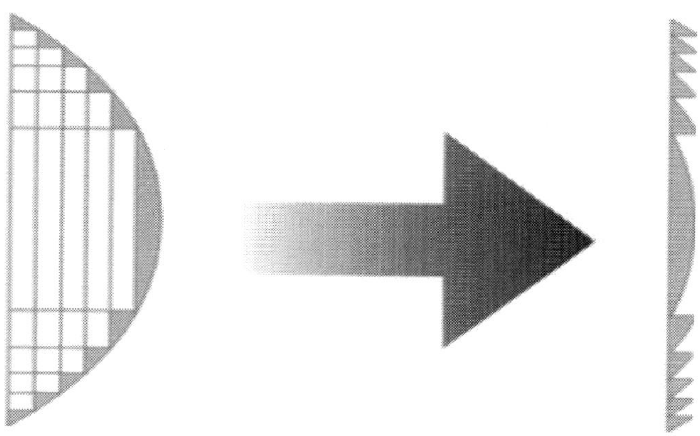

Figure 2.10
Fresnel lens.

Figure 2.11
Parabolic trough solar power plant by LUZ.

is 60−450°C. In 1985−91, the American and Israeli United Corporation (called LUZ) built nine mega solar power plants (SEGSI to SEGSIX) in the Mojave Desert in southern California, which are typical representations of parabolic trough concentrators, as shown in Fig. 2.11 [8]. The main characteristics of a parabolic trough concentrator are that it is more easily manufactured than the rotating parabolic concentrator with a low initial cost, and only one-dimensional sunlight tracking is necessary. A linear receiver is required because the sunlight will be concentrated into one line.

Another reflective solar concentration system using several heliostats is called a solar tower concentrator. The heliostats consist of an array of flat or micro-curved reflectors. From the microscopic view (the view lower than each heliostat), it is nonconcentrated, whereas from the large-scale view (the view of solar farm), it is concentrated because all

Figure 2.12
Tower type solar power plant with a heliostat concentrator.

of the sunlight reflected by the heliostats is focused on the receiver on the top of the tower, as shown in Fig. 2.12. Each heliostat is duel-axis tracking, and because the position of each heliostat is relatively different from the central tower, their tracking systems are asynchronized. Therefore, each heliostat needs to have an independent sun-tracking system, which makes the control system relatively complicated. The concentration ratio and temperature of solar tower concentrator are similar to or even higher than the dish solar parabolic concentrator. With the development of the heliostat and its control technologies, and the enhanced reduced cost of the whole system, the commercial future of the solar tower concentrator is promising [9].

There is another array Fresnel lens concentration system that is similar to the heliostats. The single reflector is nonconcentrative flat or microcurved, but when the Fresnel lens are aligned in parallel and sun is tracked using a single-tracking system, sunlight may be concentrated and reflected onto the receiver. The receiver is linear, which is similar to the trough system [10] (Fig. 2.13).

The main characteristic of a nonconcentration system is that it does not need a sun-tracking system; therefore it is normally stationary or needs to be adjusted only seasonally. It can receive not only direct beams but also diffuse ones. Its main types include: (1) a compound parabolic concentrator (CPC), which has a low concentration ratio (normally around 5), and a temperature range from 60 to 150°C; (2) a flat-plate thermal collector; (3) a vacuum tube thermal collector; (4) a solar pool; and (5) another collector called a luminescent solar concentrator (LSC), which is a transparent plate (e.g., acrylic) pasted with luminescent material. The luminescent materials can absorb incoming sunlight with a wide frequency range and reemit the energy in the form of light in a narrow frequency range. The light can be totally reflected and guided to the boundary of the plate owing to the difference between the plate and the surrounding medium. Its concentration ratio is

Figure 2.13
Solar power plant with linear Fresnel reflectors.

determined by the ratio of the flat-plate area and the boundary area, which is easily 10:100. This flat plate can absorb incident light from all directions and can also absorb diffuse light without a solar-tracking system.

Research on solar concentrators is mainly focused on methods to improve the efficiency of the solar concentrator, such as improving the manufacturing precision of the reflective surface, developing materials with high reflectivity. Correspondingly, reducing the cost for manufacturing is also a research focus.

2.3 Flat-Plate Solar Collector

The flat-plate solar collector is the most widely used solar use device. This kind of collector can provide only low-temperature heat because a concentration mirror is not applied. A typical flat-plate solar collector is composed of a transparent cover, heat absorber plate, and heat-insulated backing; its cross-sectional structure is illustrated in Fig. 2.14

A flat-plate solar collector has following characteristics, compared with a collector with a solar concentrator:

1. It has no tracking system and a simple structure, it is reliable, and it has a low cost;
2. Direct and diffused solar radiation are both used, whereas the solar concentrator can use only direct radiation;
3. For low heat flux, it has a low-temperature working fluid and safe operation. The temperature normally ranges between 50 and 70°C, and reaching a maximum of 90°C.

2.3.1 Structure of Flat-Plate Solar Collector

The working process of a flat-plate solar collector is that sunlight penetrates the transparent cover and strikes the absorber plate painted with a special coating. The absorber is heated by absorbing solar energy. Meanwhile, the absorbed heat is transferred to the heat-transport fluid within the thermal collector. Useful heat output is generated when the temperature of heat-transport fluid is increased; on the other hand, the heat-transport

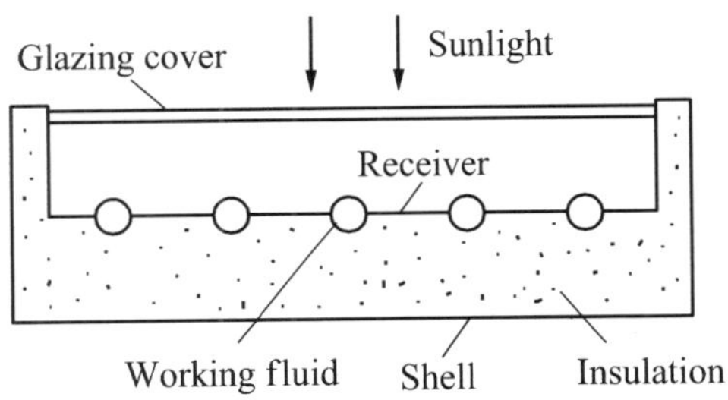

Figure 2.14
Schematic diagram of the structure of a flat-plate solar collector.

fluid can also dissipate heat to the surroundings. As a result, the greenhouse effect is formed and more heat is transferred to the heat-transport fluid, and thus the efficiency of the solar collector is enhanced. Main components of the flat-plate solar collector are discussed next.

2.3.1.1 Transparent Cover

The purposes of a transparent cover are to make solar radiation pass through and to protect from heat dissipation. Although air between the heat absorber and transparent cover is a good insulator, heat transfer between the absorber and the transparent cover could still happen owing to natural convection and heat radiation. Afterward, heat will be lost from the transparent cover to the surroundings. The transparent cover may be single, double, or multiple layers. Normally, double or multiple layers would be used only when the temperature difference between the absorber and ambient is large.

1. Function of Transparent Cover
   There is heat dissipation into the environment when received solar radiation is converted into heat by the heat absorber, which makes it impossible to convert all solar radiation into effective heat energy. This is especially obvious for a flat-plate collector, whose heat absorption area equals the light transmission area as well as the heat dissipation area. To reduce energy loss in this part, it is necessary to place a cover above the heat absorber; its requirement is to transmit visible light but block far-infrared rays, which ensures that incoming solar radiation energy is larger than heat dissipation energy, so that the rise in temperature rise on the heat absorber can be improved. This effect is called the greenhouse gas effect.
2. Basic Requirements for the Cover of a Thermal Collector
   a. High full light transmittance: The cover should has high full light transmittance for the sunlight spectrum; thus the stronger the light is on the heat absorber, the larger the received heat is and the higher the collector efficiency is.
   b. High impact strength: The covering should have high impact strength so that the external forces will not damage the system.
   c. Good weather fastness: The collector is usually installed outdoors and facing the sun. It will be eroded by cold, heat, light, wind, rain, and snow in the long term. If the covering has poor weather fastness, its light transmittance and strength may drop dramatically and seriously influence the collector and even the whole system efficiency. Moreover, the lifetime of the system may be largely shortened.
   d. Low thermal conductivity: When the collector is functioning, its water temperature within the absorber tube is 80–90°C, whereas the air temperature between the heat absorber and the covering is 50–70°C, and there must be some heat dissipation through the covering. The amount of heat dissipation is directly proportional to the thermal conductivity of the covering: The lower the conductivity is, the better the

Table 2.3: Comparison of comprehensive performance for three covering materials [11].

Material	HSG	MMA	FRP
Light transmittance	High	High	High
Impact strength	Medium	Poor	High
Heat resistance	Good	Poor	Fine
Thermal insulation	Fine	Good	Good
Antiaging	Good	Fine	Fine
Specific weight	High	Poor	Medium
Machinability	Poor	Fine	Good
Cost	High	High	Medium

thermal insulation is, and the less heat loss there is, the higher the collector efficiency is. In a real application, a dual-layer or even multiple-layer covering is used.

e. Good machinability: The covering needs to have good tailing capacity and one-way bending ability, which are convenient for the manufacturer to process the product.

In our country, normal flat glass is usually used as a covering whose transmittance is low because of the high iron content. The transmittance is below 80% with a thickness of 3 mm, which is much lower than that of foreign glass (91%). In addition, because it has low strength, it is easily broken, leading to inconvenience for collector manufacturers and users. Stalinite glass and Solar-E solar plate is used to replace conventional glass material. The innovative Solar-E plate is composed of high-energy resin and high-performance fiber, and its surface is covered with a special protection layer. This plate is heat resistant, antiaging, and highly transmittant. Owing to its advantages of light weight, high strength, corrosion resistance, and high light transmittance, it is widely used in China. Main materials for this innovative covering are high strength heat-resistant glass (HSG), methyl methacrylate plate (MMA), and glass fiber reinforced plastics (FRP). The comprehensive performance is shown in Table 2.3.

2.3.1.2 Heat Absorber

The absorber area for a flat-plate collector is as same as for the transparent covering; the heat absorber should be designed to absorb as much solar radiation as possible, and as rapidly as possible, to deliver the received solar radiation to the working fluid. At the same time, the heat dissipation to the ambient needs to be minimized.

The absorbers are all painted with a matt black coating to increase solar radiation absorptivity. A good coating has good absorptivity for incident light with all incidence angles; normally absorptivity can be above 95%.

To deliver the absorbed heat rapidly to working fluid, the absorber plate and the tubes containing the working fluid are all made of a high-conductivity metal material. Heat resistance between the heat absorber and the working fluid should be small enough so that heat loss will increase owing to the large temperature difference between the heat absorber and the working fluid. The material for heat absorber is basically steel and stainless steel.

The heat absorber has the following structures:

1. Tube-plate type: There are various structural forms to a tube-plate type collector, including the tube-plate trough, fin, and tube-plate colligation. The cross-sectional structure of various heat absorbers is shown in Fig. 2.15. There are various methods of combining the tube and plate. The old technology is relatively backward; for example, the tube and heat absorber are welded onto the connection area along the direction of the tube length, or they fill the gap with heat-conductive silica gel or tightly connect the tube and plate using crafts such as copper brazing, electric resistance welding, ultrasonic welding, and high-frequency welding. To reduce contact heat resistance between tube and plate, Beijing Solar Research Institute imported a Canadian production

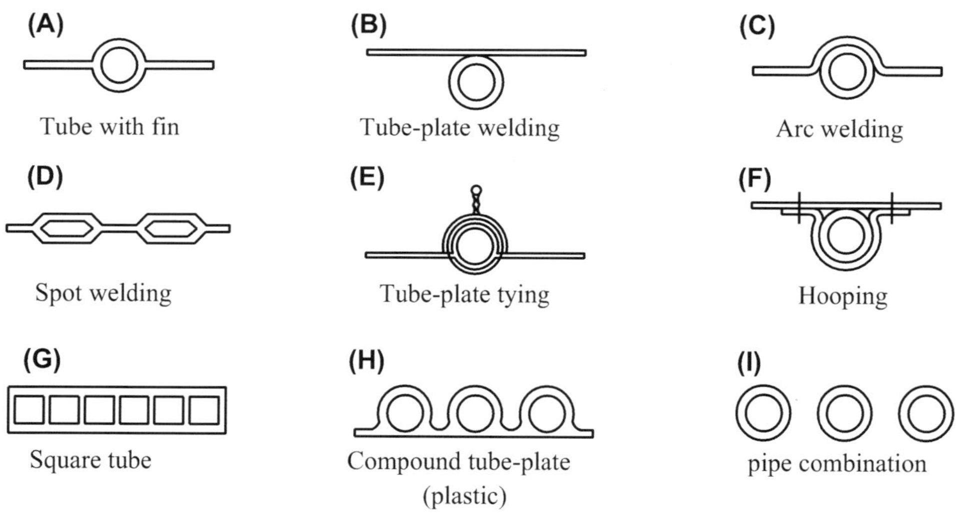

Figure 2.15
Cross-structure of the receiver.

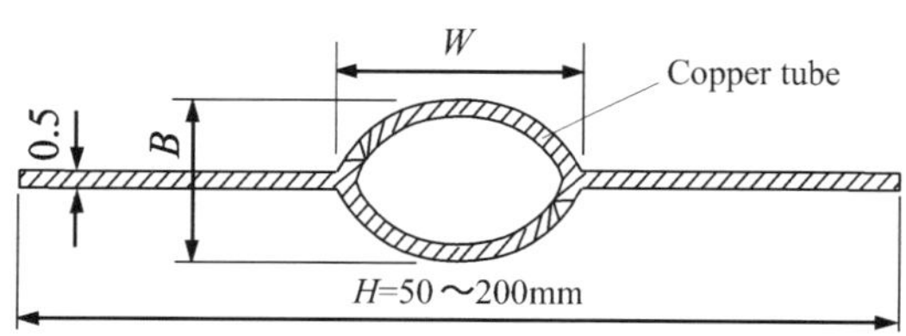

Figure 2.16
Schematic diagram of copper–aluminum composite receiver.

line to produce an international-level solar copper–aluminum compound heat absorber tape (the cross sectional structure is shown in Fig. 2.16), which largely improved the solar collector quality. This method is to roll one copper tube between two aluminum sheets and blow it within a high-pressure gas device. Features of this type of material are: (1) high thermal efficiency: rolling of copper and aluminum could reduce contact electric resistance; (2) pure water: because the water is in contact with copper, copper will not rust; (3) saving of copper material, low cost, mechanical and large-scale production, and quality assurance; (4) since the absorber is blown by high-pressure gas, it has good pressure resistance. On the basis of the copper–aluminium compound heat absorber, an all-new copper heat absorber that could further improve the reliability of the absorber seal has been developed.

2. Flat-box type: After the metal plate is molded to appropriate flow tubes, the flow tubes are welded together with upper and bottom plates and the connection between flow tubes is welded by point or rolling welding, whereas the surrounding heat absorber uses rolling welding. Materials are antirust aluminum alloy, galvanized plate, stainless-steel plate, and so forth. A feature of this heat absorber is that it is a combination of tube and heat absorber, which could increase collector efficiency. Disadvantages are that it has a special requirement for welding technology: there is a limitation in pressure bearing; there is a relatively high cost; when galvanized plate is used, the welding point might erode or begin punching; apart from these issues, the cross-sectional area of the tube is large, as is the water capacity, but the thermodynamics performance of the collector is reduced.
3. Box-tube type: The abrasive tool is used to squeeze, stretch, and mold the tube so that the fin or heat absorber bar is placed on its side; then the upper and lower collector tube is welded into the heat absorber. Its feature is good pressure bearing because an antirust aluminum alloy is used. A disadvantage is that it is easy to erode; thus the water quality is a special requirement. Because of the limitation of the craft technology, the wall thickness of the box-tube type is larger than that of the tube-plate type, and material consumption is also large. The high thermal capacity of the collector could lead to a large water capacity and reduced thermodynamic performance.

2.3.1.3 Selective Coating on Heat Absorber

The wavelength of solar radiation reaching the earth is in the range of 0.3–2.5 μm, whereas heat radiation generated by collector absorbers mainly focuses on the wavelength range of 2–20 μm. Therefore, using a selective coating with special solar radiation features in different ranges of wavelength could increase the absorption of solar radiation and at the same time reduce heat radiation loss on the absorber. The surface reflectivity of a metal absorber is high and its absorbability is low, only 20–40%. To increase the absorption effect, the metal surface is coated with special paint. The types of coating are selective and nonselective. The selective coating has high absorbability α for short-wavelength solar radiation but low long-wavelength emissivity ε_T, which could be expressed by: $\alpha/\varepsilon_T > 1$. If $\alpha/\varepsilon_T = 1$. It is called neutral or nonselective coating. The nonselective coating may be used on a heat absorber when the operating temperature is between 40 and 60°C. It has high absorptivity $\alpha = 0.92-0.96$, but the emissivity is also large, which may increase heat loss. The coating can be thin and even.

The selective absorption coating has various material and painting technologies. It is normally easy to improve the absorptivity of selective coating α, but when the absorptivity increases, the emissivity ε_T may also increase. Therefore, the purpose of various material technologies is to improve absorptivity with low emissivity. In the late 20th century, the selective absorption coating used for solar collectors in our country had solar absorptivity $\alpha = 0.90-0.94$ and emissivity $\varepsilon_T = 0.29-0.50$. Collector efficiency improved by 3% compared with black painting. To reduce emissivity further, the electrochemical method is used to fabricate selective coating whose absorptivity $\alpha = 0.94$, and emissivity $\varepsilon_T = 0.13-0.15$. Currently some manufacturers in our country have used magnetron sputtering to fabricate a selective coating with an absorptivity of some coating $\alpha = 0.96$, and emissivity $\varepsilon_T = 0.08$. Thus the product thermal performance has been largely improved [12].

2.3.1.4 Thermal Insulation Layer and Its Shell

The function of the insulation layer is to reduce heat dissipation to the surroundings so that the thermal efficiency of the heat collector can be improved. Requirements for the insulation layer material are good heat insulation, i.e., small heat conductivity, low deformability, no volatility, no toxic gas, and no water absorption. In this way, a thinner insulation layer can achieve better insulation effects. The thickness of insulation layer on the bottom is 3–5 cm whereas that of the surrounding insulation layer is half of the bottom. The minimum thickness of the insulation layer on the bottom of the thermal collector can be calculated using Eq. (2.5):

$$\delta \geq \frac{\lambda_{100}}{1.45}(\mathrm{m}) \tag{2.5}$$

where λ_{100} is the heat conductivity of the insulation material at 100°C, W/m K.

Commonly used insulation materials include rock wool, mineral wool, polystyrene, and polyurethane. Because polystyrene shrinks at higher temperatures (the operation temperature should be no more than 70°C), a thin polyester film coated with aluminum can be inserted between itself and the heat absorber so that it can work under a low temperature, but there will still be some shrinkage after long use; thus more attention is necessary during operation. Mineral wool is mostly used in products currently; it is much better when it is foamed by polyurethane.

The shell of the thermal collector, i.e., back baseboard and sideboard, is not only for appearance but also to combine the heat absorber, covering, and thermal insulation material, which is convenient to install. Currently used shell materials include steel, color sheet steel, embossed aluminum sheet, aluminum sheet, stainless steel, plastic, and fiberglass; cement can also be used.

2.3.2 Performance Parameters of Flat-Plate Solar Collector

2.3.2.1 Fundamental Energy Conservation Equation for Flat-Plate Solar Collector

The flat-plate solar collector is a heat exchanger that converts solar radiation energy into heat energy and is energy conservative, as shown in Fig. 2.17:

1. Rate of solar radiation received by a solar collector (Eq. (2.6)):

$$Q_A = A_C H(\tau\alpha) \tag{2.6}$$

2. Rate of heat loss from a solar collector (Eq. (2.7)):

$$Q_L = A_C U_L(t_p - t_a) \tag{2.7}$$

3. Rate of useful output from a solar collector (Eq. (2.8)):

$$Q_u = A_C M C_p\left(t_{f,o} - t_{f,i}\right) \tag{2.8}$$

4. Rate of heat capacity gradient within a solar collector (Eq. (2.9)):

$$Q_S = C_C(dt/d\tau) \tag{2.9}$$

In Eqs. (2.6)–(2.9), A_C is the area of the solar collector, m^2; H is solar radiation, W/m^2; $(\tau\alpha)$ is the product of transmittance and the covering and absorptivity of the heat absorber, %; U_L is the collector overall thermal loss coefficient, $W/m^2 \cdot °C$; t_p and t_a are the temperature of the heat absorber plate and ambient temperature, °C; $(t_{f,o} - t_{f,i})$ is the inlet and outlet temperature difference of the working fluid, °C; M is the mass flow rate of the working fluid per unit area, kg/m^2 h; C_p is the constant pressure-specific heat of the

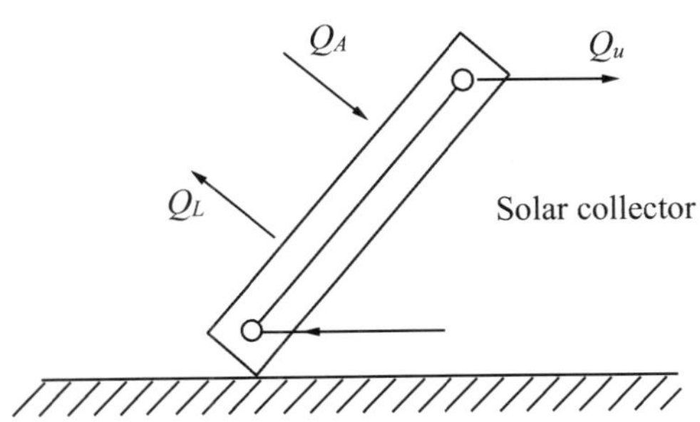

Figure 2.17
Schematic of energy balance of flat-plate solar collector.

working fluid, kJ/kg °C; C_C is the constant volume-specific heat of the working fluid, kJ/kg °C; τ is time, s; and t is temperature, °C.

Under stable working conditions, when $dt/d\tau = 0$, $Q_S = 0$, and no heat absorption or heat dissipation happens in any part of the solar collector. Under unstable working conditions, when $dt/d\tau > 0$, and continuous heat absorption happens in every part of the solar collector, then $Q_S > 0$; when $dt/d\tau < 0$, and continuous heat dissipation happens in every part of the solar collector, then $Q_S < 0$.

2.3.2.2 Instant Efficiency Equation and Efficiency Curve

When sunlight perpendicularly irradiates onto the solar collector surface and the solar collector is under stable conditions, the instant efficiency equation can be deduced from the energy conservation equation under a steady state (Eq. (2.10)):

$$\begin{aligned} \eta &= \text{useful energy}/\text{total solar radiation} \\ &= Q_u/(A_C H) = (Q_A - Q_L)/(A_C H) = (\tau\alpha) - U_L(t_p - t_a)/H \end{aligned} \tag{2.10}$$

Because it is hard to measure the heat absorber temperature but easy to measure the inlet and outlet water temperatures, the average water temperature of solar collector t_{fv} is used to express the instant equation (Eq. (2.11)):

$$\begin{aligned} \eta &= F'[(\tau\alpha) - U_L(t_f - t_a)/H] \\ t_f &= \left(t_{f,i} - t_{f,o}\right)/2 \end{aligned} \tag{2.11}$$

where F' is the efficiency factor of the solar collector, %.

In the measurement of instant efficiency of the solar collector, the intercept point on the transverse is an important feature point; the accuracy of this measurement point will influence the accuracy of the solar collector curve. This point corresponds to the maximum temperature that the water within the solar collector can reach; it is also called the restrictive temperature or stagnation temperature. The point on the vertical axis is another important feature point that corresponds to the maximum efficiency of the solar collector, in which the average water temperature equals the ambient temperature. It is easy to measure the inlet and outlet water temperatures (and average temperature), but it is hard to measure it accurately because under the constant inlet water temperature, the outlet water temperature varies with the amount of solar irradiance. It is much harder to adjust the inlet water temperature so that the average water temperature can equal the ambient temperature. Therefore, there is another instant equation that is expressed by the inlet water temperature. With this equation, instant efficiency can be measured just by equalizing the inlet water temperature and ambient temperature (Eq. (2.12)):

$$\eta = F_R\left[(\tau\alpha) - U_L\left(t_{f,i} - t_a\right)/H\right] \tag{2.12}$$

where F_R is the heat transfer factor, %.

These three methods use different temperatures to describe the instant efficiency equation from the view of measurement, which could also be described from the view of energy conservation: total solar radiation minus the heat dissipation of the solar collector is useful energy. The heat dissipation of the solar collector results from the solar collector temperature that is higher than the ambient temperature, whereas temperatures in difference parts of the solar collector are different; usefulness calculated from different heat dissipations varies. The instant efficiency equation using heat dissipation based on different temperatures (assuming the collector overall thermal loss coefficient is a constant) [13] may be:

1. Instant efficiency equation based on heat absorber temperature t_p (Eqs. (2.13) and (2.14)):

$$Q_u = A_C(\tau\alpha)H - A_C U_L(t_p - t_a) \tag{2.13}$$

$$\eta = Q_u A_C H = (\tau\alpha) - U_L(t_p - t_a)/H \tag{2.14}$$

2. Instant efficiency equation based on average water temperature t_f (Eqs. (2.15) and (2.16)):

$$Q_u = F' A_C[(\tau\alpha)H - U_L(t_f - t_a)] \tag{2.15}$$

$$\eta = F'[(\tau\alpha) - U_L(t_f - t_a)/H] \tag{2.16}$$

3. Instant efficiency equation based on inlet water temperature $t_{f,i}$ (Eqs. (2.17) and (2.18)):

$$Q_u = F_R A_C\left[(\tau\alpha)H - U_L\left(t_{f,i} - t_a\right)\right] \tag{2.17}$$

$$\eta = F_R\left[(\tau\alpha) - U_L\left(t_{f,i} - t_a\right)/H\right] \tag{2.18}$$

4. Instant efficiency equation based on useful energy taken away by water (Eqs. (2.19) and (2.20)):

$$Q_u = MC_P\left(t_{f,o} - t_{f,i}\right) \tag{2.19}$$

$$\eta = MC_P\left(t_{f,o} - t_{f,i}\right)/A_C H \tag{2.20}$$

These four methods can be expressed on orthogonal coordinates. If U_L is assumed to be constant, they are all linear, as shown in Fig. 2.18.

2.3.3 Heat Loss Coefficient of Flat-Plate Solar Collector

There is unavoidable heat dissipation to the surroundings once the heat absorber temperature is higher than the ambient temperature. The more heat loss there is in

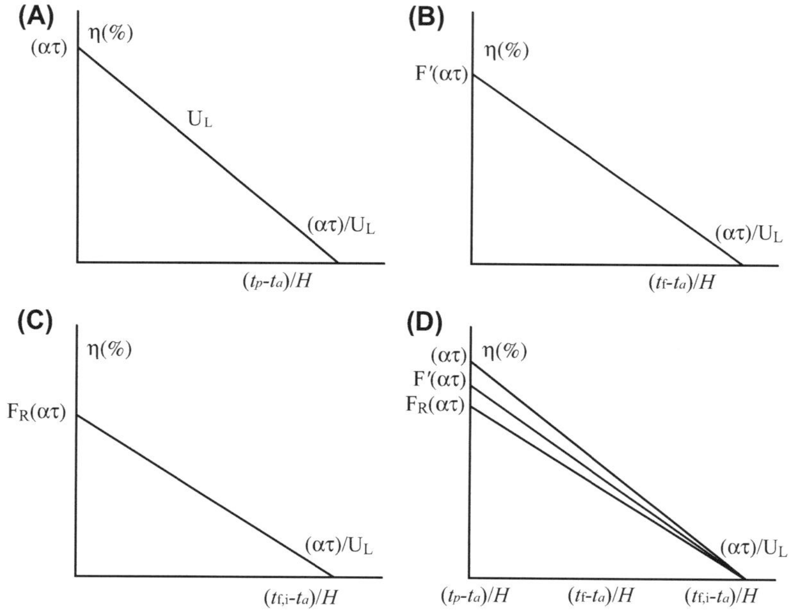

Figure 2.18
Methods to express the efficiency curve of the flat-plate solar collector: (A) Based on heat absorber temperature; (B) based on the average water temperature; (C) based on the inlet water temperature; and (D) based on useful energy taken away by water.

this area, the less energy can be used, which will directly influence the solar collector efficiency.

The area of highest temperature in the solar collector is the heat absorber, t_p, and it is where various heat loss happens. The total heat loss coefficient U_L is the sum of the heat loss coefficient of the top, U_t, edge, U_e, and bottom, U_b, as shown in Fig. 2.19.

The corresponding heat loss in three parts is (Eqs. (2.21)–(2.23)):

$$Q_t = A_t U_t (t_p - t_a) \tag{2.21}$$

$$Q_e = A_e U_e (t_p - t_a) \tag{2.22}$$

$$Q_b = A_b U_b (t_p - t_a) \tag{2.23}$$

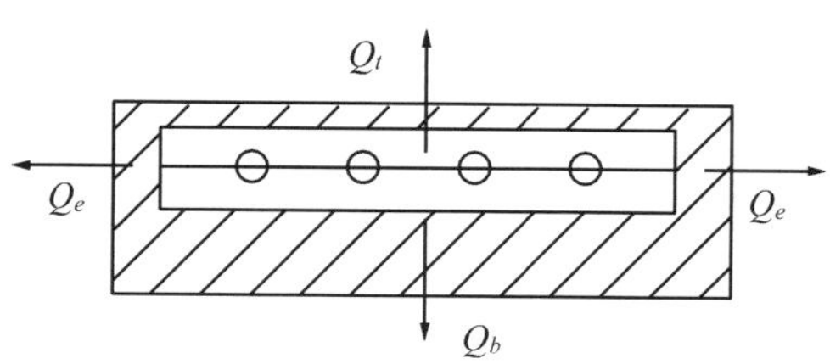

Figure 2.19
Heat loss diagram of a flat-plate solar collector.

where A_t, A_e, and A_b are the top, edge, and bottom, respectively, of the solar collector; t_p and t_a are the heat absorber and ambient temperature, respectively; and Q_t, Q_e, and Q_b are the heat loss on the top, edge, and bottom of the solar collector.

The heat loss on the top is the main part of heat loss in the solar collector, especially for the solar collector with good insulation on the bottom. Heat loss on the top differs from that the edge and bottom. Because the top part has a window for harvesting sunlight and dissipates heat, heat loss on the top is mainly through convection and radiation, which is much more complex than conductive heat transfer on the edge and bottom of the solar collector. The amount of heat loss is also much higher. It is a complicated process to calculate the heat loss coefficient on the top, where iterations are required. For the purpose of simplification, Klein concluded an experiential equation, i.e., the famous Klein equation (Eq. (2.24)):

$$U_t = \left\{ \frac{N}{(344/T_p)[(T_p - T_a)/(N+f)]^{0.31}} + \frac{1}{h_w} \right\}^{-1} + \frac{\sigma(T_p - T_a)\left(T_p^2 + T_a^2\right)}{[\varepsilon_p + 0.043N(1-\varepsilon_p)]^{-1} + (2N + f - 1)/\varepsilon_g - N} \tag{2.24}$$

where $f = \left(1 - 0.04h_w + 0.0005h_w^2\right)(1 + 0.091N)$; $h_w = 5.7 + 3.8v$; N is the number of glass layers; h_w is the convective heat transfer coefficient between wind and the glass plate, kJ/m^2 h °C; ε_g is the emissivity of glass, %; ε_p is the emissivity of the heat absorber, %; T_p is the temperature of the heat absorber, K; t_a is the ambient temperature, K; and v is the wind velocity, m/s.

For a heat absorber with a temperature of 40–130°C, Klein's (1974) experiential equation is close to the theoretical one; the error is within ±0.20 W/m^2 °C [14].

2.3.4 Factors Influencing the Efficiency of a Flat-Plate Solar Collector

The structural material, climate, and operation conditions are main factors influencing the efficiency of a flat-plate solar collector.

2.3.4.1 Structural Material of Solar Collector

1. Material for the covering should be high in light transmittance and thin, and it should have fewer layers so that its overall transmittance can be high. Certainly the heat loss coefficient can be high with fewer layers; both transmittance and the heat loss coefficient can influence the solar collector efficiency.

2. When the absorptivity α of the coating on the heat absorber is increased, $\tau\alpha$ can increase as well; when the emissivity ε of the coating on the heat absorber is decreased, the heat loss coefficient U_L will decrease.
3. The conductivity of the heat-absorber material λ needs to be high. Besides high fine efficiency F, fine bonding between the tube and the absorber are required. Under these conditions, the coefficient factor F' will increase.
4. When the conductivity of the insulation material is low, heat loss on the edge and bottom may be low; thus good insulation material is required. The thickness needs to be appropriate, not too thick.
5. Good tightness on the shell can prevent gas leakage and rainwater from entering. Normally, a solar collector with good tightness can largely decrease heat loss. A shell with poor tightness will diffuse warm air within the collector and take the useful heat, in which way heat loss will increase; sometimes, when rainwater come into the solar collector the thermal efficiency of the solar collector will become low.

2.3.4.2 Climate

With constant solar irradiation, higher ambient temperature would increases collector efficiency; thus efficiency in summer is higher than in the spring, autumn, and winter.

When the ambient is constant, higher solar irradiation can result in high collector efficiency. Thus efficiency on a sunny day with good sunlight is higher than that on a cloudy and poorly sunlit day.

When both solar irradiation and ambient temperature are constant, higher wind speed may result in lower collector efficiency. Thus less windy days are good for the solar collector efficiency.

2.3.4.3 Operation Conditions

When the flow rate is constant, increasing the inlet flow temperature will reduce the solar collector efficiency. This is the result of increased average flow temperatures such that it enhances heat loss and therefore reduces efficiency.

When the inlet flow temperature is constant while the flow rate is reduced, efficiency also decreases. This is because a reduced flow rate increases the flow temperature increment such that the heat loss will increase.

2.3.4.4 Balanced Temperature of the Flat-Plate Solar Collector

When the solar collector efficiency is 0, the flow temperature within the collector is called the balanced temperature, which is the point where the x-axis and efficiency curve meet. It represents the maximum temperature the collector can reach under certain weather conditions and is called the limit temperature of the collector.

2.3.5 Choice of the Flat-Plate Solar Collector

The choice of a flat-plate solar collector is crucial to improving its performance and reducing its cost. Normally it is chosen based on the application and operation conditions. If the required output temperature is lower than 60°C and the ambient temperature is high, most flat-plate solar collectors can meet this requirement. However the efficiency of the collector will be significantly reduced when the ambient temperature is lower than 7°C and the required output temperature is higher than 45°C.

When choosing a solar collector, make the most use of the characteristics of various collectors so that the system efficiency will be optimal, one solar thermal system can choose two different collectors: for example, the front of the system uses a flat-plate solar collector that works under low-temperature conditions while the back uses an evacuated tube collector or heat pipe vacuum tube that works under high-temperature conditions. Apart from that, building a fine solar water heating system or solar collector system needs to consider the issues described in the following sections.

2.3.5.1 Choice of Flow Direction

This main function of the solar collector is to convert solar energy into heat, and immediately to transfer the heat to the heat transfer fluid. To avoid overheating some parts of the heat transfer fluid, the working fluid needs to take away heat with no delay. Based on the method for removing heat by the working fluid, the solar collector can be classified in two ways: natural circulation and forced circulation. Natural circulation is forced by the density difference between hot and cold fluid. Cold fluid flows downward whereas hot fluid flows upward, which forms circulation. On the other hand, forced circulation is initiated by an external force such as an electrical pump. There are merits and faults to both types of circulation, and selection is based on the real application that is required. The following four aspects need attention:

1. The choice of the system needs to consider the user requirement (water temperature or water use period), local climate, building condition, and cost.
2. When designing the natural circulation system, it is not necessary to increase solely the head difference between the water tank and collector, or to use an overly large tube to increase circulation flux. These actions normally cannot increase system efficiency. On the contrast, it may increase the heat dissipation area and heat capacity so that the system efficiency will decrease. The head difference between the water tank and collector needs to be considered in combination with system resistance to ensure that the water in the tank flows through the collector only once or twice a day. To prevent the night heat dissipation caused by hot water in a storage tank flowing in reverse to the collector, the head difference between the bottom of the water tank and the top of the solar collector needs to be 0.3–0.5 m.

3. The advantage of a hot water system with discharging constant-temperature water is that it can obtain hot water with a required temperature. For this kind of system, once the temperature of the user-end hot water is constant, daily efficiency will not change much.
4. When designing a forced circulation system, under the conditions of ensuring the water temperature required, a pump needs to be selective with appropriate efficiency. In this way, both financial and electric energy cost can be saved.

2.3.5.2 Determining the Area of a Solar Collector

After determining the form of the solar collector, both technical and economic aspects need to be comprehensively considered when choosing the flat-plate or evacuated tube collector. After that, the curve equation of efficiency of the selected collector, the capacity of the hot water tank, the initial and end-use water temperatures, and local climate data are used to calculate the solar collector area in different seasons. Normally the climate data in spring or autumn are referred to when determining the collector area.

2.3.5.3 Composition Form of Solar Collector and Flow Distribution Relations

All large-scale solar collector systems are formed by several simple thermal collection components. The array of these components is especially important.

In a single solar collector (or group of them), the flow rates within upward pipes are evenly distributed, but when several collectors' components form a large-scale system, the flow distribution within the pipes are uneven. In such a way, the overall performance of the system will be reduced. This is because the temperature of the heat absorber on the sides of pipe with a low flow rate will be higher than that with a high flow rate. Because more heat loss takes place in a high temperature area, the overall thermal performance of the whole collector will be reduced. To achieve better performance of the collector, the design of an upper and lower interconnection box for a single collector or the water collection piping for a group of collectors is vitally important, especially for a large-scale forced circulation system.

The calculation of flow distribution is based on conventional hydraulics. Assuming the collector is the classic tube-plate structure with a constant distance between pipes, a uniform diameter for the upper and lower interconnection box, and isothermal flow within the pipe, the flow rate within the risers of the collector is determined by the inlet and outlet static pressure difference. For a detailed calculation, please refer to relevant professional books.

Normally, no more than 12 single collectors are allowed to connect in parallel, and maximum of 24. When more collectors are used, a parallel—series connection needs to be considered. Moreover, in every level of parallel connection, no more than 6—10 single collectors are allowed to be used, and then three to five levels are connected in series. This number can be increased for a large-scale system. To avoid excessive connections in parallel and series, collectors can be arranged in several complete separate arrays with the same water source and hot water destination. If all collectors are connected in series, it is not beneficial in a real application because the uneven flow distribution could largely reduce efficiency and the series connection array is sensitive to temporary fluctuations of solar radiation, air temperature, wind speed, and so forth.

2.3.5.4 Thermal Insulation of the System

In a solar thermal system, the thermal insulation of circulation pipes and water tank is crucial for its influence on system efficiency. Especially for hot water systems in high-latitude areas, good or bad system thermal insulation will directly affect the thermal performance of the device. Normally, it is better for the thickness of the insulation layer outside the water reservoir to be 20–50 cm; consideration of the thermal insulation of the pipe should include the critical radius of natural convection heat dissipation between the thermal insulation material and ambient temperature (Eq. (2.25)):

$$r_e = \frac{\lambda}{h_0} \tag{2.25}$$

where λ is the thermal conductivity of the thermal insulation material; h_0 is the coefficient of convective heat transfer between the outer layer of pipe and the ambient temperature. If the outer radius of the pipe is less than the critical radius of the thermal insulation layer, increasing the thickness of the insulation layer will increase the heat loss, and vice versa.

2.3.5.5 Other Issues That Need to Be Considered

1. Tilt angle of collector: To receive maximum annual daily solar irradiation, the solar collector needs to face the equator; its tilt angle should be close to the local latitude, φ. If optimum daily solar irradiation is required for winter, the tilt angle should be increased to $\varphi + 10$ degrees; the tilt angle needs to switch to $\varphi - 10$ degrees if optimum daily solar irradiation is required for summer.
2. The circulation piping needs to be compact to the greatest extent; the upper pipes should avoid up-sweeping and then down-warping. Otherwise an air plug phenomenon may happen to the rising part and block water circulation.
3. An exhaust hole needs to be installed on top of the water reservoir so that evaporating air from the heating water can exhaust from the top.
4. The open position of the upper circulation pipe to the water reservoir should be considered with the water top-up method. If cold water is filled from the make-up water tank to the bottom of water reservoir, to use hot water fully above the opening of the circulation pipe, the opening position can be a little higher, which is one-fifth to one-sixth the water tank height. If there is no make-up water tank and the hot water

pipe on the bottom of the tank is used as the cold filling water pipe, the opening position can be a little lower, best at one-half to one-third the water tank height. In this way, the water level will not be lower than the opening position and the system will not stop circulating. No overheating will occur to the thermal collector.

2.3.5.6 Antifreezing Measures of the System

When the solar collector is used in high-latitude areas, especially when water is used as a heat transfer fluid, some measures need to be taken to prevent water from freezing at night, which will cause cracks on the heat absorber. Therefore, antifreezing measures need to be considered when designing and operating the collector. The evacuated tube collector does not need to consider antifreezing measures (the connecting pipe needs antifreezing):

1. During the freezing season, especially during nighttime, the collector needs to be evacuated so that the system can stop working. This is a passive measure that will reduce the year-round use ratio. However, extra investment is not necessary.
2. Install an auto-evacuation device in thermal collector. When the temperature of water within the heater is lower than 0°C, the device will turn on automatically to evacuate the water.
3. Put an antifreezing cover on the collector in winter night. This is suitable for a small-scale system.
4. For a forced circulation system, a so-called "sinking" system can be used; i.e., the water reservoir is put under the collector and the system is open system and in contact with the atmosphere. When the pump stops operating, the water within the collector can flow back into the water reservoir, as shown in Fig. 2.20.
5. The dual-loop system uses an antifreezing heat transfer medium, shown in Fig. 2.21. The solar collector and heat exchanger form the first loop, which uses the antifreezing heat transfer medium. The water reservoir itself forms the water supply system, which heats the water in the tank by effective heat from the heat exchanger. The operation method of the dual-loop system can be forced or natural circulation can be used. The

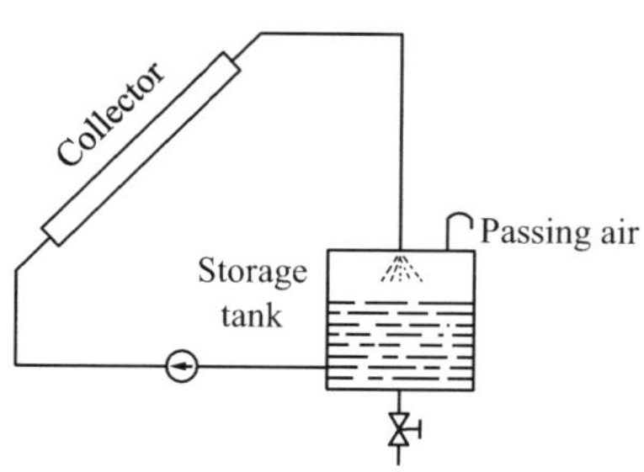

Figure 2.20
Falling water antifreeze solar system.

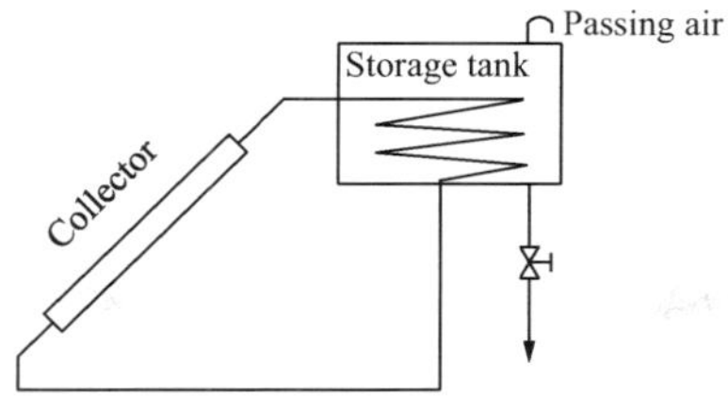

Figure 2.21
Compound circulation solar system with antifreeze medium.

latter method can be used for a small-scale system. This antifreezing hot water system is safe, reliable, and easy to use. However, because of the antifreezing heat transfer medium and heat exchanger, the system cost may increase; moreover, the daylight efficiency of the system is relatively low. Currently this type of system is widely used.

2.4 Vacuum Tube Solar Collector

To reduce heat loss in the heat absorber so that the solar collector would have higher thermal efficiency and provide high-temperature heat energy, the concept of a vacuum tube solar collector was developed at the beginning of the last century. Normally the vacuum tube collector consists of several evacuated collector tubes; each tube is formed of a transparent outer tube and endothermic inner tube. The latter is a metal tube, but glass is also applicable. The space between the inner and outer tubes is vacuumed to 10^{-3} to 10^{-4} mmHg evacuation. Therefore the cross-ventilation of air is completely restrained and heat loss can be reduced to near zero. The remaining heat loss is radiative heat transfer between the heat-absorbing tube and the outer tube. Low-emissivity coating is painted on the outer surface of heat-absorbing tube so that radiative heat transfer can be reduced. Under a similar ambient environment and operation temperature, the heat loss of the vacuum tube solar collector is one-eighth the flat-plate solar collector

Currently, there are two main types of vacuum tube solar collectors: all-glass and metal-glass. Each has its own advantages and disadvantages and needs to be selected based on different occasions.

2.4.1 All-Glass Vacuum Tube Solar Collector

The all-glass vacuum tube solar collector is formed by inserting several all-glass vacuum tubes into the header.

The all-glass vacuum tube solar collector is composed of inner and outer glass tubes. The outer surface of the inner tube has a selective absorber layer with high absorptivity and low emissivity. The interlayer between tubes is highly vacuumed; its shape is like a slim

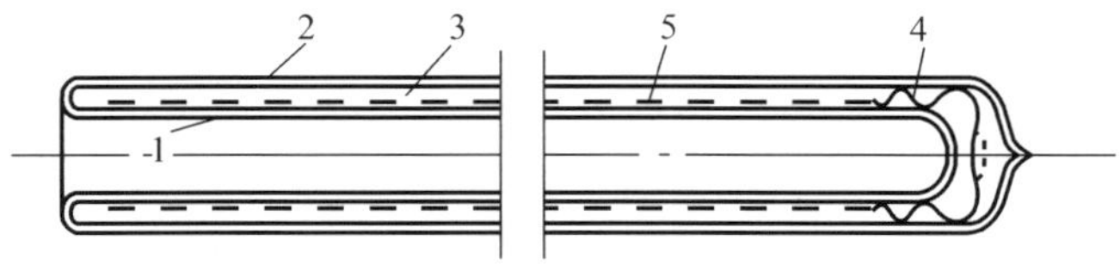

Figure 2.22
Whole-glass vacuum tube solar collector: 1, inner glass tube; 2, outer glass tube; 3, vacuum region; 4, support with getter; 5, selective absorbing surface.

Thermos core, as shown in Fig. 2.22. It uses a single-end opening. One end of the inner and outer tube opening is sealed annularly whereas the other end is a closed hemispherical round head. The inner and outer tubes are supported by a spring clinch that can be freely stretched to buffer the stress of the tubes caused by a contraction. A getter is stalled on the spring clinch that can absorb gases produced by the operating vacuum so that the degree of vacuum is maintained.

The quality of the all-glass vacuum tube solar collector is related to the selected glass material, vacuum performance, and absorption layer.

Glass with high light transmittance and heat capacity, low thermal expansivity, good thermal shock resistance, high mechanical strength, and good chemical resistance and processing suitability is preferred. Currently, borosilicate glass-3.3 is preferred as the material for a vacuum tube solar collector. The glass's thermal expansivity is $3.3 \times 10^{-6}/°C$; the iron oxide content is 0.5%; its heat resistance temperature difference is larger than 200°C. Besides, it has high strength, which meets basic requirements for manufacturing a vacuum tube solar collector [15].

The guarantee of the degree of vacuum is an important indicator of the quality and service life of a vacuum tube solar collector. According to national standards, the vacuum degree level of the collector tube should be no more than 5×10^{-2} Pa.

Special requirements for the selective absorber layer in the vacuum tube of a solar collector need to be considered: one is performance in a vacuum condition; another is heat resistance. When it is under operation, the degree of vacuum within the tube should not be influenced. The other performance indicator needs to be maintained. Normally, the manufacture of a selective absorber layer for the vacuum tube solar collector requires special devices. There are various processing technology routines, such as chemical conversion, electroplating, spray pyrolysis, heat tinting, vacuum evaporation, and magnetron sputtering. The technology used to process the selective absorber layer for a vacuum tube solar collector in our country is magnetron sputtering; the product quality and effects are all relatively ideal.

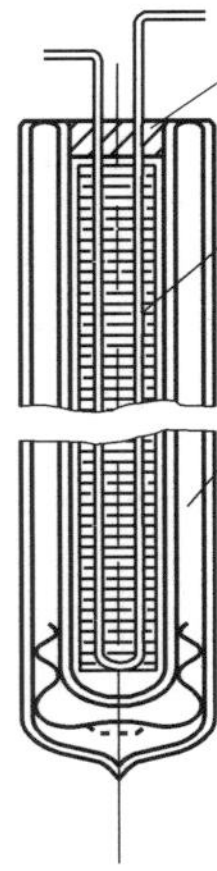

Figure 2.23
Whole-glass vacuum tube solar collector with a U loop receiver.

An all-glass vacuum tube solar collector has many advantages, but because there is water in the tubes, if one is broken the whole system will stop working. To solve this defect, two improvements were developed based on the original all-glass vacuum tube solar collector. One is to insert a heat pipe into the tube; another is to apply a U-shape tube heat absorber into the tube, as shown in Figs. 2.23 and 2.24. Therefore, water is not used and will not influence the operation of the system. In fact, these developments have been incorporated into the metal-glass vacuum tube solar collector.

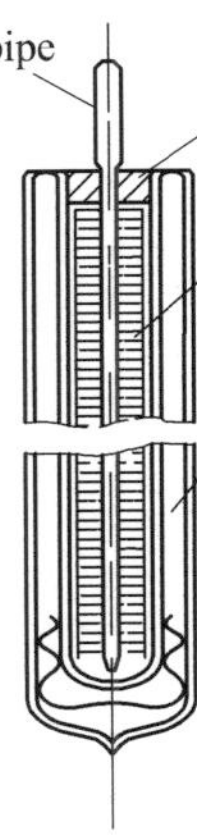

Figure 2.24
Heat pipe—type vacuum tube solar collector.

2.4.2 Metal-Glass Vacuum Tube Solar Collectors

The metal-glass vacuum tube solar collector also consists of several similar vacuum collector tubes. Based on the different structures for heat collection and heat absorption of tube collectors, they can be divided into the U-shape tube, down-hole coaxial tube, heat pipe, inner-light concentration, straight-through, and thermal storage types, which will be introduced separately.

2.4.2.1 U-Shape Tube Vacuum Solar Collector

The U-shape tube vacuum solar collector is shown in Fig. 2.25. There are two different shapes of heat absorber inserting into the tube: the plate fin and the cylindrical fin. The metal fin is welded with the U-shape tube, the surface of the heat-absorbing fin is painted with selective coating, and the tube is vacuumed. The tube (normally copper) is heat sealed with glass, or the U-shape tube is combined with the thermal insulation cover that can be guided out of the collector as the inlet and outlet of the heat transfer fluid (normally water).

2.4.2.2 Heat Pipe Vacuum Solar Collector

As shown in Fig. 2.26, the heat pipe vacuum solar collector can be divided into two types based on different heat absorbers: one is a heat pipe plate fin structure and the other is a heat pipe cylindrical fin. The heat pipe vacuum solar collector consists of a heat pipe, heat absorber, and vacuum glass tube. The working principle is that sunlight penetrates the glass and hits the heat absorber. Then the absorbed heat rapidly vaporizes the working

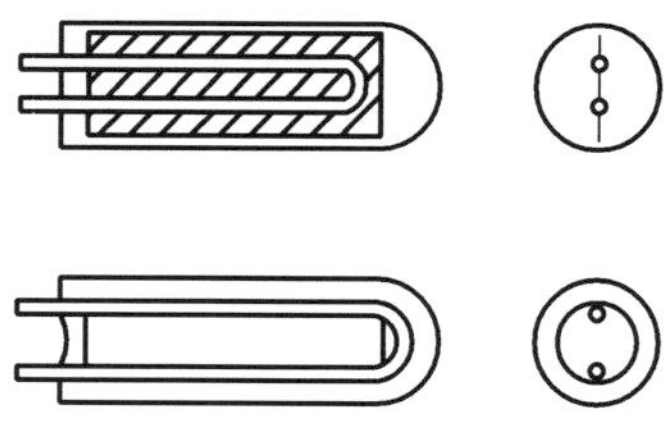

Figure 2.25
U-shape tube type.

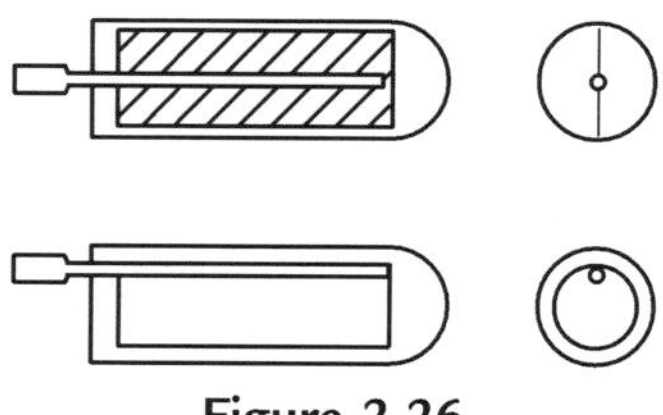

Figure 2.26
Heat pipe type.

fluid; the vaporized fluid will rise to the condensation end of the heat pipe, which will heat water in water tank or header. At the same time, the vaporized working fluid releases the latent heat and condenses into liquid. Finally, the liquid flows back to the lower end of the heat pipe owing to gravity. This repeating process continually absorbs solar radiation and rapidly delivers heat to the medium (water). This one-way heat transfer process is determined by the property of the heat pipe. To ensure the normal operation of the heat pipe, the tilt angle between the heat pipe vacuum solar collector and ground surface needs to be more than 10 degrees.

2.4.2.3 Down-Hole Coaxial Tube Vacuum Solar Collector

The down-hole coaxial tube vacuum solar collector is also called a straight-flow vacuum solar collector, shown in Fig. 2.27. The heat-absorbing tube is two metal tubes with an outer tube covering the inner one. The outer tube is welded with the heat absorber and the bottom of the outer tube is sealed. The outer tube is sealed by heat with the glass and stretches out of the vacuum tube, which is vacuumed. When it is under working conditions, cold water enters from the inner tube and is heated by the heat absorber. Then warmed water flows out through the interlayer between the inner and outer tubes. The main characteristic is high thermal efficiency and reliable operation.

2.4.2.4 Inner-Light Concentration Vacuum Tube Solar Collector

An inner-light concentration vacuum tube solar collector is shown in Fig. 2.28. This is a solar collector with a light concentration reflector in the vacuum tube. It consists of a concentrating reflector, a heat-absorbing tube, and a vacuum glass tube. The aluminum painting on the lower semicircumference or the compound parabolic cylindrical surface can be used as a reflective surface. The heat-absorbing tube can be either a heat pipe or a concentric tube, but its surface has to deposit a selective absorbing coating with high temperature resistance, which can be 150°C. A solar tracking system is not required, but the manufacturing process is relatively complicated.

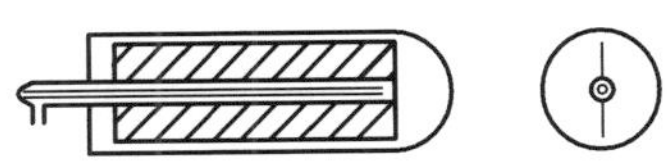

Figure 2.27
Coaxial casing type.

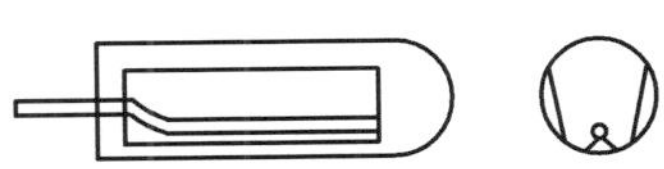

Figure 2.28
Inner-light concentration type.

2.4.2.5 Straight-Through Vacuum Solar Collector

A straight-through vacuum solar collector is shown in Fig. 2.29. According to the difference in heat-absorbing structures within the vacuum tube, it can be divided into a heat-absorbing plate and heat-absorbing tube. No matter what the structure is, both the heat-absorbing plate and tube need a selective absorbing coating with high temperature resistance. The heat transfer fluid enters the heat-absorbing tube from one end, is heated within the vacuum tube, and then exits the end. Therefore, it is called straight-through. Because both ends of the interlayer between the metal absorber and the glass tube need to be sealed, considering the difference in the contraction of the metal and glass, a metal corrugated pipe is used at the seal position as a transition. Main advantages of this type of tube are its high operation temperature and the ease of assembly, which is especially suitable for a large-scale solar thermal system. Combined with a light concentration reflector, the temperature can reach 300−400°C, which can also be used for solar power generation. The KJC solar power plant in California, United States, uses this type of product; the overall output power can be 150 MW. It uses a cylindrical reflector and single-axis solar tracking. The working fluid within the vacuum tube can reach higher than 400°C.

2.4.2.6 Thermal Storage Vacuum Solar Collector

A thermal storage vacuum collector is shown in Fig. 2.30. This is a kind of vacuum solar collector that combines a large-diameter vacuum collector tube with a thermal storage water tank. It is also called a vacuum integral solar water heater. This product is composed of a heat-absorbing barrel with a metal covering on the end of a glass tube, a supporting frame, a getter, etc. The heat-absorbing barrel stores water; it outer surface has a selective absorbing coating. The interlayer of the vacuum tube and the outer surface of the barrel have to be vacuumed; the frame is used to support the barrel and vacuum tube. The diameter of this type of thermal collector tube should normally be larger than 100 mm.

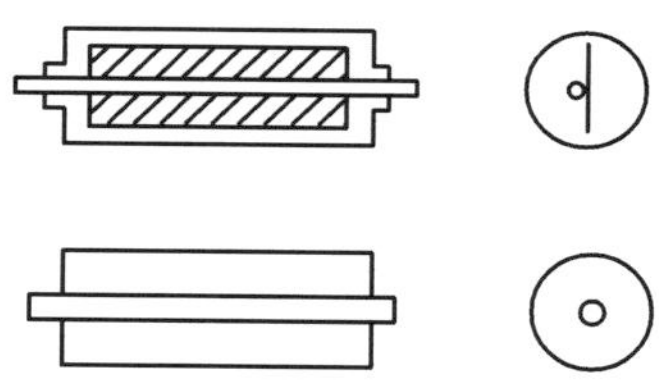

Figure 2.29
Direct-flow vacuum tube solar collector.

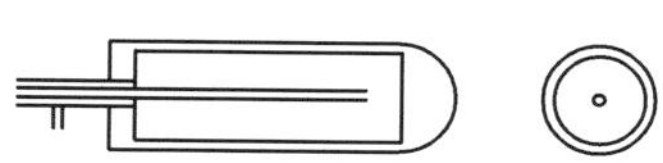

Figure 2.30
Vacuum tube solar collector with heat storage.

In the daytime, the heat-absorbing barrel can convert solar radiation into heat energy and directly heat water within the barrel. When in operation, cold water is slowly injected through the insert pipe so that hot water can be ejected for use. During nighttime, because the vacuum is thermally insulated, the hot water temperature within the barrel drops very slowly. Main features are that no water tank is required, and its structure is compact and convenient to use. The number of thermal storage vacuum tubes relies entirely on the volume of consumed water. After that, it can rapidly assemble the thermal storage vacuum solar collector or the hot water system based on the user's requirements.

2.4.3 Modules of Vacuum Solar Collector

Regardless of the vacuum tube solar collector or vacuum tube solar heating system, both use a header box to assemble several vacuum tube collectors into the solar collector module. Then the modules are grouped into a whole device or system according to the requirements.

2.4.3.1 Two Basic Structures of All-Glass Vacuum Tube Collectors

The all-glass vacuum tube collector module normally consists of a collector tube, reflective plate (sometimes not used), header box, tail stock, and supporting frame. Based on the orientation of the collector, it can be divided into south–north and east–west placement, as shown in Figs. 2.31 and 2.32, respectively. Both structures of collectors should have a tilt angle from the earth.

2.4.3.2 Header Box

The header, also called the connection tube, is an important component to connect the vacuum tube collectors into a single module unit, as shown in Fig. 2.33.

The header can be designed and manufactured based on whether it is pressure-bearing or non–pressure-bearing. A pressure-bearing header should bear a pressure of 0.6 MPa,

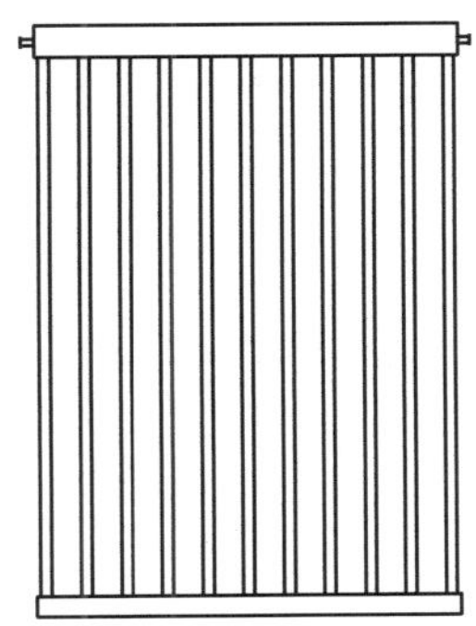

Figure 2.31
South–north-oriented vacuum tube collector.

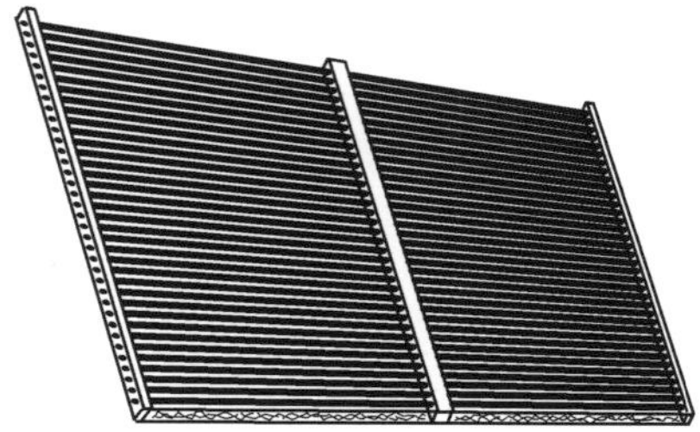

Figure 2.32
Vacuum tube solar collector placed in east–west direction.

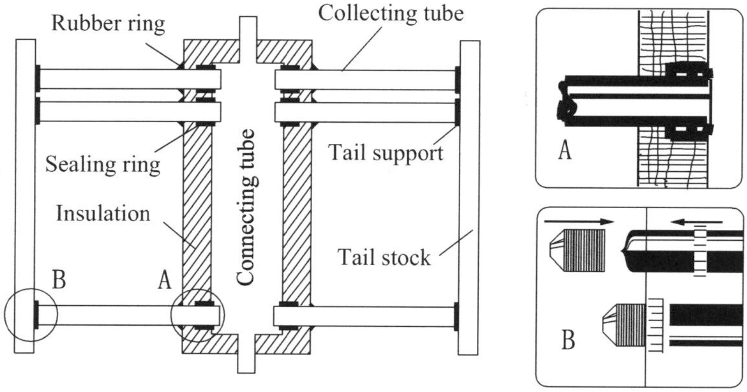

Figure 2.33
Combination of various reflectors and vacuum tube.

whereas a non–pressure-bearing header also has pressure-bearing requirements owing to the requirement of the operation and system; it can be designed to be 0.05 MPa.

The header can be designed as either tetragonal or tubular. The hole spacing may be designed as one- or two-sided. The two ends of the header are welded with an inlet and outlet tube covered with an insulation layer and shell. The opening end of the vacuum tube inserts directly into the header through a Silastic seal ring. The spacing between the tubes should refer to design requirements. For example, reasonable tube spacing for a 47-mm-diameter standard tube is around 75 mm.

2.4.3.3 Design of Reflector for Vacuum Tube Solar Collector

Because the cost of vacuum collectors tube is high, if they are arranged intensively, the cost will be high and it is difficult to process them. To meet requirements for system thermal efficiency and water yield, as well as to reduce cost, various reflectors are designed, as shown in Fig. 2.34.

Fig. 2.34A shows a flat diffuse reflector. Generally, an aluminum or matt white–painted plate can be used as a reflector whose structure is simple; thus the cost will be low.

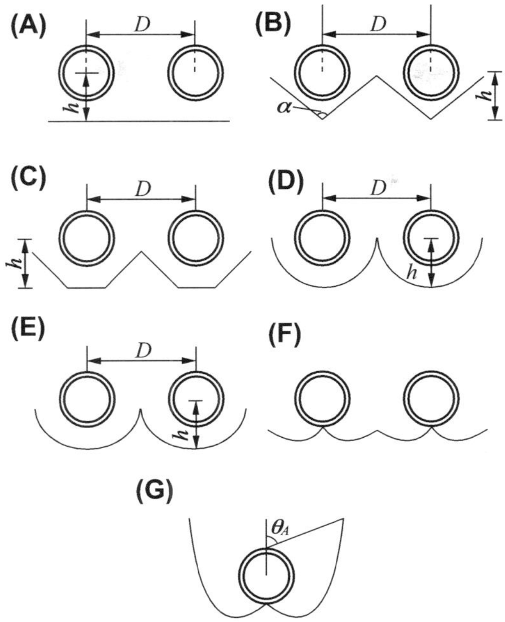

Figure 2.34

Schematic diagram of connection of collecting header and vacuum tube receivers: (A) Flat-plate reflector; (B) V-shaped reflector; (C) V-shaped and flat-base combined trough diffuse reflector; (D) cylinder reflector; (E) cylinder parabolic reflector; (F) compound mirror concentrator; (G) compound parabolic concentrator.

According to optical calculation, when light is in perpendicularly incidence, the tube spacing is double that of the tube diameter, and when a matt white–painted diffuse reflector is used, received solar radiation at the back half of the collector tube is 43% of the front half. Fig. 2.34B shows a V-shape reflector. The geometric parameters are as follows: $D/d = 1.6–17$, $\alpha = 90–103$ degrees. Fig. 2.34C shows a V-shape and flat-base combined trough diffuse reflector whose effect is similar to that of a flat diffuse reflector when D and h are equal. Fig. 2.34D–G is concentrator reflectors whose applications are mainly middle- or high-temperature solar collector systems.

Dust and dirt influence the reflection effect when it is exposed in the air for a long time; therefore regular cleaning is necessary if conditions allow.

2.5 Solar Pond

2.5.1 Basic Concept of Solar Ponds

When solar radiation enters a natural pond and spreads deeply, its energy is absorbed by water along a path; heated water rises owing to the smaller density, which will cool down on the water's surface and sink again. Therefore, natural water cannot reserve solar energy.

However, for a salty pond, if the brine strength increases with the depth of the pond, the natural convection described can be suppressed, and it can reserve solar energy. This is a so-called gradient solar pond, or a solar pond in short.

A solar pond normally consists of three zones. The upper layer is a convective zone with low and thin saltiness, the middle layer is a nonconvective gradient saltiness zone with a thickness of 1–1.5 m, and the lower layer is a heat reservoir area with high and even saltiness whose thickness is 1–3 m. For the normal operation of a solar pond, the most important thing is to keep the stability of the gradient saltiness zone in the middle layer; no convection happens in that layer. The solar pond can provide heat with a temperature of more than 90°C. Because the heat capacity of the solar pond is large, it can be used as a solar collector and heat reservoir. It can even collect and reserve solar energy in the summer and be used in the winter.

The concept of solar ponds traces back to the beginning of the last century. In 1902, Kalecsinsky observed that in a naturally layered salty pond, the temperature at a depth of 1.3 m could reach 70°C in Hungary. Therefore, he was the first to come up with the idea of using the salty gradient pond as a solar collector. In 1948, Bloch suggested that Israeli National Research Committee use an artificial layered salty pond as a solar collector. In 1959, under the guidance of Tabor, a large amount of pioneering research was undergone in the Israeli National Physics Laboratory, which founded the development and application of solar ponds. Afterward, Tabor, Weinberger, Elata, Levin, and Matz conducted some in-depth theoretical research on the fundamental physics of nonconvective solar ponds. They also carried out indoor and field experimental research to discuss the influence of various factors on solar pond performance.

2.5.2 Basic Structure of Solar Ponds and Basic Conditions to Maintain Stability

For liquid with a uniform density, if it is heated from the bottom, its density in the bottom layer may decrease with the rising temperature. This density gradient and buoyancy caused by the temperature difference will inevitably cause natural convection and rapidly transfer heat from the bottom to the top. As a result, overcoming heat convection in a pond is the key to heating its reservoir. The gradient solar pond just uses the liquids in the pond with different densities to overcome the density variance and buoyancy caused by the temperature difference. A cross-section figure of an ordinary gradient solar pond is shown in Fig. 2.35, and the distribution of saltiness is shown in Fig. 2.36.

In a salty gradient solar pond, the saline solution itself has a certain concentration gradient (increasing from the bottom to the top). Assuming the density gradient is caused by a concentration difference that could be offset by that caused by the temperature difference, the thermostability of the solar pond could be guaranteed, i.e., the bottom temperature is

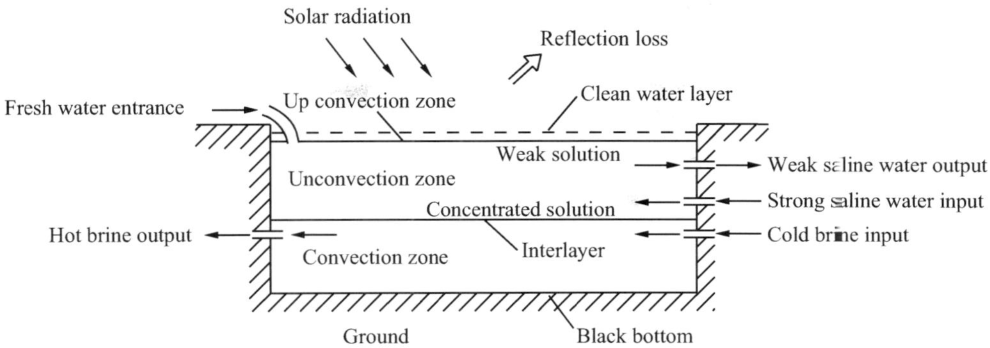

Figure 2.35
Schematic diagram of a nonconvection solar pond.

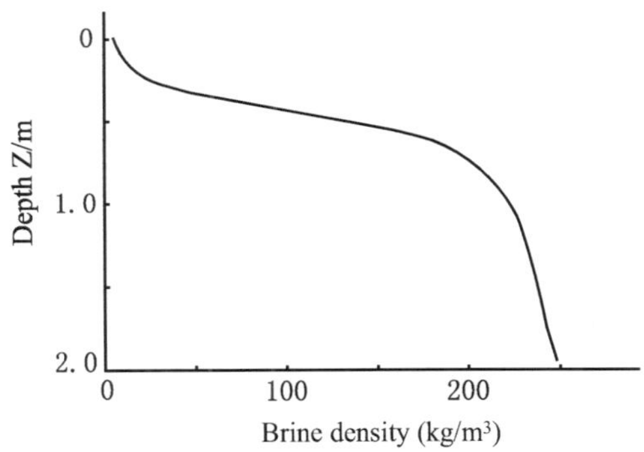

Figure 2.36
Diagram of concentration distribution in a solar pond.

higher than the top temperature and no natural convection happens. Therefore, the thermostability condition of a solar pond could be expressed as Eq. (2.26)

$$\frac{d\rho}{dz} = \frac{\partial \rho}{\partial s}\frac{ds}{dz} + \frac{\partial \rho}{\partial T}\frac{dT}{dz} \geq 0 \tag{2.26}$$

where s stands for saltiness, ρ stands for the density of solution, T stands for temperature, and z stands for the depth of pond water from its surface.

2.5.3 Solar Pond Liquid and Absorption of Solar Radiation at Pond Bottom

After solar radiation enters the pond, because of the scattering and absorption effect, the strength of the solar radiation will die out along the spread path. In a semitransparent medium, the radiation absorption follows Bouguer's law: the amount of damping dq_λ

incidence radiation with a wavelength of λ by a unique absorbent with a thickness of dl is normal to strength q_λ and thickness dl, i.e., (Eq. (2.27)):

$$dq_\lambda = -k_\lambda q_\lambda dl \tag{2.27}$$

where, k_λ is called a monochromatic absorption coefficient. If Eq. (2.26) is integrated, the transparent radiation strength needed for q_λ^t to traverse the medium with a thickness of t is (Eq. (2.28)):

$$q_\lambda^t = q_\lambda e^{-k\lambda t} \tag{2.28}$$

A relatively easy way to express the radiation strength in any depth z is to use several extinction coefficients with every extinction coefficient response to a specific spectral band. Then the following equation could be obtained (Eq. (2.29)):

$$H = H_0 \sum_{j=1}^{m} \beta_j e^{-\eta_j z} \tag{2.29}$$

where H_0 is the radiation in $z = 0$ (pond water surface), and β_j is the radiation components, which meet the following condition (Eq. (2.30)):

$$\sum_{j=1}^{m} \beta_j = 1 \tag{2.30}$$

where η_j is the extinction coefficient of the jth spectral band. This method is similar to integration of the absorption coefficient K. In 1975, A. Rabl and C.E. Nielsen gave typical numbers of β_j and η_j for water, which are listed in Table 2.4.

The data listed in Table 2.4 are for pure water. Someone conducted experiments using pure NaCl with different concentrations; results showed that a 5% difference between that and pure water. Li [12] introduced a more simple approximate equation (Eq. (2.31)):

$$H = 0.36 - 0.08 \ln x \tag{2.31}$$

There is a certain selectivity of solar radiation absorption for pond water. In water, long-wave radiation ($\lambda > 1.2$ μm) can be strongly absorbed and rapidly damps. Its energy is reduced two orders of magnitude 1 cm under water's surface, whereas radiation with a

Table 2.4: Extinction coefficient of water [16].

j	Spectral band (μm)	β_j	η_j (cm^{-1})
1	0.2–0.6	0.237	3.2×10^{-4}
2	0.6–0.75	0.193	4.5×10^{-3}
3	0.75–0.9	0.179	0.03
4	0.9–1.2	0.167	0.35

Table 2.5: Percent energy of solar radiation with different wavelengths when transversing different depths of water [16].

	Distance sunlight travels in water (m)				
Wavelength (μm)	0	0.01	0.1	1.0	10
0.2–0.6	23.7	23.7	23.6	22.9	17.2
0.6–0.9	36	35.3	30.5	12.9	0.9
0.9–1.2	17.6	12.3	0.8	—	—
>1.2	22.4	1.7	—	—	—
Total	100	73.0	54.9	35.8	18.1

wavelength of 0.3–1.2 μm can also be absorbed by water. The effects of different wavelengths on the absorption of solar radiation are shown in Table 2.5.

This discussion shows that because of the reflection, diffusion, and absorption effects of pond water, actual radiation reaching the pond bottom is dramatically reduced. For example, in a 1-m-deep pond, received radiation on the pond bottom is only 35% of incident radiation. However, under certain conditions, absorbed solar radiation on the upper part of water can just supplement energy flowing from the inside part of the pond to the outside part. In other words, when heat is extracted from the bottom, an isothermal layer is formed in the convection zone that makes the temperature gradient on the upper boundary almost zero; it is equivalent to covering an ideal insulation layer and a heat reservoir in a solar pond can be achieved.

2.5.4 Thermal Efficiency of Solar Pond and Its Characteristics in Steady State and Transient State

In an ideal solar pond system as shown in Fig. 2.35, assuming that its lateral dimension is small compared with the depth of a solar pond (this assumption is valid for a large-scale solar pond), the nonconvection zone could be described by the one-dimensional heat transfer equation with an internal heat source (Eq. (2.32)):

$$\frac{\partial}{\partial z}\left(K\frac{\partial T}{\partial z}\right) = \rho c\frac{\partial T}{\partial t} + \frac{\partial H}{\partial z} \tag{2.32}$$

where t is the time variable; and K, p, and c are the heat transfer coefficient, density, and specific heat, respectively, of the solution within the pond. Note: Eq. (2.32) ignores the influence of the vertical movement of pond water on pond temperature, which is allowed in a stable solar pond.

Because the heat capacity of a convection zone under the solar pond is large, its change on the status is slow even under direct sunlight; thus the problem could be simplified so that the solar pond is operating under stability, so that the degree of influence of various

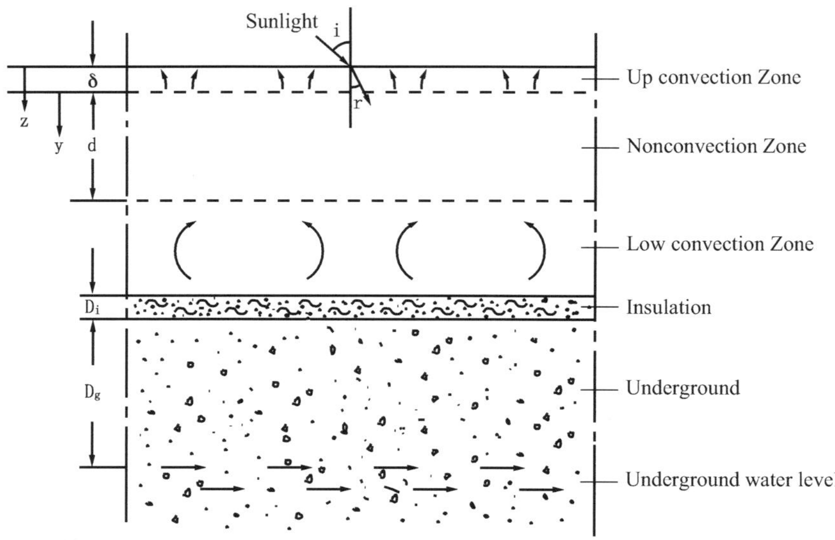

Figure 2.37
Model of steady-state solar pond.

parameters for a solar pond can be clearly observed. When a solar pond reaches stability, the temperature in any location will not change with time (Fig. 2.37). The average solar radiation is used for calculation, which is also unchangeable with time. Thus, Eq. (2.32) can be simplified as [17] (Eq. (2.33)):

$$-\frac{\partial}{dz}\left(K\frac{dT}{dz}\right)+\frac{dH}{dz}=0 \tag{2.33}$$

For convenience in calculating and wider universality, Eq. (2.31) can be rewritten in normal form (Eq. (2.34)):

$$h=a-b\ln x \tag{2.34}$$

It also can be written as (Eq. (2.35)):

$$h=b\ln\frac{x_0}{x} \tag{2.35}$$

where $x_0=e^{\frac{a}{b}}$, a, and b are all constants. It is considered that the pond surface will also reflect sunlight; thus the average solar radiation reaching the depth of z within pond H can be expressed as (Eq. (2.36)):

$$H=\overline{p}\cdot\overline{H}\cdot b\ln\left[\frac{x_0}{z/\cos\gamma}\right] \tag{2.36}$$

where $\overline{p}$ is the average transmission coefficient of the pond surface, as an estimation, in regions whose latitude is less than 40°, and the value is between 0.89 and 0.97; the larger value is used for summer and low-latitude areas, whereas the smaller value is used for winter and high-latitude areas. γ is the refraction angle of sunlight on the pond surface.

Defining the surplus temperature (Eq. (2.37)):

$$\theta = T - T_a \tag{2.37}$$

Then, integrating Eq. (2.33), we get (Eq. (2.38)):

$$-K\frac{d\theta}{dz} + H = C_1 \tag{2.38}$$

where, C_1 is the integral constant. The value of C_1 can be determined by the heat balance equation on the lower boundary ($y = d$) of the nonconvection zone; i.e., the solar radiation that travels through the lower boundary of the nonconvection zone should be equal to the sum of effective heat output Q and upward and downward heat transfer loss (Eq. (2.39)):

$$-K\frac{d\theta}{dy}\bigg|_{y=d} + \overline{p}\cdot\overline{H}\cdot b\ln\left[\frac{x_0}{(d+\delta)/\cos\gamma}\right] = Q + \frac{\theta_d}{\frac{D_i}{K_i} + \frac{D_g}{K_g}} \tag{2.39}$$

Thus, the value of C_1 is (Eq. (2.40)):

$$C_1 = Q + \frac{\theta_d}{\frac{D_i}{K_i} + \frac{D_g}{K_g}} \tag{2.40}$$

After further deduction and considering the change of heat transfer coefficient of pond water with temperature and concentration, the average thermal efficiency η of solar pond is (Eq. (2.41)):

$$\eta = \frac{Q}{\overline{H}} = \frac{\overline{p}\cdot b}{d} \times \ln\left[\delta^{\delta}\frac{(x_0\cos\gamma)^d}{(d+\delta)^{d+\delta}}\right] + \overline{p}\cdot b - \frac{\theta_d}{\overline{H}}\left(\frac{1}{\frac{D_i}{K_i} + \frac{D_g}{K_g}} + \frac{1}{\frac{a}{K}}\right) \tag{2.41}$$

It is clear that there is a linear relationship between η and $\theta_d/\overline{H}$, which is similar to the well-known thermal efficiency equation of a flat-plate solar collector. Various factors influence the thermal efficiency of solar ponds. Apart from the working temperature (θ_d) and solar radiation strength, the solar pond structure is another important factor: for instance, the depth of nonconvection zone d, the thickness of pond floor thermo-insulation layer D (if there is any), its heat conductivity coefficient K_i, the underground condition and its heat conductivity coefficient K_g, the depth of underground depth (which determines D_g), and so forth. The influence of reflection loss on the pond surface and solar radiation absorption by pond water are reflected by $\overline{p}$, b, x_0, and $\cos\gamma$ in the equation. $\cos\gamma$ also reflects the influence of latitude. Moreover, the effect of upper convection zone thickness δ on the solar pond is clear. Figs. 2.38–2.40 are results based on Eq. (2.41) and reflect the influence of the upper convection zone thickness and underground water and soil on

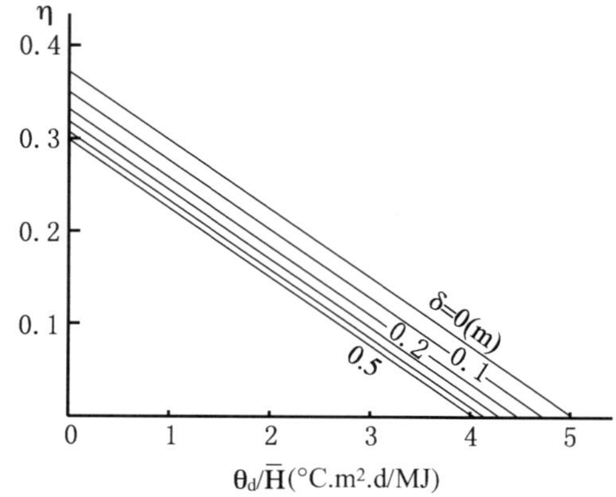

Figure 2.38
Effect of UCZ (up convection zone) thickness on solar pond efficiency.

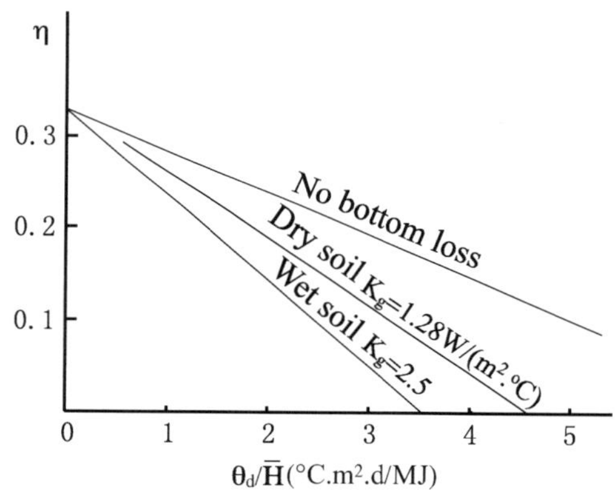

Figure 2.39
Effect of underground soil on solar pond efficiency [13,14].

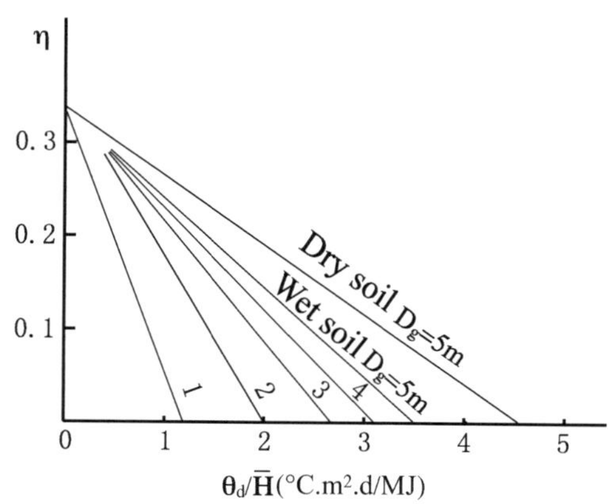

Figure 2.40
Effect of underground water level on solar pond efficiency [13,14].

solar pond performance. Except for some special indications, the calculation is based on the following conditions: latitude 38°; $\delta = 0.2$ m; $d = 1$ m; $D_i = 0$; $D_g = 5$ m; $K_g = 1.28$ W/(m °C) (for dry clay). Controlling the upper convection zone thickness has an obvious effect on solar pond efficiency.

According to this steady-state analysis method, for a solar pond built in the 38° latitude area, average annual solar radiation is 17 MJ (m^2 d) when $\delta = 0.2$ m, $d = 1$ m; the 5-m thick dry clay is used as thermal insulation, $\theta_d = 40$°C, and annual average efficiency is 16%. In other words, the solar pond can provide low-temperature (higher than ambient temperature, 40°C) heat energy with 16% efficiency. It is easy to calculate other conditions using Eq. (2.39).

2.5.5 Combination of Solar Pond and Seawater Desalination

In some areas that lack fresh water but have seawater and bitter salty water, if there are enough solar resources in that area, the use of a solar pond as a heat source of desalination seems to be promising. Normally the disk distillation device (see Chapter 9) has a low internal temperature that is only 40–50°C. Even for a complicated multiple-stage evaporative seawater desalination device, it could work with a heat source temperature of 75°C. At the low heat energy consumed for water production (about one-tenth the latency for vaporizing water), the solar pond could sufficiently reach this temperature [18]. Generally, areas that lack fresh water are vast in territory, which could provide plenty of land. The combination of solar pond and a seawater desalination device could provide necessary conditions for developing the natural resources of these areas. Therefore, using a solar pond for seawater and salty bitter water desalination meets actual requirements.

A disadvantage of solar ponds is that the pond location is limited to salt-producing areas or areas with mass and cheap salt. On the other hand, the environment needs to be protected from saline pollution.

To avoid the disadvantage of solar ponds, the design concept of a pure water solar pond is promoted whose feature is to try different measures to restrain heat loss on the pond water surface.

2.6 Concentrating Solar Collector and Its Characteristics

With the development of solar desalination technology, the variety of solar desalination systems has increased, needing various forms of heat sources. There has been a tendency to combine solar collectors and solar desalination devices because conventional seawater desalination technology has demonstrated mature experience and good economic efficiency; in addition, it has high heat transfer efficiency, a compact structure, and relatively low cost. From the viewpoint of thermodynamics, seawater desalination devices driven by heat have a higher coefficient of performance when operated under higher temperatures. Therefore, solar desalination systems also have medium- or high-temperature operations.

Flat-plate or vacuum tube solar collectors are simple in structure and convenient to use, but their heat-collecting temperature is relatively low, normally under 80°C, and efficiency drops dramatically when it the collectors are operated at a higher temperature; thus the application range is limited. To meet the requirement of a higher application temperature, a concentrating solar collector is necessary. In addition, with significant progress in concentrating solar collector technology, various concentrating solar collectors have been produced and applied to the industry.

2.6.1 Focusing Solar Collector System

Focusing solar collector systems can be seen as optical systems consisting of a light source (moving sun), concentrator, and receiver. As shown in Fig. 2.41, the concentrator concentrates solar radiation on an unthreaded hole to a focal plane on the receiver through refraction and reflection. The medium on the receiver removes heat converted from solar energy. Therefore, for a focusing solar collector system, the energy flux density on the receiver is much higher than that on unthreaded hole; the temperature on the focal plane is much higher than that on the flat-plate collector.

2.6.1.1 Optical Efficiency of Solar Concentration System

A nonideal concentration system cannot completely transfer concentrated light into the next assembly unit; therefore, there is an issue regarding concentration efficiency. Causes of low concentration efficiency include the device structure, transmittance, reflectivity, and workmanship.

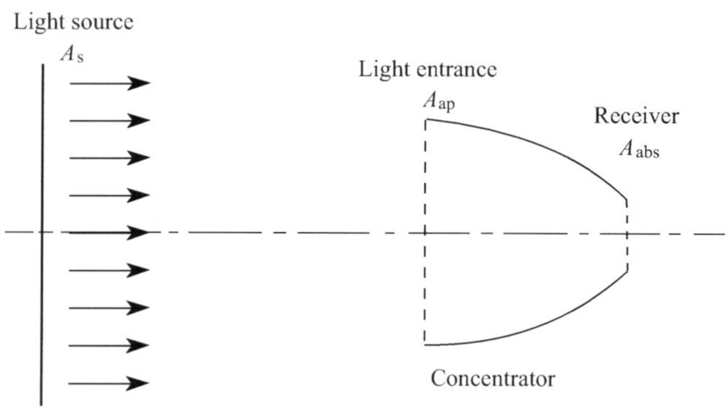

Figure 2.41
Diagram of a general concentrator.

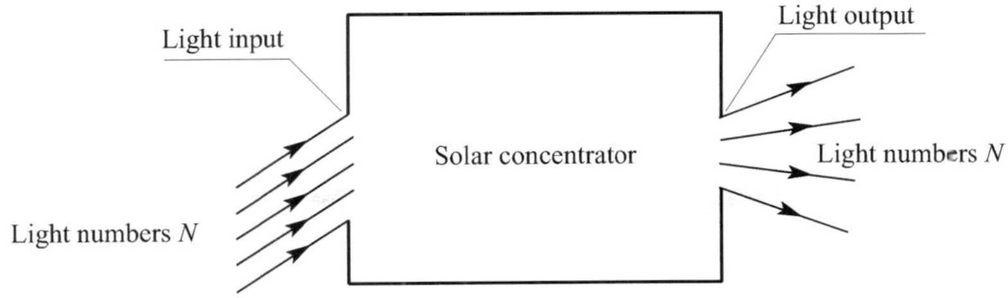

Figure 2.42
Optical efficiency of a solar concentrator.

For an abstract concentrator as shown in Fig. 2.42 it is assumed that N parallel lights enter the concentrator at an angle of θ, and after reflecting within the concentrator, only N' parallel lights escape the exit; the system energy transmittance is N'/N, i.e., optical efficiency, sometime called transmission.

In a solar energy system, the concentrator optical efficiency, η_0, may be defined as the fraction of total solar radiation I that can reach the receiver (Eq. (2.42)):

$$\eta_0 = \frac{q_{abs}}{I} \tag{2.42}$$

where q_{abs} is the received energy. Optical efficiency η_0 represents the optical performance of the focusing solar collector, which reflects the degree of nonidealization of the collector (such as the shape, optical precision, and reflectivity) when it is in the optical process of concentrating solar radiation; the optical loss on the receiver surface could also cause the nonidealization of the collector. Because optical efficiency is mostly related to a geometrical relationship, for most normal solar collectors or concentrations, it is easy to deduce the approximate optical efficiency calculation equation. For instance, for a flat-plate collector (single covering), the transmittance of a covering is set to be τ, the absorptivity of absorber is α, and thus the optical efficiency is (Eq. (2.43)):

$$\eta_0 = \tau\alpha \tag{2.43}$$

For the parabolic trough concentrator, if the reflectivity of trough is ρ, the transmittance of glass tube absorber is τ, the absorptivity of absorber is α, and the approximate system optical efficiency for the direct solar radiation is (Eq. (2.44)):

$$\eta_0 \approx \rho\tau\alpha \tag{2.44}$$

In these equations the multiple reflection between the receiver and covering plate is ignored, but on some occasions, it cannot be ignored, which may increase the optical efficiency by around 1%.

2.6.1.2 Concentration Ratio of Solar Concentrator

The solar concentration ratio is an important concept for a focusing solar collector. As mentioned, the energy flux density is only 800–1000 W/m^2. Therefore, it is necessary to concentrate light to obtain higher solar collecting temperatures. The concept of a concentration ratio is introduced to evaluate and compare the quality of solar collectors with a concentration function.

The definition of a concentration ratio of solar concentration is the ratio of solar radiation entering the collector to solar radiation received by the receiver. It represents the system's ability to concentrate solar energy (Eq. (2.45)):

$$C_{flux} = \frac{I_{ap}}{I_{abs}} \tag{2.45}$$

where C_{flux} is called the energy flux density ratio. It is apparent that the larger C_{flux} is, the higher is the system's degree of concentration, and correspondingly, the higher the solar concentration temperature is that the system can collect. However, because the intensity of solar radiation varies throughout the day and is hard to measure, it is not easy to obtain the concentration ratio using that equation. To simplify the research problem, the concept of the geometrical concentration ratio was introduced, which is defined by the ratio of the light aperture area to the receiver area, i.e., (Eq. (2.46)):

$$C = C_{geom} = \frac{A_{ap}}{A_{abs}} \tag{2.46}$$

According to this theory, the larger the geometrical ratio is, the higher will be the concentration degree of the system, as well as the concentration temperature. The concentration ratio is a geometrical parameter that does not change once the system is manufactured; thus it is convenient for application in practical engineering.

2.6.1.3 Acceptance Angle of Solar Concentrator

For the solar concentrator shown in Fig. 2.43, because the receiver on the focal plane cannot be infinite, it can receive sunlight only within a certain range. When sunlight enters the concentrator in a certain direction, it will be reflected or refracted by the concentrator and projected onto the receiver, but when sunlight diverges from its original direction, the reflected light might not be able to reach the receiver and may not be effectively accepted. This maximum allowed deviation angle is called the acceptance angle of the concentrator. In fact, this process is easily understood: the sun moves in the sky and the incidence angle of sunlight to any point or surface change with time, and therefore any solar collector or concentration will not be able to receive all incident sunlight on its aperture surface if it is stationary, unless it can track the sun accurately. If it is stationary, it can receive only a

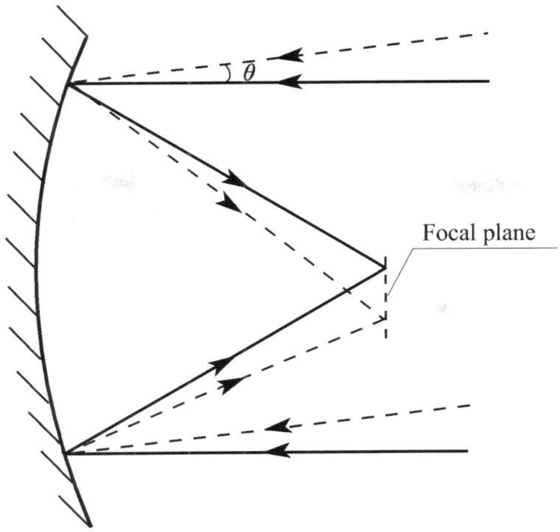

Figure 2.43
Imaging focusing principle of mirror by reflection.

part of sunlight, whereas the rest of sunlight cannot reach the receiver. Therefore, for the concentration solar collector system, the acceptance angle $2\theta_a$ relates well to the concentration process. The definition of an acceptance angle is: when the whole or partial collector stays stationary or remains its relative position unchanged, the angle range of sunlight reaching the device and received by receiver is called acceptance angle.

The acceptance angle is an important parameter for a solar concentration system; it is well related to the performance, especially the economic performance of a concentration system. A larger acceptance angle means that more sunlight can be received by a system when it is not moving; when a sun-tracking system is used, low tracking accuracy is required. Therefore, a larger acceptance angle is requested for a solar collector in most cases. However, the acceptance angle and concentration ratio are contradictory parameters. The larger the acceptance angle is, the smaller the concentration ratio is. Therefore, in practical engineering, the requirements of the user need to be considered when designing an acceptance angle so that the best economic performance may be achieved.

According to the second law of thermodynamics, it is easy to deduce the relationship between the concentration ratio and the half-acceptance angle of system θ_a for a two-dimensional concentration system such as a trough focusing system [19] (Eq. (2.47)):

$$C_{ideal,2D} = \frac{1}{\sin\theta_a} \tag{2.47}$$

For three-dimensional point focusing system such as a dish paraboloid, the relation can be expressed as (Eq. (2.48)):

$$C_{ideal,3D} = \frac{1}{\sin^2\theta_a} \tag{2.48}$$

Because the sun itself is not a point whereas a circular disc with angular diameter $\Delta_s = 4.7$ mrad, the thermodynamic limitation for the geometrical concentration ratio of a two-dimensional line-focusing system is 213, whereas for a three-dimensional focusing system the maximum geometrical concentration ratio is 45,300. However, most of practically used concentrators, especially a paraboloid focusing or lens focuser, the actual concentration ratio is two to four times lower than the limit thermodynamic concentration ratio.

2.6.1.4 Acceptance of Diffuse Solar Radiation for Solar Concentration System

If little or no sun tracking is needed for a solar collector, there has to be a large acceptance so that the collector can collect more diffuse solar radiation. In fact, the acceptance of the diffuse solar radiation is also not ignorable, because under normal weather conditions, the fraction of diffused sunlight is about 50%.

In most solar concentrators such as a parabolic trough or disk concentrator, V-shaped trough concentrators, and CPCs, almost all radiation from the receiver can reach the light aperture. Therefore the visual quotient of the receiver to the light aperture $\Im_{abs-ap} = 1$; thus it can approximately obtain the fraction of diffuse solar radiation on the solar collector with a concentration ratio of C (Eq. (2.49)):

$$\Im_{ap-abs} = \frac{1}{C} \tag{2.49}$$

This equation applies to most solar collectors. In fact, the visual quotient of the receiver to the light aperture $\Im_{abs-ap}$ is less than 1 for most solar collectors. Sometimes it may be low to 0.5, in which the time to the receiver receives little diffuse radiation, and thus sun tracking is necessary or the effect is not good. In fact, it is meaningful to consider the contribution of diffuse solar radiation only when the concentration ratio is less than 10. The amount of contribution from diffuse solar radiation is determined by the geometrical structure. Normally, the larger the acceptance angle is, the more solar radiation can be accepted.

2.6.1.5 Truncation Ratio of Receiver

The optical property of concentrating solar collection systems is largely different from that of flat-plate collectors. First, in concentrating solar collection systems, the reflector and receiver are two different and independent components. Second, concentrating solar collectors need a sun-tracking system because its acceptance angle is much smaller than that of a flat-plate collector. In other words, the receiver of a concentrating solar collector cannot accept all solar radiation entering the light aperture. There is a truncation ratio or truncation factor, γ. After considering these factors, the amount of solar radiation incident on the aperture area of collector can be calculated, as well as the solar radiation intensity accepted by the receiver (Eq. (2.50)):

$$S = I_b\rho(\gamma\tau\alpha)_n K_{\gamma\tau\alpha} \tag{2.50}$$

where I_b is the solar radiation intensity on the light aperture of concentrator, apart from a concentration system with a low concentration ratio, in which the diffuse solar radiation is normally not considered; and ρ is the reflectivity of reflector. If the concentrator is a refractive concentration system, ρ is its transmittance. $K_{\gamma\tau\alpha}$ is called the beam angle correction factor, which is set to account for the deviation of incident light on the aperture.

γ, τ, and α are all functions of the sunlight incidence angle. The influence of the incidence angle on them is individual or mutual, which needs to be considered according to different conditions. τ is the transmittance of the receiver outer covering or circular tube; α is the acceptance rate of the receiver; and γ is the truncation ratio, which is defined by the amount of solar radiation accepted by the receiver to that reflected by the reflector. For example, for a linear focusing concentrator, its concentrated energy distribution on the focal plate is shown in Fig. 2.44. The receiver can accept only sunlight within the range of A to B, and then the truncation ratio can be defined as (Eq. (2.51)):

$$\gamma = \frac{\int_A^B I(x)dx}{\int_{-\infty}^{\infty} I(x)dx} \tag{2.51}$$

As shown in Fig. 2.44, to receive as much as sunlight from the reflector, the truncation ratio must be large enough: that is to say, the tube diameter or superficial area needs to be large enough. However the large receiver would on reduce the concentration ratio and

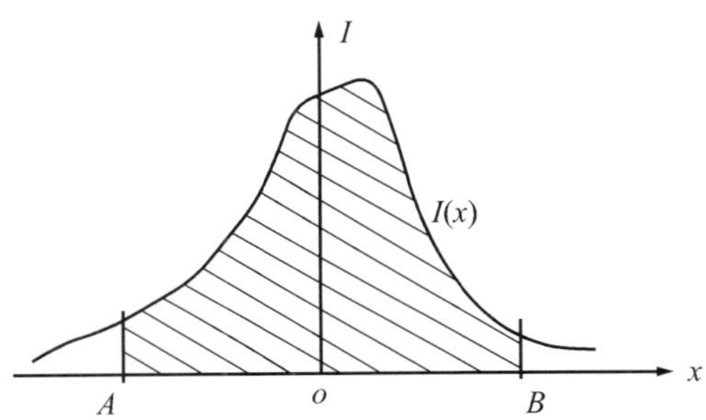

Figure 2.44
Concentrated energy distribution on the focal plate for a linear focusing concentrator.

increase the cost of the material; therefore it should not be expected that 100% of sunlight could be accepted. Mostly a truncation ratio of around 0.9 is appropriate [20].

2.6.2 Basic Structure of Concentrating Solar Collector

A solar concentrator can improve the optical energy flux density on a receiver and thus increase the receiver temperature. With the investment in technology, research into concentrating solar collectors in our country has developed significantly, various innovative concentration systems have continuously arisen, and there has been an especially big step in technological and control levels. Currently, solar concentrators have several categories: the first one is reflective concentrators such as parabolic reflectors and CPC reflectors. The second one is refractive concentrators such as Fresnel lenses, convex lenses, and so forth. The forms of concentration include linear concentration such as trough parabolic concentrators, and point concentration such as disk parabolic concentrators. As to the receiver itself, its structure also varies; it could be cylindrical, a flat-plate shape, a cavity shape, and so forth. Fig. 2.45 illustrates some common forms of concentrating solar collectors [20].

In Fig. 2.45, the concentration ratios of the (A), (B), and (C) concentration systems are relatively low and do not exceed 10. Thus the latter concentration systems are selected for their high collecting temperature. For example, a parabolic concentration system can obtain a high temperature exceeding 3000°C but a concentrator with a low concentration ratio is normally simple in structure and without a sun-tracking system. It can use some part of diffused light and thus it is widely used in medium-low temperature solar collector systems. The collecting temperature (for a trough linear concentrator) ranges from 100°C to 500°C for focal concentration systems; it could be 300−1500°C or even 3000°C at maximum. The concentrator in Fig. 2.45J uses a small-scale disc parabolic to concentrate lights. Then the light is directed into optical fiber through which solar energy travels into a small space to achieve high-temperature concentration. This type of solar concentrator can collect a temperature of more than 800°C, which is an innovative developing concentration technology. Fig. 2.45N shows a compound multiple-curve concentrator consisting of several paraboloids or evolents, or several combinations of flat plates and paraboloids. The receiver is normally flat-plate. The circular vacuum tube is also allowed under certain circumstances. Advantage of this type of concentrator are low sun-tracking requirements and good thermal insulation, but disadvantages are its complicated structure and high cost. With the diversification of the market, this compound solar concentration collector has been increasingly viewed by researchers.

Generally, for concentration collectors, especially high concentration collector systems, its sectional area on the receiver is always much smaller than the light aperture of the concentrator. Therefore, to collect maximal sunlight, real-time sun tracking is necessary.

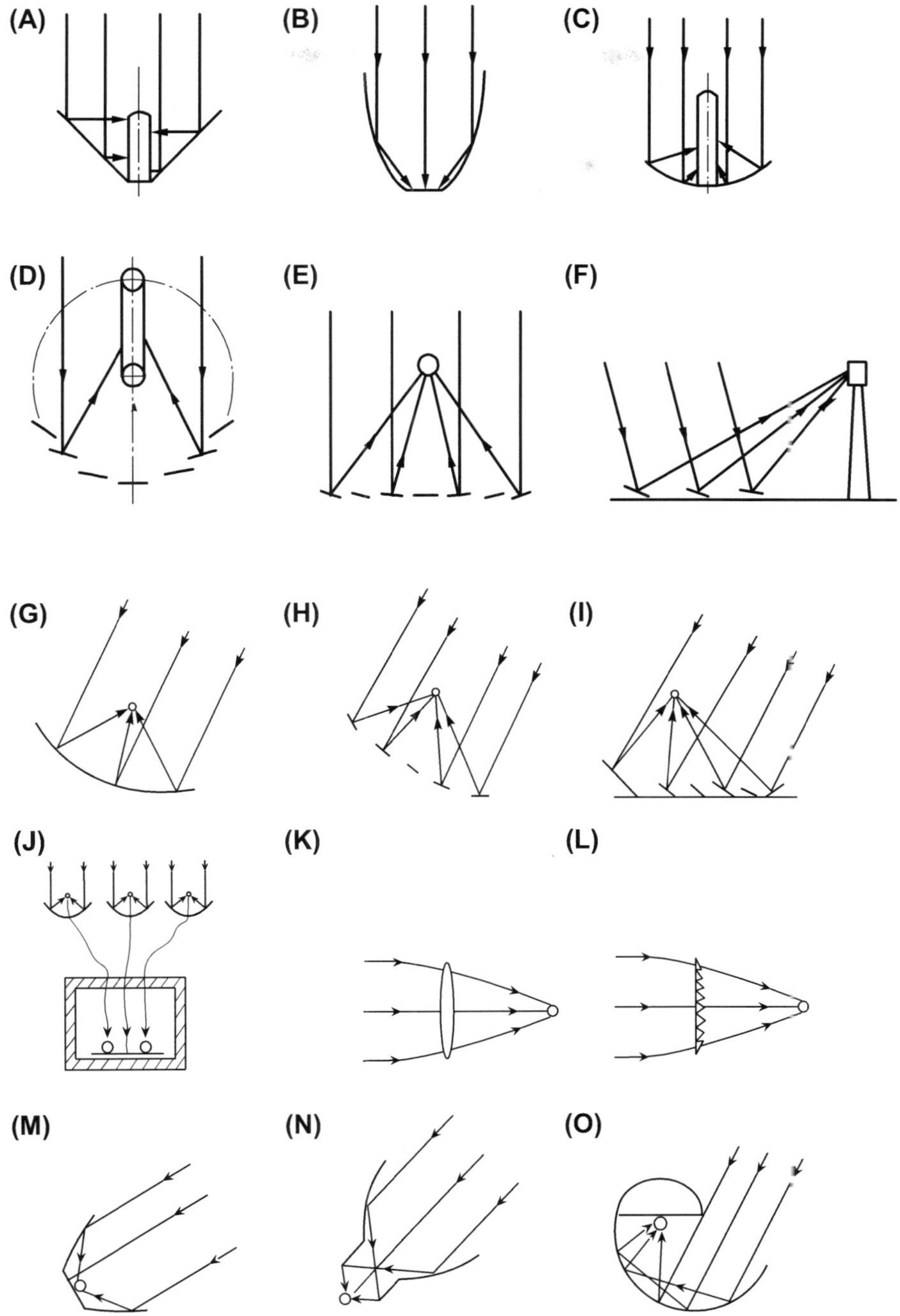

Figure 2.45

Various kinds of concentrating solar collector structures: (A) conical concentrator; (B) CPC concentrator; (C) sphere concentrator; (D) cylinder array concentrator; (E) array Fresnel lens concentrator; (F) heliostat tower concentrator; (G) parabolic concentrator; (H) array parabolic concentrator; (I) reflective Fresnel lens concentrator; (J) small disk reflective concentrator; (K) convex lens concentrator; (L) transmittance Fresnel lens concentrator; (M) trough conical concentrator; (N) mirror dual-focus concentrator; (O) multicurved compound concentrator.

This will have an influence on the positive aspect, because the concentration aperture is always facing the sun so that it can receive maximum sunlight intensity, which is 10–40% higher than sunlight on the flat plate. However, on the negative aspect, when moving components are added to the system, the system itself will consume some power, and the system will become complicated with a higher failure rate.

Generally, for linear focusing solar concentration systems such as trough parabolic concentration systems, array Fresnel lens concentrators use linear symmetry, tube receiver and single-axis sun-tracking.

For focal solar concentrators such as disc parabolic concentrators or other circular symmetrical concentrators, the symmetry axis needs to remain parallel to sunlight; thus it should be able to move in two independent directions using a dual-axis sun-tracking system. In other words, direct solar radiation entering the concentrator's aperture has to be perpendicular to the aperture plane so that solar radiation entering the concentrator can reflect onto the receiver maximally.

2.6.3 Influence of Concentration Ratio on System Collecting Temperature

The previous section showed that the concentration ratio is one of the most important indicators for a concentration solar collector. The concentration ratio has two categories: radiation flux and geometric. The radiation-flux concentration ratio is defined by the ratio of average radiation flux on the receiver to that on the light aperture; it is sometimes called the energy flux density ratio and represents the concentrated magnification of solar energy. However, the geometrical concentration ratio, or concentration ratio in short, is mostly used. Its definition is the ratio of the aperture area to the receiver area, and represents the degree of possibility that a collector could concentrate energy, which is a geometrical characteristic parameter for the concentrating collector. The geometrical concentration ratio is defined as (Eq. (2.52)):

$$C = \frac{A_a}{A_{abs}} \tag{2.52}$$

where A_a is the light aperture area and A_{abs} is the total receiver area. The concentration ratio is an important parameter for concentrating solar collectors; it determines the collecting temperature. Generally, the higher the required collecting temperature is, the larger the system concentration ratio is. Certainly, the higher requirement is expected for concentration systems and tracking accuracy.

Fig. 2.46 illustrates the calculated relationship between the concentration ratio and the receiver surface temperature for various solar collectors. Properly speaking, it is the relationship between average solar radiation intensity on the receiver surface and the working temperature [13]. Higher receiver surface temperatures can increase heat loss.

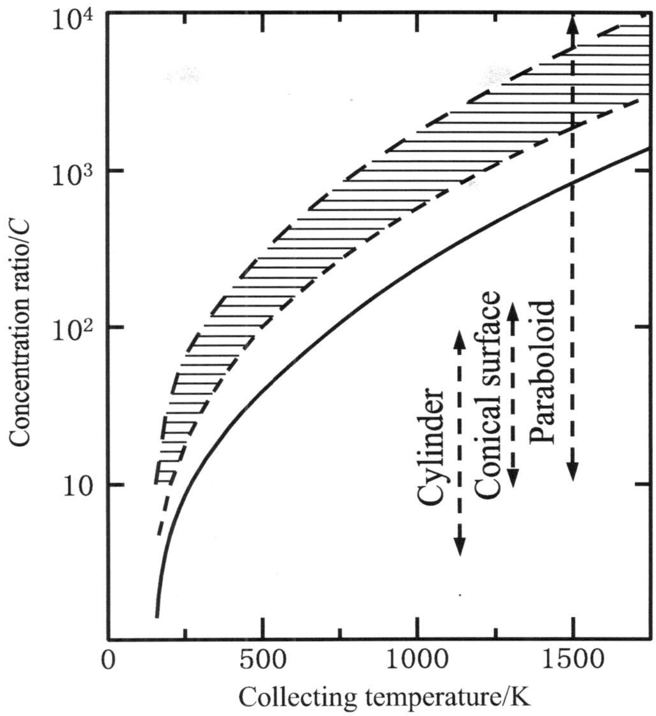

Figure 2.46
Relationship between the concentration ratio and collecting temperature [13].

When the received heat is balanced with heat loss, the relationship between the balance temperature and radiation intensity on the focal plane or concentration ratio is represented by a solid line in the figure. If there is useful heat output under a certain temperature, the concentration ratio needs to be improved. The shaded zone represents collector efficiency within a 40–60% working range; this is also a feasible system working range. The figure also gives the approximate range for a concentration ratio of several reflective concentrators for reference purposes.

The concentrator is normally divided into two categories: nonimaging concentration and imaging concentration. The nonimaging concentrator could not make the sunlight converge to the focal point or line but the face on the receiver. Therefore, the clear image of sun could not be generated on receiver of concentrator. The concentration ratio of linear nonimaging collectors is normally low and within 10. The imaging concentrator is similar to camera lens, and it will form sun image on receiver. This type of solar concentration has a higher concentration ratio and may be as high as hundreds or thousands. It can achieve a high solar collection temperature, which is the main method of using sunlight to capture a high temperature. Disadvantages are that an ideal reflector and higher tracking

Table 2.6: Performance of several common concentrators [13].

Concentrator type		Concentration ratio range	Maximum operating temperature (°C)
Three-dimensional concentrator	SRTA	50–150	300–500
	Fresnel lens	100–1000	300–1000
	Dish paraboloid	500–3000	500–2000
	Tower	1000–3000	500–2000
Two-dimensional concentrator	Compound paraboloid (seasonal adjustment)	3–10	100–150
	Fresnel lens	6–30	100–200
	Paraboloid and Fresnel lens reflector	15–50	100–300
	FMSC	20–50	300

SRTA, spherical surface concentrating collectors; *FMSC*, strip surface concentrating collectors.

accuracy are required, and thus the construction cost is high. Therefore, in practical engineering, the choice of an imaging or nonimaging concentrator should be based on user requirements instead of a higher collecting temperature concentrator; otherwise the economic performance of the system will be influenced. Table 2.6 lists the concentration ratio and operating temperature range for several solar concentration collection systems.

2.6.4 Thermal Efficiency of Solar Concentration System

Solar concentration systems collect sunlight, which is converted into high-temperature heat through a receiver, but the receiver cannot deliver all of the received heat to users because of heat dissipation. For the purpose of describing heat dissipation, system thermal efficiency is introduced as an evaluation indicator.

Assuming the effective useful heat of the concentration solar collector system to be Q_u, the energy balance equation is (Eq. (2.53)):

$$Q_u = Q_{abs} - Q_l \tag{2.53}$$

where Q_{abs} is the heat received by the receiver and Q_l is the receiver heat lost to the ambient atmosphere.

The instant collector efficiency η_c could be expressed as (Eq. (2.54)):

$$\eta_c = \frac{Q_u}{IA_a} = \eta_o - \frac{Q_l}{IA_a} \tag{2.54}$$

where I is the solar radiation intensity perpendicular incident on the light aperture; A_a is the light aperture; and η_0 is the optical efficiency of the collector.

The optical efficiency η_0 in this equation is (Eq. (2.55)):

$$\eta_0 = \frac{Q_{abs}}{IA_a} \tag{2.55}$$

where η_0 is the ratio of the total received energy on the receiver to that on the aperture.

The optical loss of the concentrating solar collector is more obvious than for the flat-plate one, and only direct solar radiation is used for the concentrating solar collector. Some portion of diffuse solar radiation could be used only on a low-concentration ratio collector. Therefore, the loss in a diffuse component and optical loss should be considered in the energy balance of a concentrating solar collector.

The value of Q_l is related to the system working temperature. When the receiver temperature is higher than the ambient temperature, heat loss might be generated in the collection system. The higher the temperature is, the more heat loss there is. Assuming the heat loss on per unit receiver area is $\dot{q}_l(W/m^2)$, the system thermal efficiency turns to (Eq. (2.56)):

$$\eta = \eta_0 - \frac{\dot{q}_l}{I} \cdot \frac{1}{C} \tag{2.56}$$

This equation tells us that the smaller receiver is beneficial for thermal insulation; a smaller receiver area means a high concentration ratio. A solar collection system with a high concentration ratio requires not only high optical precision in the reflection mirror but also high accuracy in the sun-tracking system. To show the relationship more intuitively between heat loss and the operation temperature, the heat loss coefficient of a solar collection system is defined as U, which is (Eq. (2.57)):

$$U = \frac{\dot{q}_l}{(T_{abs} - T_a)} \tag{2.57}$$

where T_{abs} is the receiver surface temperature and T_a is the ambient temperature.

Research showed that the relationship between the value of U and the operation temperature is not obvious; thus in most cases U may be regarded as a constant. Then the linear efficiency equation is (Eq. (2.58)):

$$\eta = \eta_0 - U(T_{abs} - T_a)/I \cdot C \tag{2.58}$$

This processing method is convenient because the collector efficiency is determined by only two parameters: η_0 and U. If the $(T_{abs} - T_{amb})/I$ is the horizontal axis, η is the vertical axis and shows the efficiency variance in the figure. It will be a line with an intercept of η_0 and a slope of $-U$, as shown in Fig. 2.47A.

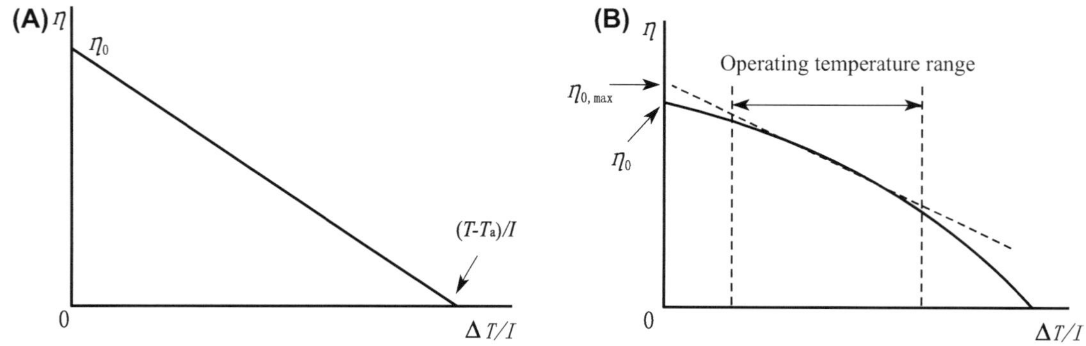

Figure 2.47
Relationship between efficiency and the operation temperature.

Strictly speaking, U is not a constant while a coefficient that is related to parameters such as receiver temperature, ambient temperature, sky temperature, wind speed, and collector inclination. However much study has shown that in most cases, this relationship is not obvious and its influence is small. Thus it should only be given due consideration in real practice.

For example, when the value of U increases with rising temperature, the correction between efficiency and the operation temperature is shown in Fig. 2.47B. The solid line is the actual variation curve whereas the dotted line is a linear line when U is nearly constant.

For a high-temperature solar collector, because the range of operation temperature crosses hundreds of degrees Celsius, it seems to be necessary to consider the influence of temperature on the value of U. This time, U may be expanded to the sum of a constant plus a quadratic correction term, i.e., (Eq. (2.59)):

$$U = U_0 + U_1(T_{abs} - T_a) \tag{2.59}$$

Sometimes for the convenience of calculation, the value of U can be summarized as an equation that varies with temperature (Eq. (2.60)):

$$U = U_0(T_{abs} - T_a)^p \tag{2.60}$$

where U_0 and p are two experimental constants that are determined by measurement and U_1 represents the changing scope of heat loss coefficient with temperature. Some researchers also use Eq. (2.61) to calculate the value of U:

$$U = U_0 + U_1(T_{abs} - T_a)/I \tag{2.61}$$

The heat transfer coefficient is independent of the solar radiation intensity.

2.7 Nonimagining Concentrator

A conventional solar collector can be divided into two categories: concentration and nonconcentration. Typical representatives of nonconcentration collectors are flat-plate and all-glass vacuum tube collectors. There are various types of concentration collectors, which can also be divided into two categories based on the form of concentration form: strictly concentration collectors and nonstrictly concentration collectors, or imaging and nonimaging concentrators. Imagining concentrators can strictly concentrate and form a sun image whereas nonimaging concentrators cannot strictly generate a focal point and roughly concentrate sunlight. If the concentration collector is divided by the tracking form, there will be tracking and nontracking concentrators. As is indicated by its name, a nontracking concentrator does not need to track the sun with time; it only needs to be fixed in an optimal position, or its tilt angle only needs to be adjusted at a fixed period. The advantage of this kind of device is its simple operation, but its collection temperature is low and can only reach 100−150°C, and in some period such as the morning and evening, its efficiency will be significantly reduced. The tracking concentration solar collector needs real-time solar tracking; thus its efficient is high, but the system is complicated and its maintenance is inconvenient. Comprehensive consideration is needed when making a choice.

A CPC is a typical nontracking, nonimaging concentrator; its most notable advantages are its large concentration angle, no real-time sun-tracking, high reliability, and its simple structure. Therefore, it has a good market to some extent.

2.7.1 Structural Characteristics of Compound Parabolic Concentrators

CPCs are composed of two curved surfaces. These structures are is clearer especially for trough CPCs. The structure and design concept of CPC are shown in Fig. 2.48.

In designing a CPC, the focal point of paraboloid A lies on the lower end of paraboloid B; similarly, the focal point of paraboloid B lies on the lower end of paraboloid A. The maximal concentration angle θ_{max} may be simply defined as half of the angle between the connection lines of the upper and lower ends of the two paraboloids. If the angle between the incident light and the symmetric axis is defined as incidence angle θ_i, as shown in Fig. 2.49A, $\theta_i < \theta_{max}$, and the incident light could be reflected to a light exit. When $\theta_i = \theta_{max}$, i.e., the incident light is parallel to the axis of paraboloid A, the incident light would be reflected to the focal point of paraboloid A, which is also the lower end of paraboloid B; as shown in Fig. 2.49B. When $\theta_i > \theta_{max}$, the incident light cannot escape from the light exit. However, it can be reflected back to the light aperture. Therefore, the maximal concentration angle is the key factor in designing CPCs.

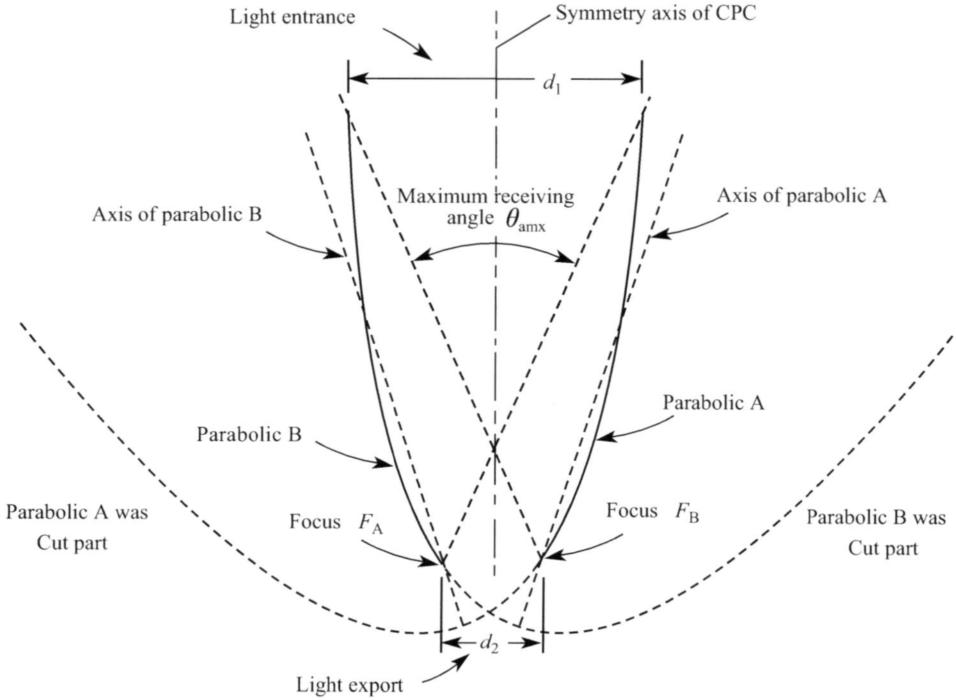

Figure 2.48
Structure of a CPC solar concentrator.

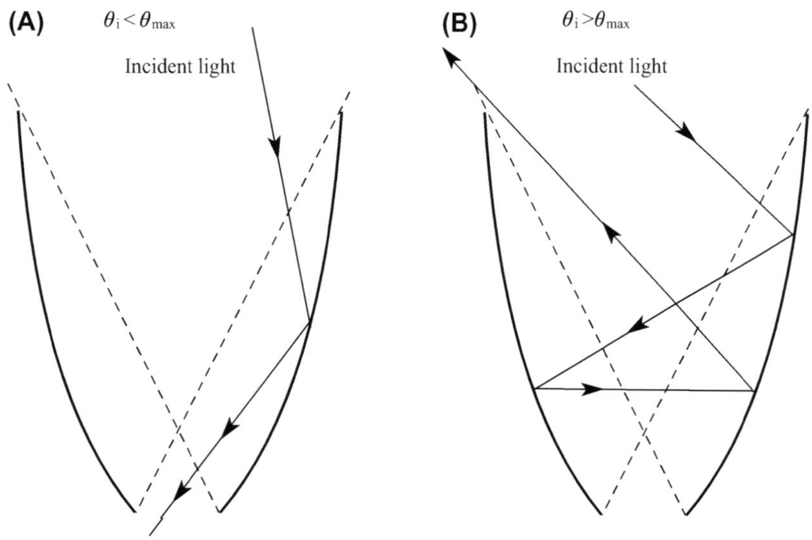

Figure 2.49
Optical path of a single CPC.

2.7.1.1 Maximum Concentration Angle of Compound Parabolic Concentrator

According to the optical path within a CPC, only if the angle between the incident light and the asymmetrical axis is smaller than the maximum concentration angle θ_{max} can light transmit through a CPC. So what are the influential factors for the maximal concentration angle? Apparently it is related to the structural parameters, and one of the most important parameters is concentration ratio C. This concentration ratio is the system geometrical concentration ratio, or simply, it is the ratio of the light aperture area and light exit area. For a two-dimensional trough concentrator, it is the ratio of aperture width d_1 to exit width d_2.

After geometrical calculations, the correlation between concentration ratio C and maximum concentration angle θ_{max} for a CPC could be expressed as (Eq. (2.62)):

$$\sin\theta_{max} = \sin\alpha = \frac{1}{C} \tag{2.62}$$

To obtain the maximal concentration angle through the concentration ratio, Eq. (2.63) is used:

$$\theta_{max} = \alpha = \arcsin\left(\frac{1}{C}\right) \tag{2.63}$$

According to the geometrical relationship, it is easy to get the height of s full CPC (Eq. (2.64)):

$$h = d_2(C+1)\sqrt{C^2-1} \tag{2.64}$$

where $d_2$is the light exit width. As we know, for a concentrator with a half-acceptance angle of θ_a, its thermodynamic limitation for concentration ratio is (Eq. (2.65)):

$$C_{2-D} = \frac{1}{\sin\theta_a} \quad \text{or} \quad C_{3-D} = \frac{1}{\sin^2\theta_a} \tag{2.65}$$

This is same as the relationship between the geometrical concentration ratio and the concentration angle. Because a CPC has larger half-acceptance angle, for example, the half-acceptance angle of a trough CPC collector ranges from 15 to 20 degrees, sometimes even reaching 30 degrees, its concentration ratio is only 4, for a three-dimensional CPC, its concentration ratio is also only in the range of 10–15. Therefore, the concentration ratio of CPC is largely limited by the concentration angle. When a large acceptance angle is required, the concentration ratio would be low, and thus it is impossible for the collector temperature to be high.

Common structures of CPC combined with other collectors are shown in Fig. 2.50.

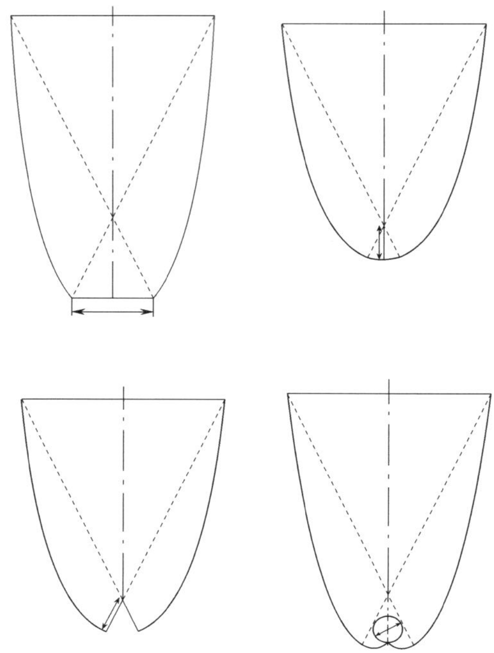

Figure 2.50
Combination of CPCs with other forms of collectors.

2.7.1.2 Truncation of Compound Parabolic Concentrators

The CPC is normally high and needs a large amount of reflective material, which is regarded as a disadvantage of CPC. However with the increase of height in the upper half of a CPC, the increase in aperture width is small, and sometimes there is no addition. Therefore, a consideration is to use a non—full size CPC or to truncate the upper part of a full-size CPC. The practice or calculation shows that when the upper part of a CPC is truncated, the loss in concentration ratio is small or the decrease of incoming light into the device is small. For example, when CPC is truncated by one-third, the decrease in concentration ratio is only 10% [21]. Fig. 2.51 compares the geometrical relationship before and after truncation.

The CPC structure shows that the focal length of parabolic-forming CPC is (Eq. (2.66)):

$$f = a'(1 + \sin\theta_a) \tag{2.66}$$

It is easy to derive the light aperture width after truncation from the geometrical relationship in Fig. 2.51 (Eq. (2.67)):

$$a_T = \frac{f\,\sin(\theta_T - \theta_a)}{\sin^2(\theta_T/2)} - a' \tag{2.67}$$

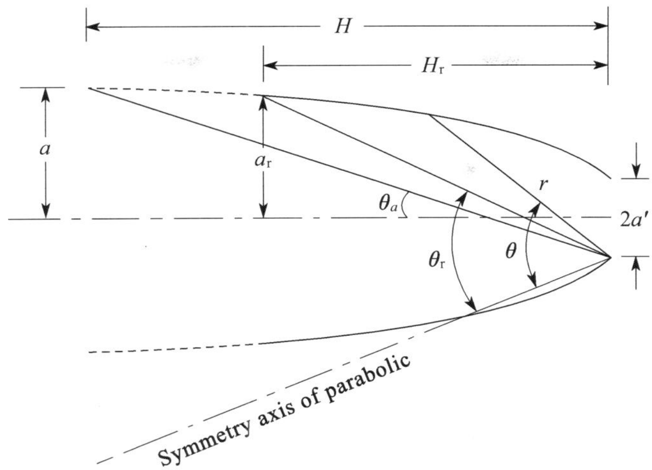

Figure 2.51
Geometry relation of truncated type CPC.

The light aperture width before truncation is (Eq. (2.68)):

$$a = \frac{a'}{\sin \theta_a} \tag{2.68}$$

The height of the concentrator after truncation is (Eq. (2.69)):

$$H_T = \frac{f \cos(\theta_T - \theta_a)}{\sin^2(\theta_T/2)} \tag{2.69}$$

The height of the concentrator before truncation is (Eq. (2.70)):

$$H = \frac{f \cos \theta_a}{\sin^2 \theta_a} \tag{2.70}$$

Thus the height–width ratio after truncation of the CPC is (Eq. (2.71)):

$$\frac{H_T}{a_T} = \frac{(1 + \sin \theta_a)\cos(\theta_T - \theta_a)}{\sin(\theta_T - \theta_a)(1 + \sin \theta_a) - \sin^2(\theta_T/2)} \tag{2.71}$$

The height–width ratio before truncation of the CPC is (Eq. (2.72))

$$\frac{H}{a} = \frac{(1 + \sin \theta_a)\cos \theta_a}{\sin \theta_a} \tag{2.72}$$

Fig. 2.52 shows the influence of truncation on the height–width ratio of a CPC under a different half-acceptance angle.

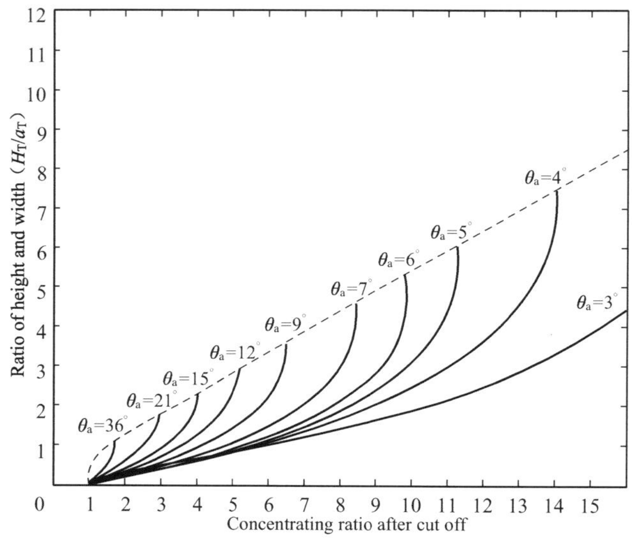

Figure 2.52
Effect of a truncated CPC on the ratio of height and width [21].

It is easy to obtain the concentration ratio of a CPC after truncation using the geometrical relationship in Fig. 2.51 (Eq. (2.73)):

$$C_T = \frac{a_T}{a'} = \frac{f \sin(\theta_T - \theta_a)}{a' \sin^2(\theta_T/2)} - 1 = \frac{(1 + \sin \theta_a)\sin(\theta_T - \theta_a)}{\sin^2(\theta_T/2)} - 1 \tag{2.73}$$

The CPC concentration ratio after truncation is related to the concentration angle after truncation, θ_T, and the concentration angle of the original full-size CPC, θ_a. The analysis shows that when the truncation ratio is less than 0.5, the change in concentration ratio is small, especially when the original concentration angle is large. When the original concentration angle is small, the concentration ratio will decrease dramatically when the CPC is truncated by 0.5. The larger is truncation ratio is, the faster the drop is in the concentration ratio. When the original concentration angle is large, as mentioned, the reason for the slight influence of CPC truncation on the CPC concentration ratio is that the increase in aperture width and height is not obvious. Apart from that, there are two other reasons: (1) after the CPC is truncated, the light that is originally outside the acceptance angle can be reflected to the exit; the incident angle of light $|\theta| > \theta_a$, θ_a, is the original CPC half-acceptance angle, whereas the original light with an incident angle less than $|\theta_a|$ can still reach the light exit; (2) when the CPC is truncated, the reflection number of some light within the CPC is reduced, which is also beneficial to the receiver. Therefore, for a combination of several reasons the CPC truncation has little influence on collector performance. In real practice, most CPC collectors are truncated.

The reflective area represents the material needed to manufacture a CPC device, at the same concentration ratio, but a smaller reflective area is required to save material. The reflective area of the truncated CPC will be significantly reduced, which is good for cost savings.

The maximum concentration angle of CPC is an important performance indicator. Only incident light within the maximum concentration angle can escape from the light exit. For a concentrator whose height to light aperture width ratio is larger than 1, its maximal concentration angle is less than 15 degrees, which is not beneficial for all-day solar collection. Therefore it is necessary to investigate the structural dimension of a CPC to improve its collection performance.

For a full-size CPC with a given concentration ratio, there are four methods for improve its maximum concentration ratio and concentration performance:

1. Truncate from the top of the CPC.
2. Truncate from the bottom of the CPC.
3. Translate the reflector outward using the symmetry axis as the center.
4. Rotate the reflector toward the central axis for a certain angle.

Among these methods, truncation from the top is the most commonly used. In real design practice, it can be truncated by one-third or even half its height to reduce the material cost and shrink the body size, although this process has little influence on CPC performance. The calculation shows that when trough CPC is truncated by 50% from its top, i.e., the height light aperture ratio is reduced by approximately 50%, the concentration ratio is reduced by only 10%, whereas the average reflection number is reduced by 20%, which partly improves the concentrator performance.

2.7.1.3 Reflection Number of Light Within Compound Parabolic Concentrator and Its Transmittance

The light within the maximum concentration angle can be reflected to the light exit, but the reflection of light could can optical loss; the higher the reflection number is, the more optical loss there is. Radiation loss caused by reflection is expressed as (Eq. (2.74)):

$$\text{The reflective loss} = 1 - \rho^{\overline{n}} \tag{2.74}$$

where ρ is the reflectivity of the reflector and $\overline{n}$ is the average reflection number, which is determined by the incidence angle, concentrator height, and concentration ratio.

The reflection number of light within the CPC is determined by the incidence angle, θ. It also determines the position of light on the light aperture. The proportion of light traveling through the CPC varies with the reflectivity of the wall surface, which can be estimated by Eq. (2.75):

$$\tau_{\text{CPC}} = \rho^{<n>} \tag{2.75}$$

where $<n>$ is the average reflection number of light on the reflector. Apparently, for light with an incidence angle of θ, its reflection number must be related to the CPC half-acceptance angle, and it also relates to the position of light on the CPC light aperture. If the CPC is truncated, it should also be related to the size of the truncated part. Generally, when $\theta_a = 5$ degrees, the $<n>$ ranges from 1 to 2; when $\theta_a = 11.5$ degrees, the $<n>$ ranges from 0.76 to 0.86 and the average value is 0.82. In conclusion, the larger the acceptance angle is, the less the reflection number is [21]. Different shapes of CPC collectors may also influence the average reflection number.

The transmittance of the concentrator can be defined as the number of light on the light exit divided by that on the light aperture, or the amount of light that can reach the light exit among the lights collected by the concentrator. This definition is similar to optical efficiency.

The analysis of a CPC needs to use the concept of light transmittance, especially for three-dimensional CPC. This kind of concentrator looks like a funnel whose light inlet and outlet are very clear. In the study of CPC concentration performance, one of the most important conclusions is the concentration ratio (Eq. (2.76)):

$$C = \frac{a}{a'} = \frac{1}{\sin \theta_a} \tag{2.76}$$

where θ_a is the maximal concentration angle of the CPC. This means that all incidence light within θ_a can reach the light exit, which is easy to understand for a two-dimensional CPC. However, the situation is different for a three-dimensional CPC, because there is a nonnormal incidence multireflection in a three-dimensional CPC and the sunlight is not exactly parallel light but nonparalleled light with an angle of $\frac{1}{2}\Delta_s \approx 16'$.

When a beam of parallel light enters the aperture of a three-dimensional CPC along the direction of its symmetry axis, light can completely arrive at the light exit after reflection on the reflective surface of the CPC. When the parallel light does not enter the CPC along the direction of the CPC symmetry axis but with an angle from the symmetry axis, θ, some part of light will go through several reflections on the reflective surface and transmit downward in a rotary type, forming the phenomenon of "optical rotation," as shown in Fig. 2.53.

Will the existence of "optical rotation" influence the light transmittance of a CPC? The answer is yes. Research showed that the change in light transmittance from 0.1 to 0.9 happens in the range of $\Delta\theta < 3$ degrees. However there will be optical loss around $\theta \approx \theta_a$. When $\theta > \theta_a$, there will still be some light transmitting through the CPC.

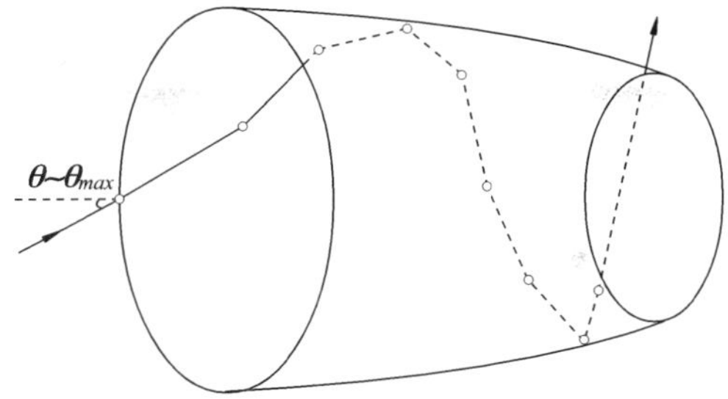

Figure 2.53
Transmission of nonnormal incidence light within a three-dimensional CPC.

2.7.2 Orientation of Compound Parabolic Concentrator and Its Received Energy

The biggest advantage of a CPC is its low requirement for sun-tracking accuracy owing to its large acceptance angle. There will still be some concentration gains even without a sun-tracking system. However, optimal orientation needs to be selected to achieve maximal concentration gains.

For a trough CPC system, if the sun is tracked continually or discontinuously, it appears that the east–west orientation would be an optimal option; under that condition, the opening of the CPC faces the equator and its tilt angle, β, is related to the local geographical latitude. A higher-frequency adjustment of the CPC is required for a smaller acceptance angle.

Sometimes the reflected sunlight from ground is still valid for a CPC, i.e., if the actual concentration angle of a CPC is θ_c, when $\beta + \theta_c > 90$ degrees, the CPC can receive part of the reflected light from the ground, which means under that condition, the receiver can "see" the ground. The relevant angles are given in Fig. 2.54, which gives

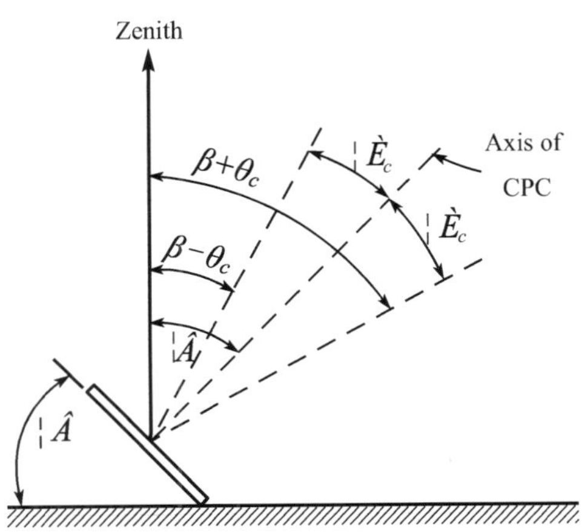

Figure 2.54
Projection of CPC acceptance angle for a CPC on the south–north plane.

the projection relationship of incident light on the south—north plane for a CPC system that is oriented east—west, south-facing (north-facing for the southern hemisphere), and tilted by β.

The figure shows that incident light whose incidence angle θ is within the range $\beta - \theta_c < \theta < \beta + \theta_c$ can be reflected with the receiver. If θ_z and γ_s are the azimuth and zenith angle of sunlight, respectively, only when θ_z and γ_s, respectively, meet certain conditions can sunlight be reflected to the receiver. Mitchell (1997) gave this condition [22] (Eq. (2.77)):

$$(\beta - \theta_c) \leq \tan^{-1}(\tan \theta_z \cos \gamma_s) \leq (\beta + \theta_c) \tag{2.77}$$

Sometimes some transparent cover will be added to the CPC; the benefit of adding a cover is to protect reflective and receiver surfaces from the erosion of ambient dust. It reduces heat loss. Direct and indirect sunlight entering a CPC could be influenced by the cover, and the degree of influence relies on the transmittance of cover τ_c.

For diffuse radiation, only some part of diffuse light entering the CPC can reach the receiver, called effective diffuse radiation. The amount of effective diffuse radiation is also related to the system acceptance angle. Assuming that the diffuse light is isotropy, the rough relationship between the average incidence angle of effective diffuse light from sky θ_e and the CPC acceptance angle θ_c is [13] (Eq. (2.78)):

$$\theta_e = 44.86 - 0.0716\theta_c + 0.00512\theta_c^2 - 0.00002798\theta_c^3 \tag{2.78}$$

For example, for a CPC with $\theta_c = 20$ degrees, its average incidence angle of effective diffuse radiation is 45 degrees.

2.8 Paraboloid Solar Concentrator

The paraboloid solar concentrator is currently the most widely used solar concentrator; it has the most mature technology and is applied on the market on a large-scale. There are two types of paraboloid concentrator: focal point and linear focus. They are important devices for humans to obtain high-temperature heat from sunlight. The focal point paraboloid concentrator is also a commonly used method to achieve super—high temperature heat from solar energy in manufacturing and research. From a thermodynamic view, a focal point concentration collector can theoretically collect solar energy to achieve above 5000°C; in reality, the current highest temperature collected on earth from solar energy is around 4000°C. A linear focus concentrator can also achieve 500°C, and is currently the preferred solar concentration device for large-scale solar power systems.

2.8.1 Dish Paraboloid Concentrator System

2.8.1.1 Structure and Concentration Principles for Dish Paraboloid Concentrator System

The dish paraboloid solar concentration system is also called a dish system. Its main features are that it uses a dish-shape rotational paraboloid reflective mirror to concentrate light and its structure is similar to a large-scale paraboloid radar antenna. The dish paraboloid solar concentrator is a typical focal point concentration solar collector whose concentration ratio could be hundreds or even thousands; thus it could generate a super-high temperature. The opening diameter of a collecting mirror depends on various conditions. It could be as small as a few centimeters, such as a concentrator that uses collected sunlight for optical fiber illumination; it could also be as large as a few meters, such as a solar cooker to achieve high-temperature cooking. Larger ones could be as much as several decameters, such as a large-scale concentrator to achieve high-temperature power generation. The diameter is ideally limited to between 10 and 15 m. The main reason is that it is limited by the mirror structure and self-weight. If the actual mirror deviates too much from the ideal shape, it would be easy to exceed the focal spot diameter, which is not beneficial to a high-temperature or high-efficiency solar collector.

To avoid manufacturing difficulties for large-scale reflectors, in reality, large-scale dish paraboloid concentrators consist of several dish systems. Multiple- and single-dish systems have their own features. Multiple-dish systems have low wind resistance, easy installation, low cost, and low self-consumption power; single-dish systems have a compact structure, high concentration efficiency, and simple analysis.

Fig. 2.55 shows the cross-section structure of a paraboloid concentrator. The sunlight that is parallel to the optical axis and projects on the edge of the paraboloid is called the

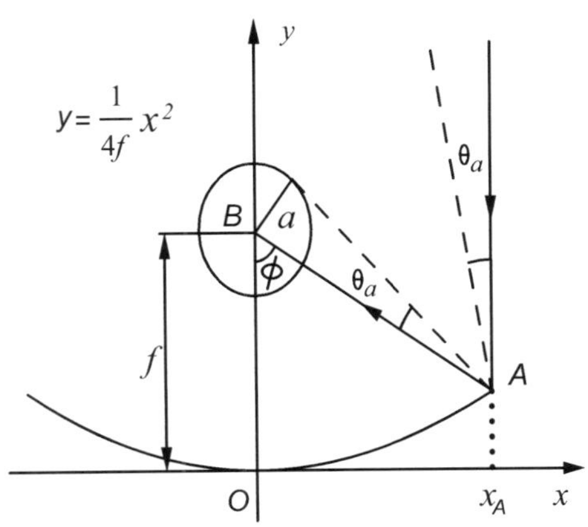

Figure 2.55
Relationship between receiver radius a and acceptance angle θ_a.

boundary incident light. Light that is reflected from the edge of the paraboloid to a focal point is called the boundary reflected light. The angle between the boundary reflected light and asymmetric axis is called the boundary angle, which is expressed by ϕ in Fig. 2.55 (i.e., angle $\angle ABO$).

An ideal paraboloid can concentrate parallel light into an ideal point; thus it is an identical reflector shape and is applied in various areas. Certainly, only light that is parallel to an optical axis (or symmetric axis) could ideally be concentrated to one point by a paraboloid.

For sunlight, although incident light travels alone the optical axis, the paraboloid still is not able to concentrate the light completely into one point, because sunlight is not completely parallel light; it has an angle, $\Delta_s = 32'$. Thus a focal area or focal spot will form around the focal point, and there is a certain energy distribution within the focal area. The size of the focal area is not caused only by the nonparallelism of sunlight but also by the geometrical error of the paraboloid. The focal area size for a paraboloid solar concentrator is much larger than its maximal concentration ratio. To make the solar collector around the focal area receive all lights, the solar collector needs to be big enough. However, there is an economical loss for a large receiver, and sometimes it is not allowed by the structure. Therefore, the receiver size is determined by the choice of an appropriate collection ratio.

As shown in Fig. 2.55, the opening diameter of the paraboloid concentrator is expressed by D; thus the collection area if sunlight is $\frac{1}{4}\pi D^2 = \pi x_A^2$. If the receiver on the focal point is spherical and its radius is a, then its total collection area is $4\pi a^2$, and then the concentration ratio of the paraboloid concentrator could be written as approximately (Eq. (2.79)):

$$C_{3D,parab} = \frac{\frac{1}{4}\pi D^2}{4\pi a^2} = \frac{\sin^2\phi}{4\sin^2\theta_a} \tag{2.79}$$

where θ_a is the half divergence angle of the boundary point to the receiver, shown as a dotted line in Fig. 2.55. If it is a two-dimensional linear focus paraboloid concentrator, i.e., the traditional trough solar concentrator, the concentration ratio could be expressed as (Eq. (2.80)):

$$C_{2D,parab} = \frac{2x_A}{2\pi a} = \frac{\sin\phi}{\pi\sin\theta_a} \tag{2.80}$$

The opening width of a paraboloid is $D = 2x_A$. According to the geometrical relationship, the relationship between the boundary angle, the opening width of the paraboloid, and the focal length is (Eq. (2.81)):

$$\tan\left(\frac{\phi}{2}\right) = \frac{D}{4f} \tag{2.81}$$

The width—height ratio of the concentrator is (Eq. (2.82)):

$$\frac{D}{H} = \frac{8f}{x_A} = \frac{8f}{r \sin \phi} \tag{2.82}$$

where r is the distance between the paraboloid boundary and the focal point and H is the vertical height of the concentrator. It could be seen from Eqs. (2.79) and (2.80) that the maximal concentration ratio happens when $\phi = 90$ degrees. When $\phi > 90$ degrees, the concentration ratio decreases instead, and the width—height ratio of the concentrator increases, which is not beneficial to the cost of the concentrator surface material.

2.8.1.2 Energy Distribution on Focal Area of Paraboloid

A paraboloid reflector is an old solar concentration device. Investigations of this kind of device have been thorough and there have been some beneficial conclusions. With the development of computer technology, many researchers have used computational programs to simulate the concentration properties of paraboloids, energy distribution on focal spots, and incidence angle correlations, and some conclusions that used to be unknown have been determined. Ray tracing is the most visualized and commonly used method that also gives the most information.

Knowing the energy distribution around the focal point is an important reference to determining the size and shape of the receiver. Therefore, in research on paraboloid concentrator performance, analyzing the energy distribution on a focal spot is an important task. Because sunlight itself is not a completely parallel light, it comes from a solar "dish" and thus it is not a truly parallel light. Therefore, despite an ideal paraboloid concentrator, its focal point is not a point but an area.

When light is reflected by a concentrator, it is impossible for the reflector surface to be an ideal reflective surface. Considering all of these factors, sunlight reflected to the focal point will scatter. This light diffusion is caused by two factors: the nonidealization of the reflector surface and the nonparallelism of sunlight.

Both theoretical and experimental studies have shown that for a solar concentrator that does not deviate from an ideal paraboloid too much, energy distribution around its focal area may be illustrated by the distribution curve in Fig. 2.56. In the central area of a solar image, the energy distribution is even, whereas on the boundary of the solar image, the energy density drops significantly. The rate of drop on the solar image boundary is related to the nonidealization of the reflector surface, whereas the size of the image is determined by the boundary angle, aperture width, focal length, and so forth.

Fig. 2.57 demonstrates a simulated curve showing the influence of the boundary angle on the energy flux density distribution, in which the horizontal axis R is the distance from

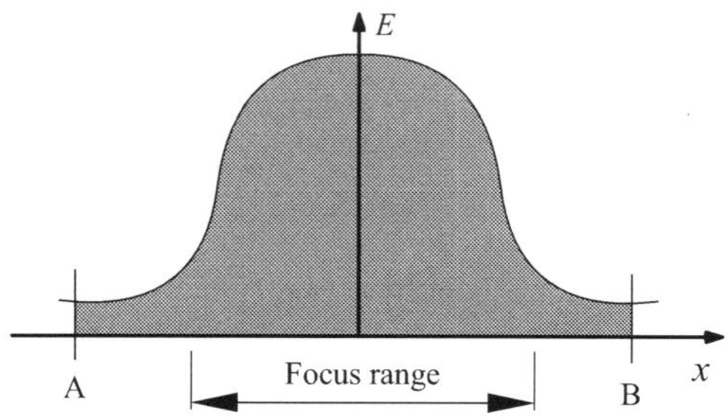

Figure 2.56
Energy distribution within a focal spot.

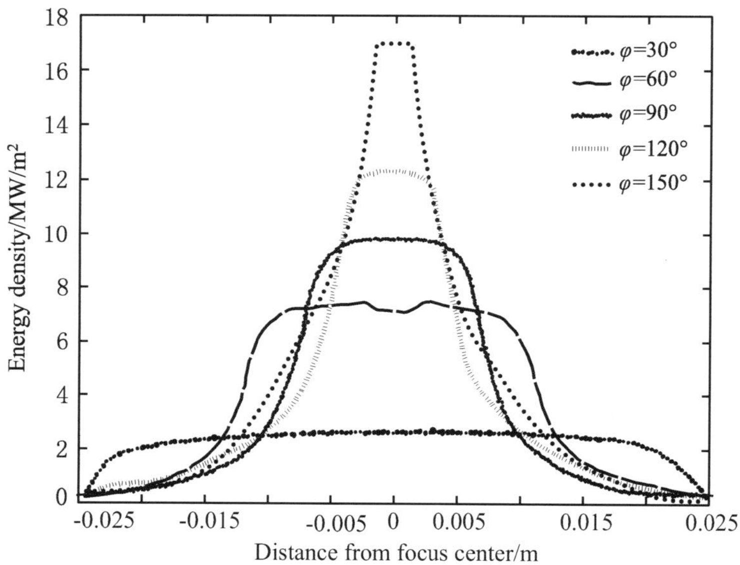

Figure 2.57
Influence of the boundary angle on the energy flux density distribution [23].

the focal spot. The larger the boundary angle is, the more centralized the energy density is within the focal spot.

2.8.1.3 Energy Flux on the Receiver and Intercept Factor

There are many receivers for a paraboloid concentrator, including flat-plate, sphere, cavity, and so forth. The cavity type has become popular owing to its low heat dissipation. It is also conveniently connected with heating devices. Fig. 2.58 illustrates some structures for cavity-type receivers. Fig. 2.59 shows a paraboloid concentrator and cavity receiver combination [19].

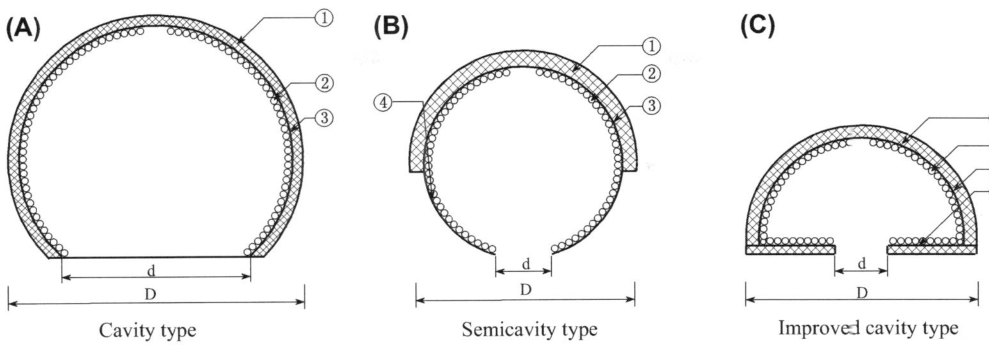

Figure 2.58
Structure of a cavity-type solar receiver.

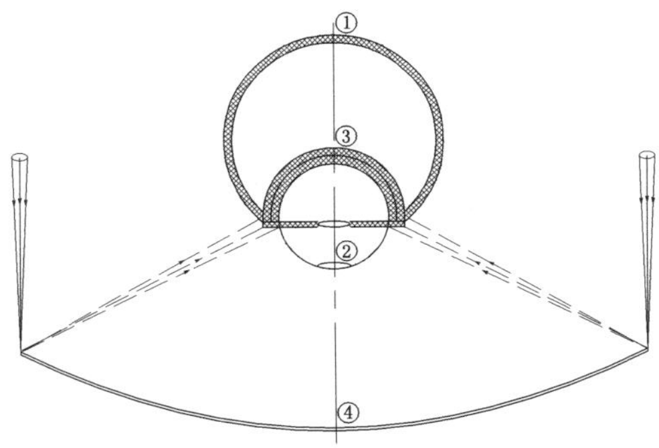

Figure 2.59
Combinations of paraboloid concentrator and cavity: 1, cavity absorber; 2, semicavity absorber; 3, improved cavity absorber; and 4, plate light collector.

It is impossible to receive all reflected light completely around a focal point no matter what receiver is used. Therefore it is necessary to investigate the collecting efficiency of the receiver. $B_{eff}(\theta)$ is set to be the radiation intensity from the direction of θ (W/m^2 mrad); $f(\theta)$ is the acceptance function with the angle, i.e., the amount of radiation that can be delivered to the receiver. Therefore the obtained energy flux is the product of $B_{eff}(\theta)$ and $f(\theta)$, and integrating the product by the incidence angle, the energy flux intensity on receiver is (Eq. (2.83)):

$$I_b = \int_{-\infty}^{\infty} d\theta f(\theta) B_{eff}(\theta) \tag{2.83}$$

This equation considers the size of the receiver because the size is contained in concentration ratio C, and C is also contained in the angular acceptance function. If the receiver can collect lights from all directions, it is not related to the angular acceptance function, or the function equals 1; thus the total energy flux intensity on the receiver turns to be (Eq. (2.84)):

$$I_b = \int_{-\infty}^{\infty} d\theta B_{eff}(\theta) \tag{2.84}$$

In fact, the I_b here is the sum of all lights projecting around the focal point. Thus, the intercept factor is as follows (Eq. (2.85)):

$$\gamma = \frac{I_{in}}{I_b} = \int_{-\infty}^{\infty} d\theta \frac{f(\theta)B_{eff}(\theta)}{I_b} \tag{2.85}$$

where I_{in} is the sum of intercepted lights by the receiver that project around the focal point. This equation is useful in a real application because it is an independent magnitude that is not related to the radiation intensity. The system's optical efficiency can be obtained after the intercept factor is obtained (Eq. (2.86)):

$$\eta_o = \rho\tau\alpha\gamma \tag{2.86}$$

where ρ is the reflectivity of the reflector, $\tau\alpha$, the product of receiver transmittance and absorptivity; and γ is the intercept factor of the receiver. It is convenient to use this equation because the relationship between relevant parameters is clear. If the given direct solar radiation intensity is I_b, the collector area is A, and then the power provided by receiver is $W = \rho\tau\alpha\gamma A I_b$.

2.8.2 Trough Paraboloid Solar Concentrator

2.8.2.1 Structure of Ideal Trough Paraboloid Solar Concentrator

The trough paraboloid solar concentrator is a typical linear focus concentrator. This is also one of the most widely used and investigated solar concentrators. Previous massive theoretical analysis and experimental measurements have proved that it is an effective solar collection device.

To deeply understand the working principles of this type of thermal collector, it is necessary to acknowledge its optical and concentration properties, especially solar radiation energy flux distribution in the focal area. The light concentration principle of a trough paraboloid concentrator is shown in Fig. 2.60. Sunlight enters perpendicularly to the light aperture and projects on the paraboloid concentrator, and it is reflected onto the receiver by the reflector and through the receiver so that high-temperature heat can be collected. Finally, the heat is transferred by the working fluid within the receiver.

According to the concentration ratio of a trough paraboloid concentrator, its working temperature ranges from 100 to 500°C. Circular tube collectors are normally used for this type of solar collector. To reduce heat dissipation, the collector is covered by a glass tube around its outside and the glass tube is vacuumized. The flat-plate collector is also

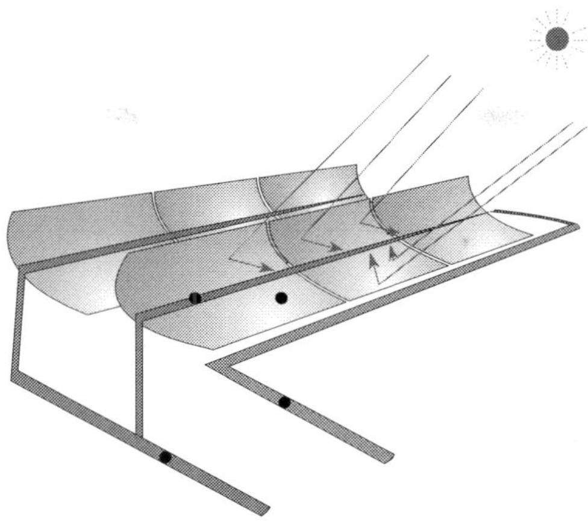

Figure 2.60
Light concentration process for a trough paraboloid solar concentrator.

applicable; it is normally called a fin tube collector. Advantages of this type of collector are that the working fluid flows only within a very small tube and the fin (flat-plate) is tightly attached to the small tube to collect sunlight. This method can reduce the weight of the heat transfer fluid and increase the temperature, which equivalently reduces the material consumption of the collector. To reduce heat dissipation from the fin, sometimes the fin is also placed in a vacuumized glass tube.

Fig. 2.61 shows several collectors used for trough concentrators. Some large-scale concentration systems might need a second concentration to improve the concentration

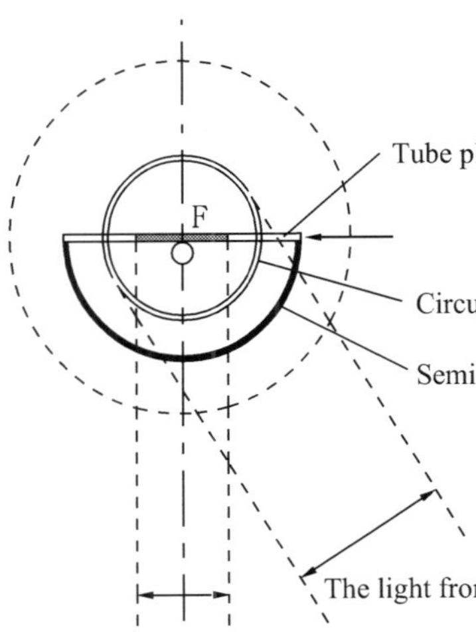

Figure 2.61
Collectors used for a trough concentrator [19].

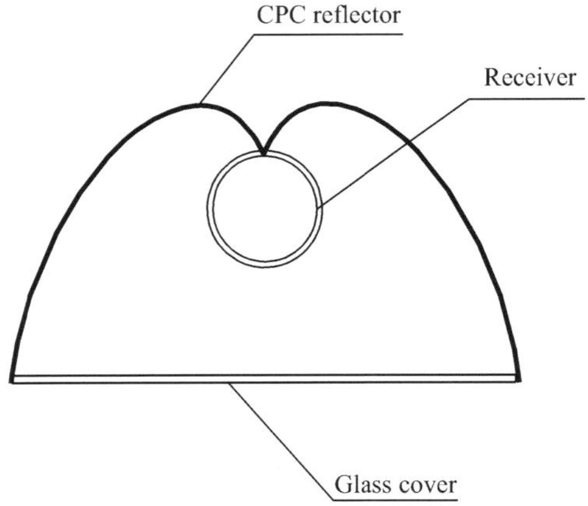

Figure 2.62
A CPC concentrator and tubular collector.

ratio further and the collected temperature, as shown in Fig. 2.62, whereas a simple collector is used for a simple paraboloid concentrator, as shown in Fig. 2.63.

For this kind of concentration collector, the provided solar radiation intensity on the unit light aperture area is (Eq. (2.87)):

$$S = I_b\rho(\gamma\tau\alpha)_n K_{\gamma\tau\alpha} \tag{2.87}$$

To calculate S, it is necessary to know the image features formed by reflected light. Theoretically, the focal spot for an ideal trough concentrator is an ideal straight line.

The geometrical structure of a paraboloid concentrator with a linear focus is shown in Fig. 2.64. The concentrator width is a and the focal length is f. The parabolic equation on a coordinate system is (Eq. (2.88)):

$$y = \frac{x^2}{4f} \tag{2.88}$$

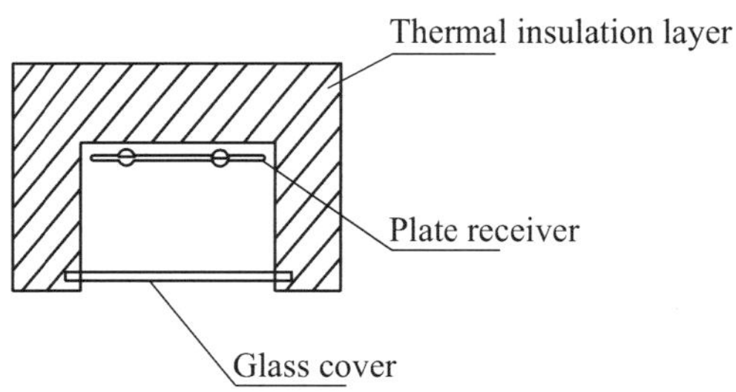

Figure 2.63
A single-side flat-plate collector.

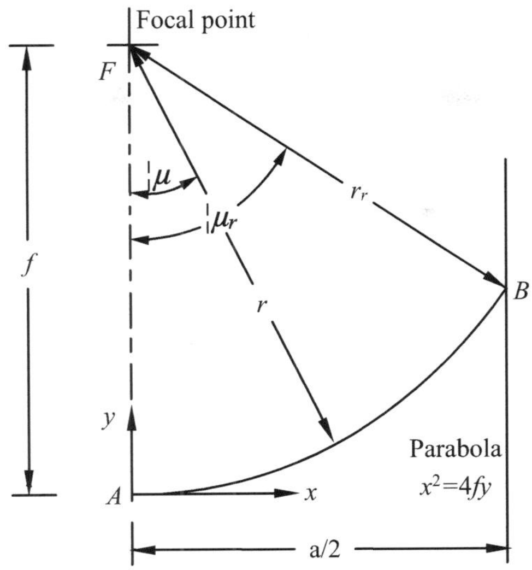

Figure 2.64
Geometrical structure of a paraboloid concentrator.

Fig. 2.65 gives the concentration principle of a paraboloid concentrator. The direct radiation incident on point B of a paraboloid concentrator, point B, is the boundary point of the paraboloid concentrator, and the distance to focal point is $r_{r,1}$, which is also called the maximum mirror radius. ϕ_r is called the boundary angle, i.e., $\phi = \angle AFB$. It is easy to

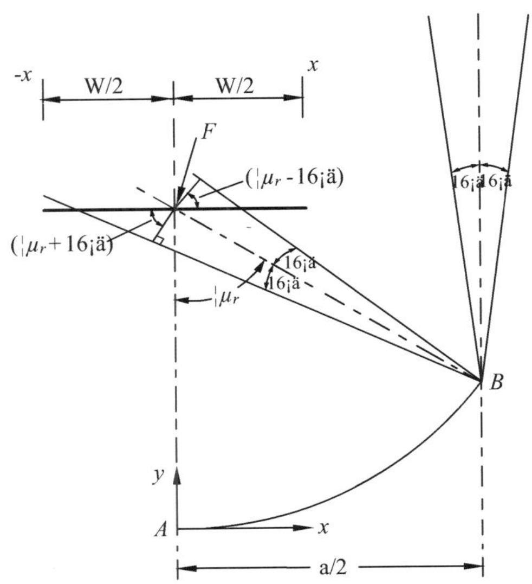

Figure 2.65
Concentration principle of a paraboloid concentrator.

calculate the relationship of the boundary angle, focal distance, and light aperture width (Eq. (2.89)):

$$\phi_r = \tan^{-1}\left[\frac{8(f/a)}{16(f/a)^2 - 1}\right] = \sin^2\left(\frac{a}{2r_r}\right) \tag{2.89}$$

The mirror radius on any point of parabolic is (Eq. (2.90)):

$$r = \frac{2f}{1 + \cos\phi} \tag{2.90}$$

In fact, direct solar radiation is not strictly parallel light, whereas cone-shaped light with an angle of $\Delta_S = 0.53$ degree; its half-angular breadth is 0.267 degrees, or 16′. For the convenience of discussion, it concentrator is assumed to be symmetrical, the incident light is perpendicular to light aperture, and direct radiation enters the concentrator along the central axis plane and then a focal zone or area will form on the focal point position. The focal zone is not an exact point but a focal spot with a certain size. The size of the focal spot and energy distribution on the focal spot is determined by idealization of the reflective surface: the more ideal the reflective surface is, the smaller the distribution focal spot is. The range of the focal spot is the smallest for an ideal paraboloid. The width of the focal zone or area is determined by the reflected light from the paraboloid boundary. It is called the minimum width of the focal zone, and also the width of the solar image, which apparently increases with the increase in boundary.

Figs. 2.61–2.63 show several shapes of collectors. Their minimum dimension all should be larger than the real focal spot area. For an ideal paraboloid reflector, the minimum intercepted solar image by the collector could be calculated, and the diameter for cylinder collector is (Eq. (2.91)):

$$d = 2r_r \sin 0.267 = \frac{a \sin 0.267}{\sin\phi_r} \tag{2.91}$$

For a flat-plate collector on a focal plane, width W is (Eq. (2.92)):

$$W = \frac{2r_r \sin 0.267}{\cos(\phi_r + 0.267)} = \frac{a \sin 0.267}{\sin\phi_r \cos(\phi_r + 0.267)} \tag{2.92}$$

Note: W is also the diameter of a semicircle collector.

2.8.2.2 Positioning of Trough Paraboloid Solar Concentrator

In practice, a trough paraboloid concentrator has an issue of positioning: no matter whether it has an east–west or south–north alignment, or a horizontal or tilt alignment, it can track the sun in one only direction. It is impossible for sunlight to enter the light aperture perpendicularly while from an angle of θ with the normal light aperture plane.

Fortunately, for a trough concentrator with very high tracking accuracy, in principle sunlight is always on a $y-z$ plane. The z-axis is parallel to the focal line (or tracking rotary axis) of the trough concentrator.

For a trough paraboloid concentrator with a good tracking system, when sunlight enters the light aperture with an incidence angle of θ, after reflection, the distance to the collector is increased by $1/\cos\theta$ compared with the normal incidence. It is apparent that the solar image on the focal line will increase with it. The solar view becomes 0.267 degree/cos θ at that time; thus the minimal image width becomes (Eq. (2.93)):

$$W = \frac{2r_r \sin(0.267/\cos\theta)}{\cos(\phi_r + 0.267/\cos\theta)} \tag{2.93}$$

The influence of incidence angle θ to the image width may also be expressed by the influence of θ to W/W_0, where W_0 is the image width when $\theta = 0$ (Eq. (2.94)):

$$\frac{W}{W_0} = \frac{\sin(0.267/\cos\theta)}{\sin 0.267} \cdot \frac{\cos(\phi_r + 0.267)}{\cos(\phi_r + 0.267/\cos\theta)} \approx \frac{1}{\cos\theta} \tag{2.94}$$

Note that the inequality sign in Eq. (2.94) is valid only when ϕ is small.

For a trough paraboloid concentrator, if it is aligned in the east–west direction, because the incidence angle of sunlight θ in the morning and late afternoon is large, the solar image on the focal line would be large in the morning and afternoon. In addition, the solar image becomes at noon. Therefore, the highest efficiency and most obvious collection effect occur at noon. If it is aligned in the south–north direction, the incidence angle in summer does not vary much, and thus the size of the solar image does not obviously change. In fact, apart from the size of solar image, the incidence angle θ also influences the total incident energy on the concentrator and the boundary effect of the solar concentrator. On the other hand, for a concentrator whose position is adjustable, it could be adjusted for the season, time of the day, or geometrical latitude. For a low-latitude area, it is better to be aligned in the south–north direction. Alternatively, it could be aligned to east–west but 10–15 degrees southward, which is good for heat use at night.

2.8.2.3 Receiver and Acceptance Angle of Trough Paraboloid Concentrator

For a trough paraboloid concentrator with a boundary angle of ϕ, if the circular tube is used as a collector, its angular acceptance function can be calculated by the geometrical relationship shown in Fig. 2.66. The focal length of the paraboloid is f, the opening width of the collector is a, the receiver diameter is d, and then the system concentration ratio is (Eq. (2.95)):

$$C = \frac{a}{\pi d} \tag{2.95}$$

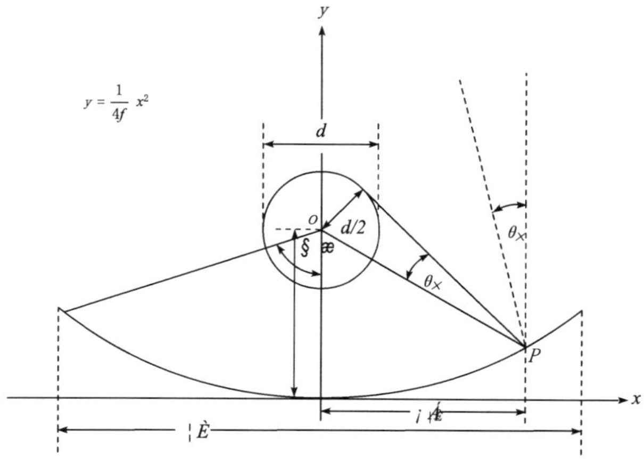

Figure 2.66
Geometrical relationship of a trough paraboloid concentrator.

The given parabolic equation is (Eq. (2.96)):

$$y = \frac{x^2}{4f} \tag{2.96}$$

If the light incident point stays $P = (x,y)$ and light can reach the center of the receiver after reflection by the reflector, then all of the light reaches point P. As long as its angle with normal incident light θ is smaller than the maximum allowed angle of deviation θ_x, incident light can reach the collection tube after reflection. This θ_x is the so-called "minimum tracking accuracy requirements" for point P, i.e., the reflected light on point P can exactly reach the boundary of the collection tube. The light of the angle that is larger than θ_x cannot be collected by the collection tube. According to this geometrical relationship, the result is (Eq. (2.97)):

$$\sin\theta_x = \frac{d/2}{\sqrt{x^2 + (f-y)^2}} = \frac{d}{2f\left[1 + \left(\frac{x}{2f}\right)^2\right]} \tag{2.97}$$

It is apparent that the value of θ_x changes with the position of P; when x is reduced, the θ_x increases, otherwise θ_x will decrease. Especially when $x = \frac{a}{2}$, θ_x reaches the minimum value θ_1, which means that all light within the deviated normal incident angle θ_1 can be collected. This angle θ_1 is also called the half-acceptance angle of the trough paraboloid. The half-acceptance angle θ_1 can be rewritten by the boundary angle ϕ (Eq. (2.98)):

$$\sin\theta_1 = \frac{\sin\phi}{\pi C} \tag{2.98}$$

When $|\theta| < \theta_1$, all light can be collected. At this time, the system acceptance function is $f_{PT,cyl}(\theta) = 1$ (for a trough parabolic and circular tube collector), but it does not mean that all light that is larger than θ_1 will not be accepted. There is still some light that can be received, as long as its reflection point is much closer to the central or original point of the paraboloid. At this time, a typical offset angle θ_2 has to be mentioned (Eq. (2.99)):

$$\sin\theta_2 \approx \frac{d}{2f} \quad (\theta_2 > \theta_1) \tag{2.99}$$

If the deviation angle is large than θ_1 but smaller than θ_2, there will still be some part of light reflected to the collector. This part of light is in the range of $-x_\theta$ to x_θ (the meaning of x_θ is given in Fig. 2.66), and (Eq. (2.100)):

$$x_\theta = 2f\left(\frac{d}{2f\sin\theta} - 1\right)^{1/2} \tag{2.100}$$

The corresponding acceptance angle is $\theta = \frac{2x_\theta}{a}$. For light whose incidence angle is larger than θ_2, only the light that directly projects onto the collector can be received, but for a high concentration ratio system, this type of light makes no sense.

2.9 Array Fresnel Lens Reflective Concentration System

The trough paraboloid system is the most mature high-temperature solar collection technology. Its advantages are the maturity of its technology, its high operation temperature of 400°C, that it can easily be tracked and controlled, and its large-scale production. Its disadvantages are also obvious. Primarily, the trough collector is very wide and long, and thus its wind resistance is high. A large support structure is necessary to protect it from wind damage and device transformation, and therefore the cost is largely increased. At the same time, the power for a driven device becomes huge, consuming power generation from itself. In addition, the paraboloid is large. When it is placed upward, it easily accumulates dust and is hard to clean, which reduces system reflectivity.

To overcome the disadvantages of the trough paraboloid concentration system, the concept of the large-scale array Fresnel lens high efficiency concentration system was introduced. It uses several flat mirrors with different tilt angles as reflective mirrors. These reflective mirrors reflect sunlight onto the receiver; the principle is shown in Fig. 2.67.

The array Fresnel lens reflective concentration system uses several flat mirrors to replace the concentration concept of a single curved surface for a trough paraboloid concentration system so that the disadvantages can be avoided.

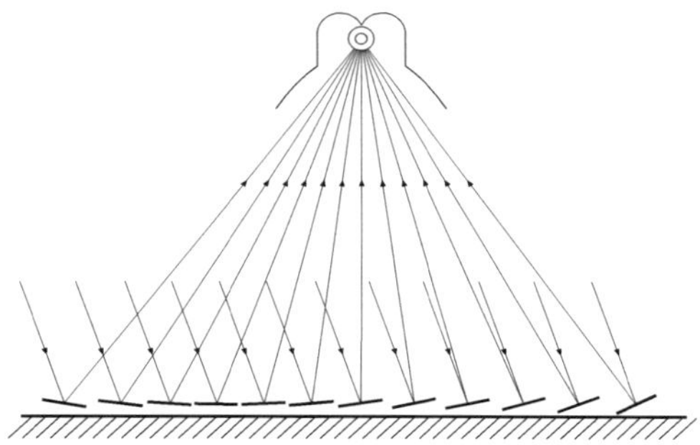

Figure 2.67
Array Fresnel lens reflective concentration system.

Advantages of an array Fresnel lens reflective concentration system are that:

1. The wind capacity is largely enhanced because the reflective mirror can be placed low and there are gaps between reflectors, which can reduce wind resistance.
2. The reflector is a flat mirror whose reflective accuracy is low. The flat mirror is available on the market. It is not hard to process as a paraboloid and bending hot glass is avoided, which thus reduces the cost of the reflector. At the same time, the weight of the single-sided glass is reduced and a huge supporting structure is not necessary. It can be driven by a normal electrical machine, which reduces the power consumption of the tracking system. Finally, cleaning and maintenance are easy.
3. Because the distance between the reflector and receiver is only 3–6 m, it is not the same as a tower power plant, whose distance is around 100 m; thus tracking accuracy is relatively low compared with a tower light collector. The investment for a tracking system is reduced.
4. The heat-transfer loop for the whole system is stationary and is separate from the reflective concentration system. The connection between moving and stationary parts of the system can be avoided, which needs to be solved for a conventional trough or dish concentration system; thus the cost of the tube is reduced.
5. The second reflector is designed for this type of system. It can collect a larger range of sunlight reflected from a Fresnel lens for the same vacuum collector tube, which increases the concentration ratio and collector efficiency. The phenomenon of heat bending is also avoided.

The design concept of a large-scale reflective Fresnel lens concentration system is to use a flat mirror to replace a trough paraboloid reflector, reduce cost; reduce wind resistance using gaps between flat mirrors, and make the operation easy. This system consists of four major parts: a group of flat mirrors, a receiver, a second concentrator, and a sun-tracking system.

2.9.1 Structure of Array Fresnel Lens Reflective Concentration System

There are various forms of array Fresnel lens reflective concentration structures. The most typical is shown in Fig. 2.67. Its reflective mirror is symmetrically distributed along two sides of the receiver. This system has good integrality and appearance, and it is also easy to design and control. However it has one shortcoming, because the reflective surface of this type of Fresnel lens concentrator is single-axis tracking; when solar altitude angle is low, the front reflective mirrors will shade the back ones such that the concentration efficiency is influenced. This phenomenon is called the self-shading effect. Moreover, regardless of whether it is morning or late afternoon, both the receiver and second concentrator will shade the reflective mirror, which will reduce the reflective area. In addition, if the site is limited, the development in the system "length" direction will be limited but only in the "width" direction, and then installing one receiver is not adaptive to environmental change. It is a wise choice to design two or more receivers [25].

To resolve these issues, it is appropriate to design a Fresnel lens reflective collector with two or more receivers, as shown in Fig. 2.68. This structure eliminates the system's self-shading that reduces collector efficiency by placing the receiver at two sides of the concentrator. This method largely reduces the negative effect from the receiver and improves the concentration effect of the Fresnel lens concentrator when the solar latitude angle is low. However, this method also increases the incidence angle on the receiver surface and increases the light acceptance area; thus the concentration ratio will decrease and the required vacuum tube collector will increase.

Installation of an array Fresnel lens solar concentration system will meet the same issues as for a trough paraboloid concentrator, which are the cosine effect and an increased focal spot. In an array Fresnel lens system, the cosine effect will be serious, because the flat mirror is used as a reflector to reflect sunlight onto the receiver, whereas under most

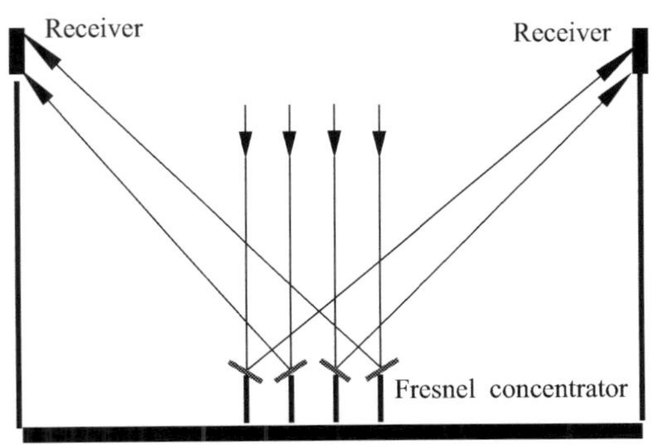

Figure 2.68
One type of Fresnel lens collection system [24].

conditions, the reflector does not face right to the sun. Thus its effective receiving area will decrease, which is called the cosine effect. On the other hand, when sunlight enters with a certain tilt angle, the light will diverge more seriously owing to the long distance between the receiver and reflector. Therefore, a second concentrator is necessary.

In a high-altitude area, east–west alignment is ideal. However, for a low-altitude area, south–north alignment is better. Because the degree of concentration is far less than for a trough paraboloid, this system has a better effect in a low-altitude area. The tracking system for a Fresnel lens concentrator is single-axis. In the tracking process, the tracking angles for each piece of reflector are different and change with the season; the value of angular change for each mirror is also different. An independent tracking system is needed for each reflector, which increases the cost of the tracking system. To reduce the cost, it is useful to add a mechanical transmission part between individual reflectors to form a linkage such that the number of independent tracking devices is reduced as well as the cost. Experimental data showed that the thermal efficiency of the whole concentration system is reduced by only 0.2% with a linkage device, and the cost can be reduced by 50% [26].

2.9.2 Structure and Choice of Receiver

In an array Fresnel lens reflective concentration system, because sunlight is reflected by a flat mirror, the reflected solar spot has a large area, which is different from the paraboloid whose reflected light is located in a line. Therefore a second concentration is necessary for this system. Some researchers have come up with the idea of using mechanical technology to bend the glass mirror slightly such that the reflected light spot can be within a smaller zone. However, this method cannot completely resolve the problem; a second concentrator is still necessary.

The design for a second concentrator has many concepts, but the most popular one is still the CPC owing to its large acceptance angle, which is easy to manufacture and connects to receiver. The design shown in Fig. 2.69 is also a good choice.

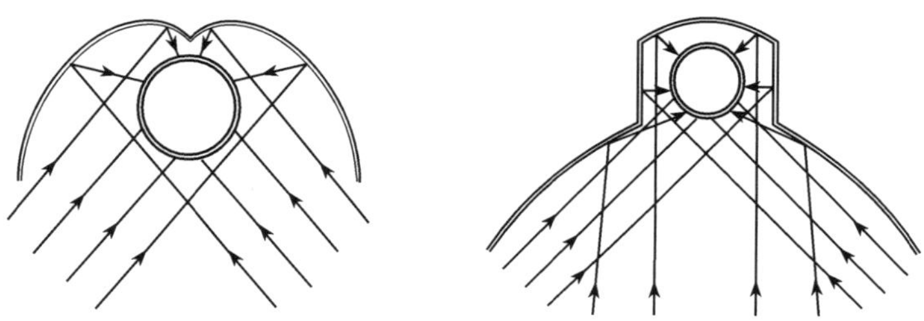

Figure 2.69
Several innovative structures of a second concentrator.

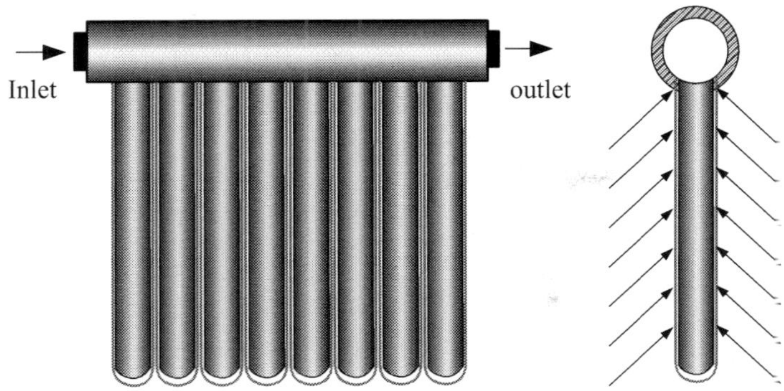

Figure 2.70
Compound vacuum tube collector.

Although the second concentrator is used, the acceptance area is still large and a single vacuum tube is not large enough to collect all light. Thus in most situations, the compound vacuum tube is used as a light collector, as shown in Fig. 2.70. To keep the light from traveling through gaps, sometimes a double-layer vacuum tube collector is necessary. However some parts of a vacuum tube cannot receive sunlight if the double-layer vacuum tube collector is used. The cost of a compound vacuum tube collector is more expensive than other parts, to reduce waste. The following improvements are used:

1. Change the gap between tubes from the original intensive alignment (as shown in Fig. 2.71A) to an interlacing alignment (as shown in Fig. 2.71B). In an interlacing alignment, the gap between two adjacent tubes is an important parameter. A large gap will result in reduced light absorptivity and too much light traveling through the gap between collection tubes; but too small a gap will result in vacuum tube waste.

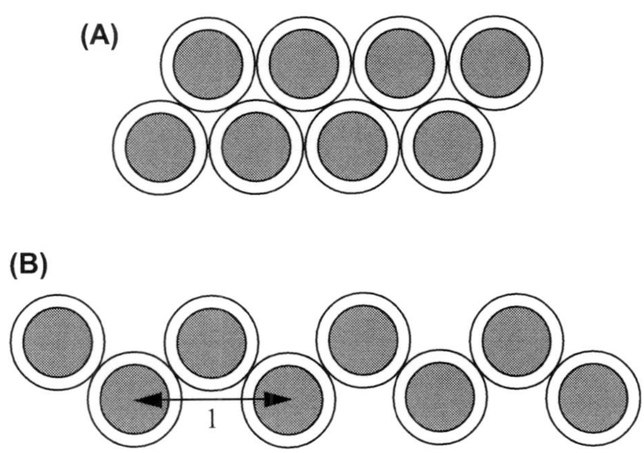

Figure 2.71
The combination pattern of a vacuum tube.

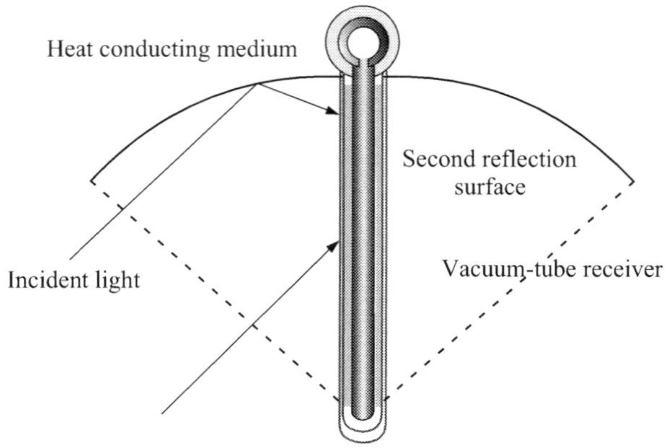

Figure 2.72
Combination of receiver and second reflector.

Experimental results showed that the gap distance should be 1.4–1.6 times that of the vacuum tube diameter, but detailed dimensions have to be decided on by both the structural dimension of the concentrator and the installation height of the collector.

2. Install a second reflector on top of the vacuum collector, as shown in Fig. 2.72. In this method, the light from the flat mirror can be divided into two parts. One part will project directly onto the vacuum tube collector and be absorbed; another part of light will also project onto the vacuum tube after the second reflection. A structure of this kind of receiver can respond to a single Fresnel lens reflector with a large area, which can reduce the number and cost of Fresnel lens reflectors. On the other hand, light with a wider angular range can be received by a collector if a second reflector is installed, which could widen the width of the Fresnel lens reflector array and improve the solar collection area. However, some light would unavoidably be absorbed into the reflective surface or would reach the area that is not covered by the vacuum collector after the second reflection, so overall collection efficiency would be decreased.
3. Installed the vacuum tube collector horizontally. Correspondingly, the position of second reflective concentration has to be changed, as shown in Fig. 2.73. A comparison experiment for a horizontal and vertical collector showed that the vertical collector had better light collector performance. Under the same experimental conditions (receiver height, primary reflector dimension, vacuum tube collector model, and weather conditions), the horizontal vacuum tube collector collected 17% more energy than did the vertical collector [27,28].

The purpose of a Fresnel lens concentration system is to generate high-temperature and high-pressure heat. The all-glass vacuum tube collectors in this structure are completely incapable of bearing pressure. They are metal tubes placed within the glass vacuum tube, through which the heat could be transferred to a tube containing heat transfer fluid. The tube has a thick wall that can bear 5 MPa pressure.

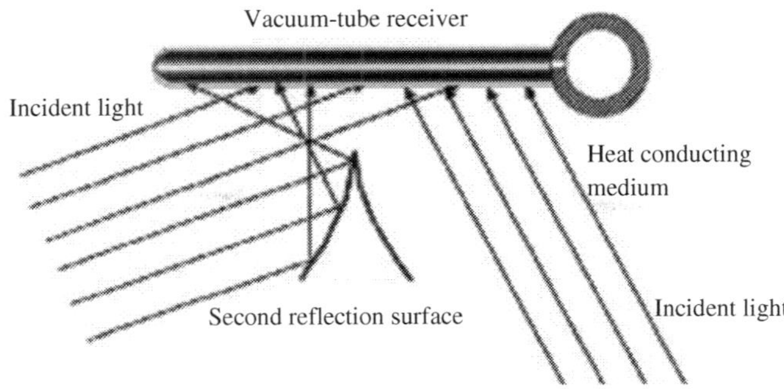

Figure 2.73
Combination of receiver and second reflector.

2.10 Transmission-Type Line-Focus Fresnel Solar Concentrator

2.10.1 Concentration Principle of Fresnel Lens

There are two kinds of Fresnel lenses for solar concentration, including point focus and line focus. The Fresnel lens was invented by a French physicist, Augustin Jean Fresnel, in 1822 [29]. The original idea was to deduce the thickness of a large lens to decrease its weight. He found that energy refraction happens only on the interface of the lens. The imaging characteristic is determined by the curvature of the optical surface and the thickness along the optical axis barely has an effect on it. Thus, chipping off the extra material and keeping the curvature unchanged can realize the same concentration as before. It can be explained by Fig. 2.74, in which a spherical lens is considered an arrangement consisting of a series of discrete lenses. Therefore the redundant materials can be cut off and the curvature kept unchanged to ensure that it does not change the refraction. Every element is then moved onto the common plane perpendicular to the optical axis. The new lens has the same concentration compared with the original lens, which is called a Fresnel lens. Fresnel lenses are often designed as different structures to match various applications. General types are spherical, flat plate, dome-shaped, etc. Its concentration principle is shown in Fig. 2.75.

With the development of materials science and manufacturing technology, some kinds of translucent plastics with high transmittance have been found, such as polycarbonate (PC), polymethyl methacrylate (PMMA), etc., which make Fresnel lenses with less volume, light weight, and better concentration. Thus Fresnel lens has gradually become recognized in the solar energy field [30].

2.10.2 Optical Loss of Fresnel Lens

Optical efficiency is an important parameter for illustrating the performance of a Fresnel lens. Optical loss is directly influenced by optical efficiency. Several factors lead to optical

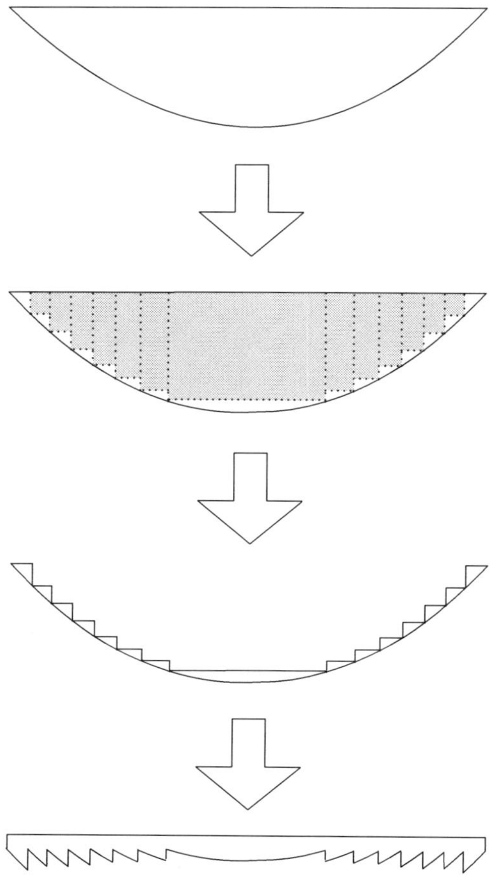

Figure 2.74
Changes in Fresnel lenses.

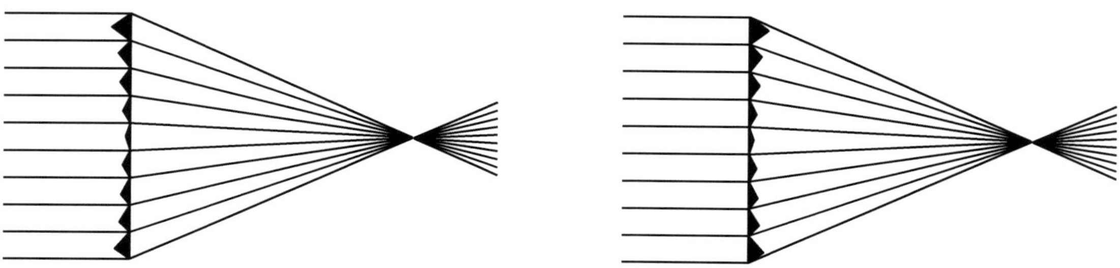

Figure 2.75
Concentration principle of Fresnel lens.

loss: reflection, adsorption, technological defects, and structural defects. Technological defects illustrate unavoidable modifications on the outlines of a Fresnel lens in manufacturing. The difference between the original and modified lens leads to optical loss. A structural defect loss is caused by scattering of the joint faces of discrete lenses. The scattered light cannot reach the focus and thus causes loss.

As for the limited size of the prism, the optical process of one prism is shown in Fig. 2.76. Parallel lights converge when going through the Fresnel lens. However the focuses may be not on one point. Fig. 2.76 shows four lights, A, B, C, and D, reaching different positions; they will generate different optical losses. Light A shoots in the joint point of two prisms where there is a small uncertain surface caused by manufacturing. Because the joint point is not a sharp connection, light A will be refracted to somewhere instead of the focus. Light B shoots into the prism from its middle position. Light B will be reflected a certain percentage on two interfaces. In addition, the material will absorb a certain percentage of energy during the light going through. When light C reaches the peak of the prism, the optical loss is similar to the loss of light A. Light D reaches the cliff surface of the prism and will generate a large percentage of reflection because the incidence angle is approaching 90 degrees. Moreover, the transmitted light will not reach the focus because it does not match the refraction conditions.

According to this analysis, the optical loss of a Fresnel lens can be caused by many factors. Because a Fresnel lens is transformed from a smooth optical lens to discrete lenses, its focus also changes to a nonideal point, or a focal area, actually. However, if the focal area is small enough, it can still meet application requirements.

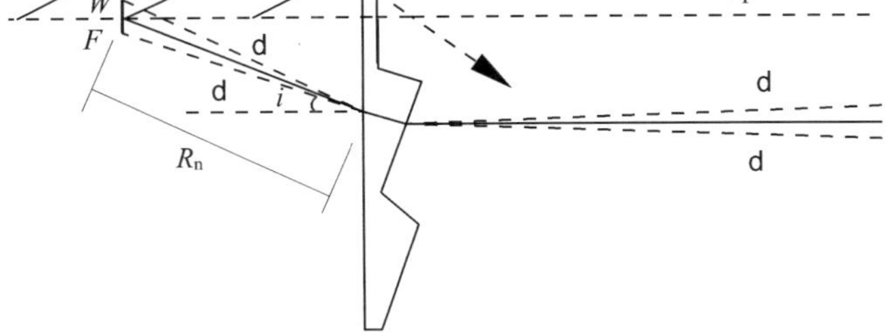

Figure 2.76
Optical loss and concentration ratio of a Fresnel lens.

2.10.3 Concentration Ratio of Fresnel Lens

If a receiving plane is set on the focal area of a Fresnel lens, the size of the focal area and its energy distribution are relative to the focal length, the diameter of aperture, and the size of the prisms' elements. Generally the middle part of the focal area has the highest energy density and will decline when moving from the middle area. As for the point focus Fresnel lens, the focus is a circular area. Similarly, a line-focus Fresnel lens has a rectangular focal area with a small width. It can be explained as follows: if one prism is defined as the refractive surface, there is usually one light that goes through a point on the surface that can reach focus *F*. This point is generally the middle point of the surface (such as light B shown in Fig. 2.76). Lights A and C going through Fresnel lens cannot reach *F* and will reach the adjacent area of *F*. The size of the adjacent area is defined as the spherical difference. The larger the size of prism is, the bigger the spherical difference is. Thus, the spherical difference can be deduced by shrinking the size of prism, but it cannot be eliminated. If the size of the prisms is relative large, the surface of the prism can be designed to be curve-shaped to deduce the spherical difference. For example, change the plane between lights A and C to be a curved surface in Fig. 2.76; all light between A and C will reach *F*.

The smallest size of s prism is limited by Fraunhofer diffraction. The single-slit Fraunhofer diffraction formula with infinity light source is (Eq. (2.101)):

$$b_{\min} = \frac{m\lambda}{\sin \delta} \tag{2.101}$$

where $b_{\min}$ is the minimal size of a prism that is the condition of occurrence of Fraunhofer diffraction. As for a line-focus Fresnel lens, $m = 1.0$ and λ is the wavelength of light that goes through the single slit. The sun is a huge sphere that has a 32 degrees opening angle relative to the earth. The opening angle makes the sunlight that reaches to the earth become not absolute parallel light. δ is the half-opening angle, so it is 16 degrees. It can be easily figured out that $b_{\min} = 0.21$ mm. In other words, if the size of prism is defined to be larger than 0.21 mm, diffraction can be avoided.

The difference between this analysis and the actual process is that if all refracted light goes through the Fresnel lens and can converge on the focal area, it should meet the follow relation (Eq. (2.102)):

$$\mathrm{W} \geq 2R_n \tan \delta / \cos i \tag{2.102}$$

where R_n is the light path length from the surface of the lens to the focal area. According to Fig. 2.76, the geometric concentration ratio is (Eq. (2.103)):

$$C_g = \frac{b}{W} = \frac{b \cos i}{2R_n \tan \delta} \tag{2.103}$$

where b represents the width of the prism. Therefore the opening angle of the sun to the earth is a factor of the focal area's generation.

Lorenzo [31–33] analyzed and calculated the Fresnel lenses with various structures using geometric optics. He found that the flat-plate Fresnel lens has the smallest concentration ratio and the spherical-shaped Fresnel lens has the largest concentration ratio and can even be beyond the parabolic trough. The concentration ratio is proportional to the refraction index of the material. When considering the influence of color difference, the design of a Fresnel lens with a high concentration ratio must reckon the effect of color difference.

2.10.4 Curve-Shaped Fresnel Lens

A flat-plate Fresnel lens is easily designed and fabricated because its one side is a completely flat plane. Therefore it is applied to a solar concentration system much earlier than it is in others. However, curve-shaped Fresnel lenses have many more advantages than do flat-plate Fresnel lenses, especially regarding transmittance loss. Experimental results showed that flat-plate Fresnel lenses have higher transmittance in their middle part. When deviating from the middle part, transmittance quickly decreases and the uniformity of the focal area rapidly becomes worse.

Krichman et al. [34] developed a new kind of convex-shaped line-focus Fresnel lens, which provided a new idea for studying shaped-Fresnel lenses and proved that the geometrical concentration ratio (GCR) is close to the theoretical value. In 1980, Krichman took the influences of color and phase differences of solar light into account and designed a convex-shaped Fresnel lens whose GCR reached 80 [35]. In 1987, Mijatovic et al. [36] designed a solar concentrating system with a cylindrical Fresnel lens. Experiment results showed that the photoelectric conversion efficiency was higher than 30% after considering the homogeneity of energy distribution. Thus it could be successfully applied to high-power photovoltaic (PV) power generation. Leutz et al. [37] designed and manufactured a nonimaging dome-shaped line-focus Fresnel lens based on the edge light principle. This lens had several advantages such as a short focal length, wide receiving angle, and high GCR. The average optical efficiency of the lens was 72%.

In the actual use of Fresnel lenses, when approaching the flanks, the converted angles of light become larger; correspondingly, the reflectance becomes higher. To keep the entire transmittance of a Fresnel lens, the transmittance everywhere and in the lowest part should be concerned. The flat-plate Fresnel lens is used as an example to illustrate the changes.

The structure shown in Fig. 2.77A is the traditional structure of a flat-plate Fresnel lens. To avoid the weakness, the modified structure is given in Fig. 2.77B. Each prism-like element is modified according to the optimal transmittance condition of the prism so that the transmittance of the flat-plate Fresnel lens will be optimal on the whole [38].

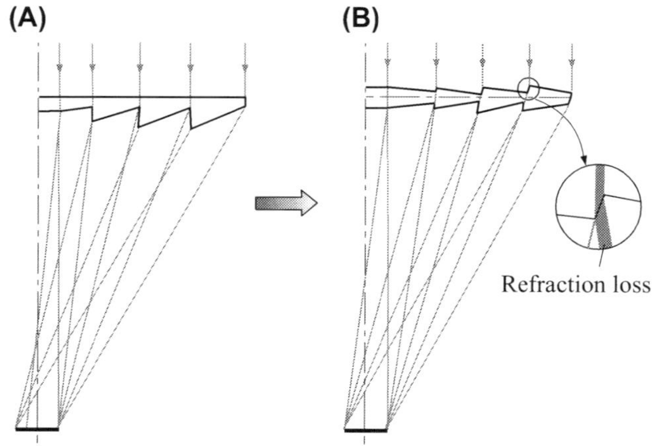

Figure 2.77
Traditional flat-plate Fresnel lens (A) and its modified structure (B).

However a defect appearing in the modified structure is that a little slope is added to the incident surface in the large section of each prism-like element. The light shooting into these slopes cannot reach the focal area. Meanwhile, the new structure is too complex to fabricate. If all incident interfaces of structure (B) could be moved to one surface, the difficulties will be cut down to a great extent. Hence a transformation from a flat-plate Fresnel lens to a curved-Fresnel lens will be investigated.

The transformation process is revealed in Fig. 2.78. The step from Fig. 2.78A to B is the optimization of a traditional flat-plate Fresnel lens. Beginning with the outermost element, move them down one by one and make the arrays of the element's large section collinear with the arrays of the next element's small section; then structure (C) is obtained. The incident surface of structure (C) is a broken surface and reflectivity is improved because it has same incident angles as structure (B), but the slopes disappear. Structure (C) has great advantages such as that it can easily be manufactured, it has less wind drag, and it does not easily accumulate rain, dust, etc. To enhance the advantages, the broken surface can be replaced with a curved surface such as one that is cylindrical, parabolic, or a fitting, that approximately passes the nodes. The curved Fresnel lens, structure (D), can be obtained. Compared with structure (C), (D) has a nearly equivalent transmittance and more prominent advantages.

This analysis shows that the curve-shaped Fresnel lens has better transmittance than a flat-plate Fresnel lens. Considering how simple it is to manufacture, single regular curves should be chosen to design Fresnel lens, including circles, ovals, parabolas, etc. Which curve is the optimal choice? It can be known by an optical simulation.

To ensure that the comparison uses the same concentration ratios, assume that the chord of concentrators is equal to 0.64 m. The focal lengths are equal to 1 m. The radius of circular

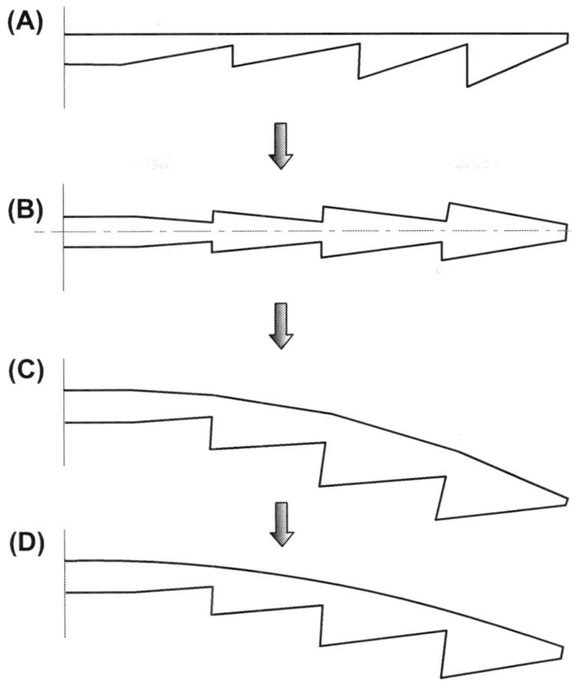

Figure 2.78
Transformation of a Fresnel lens.

Fresnel lens is 0.5 m. The minimum thickness of Fresnel lenses is 2 mm and the maximum is 7 mm. Set a flat-plate plane as the receiver that has 100 mm width and the light reaching the receiver is entirely collected.

The optical simulation on these three kinds of Fresnel lenses illustrates the influence of tracking error on the collection efficiency. The results are shown in Fig. 2.79.

Referring to Fig. 2.79, the tendencies of collection efficiencies of three kinds of Fresnel lenses change similarly with the tracking error. When the tracking error is less than 1 degree, the three all have good collection, and when it is larger than 3 degrees, the collection efficiency is 0. However when the tracking error is less than 1 degree, the parabolic Fresnel lens has better collection efficiency than the others. Although tracking error is larger than 1.5, the cylindrical Fresnel lens is better. Thus, for different concentration systems, the type of Fresnel lens can be chosen by the simulated result.

This analysis of tracking precision is under the condition in which the incidence light is perpendicular to the axis (namely, incidence light is parallel to the generatrix of every Fresnel lens). Define the line that is perpendicular to the axis as the normal line and the angle between the normal line and incidence light as the tilt angle. If the inclination angle

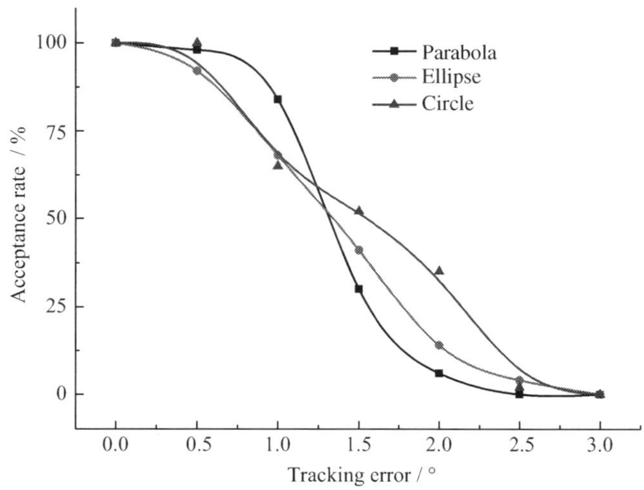

Figure 2.79
Influence of tracking error on collection efficiency.

changes, the collection efficiency will also be different. Fig. 2.80 shows the different refractions in different inclination angles of a cylindrical Fresnel lens. The large tilt angle will lead to large refraction and cause light diffusion. In addition, the focal area deviates from the initial position and the focal length become shorter, as shown in Fig. 2.81.

To compare the changes of concentration effect of different kinds of Fresnel lens with the inclination angle, a simulation result is given in Fig. 2.82. The curves show the percentage of received light to the total incident light.

When the tilt angle less than 5 degrees, all three have 100% collection. When the tilt angle equals 5 degrees, the collection efficiency of the oval Fresnel lens declines first.

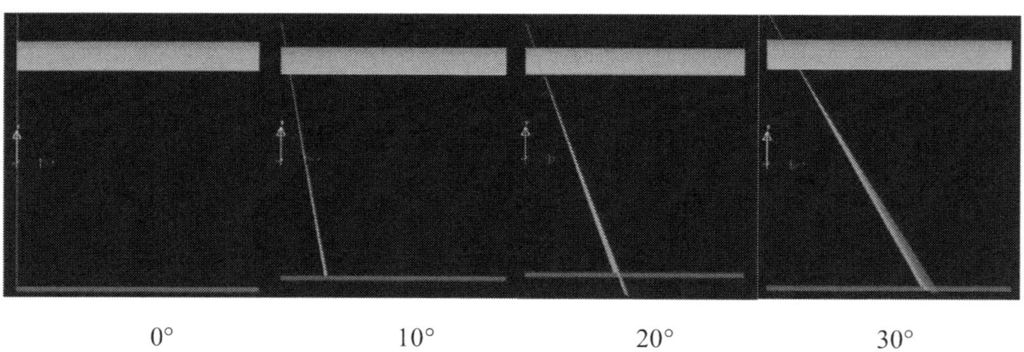

Figure 2.80
Refraction in different tilt angles.

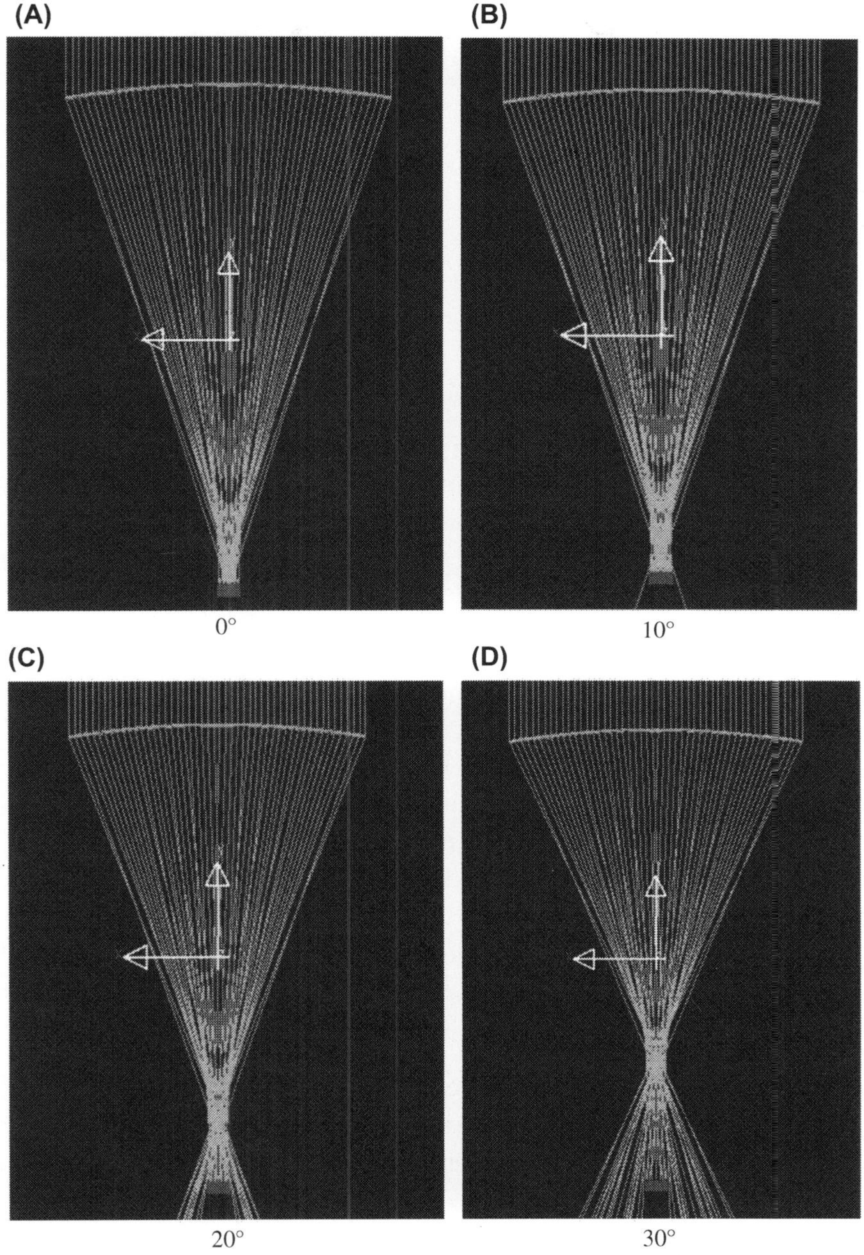

Figure 2.81
Changes in focal length and focal area with tilt angle: (A) 0°, (B) 10°, (C) 20°, (D) 30°.

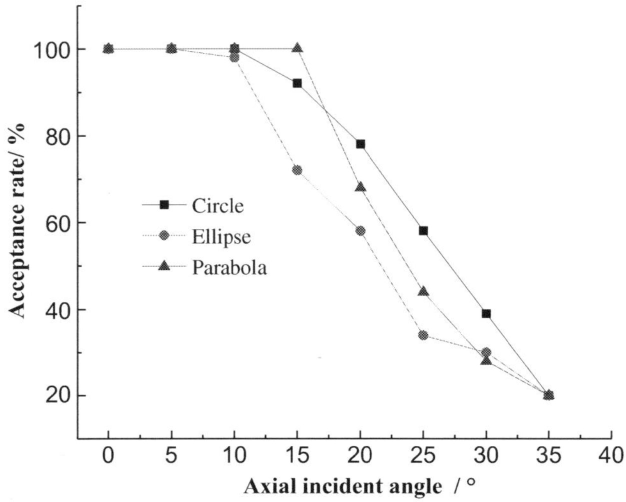

Figure 2.82
Percentage of received light to the total incident light of three kinds of Fresnel lens.

When the tilt angle increases to 10 degrees, the collection efficiency of the cylindrical Fresnel lens goes down after the oval Fresnel lens. In this condition, the collection efficiency of the oval Fresnel lens declines substantially, whereas the parabolic Fresnel lens keeps high collection efficiency. However, when the tilt angle increases to 15°, the collection efficiency of parabolic Fresnel lens decline suddenly. Comprehensively speaking, the tendency of a cylindrical Fresnel lens changes relatively slower than for the two others and the oval Fresnel lens is second. As for the parabolic Fresnel lens, although the high efficiency interval is the widest, its declination after the high efficiency interval is much faster than for the others. Therefore, the cylindrical Fresnel lens has better collection efficiency than for the others overall. Fig. 2.83 shows the variations in focal length with an increase in the tilt angle.

Fig. 2.83 shows that three kinds of Fresnel lens have similar tendencies when the tilt angle increases. For a certain tilt angle, the cylindrical Fresnel lens has the smallest change. Therefore the cylindrical Fresnel lens has the best collection efficiency.

2.10.5 Fresnel Lens Used for Solar Energy Field

The parabolic trough is a typical line-focus solar concentrator. The traditional parabolic trough solar concentrator is widely used in the solar collection field, especially in a solar thermal power plant, because it has the most mature technology. Under the condition of accuracy tracking by a precise mechanism, it can achieve heat at a temperature higher than

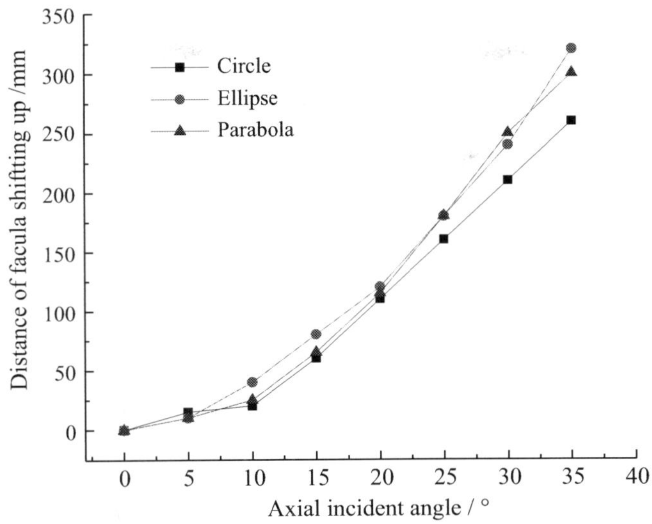

Figure 2.83
Variations in focal length with increase in tilt angle.

400°C. However, although the parabolic trough has perfect theoretical concentration performance, it has fatal defects on the engineering list:

1. To obtain enough solar energy to reach a high temperature, a parabolic trough usually has a large aperture. The width of a trough generally approaches 6–8 m. Parabolic reflectors are glass mirrors, which are heavy. Besides, it must have enough strength and stiffness to resisting deformation, especially wind load, or the reflected lights will not reach the receiver. Meanwhile, there also needs to be a strong supporting frame to support the heavy reflectors. Unavoidably, the supporting frame is much too heavy itself. Thus a large torque needs to drive the support to track the sun, which consumes a large amount of electricity. Generally, energy consumed by the driving mechanism is 25% of the total electricity the parabolic trough can obtain, which makes the parabolic trough system have lower efficiency.
2. The aperture of the trough is usually upward to the sky so that it looks like a big groove where dust and snow are easily accumulated. In addition, it is not easily cleaned. All of this leads to reduced efficiency. Besides, the large aperture has low resistance to rain and snowstorms. It must turn its convex surface (the back of it) to the sky, which also consumes electricity and decreases energy production.
3. The receiver of parabolic trough is an evacuated tube receiver. During the working time, the light is received by the bottom wall of the tube so that the heat between the bottom and the top is different, which leads to hot bending of the tube and will even break the evacuated tube.

4. The evacuated tube receiver is set up in the sky. When it works at a high temperature, the radiation heat transfer will generate between the tube and the space, which will cause large energy losses.
5. The parabolic trough needs a high-precision tracking mechanism, especially for a large concentrator, and tracking error needs to be no more than 0.1 degrees. If sunlight shoots in at an angle, tracking error is required to be even higher. Generally, a large-scale system is tracked by a hydraulic mechanism, which increases the cost.

These defects limit the applications of a parabolic trough on a solar system. In addition, the great cost of a solar thermal power system is unacceptable to users. Fresnel lens adopts a reverse solar concentration strategy compare with a parabolic trough. A Fresnel lens uses refraction instead of reflection to change the direction of light to converge all of it. This kind of system mainly consists of four elements: a solar collector, energy receiver, secondary reflector, and tracking mechanism, as shown in Fig. 2.84. The Fresnel lens is the main optical element used for light collection. The other elements are similar to the parabolic system. The receiver is set behind the Fresnel lens so that a secondary reflector can be added under the receiver. Therefore the secondary reflector can reflect the deviated light onto the receiver again. This is the particular advantage of a Fresnel lens system.

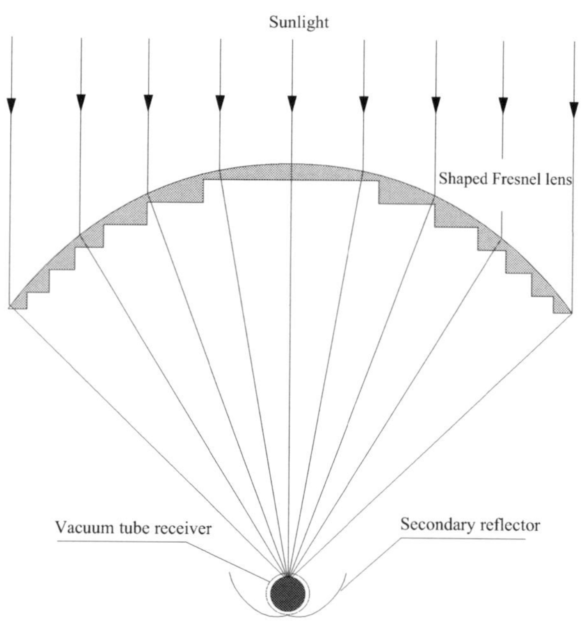

Figure 2.84
Concentrating system of a line-focus Fresnel lens system.

If the convex surface for a Fresnel lens faces upward toward the sky, the system has a simple structure and the whole structure can roll along with the center of gravity. Compared with a parabolic system it has several advantages:

1. The resistance of the wind load is much better because the Fresnel lens is convex-shaped and the system looks like a cylinder. In addition, the system can be set close to the land, or the upper part of it can even be set underground.
2. The addition of a secondary reflector decreases tracking precision. It differs from a parabolic trough, which has only one tube receiver. A Fresnel lens system can gather light that reaches the secondary reflector, which is wider than the receiver is. Therefore the required tracking precision is about 1.5 degrees. It is seven times smaller than a parabolic trough. Thus, the tracking mechanism can use a stepping motor to take the place of the hydraulic mechanism, which is much cheaper. Meanwhile, electricity consumption is decreased. Of course, the driving torque is much smaller when the system rolls along with its center of gravity.
3. In a Fresnel system, the convex surface faces upward to the sky and receiver. The secondary reflector is set under the lens so that wind resistance is reduced. Meanwhile, it resists the accumulation of rain and dust, which give it good performance when it is places in a surrounding for a long time. This structure has good resistance to rainstorms and snowstorms.
4. Because a Fresnel system takes secondary reflectors, the tube is put in a space similar to a greenhouse so that heat transfer loss is reduced. Meanwhile, the upside and downside of the receiver are both heated so that collection efficiency is improved and the hot bending of the receiver tube can be avoided.
5. The material of a Fresnel lens is made of organic transparent plastic, which is easily shaped and even produced by an assembly line. The costs of manufacture, assembly, and maintenance are greatly decreased.

Fig. 2.85 shows a solar thermal trial project consisting of a line-focus cylindrical Fresnel lenses. Experimental results illustrate that for a cylindrical Fresnel lens with a chord length of 640 mm, the highest temperature can easily reach 240°C. When the tilt angle of incidence is within 30 degrees, the collection efficiency approaches 55%. When tilt angle is 30−45 degrees, the collection efficiency is about 40%.

2.11 Tower Solar Concentration System

A tower solar concentration system is also called a centralized collection solar concentration system. The basic form of this kind of solar collection system is a group of heliostats that independently track the sun to concentrate sunlight onto the receiver placed on top of a tower. The concentrated light is used for high-temperature heat.

Figure 2.85
Solar thermal trial project with Fresnel lenses.

2.11.1 Working Principle of Tower Solar Concentrator

Fig. 2.86 shows a working principle diagram of a tower solar concentrator. This system contains four main components: a central receiver, reflective mirror (heliostat), high tower, and sun-tracking system. The working principle is to use heliostats to concentrate sunlight onto the receiver placed on top of a tower. Solar energy is converted to high-temperature heat through a receiver, and thus high-temperature power generation is achieved. The central receiver (on the top of high tower) is surrounded by heliostats. In addition, each piece of the heliostat precisely tracks the sun's position with the help of a computer to achieve maximal solar concentration efficiency.

For concentration purposes, the heliostat cannot face the sun directly; thus the effective interception area is far smaller than its real area. To protect the shading effect from the nearby reflective mirror, there has to be some distance between adjacent reflective mirrors,

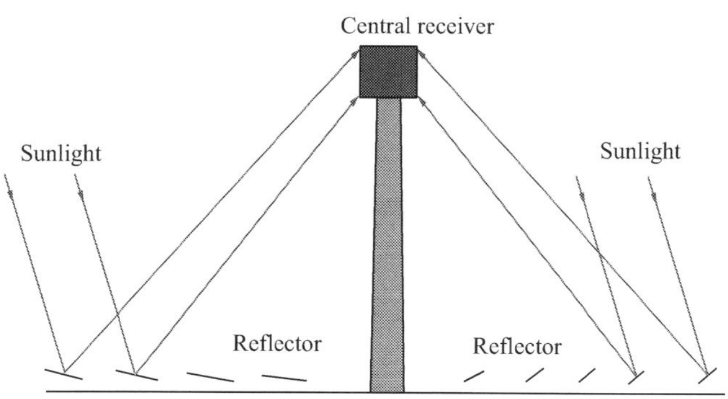

Figure 2.86
Alignment of reflective mirror and central receiver.

which increases use of the land area. The larger the scale is of this solar concentration system, the more land is used for the reflective mirror array and the higher the receiver tower is. There are various shapes of receivers on the top of tower: tubes, cylinders, dishes, cavities, and others.

The current application for the tower solar concentration system is high-temperature solar power generation. A high concentration ratio is one advantage of the system; thus it can achieve a high vapor temperature such as 500°C. Therefore in this type of solar power system, developed technologies that are used in thermal power plants can be used. For example, a solar tower power plant using a vapor Rankine cycle and high-temperature, high-pressure thermal power plant can share same parameters. In this way, not only can the solar power plant achieve high thermal efficiency, it is also easy to obtain the corollary equipment. What is more, this can reduce the construction cost of the solar power plant.

The tower concentrator can generate 500°C high-temperature heat. Of course, it can also generate 200°C heat for solar desalination. Some scholars demonstrated that using a tower concentrator to achieve solar desalination has more economic benefits than tower power generation; the technology is more developed.

2.11.2 Heliostats for Tower Solar Concentration System

The heliostat is the key device for tower solar power systems. In a tower solar concentration system, hundreds and even thousands of heliostats are used; heliostats can continuously track the sun by an independent control system and concentrate solar radiation on the central receiver placed on top of a tower for heat conversion. Therefore the design of a heliostat field is one of the most important parts of designing a tower solar power plant. Good heliostat design can reduce power generation costs and lay a solid foundation for the functionalization of solar power generation.

A heliostat is a reflective mirror that is installed on the metal construction and can automatically track the sun on a dual axis. Its latitude and azimuth angle are adjusted by a control system based on the solar position such that sunlight can be concentrated precisely on the receiver placed on the top of a high tower. To ensure the normal, steady, safe, and highly efficient operation of a tower solar power plant, heliostat performance has to meet these basic requirements: high mirror reflectivity; small flatness error; high structure mechanical strength; good ability to resist over level-8 typhoon attack; stable operation; high concentration positioning accuracy; flexible handling; automatic turnover and collapse; all-day operation; mass production; and easy installation, maintenance, and cleaning. The heliostat surrounding the high tower should be placed in a certain pattern; the coverage area is related to generated power, and the layout plan should regard optimal annual average solar thermal use as the objective function.

Sunlight is not a completely parallel light. Its direct radiation reaches the earth's surface in a field angle of $\Delta_s = 32'$. Because of the solar field angle, the solar image generated by any optical system is a limited image, and theoretically the size of the solar image (diameter d) is determined by the focal distance of the optical system; the function is $d = f \cdot \Delta_s$. For example, if the focal distance of concentrator is 50 m, the theoretical solar image size is 465 mm; if the focal distance is 100 m, the theoretical solar image size is 930 mm. The actual solar image size is larger than theoretical ones for to various reasons.

To reflect light smoothly onto the high tower, it has to be a certain focal distance to ensure high optical efficiency. Therefore, to ensure concentration quality, the surface of the heliostat is a slight curved surface [39]. From the view of a mirror surface shape, currently used heliostats are mainly concaved mirrors, curved mirrors, spherical mirrors, and so forth. In a solar tower power plant, because the distance between the heliostats and the central receiver on top of the tower is large, defocus seems to be highly possible after reflection by heliostats. To concentrate more than 95% reflected sunlight on the receiver, currently the most widely used heliostat is a concave mirror with a small radian (16′). From the view of the material used, heliostats have two types of reflective mirror:

1. Strain mental layer reflector
   Its mirror surface is made of 0.2- to 0.5-mm-thick stainless-steel material; the curvature of strain mental layer can be adjusted by adjusting the internal pressure of the reflective mirror. The advantage of this kind of heliostat is that its mirror surface consists of a whole piece of continuous metal layers so that the focal point of the heliostat can by adjusted by changing the internal pressure of the heliostat. The large disadvantage of this kind of heliostat is its low reflectivity and complex structure.
2. Glass reflective mirror
   The heliostats used in current completed, operating, and plan tower solar concentration systems are all-glass reflective mirror. Their advantages are high antideformation ability, high reflectivity, ease of cleaning, and so forth. Currently, the glass reflective mirror uses mostly glass with a film coating on its back. The solar absorptivity of silver is low and its reflectivity can be as high as 97%; thus silver is the most appropriate solar reflection material, but it has to be protected because it will degenerate in an outside environment.

Currently used silver coating glass mirrors are mostly four-layer second mirrors made by wet chemistry or magnetron sputtering. The 3- to 6-mm-thick glass is used as a basement to place the mirror, and it can provide a hard surface for mechanical cleaning. It would be much better if the gap between the mirror and sliver coating were filled or painted with a multiple-layer protection.

The mirror supporter and basement are the basis for supporting the heliostat. For reasons of seasonal resistance and mechanical strength, the mental heliostat supporter is globally and widely used for most tower solar power stations. There are two types of heliostat supporters: one is a steel plate structure supporter, which has good sand resistance and protection to the mirror surface, and thus the mirror itself can be thin, which is good for a smooth reflective curved surface. Another is a steel frame structure supporter; this structure reduces the mirror's weight, i.e., energy consumption to operate the heliostats. However, this steel frame structure creates a new problem in that the connection between the supporter and mirror should avoid damage to the mirror coating, soundness of the combination of the supporter and mirror, and ejection of rainwater. To solve these problems, some measures are taken as follows: past ceramic substrates on the outer layer of the protection to connect with the supporter; use gel to paste supporter and mirror; fix supporter and mirror by rivet.

The construction of a heliostat for a tower solar concentration system has to consider the following issues [39,40]:

1. Mechanical collision
   The heliostat is a kind of mirror (reflective mirror). It is conventionally rectangular in shape. Imitating the sun-facing feature of a sunflower, each heliostat can be controlled by a two-dimensional independent device. It rotates around one fixed axis and one rotation axis perpendicular to the fixed axis such that it can track the solar position change all of the time and solar radiation can be reflected to a fixed target. According to the difference in the rotational axis, the tracking system for a heliostat can be divided into two categories: rotating with a vertical axis and horizontal axis. Therefore in the process of designing a heliostat field, the space needed for the free rotation of a heliostat under different tracking methods has to be considered to avoid mechanical collision between adjacent heliostats.
2. Optical issue
   In the process of receiving and reflecting sunlight, the heliostat has cosine loss, shading and obstruction loss, attenuation loss and overflow loss. Therefore, in designing heliostat field, the causes of these losses should be considered and avoided to collector more solar radiation.
3. Tracking and transmission device for heliostat
   There are two methods to track the sun for a heliostat: azimuth-elevation angle tracking and self-rotation—elevation angle tracking. Azimuth-elevation angle tracking uses the rotation basement or rotation device above the rotation basement to adjust the position of the heliostat when it is in operation; at the same time, the elevation angle will also be adjusted. Self-rotation—elevation angle tracking uses mirror self-rotation and adjusts the mirror elevation angle to achieve solar tracking for the heliostat.
   The transmission method for a heliostat may take mostly gear transmission, hydraumatic transmission, or a combination of gear and hydraumatic transmission. Because the slight change in heliostat position can cause obvious deviation for reflected light in a large range, no clearance gear transmission or hydraumatic transmission is currently

used. In the design and research of a heliostat, sealing for sand protection and lubricating oil leakage protection are key steps. The main basis for the choice of transmission method is as follows: low power, high-tracking accuracy, low manufacturing cost, and ability to withstand a windy and sandy environment.

2.11.3 Central Receiver for Tower Solar Concentration System

The central receiver (also called the heat absorber or solar cooker) is placed on the top of high tower, which is the key component for light-thermal conversion. Its function is to convert concentrated light from the heliostat's sunlight into heat and to heat the working fluid to 500°C for high-temperature heat provision.

Because a solar receiver is the core technology to achieve tower solar concentration power generation, many countries have invested massive resources to study this technology. Pioneer countries such as European countries, America, and Israel have achieved breakthrough results. Current types of receivers for tower solar concentration are mainly indirect and direct irradiation. The main feature of an indirect irradiation solar receiver is that heat transfer to the heat-bearing tool is indirect; concentrated light will heat the metal surface first, and then heat is transferred to the working fluid when the metal surface is heat. The main feature of a direct irradiation solar receiver is that incident sunlight can directly heat the working fluid, or the receiver and working fluid heating happen on the same surface [41].

1. Indirect radiation receiver
   An indirect radiation receiver is the most commonly used early-stage solar power device. Typical examples are a straight tube, multiple-stage tubulation, and so forth. Solar radiation is directly incident on the outer wall of tabulation and heats the tube wall, such that the working fluid within the tube is also heated to form high-temperature heat. The advantage of this type of receiver is that it can use several types of working fluid, such as water, fused salt, and air. The fluid temperature is between 100 and 600°C. Another advantage is that it has high-pressure bearing, which can be above 12 MPa, and solar radiation intensity can reach 1000 kW/m^2.
2. Direct radiation receiver
   A direct radiation receiver is also called a cavity receiver. Because a cavity receiver is similar to an empty cavity with a small hole on its surface, it has the feature of a blackbody; thus its reflectivity is low and it can effectively absorb received solar energy, such that the issue of selective absorption coating can be avoided because the normal selective absorption coating has the issues of heatproof and aging. If this type of receiver is used, sunlight is incident only from its window direction; thus its acceptance angle will decrease, which will limit the layout of the heliostat field. The working temperature of a cavity receiver can reach 500–1300°C and its working pressure is 3 MPa. It is one of the most researched receivers.

2.11.4 Heat Storage System of Tower Solar Power Generation

For solar thermal power generation, the functions of a storage system are to adjust loading, reduce the device capacity and investment cost, further improve solar resources and device use ratio, and improve solar thermal power system reliability and economic performance.

Objectively speaking, energy storage is the weakest part of the solar use technology. Solar-sensible heat storage technology is relatively developed and has practical value. Materials for sensible heat storage system are mainly fused salt, oil, and ceramics. Some countries such as America and France have achieved substantial progress in sensible heat storage systems.

Currently the most used method is single-phase heat storage. To improve the generated steam temperature for heat storage subsystems, its maximum operation temperature for heat storage fluid is continuously increasing, but the exergy loss during this heat storage process is large and not beneficial [42]. To extend the system operation period and satisfy the need for a large scale, a significant increase in high-temperature heat storage working fluid is necessary, which increases investment in a single-phase heat storage subsystem and the total system. To overcome these difficulties, the tower power cycle with dual-phase energy storage was proposed. This cycle uses a high-temperature and medium-temperature phase-changing material as the heat storage medium to achieve a dual-phase heat-storage operation model, which solves or relieves the problem of large energy loss in single-phase solar heat storage when storing or releasing energy.

Solar energy storage can also use latent heat storage and chemical reaction heat storage. Chemical reaction heat storage has maximal heat storage density and can save device cost; it is regarded as the most promising heat storage method.

2.11.5 Optical and Thermal Efficiencies of Tower Solar Concentration System

The tower solar concentration system is a complex system influenced by lots of factors. From the large aspect, the quality of the heliostat, receiver, and heat-driven power conservation system are key factors influencing the total system efficiency. Besides, three influence factors also contain and relate to each other.

The quality of the heliostat mainly includes mirror reflectivity, mirror transmittance (for a reflective mirror with a thin film coating), concentration ratio, and sun-tracking accuracy.

The mirror reflectivity of the reflectivity of reflective surfaces that forms the heliostat and reflects sunlight is expressed by ρ. For tension metal reflective surfaces, its reflectivity can reach 92%; whereas for glass reflective surfaces, owing to the silver coating reflective layer, reflectivity can be 95–97%. Therefore, the reflectivity is high and meets the requirement for a heliostat.

The mirror transmittance is for heliostat made of glass. The reflective mirror for a glass heliostat is a coating on the back of glass, which can prevent the mirror surface from aging and can easily be mechanically cleaned. Light from the sun has to penetrate the glass layer to each reflective mirror, and reflected sunlight has to pass through the glass to reach the receiver. This sunlight has to penetrate twice within the glass layer. If the transmittance of the glass is τ after two penetrations, only τ^2 of sunlight is reached. Although a high-transmittance float glass has been developed whose transmittance can reach more than 92%, owing to double penetration, the glass transmittance still has a large influence on heliostat performance [43].

The concentration ratio is the percentage of reflected light reaching the receiver to the total incident sunlight; it is expressed by η_f. This means that not all reflected light can arrive at the receiver on the high tower. This is one reason why the receiver area is limited; on the other hand, there is error in the mirror, which causes defocus of light. All of these errors make some part of light reflect to the sky instead of to the receiver.

Tracking error is caused by the tracking system because the tower receiver is far from the heliostat, which could be dozens of or hundreds of meters away. For a distance of 100 m and receiver diameter of 2.5 m, tracking accuracy should be within 1.4 degrees; otherwise there will be no light spot in the receiver. Adding that the heliostat area is large, which is on a magnitude of dozens or hundreds of square meters, the requirement for a tracking system is high.

The optical efficiency of the system is (Eq. (2.104)):

$$\eta_o = \rho \cdot \tau^2 \cdot \eta_f = 0.95 \times 0.92^2 \times 0.95 = 76.4\% \tag{2.104}$$

This an ideal condition; in most cases, it is impossible for η_f to be 95%. Thus the system optical efficiency should be lower, especially when there is dust on the mirror. This optical efficiency is for direct solar radiation. There will still be a large number of diffuse light under a cloudy sky, at which time the diffuse light is stronger than direct light. For a tower solar collector system, the collection of diffuse light has no efficiency, which is one of the problems with a tower solar collector system.

Once the optical efficiency is obtained, system thermal efficiency is mainly based on the quality of the receiver, which includes the acceptance rate of the receiver, heat emissivity, and heat transfer ratio.

Sunlight reaching the receiver cannot be absorbed completely because the receiver surface is not a total blackbody and has reflectivity. For a cavity receiver, it has an issue of the acceptance angle. When the incidence angle is larger than a certain degree, the incident light cannot be accepted by reflected light; and the η_c is used for the collection rate.

Heat emissivity is the result of the heat radiation effect caused by the high-temperature surface; although there is a selective absorption coating, it is still impossible to control the heat emission to the outside. Large conventional or conductive heat transfer can also happen to some receivers. These overall heat dissipation rates can be expressed by heat emissivity η_r.

The heat transfer ratio stands for the fraction of received heat that is transferred to the working fluid to generate work. Some part of energy is dissipated during the transporting process.

Based on the rough analysis of these two parts, the system thermal efficiency can be expressed simply as (Eq. (2.105)):

$$\eta_h = \eta_o \cdot \eta_c (1 - \eta_r) \cdot \eta_t \tag{2.105}$$

If $\eta_c = 85\%$, $\eta_r = 8\%$, $\eta_t = 95\%$, the system thermal efficiency is (Eq. (2.106)):

$$\eta_h = 56.8\% \tag{2.106}$$

The calculated thermal efficiency is high, which is unachievable in real practice. This is because of various uncertain factors in an actual system, which will have a negative effect on the system, such that system overall efficiency will be reduced.

2.12 Compound Solar Concentrator With Multiple Surfaces and Multiple Elements

2.12.1 Introduction

Solar concentration is the most general way to use solar energy. Reflectors with parabolic surfaces and lenses with convex-shaped surfaces (hyperbolic surfaces, actually) are all provided with the property of converging light. Thus, they were used as solar collectors much earlier. In the 1980s in Europe and America, countries built several solar power plants with parabolic trough collectors [44]. However, parabolic trough collectors had major weaknesses such as large wind resistance, large accumulation of dust and snow, high requirements of tracking precision, and huge energy consumption for driving and operation. Naeenihad analyzed the wind load of large parabolic trough collectors taken using collectors with a size of 3.4 × 25 m as an example; the results showed that the wind load under the conditions of the aperture perpendicular to the direction of wind was 10 times larger than the condition when parallel [45].

Fresnel lens was invented by Augustin Jean Fresnel in 1822 with a glass material and applied as a beacon [29,46]. It has several advantages including small volume, light weight, low material cost, and small *f* number, and it effectively increases energy density. With the development of material technology, plastics such as PMMA and PC have gradually taken the place of glass. The plastic Fresnel lens has brought the advantages of Fresnel lenses to a higher level. In 1951, Mill OE designed and fabricated the world's first plastic Fresnel lens by injection molding [47]. Nevertheless the Fresnel lens was not applied in solar concentration widely until today owing to its intrinsic properties. First, the optical efficiency of the Fresnel lens is less than that of a simple mirror because of the reflection loss on the interfaces and the adsorption loss of the material. Second, the plastic material is prone to aging and is easily scratched. Third and most important, the focal length of the Fresnel lens is not stable. M. Collares-Pereira et al. found that the focal length of the Fresnel lens became shorter with the off-axis incident angle [48]. Hongfei Zheng et al. [49] also found that the focal length of the line-focus Fresnel lens became shorter when incident light deviated from normal incidence although incident light shot along the symmetrical plane of the Fresnel lens.

Because of these problems, solar collectors with single concentrating elements always have demerits. Therefore, compound solar concentrators with multiple surfaces and multiple elements have been put forward. The CPC offers a good start.

2.12.2 Compound Solar Concentrator With Multiple Elements and Multiple Surfaces

The compound solar concentrator is generally composed of two or more optical elements. It is also provided with complex-shaped surfaces.

2.12.2.1 Compound Parabolic Concentrator

The most popular nonimaging concentrator is the CPC first used for solar energy by Winston and Hinterberger (1975), and was composed of two pieces of parabolic reflectors. As shown in Fig. 2.87, a whole parabolic line is divided into two parts. Half of it rotates onto the position of 2, and then mirror two rotates to get part three. Parts two and three consist of CPC.

CPCs are relevant for solar energy collection because they achieve the highest possible concentration for any acceptance angle (tracking requirement) and they have a constant acceptance angle over the entire entrance aperture [50]. Because it is composed of two parabolic lines and has two focuses, the size of the receiver must big enough to receive the entire incidence. The structure of the receiver is designed according to the edge-ray principle [51].

In 1976, A. Rabl et al. introduced an ideal solar concentrator that consisted of two parts of parabolic reflectors, as shown in Fig. 2.88 [52]. This concentrator could collect the coming incident rays completely within the range $-\theta_{in}$ to θ_{in}.

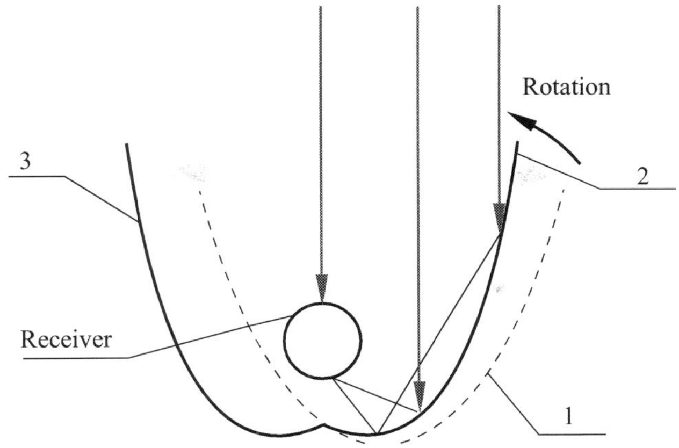

Figure 2.87
Structure of CPC.

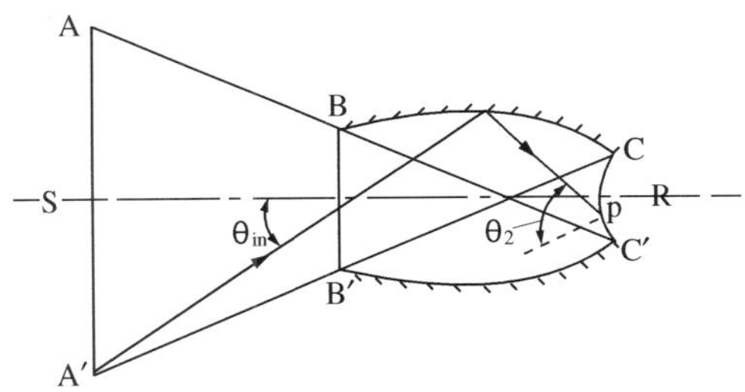

Figure 2.88
Ideal CPC.

J.M. Gordon et al. [53] thought that the CPC was ideal only for a uniform incident solar flux averaged over the energy collection period, but a solar energy flux map provided by nature is not uniform. He put forward a modified CPC collector, a tailored edge-ray concentrator (TERC), which could be used to determine optimal reflector contours. The totally stationary TERCs have around three times the geometric concentration of corresponding optimized stationary CPCs and greater energy delivery.

1. Multiple Surface Trough
   To widen the acceptance angle, Hongfei Zheng et al. [54] put forward a multiple surface trough, shown in Fig. 2.89. It is also a CPC-type collector, but it is different from the traditional CPC shown in Fig. 2.87. The inner parts of two parabolic lines were selected as the reflectors and two flat mirrors make the focus exist in the symmetrical line of the trough. It is one kind of imaging concentrator and thus it can receive solar energy by focusing sunlight into a small area. However, this kind of collector has a small aperture and high height. Hongfei Zheng improved the structure of the multiple surface trough and designed a solar funnel used for a high-temperature solar stove [55–57]. The highest temperature is about 250°C and the collection efficiency is about 43%.

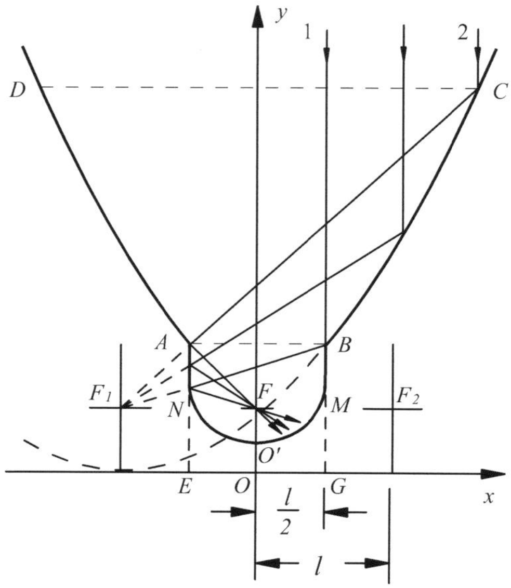

Figure 2.89
Multiple surface trough.

2. Asymmetric Compound Parabolic Concentrator Solar Collectors
   Baum et al. [58] presented a geometric analysis with optimal configurations of asymmetric CPC (ACPC) solar collectors that opened up a new way to study CPCs. ACPCs are mostly obtained by rotating and tailoring the symmetric CPC, demonstrated in Fig. 2.90.

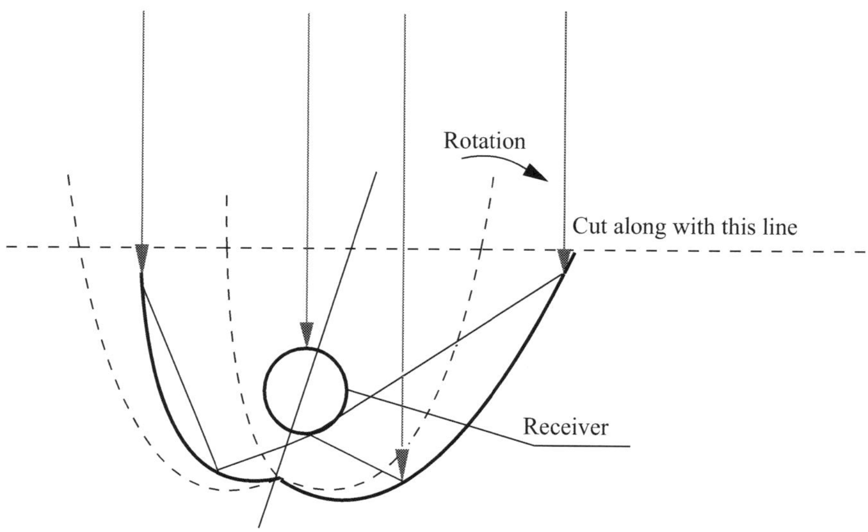

Figure 2.90
Structure of an ACPC.

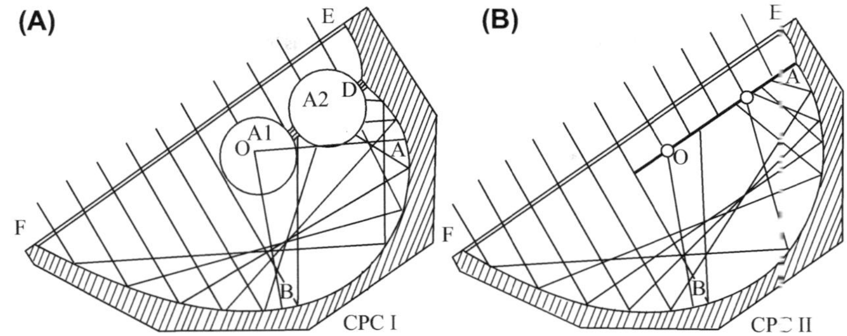

Figure 2.91
ACPC with different receivers.

Y. Tripanagnostopoulos et al. [59] designed and tested a solar collector consist of two separate absorbers horizontally incorporated into a stationary ACPC mirror, shown in Fig. 2.91A. This kind of ACPC has a large receiving area because it has two receivers; thus it has a low geometrical concentration ratio and can be used only for low-temperature fields. Y. Tripanagnostopoulos also designed an integrated collector storage solar system (ICS) with a parabolic-involute compound ACPC [60]. The new ACPC has only one receiver, shown in Fig. 2.91B. A comparison between the new ACPC and symmetric CPC was investigated. Results showed that the new ACPC had lower mean daily efficiency and contributes to better water heat preservation during the night compared with the system with a symmetric CPC reflector. M. Souliotis et al. [61] studied the concentration performance of the ACPC shown in Fig. 2.91B. The optical simulation illustrates that the 100% acceptance angle of ACPC approached ±45 degrees, which means it could be set stationary in a year.

HamdiKessentini et al. [62] designed an ACPC–CPC combined integrated solar collector showed in Fig. 2.92, which was applied on an ICS. The ACPC–CPC combined collector has a larger aperture and the advantage of not needing tracking. However, it cannot provide a high temperature because of its low geometrical concentration ratio.

3. Dielectric Totally Internal Reflecting Concentrator

   A totally internal reflecting concentrator uses total reflection on the medium's interface to take the place of a mirror's reflection. Generally, it could be a CPC-shaped vessel with reflecting sidewalls filled with a transparent high-index liquid or a solid transparent dielectric in which the mechanism for reflection off the sidewalls is totally internal reflection. In the latter case, such devices are referred to as dielectric totally internal reflecting concentrators [63,64].

   Xu Yu et al. [65] designed a low concentration solid dielectric compound parabolic concentrator (dCPC), shown in Fig. 2.93. This kind of dCPC has a large acceptance angle but light with an incident angle that a PV cell could receive must be limited

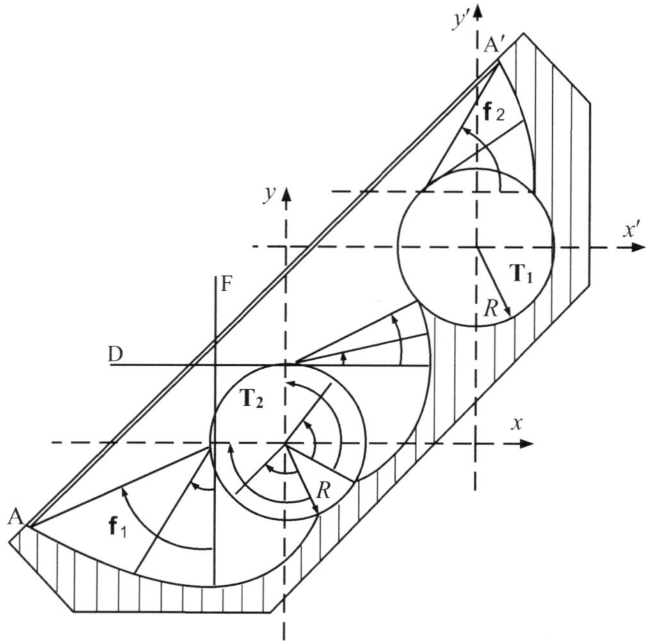

Figure 2.92
ACPC–CPC combined integrated solar collector.

to ±30 degrees, which seems to be a demerit. However, Chao qingFeng et al. [66] used this disadvantage to design a PV/D/T system used for green-building design. A series of dCPCs was ranked with each other and set in a plane. It was used as the roof of a building. At noon, solar radiation is intense and the incident angle is smaller; thus the dCPC can generate electricity. In the morning and afternoon, the building needs light and coincidently, the transmittance of dCPC is larger than at noon.

2.12.2.2 Transmission-Type Concentrators

Transmission-type concentrators have many advantages such as a compact structure, easy installation, and low wind resistance. Pereira et al. [67] described lens–mirror combinations with maximal concentration, which was a modified type of reference [53]. The two-stage compound concentrator was composed of a lens and a couple of elliptic mirrors, as shown in Fig. 2.94A; it can collect all radiation incident on the lens aperture AA with incident angles no more than θ. Similarly, another structure is shown in Fig. 2.94B that consisted of a flat-plate Fresnel lens and a couple of flat mirrors, which was mainly researched in the literature [48]. Pereira found that the focal length became shorter when the incident angle deviated from normal incidence. Compared with a CPC, this lens–mirror combination concentrator had a depth–aperture ratio of approximately 1, much smaller than a single CPC. Thus, it had a higher geometric concentration.

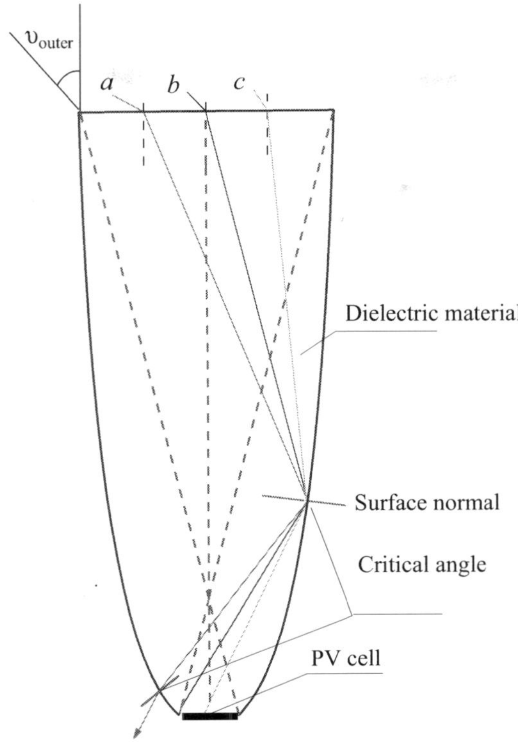

Figure 2.93
Structure of the totally internal reflecting concentrator.

Hongfei Zheng presented a cylinder-shaped solar concentrator (Fig. 2.95), which is composed of an arched Fresnel lens, a Fresnel mirror, and a secondary reflector. An evacuated tube or a concentrating PV cell can be used as the receiver being placed at the focus of the concentrator. This cylindrical compound Fresnel solar concentrator may have the advantages of good weather resistance, low driving force during tracking, good appearance, and easy modularization. The simulation and experiment showed that about 90% of incident sunlight could still be gathered by the absorber when the tracking error is within 1.5 degrees. There is also discussion about the influence of the solar azimuth angle on optical efficiency.

2.12.2.3 Multiple-Element, Complex-Shaped Compound Concentrator

Convectional regular structure concentrators always cannot match the rigorous requirement of concentration, or they need the high precision of a tracking mechanism. Thus some researchers tried to solve these puzzles by using a multiple-element or complex-shaped concentrator. Generally, multiple-element concentrators and

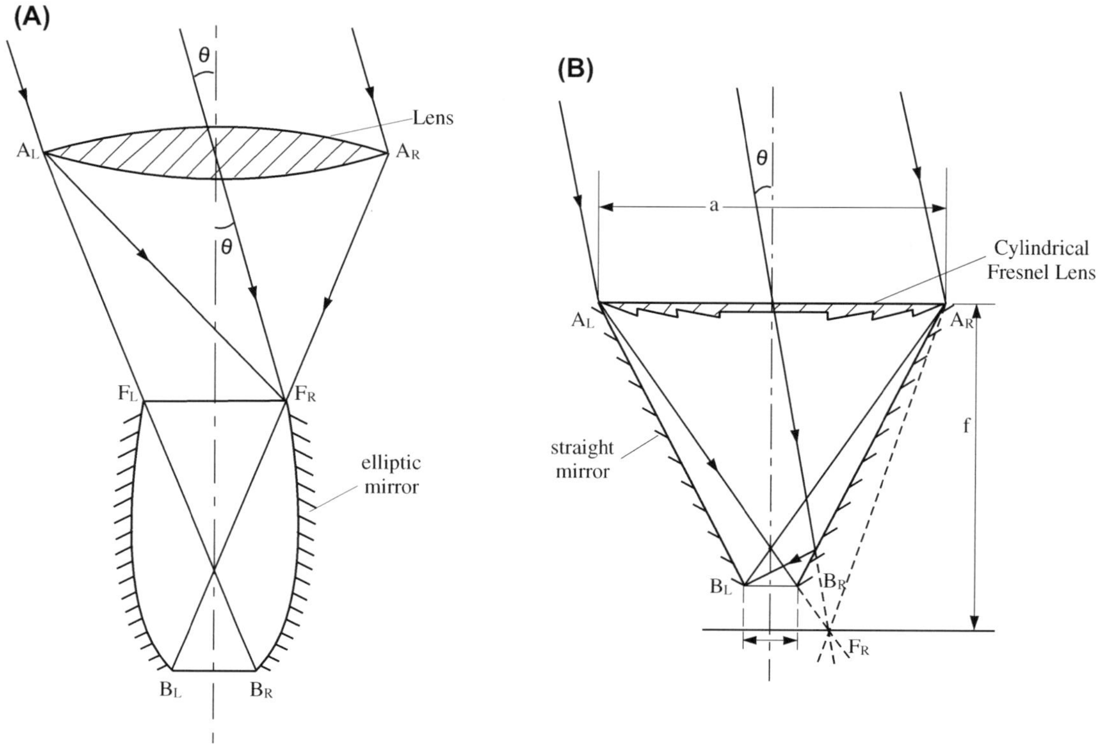

Figure 2.94
Lens–mirror combination concentrator.

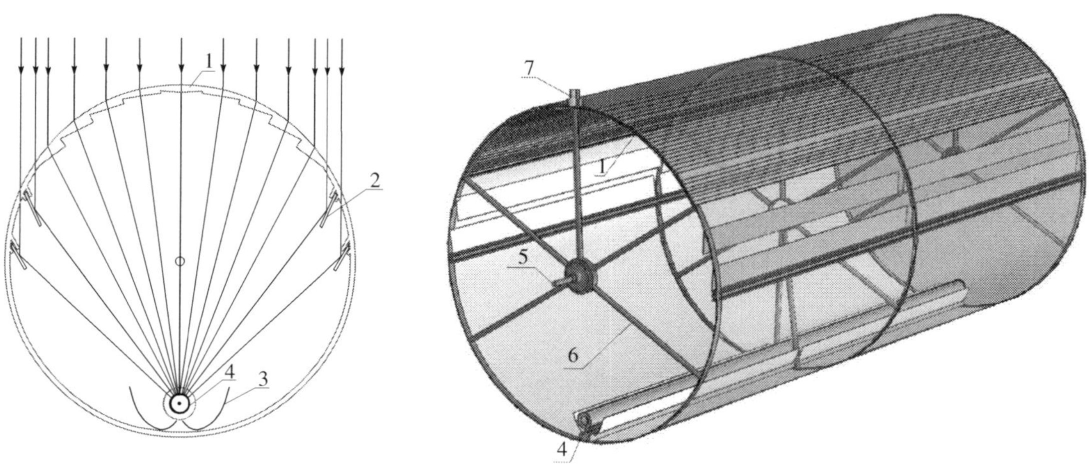

Figure 2.95
Cylinder-shaped solar concentrator: 1, Fresnel lens; 2, Fresnel reflector; 3, secondary reflector; 4, evacuated-tube absorber; 5, shaft; 6, bracket; 7, tracking signal sensor.

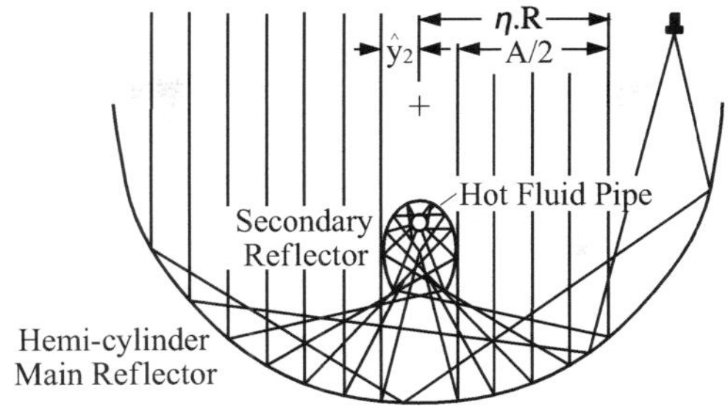

Figure 2.96
Linear focusing solar concentrator with two reflecting troughs.

complex-shaped concentrators are jointly used. The earlier research was done in 1996. John et al. [68] presented a linear focusing solar concentrator with two reflecting troughs, which is shown in Fig. 2.96. The concentrator has a large stationary hemicylinder reflector and a small motive reflector that tracks the sun. Energy is carried to the job by solar-heated fluid in a pipe at focus in the secondary reflector assembly. The deep secondary reflector can reduce convective heat loss because wind on the hot fluid pipe is small. Compared with parabolic reflectors, the two-trough structure has better performance in actual weather owing to its stationary main reflector. However, sun tracking is more complex because the secondary reflector moves against to the main reflector.

A two-dimensional compound parabolic concentrator (2D-CPC) designed by Winston et al. is well-known as a typical static concentrator, but it has the problem that its height is too high to use in a PV module application. In 1999, Yoshioka et al. [69] designed a two-dimensional compound elliptic lens (2D-CEL) for a PV static concentrator shown in Fig. 2.97. The 2D-CEL has only a small reflective loss at a boundary surface between materials with different refractive indices because the 2D-CEL is formed of only a transparent resin. Besides, it has high acceptance angle that reaches 78 degrees and performs better than a 2D-CPC. The maximum yearly averaged optical concentration ratio of 1.75 was obtained for global radiation when the 2D-CEL was installed at certain tilt angle equal to the latitude of Tokyo (N35°).

In 2000, Julio Chaves et al. [70] design an ultraflat compact concentrator, which is composed of a large number of small structures placed side by side, as shown in Fig. 2.98. These compact designs are much more compact than traditional ones such as lens–mirror combinations or parabolic primaries with nonimaging secondaries. The acceptable angle can be designed freely. However, the complicated structure makes it hard to fabricate and repeated reflections lead to a large decrease in optical efficiency.

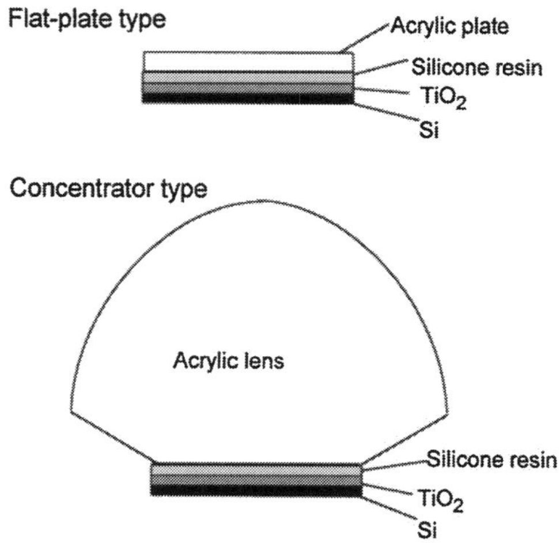

Figure 2.97
Two-dimensional compound elliptic lens.

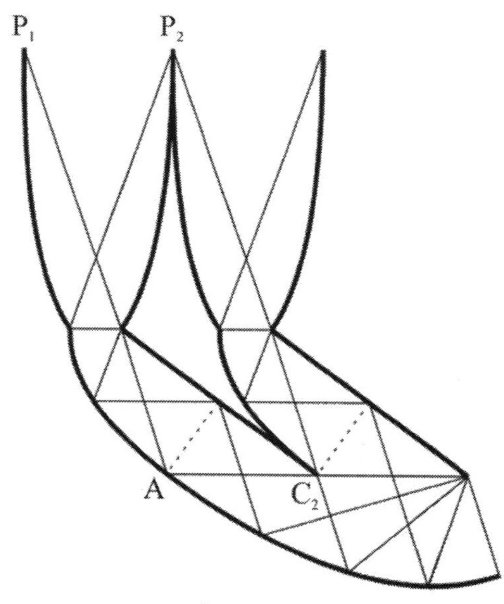

Figure 2.98
Ultraflat compact concentrator.

Jung Min Kim et al. [71] presented an analysis of a concentrator based on an optically transparent planar waveguide with a diffused reflector on the rear surface and PV devices placed at the edge of the waveguide. It can be regarded as a dielectric total inner reflection linear Fresnel reflector. The concentrator can be stationary and is easily made. When in a

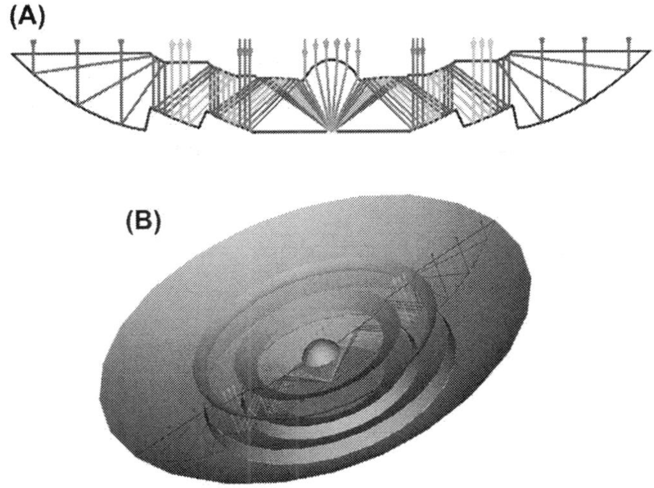

Figure 2.99
A thin profile solar concentrator.

low geometric concentration ratio, the optical efficiency can reach 80% and the acceptable angle is about 60 degrees. Michael et al. [72] conducted a performance investigation on a wedge-shaped LSC. The concentrator is designed based on total inner reflection and thus it has an abnormal property. It is different from the convectional solar concentrator in that under the conditions of summer, when the sun is high in the sky, the average wedge LSC concentrator efficiency is 3.5%, whereas the planar LSC achieves an efficiency of 6.3%. However, when the sun stays low in the sky, such as during early winter, the wedge LSC concentrates light with a maximum efficiency of 32.8%, which is more than four times greater than that of the planar LSC at 7.6%. Therefore, it is suitable for a high-latitude area.

In 2015, Chao-Wen Liang et al. [73] designed a thin profile solar concentrator using a toroidal confocal relay, as shown in Fig. 2.99. It has a high concentration; in particular, the focus is on the center of the bottom surface, which means that the focal length is zero. A great accompanying advantage is the small volume of concentrating system when it is used for a concentrated PV. However, the toroidal confocal lens need high tracking precision that is smaller than 0.46 degree.

In 2015, Milorad Bojic et al. [74] designed a seashell solar concentrator. It is a nonimaging, asymmetric, stationary concentrator that consists of large parabolic reflectors and a small semicircle reflector, as shown in Fig. 2.100. A heat-adsorptive fin is assembled on the hemicircle reflector. Experimental results showed that the optical efficiency was about 80% with an acceptance angle of 50 degrees.

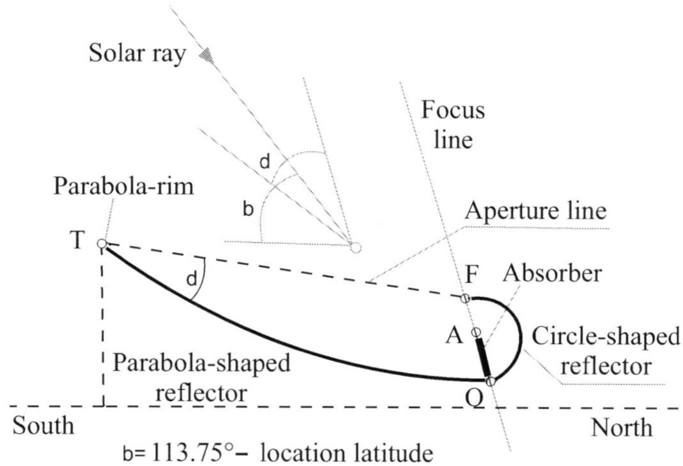

Figure 2.100
Structure of seashell solar concentrator.

References

[1] Johnson FS. J Meteorol 1954;11:431 [The Solar Constant].
[2] NASA SP-8055. Solar electromagnetic radiation. National Aeronautics and Space Administration; May 1971.
[3] Willson RC, Gulkis S, Jamssen M, Hudson HS, Chapman GA. Science 1981;211:700 [Observation of Solar Irradiance Variability].
[4] Duncan CH, Willson RC, Kendall JM, Harrison RG, Hickey JR. Sol Energy 1982;28:385 [Latest Rocket Measurements of the Solar Constant].
[5] Spencer JW. Search 1971;2(5):172 [Fourier Series Representation of the Position of the Sun].
[6] Iqbal M. An introduction to solar radiation. Toronto: Academic; 1983.
[7] Chou Q, Ge X. Solar energy power technology. J Nature (in Chinese) 1996;18(6):344–8.
[8] He H. Progress of Solar power plant. Sol Energy (in Chinese) 2004;3:43–4.
[9] Schmitz M, Schwarzbozl P, Buck R, Pitz-Paal R. Assessment of the potential improvement due to multiple apertures in central receiver system with secondary concentrators. J Sol Energy 2006;80:111–20.
[10] Ausra, Inc.. An introduction to solar thermal electric power[r]. Palo Alto. 2007.
[11] Wang C, Zheng R. New energy applied in architecture (in Chinese). Beijing: China Electric Power Press; 2003.
[12] Li S. Solar energy physics (in Chinese). Capital Normal University Press; 1996.
[13] Duffie JA, Beckman WA. Solar engineering of thermal processes. John Wiley & Sons, Inc.; 1980.
[14] Klein SA, Cooper PL, Beckman WA, Duffie JA. TRNSYS, a transient simulation program. University of Wisconsin; 1974. Engineering Experiment Station Report.
[15] Yin Z. Solar energy application (in Chinese). Beijing: People's Education Press; 1995.
[16] Ge X, Wang Y, Weide L, Gong B. Solar energy engineering – principle and application (in Chinese). Beijing: Academic Journal Publishing House; 1988.
[17] Kooi CE. The steady state salt gradient solar pond. Sol Energy 1979;23:37–41.
[18] Kreider JF, Kreith F. Solar energy handbook. McGraw Hill; 1979.
[19] Rabl A. Active solar collectors and their applications. 1985. p. 170–1.

[20] Zheng H, He K, Tao T. Solar energy concentrating and collecting technology (in Chinese). Beijing: Weapon Industry Press; 2010. p. 8.
[21] Welford WT, Winston R. Non imaging optical concentrator (in Chinese). Translated by Guoqiang Wang. Beijing: Science Press; 1987.
[22] Rabl A, Bendt P. Effect of circumsolar radiation on performance of focusing collectors. ASME J Sol Energy Eng 1982;104:237−41.
[23] Liu Y. Theoretical and experimental study of focal spot energy density distribution of solar concentrator (in Chinese). Doctor paper of Harbin Institute of Technology. 2008.
[24] Neumann A, et al. Representative terrestrial solar brightness profiles. Trans ASME 2002;124:198−204.
[25] Sharan SN, Kandpal TC. Testing of a prototype combined photovoltaic -thermal linear Fresnel concentrator-receiver system. J Energy Convers Manag 1992;33(1):37−9.
[26] Haberle A, Berger M, Luginsland F, Zahler C, Baitsch M, Henning H-M, et al. Linear concentrating Fresnel collector for process heat applications[r]. Freiburg, Germany. 2006.
[27] Yang L, Ying X, Chen Q. Design and making of large scale Fresnel lens (in Chinese). J Opt Technol 2001;27(6):499−502.
[28] Morrison GL, Mills DR. Compact linear Fresnel reflector solar thermal power plants. J Solar Energy 2000;68(3):263−83.
[29] Robert MC. The linear Fresnel lens solar concentrator: transverse tracking error effects [M]. Washington D.C.: National Aeronautics and Space Administration; 1977.
[30] Wu H. Fresnel solar concentrator (in Chinese). Wu Han, Doctor paper of Wuhan Institute of Technology. 2010.
[31] Lorenzo E, Luque A. Fresnel lens analysis for solar energy application. J Appl Opt 1981;20:2941−5.
[32] Lorenzo E. Chromatic aberration effect on solar energy systems using Fresnel lens. J Appl Opt 1981;20:3729−32.
[33] Lorenzo E, Luque A. Comparison of Fresnel lenses and parabolic mirrors as solar energy concentrators. J Appl Opt 1982;21:1851−3.
[34] Krichman EM, Friesem AA, Yekutieli G. Highly concentrating Fresnel lens. Appl Opt 1979;18(15):2688−95.
[35] Krichman EM. Color-corrected Fresnel lenses for solar concentration. Opt Lett 1980;5(1):35−7.
[36] Mijatovic M, Dimitroviski D, Veselinovic V. Fresnel lens absorber system with uniform concentration and normal incoming lights to the absorber. J Opt Paris 1987;18(5−6):261−4.
[37] Leutz R, Suzuki A, Akisawa A, Kashiwagi. Design of a nonimaging Fresnel lens for solar concentrators. Sol Energy 1999;65(6):379−87.
[38] Ma X, Zheng H, Tian M. Optimize the shape of curved-Fresnel lens to maximize its transmittance. Sol Energy 2016;127:285−93.
[39] Zhang Y, Wang J, Zhang W. Tower and trough solar power plants (in Chinese). J Sol Energy 2006;2:29−32.
[40] Kistler BL. A user's manual for DESOL3: a computer code for calculating the optical performance and optimal system design for solar thermal central receiver plants[a]. New Mexico: The United States Department of Energy; 1986. p. 39−122.
[41] Sassi G. Some notes on shadow and blockage effects. J Sol Energy 1993;31(3):331−3.
[42] Siala FMF, Elayeb ME. Mathematical formulation of a graphical method for a no-blocking heliostat field layout. J Renew Energy 2001;23:77−92.
[43] S'anchez M, Romero M. Methodology for generation of heliostat field layout in central receiver systems based on yearly normalized energy surfaces. J Sol Energy 2006;80:861−74.
[44] Price H, Lupfert E, Kearney D, Zarza E, et al. Advances in parabolic trough solar power technology. J Sol Energy Eng 2002;124(5):109−25.
[45] Naeeni N, Yaghoubi M. Analysis of wind flow around a parabolic collector: fluid flow. J Renew Energy 2007;32:1898−916.

[46] Leutz R, Suzuki A. Nonimaging Fresnel lenses: design and performance of solar concentrators [M]. Heidelberg: Springer Verlag; 2001.
[47] Miller OE, Mcleod JH, Sherwood WT. Thin sheet plastic Fresnel lenses of high aperture. J Opt Soc Am 1951;41(11):807–15.
[48] Collares-Pereira M, Rabl A, Winston R. Lens–mirror combinations with maximal concentration. Appl Opt 1977;16(10):2677–83.
[49] Zheng H, Feng C, Su Y, Dai J, Ma X. Design and experimental analysis of a cylindrical compound Fresnel solar concentrator. Sol Energy 2014;107:26–37.
[50] Rabl A. Optical and thermal properties of compound parabolic concentrators. Sol Energy 1976;6(18):497–511.
[51] Winston R. Principles of solar concentrators of anovel design. Sol Energy 1974;16:89–94.
[52] Rabl A, Winston R. Ideal concentrators for finite sources and restricted exit angles. Appl Opt 1976;15:2880.
[53] Gordon JM, Lasken M, Ries H. Upper bounds for the yearly energy delivery of stationary solar concentrators and the implications for concentrator optical design. Sol Energy 1996;4(58):197–202.
[54] Tao T, Hongfei Z, Kaiyan H, Mayere A. A new trough solar concentrator and its performance analysis. Sol Energy 2011;85:198–207.
[55] Zheng H, Tao T, Ma M, Kang H, Su Y. Experimental test of a novel multi-surface trough solar concentrator for air heating. Energ Convers Manage 2012;63:123–9.
[56] Kaiyan H, Hongfei Z, Tao T, Xiaodi X. Experimental investigation of high temperature congregating energy solar stove with sun light funnel. Energ Convers Manage 2009;50:3051–5.
[57] Xiaodi X, Hongfei Z, Kaiyan H, Zhili C, Tao T, Guo X. Experimental study on a new solar boiling water system with holistic track solar funnel concentrator. Energy 2010;35:692–7.
[58] Baum HP, Gordon JM. Geometric characteristics of ideal nonimaging (CPC) solar collectors with cylindrical absorber. Sol Energy 1984:33.
[59] Tripanagnostopoulos Y, Yianoulis P, Papaeflhimiou S, Souliotis M, Nousia Th. Cost effective asymmetric CPC solar collectors. Renew Energy 1999;16:628–31.
[60] Tripanagnostopoulos Y, Souliotis M. Integrated collector storage solar systems with asymmetric CPC reflectors. Renew Energy 2004;29:223–48.
[61] Souliotis M, Quinlan P, Smyth M, Tripanagnostopoulos Y, Zacharopoulos A, Ramirez M, et al. Heat retaining integrated collector storage solar water heater with asymmetric CPC reflector. Sol Energy 2011;85:2474–87.
[62] Kessentini H, Bouden NC. Numerical and experimental study of an integrated solar collector with CPC reflectors. Renew Energy 2013;57:577–86.
[63] Ning X, Winston R, O'Gallagher. Dielectric totally internally reflecting concentrators. Appl Opt 1987;26:300–5.
[64] Welford WT, Winston R. High collection non-imaging optics. 1989. p. 147–63.
[65] Yu X, Su Y, Zheng H, Riffat S. A study on use of miniature dielectric compound parabolic concentrator (dCPC) for day lighting control application. Build Environ 2014;74:75–85.
[66] Feng C, Zheng H, Wanga R, Yu X, Su Y. A novel solar multifunctional PV/T/D system for green building roofs. Energy Convers Manag 2015;93:63–71.
[67] Pereira M, Rabl A, Winston R. Lens–mirror combinations with maximal concentration. J Opt Soc Am 1976;66:1068.
[68] Richter JL. Optics of a two-trough solar concentrator. Sol Energy 1996;2(56):191–8.
[69] Yoshioka K, Suzuki A, Saitoh T. Performance evaluation of two-dimensional compound elliptic lens concentrators using a yearly distributed insolation model. Sol Energy Mater Sol Cells 1999;57:9–19.
[70] Chaves J. Ultra flat ideal concentrators of high concentration. Sol Energy 2000;4(69):269–81.

[71] Kim JM, Dutta PS. Optical efficiency—concentration ratio trade-off for a flat panel photovoltaic system with diffuser type concentrator. Sol Energy Mater Sol Cells 2012;103:35—40.

[72] Hughes MD, Maher C, Borca-Tasciuc D-A, Polanco D, Kaminski D. Performance comparison of wedge-shaped and planar luminescent solar concentrators. Renew Energy 2013;52:266—72.

[73] Liang C-W, Lin J-S. High concentration thin profile solar concentrator utilizing toroidal confocal relay. Sol Energy 2015;122:264—70.

[74] Bojic M, Marjanovic N, Miletic I, Bojic L. Comparison of optical performances of sea-shell trough solar concentrators. Energy Build 2015;98:144—50.

CHAPTER 3

Fundamental Relationships of Heat and Mass Transfer in Solar Seawater Desalination Systems

Chapter Outline

Solar Energy Desalination Technology. http://dx.doi.org/10.1016/B978-0-12-805411-6.00003-8

3.1 Physical Properties and State Parameters of Moist Air

3.1.1 Physical Properties of Moist Air

The seawater desalination systems, especially the ones based on distillation either through multistage flash distillation or atmospheric distillation, are affected by air due to air leakage. Air is a noncondensable gas, which has a significant influence on water evaporation and condensation. In reality, the air in the system is the combination of air and water vapor, or moist air in other words. Therefore it is important to understand the properties of moist air for improving the efficiency of the desalination system.

Moist air refers to the air with water vapor, and dry air is the air without water vapor. Water vapor changes state to gas or solid under certain conditions. Dry air is a mixture of gases, which consists of nitrogen (N_2, 78.084%), oxygen (O_2, 20.948%), argon (Ar, 0.934%), carbon dioxide (CO_2, 0.0314%), neon (Ne, 0.0018%), and helium (He, 0.00052%) by volume fraction and various other components. The scientific test presents that the combination of dry air is relatively stable under normal circumstances. Hence it is reasonable to consider the dry air as a whole when investigating the physical properties of moist air. Water vapor takes a small part in moist air at a normal pressure but changes

Table 3.1: Physical properties of saturated air at different temperature (reference temperature $T_{reference} = 0°C$) [1].

Properties of water and water vapor				Physical properties of 1 m^3 saturated air at temperature T				
T (°C)	p_w (kPa)	h_w (kJ/kg)	h_{fg} (kJ/kg)	h_{vap} (kJ/kg)	m_v (kg/m^3_{air})	$H_{v,a}$ (kJ/m^3_{air})	H_{air} (kJ/m^3_{air})	$H_{a,s}$ (kJ/m^3_{air})
100	101.3	418.7	2255.2	2674.0	∞	∞	131.4	∞
95	84.5	397.6	2268.1	2666.0	3.774	10,060	124.7	10,185
90	70.1	376.6	2281.5	2658.1	1.684	4476.5	118.0	4594.5
85	57.8	355.6	2294.5	2650.1	1.007	2668.5	111.3	2779.9
80	47.3	334.6	2307.0	2641.7	0.634	1675.3	104.4	1779.7
75	38.5	313.8	2319.6	2633.4	0.443	1168.2	97.9	1266.1
70	31.1	292.8	2332.2	2625.0	0.321	843.5	91.4	934.9
65	25.0	271.9	2344.3	2616.3	0.237	621.3	84.5	705.8
60	19.9	251.0	2357.0	2607.9	0.177	461.9	78.2	540.1
55	15.7	230.1	2369.0	2599.1	0.133	346.0	71.8	417.8
50	12.3	209.2	2381.1	2590.3	0.100	259.8	65.3	325.1
45	9.58	188.3	2394.0	2581.1	0.076	195.1	58.7	253.8
40	7.38	167.4	2405.0	2572.3	0.057	146.2	52.2	198.4
35	5.62	146.4	2417.1	2563.5	0.042	108.9	45.7	154.6
30	4.25	125.5	2428.9	2554.3	0.032	80.8	39.2	120.0
25	3.16	104.6	2440.5	2545.1	0.023	59.3	32.6	92.0
20	2.34	83.6	2452.7	2536.3			26.1	

Note: Dry air is at 25°C, 101.3 kPa.

with temperature and pressure. Table 3.1 demonstrates the variation of water vapor content in saturated humid air.

In thermodynamics, dry air can be assumed as an ideal gas at standard ambient temperature and pressure (some seawater desalination system belongs to this area). When the temperature is low, the water vapor in moist air can be also assumed as an ideal gas approximately because the small portion of water vapor is sometimes overheated so that it has low partial pressure and huge specific volume. Meanwhile, the moist air consisting of air and water vapor follows the principle of ideal air approximately. Hence the relationship of state parameters can be expressed by the ideal gas equation of state:

$$pv = RT \tag{3.1}$$

$$pV = mRT \tag{3.2}$$

where p is the pressure of moist air in Pa; v is the specific volume of moist air in m^3/kg; R is the gas constant, which is determined by gas properties in J/(kg K); V is the total volume in m^3; T is the thermodynamic temperature of gas in K; and m is the total mass of moist air in kg.

Table 3.2: Constants of Van der Waals equation of state.

Substance	$a/\left(\frac{m^6\ MPa}{kmol^2}\right)$	$b/\left(\frac{m^3}{kmol}\right)$
Nitrogen N_2	0.136788	0.03864
Oxygen O_2	0.138105	0.03184
Argon Ar	0.136687	0.03227
Carbon dioxide CO_2	0.365377	0.04280
Water vapor H_2O	0.552626	0.03042

When the total mass of moist air is in kmol, the ideal gas equation of state is expressed by $pV_m = R_0T$, where V_m is the volume of 1 kmol gas in m^3/kmol; R_0 is the universal gas constant in J/kmol K.

But at high temperature and high pressures, the moist air should be calculated as real gas. The common gas equations of state are Van der Waals equation of state and Virial equations of state, which are:

$$\left(p + \frac{a}{v}\right)(v - b) = RT \tag{3.3}$$

and

$$pv = RT\left(1 + \frac{b - a/RT}{v} + \frac{b^2}{v^2} + \frac{b^3}{v^3}\right) \tag{3.4}$$

where a(m^6 Pa/kg^2) and b(m^3/kg) are Van der Waals constants. Table 3.2 demonstrates some common values of them.

If water vapor occupies a large proportion in moist air, the physical properties of moist air can be obtained by searching the tables and diagrams of water vapor directly, or calculating the related parameters and obtaining by weighted sum of relative dry air and water vapor content. The physical properties of dry air and dry steam vapor are demonstrated in Appendix Tables 1 and 2.

3.1.2 State Parameters of Moist Air

The state parameters of moist air are critical to determine the state and physical properties of moist air. The calculation of moist air is commonly used. Hence it is necessary to discuss the state parameters in the calculation of seawater desalination in industry.

1. Pressure
   Daltons Law states that the total pressure p of moist air is the sum of partial pressures of dry air p_a and water vapor p_v: $p = p_a + p_v$.

At a certain temperature, the more water vapor in air, the wetter the air and the larger the partial pressure of water vapor. The extra water vapor will dissolve out when it exceeds a certain value. Therefore when the air contains a maximum amount of moisture that it can hold at a particular temperature, the air is in a state of saturation and is called "saturated air". The corresponding partial pressure of water vapor is referred as "partial pressure water vapor" by $p_{v,s}$. It can be found in a common saturated steam table, as shown in Appendix Table 3
The internal pressure of a seawater desalination system is tested by pressure gauge or vacuum gauge. The pressure gauge is used to test the operating pressures under positive pressures, and the vacuum gauge is for the working pressures under negative pressures or vacuum conditions. The operating pressure is not the actual or absolute pressure in the system but the difference with local barometric pressure. The relationship between them is:

$$\text{Absolute pressure} = \text{local barometric pressure} + \text{operating pressure}$$
$$\text{(under positive pressure)}$$

$$\text{Absolute pressure} = \text{local barometric pressure} - \text{operating pressure}$$
$$\text{(under negative pressure)}$$

Only the absolute pressure is the state parameter of moist air. When the operating pressure is not indicated, the pressure should be considered as absolute pressure. The local barometric pressure can be obtained by barometer.

2. Temperature
The temperature of moist air represents its hotness and coldness. The common international criteria are absolute temperature scale (or Kelvin scale) in K, represented by T; Celsius temperature scale in °C represented by t; and Fahrenheit temperature scale in $°F$, represented by t. The relationship between the three scales is:

$$t = T - 273.15 \approx T - 273 \tag{3.5}$$

$$t°C = \frac{5}{9}(t°F - 32) \tag{3.6}$$

3. Density and specific volume
Density is the mass of a unit of volume for moist air and referred by ρ

$$\rho = \frac{m}{V}\ \text{kg/m}^3 \tag{3.7}$$

where m is the mass of moist air in kg; V is the volume occupied by moist air in m^3. Specific volume is the volume occupied by a unit of mass of moist air and represented by v:

$$v = \frac{V}{m} = \frac{1}{\rho}\ (\text{kg/m}^3) \tag{3.8}$$

Because the content of water vapor changes during the transmission of moist air, the benchmarks of moist air are different at initial and final states. However, the quantity of dry air is stable, resulting in applying 1 kg dry air as the standard in the calculation of moist air, which is represented by the subscript of a. Therefore it is important to notice the units and meanings in calculations of moist air.

4. Absolute humidity and relative humidity
Humidity is the amount of water vapor in the air, which can be signified by absolute humidity and relative humidity. Absolute is the total mass of water vapor present in a given volume of air, which is the density of water vapor ρ_v

$$\rho_v = \frac{m}{V} = \frac{p_v}{R_v T} \tag{3.9}$$

where R_v is the gas constant of water vapor.
Relative humidity (RH) is the ratio of the absolute humidity ρ_v and the maximum absolute humidity (absolute humidity of saturated air) $\rho_{v \cdot s}$ of moist air at the same temperature:

$$\varphi = \frac{\rho_v}{\rho_{v \cdot \max}} = \frac{\rho_v}{\rho_{v \cdot s}} \tag{3.10}$$

According to Eq. (3.9),

$$\varphi = \frac{p_v}{p_{v \cdot s}} \tag{3.11}$$

where $p_{v \cdot s}$ is the maximum partial pressure of water vapor in moist air. Therefore the RH is also the partial pressure of water vapor to the equilibrium vapor pressure of water at the same temperature. The smaller the RH, the drier the moist air and vice versa. When the RH $\varphi = 100\%$, the moist air is saturated and cannot hold more moisture.

5. Moisture content (humidity ratio)
Moisture content d is the mass of water vapor present in a unit of mass for dry air, which can be expressed as:

$$d = \frac{m_v}{m_a} \text{ kg/kg(a)} \tag{3.12}$$

where m_v and m_a are the mass of water vapor and mass of dry air in moist air; the unit kg(a) refers to per kilogram of dry air.
Then the following equation can be driven based on the ideal gas equation of state

$$d = 0.622\frac{p_v}{p_a} = 0.622\frac{p_v}{p_b - p_v} = 0.622\frac{\varphi \cdot p_{v \cdot s}}{p_b - \varphi \cdot p_{v \cdot s}} \tag{3.13}$$

where p_b is barometric pressure.

6. Dew point
   The water vapor is overheated in unsaturated air. Dew point t_d is the temperature to which moist air must be cooled to make the water vapor from overheating to saturation and maintain a constant partial pressure of water vapor p_v in moist air at the same time. For an air mixture with constant moisture content d, when the temperature decreases to t_d, the moist air is saturated ($\varphi = 100\%$), and water will leave the air and condense into "dew" if the temperature continues to fall. Therefore dew point is the critical point of water condensation. In addition, dew point depends on the moisture content in moist air only, which means t_d is constant when the moisture content is stable.
   In seawater desalination, the condensing temperature must be lower than the dew point if condensing fresh water from moist air. The lower the condensing temperature compared with the dew point, the fresher water will be condensed.
7. Enthalpy of moist air
   Moist air is a mixture of dry air and water vapor. The enthalpy of moist air is the total of the enthalpy of the dry air and the enthalpy of the water vapor:

$$H = H_a + H_v = m_a h_a + m_v h_v \tag{3.14}$$

Because only the quantity of dry air is constant in the thermodynamic process of moist air, specific enthalpy is for a unit mass of dry air, which can be expressed as:

$$h = \frac{H}{m_a} = h_a + d \cdot h_v \ \text{kJ/kg(a)} \tag{3.15}$$

As the enthalpy of dry air is 0 at 0°C, the enthalpy of dry air at temperature t is:

$$h_a = C_{p,a} \cdot t \tag{3.16}$$

where $C_{p,a}$ is the specific heat of air at constant pressure and $C_{p,a} = 1.005$ kJ/(kg K). The enthalpy of water vapor can be calculated by:

$$h_v = h_c + C_{p,v} \cdot t \tag{3.17}$$

where h_c is the enthalpy of vaporization at 0°C, which is $h_c = 2501$ kJ/kg; $c_{p,a}$ is the specific heat of air at constant pressure and equals $c_{p,v} = 1.86$ kJ/(kg K). The enthalpy of moist air can be expressed as:

$$h = 1.005t + d(2501 + 1.86t), \ \ \text{kJ/kg(dry air)} \tag{3.18}$$

3.2 Psychrometric Chart and the Basic Thermodynamic Process of Moist Air

3.2.1 Psychrometric Chart (h-d Diagram)

For convenient calculation, the coordinate graphs with state parameters for moist air are applied to determine the states and parameters of moist air and analyze its thermodynamic process. The most common one is the psychrometric chart (h-d diagram).

There are only two independent coordinates in a two-dimensional coordinate plane normally. However, the state of moist air depends on temperature t, moisture content d, and barometric pressure p_b. Therefore three independent coordinates are required. If the atmospheric pressure is provided, t and d can be drawn in a 2-D coordinate plane. Different atmospheric pressures p_b correspond to different diagrams. A suitable h-d diagram should be selected according to local atmospheric pressure.

In China, the psychrometric chart in use has enthalpy as the vertical axis and moisture content as the horizontal axis, also called the h-d diagram, as shown in Fig. 3.1 and Appendix Figure 1. The plotting process of it is introduced in the following part.

1. Isenthalpic lines and lines of constant moisture content
   In order to make the diagram clearer, isenthalpic lines are parallel with the 135° tilt angle relative to the vertical axis. The lines of constant moisture content are parallel to the vertical axis. The origin is the state of dry air at $t = 0$ and $d = 0$.
2. Isothermal lines
   Isothermal lines are plotted based on the formula $h = 1.01t + d(2501 + 1.86t)$, and the linear relationship between h and d can be found when the temperature is constant, as a

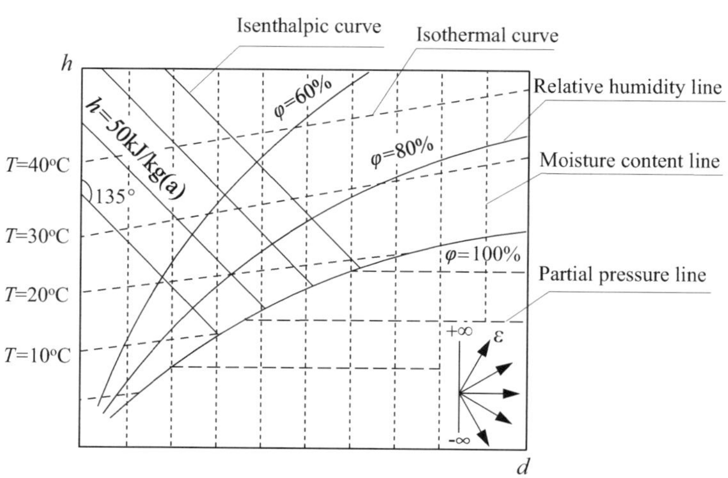

Figure 3.1
Psychrometric chart.

result of which the isothermal lines in the h-d diagram is a series of straight lines. According to the equation, 1.01t is y-intercept and $(2501 + 1.86t)$ is the slope; when t values are different, the slopes of each isothermal lines are different. Hence, isothermal lines are not parallel to each other, but they can be assumed as parallel lines because $1.86t$ is much less than 2501, causing less obvious effects on slope by t.

3. Lines for constant relative humidity

   The lines for constant RH are determined by formula $d = 0.622\frac{\varphi \cdot p_{v \cdot s}}{p_b - \varphi \cdot p_{v \cdot s}}$. Under a certain atmospheric pressure p_b, d depends on $p_{v,s}$ only when RH φ is constant. $p_{v,s}$ is the single-valued function to temperature t, and the value can be found in the tables of water vapor properties. Therefore the value of d can be obtained at different temperature t, and the line of constant φ can be plotted by connecting the points on the i-d diagram according to the values of t and d. The lines of constant φ are a series of upward-concave divergent curves. The y-axis is the line of constant φ at $\varphi = 0\%$. The line of constant φ at $\varphi = 100\%$ also represents the state of saturated moist air; the top left of the line is the region for moist air (also called the "unsaturated region"), and the area at the bottom right is the region of oversaturated moist air. Due to the instability of oversaturation, condensation happens easily. The water droplets are suspended in midair and form fog, and this region is also called the "foggy region". In the region of moist air, water vapor is overheated and has a stable state.

4. Lines of partial pressure of water vapor

   The formula $d = 0.622\frac{p_v}{p_b - p_v}$ can be expressed as $p_v = \frac{p_b \cdot d}{0.622 + d}$. When the barometric pressure B is constant, the equation is in the form of $p_v = f(d)$, which means the partial pressure of water vapor p_v only depends on moisture content d. Therefore the lines of partial pressure of water vapor can be obtained by drawing a horizontal line above the line of d and labeling the values of p_v corresponding to the values of d.

5. Lines of heat humidity ratio

   In the calculation of moist air, the changes of enthalpy and moisture content are important. In the thermodynamic process, the ratio of change in enthalpy to the change in humidity is called heat humidity ratio and is represented by ε, which is:

$$\varepsilon = \frac{\Delta h}{\Delta d} \tag{3.19}$$

   The unit of Δd is g in most cases, so the ε in the h-d diagram should be $\varepsilon = 1000\,\frac{\Delta h}{\Delta d}$. The line presenting the process of moist air with constant ε in the h-d diagram is straight, and the unit of ε is kJ/kg or kJ/g, so ε is the slope of it. Because the slope is not related to the start position, a series of lines with different ε can be plotted in the h-d diagram by applying any point as reference. In practice, the line for the phase-change process of air can be drawn by translating the referenced line with the same ε to the point of initial state.

3.2.2 Basic Thermodynamic Process of Moist Air

The complex thermodynamic processes in industry are the combination of several typical thermodynamic processes. In the thermodynamic process of moist air, the phase change of water vapor in moist air makes the changes on the mass of moist air. Therefore both the energy equation and the mass conservation equation should be applied in calculations. The analysis of the thermodynamic process on moist air is an important part in analyzing the psychrometric chart.

1. Heating or cooling process
   The feature of simple heating or cooling on moist air is used to maintain moisture content, which means the process follows the line constant ω as $\varepsilon = \pm\infty$. The enthalpy increase in the heating process is shown by 0–1 in Fig. 3.2. With the introduction of heat, the temperature of moist air rises and RH decreases. The cooling process is the reverse of the heating process, as shown by 0–5. For a unit mass of dry air, the heat added or ejected is:

$$q = h_2 - h_1 \tag{3.20}$$

where h_2 and h_1 are the enthalpies of initial and final state, respectively. The changes of macroscopic kinetic energy and gravitational potential energy are ignored.

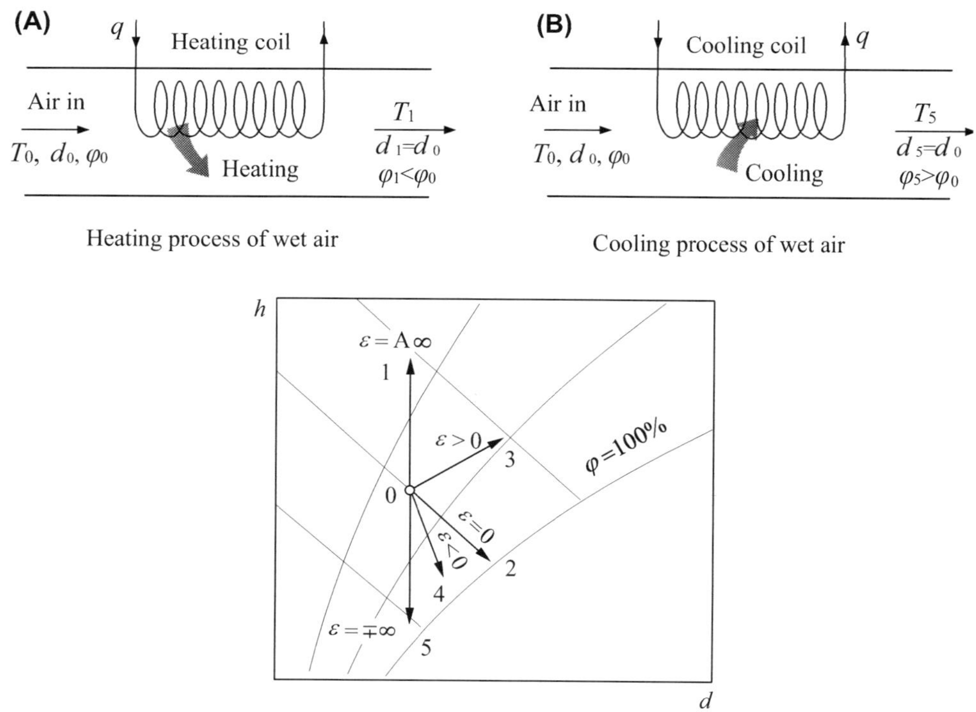

Figure 3.2
Heating and cooling processes of moist air: (A) heating process of wet air (B) cooling procss of wet air.

2. Adiabatic humidification process
Adiabatic humidification is introducing water into air to improve the moisture content without the contribution of thermal energy from an external source. The latent heat required in evaporation is from air itself in the adiabatic process. Because the temperature of air decreases after humidification, this process is called the evaporative cooling process. Ignoring the changes of macroscopic kinetic energy and gravitational potential energy, for a unit mass of dry air, the energy balance equation is:

$$h_1 + (d_2 - d_1)h_w = h_2 \tag{3.21}$$

where $d_2 - d_1$ is the increase of moisture content of the air, which means the added moisture for a unit mass of dry air; h_w is the enthalpy of water. Compared with the initial and final enthalpy h_1 and h_2, $(d_2 - d_1)h_w$ is always small, so it can be ignored in calculation, which derives:

$$h_1 \approx h_2$$

It also indicates that the enthalpy of moist air can be assumed as constant approximately in the adiabatic humidification process in which $\varepsilon = 0$. The process tends to the increase of φ and d and the decrease of t following the line of constant h, as shown by 0–2 in Fig. 3.2.

3. Heating and humidifying process
With the introduction of moisture and heat, both the enthalpy and moisture content of moist air will increase and $\varepsilon > 0$. The line of this process is shown by 0–3 in Fig. 3.2, which is between the line 0–1 and line 0–2. The air-continuing heat and moisture ejection in the workshop and mine follows this process.
If the air is cooled to eject heat when the moisture is added, $\varepsilon < 0$ in this humidification and cooling process. The process line is located between line 0–2 and line 0–5, as shown by line 0–4 in Fig. 3.2.

4. Cooling and dehumidification process
In the cooling process of moist air, if cooling is continued when it is cooled to saturation, the water vapor will be condensed. At this moment, the air is always saturated, and the moisture content and temperature decrease following the line of saturated air, as shown by 1–2 in Fig. 3.3.
In the cooling and dehumidification process, the condensed moisture from a unit mass of dry air is the decrease of moisture content of moist air $(d_1 - d_2)$. The energy q taken away by the cooled water is:

$$q = (h_1 - h_2) - (d_1 - d_2)h_w \tag{3.22}$$

where h_w is the enthalpy of condensed water; $(d_1 - d_2)h_w$ is the energy taken by the condensed water.

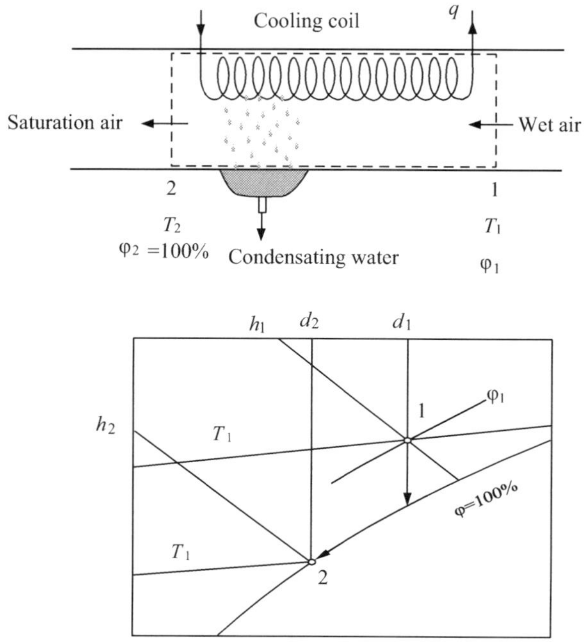

Figure 3.3
Cooling and dehumidification process.

5. Adiabatic mixing process
   In air conditioning units, the air with required temperature and humidity is achieved by mixing several streams in different conditions. If there is no heat transfer between this thermodynamic system and its surroundings, this mixing process is called adiabatic mixing. The obtained moist air condition is determined by the condition of each stream and their proportions of flow.
   Fig. 3.4 presents the adiabatic mixing of streams in state 1 and state 2 with the mass flow rate of dry air of $\dot{m}_{a1}$ and $\dot{m}_{a2}$.The condition of mixed air is represented by 3 and its mass flow rate of dry air is $\dot{m}_{a3}$. According to the principle of mass balance, for dry air:

$$\dot{m}_{a1} + \dot{m}_{a2} = \dot{m}_{a3} \tag{3.23}$$

For the mass of water vapor in streams:

$$\dot{m}_{a1}d_1 + \dot{m}_{a2}d_2 = \dot{m}_{a3}d_3 \tag{3.24}$$

In the process of adiabatic mixing, the equation of energy balance when ignoring the macroscopic kinetic energy and gravitational potential energy is:

$$\dot{m}_{a1}h_1 + \dot{m}_{a2}h_2 = \dot{m}_{a3}h_3 \tag{3.25}$$

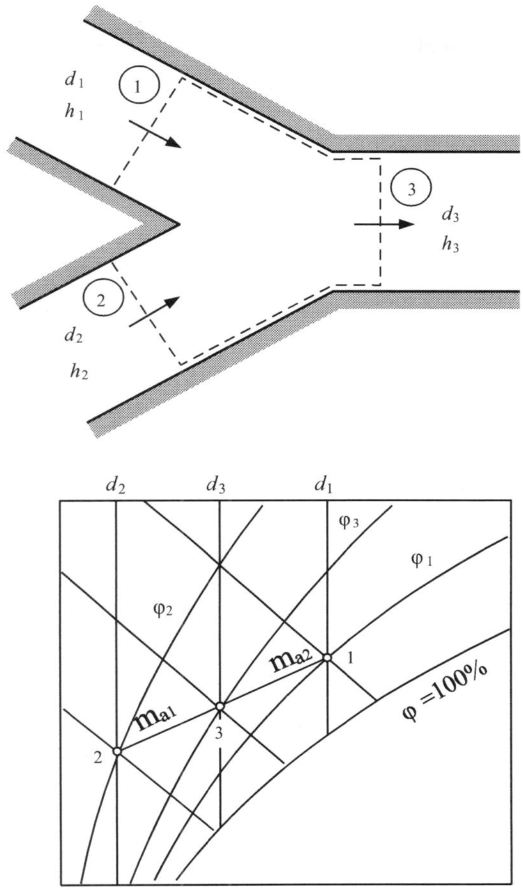

Figure 3.4
Mixture process of wet air.

The specific enthalpy is for a unit mass of dry air.

The three equations are the basic equations of the adiabatic mixing process. If the condition and flow rate of each stream are given, the mass flow rate, moisture content, and enthalpy of mixed air can be calculated based on them, by which the condition of mixed air can be determined.

The state point 3 of mixed air can also be obtained by graphic method in the psychrometric chart. By simultaneous Eqs. (3.23) and (3.24), it can derive:

$$\frac{\dot{m}_{a1}}{\dot{m}_{a2}} = \frac{d_3 - d_2}{d_1 - d_3} \tag{3.26}$$

Similarly, using Eqs. (3.23) and (3.25), the following can be obtained:

$$\frac{\dot{m}_{a1}}{\dot{m}_{a2}} = \frac{h_3 - h_2}{h_1 - h_3} \tag{3.27}$$

Hence,

$$\frac{\dot{m}_{a1}}{\dot{m}_{a2}} = \frac{d_3 - d_2}{d_1 - d_3} = \frac{h_3 - h_2}{h_1 - h_3} \tag{3.28}$$

It is shown that the state point 3 of mixed air is on the line connected by state points 1 and 2 of the streams before mixture. The ratio of the distance $\overline{13}$ from point 3 to 1 and the distance $\overline{32}$ from point 3 to 2 is in inverse proportion of the mass of two air streams, and the point of mixed air is close to the point referring to the stream with larger mass.

3.3 Heat and Moisture Transfer in Direct Contact Between Air and Water

3.3.1 Heat and Moisture Transfer Between Air and Water

When air meets the open surface of water or water splashes, heat and moisture exchanges occur on the water surfaces. Only sensible heat exchange or moisture exchange (mass exchange) and sensible heat change together with latent heat exchange happen depending on the temperature of water. Sensible heat exchange is caused by the temperature difference between air and water, and heat transfers by conduction, convection, and radiation. Latent heat transfer comes from the condensation (or evaporation) of water vapor in moist air, which results in ejection (or absorption) of latent heat of vaporization. The total heat exchange is the algebraic sum of sensible heat exchange and latent heat exchange [3].

According to the principles of mass transfer, when air contacts water directly, a boundary layer of saturated air (Fig. 3.5) is formed in which temperature equals the temperature of the water surface near water surface or water droplets due to the random movement of water molecules. In addition, the concentration of water molecules or partial pressure of water vapor in the boundary layer is determined by the temperature of saturated air in this layer. The water molecules stay random movements in the surroundings of this boundary layer, which causes some water molecules to come into the boundary layer, and some go out and return to the water.

If the temperature of the boundary layer is higher than the temperature of the surroundings, heat will be ejected from the boundary layer to the surroundings, and vice versa.

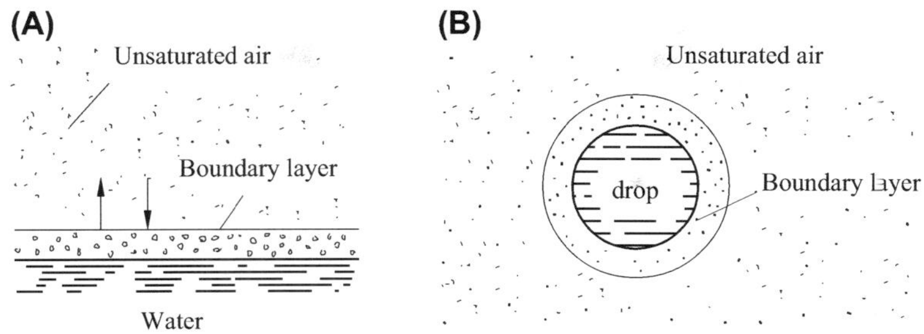

Figure 3.5
Heat and moist transfer between water and air: (A) open water surface and (B) spraying water drop.

If the concentration of water molecules in the boundary layer is larger than that in the surrounding air (which also means the partial pressure of water vapor in boundary layer is larger than that in surrounding air), the water molecules moving from the boundary layer to the surroundings are more than the molecules from the surroundings to the boundary layer, which results in the increase of molecules in the surroundings and vice versa. In evaporation, the reduced water molecules in the boundary layer are supplied by the water molecules coming out from water surfaces; in condensation, the exceeded water molecules in the boundary layer will go back to the water surface.

According to the earlier discussion, mass transfer will occur by water molecule movements from a higher concentration region to a lower concentration region, if there is concentration difference (or partial pressure difference) of water vapor between unsaturated air and boundary layer. It is also indicated that the concentration difference drives mass transfer, like the temperature difference drives heat transfer.

Sensible heat transfer between air and water depends on the temperature difference between the boundary layer and the surroundings, and latent heat transfer from the moisture exchange is determined by the concentration difference (or partial pressure difference) of water molecules between them.

When air contacts water on a small surface with area $dA(\text{m}^2)$, the amount of sensible heat transfer is:

$$dQ_x = a(t - t_b)dA(W) \tag{3.29}$$

where a is sensible heat flux transfer coefficient between the air and water surface, W/(m^2 °C); t is air temperature of surroundings, °C; and t_b is temperature of boundary layer, °C.

The amount of moisture transfer is:

$$dW = a_D(C_a - C_b)dA \quad \text{kg/s} \tag{3.30}$$

where a_D is the moisture transfer coefficient calculated by the concentration difference between air and water surface, m/s; C_a is the concentration of water molecules of air in the surroundings, kg/m^3; and C_b is the concentration of water molecules of air in the boundary layer, kg/m^3.

It can also be expressed as:

$$dW = \beta(p_v - p_{v,b})dA \quad \text{kg/s} \tag{3.31}$$

where β is the moisture transfer coefficient calculated by the partial pressure difference between air and water surface, kg/(N s); p_v is partial pressure of water vapor in surroundings, Pa; and $p_{v,b}$ is partial pressure of water vapor in the boundary layer, Pa.

Because the difference of partial pressure for water vapor can be represented by the difference of moisture content in a small range of temperatures, the moisture transfer can be expressed as:

$$dW = \sigma(d_a - d_b)dA \quad \text{kg/s} \tag{3.32}$$

where σ is the moisture transfer coefficient calculated by the difference of moisture content between air and water surface, kg/(m^2 s); d_a is the moisture content of air in the surroundings, kg/kg$_{\text{dry air}}$; and d_b is the moisture content of air in the boundary layer, kg/kg$_{\text{dry air}}$.

The latent heat exchange occurring with moisture transfer is:

$$dQ_q = h_{fg} \cdot dW = h_{fh} \cdot \sigma(d_a - d_b)dA(W) \tag{3.33}$$

where h_{fg} is latent heat of vaporization at temperature t_b, J/kg.

According to the total heat exchange $dQ_z = dQ_x + dQ_q$, it can be derived that:

$$dQ_z = [a(t - t_b) + h_{fg} \cdot \sigma(d_a - d_b)]dA(W) \tag{3.34}$$

Where dA is the actual area of water surface touching the air. If the equation can be integrated by area, the total amount of heat exchange can be obtained. However, the area of water surface is difficult to determine sometimes, so there may be difficulties in practical calculations.

3.3.2 Phase Change in Direct Contact Between Air and Water

When air flows on water surfaces or around water droplets, it will take some saturated air away and supply some new air to achieve the saturation. Hence the saturated air will mix with the passed unsaturated air to make changes on the state of air. The moisture exchange between air and water can be assumed as an air mixture of saturated and unsaturated air. On the i-d diagram, the state point of mixed air should be on the line connecting the points of initial air state and saturated air state at the same temperature. It is obvious that the point of final air state is closer to the point of saturation if saturated air becomes more. Therefore if the amount of water contacting air is finitely great and the time of contacting is finitely long, which means under the assumed conditions, all of the air can reach saturation conditions and has the same temperature as water; the point of final air state should be located above the curve of saturation on the i-d diagram, and the final temperature equals the water temperature. If the temperatures of water are different, the change of air states will be different. Hence under these conditions, seven typical processes of air state changes can be obtained at different water temperatures, as shown in Fig. 3.6 and Table 3.3.

In the seven processes, line *A*-2 is the boundary of humidification and dehumidification, line *A*-4 is the boundary of increase and decrease enthalpy, and line *A*-6 is the boundary of the heating and cooling of air.

3.3.3 Lewis Relationship and Its Application

In the research of adiabatic humidification, the American scholar Lewis introduced that when air contacts water directly, there is a relationship between heat and mass transfer, which the ratio between heat transfer coefficient and mass transfer coefficient is constant, and it equals the specific heat of moist air. It is also called the Lewis number.

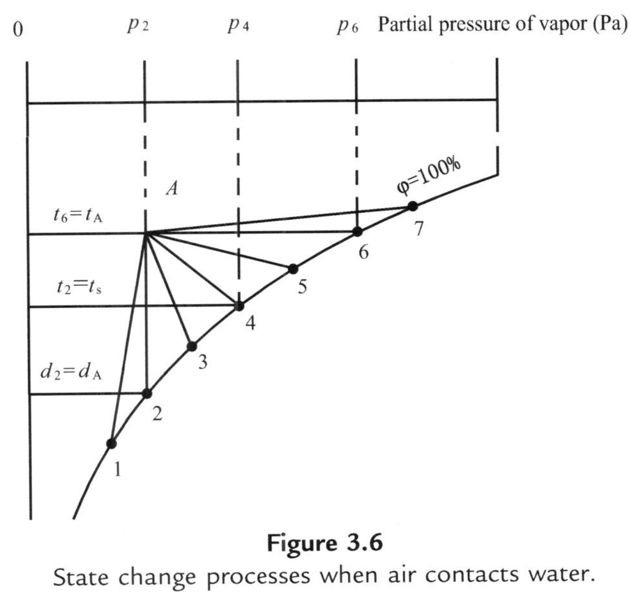

Figure 3.6
State change processes when air contacts water.

Table 3.3: Properties of difference processes in direct contact between air and water.

Progress line	Water temperature	t or Q_x	d or Q_q	h or Q_z	Type of process
A-1	$t_\omega < t_l$	Decrease	Decrease	Decrease	Cooling and dehumidification
A-2	$t_\omega = t_l$	Decrease	Stay constant	Decrease	Sensible cooling
A-3	$t_l < t_\omega < t_s$	Decrease	Increase	Decrease	Minus enthalpy humidification
A-4	$t_\omega = t_s$	Decrease	Increase	Stay constant	Isenthalpic humidification
A-5	$t_s < t_\omega < t$	Decrease	Increase	Increase	Increase enthalpy and humidification
A-6	$t_\omega = t$	Stay constant	Increase	Increase	Isothermal humidification
A-7	$t_\omega > t$	Increase	Increase	Increase	Heating and humidification

This theory is discussed as follows:

For adiabatic humidification, the sensible heat loss of air is equal to the sensible heat required by water vaporization on the interface with area dA, which is:

$$dQ_x = dQ_q \tag{3.35}$$

or

$$\alpha(t - t_b)dA = h_{fg} \cdot \sigma(d_a - d_b)dA \tag{3.36}$$

Rearrange,

$$d_a - d_b = \frac{\alpha}{\sigma \cdot h_{fg}}(t - t_b) \tag{3.37}$$

In addition, the sensible heat ejected by air comes back to air in the form of latent heat in adiabatic humidification. For 1 kg of moist air, the heat balance equation can be expressed as:

$$c_p \cdot (t - t_b) = h_{fg} \cdot (d_a - d_b) \tag{3.38}$$

Rearrange, $(d_a - d_b) = \frac{c_p}{h_{fg}}(t - t_b)$

When comparing Eqs. (3.37) and (3.38), one can obtain: $c_p = \frac{\alpha}{\sigma}$, which is the famous equation of Lewis relationship, and the mass transfer coefficient can be calculated by heat transfer coefficient according to it.

3.4 Water Vapor Generation Under Constant Pressure

3.4.1 Three Stages of Water Vapor Generation Under Constant Pressure

In common seawater desalination systems, especially in traditional solar distillators, water vapor is generated under constant pressure normally. It is necessary to discuss the stages of water vapor generation under constant pressure.

At any temperature, water molecules can go into the vapor phase, but this macrophenomenon will stop under constant pressure. If a certain amount of water is heated continuously, it will totally become water vapor. Generally, this process can be divided into three stages:

1. Heating
   Assuming the pressure of water is 0.1 MPa and temperature is 0°C at the beginning, it is presented in a *p-v* diagram and a *T-s* diagram in Fig. 3.7A and B by 1°.
   While heating water and maintaining constant pressure at the same time, there will be a small quantity of increase in volume, specific volume, and entropy. When the temperature reaches 99.634°C, it is boiling. If it is heated continuously, the boiling water will generate water vapor. This boiling point (99.634°C corresponding to 0.1 MPa) is also called saturation temperature t_s. The water at saturation temperature is called saturated water, and the state parameters v,h,s are expressed as v',h',s' with the superscript "′" to distinguish from other states. The water with temperature lower than saturation temperature is called unsaturated water (or subcooled water). The heat required for heating a

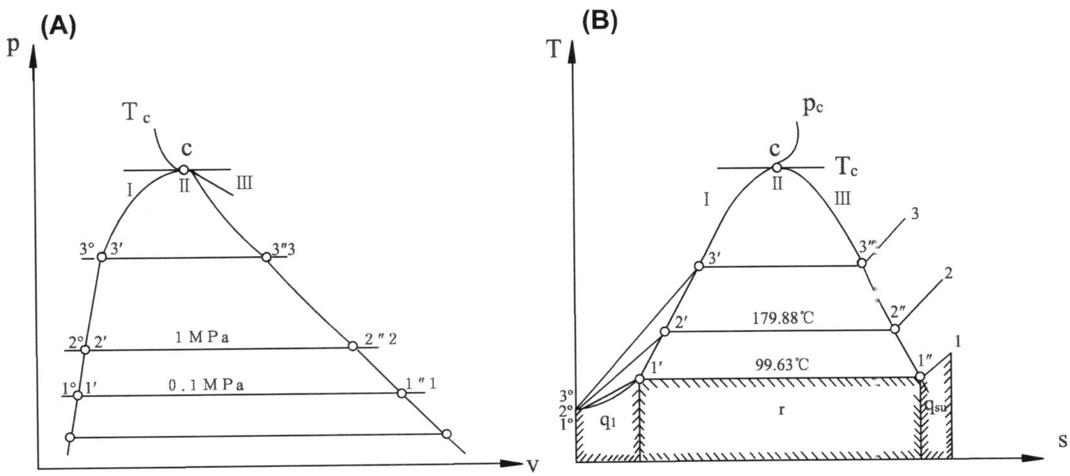

Figure 3.7
Water vapor generating processes under constant pressure [2].

unit mass of water from unsaturation to saturation at temperature 0.01°C is called the heat of liquid referred by q_1. According to the first law of thermodynamics:

$$q_1 = h_{1''} - h_{1'} \tag{3.39}$$

It can be presented by the shaded area below 1°-1′ on a *T-s* diagram, as shown in Fig. 3.7B. 1′ is the corresponding state of saturated water.

2. Vaporization
If heating saturated water under constant pressure, water will boil, and phase transition occurs from water to vapor. The temperature stays constant at saturation temperature t_s during boiling. During the phase change process, water and vapor exist at the same time, which is called wet saturated vapor. With the heat being added continuously, water decreases and vapor increases until all of the water becomes vapor with a state of 1″. It is called dry saturated vapor. The parameters of it should be presented with superscript "″", referring to its specific volume, enthalpy, and entropy by, respectively, v'',h'',s''. The process of heating saturated water to dry saturated vapor under constant pressure is presented by horizontal line 1′-1″ on *p-v* and *T-s* diagram because the specific volume increases due to the increase of vapor, and entropy rises due to absorbing heat, although the pressure and temperature are constant during this process. The heat absorbed in this process is called latent heat of vaporization and is expressed by h_{fg}. Therefore:

$$h_{fg} = h'' - h' \quad \text{or} \quad h_{fg} = T_s(s'' - s') \tag{3.40}$$

It is presented by the shaded area under 1′-1″ on the *T-s* diagram.

3. Superheating
If heating the dry saturated vapor, the temperature, specific volume, and entropy of it will rise, as shown by 1″-1 in Fig. 3.7. Because the temperature in this stage is higher than the saturation temperature at the same pressure, it is called superheated steam. The temperature difference between superheated steam and saturation temperature at the same pressure is expressed as:

$$D = t - t_s \tag{3.41}$$

D is called degree of superheat. The heat absorbed in this process is called heat of superheat:

$$q_{su} = h_1 - h_{1''} \tag{3.42}$$

which can be expressed by the shaded area under line 1″-1 in the *T-s* diagram of Fig. 3.7.
If changing the pressure *p*, for example, to increase the pressure and investigate the vaporization process under constant pressure, three similar stages can be obtained.

In Fig. 3.7, lines 2°-2′-2″-2 are the corresponding vaporization process at constant pressure $p = 1$ MPa. Although the three stages are comparable, the saturation temperature rises with the increase of pressure. The saturation temperature at 1 MPa is 179.916°C rather than 99.634°C. The saturation temperature is constant with the pressure staying constant, and vice versa. There is one-to-one correspondence between them. The pressure for a corresponding saturation temperature is called saturation pressure and is expressed by p_s, which is:

$$p_s = p_s(t_s) \tag{3.43}$$

Except the increase of saturation temperature after rising pressure, the vaporization process speeds up, which means $(v'' - v')$ and $(s'' - s')$ decrease and the latent heat of vaporization reduces. When the pressure is increased to 22.064 MPa, $t_c = 373.99$°C, $v'' = v'$ and $s'' = s'$ under which condition there is no difference between saturated water and saturated vapor. This is a point of the phase called the critical point or critical state, as shown by point c in Fig. 3.7. The state parameters of critical state are called critical state parameters, as critical pressure p_c, critical temperature t_c, and critical specific volume v_c. When $t > t_c$, the steam cannot be liquefied no matter how big the p is. It is called permanent gas.

The saturated water line is obtained by connecting all of the points of saturated water $1',2',3',\ldots$ at different pressures and critical point c. Connecting the points of dry saturated vapor $1'',2'',3'',\ldots$ and critical point c can determine the saturated vapor line. They are named saturation lines. The saturation lines divide p-v and T-s diagrams into three regions: subsaturated water (region on the left of saturated water line), wet steam (also called two-phase region, saturation region) in the middle of two saturation lines, and superheated steam (region on the right of saturated vapor line). The water and vapor in three regions and one saturation line present five states: unsaturated water, saturated water, wet steam, dry saturated vapor, and superheated steam.

It is important to mention that wet steam is the mixture of water and saturated vapor. Although the pressure (saturated pressure) and temperature (saturated temperature) of wet steams with different amounts of water vapor (or water) are the same, the states of the wet steams are different. To describe the vapor content in wet steam to determine the state of it, the dryness fraction x is applied, which is the mass percentage of saturated vapor in wet steam:

$$x = \frac{m_v}{m_v + m_w} \tag{3.44}$$

where m_v and m_w are the mass of saturated vapor and saturated water in wet steam. It is obvious that the dryness fraction of saturated water is $x = 0$ and $x = 1$ for saturated dry vapor.

3.4.2 Table of Thermodynamic Properties for Steam (Steam Table)

The steam table is an important tool for thermodynamic calculations. Because the popularization of water vapor applied in industry, the steam table used currently is uniform all over the world. It is produced based on *1968 IFC Formulation for Scientific and Industrial Use* by the International Formula Council (IFC). The diagram set the point of water in the liquid phase for triple point as reference point and ruled that the internal energy and entropy of saturated water at triple point is zero.

Normally, the thermodynamic properties of water and vapor are expressed by setting pressure p and temperature t as independent variables and specific volume v, enthalpy h, and entropy s as dependent variables. Internal energy can be calculated by $u = h - pv$. Because there is only one independent variable between pressure and temperature under the states on saturation lines and wet steam, and both temperature and pressure are independent variables for unsaturated water and superheated steam, the steam table is divided into "saturation water and steam table" and "unsaturated water and superheated steam table".

According to the requirements in engineering calculations, the saturation water and steam table is divided into two types: one is arranged by temperature t_s (as shown in Appendix Table 3), and the other is arranged by pressure p_a (as shown in Appendix Table 4) to show the different saturation pressure p_a (or saturation temperature t_a), v',v'',h',h'',r,s', and s'' under different t_a or p_a. For the wet steam with dryness fraction x, the parameters can be calculated by the following formulae according to the state parameters of saturated water and steam with the same t_a (or p_a):

$$v = xv'' + (1 - x)v' \tag{3.45}$$

$$h = xh'' + (1 - x)h' \tag{3.46}$$

$$s = xs'' + (1 - x)s' \tag{3.47}$$

The table of unsaturated water and superheated steam demonstrates the specific volume v, enthalpy h, and entropy s under different pressure and temperature of unsaturated water and superheated steam. The thick black line in this table separates the unsaturated water and superheated steam. The region above the line represents unsaturated water and below shows superheated steam. The parameters of saturation water and steam on heat of this table can be applied as a reference.

3.5 Liquid Boiling Under Constant Pressure

Boiling is the vaporization of liquid on the interface of solid−liquid. It happens when heating the surface temperature T_w to the temperature above the saturation temperature T_s

corresponding to the pressure of liquid. The heat transfer from solid to liquid is based on Newton's law of cooling:

$$q = h(T_w - T_s) = h\Delta T \tag{3.48}$$

Where $\Delta T = T_w - T_s$ is the excess temperature superheat; h is the heat transfer coefficient between heating surface and liquid, W/(m^2 °C).

The features of boiling are in the form of bubbles. The small bubbles generate, become bigger, and finally leave the surface. The complex process of the bubble transformation is related to the excess temperature, surface properties, and thermophysical properties (e.g., surface tension). Therefore the forming process of bubbles affects the fluid motion near the surface, which also influences heat transfer coefficient significantly.

Boiling can occur under any conditions. There are two typical types: pool boiling (boiling happens when a heating surface is submerged in a large body of stagnant liquid) and forced convective boiling in tubes.

3.5.1 Pool Boiling

In pool boiling, liquid is stagnant, and the liquid motion near the surface is caused by the free convection and mixing of liquid due to the generation and escape of bubbles. On the contrary, for the forced convective boiling, the motion of liquid comes from the external influences and the mixing due to the free convection and bubbles. Boiling can be categorized into subcooled boiling and saturated boiling (bulk boiling). In subcooled (partial) boiling, the liquid temperature is lower than the saturation temperature; bubbles generate on the surface and condense in liquid. However, in saturated boiling, liquid temperature exceeds saturation temperature; bubbles generate on the surface and go through the liquid due to the buoyancy. They escape from the free surface finally, as shown in Fig. 3.8. Under this condition, the liquid temperature decreases rapidly near the

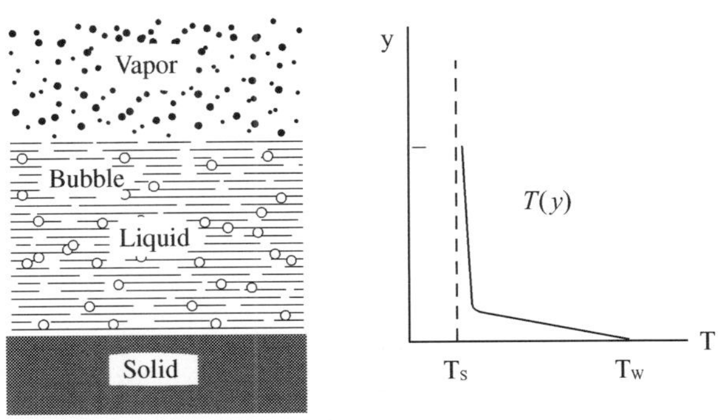

Figure 3.8
Temperature distribution of saturated pool boiling with liquid–vapor interface.

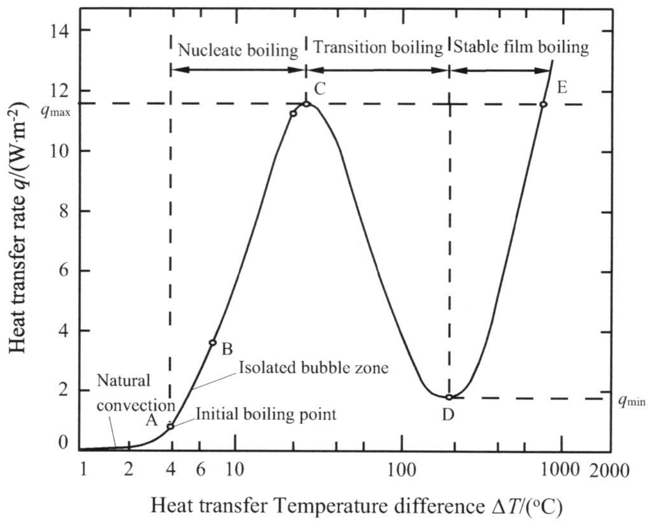

Figure 3.9
Boiling curve of saturated water under atmospheric pressure.

solid surface, but the temperature of most liquid is still a little higher than saturation temperature. The bubbles generating on the liquid–solid interface can rise and pass through the liquid–vapor interface.

When the liquid contacts the surface with high temperature and the temperature rises up to a certain value, boiling heat transfer will occur. Experiments indicate that four regions with different heat transfer principles will occur as the increase of the difference Δt (excess temperature) between the temperature of heating surface t_w and the corresponding saturation temperature at the same pressure t_s in a big container. Fig. 3.9 shows the change of heat flux with the excess temperature of surface Δt in pool boiling of water under atmospheric pressure.

1. The region before point A on the curve is called the free convection region. At the beginning, the excess temperature of the surface is low and boiling has not started. There are no bubbles generating on the heating surface, so only heated fluid floats upward. Therefore heat transfer follows the law of unidirectional flow.
2. The area between curve AC is called the nucleate boiling (spray boiling) region. From point A, bubbles will generate on some specific points (called nucleation sites) on the heating surface. At the beginning, the generated bubbles on nucleation sites do not interfere each other, which is called the region of isolated bubbles, as shown by section AB on the curve. With the increase of excess temperature, the number of nucleation sites rises and bubbles interfere to merge into air slugs and air columns as section BC on the curve. In the regime between A and C, the massive interference and coalescence happen, and heat transfer coefficient, e.g., heat flux, increases rapidly. Because

nucleation sites play a critical rule in this process, the boiling in this regime is called nucleate boiling (or spray boiling). The range of excess temperature is around $5°C \le \Delta t \le 50°C$.

3. The regime between curve CD is called the transition boiling region. The heat transfer principle changes when increasing Δt from the peak. Because bubbles gather on the heating surface and the exclusion of vapor becomes worse, heat flux reduces with the increase of Δt. This circumstance will continue until the heat flux reaches the lowest q_{min}. This regime is called transition boiling, which is an unstable process. In this regime, the excess temperature range is $50°C \le \Delta t \le 150°C$.
4. The regime between curve DE is a stable film boiling region. The heat transfer principle changes again from point q_{min}. At this moment, a stable vapor layer is already formed on the heating surface. The generated vapor escapes from the film layer regularly and q rises with the increase of Δt. This regime is called stable film boiling.

In practice, the peak of heat flux introduced earlier is significant. It is called critical heat flux, and point C is the critical point. For the heating equipment with a certain heat flux, e.g., electric water heater, water wall boiler, etc., the heat flux must be controlled below the critical heat flux q_{max} or take actions on the region where film boiling may occur. Otherwise, once the heat flux exceeds the peak and film boiling happens, the temperature of walls will increase rapidly or the equipment will burn out due to the sharp decrease of heat emission coefficient. Therefore q_{max} is called burnout point.

Because the high heat transfer rate and heat transfer coefficient occur when the excess temperature is small, especially for a seawater desalination system, most equipment are desired to operate in the regime of nucleate boiling. Most of the heat transfer coefficients in this regime are beyond 10^4 W/(m^2 K), which is much bigger than those in normal convective heat transfer without a phase change. Hence it is significant to understand the heat transfer in detail of this regime. The following equations show the heat transfer coefficient of nucleate boiling in a pool boiling under $(0.2 \sim 101) \times 10^5$ Pa [4].

$$h = 0.142 \Delta t^{2.33} p^{0.5} \quad \mathrm{W/(m^2\ °C)} \tag{3.49}$$

Because $q = h \cdot \Delta t$, it can be expressed as:

$$h = 0.55 q^{0.7} p^{0.15} \quad \mathrm{W/(m^2\ °C)} \tag{3.50}$$

where Δt is excess temperature of the wall, $\Delta t = t_w - t_a$, °C; q is wall heat flux, W/m^2; and p is absolute pressure of boiling, Pa.

Eq. (3.49) is suitable when the excess temperature of the wall is known; Eq. (3.50) is applied when the heat flux is given.

We know that the critical heat flux q_{max} is an important point on the boiling curve. It is expected that the boiling process happens close to that point, but the risk of exceeding it should be considered as well. Another equation is suggested for calculating critical heat flux by [5]:

$$q_{max} = 0.18 h_{fg} \rho_v \left[\frac{6g(\rho_l - \rho_v)}{\rho_v^2} \right]^{1/4}$$

where h_{fg}(kJ/kg) is the latent heat of liquid vaporization. The subscripts l,v refer to the state of liquid and vapor, respectively. σ(N/m) is the surface tension of liquid, g(m/s^2) is the local gravitational acceleration, and ρ(kg/m^3) is the density of the substance.

3.5.2 Heat Transfer Coefficient of Convective Boiling in Tubes

When liquid boils in a large space, bubbles can leave and float up from the heating surface without limitations. The proportion of gas contained in a liquid–gas mixture is small, so the effects on flow by bubbles belong to the natural convection. But the conditions are different when liquid flows and boils in a small space (in a tube). When liquid is heated by tube, the gas content and flow of the mixture change along the tube. The flow patterns of liquid and gas change as well as the heat exchange types. The heat transfer type affects flow pattern, and flow type changes heat transfer type.

Fig. 3.10 shows a flow process in a heated pipe. At the beginning, forced convection heated the subcooled liquid coming into the pipe. The heat transfer of forced convection

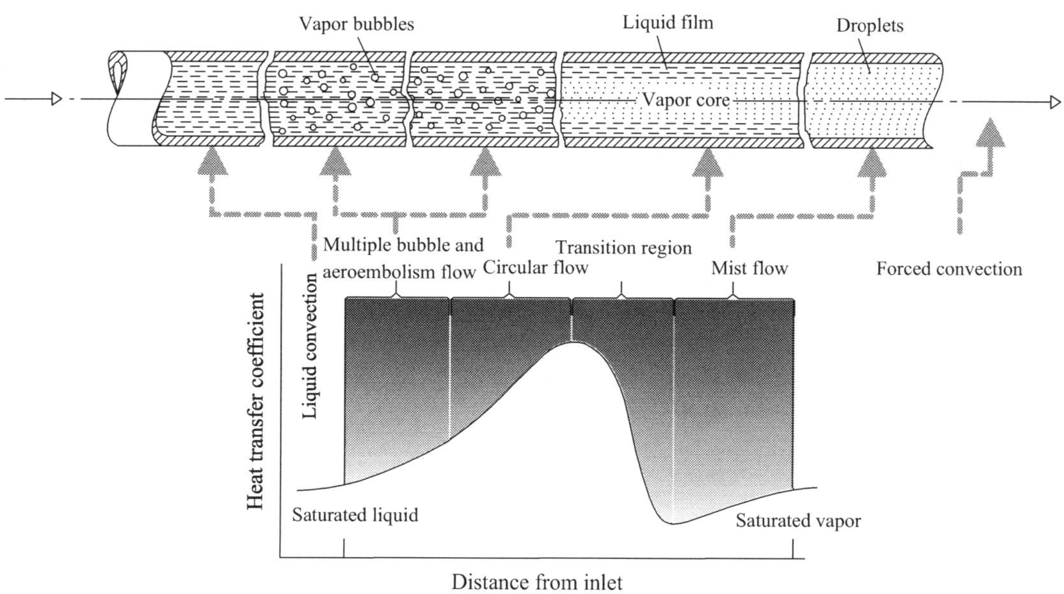

Figure 3.10
Flow conditions of forced convection boiling in a tube [5].

here can be calculated by traditional heat transfer equations of force convection. Boiling will occur shortly, and bubbles generate and grow to be taken into the mainstream. This kind of flow with many bubbles causes the obvious increase of heat transfer coefficient of convection. With the rising of vapor volume percentage, bubbles will merge to aeroembolism or air mass of vapor. There is a circular flow region after the aeroembolism region. A thin film will be formed in this region and moves along the internal surface. Vapor flows along the center axis of the pipe with higher speed. In the regions of bubble flow and most circular flow regions, heat transfer coefficient increases continuously. But there will be some drypoint on the internal surface where the heat transfer coefficient begins to decrease in the end. The features in transition regions are that the number of drypoints grows until the surface is completely dry, and the liquid left will stay in the mainstream in the form of liquid droplets. The heat transfer coefficient reduces along this region. In the mist flow region, heat transfer coefficient hardly changes. The mist flow continues until all the droplets become vapor. In addition, the forced convection on the surface makes the vapor overheat. This kind of heat transfer and two-phase flow interact as both cause and effect, which make the boiling in pipe be too complicated to figure out its physical processes completely until now. None of the appropriate methods to calculate that can be generally accepted. Hence only empirical equations under certain conditions can be used.

The average heat transfer coefficient h_b of boiling in a tube can be calculated as follows:

$$h_b = cq^b \tag{3.51}$$

where c and b are the coefficients related to refrigerant type and heat flux density; q is heat flux density.

The average heat transfer coefficient of boiling in a horizontal tube can be expressed as follows:

$$h_b = Aq^{0.6}v_m^{0.2}d_i^{-0.2} \tag{3.52}$$

where v_m is flow rate of refrigerant, A is coefficient, and d_{i} is diameter of the tube.

Eq. (3.52) is suitable when:

$$q = 0.6{\sim}25000\ \mathrm{W/m^2}$$
$$v_m = 50{\sim}600\ \mathrm{kg/(m^2\ s)}$$

It was put forward that it can be calculated by a piecewise function according to the research data from foreign countries.

At the inlet of the tube, the dryness of vapor is low and the flow rate of the refrigerant is small. The heat flux density is bigger compared with that on the internal surface. The energy transferred from the exterior can make bubbles generate and boil. During this time, the heat emission factor is mainly determined by heat flux density, but the influences from the flow rate of the working substance are small.

With the heating effects on the tube wall, vapor dryness increases and the flow rate of the working substance rises. It can be predicted that the heat emission intensity will be determined by the flow rate of the working substance rather than heat flux density. Heat transfer coefficient h only depends on flow rate and dryness x at this time. This kind of heat transfer is called convective heat transfer. According to the data by B. Slipcevic and Chawla, the formulae for different heat transfer regions are summarized as follows:

1. Convective heat transfer region

$$h = y\frac{v_m^{1.4}}{d^{0.5}} \tag{3.53}$$

$$y = \frac{0.0115(\lambda')(\rho')^{0.06}}{g^{0.3}(\rho'')^{0.66}(\mu')^{0.225}(\mu'')^{0.575}} \tag{3.54}$$

   where d is inner diameter of the tube, superscript "$'$" is the value for saturated liquid, and superscript "$''$" is the value for saturated gas.
2. Nucleate heat transfer region

$$h = \frac{Bv_m^{0.1}\varphi^{0.7}}{\varphi^{0.7}d^{0.5}} \tag{3.55}$$

   where B is the physical parameter of the working substance when nucleate boiling happens in a tube, which is the function of temperature of vaporization. φ is the ratio of internal and external surface area.

The boundary of the two heat transfer regions depends on the relationship between mass flow rate and v_m heat flux density q.

When $v_m > (B/y)^{0.769}q^{0.538}$, it should be calculated based on convective heat transfer;

when $v_m < (B/y)^{0.769}q^{0.538}$, it should be calculated based on the heat transfer of nucleate boiling.

It can be seen that the main factor affecting heat transfer in boiling is heat flux density, physical parameter, and the correlation between gas and liquid flow.

3.5.3 The Influence of Surface Roughness

The influence of surface roughness (by machining, grooving, scoring, or sandblasting) on the maximum and minimum heat fluxes and on film boiling is negligible. However, as demonstrated by Berenson [14], increased surface roughness can cause a large increase in

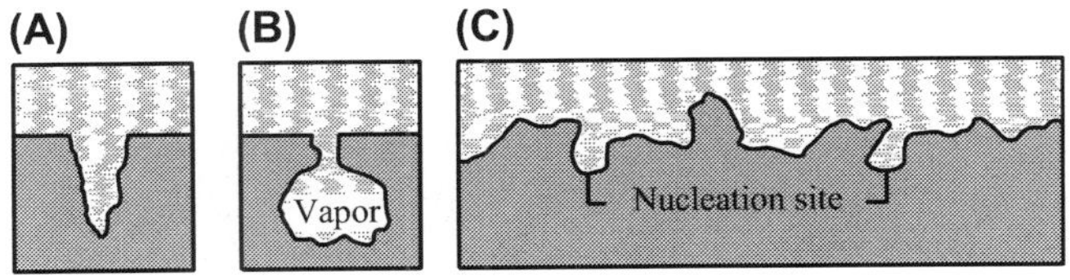

Figure 3.11

Formation of nucleation sites: (A) wetted cavity with no trapped vapor, (B) reentrant cavity with trapped vapor, and (C) enlarged profile of a roughened surface.

heat flux for the nucleate boiling regime. As Fig. 3.11 illustrates, a roughened surface has numerous cavities that serve to trap vapor, providing more and larger sites for bubble growth. It follows that the nucleation site density for a rough surface can be substantially larger than that for a smooth surface. However, under prolonged boiling, the effects of surface roughness generally diminish, indicating that the new, large sites created by roughening are not stable sources of vapor entrapment.

Special surface arrangements that provide stable augmentation (enhancement) of nucleate boiling are available commercially and have been reviewed by Webb [15]. Enhancement surfaces are of two types: (1) coatings of very porous material formed by sintering, brazing, flame spraying, electrolytic deposition, or foaming and (2) mechanically machined or formed double-reentrant cavities to ensure continuous vapor trapping (see Fig. 3.12). Such surfaces provide for the continuous renewal of vapor at the nucleation sites and heat transfer augmentation by more than an order of magnitude. Active augmentation techniques, such as surface wiping–rotation, surface vibration, fluid vibration, and electrostatic fields, have also been reviewed by Bergles [16,17]. However, because such techniques complicate the boiling system and, in many instances, impair reliability, they have found few practical applications.

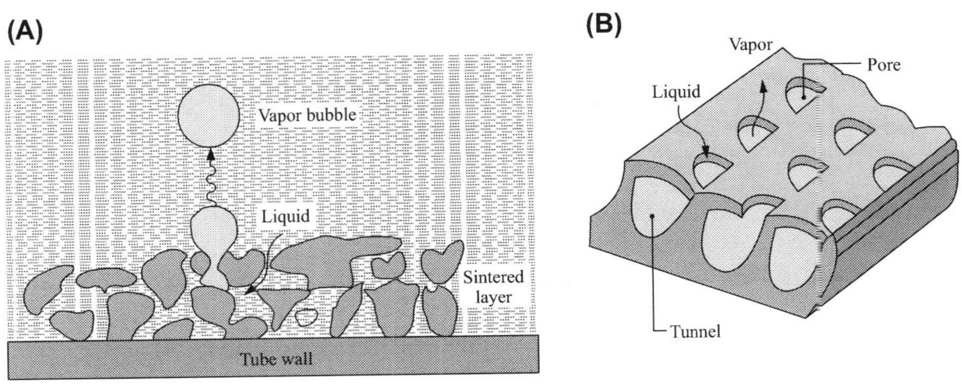

Figure 3.12

Typical structured enhancement surfaces for augmentation of nucleate boiling: (A) sintered metallic coating and (B) mechanically formed double-reentrant cavity.

3.5.4 Factors Affecting Heat Transfer of Liquid Boiling

Boiling begins from the bubble generation from the heating surface. The amount and generating frequency of bubbles determine the boiling extent. There are two criteria for bubble generation: first, the liquid must be overheated; second, there must be a nucleation site.

According to the Clausius–Clapeyron equation, the relationship between the radius R of bubbles and levels of overheating can be derived as follows:

$$R = \frac{2\sigma T_s}{r\rho_g(T_L - T_s)} \tag{3.56}$$

where σ is surface tension, r is latent heat of vaporization, ρ_g is vapor density, and $(T_L - T_s)$ is temperature difference between liquid temperature and saturation temperature, which is the degree of superheat.

It is shown that the radius R of bubbles is inversely proportional to the degree of superheated $(T_L - T_s)$. The higher the superheated degree, the more bubbles will generate from bubble cores.

The nucleation site is the sufficient condition of boiling. Theoretical and experimental results show that bubbles only occur at the locations of some nucleation sites. Currently, there are two statements of nucleation sites: one is homogeneous nucleation, and the other is heterogeneous nucleation. The first believes that the size of homogeneous nucleation is as small as the order of magnitude of molecular volume. They are generated from the fluctuation of overheated liquid. High-energy molecules gather to a molecular beam with a certain size according to a certain presumption and form a bubble embryo with a certain size. They will become homogeneous nucleation that can generate bubbles when the superheat degree is high enough. However, the heterogeneous nucleation reckons that bubbles generate from the crack and cavity on the wall that is filled with solid particles or noncondensable gases, or from the depressions on the wall due to corrosion and incrustation, etc. In all, the temperature difference, heat flux intensity, and the condition of the heating surface are the main factors that influence liquid boiling.

3.6 Condensation of Water Vapor Under Constant Pressure

When saturated vapor contacts the vertical surface whose temperature is lower than the saturation temperature, it will release the latent heat of vaporization to condense to a liquid and stick on the wall. Because of the capability of condensate wetting the wall, the vapor condensation may become two forms: film-wise condensation and drop-wise condensation. If the condensate wets the wall very well (as shown in Fig. 3.13A, contact

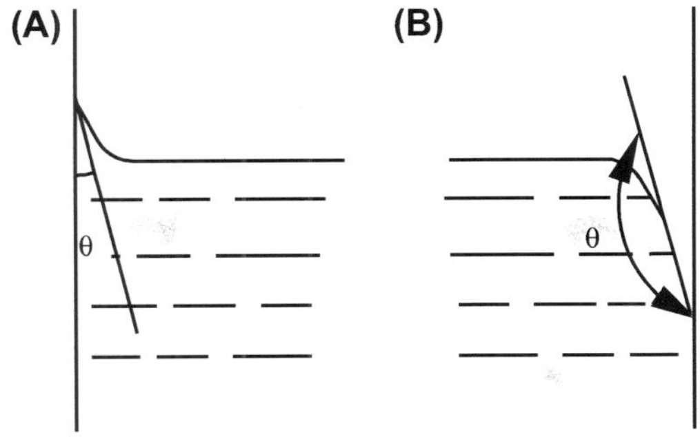

Figure 3.13
Contact angle: (A) wet capacity strong and (B) wet capacity weak.

angle $\theta < 90°$), there will be a complete film forming on the surface, which is called film-wise condensation. During this time, the wall is covered by the film all the time, so that the latent heat released from condensation must transfer onto the cooling wall through this film. The liquid film becomes the main thermal resistance. If condensate cannot wet the surface well (as shown in Fig. 3.13B, contact angle $\theta > 90°$), condensates form small droplets rather than film, which is called drop-wise condensation. On a nonhorizontal wall, droplets roll down along the surface due to gravity when they grow to a specific size. During the rolling down, they will take away the droplets along the road to "clean" the surface. More surface will be exposed to vapor in order to greatly reduce the thermal resistance. Therefore the heat transfer coefficient of drop-wise condition is much bigger than that of film-wise coefficient, which can reach 5 to 10 times higher normally.

Experiments show that almost all of the common vapors, including water vapor, can have film-wise condensation on clean surfaces of popular engineering materials. In engineering, film-wise condensation is used as a reference for design in engineering usually. The following sections will present the heat transfer calculations of vapor film-wise condensation on the vertical plate and external surface of the horizontal tube, which are commonly used.

3.6.1 Heat Transfer of Film-Wise Condensation on Vertical Plate

The vapor film on the top of the vertical plate is thinner, and the film is in laminar flow. During the process that film flows down along the gravity direction, vapor condensation continues. The thickness of the film increases, and the film flows faster. When the thickness reaches a specific value, its flow type changes from laminar flow to turbulent flow. The changes of partial heat transfer coefficients are shown in Fig. 3.14.

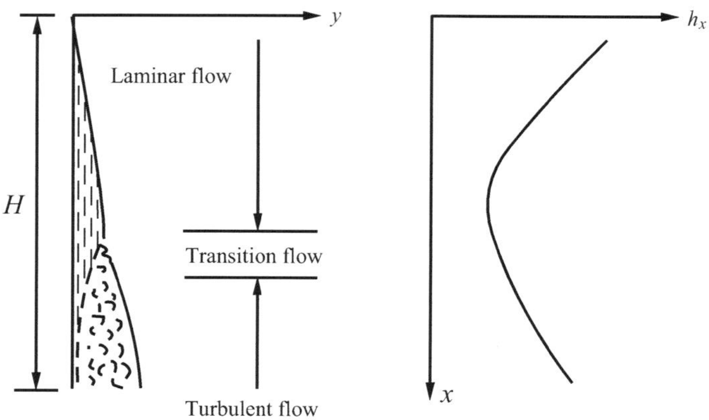

Figure 3.14
Vapor film condensation on the vertical plate and its heat transfer coefficient variation with plate length.

Experiment results demonstrate that the partial heat transfer coefficient of laminar flow (Re < 1600) on the vertical plate can be expressed as [5]:

$$h_x = \frac{\lambda_l}{\delta(x)} = \left[\frac{g h_{fg} \rho_l (\rho_l - \rho_v)}{4\mu_l (T_{sat} - T_w) x}\right]^{\frac{1}{4}} \tag{3.57}$$

The average convective heat transfer coefficient of the whole plate is:

$$\bar{h} = \frac{1}{L}\int_0^L h(x)dx = \frac{4}{3}h_x\big|_{x=L} \tag{3.58}$$

or

$$\bar{h} = C\left[\frac{g h_{fg} \rho_l^2 \lambda_l^3}{\mu_l (T_{sat} - T_w) L}\right]^{\frac{1}{4}} \quad \mathrm{W/(m^2\ ^\circ C)} \tag{3.59}$$

This is the famous Nusselt's equation. The C in the equation is a constant shown in Table 3.4;

g is gravitational acceleration, m/s^2;
h_{fg} is latent heat of vaporization, which can be looked for by t_a, J/kg;
L is height of vertical plate, m;
T_{sat} is saturation temperature under corresponding pressure, °C;
T_w is temperature of surface, °C;
ρ_v is vapor density (kg/m^3); and
ρ_l,λ_l,μ_l is density of condensate (kg/m^3), thermal conductivity [W/(m °C)], and dynamic viscosity [W/(m s)].

Table 3.4: Values of coefficients C, C', and dimensions L in film condensation of laminar flow.

		C	C'	L
Vertical plate (or tube)	Laminar flow	0.943	0.925	Height $H = L$
	Wave	1.130	1.180	
Horizontal tube		0.725	0.650	Outer diameter $d_0 = L$

The values of thermophysical properties for the condensate in the earlier equations is based on the temperatures in the following equation:

$$T' = T_{sat} - 0.75(T_{sat} - T_w) \tag{3.60}$$

Nusselt's equation can also be applied for the pipe wall (for both vertical and horizontal). But the differences of coefficients and dimensions should be noticed. For easier application, the temperature difference ΔT is replaced by the heat flux density q (unit: W/m^2); the parameter of physical property is referred by β:

$$\beta = \frac{\lambda_l^3 \rho_l^2 h_{fg} g}{\mu_l} \tag{3.61}$$

Hence the average heat transfer coefficient can also be expressed as:

$$\overline{h} = C'\left(\frac{\beta}{\rho_l q}\right)^{1/3} \tag{3.62}$$

It is important to mention that Nesselt's equation is derived under the condition that the flow rate of vapor is zero. If the flow rate of vapor is not zero, its effects must be considered. In addition, corrections are required when it is for the pipes with multiple rows or ribs outside the pipe.

Therefore the total heat flux of the whole surface is:

$$q_t = \overline{h}A(T_{sat} - T_w) \tag{3.63}$$

where A (m^2) is the area of the surface. The total mass flow rate of condensate is:

$$m_e = \frac{q}{h_{fg}} = \overline{h}A(T_{sat} - T_w)/h_{fg} \tag{3.64}$$

When Re > 1600, the film changes from laminar flow to turbulent flow. The average heat transfer coefficient of the whole vertical surface is suggested to be calculated by the equation:

$$\overline{h} = 0.0077\left[\frac{g\rho_l^2 \lambda_l^3}{\mu_l}\right]^{\frac{1}{3}} \mathrm{Re}_\delta^{0.4} \tag{3.65}$$

Re_δ is the Reynolds number, where the thickness of film is δ. Normally, it is calculated by:

$$Re_\delta = \frac{4\Gamma}{\mu_l} \tag{3.66}$$

where Γ is the mass flow rate of a unit width on the surface in kg/(s m). The top of the horizontal surface $\Gamma = 0$. It rises with the increase of x. The total heat flux and mass flow are still calculated by Eqs. (3.63) and (3.64). For the vertical tube, the formula for the vertical plate can still be utilized in the same way when its diameter is far less than its height.

3.6.2 Film Condensation Outside the Horizontal Tube

The film of film condensation on the external surface of the tube (as shown in Figs. 3.15 and 3.16) is in laminar flow often. The average heat transfer coefficient of condensation is [6]:

$$\overline{h} = 0.725\left[\frac{g h_{fg} \rho_l^2 \lambda_l^3}{\mu_l d (T_{sat} - T_w)}\right]^{\frac{1}{4}} \text{ W/(m °C)} \tag{3.67}$$

Where d is the outer diameter of the tube in m.

The form of equations for heat transfer coefficient of film condensation on the exterior of the horizontal tube $\overline{h}$ is similar to that on the vertical plate $\overline{h}_H$. The only differences are changing H to d and changing coefficient 1.13 to 0.725. Because the length of the tube H is far larger than the outer diameter of the tube d in engineering, it is positioned horizontally most of the time. The bigger heat transfer coefficient of condensation can be obtained under the same conditions.

In industry, the condenser consists of pipeline bundles commonly. There are no simple and precise equations to calculate the average heat transfer coefficient of vapor condensing outside pipe bundles. Generally, Eq. (3.67) is used with the replacement of d with $N \cdot d$.

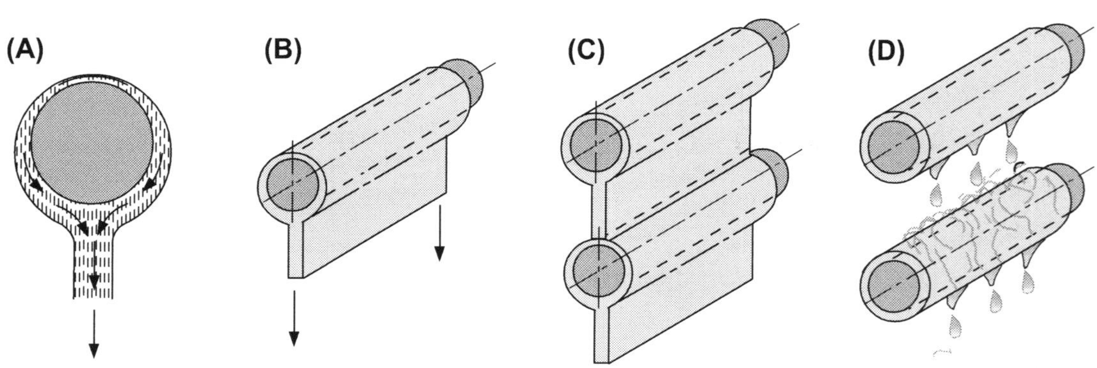

Figure 3.15
Film condensation on (A) a sphere, (B) a single horizontal tube, and (C) a vertical tier of horizontal tubes with a continuous condensate sheet and (D) with dripping condensate.

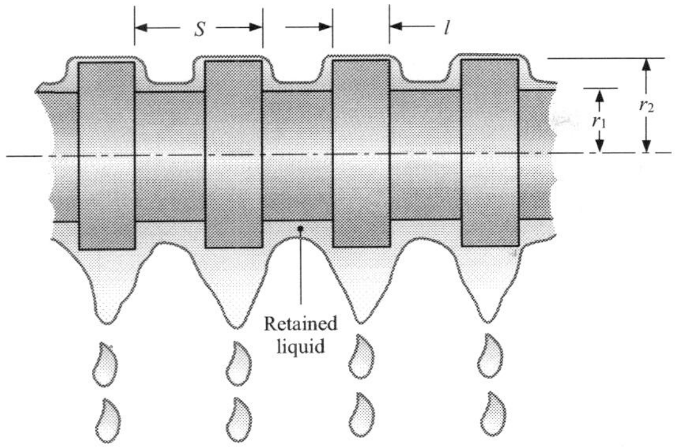

Figure 3.16
Condensation on a horizontal finned tube.

Therefore the average heat transfer coefficient of condensation outside horizontal pipe bundles is:

$$\overline{h} = 0.725\left[\frac{gh_{fg}\rho_l^2\lambda_l^3}{\mu_l d \cdot N(T_{sat} - T_w)}\right]^{\frac{1}{4}} \text{ W/(m °C)} \tag{3.68}$$

where N is the average number of rows in vertical direction.

3.6.3 Film Condensation Inside the Horizontal Tube

The condensation inside the horizontal tube is complex normally, and it is significantly affected by the vapor flow rate inside the tube. If the flow rate is small, the condensation in the horizontal tube can be described approximately, as shown in Fig. 3.17A. The condensate flows from the upper half to the lower half. It accumulates at the lower half and flows with the vapor along the pipe finally. For the vapor with low flow rate, that is,

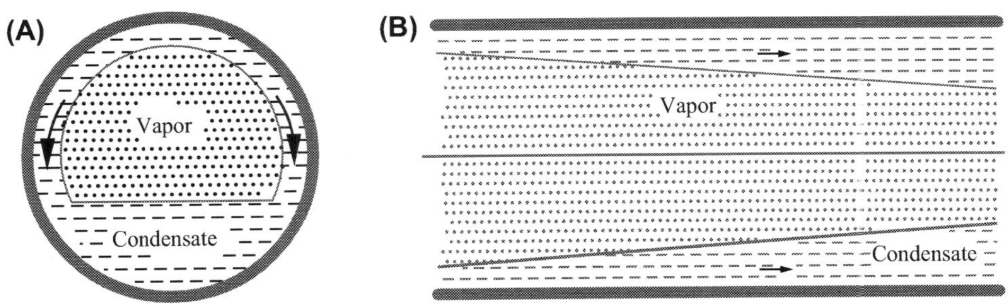

Figure 3.17
Film condensation in the horizontal tube: (A) condensate flow in a cross section under low vapor velocity and (B) axial flow of condensate under high vapor velocity.

the Reynolds number of vapor at the inlet is Re < 3500, and the following equation can be used for calculating heat transfer coefficient [5,6]:

$$\overline{h}_L = 0.555\left[\frac{g\rho_l^2\lambda_l^3 h'_{fg}}{\mu_l(T_{sat} - T_w)d}\right]^{1/4} \tag{3.69}$$

in which,

$$h'_{fg} = h_{fg} + \frac{3}{8}c_{p,l}(T_{sat} - T_w) \tag{3.70}$$

$c_{p\cdot l}$ is the specific heat of liquid at constant pressure, W/(kg K); d is the tube diameter in m.

Under the higher flow rate of vapor, the two-phase flow changes to a circular flow, as shown in Fig. 3.17B. Vapor occupies the center of the circular ring. The diameter of the vapor in the center reduces with the increase of condensate thickness along the flow direction.

There is no unified view and calculating method due to its complexity. The approach introduced below is appropriate and more accurate when the following conditions are satisfied: gas velocity is $\upsilon_v = 3 \sim 300$ m/s, the ratio of operating pressure and critical pressure is $\varepsilon_p = 0.0019 \sim 0.44$, and the Reynolds number of total condensate is $\text{Re}_L = 386 \sim 6.29 \times 10^4$.

$$h_x = h_L(1-x)^{0.8} + \frac{3.8x^{0.76}(1-x)^{0.04}}{\varepsilon_p^{0.38}} \tag{3.71}$$

where h_x is heat transfer coefficient of partial condensation; h_L is heat transfer coefficient of liquid when the vapor condensates completely.

$$h_L = 0.023\text{Re}_L^{0.8}\text{Pr}_e^{0.4}\lambda_L/d_i \tag{3.72}$$

where λ_L is thermal conductivity of condensate, d_i is the inner diameter of the tube, and x is dryness of vapor on the cross section.

When the change of x is smaller than 40%, x is close to linear distribution.

Hence $x = \frac{1}{2}(x_1 + x_2)$

When the variation range of x is bigger, it should be calculated in the segment.

The heat transfer coefficient of condensation for Freon is:

$$h = 0.555\left(\frac{\beta}{d_i\Delta T}\right)^{1/4} = 0.455\left(\frac{\beta}{d_i\psi}\right)^{1/3} \tag{3.73}$$

where ΔT is the temperature difference between condensation temperature and surface temperature; ψ is heat flux density on the inner surface of the tube.

The heat transfer coefficient of condensation for ammonia in a tube can be calculated as follows:

$$h = 2116\Delta T^{-0.167} d_i^{-0.25} = 86.88\psi^{0.2} d_i^{-0.33} \quad (3.74)$$

3.6.4 Dropwise Condensation

Typically, the heat transfer coefficient for dropwise condensation is an order of magnitude larger than those for film condensation. In fact, in heat exchange applications for which dropwise condensation is promoted, other thermal resistances may be significantly larger than that due to condensation, and therefore reliable correlations for the condensation process are not needed.

Of the many surface-fluid systems studied, most of the data are for steam condensation on well-promoted copper surfaces (that is, surfaces for which wetting is inhibited) and are correlated by an expression of the form [18].

$$\overline{h}_{dc} = 51104 + 2044 T_{sat}(^\circ\mathrm{C}) \quad 22^\circ\mathrm{C} \le T_{sat} \le 100^\circ\mathrm{C} \quad (3.75)$$

$$\overline{h}_{dc} = 255510 \quad T_{sat} \ge 100^\circ\mathrm{C} \quad (3.76)$$

where the heat transfer coefficient has units of ($\mathrm{W/m^2\ K}$). The condensation rate can be calculated by,

$$\dot{m} = \frac{q}{h_{fg}} = \frac{h_{dc} A (T_{sat} - T_s)}{h_{fg}} \quad (3.77)$$

where h_{fg} is the latent heat of vaporization of water. The properties are evaluated at the film temperature $T_f = (T_{sat} + T_s)/2$. The vapor density ρ_v and latent heat of vaporization h_{fg} should be evaluated at T_{sat}. The effect of subcooling, $T_{sat} - T_s$, on h_{dc} is small and may be neglected.

3.6.5 Analysis of the Factors Influencing Condensation Heat Transfer

As discussed in the previous section, the heat transfer coefficient of film is $h = \lambda/\delta$, where λ and δ is thermal conductivity and thickness of the film. Obviously, the factors that help to reduce the thickness of film δ are a benefit to the condensation heat transfer; otherwise they are negative factors.

3.6.5.1 Effects of Flow Regime

1. The flow regime of condensate in liquid film is referred by Re. It is in laminar flow when Re < 400, which was introduced before. Within the laminar region, the flow of condensate increases with the rising of Re. Hence the thickness of film δ rises, and the

heat transfer coefficient of film h decreases. When Re > 400, it is in turbulent flow. With the increase of Re, δ reduces so that h rises.

2. The flow regime of refrigerant is represented by v_w. When v_w increases, the surface temperature T_w drops, the flow rate of refrigerant rises, and the thickness of laminar flow film increases with the rising of v_w, which causes the reduction of h. For the film of turbulent flow, h increases with the enhancement of v_w.
3. The flow regime of vapor is referred by its flow rate v_v. When vapor flows to the same direction with liquid film, v_v increases and δ decreases so that h rises. When the flow direction of vapor is opposite to that of liquid film, h reduces with the increases of v_v. The conditions in laminar flow film was introduced earlier. If the film is in turbulent flow, the conditions are the opposite.

3.6.5.2 Effect of Noncondensing Gas

If a small amount of noncondensing gas is contained in condensed gas, the diffusion of the gas mixture will be a component that diffuses through another static component. During diffusion, noncondensing gas is taken to and accumulates on the condensation surface by steam. Fig. 3.18 shows the heat transfer model of film condensation when noncondensing gas exists. The total heat flux q passing through the cooling surface is the sum of the sensible heat q_g emitted by noncondensing gas when it is cooled down and the latent heat $h_{fg} \cdot m$ by vapor condensation where m is the mass of condensed vapor. The partial pressure of noncondensing gas is the biggest on the surface. Because of the accumulation of noncondensing gas on the interface, the mainstream vapor must pass through this gas layer by diffusion in order to reach the gas–liquid interface to condense, which results in the partial pressure of vapor near the interface decreasing, and the saturation temperature of vapor decreasing. The temperature of the gas–liquid interface is unknown rather than the saturation temperature of vapor mainstream.

On the condensing surface, the partial pressure of noncondensing gas increases up to which is higher than the partial pressure of noncondensing gas in a gas mixture. Then a force

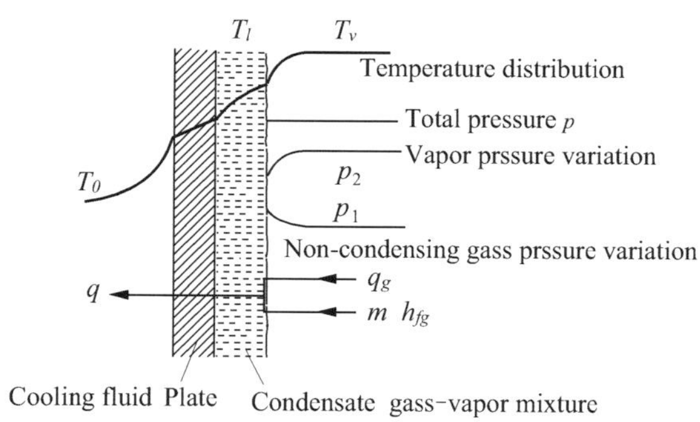

Figure 3.18
Film condensation when noncondensing gas exists [7].

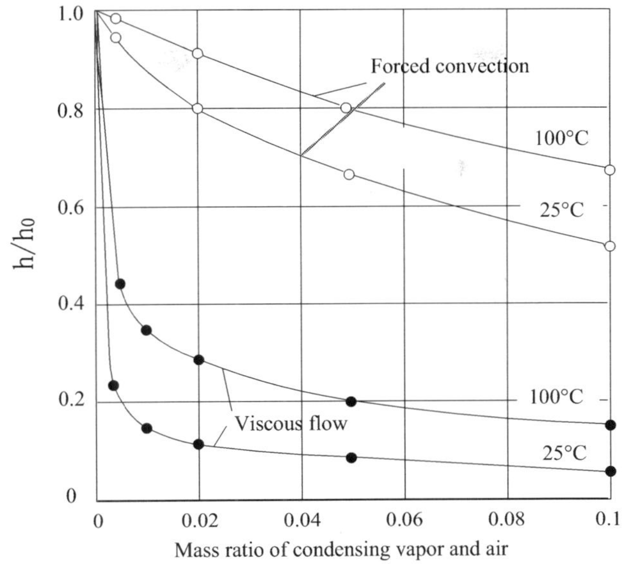

Figure 3.19
The influence of noncondensable gas on the condensation process.

makes noncondensing gas diffuse to the opposite direction of the interface, which offsets a part of diffusion to the interface of the vapor–gas mixture. Because the accumulation and reverse diffusion of noncondensing gas on interface, the thermal resistance of the gas–liquid interface increases obviously so that it affects heat transfer significantly. Fig. 3.19 presents the effects of the proportion of noncondensing gas content on heat transfer rate. ρ_a and ρ_v refer to the mass content of air and vapor in a unit volume of the gas mixture in kg/m^3. h and h_0 represent the average heat transfer coefficient with and without air, respectively.

It can be seen that the heat transfer coefficient can be reduced by 60% with 1% air content. The reduction is 50% when air content is 0.5%. For the condenser utilized in industry, the content of noncondensing gas is restricted within 0.5%.

It should be mentioned that the condensation of vapor in the condenser is continuous. The content of noncondensing gas is low at inlet and high at outlet, due to which the heat transfer coefficients of inlet and outlet differ many times.

Actually, in most of the facilities of falling film evaporation and condensation, the existing noncondensing gas is inevitable. For example, in seawater desalination, the vapor of falling film evaporation is surrounded by a mixture of air and vapor rather than pure vapor. The air in it will prevent the evaporation and condensation. Removing the noncondensing gas must be taken into consideration.

Except flow type and noncondensing gas, the factors that affect condensation heat transfer are fluid physical properties, surface conditions, and position of placement etc.

3.6.5.3 Enhancement of Condensation Heat Transfer

The best way of enhancing condensation heat transfer is to change the film-wise condensation to drop-wise condensation somehow. Drop-wise condensation can reduce the liquid film to a minimum so that the thermal resistance will reduce to a minimum. With the influence of surface tension, circulation will occur inside the droplet to enhance transfer rate. Other technologies of enhancing condensation heat transfer will be discussed in the following sections.

3.7 Falling Film Evaporation of Water Under Constant Pressure

When the temperature of vapor is cooled below its saturation temperature T_{sat} on a surface, condensation will happen on it. On the contrary, if the liquid film is heated to its saturation temperature, evaporation will happen. Hence evaporation and condensation are two generating processes that are reversed.

3.7.1 Falling Film Evaporation on Vertical Plate

Falling film evaporation is the same as the process of falling film condensation essentially, but the different generating direction is reversed. However, they are different in practical industrial application. The difference is that the thickness of the film begins from a limited thickness for falling film evaporation. The Reynolds number (Re) decreases with the falling of the film. In a seawater desalination system, the changes of Re is not big because the evaporation rate is much smaller than the rate of film falling.

In falling film evaporation, the liquid close to the wall may be overheated due to $T_w > T_{sat}$. Boiling will occur when overheating is too much. At this time, the heat transfer process becomes complicated. Fig. 3.20 shows the temperature distribution line along the turbulent evaporating film.

In a normal evaporating process, the most common types are wavy laminar film evaporation and turbulent film evaporation. If ignoring the disturbing influences on film falling from vapor, Eqs. (3.78)–(3.80) can still be applicable to film falling evaporation calculation. Hence for the film falling evaporation on a vertical plate, there is [8]:

$$\mathrm{Nu} = \left(\frac{3}{4}\mathrm{Re}\right)^{-\frac{1}{3}}, \quad \mathrm{Re} < 30, \quad \text{for laminar flow.} \tag{3.78}$$

$$\mathrm{Nu} = 0.822\mathrm{Re}^{-0.22}, 30 < \mathrm{Re} < \mathrm{Re}_{tr}, \text{for wavy condensation} \tag{3.79}$$

$$\mathrm{Nu} = 3.8 \times 10^{-3}\mathrm{Re}^{0.4}\mathrm{Pr}_l^{0.65}, \mathrm{Re}_{tr} < \mathrm{Re}, \text{for turbulent flow.} \tag{3.80}$$

where $\mathrm{Re}_{tr} = 5800\mathrm{Pr}_l^{-1.06}$, which is the Reynolds number of the place where turbulence occurs; $\mathrm{Pr}_l = c_{pl}\mu_l/k_l$ is the Prandtl number of the liquid. The energy balance equation

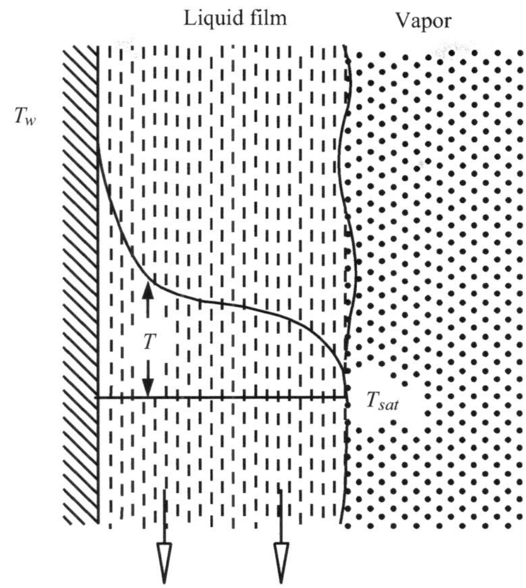

Figure 3.20
Temperature distribution along the turbulent evaporating film.

inside the film can be applied to figure out the relationship between the average Nusselt number (Nu) and film Reynolds number. On the both sides for the film, there is [8]:

$$k_l \frac{\partial T}{\partial y}\bigg|_{y=0} = h_x(T_{sat} - T_w) = h_{fg}\frac{d\Gamma}{dx} \tag{3.81}$$

Then [8],

$$\frac{dx}{\left(\nu_l^2/g\right)^{\frac{1}{3}}} = \frac{\mu_l h_{fg}}{4k_l(T_{sat} - T_w)\text{Nu}} d\text{Re} \tag{3.82}$$

Applying the generally defined Prandtl number and Jakob number (Ja), the equation above can be derived to:

$$\frac{dx}{\left(\nu_l^2/g\right)^{\frac{1}{3}}} = \frac{\text{Pr}_l}{4\text{Ja}_l}\frac{d\text{Re}}{\text{Nu}} \tag{3.83}$$

Assuming Nu is constant on the whole plate and it equals its average value. Rearranging the equation above and integrating along the whole vertical plate, the average Nusselt number is:

$$\overline{Nu} = \frac{\text{Pr}_l}{4Ja_l}\frac{\left(\frac{\nu_l^2}{g}\right)^{\frac{1}{3}}}{L}(\text{Re}_0 - \text{Re}_L) \tag{3.84}$$

where Re_0 is the Reynolds number when film is at $x = 0$; Re_L is the Reynolds number when the film is at $x = L$.

A study [9] introduced a series of empirical equations of heat transfer coefficient (when $q < 0.7 \times 10^5$ W/m^2) as follows:

$$h' = 0.9\mathrm{Re}^{-0.22} \quad \text{(Laminar flow, evaporation)}$$
$$h' = 0.006\mathrm{Re}^{0.4} \quad \text{(Turbulent flow Re} > 3200\text{, evaporation)}$$

h' is dimensionless heat transfer coefficient, and $h' = h\left(\frac{v_l^2}{k_l^3 g}\right)^{\frac{1}{3}}$.

3.7.2 Falling Film Evaporation on the Horizontal Tube

Falling film evaporation on the horizontal tube is a new technology for seawater desalination. Because the phase changes occur on both sides of film evaporation and the film flow of the liquid outside the tube contributes to speed up the separation of vapor and liquid, its heat transfer coefficient is high. Generally, for a smooth tube, the heat transfer coefficient for falling film evaporation on the horizontal tube is equal to the 1.5–2 times of flash evaporation and evaporation on the vertical tube. The area required for heat transfer can be greatly saved under the same conditions. Meanwhile, the scaling on tubes can be improved because heat can transfer when the temperature difference is low, leading to the reduction of overheating on the surface. The falling film evaporation on the horizontal tube is taken into consideration by more and more scholars due to the reasons mentioned earlier. They think it is an energy-saving technology for evaporating desalination, which has a lower energy consumption and its temperature requirements for heat supply is low. It is better for utilizing low-grade heat sources and solar energy.

The total heat transfer coefficient of falling film evaporation on the horizontal tube can be divided into two parts: film (heat transfer) coefficient of evaporation outside the tube and film coefficient of condensation inside the tube, as shown in Fig. 3.21. The film coefficient of evaporation outside the tube is mainly determined by the flow type of falling film outside the tube. U refers to the overall heat transfer coefficient of falling film evaporation on the horizontal tube in W/m^2 °C; h_1 and h_2 represent the heat transfer coefficient inside the tube and outside the tube in W/m^2 °C. Hence,

$$\frac{1}{U} = \frac{1}{h_1} + \frac{1}{h_2} \tag{3.85}$$

Research shows that may factors influence the heat transfer coefficient h_1 of condensation inside the tube, especially the flow rate of vapor inside the tube, tube diameter, and temperature difference of heat transfer, etc. The process of condensation inside the tube is

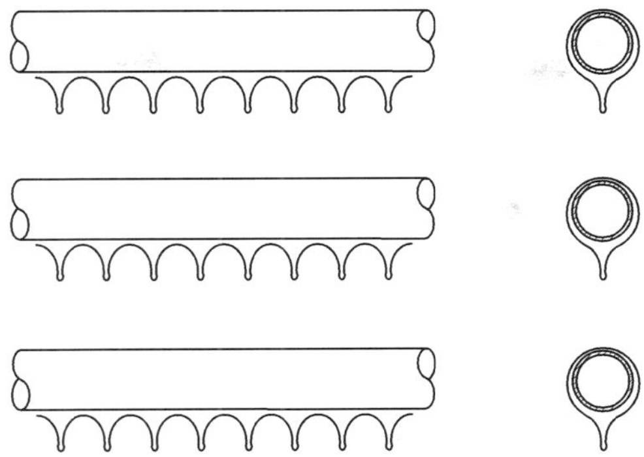

Figure 3.21
Horizontal tube falling film evaporation.

discussed in previous sections, and the calculating equations under certain conditions are provided. The following equations are based on the experiments [10] that can be compared with previous formulae:

$$h_1 = K^+ \left\{ \frac{g\rho_{1L}(\rho_{1L} - \rho_{1G})k_L^3\left[h_{fg} + \frac{3}{8}c_{p1L}(T_1 - T_w)\right]}{\mu_{1L}d(T_1 - T_w)} \right\}^{\frac{1}{4}} \tag{3.86}$$

where $K^+ = 0.3508\left(\frac{L}{d}\right)^{-0.1929}(\mathrm{Re}_\tau)^{0.0643}$

$\mathrm{Re}_\tau = \frac{v_s\rho_{1L}d}{\mu_{1L}}$ is called the cut-off Reynolds number.

Much research has introduced the calculation about the heat transfer coefficient h_2 outside the horizontal tube. Sernas (1979) summarized a heat transfer equation of falling film evaporation on the horizontal tube based on his own experimental data [11]:

$$\overline{h}_2 = Ck_{2L}\left(\frac{g}{\nu_{2L}^2}\right)^{\frac{1}{3}}\left(\frac{4\Gamma}{\mu_{2L}}\right)^{0.24}\left(\frac{\nu_{2L}}{\alpha_{2L}}\right)^{0.66} \tag{3.87}$$

where $\overline{h}_2$ is the average heat transfer coefficient of evaporation outside the tube and C is a constant. When the tube diameter is $d = 25$ mm, $C = 0.01925$; when $d = 50$ mm, $C = 0.01729$.

An experimental research [10] put forward an equation to calculate the heat transfer coefficient h_2 of evaporation outside the tube when the following three assumptions are satisfied for heat transfer on a single tube: (1) The liquid is saturated and with homogeneous temperature before dropping onto and after leaving the tube wall, which means the liquid temperature T_2 is the boiling point corresponding to the saturated vapor pressure; (2) the surface temperature T_w on the wall of heat transfer is uniform; and (3) the effects by liquid splash and the heat resistance of the tube wall are ignored:

$$h_2 = 5.169 \times 10^{-11} \frac{h_{fg} g (\rho_{2L})^2 R^2}{(T_w - T_2)\mu_{2L}} \left(\frac{\bar{\delta}}{R}\right)^{-0.422 \Delta T_2^{0.503}} \times \left(1 + \frac{\delta_{\max} - \delta_{\min}}{\bar{\delta}}\right)^{5.708} \quad (3.88)$$

Experimental results show that when the load of film flow is $\Gamma = 0.05 \sim 0.35$ kg/(m s), the temperature difference is $\Delta T = 0.5 \sim 7$°C and the diameter of the tube is in 20 ~ 40 mm. When the vapor flow rate at the end of tube is in 0.1 ~ 1.0 m/s, the overall heat transfer coefficient of falling film evaporation on the horizontal tube is around 4.0 kW/(m^2 °C), and it will decrease rapidly only in blind gut, at the end of which the flow rate is zero. The coefficient can drop from 4.0 kW/(m^2 °C) to 1.0 kW/(m^2 °C) and even lower within 10 min, which should be noticed.

The symbols used in Eqs. (3.85)–(3.88) are described as follows:

c_p, specific heat at constant pressure, kJ/kg °C;
d, diameter of heat pipe, m;
g, gravitational acceleration, m/s^2;
h, heat transfer coefficient on the surface, W/(m^2 °C);
U, overall heat transfer coefficient, W/(m^2 °C);
L, tube length, m;
R, radius of heat pipe, m;
h_{fg}, latent heat of vaporization for water, kJ/kg;
T, temperature, °C;
v_s, vapor flow rate at the end of heat pipe, m/s;
Γ, liquid load, kg/(m•s);
δ, thickness of film, m;
μ, dynamic viscosity, Pa s;
ν, kinematic viscosity, m^2/s;
ρ, density, kg/m^3

Subscript:

G, component of gas phase;
L, component of liquid phase;
1, component of condensation inside tube;
2, component of evaporation outside tube.

3.7.3 Falling Film Evaporation in Vertical Tube

The falling film evaporation in the vertical tube is one of the novel seawater desalination technologies, which is earlier than the falling film evaporation on the horizontal tube. It has the remarkable advantages of fast heat transfer, small temperature difference of heat transfer, etc. Hence it has been widely used in seawater desalination for a long time. However, it has some fixed defects, which are difficult to be accepted by people. First, the structure of the vertical tube for the device is difficult to be miniaturized because the heat transfer will not be sufficient if the tube is short. The device is very high usually so that it goes to be utilized in some restricted area. In addition, the falling film liquid in a vertical tube is difficult to distribute homogeneously with different thickness. The specific treatments are required most of the time.

Falling film evaporation still plays an important role in the seawater desalination area although it has various defects. Chun and Seban (1971) introduced the famous heat transfer equations for calculating the heat transfer coefficient of falling film evaporation on a smooth vertical tube:

$$\text{For laminar flow:} \quad h_c = 0.821\left(\frac{\nu_L^2}{k_L^3 g}\right)^{-\frac{1}{3}}\left(\frac{4\Gamma}{\mu_L}\right)^{-0.22} \tag{3.89}$$

$$\text{For turbulent flow:} \quad h_c = 3.8\times 10^{-3}\left(\frac{\nu_L^2}{k_L^3 g}\right)^{-\frac{1}{3}}\left(\frac{4\Gamma}{\mu_L}\right)^{0.4}\left(\frac{\nu_L}{\alpha_L}\right)^{0.65} \tag{3.90}$$

The symbols used in the two equations are the same to those in Eqs. (3.85)–(3.88). h_c is the heat transfer coefficient of falling film evaporation on the smooth vertical tube.

3.8 Enhancement of Falling Film Evaporation and Condensation

Heat transfer enhancement is an important area that develops rapidly in heat transfer studies, and it has become the second generation of heat transfer technology. It is widely used in industry, agriculture, space, and biology, especially in the development, application, and saving of energy.

In a seawater desalination system, heat transfer enhancement has been investigated and utilized. Some researchers pointed out that the heat transfer rate can be improved up to 300% when enhancing the condensation and evaporation on the vertical surfaces [12,13], which shows the effects of heat transfer enhancement are obvious. In seawater desalination, heat transfer enhancement is mainly applied in the enhancement of the evaporation and condensation process, which can be mainly divided into two types:

1. Passive enhancement technology
   Passive enhancement technology is the technology on heat transfer enhancement without direct external forces, which includes treated surface, rough surface, extended surface, mechanical insert, spiral tube, and various additives in liquids or solids, etc. Each of them has different features and can be selected in accordance with specific conditions. Fig. 3.22 shows several common passive enhancement surfaces.

2. Active enhancement technology
   Active enhancement utilizes the enhancements from external forces, including jet impingement, mechanical vibration, impressed electromagnetic field, and suction. These technologies require external energy and are more complex. But they have better effects, so it is worth to be applied to enhancement.
   In theory, condensation is a balance process with zero entropy increase, which means it is a process that may have no energy losses. But actually, the heat transfer in condensation is not ideal because there is temperature difference all the time. The condensation in industry is film condensation usually. The heat transfer coefficient of film condensation is relatively small, and its thermal resistance comes from condensate film. When the condensate surface is covered with condensation film, the heat ejected from vapor condensation on the gas–liquid interface must be transferred by condensation film to the condensate surface. The heat transfer resistance is determined by the thickness of the film. If the condensate is poorly drained or accumulates somewhere, the heat transfer resistance will increase a lot. Hence enhancing the heat transfer of condensation is used for draining condensate with different solutions and reducing the thickness of condensation film.
   The vapor condensates on vertical plates and the exterior of vertical tubes are common condensation types in seawater desalination. For a normal plate, the

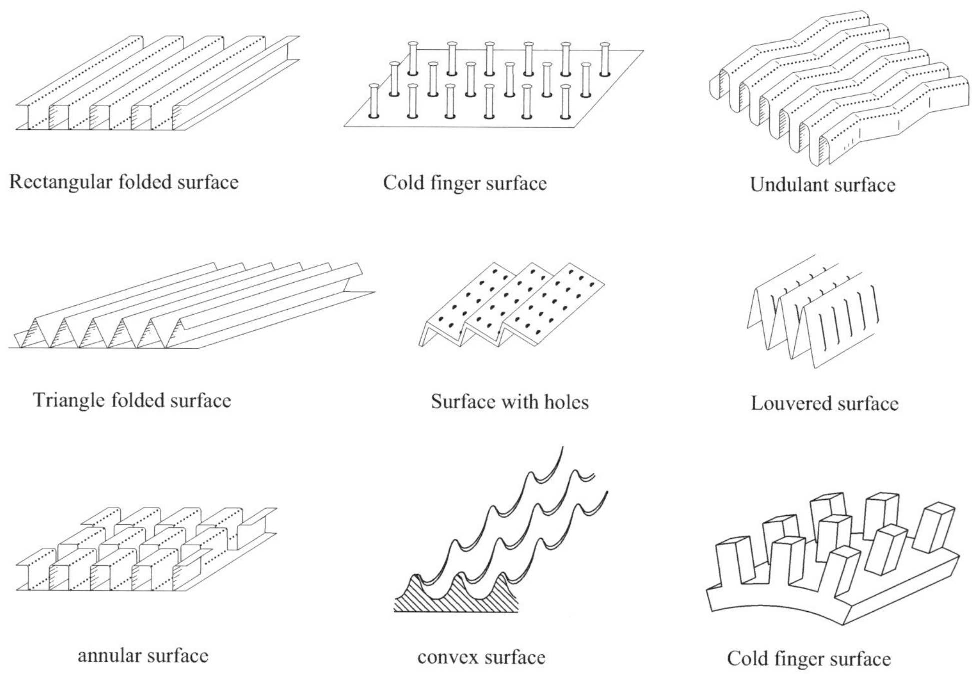

Figure 3.22
Several passive enhancement surfaces [13].

condensate flows downward, owing to gravity. Condensation film becomes thicker during this process, which results in worse heat transfer of condensation on the lower part. If adding enhancements like longitudinal rib, rectangular fin, longitudinal groove, and spiral groove, etc., the flow of condensate will be influenced by not only gravity, but also surface tension of the film. With the effects of surface tension, condensate film flows horizontally to the trough of the wave or the base of the groove in order to reduce the thickness of condensation film on the wave crest or embossment of fins. In addition, vapor condensates on top of the groove and condensate drains along the fins or trough so that the heat transfer of condensation increases. Fig. 3.23 presents the section view of several enhancement condensations on vertical plates.

The enhancements of film evaporation and film condensation are different. For film evaporation, heat resistance is mainly caused by the saturation temperature and vapor partial pressure on the gas–liquid interface, as well as the film thickness. If vapor gathers near the interface continuously, the partial pressure and corresponding saturation temperature there will increase, which will result in the reduction of the force of evaporation ($T_w - T_{sat}$). When the amount of gathered vapor reaches a certain value, evaporation stops finally. In order to enhance the heat transfer for evaporation, not only the film thickness should be reduced but the evaporated vapor should be exhausted in time, which can decrease the saturation temperature on film and

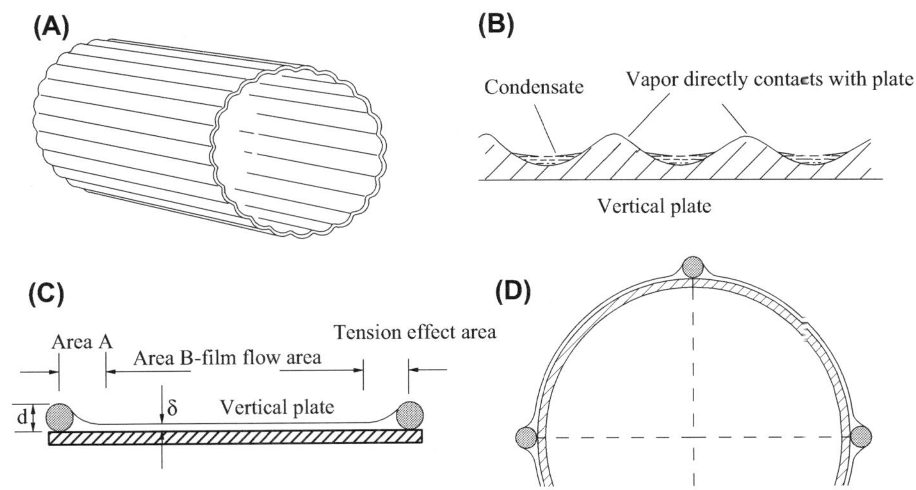

Figure 3.23

Section view of the vertical plate of several enhancement condensation [13]: (A) groove on vertical cylinder, (B) groove on vertical plate, (C) hanging wires on vertical plate, and (D) hanging wires on vertical cylinder.

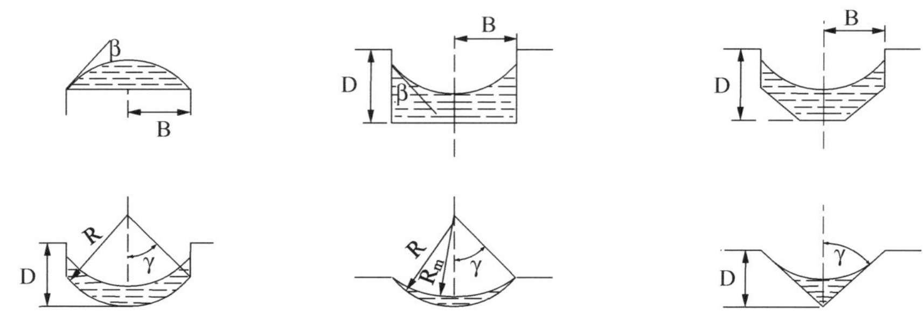

Figure 3.24
Several surface troughs for evaporation enhancement.

enhance the force of evaporation. Therefore the active enhancement technology is more appropriate for the evaporation process. Suction is the main way for enhancement.

Similar to the enhancement of condensation, adding troughs with various shapes on the vertical plate and external surface of the vertical tube can enhance the evaporation process, but the effects are not as obvious as that for condensation. Fig. 3.24 shows several common surface troughs for evaporation enhancement. Active enhancement technologies can bring better results; for instance, the evaporation efficiency can be achieved one or several times higher than the initial value by applying mechanical vibration, suction, and supersonic vibration, etc. [13].

3.9 Heat Transfer on Plate and Tube Wall in Convective Boundary Layer

During the operation of a seawater desalination system, there is a thermal process of the interior with the operating temperature that is much higher than the surrounding temperature. This may cause the energy losses from the interior to the exterior. Hence it is significant to prevent the heat losses to improve the heat transfer efficiency of the whole system.

A normal seawater desalination system consists of partitions of the external case and tubes. The main part of heat losses comes from the heat losses from the case and tube wall to the exterior. It is important to calculate the heat transfer in detail of the case and tube wall.

There are three processes in a seawater desalination system. First, the liquid (desalinated seawater, original seawater, or produced fresh water) or gas (vapor or air) in the system transfer the heat to the internal surface of the partition on the case by convection. Then heat goes through the internal surface to the external surface by conduction, or goes through the insulating layer as well, if there is one. Finally, heat goes to the surroundings from the external surface of the case or the insulating layer by convection and radiation. If there is an insulating layer with appropriate thickness and low temperature on external surfaces, the heat transfer is mainly caused by convection, and the effects by radiation can be ignored. Therefore the calculation of heat transfer in the convective boundary layer is important.

Fig. 3.25 demonstrates a working fluid with a higher temperature on the left transferring heat to the right working fluid with a lower temperature on a plate. Assume the temperatures of the left and right working fluids are $T_{\infty,1}$ and $T_{\infty,2}$, respectively; the heat exchange coefficient of both sides are h_1 and h_2, and the temperatures on the plate on two sides are T_{w1} and T_{w2}. The thermal conductivity of the plate is λ.

According to the Newton's cooling law, the rate of convective heat transfer is:

$$Q_c = hA(T_\infty - T_w) \tag{3.91}$$

where T_w is the surface temperature; A is the area of the plate. It can be also expressed as:

$$Q_c = \frac{(T_\infty - T_w)}{1/hA} = \frac{(T_\infty - T_w)}{R_c} \tag{3.92}$$

where R_c is thermal resistance of convection.

$$R_c = \frac{1}{hA} \tag{3.93}$$

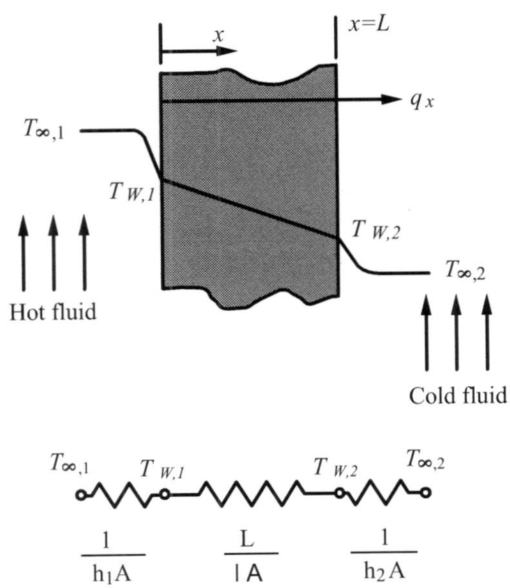

Figure 3.25
Heat transfer through a plate.

Based on the concept of thermal resistance and the thermal resistance network in Fig. 3.21, the heat flux equation of each unit can be written in the network with series connection:

$$Q_{c,1} = \frac{T_{\infty,1} - T_1}{(1/h_1 A)} \tag{3.94}$$

$$Q_k = \frac{T_1 - T_2}{(\Delta x/\lambda A)} \tag{3.95}$$

$$Q_{c,2} = \frac{T_2 - T_{\infty,2}}{(1/h_2 A)} \tag{3.96}$$

where Δx is the thickness of the plate. Because $Q_{c,1} = Q_k = Q_{c,2}$ under steady state, it is easy to derive the equation of heat flux from liquid with high temperature to the liquid with low temperature:

$$Q = \frac{T_{\infty,1} - T_{\infty,2}}{R_t} \tag{3.97}$$

where R_t is the total thermal resistance.

$$R_t = \frac{1}{h_1 A} + \frac{\Delta x}{\lambda A} + \frac{1}{h_2 A} \tag{3.98}$$

For the partition with several layers, they can be calculated in the same way, as shown in Fig. 3.26. For example, set the temperatures of working fluid on the left and right side of the partition with several layers as $T_{\infty,1}$ and $T_{\infty,4}$, and set heat transfer coefficients of them as h_1 and h_4. The heat transfer rate is:

$$Q = \frac{T_{\infty,1} - T_{\infty,4}}{\left[\left(\frac{1}{h_1 A}\right) + \left(\frac{\Delta x_a}{A\lambda_a}\right) + \left(\frac{\Delta x_b}{A\lambda_b}\right) + \left(\frac{\Delta x_c}{A\lambda_c}\right) + \left(\frac{1}{h_4 A}\right)\right]} \tag{3.99}$$

Where λ_a,λ_b,λ_c is the thermal conductivities of layers a, b, and c, respectively; Δx_a, Δx_b, Δx_c is the thicknesses of each layer.

Sometimes, the overall heat transfer coefficient U is defined in a similar way with the convective heat transfer coefficient h in Newton's cooling law to calculate the rate of heat transfer as follows:

$$Q = UA\Delta T \tag{3.100}$$

Obviously, $UA = 1/R_t$. Hence

$$U = \frac{1}{R_t A} = \frac{1}{\left[\left(\frac{1}{h_1}\right) + \left(\frac{\Delta x_a}{\lambda_a}\right) + \left(\frac{\Delta x_b}{\lambda_b}\right) + \left(\frac{\Delta x_c}{\lambda_c}\right) + \left(\frac{1}{h_4}\right)\right]} \tag{3.101}$$

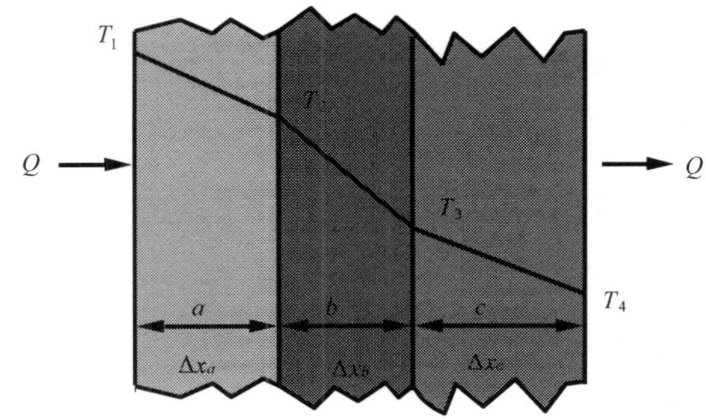

Figure 3.26
One-dimensional heat transfer through multilayer plate.

The unit of U is the same to h, which is W/(m^2 K).

The heat transfer rate through tubes in convective boundary layers can be calculated in the same way described earlier. For example, the equation to calculate the heat transfer rate for the system shown in Fig. 3.27 is:

$$Q = \frac{T_{\infty,1} - T_{\infty,3}}{\frac{1}{2\pi r_1 L h_1} + \frac{\ln(r_2/r_1)}{2\pi \lambda_A L} + \frac{\ln(r_3/r_2)}{2\pi \lambda_B L} + \frac{1}{2\pi r_3 L h_3}} \tag{3.102}$$

Where L is the total length of the tube; λ_A and λ_B are the thermal conductivities of the materials for the tube wall and insulation, respectively. The rate of heat transfer can be expressed as follows by total thermal resistance or overall heat transfer coefficient:

$$Q = \frac{T_{\infty,1} - T_{\infty,3}}{R_t} = U_1 A_1 \left(T_{\infty,1} - T_{\infty,3}\right) \tag{3.103}$$

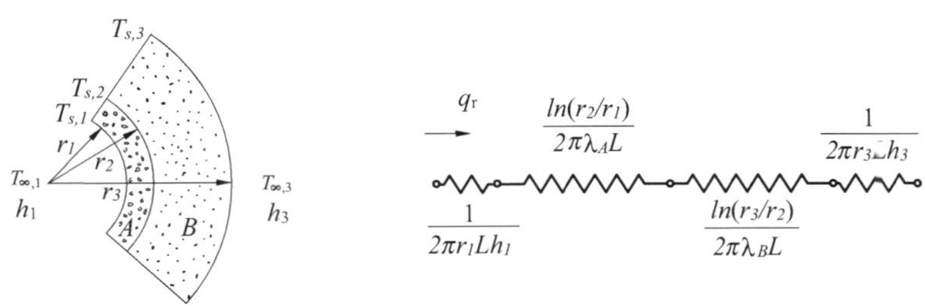

Figure 3.27
Heat transfer rate through multilayer tube.

where R_t is the total thermal resistance.

$$R_t = \frac{1}{2\pi r_1 L h_1} + \frac{\ln(r_2/r_1)}{2\pi\lambda_A L} + \frac{\ln(r_3/r_2)}{2\pi\lambda_B L} + \frac{1}{2\pi r_3 L h_3} \tag{3.104}$$

$A_1 = 2\pi r_1 L_1,$

The overall heat transfer coefficient U_1 is:

$$U_1 = \frac{1}{\frac{1}{h_1} + \frac{r_1}{\lambda_A}\ln\frac{r_2}{r_1} + \frac{r_1}{\lambda_B}\ln\frac{r_3}{r_2} + \frac{r_1}{r_3 h_3}} \tag{3.105}$$

The overall heat transfer coefficient U_1 above is defined by the area A_1 of the inside wall of the tube. It can also be defined by A_2 or A_3. No matter which cross section is applied to define U, there is:

$$U_1 A_1 = U_3 A_2 = U_3 A_3 = \frac{1}{R_t} \tag{3.106}$$

The equations of U_2 and U_3 can be derived by Eq. (3.102).

It is important to mention that we should understand four partial thermal resistances with series connection in the total thermal resistance R_t, which one plays the main role among when analyze the total thermal resistance R_t by Eq. (3.104) or the network in Fig. 3.27. In most cases, the other resistances can be ignored when comparing their effects with the main resistance unit.

It should be also noted that adding the insulating layer on the tube or increasing the thickness of the insulating layer cannot reduce heat losses under every condition. According to Eq. (3.104), when r_3 increases, the thermal resistance of convection decreases due to the increase of the heat exchange area, although the thermal resistance of conduction for the insulating layer increases.

If considering the circumstance in Fig. 3.27 again, it is simple to analyze the effects on total thermal resistance R_t by the change of r_3 with Eq. (3.104) when all the geometric parameters, thermal conductivities, and the heat transfer coefficients of convection are assumed to be constant apart from r_3. When R_t is smallest, heat losses will be the biggest. Hence the value of r_3 should be assured when R_t is smallest. Calculate the first derivative of R_t to r_3 and set its results to be zero, which is:

$$\frac{dR_t}{dr_3} = 0 \tag{3.107}$$

$$\frac{1}{2\pi k_b r_3 L} - \frac{1}{2\pi L r_3^2 h_3} = 0 \tag{3.108}$$

or

$$r_3 = \frac{k_b}{h_3} \tag{3.109}$$

It can be calculated that when $r_3 = \lambda_B/h_3$, the second derivative of R_t to r_3 is bigger than zero all the time. Hence R_t is the smallest when $r_3 = \lambda_B/h_3$. Therefore r_3 is called critical radius r_{cr} when R_t is extremely small.

$$r_{cr} \equiv \frac{\lambda_B}{h_3} \tag{3.110}$$

When $r_3 < r_{cr}$, R_t will decrease with the thickening of the insulating layer so that the heat losses increase. Increasing the thickness of the insulating layer can reduce heat losses only when $r_3 > r_{cr}$.

3.10 Heat Exchangers and Heat Transfer Rate Calculations

The solar desalination system generally includes three parts: a solar collection system, a thermal storage system, and a desalination device. Between the thermal storage system and the desalination device, the heat exchanger is often used because the seawater cannot be directionally sent into the solar collection system due to its intense corrosive. The heat exchange performance of the heat exchangers has a very important influence on the solar desalination systems. In practical engineering, the heat transfer rate of the heat exchangers is an important parameter to design the solar desalination system. Therefore it is necessary to know how to calculate the heat transfer rate of the heat exchangers and understand its performance.

3.10.1 Heat Exchanger Types

Generally, there are two kinds of different fluids flowing in the heat exchangers. The hot fluid constantly becomes cold. At the same time, the temperature of the cold fluid in another side will increase continually. According to the *flow arrangement* and *type of construction*, the heat exchangers are typically classified into four types: (1) the first type

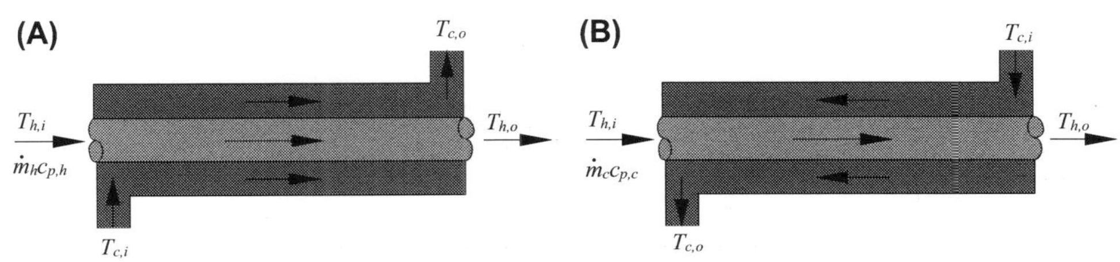

Figure 3.28
Concentric tube heat exchangers: (A) parallel flow and (B) counterflow.

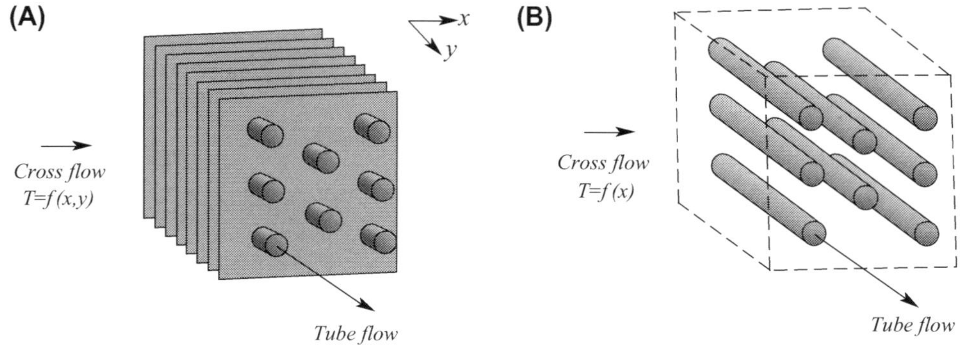

Figure 3.29
Cross-flow heat exchangers: (A) finned with both fluids unmixed and (B) unfinned with one fluid mixed and the other unmixed.

is the *parallel-flow* heat exchanger, in which the hot and cold fluids move in the same direction in a *concentric tube* (or *double-pipe*) construction, as shown in Fig. 3.28a. In the *parallel-flow* arrangement of Fig. 3.28A, the hot and cold fluids enter at the same end and leave at the same end. (2) The second type is the *counterflow* heat exchanger, in which the hot and cold fluids move in the opposite directions in a *concentric tube* (or *double-pipe*) construction, as shown in Fig. 3.28B. In the *counterflow* arrangement of Fig. 3.28B, the fluids enter at opposite ends and leave at opposite ends. (3) The third type is the *cross-flow* heat exchanger, in which the hot and cold fluids move in perpendicular directions to each other, as shown by the *finned* and *unfinned* tubular heat exchangers of Fig. 3.29. The two configurations are typically differentiated by an idealization that treats fluid motion over the tubes as *unmixed* or *mixed*. (4) The fourth type is the *complex flow* heat exchanger or *compact heat exchanger*, in which the hot and cold fluids flow in complex directions, including the *parallel-flow, counterflow, and cross-flow together*, as shown in Fig. 3.30.

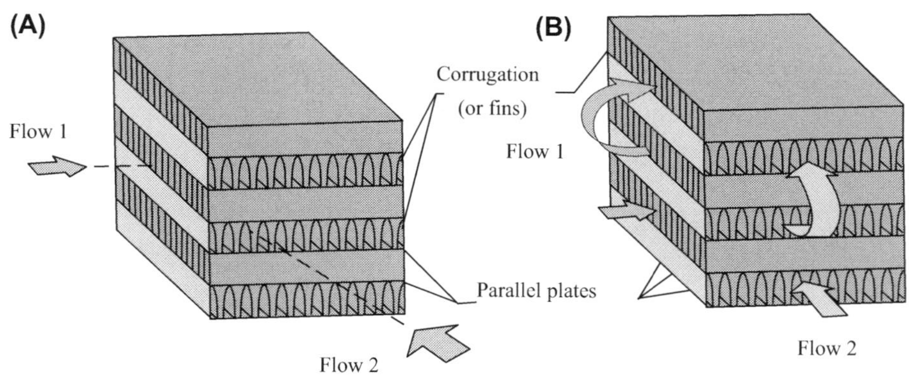

Figure 3.30
Compact heat exchanger cores [19]: (A) plate-fin (single-pass) and (B) plate-fin (multipass).

Notably, the *compact heat exchangers* can achieve a very large (400 m^2/m^3 for liquids and 700 m^2/m^3 for gases) heat transfer surface area per unit volume compared with other *construction heat exchangers,* although their configurations are becoming more complex. The *compact heat exchangers* have dense arrays of finned tubes or plates and are typically used when at least one of the fluids is a gas, and is hence characterized by a small convection coefficient. Parallel-plate heat exchangers may be finned or corrugated and may be used in single-pass (Fig. 3.30A) or multipass (Fig. 3.30B) modes of operation. Flow passages associated with compact heat exchangers are typically small (Dh 5 mm), and the flow is usually laminar.

Another common configuration is the *shell-and-tube* heat exchanger. Specific forms differ according to the number of shell-and-tube passes, and the simplest form, which involves single tube and shell *passes*, is shown in Fig. 3.31. Baffles are usually installed to increase the convection coefficient of the shell-side fluid by inducing turbulence and a cross-flow velocity component relative to the tubes. In addition, the baffles physically support the tubes, reducing flow-induced tube vibration. Baffled heat exchangers with one shell pass and two tube passes and with two shell passes and four tube passes are shown in Fig. 3.32A and B, respectively.

3.10.2 Heat Transfer Rate Calculation

Heat exchangers have various configurations with different flow ways so as to calculate the heat transfer rate becoming quite difficult. From the qualitative analysis, it is easy to understand that the heat transfer rate (the heat exchange rate between cold and hot fluids) is still relative to heat exchange area, heat transfer coefficient, and heat transfer temperature difference. Therefore in engineering for simplicity, the heat transfer rate is still expressed as follows,

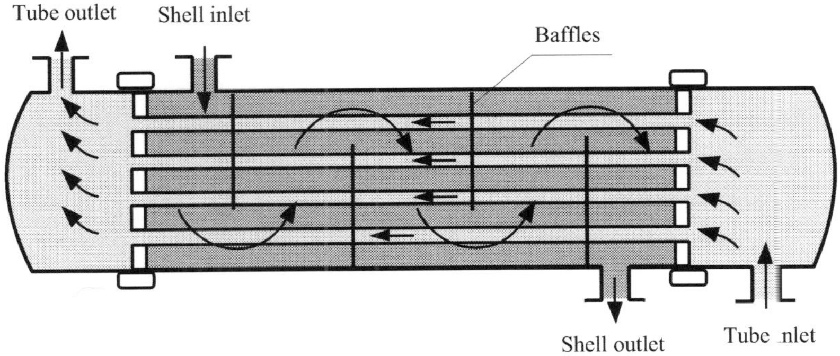

Figure 3.31
Shell-and-tube heat exchanger with one shell pass and one tube pass (cross-counterflow mode of operation).

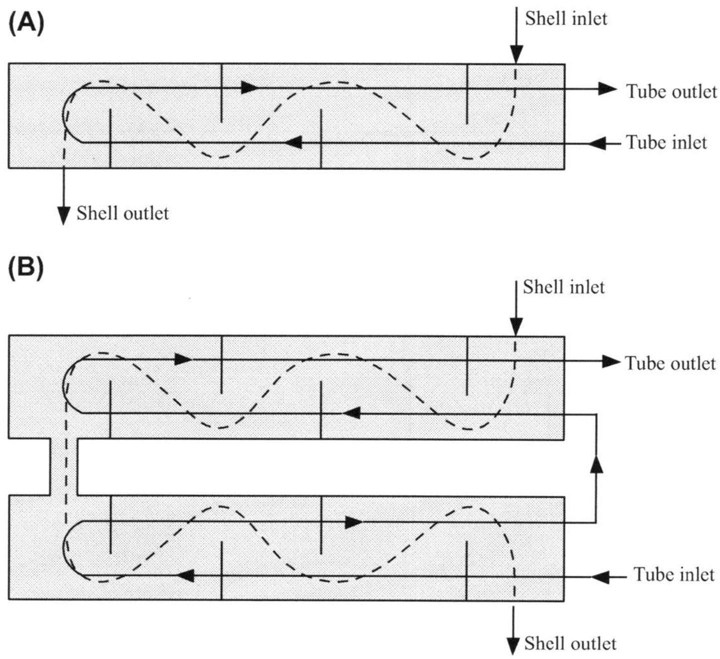

Figure 3.32
Shell-and-tube heat exchangers: (A) one shell pass and two tube passes. (B) Two shell passes and four tube passes.

$$Q = U \cdot A \cdot \Delta T_{ln} \tag{3.111}$$

where U is the total heat transfer coefficient and A is the total heat transfer area; ΔT_{ln} is the Log Mean Temperature Difference (which will be discussed in the next sections).

Note how the heat exchange area between cold and hot fluids in the heat exchanger is not just the geometrical area because the heat exchange surfaces between the cold and hot fluids generally include extended surfaces, for example fins. Therefore, only an effective area can be chosen in calculation. On the other hand, the cold and hot fluids are separated by the plate so that the heat transfer between the cold and hot fluids certainly include convection and conduction. The thermal resistances should be the sum of several thermal resistances. In the meantime, the fouling thermal resistance produced by surface corrosion during use also should be considered. Thus UA is usually called the total heat transfer coefficient. Finally, the heat transfer rate obviously is relative to the temperature difference between the cold and hot fluids. However, the fluid temperature changes with the position in the heat exchanger so that the temperature difference also changes with a different place. The next section will tell us that the temperature difference should be the Log Mean Temperature Difference.

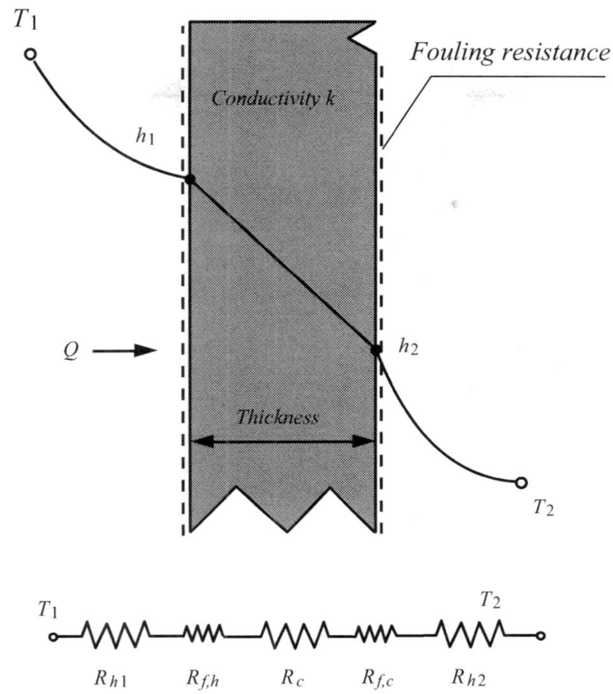

Figure 3.33
Heat transfer resistance between the cold and hot fluid.

The total heat transfer coefficient *UA* is the most basic parameter for any heat exchanger and also is most difficult to be determined. For this reason, we investigate the heat transfer model as shown in Fig. 3.33. It indicates the heat transfer process between the cold and hot fluids separated by a flat plate. Hot fluid 1 first transfers heat to the plate through convection, then the heat transfers to the side of the cold fluid by conducting, and finally to the cold fluid through convection. This is the basic heat transfer process, but in the actual heat exchanger, one still needs to consider fouling resistance on both sides of the plate due to corrosion, surface precipitation, or scale and other factors. From Fig. 3.33, we can easily get the total thermal resistance of the heat transfer process.

$$R_T = R_{h1} + R_{f,h} + R_c + R_{f,c} + R_{h2}$$

where R_{h1} is the convective resistance of the hot fluid side, $R_{f,h}$ is the fouling resistance of the hot side, R_c is the conductive resistance of the plate, $R_{f,c}$ is the fouling resistance of the cold side, and R_{h2} is the convective resistance of cold fluid side.

Taking into account that one or both sides of the fluid are often added to ribs to increase the heat transfer area and considering the formation and effects of fouling the total heat transfer coefficient of the heat exchanger is often expressed as:

Table 3.5: Representative fouling factors [20].

Fluid	Fouling factors/m^2 K/W
Seawater and treated boiler feedwater (below 50°C)	0.0001
Seawater and treated boiler feedwater (above 50°C)	0.0002
River water (below 50°C)	0.0002–0.001
Fuel oil	0.0009
Refrigerating liquids	0.0002
Steam (nonoil bearing)	0.0001

$$\frac{1}{UA} = \frac{1}{(\eta_0 hA)_h} + \frac{\xi_h}{(\eta_0 A)_h} + \left(\frac{L}{A\lambda}\right)_c + \frac{\xi_c}{(\eta_0 A)_c} + \frac{1}{(\eta_0 hA)_c} \tag{3.112}$$

where ξ_h and ξ_c are the fouling coefficient of the hot and cold side, respectively. Table 3.5 gives some typical cases of fouling coefficient; if the heat exchanger is new, the fouling coefficient is 0. $\left(\frac{L}{A\lambda}\right)_c$ is the effective conductive resistance of the plate. η_0 is the total surface efficiency of the rib. No matter whether it is the hot or cold side, it can be defined as follows for the surface without fouling.

$$Q = \eta_0 hA(T_b - T_\infty) \tag{3.113}$$

Note that T_b is the base temperature and A is the total area exposed base of the rib. Therefore as long as the total heat transfer rate, surface temperature, and heat transfer coefficient are measured, η_0 will be decided. For the rib whose shape is not particularly complex, the total surface efficiency can be estimated using the following formula.

$$\eta_0 = 1 - \frac{A_f}{A}(1 - \eta_f) \tag{3.114}$$

here, A_f is the total surface area of the rib, A is the total area of the heat transfer side, and η_f is the efficiency of a single rib, whose value for the common rib can be easily found in related books and materials.

It is worth pointing out that the fouling factors may be ignored if the heat exchanger is a new one. And in the actual situation, the separating plate is generally made of metal, and its thickness is very thin so that the conduction resistance is much less than the convection resistance. Thus under most simple situations, the simplest form of Eq. (3.112) can be used to estimate the total heat transfer coefficient UA, as follows

$$\frac{1}{UA} = \frac{1}{(\eta_0 hA)_h} + \frac{1}{(\eta_0 hA)_c} \tag{3.115}$$

Sometimes, the fin area is much bigger than the basal area so that $A_f/A \approx 1$. In this time, $\eta_0 = \eta_f$.

3.10.3 The Log Mean Temperature Difference of the Heat Exchangers

Eq. (3.111) gives the general equation for calculating the heat exchange rate of the heat exchanger, and Eq. (3.112) gives the equation of the total heat transfer coefficient. Then the remaining question is how to determine the temperature difference on both sides of the heat exchanger.

The heat exchanger is a common industrial equipment; no matter how complex the structure and form, it must meet the energy conservation law, that is, the heat released by the hot fluid must be equal to the heat obtained by the cold fluid. It is assumed that the mass flow rate of hot and cold fluids are (kg/s) and (kg/s), and their specific heat capacities are $c_{p,h}$ and $c_{p,c}$. Then the heat released by the hot fluid should be determined as follows.

$$Q = \dot{m}_h c_{p,h}\left(T_{h,i} - T_{h,o}\right) \tag{3.116}$$

The heat obtained by the cold fluid is

$$Q = \dot{m}_c c_{p,c}\left(T_{c,o} - T_{c,i}\right) \tag{3.117}$$

Where $T_{h,i}$ and $T_{h,o}$ are the temperature of the inlet and outlet of the hot fluid. $T_{c,i}$ and $T_{c,o}$ are the temperature of the inlet and outlet of the cold fluid. In order to describe conveniently, C_h and C_c are defined as the heat-capacity flow rate of hot and cold fluids, respectively. Then Eqs. (3.116) and (3.117) can be expressed as

$$Q = C_h\left(T_{h,i} - T_{h,o}\right) \tag{3.118}$$

$$Q = C_c\left(T_{c,o} - T_{c,i}\right) \tag{3.119}$$

here, $C_h = Q = \dot{m}_h c_{p,h}$, $C_c = \dot{m}_c c_{p,c}$. Note that given by Eqs. (3.118) and (3.119) is the final effect of fluid heat release or heat absorption. For a given heat exchanger, if we know its total heat transfer coefficient, we can use the Eq. (3.111) to calculate and analyze the total heat transfer rate. But in Eq. (3.111), ΔT_{ln} denotes the temperature deference, which is not the simple subtraction of hot and cold fluid temperature difference between inlet and outlet, and it is also not the difference of average temperature of the hot fluid and of the average temperature of the cold fluid, but the Log Mean Temperature Difference. Moreover, it must be paid attention that different flow forms have different Mean Temperature Difference.

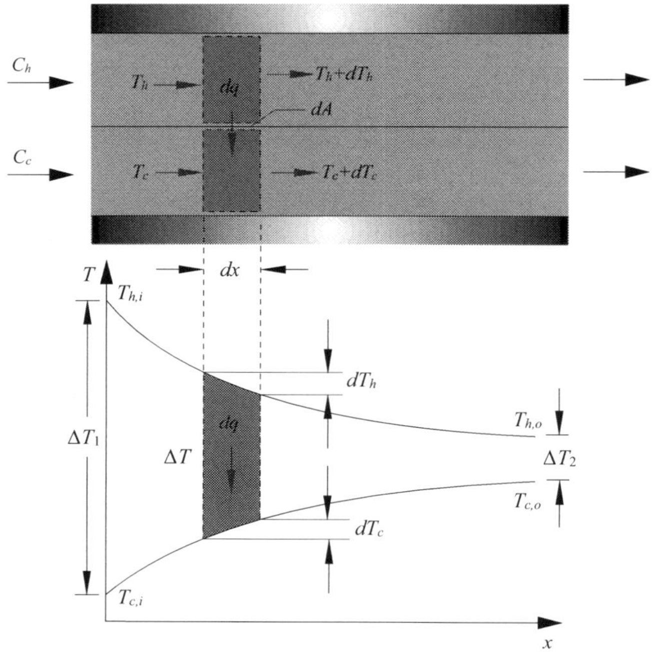

Figure 3.34
Temperature distribution in the downstream heat exchanger.

3.10.3.1 The Log Mean Temperature Difference of the Parallel-Flow Heat Exchanger

In the downstream heat exchanger, due to the fluid flow in the same direction, the temperature difference between the two fluids is the largest at the inlet of the hot fluid. With the deepening of the flow, the temperature difference between the hot and cold fluids decreases gradually. Fig. 3.34 gives the temperature change curve of hot and cold fluids. It is easy to find that the temperature difference at the inlet side is the biggest, so the heat transfer efficiency is the highest. On the outlet side, due to the decrease of temperature difference, the heat transfer rate is greatly reduced, so it is not significant to choose a long heat transfer channel. In particular, when the heat exchange channel is very long, the temperature difference between the hot and cold fluids tends to zero, and there is no heat transfer effect. Moreover, the outlet temperature of the cold fluid will never be higher than the outlet temperature of the hot fluid.

Select a short differential element of x position in the heat exchanger, as shown in Fig. 3.34, which analyzes the energy balance. It is assumed that the heat loss of the heat exchanger and the environment can be neglected, and the change of the kinetic energy and the potential energy of the fluid are neglected. It is also assumed that heat only transfers in the direction of vertical flow, and heat transfer in the other direction can be neglected. So the heat lost by the hot fluid is just the heat that gets by the cold fluid. Namely,

$$dQ = -\dot{m}_h c_{p,h} dT_h = C_h dT_h \tag{3.120}$$

and

$$dQ = C_c dT_c \tag{3.121}$$

When the heat exchange area of cold and hot fluid is dA, then the heat transfer of the differential element can also be expressed as follows, according to Eq. (3.111).

$$dQ = U \cdot \Delta T \cdot dA \tag{3.122}$$

where, $\Delta T = T_h - T_c$ is the temperature difference between the cold and hot fluid in the differential element.

Obviously, Eqs. (3.120)–(3.122) express the same heat transfer, and paying attention to that:

$$d(\Delta T) = dT_h - dT_c$$

Substituting Eqs. (3.120) and (3.121) into the formula above, one gets

$$d(\Delta T) = -dQ\left(\frac{1}{C_h} + \frac{1}{C_c}\right)$$

Then substituting the result of Eq. (3.122), the equation can be written to

$$\frac{d(\Delta T)}{\Delta T} = -UdA\left(\frac{1}{C_h} + \frac{1}{C_c}\right) \tag{3.123}$$

The Eq. (3.123) is the energy balance equation of the heat transfer element. Integrating from the inlet to the outlet of hot fluid, and considering that the heat capacities of cold and hot fluid are constant, then

$$\int_1^2 \frac{d(\Delta T)}{\Delta T} = -\mathrm{U}\left(\frac{1}{C_h} + \frac{1}{C_c}\right)\int_1^2 \mathrm{dA} \tag{3.124}$$

Upon completion of the above integrality,

$$\ln\left(\frac{\Delta T_2}{\Delta T_1}\right) = -\mathrm{UA}\left(\frac{1}{C_h} + \frac{1}{C_c}\right) \tag{3.125}$$

Now, turn around to pay attention to Eqs. (3.118) and (3.119); they can be changed into

$$\frac{1}{C_h} = \frac{(T_{h,i} - T_{h,o})}{Q} \tag{3.126}$$

$$\frac{1}{C_c} = \frac{\left(T_{c,i} - T_{c,o}\right)}{Q} \tag{3.127}$$

Substituting the results of Eqs. (3.126) and (3.127) into the Eq. (3.125),

$$\ln\left(\frac{\Delta T_2}{\Delta T_1}\right) = -\frac{UA}{Q}\left[\left(T_{h,i} - T_{h,o}\right) - \left(T_{c,o} - T_{c,i}\right)\right] \tag{3.128}$$

Further rewritten above formula, get

$$\ln\left(\frac{\Delta T_2}{\Delta T_1}\right) = \frac{UA}{Q}\left(\Delta T_2 - \Delta T_1\right)$$

Finally, the formula for calculating the total heat transfer rate is

$$Q = UA\frac{\left(\Delta T_2 - \Delta T_1\right)}{\ln\left(\frac{\Delta T_2}{\Delta T_1}\right)} = UA \cdot \Delta T_{ln} \tag{3.129}$$

Note that the $\Delta T_1 = T_{h,i} - T_{c,i}, \Delta T_2 = T_{h,o} - T_{c,o}$, and ΔT_{ln} is the Log Mean Temperature Difference, and

$$\Delta T_{ln} = \frac{\Delta T_2 - \Delta T_1}{\ln\left(\frac{\Delta T_2}{\Delta T_1}\right)} \tag{3.130}$$

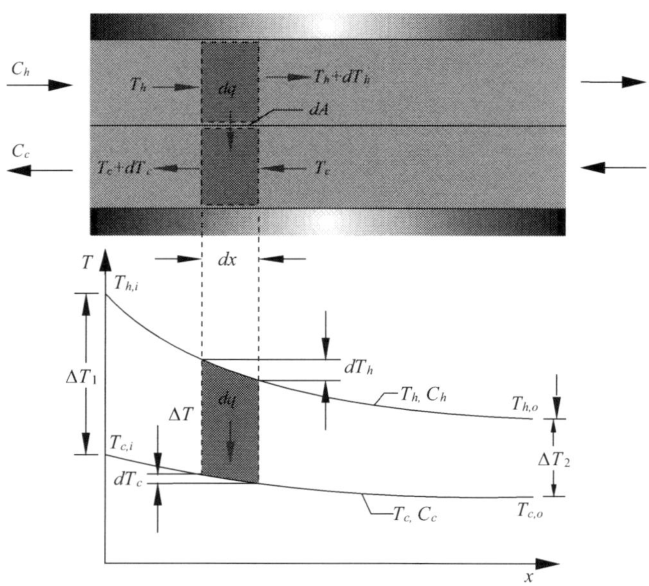

Figure 3.35
Temperature distribution in countercurrent heat exchanger.

3.10.3.2 The Log Mean Temperature Difference of the Counterflow Heat Exchanger

The discussion on the downstream heat exchanger can be similarly applied to the countercurrent heat exchanger. Fig. 3.35 shows the change curve of the temperature of the hot and cold fluids in the countercurrent heat exchanger. From the change trend of the curve, we can find that the change of temperature difference in the countercurrent heat exchanger is not as significant as the downstream heat exchanger, and the temperature difference maintains evenly in the whole heat exchanger. In particular, if the heat exchanger is long enough, the outlet temperature of the cold fluid can be higher than the outlet temperature of the hot fluid, and even the outlet temperature of the cold fluid can be close to the inlet temperature of the hot fluid, thereby obtaining a higher temperature. This is a big advantage of the countercurrent heat exchanger, but also one of the reasons for the industrial use of countercurrent heat transfer.

For the countercurrent heat exchanger, we can also choose a differential element to analyze heat transfer, as shown in Fig. 3.35. Finally, the formula of total heat transfer rate, same as Eq. (3.129), can also be obtained in the same manner as analyzing the downstream heat transfer. But at this time, the temperature difference formula of the end point can be rewritten to:

$$\Delta T_1 = T_{h,i} - T_{c,o} \tag{3.131}$$

$$\Delta T_2 = T_{h,o} - T_{c,i} \tag{3.132}$$

In order to facilitate memory and unify with the downstream heat exchanger, the downstream or countercurrent heat exchanger both substituted two temperature differences into Eq. (3.129). These two temperature differences are the subtraction of hot and cold fluid temperatures on the left and right side, respectively.

There is a special case that is worth pointing out. For the countercurrent heat exchanger, if the heat capacity flow rate of hot and cold fluids is equal, namely $C_h = C_c$, the temperature difference of hot and cold fluids in the whole heat exchanger will be constant, that is $\Delta T = \Delta T_1 = \Delta T_2$.

For cross-flow and other complex manifold heat exchangers, Eqs. (3.118) and (3.119) are still set up in principle and can also be similar to the use of Eq. (3.129) to calculate the total heat transfer rate. But the definition of the Log Mean Temperature Difference needs to be revised; there is no further expansion here, and you can refer to other specialized works.

3.11 Theoretical Power Consumption in Desalination and Minimum Heat Consumption in Distillation

Seawater desalination is a process to separate the salt and water, which can be achieved by the gas—liquid—solid phase transformation. No matter which method is to be used, the separation of salt and water is nonspontaneous and thus costs energy. How to reduce the energy consumption during the process is a question to be considered first in the desalination technology. As seen from a thermodynamics perspective, what will be the theoretical value of minimum energy consumption ($W_{\min}$) or minimum work consumption (W_{th})?

The minimum energy consumption and minimum work consumption are two different concepts. The minimum energy consumption is based on the First Law of Thermodynamics, which is to analyze the system input and output from quantitative relations of energy; the minimum work consumption is from the qualitative relations based on the Second Law of Thermodynamics.

There has been a lot of discussion among scholars on the minimum work consumption. Sideman (1980) saw the desalination as a process to separate the vapor from the seawater and calculated the theoretical W_{th} as 0.79 kW h/m^3 (at 25°C) from the relations between seawater and the entropy, enthalpy, and free energy of vapor [10].

It is assumed in the chemical calculation that the desalination process is to extract fresh water with a salt content of 500 ppm from seawater with a salinity of 34,000 ppm, and the salinity of the remaining seawater is increased by four times. Based on this assumption, the theoretical work needed in the desalination process would be [10]:

$$W_{th} = 1.41\ \text{kW h}/\text{m}^3$$

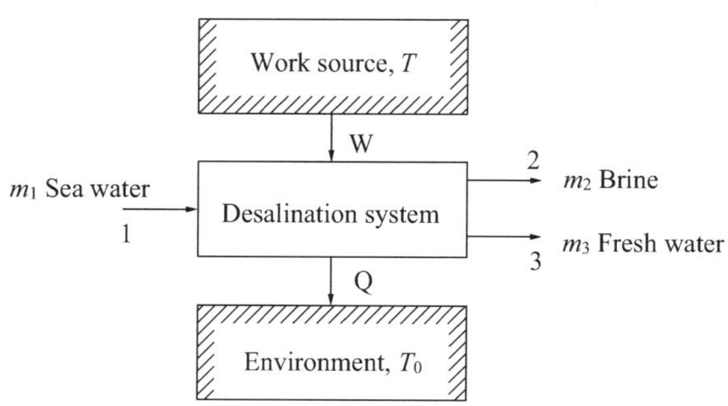

Figure 3.36
General process of desalination.

In fact, the actual desalination can hardly reach this parameter. Specific to the distillation process, in the view of Carnot Circle, the calculation of the minimum work consumption is as follows.

Fig. 3.36 shows a universal seawater desalination process, according to which, the minimum energy formula for desalting seawater is derived, and it can be applied to any mechanism [11−12]. Take the saltwater transformation system elaborated in Fig. 3.36 as the control body, and to neglect the changes of kinetic and potential energy, the steady flow process can be summarized as (based on the First Law of Thermodynamics):

$$W + m_1h_1 = m_2h_2 + m_3h_3 + Q \tag{3.133}$$

In the equation, W is the work input from the outside system; Q is the heat released to the low temperature source from the system; m_1, m_2, and m_3 are the mass flow rate of point 1, 2, and 3 respectively, also $m_1 = m_2 + m_3$; h_1, h_2 and h_3 are specific enthalpy on point 1, 2, and 3, respectively.

If this process is reversible, according to the Second Law of Thermodynamics:

$$m_2s_2 + m_3s_3 + Q/T_0 - m_1s_1 = 0 \tag{3.134}$$

where s_1, s_2, and s_3 are specific entropy on point 1, 2, and 3, respectively. The simultaneous Eqs. (3.133) and (3.134) are calculated as:

$$W = m_2(h_2 - T_0s_2) + m_3(h_3 - T_0s_3) - m_1(h_1 - T_0s_1) \tag{3.135}$$

The equation analyzed above is the minimum work needed for separating fresh water from salt water, which can be applied to any reversible processes. If the process is isothermal as well as reversible, then the T in Eq. (3.135) can be modified to T_0, then

$$W = m_2(h_2 - Ts_2) + m_3(h_3 - Ts_3) - m_1(h_1 - Ts_1) \tag{3.136}$$

By applying Gibbs Function, $g = h - T\,s$, then Eq. (3.136) can be revised as:

$$W = m_2g_2 + m_3g_3 - m_1g_1 \tag{3.137}$$

where g_1, g_2, and g_3 are the Gibbs free energy of fluid per unit mass of point 1, 2, and 3, respectively.

The actual desalination is to desalt the seawater in the concentration of C_1 to the high-salinity seawater with the concentration of C_2 and the fresh water with the concentration of C_3, that is:

$$\text{Seawater}(c_1) \rightarrow \text{Brine}(c_2) + \text{Fresh Water}(c_3)$$

The energy needed in the process is related to the concentration of the three factors. The required minimum work depends on the free energy difference between the initial and final state during the desalination process, which is Eq. (3.137). Suppose that there are n_1 mol of salt, and N_1 mol of water in m_1, g_{1s}, and g_{1w} are used to express the free energy per mol salt and water at point 1. Obviously, $m_1 g_1 = n_1 g_{1s} + N_1 g_{1w}$. Make a similar derivation of point 2 and 3, and let them be induced into Eq. (3.137), getting:

$$W = (n_3 g_{3s} + n_2 g_{2s} - n_1 g_{1s}) + (N_3 g_{3w} + N_2 g_{2w} - N_1 g_{1w}) \tag{3.138}$$

In which,

n_3, n_2, n_1, fresh water, brine, the mol number of electrolytes in seawater, mol;
N_3, N_2, N_1, fresh water, brine, the mol number of water in seawater, mol;
g_{3s}, g_{2s}, g_{1s}, fresh water, brine, the free energy per mol salt in seawater, J/mol;
g_{3w}, g_{2w}, g_{1w}, fresh water, brine, the free energy per mol water in seawater, J/mol.

Assume that all the salt in seawater is fully ionized sodium chloride, according to the universal relation between activity and free energy:

$$g = g_0 + RT \ln a \tag{3.139}$$

where a is the water activity. Let the actively be approximately replaced by the mol concentration c [2], then mol free energy of the salt and water on each point can be written as:

$$g_s = g_s^0 + 2RT \ln a \tag{3.140}$$

$$g_w = g_w^0 + RT \ln \frac{N}{N + 2n} \approx g_w^0 - 2RT \frac{n}{N} \tag{3.141}$$

where $g_s^0$and g_w^0 are the free energy of salt and water under a standard state, J/mol. The condition that N is much larger than n in the above formula should be noticed.

The total amount of the brine in final state plus the salt and water in fresh water is equal to the amount of initial state seawater, so:

$$n_3 + n_2 = n_1 \quad N_3 + N_2 = N_1 \tag{3.142}$$

Substitute these relations into Eq. (3.138) and simplify the formula, getting:

$$W = \Delta G = 2RT \left(n_3 \ln \frac{C_3}{C_1} + n_2 \ln \frac{C_2}{C_1} \right) \tag{3.143}$$

This is the calculation formula of energy consumption in the desalination theory.

Suppose that at the temperature of 25°C, salt concentration of the original state seawater, high-salinity seawater, and fresh water before and after the desalination are:

$$c_1 = 34,000 \text{ ppm} = 0.582 \text{ mol/L}$$
$$c_2 = 136,000 \text{ ppm} = 2.328 \text{ mol/L}$$
$$c_3 = 500 \text{ ppm} = 0.00855 \text{ mol/L}$$

Take 1 L of fresh water as the calculation base, then: $n_3 = 500 \times 10^{-3}$ g/L ÷ 58.5 g/mol = 0.00855 mol Salt/L. Besides, the corresponding solute mol numbers of c_1 and c_2 are 0.582 and 2.328, respectively. Meanwhile, if 1 L of fresh water produces x L of brine, then:

$$1 \times (c_1 - c_3) = x(c_2 - c_1) \tag{3.144}$$

$$x = \frac{C_1 - C_3}{C_2 - C_1} = \frac{0.582 - 0.00855}{2.328 - 0.582} = 0.3286(\text{L}) \tag{3.145}$$

Thus $n_2 = 2.328 \times 0.3286 = 0.7649$ mol.

$$\begin{aligned} W &= 2RT\left(n_3 \ln\frac{c_3}{c_1} + n_2 \ln\frac{c_2}{c_1}\right) \\ &= 2 \times 8.314 \times 298.2\left(0.00855 \times \ln\frac{0.00855}{0.582} + 0.7649 \times \ln\frac{2.328}{0.582}\right) \\ &= 5.077 \times 10^3(\text{kJ/L}) \end{aligned} \tag{3.146}$$

The density of fresh water is 1000 kg/m^3, then the minimum work needed to produce a ton of fresh water would be:

$$W_{\min} = 5.077 \times 10^3 \times 1 \times 10^3/(3.6 \times 10^6) = 1.41\left(\text{kW}\cdot\text{h/m}^3\right) \tag{3.147}$$

The theoretical work consumption (electrical or mechanical energy) is about 1.41 kW h/m^3, or 1.41 × 3600 = 5076 kJ/m^3. In practice, the energy required to produce a ton of fresh water is way larger than the number, regardless of desalination methods. The main reason would be the technology level, apart from the differences between reality and ideal conditions. This shows that there is still a large potential to develop the energy reduction of desalination. The work consumption (e.g., electrical or mechanical energy) per unit yield of the most advanced desalination unit is approximately 25,000 kJ/m^3 (fresh water) (by using reverse osmosis membrane and mechanical vapor compress distillation) [21], which is about five times the work consumption under ideal conditions. China's reverse osmosis membrane technology used in desalination has basically reached the international advanced level, and work consumption for producing unit yield of fresh water by packaged unit is no larger than 20,000 kJ/m^3, which is four times the minimum work consumption.

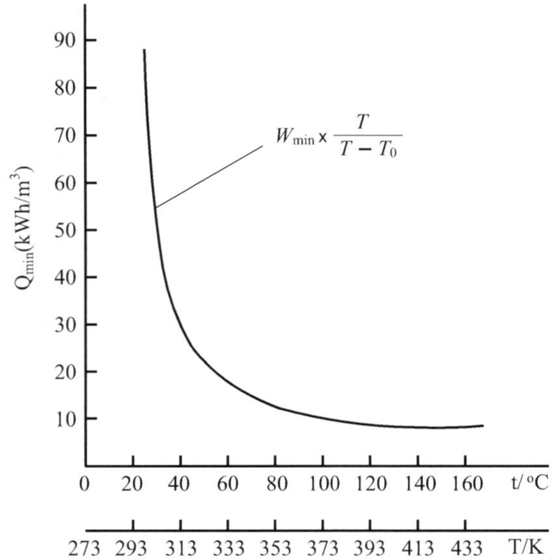

Figure 3.37
Relationship curve of Q_{min} and T.

If work is replaced by thermal energy as the power for desalination, then the minimum thermal energy calculated by the Carnot Circle Efficiency would be:

$$Q_{min} = W_{min} \frac{T}{T - T_0} \tag{3.148}$$

where T is temperature of the heat source, and T_0 is the temperature of the cold source.

For example, take the vapor at 100°C and 1 atm as the heat source; with the ambient temperature of 21°C, the minimum thermal energy needed for fresh water production per unit would be:

$$\begin{aligned} Q_{min} &= 1.41 \times \frac{373}{373 - 294} \text{ kW h}/\text{m}^3 \\ &= 6.66 \text{ kW h}/\text{m}^3(\text{fresh water}) = 23976 \text{ kJ}/\text{m}^3(\text{fresh water}) \end{aligned} \tag{3.149}$$

This roughly means the production of 1 m^3 fresh water requires 10.28 kg of 100°C vapor, because 2330 kJ heat is included in the vapor per kg.

Let both $T_0 \approx 25$°C and W_{min} be constants, and a curve can be plotted comparing Q_{min} to T, which is shown in Fig. 3.37.

As indicated in the figure, Q_{min} decreases with the increasing T, but when the temperature is above 140°C, the curve tends to be steady. So to consider from the work-

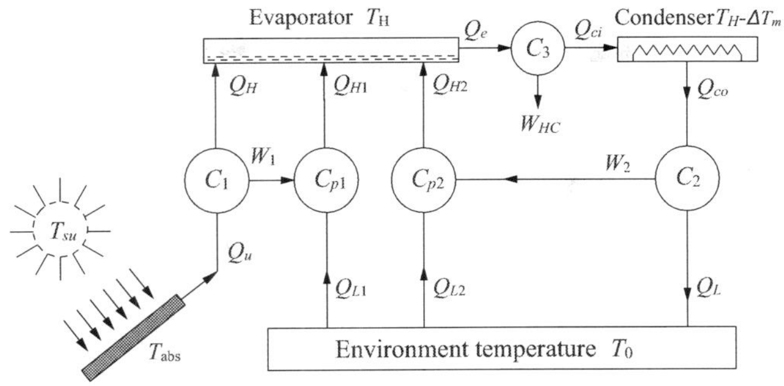

Figure 3.38
A perfect single-effect solar distillation system.

heat efficiency, the operating temperature during distillation is not necessarily above 140–150°C.

In the desalination system driven by steam, steam heating with low-pressure is deemed to be the most economical. Limited by the scale prevention method, $T_{max} < 130°C$ is proper.

As for the solar heating unit, under the present technical conditions, the maximum heat collection temperature of the solar collector can reach 120°C at most, but 75°C is the most cost-effective. At this temperature, the minimum thermal consumption in distillation is:

$$Q_{min} \approx 1.41 \times \frac{75+273}{(75+273)-(25+273)} = 9.8\left(\text{kW h/m}^3\right) \quad (3.150)$$

3.12 The Perfect Distillation Process Driven by Solar Collector

3.12.1 Perfect Single-Effect Solar Distillation Process

Utilizing solar energy for seawater desalination is a long-standing dream of mankind. The solar distillation process has been happening in the world since the Earth was born. The ocean will produce the vapor when the sunlight shines on the ocean. The vapor rises up with air and goes to the land with the wind. Finally, it becomes a cloud, rain, or snow when it meets the cool gas flow and falls down to the land to produce the fresh water, which nourishes the earth. The fresh water goes back to the ocean by way of the rivers. Ultimately, the water circulation processes are completed on the earth. That is the most primitive and the most natural solar distillation and desalination.

In that way, what is the perfect process of solar distillation when mankind uses the solar collection and engine technology to achieve the distillation process? What is its maximum potential? For answering this problem, a perfect single-effect solar distillation system driven by a solar collector and coupled with multiple engines and heat pumps is shown in Fig. 3.38. Its working principle is that sunlight shines into the collector, and then the collector provides the useful heat energy Q_u and sends it to the system. The heat energy Q_u provided by the solar collector has a higher temperature than the evaporator needs. For avoiding the irreversibility, a reversible Carnot engine could be designed between the collector and evaporator. Carnot engine C_1 provides not only heat energy to the evaporator, but also produces work for a reversible heat pump C_{p1}, which also provides the heat to the evaporator. The vapor produced in the evaporator is sent into the condenser, where the vapor is condensed into fresh water. A reversible Carnot engine C_3 is also constructed between the condenser and the environment because the working temperature T_c of the condenser is higher than ambient temperature T_0. Carnot engine C_3 releases heat to the environment and at the same time produces some work to drive another reversible heat pump C_{p2}, which also provides the heat for the evaporator through drawing heat from the environment. For a perfect distillation process, the temperature difference ΔT_m between the evaporator and condenser should be the minimum difference. However, the temperature difference must be there because the evaporating surface is seawater, while the condensing surface is fresh water. These two surfaces have different vapor partial pressure. Seawater consists of salt and water so that the vapor partial pressure on the seawater surface is smaller than that on the pure water surface. Therefore the temperature of the seawater surface has to rise a little so the vapor produced on the seawater surface goes to the pure water surface to condense.

The minimum temperature-rising quantity must satisfy the requirement that the partial pressure on the seawater surface is the same as on the pure water surface. That is to say, both the evaporation process in the evaporator and the condensation process in the condenser are irreversible processes. Exergy loss must happen in these processes. In fact, the exergy loss is translated into the pure water, assuming a reversible heat engine C_3 is installed between evaporator and condenser. It is also assumed that the work produced by the engine is the same as the exergy loss, which is transferred into pure water. Assuming Q_e represents the heat sent into the engine, Q_{ci} represents the heat sent into the condenser from the engine, and W_{HC} is the work produced by the engine, which is the same as the exergy loss in the evaporation and condensation process.

In order to analyze the maximum potential of the solar distiller, all of the other processes are assumed as the perfect processes. That is to say, any place is reversible in the solar distillation system except some irreversibility happens in the solar collector. Certainly, a little irreversibility will happen between the evaporator and the condenser. Also, it can be assumed that the water capacity in the evaporator and condenser is very small so that the sensible heat of the seawater and pure water can be ignored. At the same time, the heat capacity of the evaporator and condenser is also ignored.

According to the discussion of the fifth chapter, the heat equation of the ideal solar collector can be expressed as

$$Q_u = \left[Cf\sigma T_s^4 + (1 - Cf)\sigma T_0^4 - \sigma T_{abs}^4\right] - U_L(T_{abs} - T_0) \tag{3.151}$$

where, C is the concentrating light ratio of the solar collector; T_0 is the ambient temperature; σ is the Stefan–Boltzmann constant; T_{abs} is the temperature of the absorber of the solar collector; and U_L is the total heat transfer coefficient. $I_s = f\sigma T_s^4$ is the solar radiation value at the entrance of the solar collector. In order to simplify the discussion, assuming that the collector's concentration ratio is not very high, Cf is small and can be neglected. Eq. (3.151) can be simplified as

$$Q_u = \left(CI_s + \sigma T_0^4 - \sigma T_{abs}^4\right) - U_L(T_{abs} - T_0) \tag{3.152}$$

Eq. (3.152) implies that all other places of the solar collector are perfectly insulated except for the entrance of the solar collector. Even if in the entrance, the convection and conduction heat loss is still considered to be linear. The heat loss coefficient has no relationship with the temperature.

As a perfect Carnot engine, C_1 provides the heat to the evaporator to be

$$Q_H = \frac{T_H}{T_{abs}} Q_u \tag{3.153}$$

The work production of C_1 is,

$$W_1 = \left(1 - \frac{T_H}{T_{abs}}\right) Q_u \tag{3.154}$$

Carnot engine C_2 gives out the heat to the environment to be

$$Q_L = Q_{co} \times \frac{T_0}{T_c} \tag{3.155}$$

where $T_c = T_H - \Delta T_m$ is the temperature in the condenser. The work produced by C_2 is

$$W_2 = \left(1 - \frac{T_0}{T_c}\right) Q_{co} \tag{3.156}$$

Carnot engine C_3 works with a small temperature difference between the evaporation surface and condensation surface. It produces the work that is the exergy, which is translated into pure water. Therefore it can be thought that C_3 works between heat source temperature T_H and $T_H - \Delta T_m$. It sends the heat to the condenser as follows,

$$Q_{ci} = Q_e \times \frac{T_H - \Delta T_m}{T_H} \tag{3.157}$$

C_3 produces the work as follows

$$W_{HC} = \left(1 - \frac{T_H - \Delta T_m}{T_H}\right) Q_e \tag{3.158}$$

This work is the exergy to be transferred into pure water. It is known that heat pump C_{p1} and C_{p2} operate between T_0 and T_H. According to this condition, the total heat energy given by C_{p1} and C_{p2} can be calculated as follows

$$Q_{H1} + Q_{H2} = (W_1 + W_2)\frac{T_H}{T_H - T_0} \tag{3.159}$$

Thus the total heat energy obtained by the evaporator is

$$Q = Q_H + Q_{H1} + Q_{H2} = \frac{T_H}{T_{abs}} Q_u + (W_1 + W_2)\frac{T_H}{T_H - T_0} \tag{3.160}$$

According to the energy balance, the next equations also can be obtained

$$Q = Q_e; \quad Q_{co} = Q_{ci} \tag{3.161}$$

$$Q_{ci} = Q_e \frac{T_H - \Delta T_m}{T_H} \tag{3.162}$$

Finally, substituting Eqs. (3.154), (3.156), and (3.152) into Eq. (3.160), it is easy to find

$$Q = \frac{\frac{T_H}{T_{abs}} + \left(1 - \frac{T_H}{T_{abs}}\right) \times \left(\frac{T_H}{T_H - T_0}\right)}{1 - \left(1 - \frac{T_0}{T_H - \Delta T_m}\right) \cdot \frac{T_H - \Delta T_m}{T_H} \cdot \left(\frac{T_H}{T_H - T_0}\right)} Q_u \tag{3.163}$$

Substituting Eq. (3.152) into the above equation and doing some simplification for its denominator, Eq. (3.163) can be changed into the form as follows

$$Q = \frac{\left[\frac{T_H}{T_{abs}} + \left(1 - \frac{T_H}{T_{abs}}\right) \times \left(\frac{T_H}{T_H - T_0}\right)\right]\left[\left(CI_s + \sigma T_0^4 - \sigma T_{abs}^4\right) - U_L(T_{abs} - T_0)\right]}{\frac{\Delta T_m}{T_H - T_0}} \tag{3.164}$$

It can be found from Eq. (3.164) that as long as Q is known, the fresh water production rate can be calculated out based on the following equation

$$\mathrm{m}_e = \frac{Q}{h_{fg,T_H}} \tag{3.165}$$

where, $h_{fg,TH}$ is the latent heat of vaporization of pure water at temperature T_H.

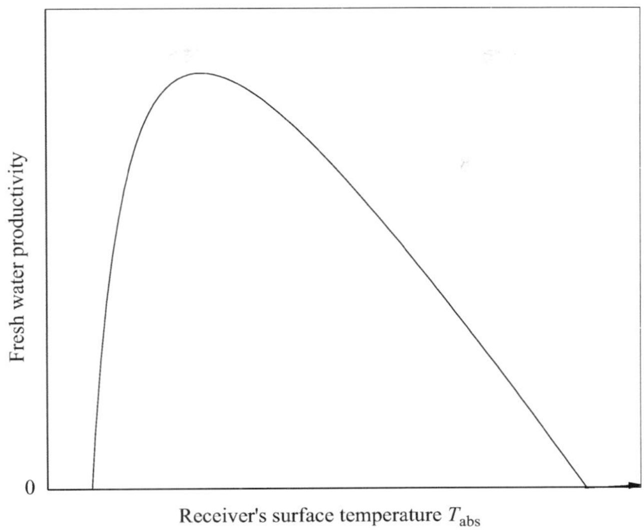

Figure 3.39
Effect of the receiver's surface temperature on the fresh water productivity.

Obviously, the total heat energy provided to the evaporator is related to the temperature T_{abs} of the receiver, evaporation temperature T_H, and heat transfer temperature difference ΔT_m. When the evaporation temperature, heat transfer temperature difference, ambient temperature, and the parameters of the collector are fixed, the variation of the productivity of the distiller with the temperature of the receiver's surface is shown in Fig. 3.39.

It is easy to find from Fig. 3.39 that the surface temperature of the receiver has a very important effect on the fresh water production. However, high temperature is not necessary. There is an optimum value. In practical work, the surface temperature of the receiver should be regulated to reach the optimum value.

For ΔT_m, it is caused by the partial pressure difference between the evaporation surface and condensation surface. The saturation vapor pressure on the pure water surface can be calculated by Antoine equation as follows

$$lgp_w = 7.07406 - \frac{1657.46}{T - 46.13} \tag{3.166}$$

where, p_w is the saturation vapor pressure of pure water surface, unit kPa at temperature T (K). This equation has very good accuracy in the range of 283K $\leq$ T $\leq$ 440K. From this equation, it easily calculates that the temperature change, which is caused by the pressure change dp_w, is as follows

$$dT = \frac{(T - 46.13)^2}{3816.44} \frac{dp_w}{p_w} \tag{3.167}$$

Supposing that seawater is only the mixture of water and NaCl salt, and the mole fraction of NaCl salt in seawater is x. Then based on the Henry's law, the saturation vapor partial pressure on the seawater surface can be calculated as

$$p_{w,s} = (1 - x)p_w \tag{3.168}$$

Therefore under a determined temperature the partial pressure difference between seawater and the pure water surface is

$$dp_w = xp_w \tag{3.169}$$

Substituting Eq. (3.169) into Eq. (3.167), someone obtains that the temperature difference caused by the pressure difference is

$$dT = \frac{(T - 46.13)^2}{3816.44} x \tag{3.170}$$

Generally speaking, the mass concentration of NaCl salt in seawater is about 3.5%. So, $x = 0.01105$. If the temperature of seawater is 50°C, the corresponding temperature change can be calculated as $dT|_{T=50} = 0.222°C$. Let Eq. (3.170) as ΔT_m substitute into Eq. (3.164), and the heat obtained by the evaporator will be

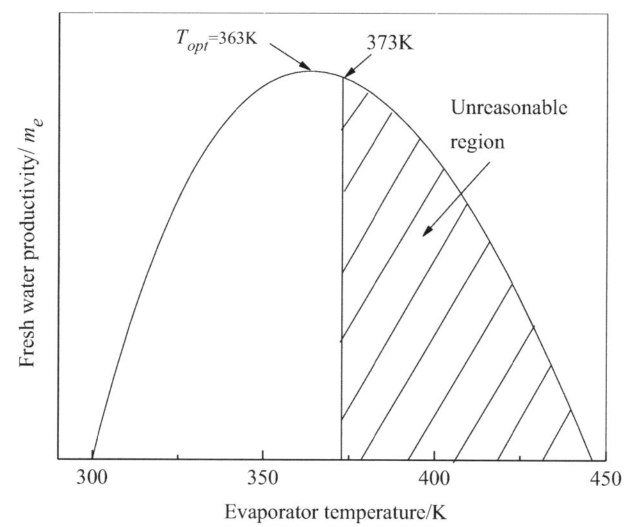

Figure 3.40
The variation of water productivity of the system with evaporation temperature.

$$Q = \frac{\left[\frac{T_H}{T_{abs}} + \left(1 - \frac{T_H}{T_{abs}}\right) \times \left(\frac{T_H}{T_H - T_0}\right)\right]\left[\left(CI_s + \sigma T_0^4 - \sigma T_{abs}^4\right) - U_L(T_{abs} - T_0)\right]}{\frac{(T_H - 46.13)^2}{3816.44(T_H - T_0)}x} \quad (3.171)$$

If the receiver's area is 1 m^2, and the concentrating ratio C is 1, $I_s = 1000$ W/m^2, $T_{abs} = 373$K, $T_0 = 300$ K, the variation of productivity of the system with evaporation temperature is shown in Fig. 3.40.

Fig. 3.40 indicates that the fresh water productivity of the system will increase with the operation temperature in the evaporator when the receiver's temperature is fixed. However, the operation temperature has an optimum value. The fresh water productivity will decrease when the operation temperature is over the optimum value. Because the operation temperature in the evaporator cannot be larger than the temperature of the receiver, the curve of the variation of the fresh water productivity with the operation temperature will produce an unreasonable region, as shown in Fig. 3.40. In this region, the productivity will continuously go down to zero. It can be seen from Eq. (3.164) that the temperature of the receiver mainly affects the heat gain of the solar collector and does not affect the shape of the curve. Therefore we always have the opportunity to choose the optimum temperature of the evaporator under any receiver temperature T_{ab}.

3.12.2 Single-Effect Ideal Distillation Process Directly Heated by Solar Collector

If the useful heat produced by the solar collector is directly sent into the evaporator other than by way of the Carnot engine, as in Fig. 3.41, the case will save the cost due to saving a heat engine and a heat pump.

Assuming the measures in the next steps keep the same, Eq. (7.160) will be changed as

$$Q = Q_u + Q_{H2} = Q_u + \left(\left(1 - \frac{T_0}{T_H - \Delta T_m}\right)Q_{co}\right)\frac{T_H}{T_H - T_0} \quad (3.172)$$

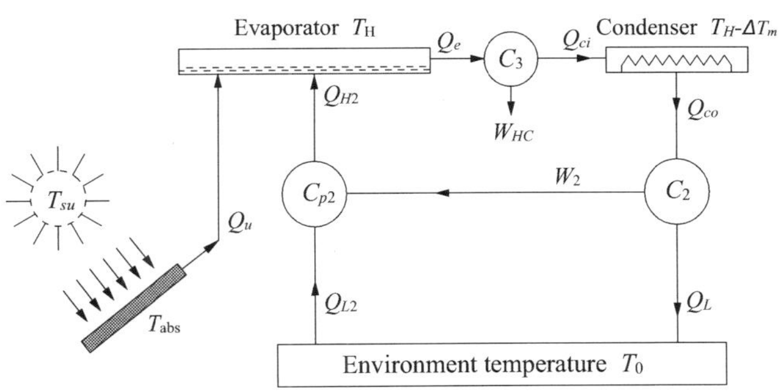

Figure 3.41
Single-effect ideal distillation process directly heated by solar collector.

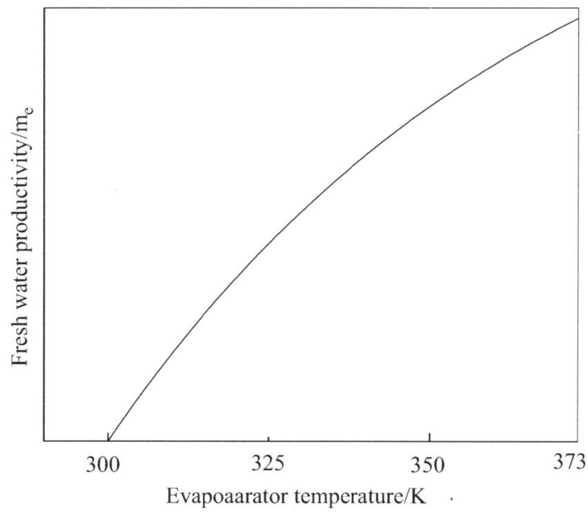

Figure 3.42
Variation of the fresh water productivity with T_H.

With further change, it can be obtained

$$Q = Q_u + \left(\left(1 - \frac{T_0}{T_H - \Delta T_m} \right) \frac{T_H - \Delta T_m}{T_H} \times Q \right) \frac{T_H}{T_H - T_0} \tag{3.173}$$

Solving the equation, Q is found as,

$$Q = \frac{Q_u}{1 - \left(1 - \frac{T_0}{T_H - \Delta T_m}\right) \cdot \left(\frac{T_H - \Delta T_m}{T_H - T_0}\right)} = \frac{Q_u}{\frac{\Delta T_m}{T_H - T_0}} = \frac{Q_u}{\frac{(T_H - 46.13)^2}{3816.44(T_H - T_0)} x} \tag{3.174}$$

Supposing the receiver's area is 1 m^2, $T_{abs} = 373$K, and $T_0 = 300$K, the variation of the fresh water productivity with T_H is easily obtained, as shown in Fig. 7.5.

Fig. 3.42 indicates that the productivity will always increase with the operation temperature in the evaporator when the receiver's temperature is determined. There is no optimum operation temperature. In the same way, the receiver's temperature mainly affects the obtaining heat of the collector but does not affect the shape of the curve in Fig. 3.42. Therefore under any receiver's temperature condition, the operation temperature of the evaporator has the same influence trend on fresh water productivity.

3.12.3 Single-Effect Ideal Distillation Process With Collector Supplying Energy by Generating Electricity

Provided the heat obtained from the solar collector is used to generate electricity directly, then the electricity drives the heat pump to heat the system, as shown in Fig. 3.43.

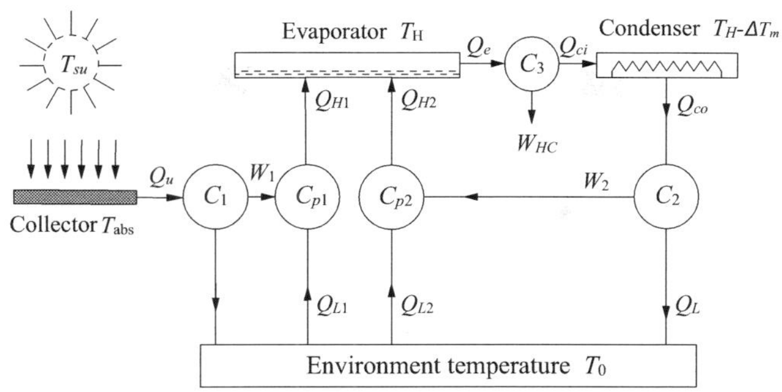

Figure 3.43
Single-effect ideal distillation process directly supplied work by solar collector.

Assuming that the subsequent devices remain unchanged, the maximum work or electricity produced by Q_u is as follows

$$W_1 = \left(1 - \frac{T_0}{T_{abs}}\right) Q_u$$

So Eq. (3.159) can be expressed as

$$Q = Q_{H1} + Q_{H2} = (W_1 + W_2)\frac{T_H}{T_H - T_0} \tag{3.175}$$

Finally, the Q can be written as

$$Q = \frac{\left(1 - \frac{T_0}{T_{abs}}\right)\frac{T_H}{T_H - T_0}Q_u}{\frac{\Delta T_m}{T_H - T_0}} = \left(1 - \frac{T_0}{T_{abs}}\right)\frac{T_H}{\Delta T_m}Q_u \tag{3.176}$$

Substituting the result of Eq. (3.170) into Eq. (3.176) gives

$$Q = \left(1 - \frac{T_0}{T_{abs}}\right)\frac{T_H}{\frac{(T_H - 46.13)^2}{3816.44}x}Q_u \tag{3.177}$$

Under the conditions of a given receiver's temperature, it is a continuous decline curve, which can be seen from Eq. (3.177). That is to say, the higher the evaporator temperature, the less energy the evaporator gets. So it is best to choose a lower temperature of the evaporator.

By carefully analyzing Eq. (3.171), (3.174), and (3.177), these equations have the same denominators; that is to say, their changes in the water production rate mainly depend on numerators. By checking, it is easy to prove that the Q in Eq. (3.171) can produce the most heat, Eq. (3.174) produces less, and Eq. (3.177) produces the least. This means that the best thermodynamic performance can be obtained by the heat of the solar collector

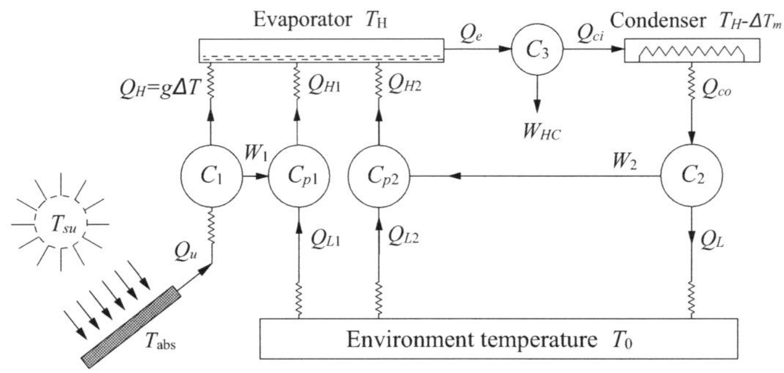

Figure 3.44
Single-effect ideal distillation process with thermal resistance.

first by heating pump one and then offering energy to the evaporator, and secondly by the heat collector directly giving energy to the evaporator. The worst thermodynamic performance is when the heat collector is firstly used to generate electricity and supply energy to the evaporator by heat pump.

3.12.4 The Finite Time Thermodynamic Process of Solar Distillation

The earlier discussion does not take into account the thermal resistance of the heating process, which cannot be ignored in the actual process. Assuming there is a thermal resistance in each heat transfer process, and the thermal resistance mainly exists in the form of convection or conduction, the actual process can be expressed as Fig. 3.44.

Due to the thermal resistance, the efficiency of the heat engine will be reduced, and the coefficient of the performance of the heat pump will be reduced too. Generally speaking, because heat engines and heat pumps exchange different heat with the hot reservoir and cold reservoir, temperature difference generated by thermal resistance is different. In order to simplify the analysis, assume the temperature difference is the same at both ends of all resistance; that is to say, increase or reduce the same ΔT between the mechanism (engines and heat pumps) and reservoir (hot reservoir and cold reservoir). Based on this, the calculation of the process of heat transfer is as follows.

The power produced by engine (C_1) can be written as

$$W_1 = \left(1 - \frac{T_H + \Delta T}{T_{abs} - \Delta T}\right) Q_u \tag{3.178}$$

The heat provided by the engine (C_1) to the evaporator can be expressed as

$$Q_H = Q_u - W_1 = \frac{T_H + \Delta T}{T_{abs} - \Delta T} Q_u \tag{3.179}$$

The heat provided by the heat pump (C_{p1}) to the evaporator can be expressed as

$$Q_{H1} = \frac{T_H + \Delta T}{T_H + \Delta T - (T_0 - \Delta T)} \cdot W_1 = \frac{T_H + \Delta T}{T_H + 2\Delta T - T_0}\left(1 - \frac{T_H + \Delta T}{T_{abs} - \Delta T}\right) Q_u \tag{3.180}$$

The power produced by the engine (C_2) can be written as

$$W_2 = \left(1 - \frac{T_0 + \Delta T}{T_H - \Delta T_m - \Delta T}\right) Q \tag{3.181}$$

So the heat provided by the heat pump (C_{p1}) to the evaporator can be expressed as

$$Q_{H2} = \frac{T_H + \Delta T}{T_H + 2\Delta T - T_0}\left(1 - \frac{T_0 + \Delta T}{T_H - \Delta T_m - \Delta T}\right) Q \tag{3.182}$$

Based on the energy balance equation, the Q can be written

$$Q = \frac{\frac{T_H + \Delta T}{T_{abs} - \Delta T} + \left(1 - \frac{T_H + \Delta T}{T_{abs} - \Delta T}\right) \times \left(\frac{T_H + \Delta T}{T_H + 2\Delta T - T_0}\right)}{1 - \frac{T_H + \Delta T}{T_H + 2\Delta T - T_0}\left(1 - \frac{T_0 + \Delta T}{T_H - \Delta T_m - \Delta T}\right)} Q_u \tag{3.183}$$

If the solar collector is the flat-plate solar energy collector, the heat generated by the collector can be expressed by using the standard formula of the plate heat collector.

$$Q_u = AF_R[I_s - U_L(T_{abs} - T_0)] \tag{3.184}$$

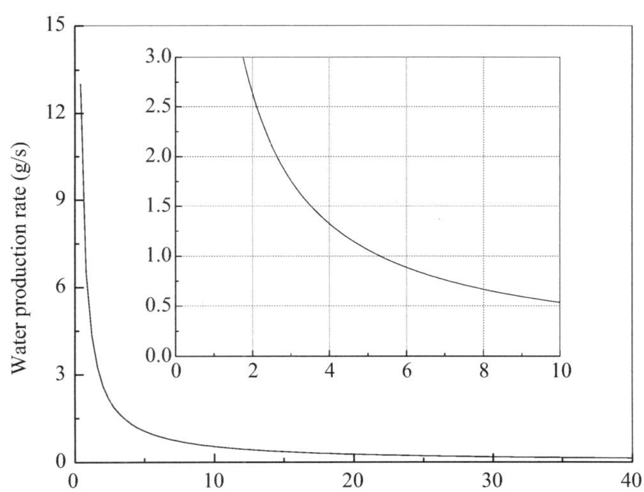

Figure 3.45
Effect of heat transfer temperature difference on water production rate.

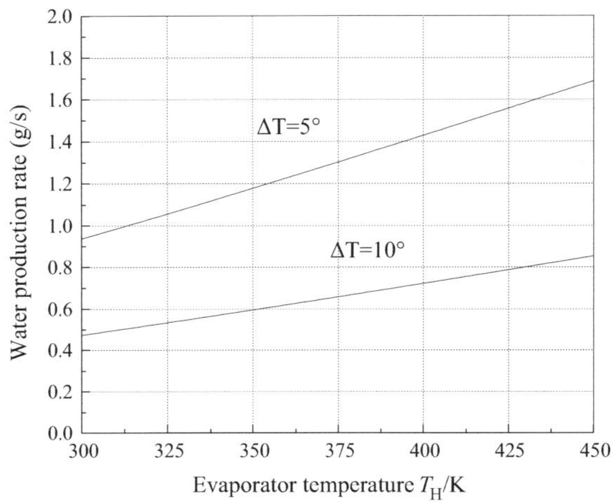

Figure 3.46
Effect of surface temperature of the collector on the water production rate based on a given surface temperature and heat transfer temperature difference.

Assuming that the collector area is 1 m^2, $U_L = 3$ W/m^2 K, I = 1000 W/m^2, $F_R = 0.8$, $T_{abs} = 470$K, $T_H = 373$K, $T_0 = 300$K, and latent heat of water is 2250 kJ/kg. The curve regarding how the ΔT influences water production rate, obtained from Eq. (3.183), is shown in Fig. 7.8.

The heat transfer temperature difference on the water production rate has a very rapid effect, especially when the temperature difference is relatively small, which can be seen in Fig. 3.45. The water production rate drops rapidly when the heat transfer temperature difference increases. This shows that it is necessary to reduce the heat transfer temperature difference of this system. The reason for this situation is that the heat transfer temperature difference directly leads to an irreversible process, which will reduce the efficiency of the heat engine and heat pump.

When $T_{abs} = 470$K, $\Delta T = 5$K, or $\Delta T = 10$K, and T_H varies between T_H and T_{abs}, the relationship between the water production rate and T_H is shown in Fig. 3.46.

Obviously, when the surface temperature of the collector and other parameters are given, the effect of evaporation temperature on the water production rate keeps a linear relationship; that is to say, the higher the evaporation temperature, the greater the water production rate. If the temperature difference is greater, the water production rate is lower at the same evaporation temperature.

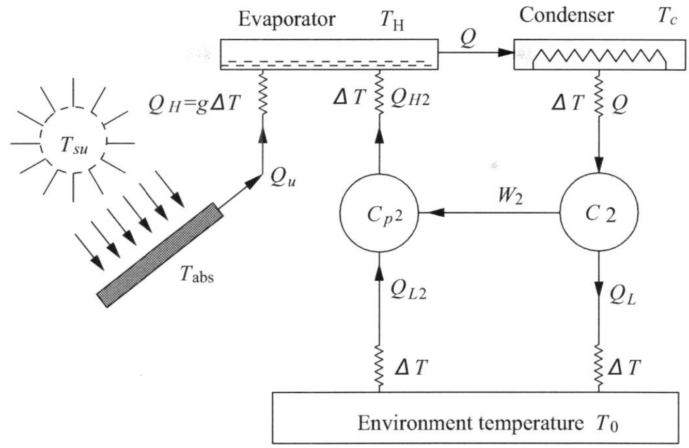

Figure 3.47
Single-effect ideal distillation process directly heated by the solar collector and containing thermal resistance.

3.12.5 Comparison of Several Operation Modes

In consideration of the heat resistance, because the ΔT_m is much smaller than the ΔT, assume $\Delta T_m = 0$ during the preliminary discussion. If the useful heat generated by the solar collector directly transports to the evaporator and does not pass through the reversible Carnot heat engine, this progress would be like Fig. 3.47.

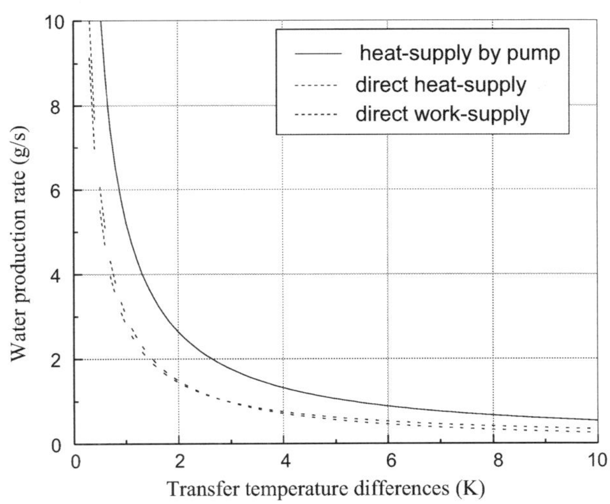

Figure 3.48
Effect of temperature difference on water production rate under different conditions of energy supply.

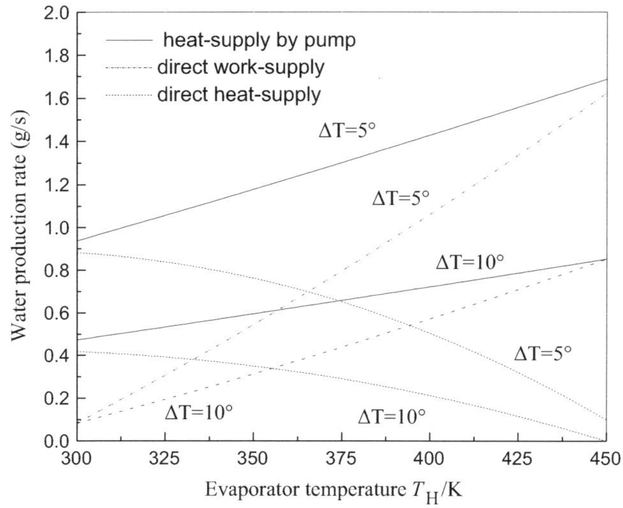

Figure 3.49
Effect of evaporator temperature on water production rate under different conditions of energy supply.

Assuming that the following measures are unchanged, Eq. (3.171) can be expressed as

$$Q = Q_u + Q_{H2} = Q_u + \left(1 - \frac{T_0 + \Delta T}{T_C - \Delta T}\right) \cdot \frac{T_H + \Delta T}{T_H - T_0 + 2\Delta T} Q \tag{3.185}$$

$$Q = \frac{Q_u}{1 - \left(1 - \frac{T_0 + \Delta T}{T_C - \Delta T}\right) \cdot \frac{T_H + \Delta T}{T_H - T_0 + 2\Delta T}} \tag{3.186}$$

For the system taking the flat plate solar collector as the heating source, there exists a linear relationship between Q and T_{abs}, which can be seen in Eq. (3.184). This kind of trend is the same as the trend that the efficiency of the flat-plate collector changes with the surface temperature of the collector. When other parameters are given (keep the same parameters in Fig. 3.45), and only considering the variation of water production rate with the ΔT, the curve of the variation of water production rate with the T curve can be seen in Fig. 3.48. The result showed that whether supplying direct heat or power, the water production rate will be reduced, but the trend is the same as the trend of supplying heat by the heat pump. When directly supplying solar energy to the system without adding a heat pump, $T_{abs} = 470\text{K}$, $\Delta T = 5\text{K}$, or $\Delta T = 10\text{K}$, and T_H varies between T_H and T_{abs}, the relationship between the water production rate and T_H can be seen in Fig. 3.49. The results in the figure show that the water production rate is increased with increasing T_H when directly supplying solar energy to the system. It has the same trend as a heat pump supplying energy, but its trend is more steep growth.

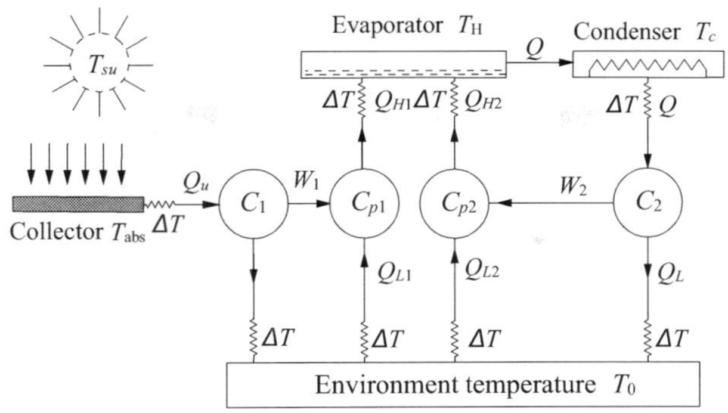

Figure 3.50
Single-effect ideal distillation process directly supplied work by solar collecto- and containing thermal resistance.

Provided that the heat produced by the solar collector is directly used to generate power, which drives the heat pump to supply heat to the system, this progress can be shown in Fig. 3.50.

Supposing that the subsequent devices remain unchanged, the maximum work or electricity produced by the Q_u is

$$W_1 = \left(1 - \frac{T_0 + \Delta T}{T_{abs} - \Delta T}\right) Q_u \tag{3.187}$$

$$W_2 = \left(1 - \frac{T_0 + \Delta T}{T_H - \Delta T}\right) Q \tag{3.188}$$

Noting that $T_H = T_C$ has been used in the earlier equations, substitute Eqs. (3.187) and (3.188) into the energy balance equation

$$Q = Q_{H1} + Q_{H2} = (W_1 + W_2)\frac{T_H + \Delta \mathrm{T}}{T_H - T_0 + 2\Delta T} \tag{3.189}$$

Finally, the Q can be written as

$$Q = \frac{\left(1 - \frac{T_0 + \Delta T}{T_{abs} - \Delta T}\right) A F_R [I_s - U_L(T_{abs} - T_0)] \frac{T_H + \Delta \mathrm{T}}{T_H - T_0 + 2\Delta T}}{1 - \left(1 - \frac{T_0 + \Delta T}{T_H - \Delta T}\right) \cdot \frac{T_H + \Delta \mathrm{T}}{T_H - T_0 + 2\Delta T}} \tag{3.190}$$

The effects of heat transfer temperature difference and evaporation temperature on water production rate are shown in Figs. 3.48 and 3.49. As can be seen from the pictures (Figs. 3.48 and 3.49), ΔT has the same effect on directly supplying work and directly supplying heat, i.e., the water production rate will be increased with the reduction of ΔT, but the trend remains the same as the heat pump supplies heat. When supplying heat by heat pump or directly supplying heat, the water production rate will be increased with the

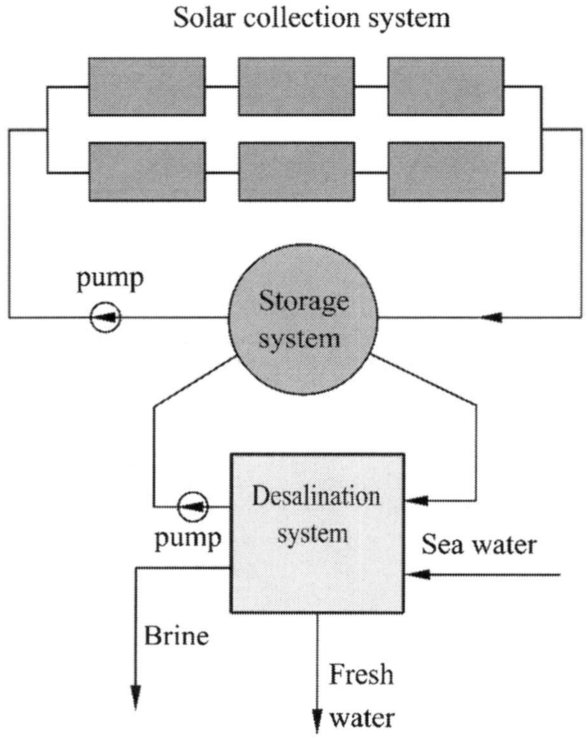

Figure 3.51
The fundamental composition of the solar desalination system.

increase of T_H, but when directly supplying power, the water production rate will be reduced with the increase of T_H, which is a completely different result in two different progressions and should be worth noting.

3.13 The Performance Evaluation of Solar Desalination System

In general, the solar desalination system may include several parts, such as collector, storage device, desalination unit, and other systems. Every part will be with its efficiency. A traditional solar desalination system is shown in Fig. 3.51. It could be seen from Fig. 3.51 that the total efficiency of the system should be the product of the efficiency of all parts. That is,

$$\eta_T = \eta_c \cdot \eta_s \cdot \eta_d \tag{3.191}$$

where, η_c is the efficiency of the solar collector, η_s is the efficiency of the storage device, and η_d is the efficiency of the desalination unit.

In fact, solar desalination has many structures and forms, as shown in Fig. 3.51. They may be directly driven by solar energy or combined by the traditional industrial desalination system. Also, they may be multiple stage devices. Hence there should be different methods to evaluate them. The evaluation index generally includes efficiency, gain output ratio (GOR), performance ratio (PR), and energy recovery ratio.

3.13.1 The Efficiency of Solar Desalination System

For simple, passive, or single-effect solar desalination systems, for example, solar stills, we can use the efficiency to evaluate the system performance. The efficiency is defined as the ratio of total latent heat of evaporation of the distillate to the solar energy incidence into the system. It can be mathematically expressed as,

$$\eta_d = \frac{m_d \cdot h_{fg}}{A \cdot I_s} \tag{3.192}$$

where, m_d is the distillate production, h_{fg} is the latent heat of evaporation, A is the receiver area of the solar desalination system, and I_s is solar radiation. Generally, we hope the solar desalination system has higher efficiency.

3.13.2 Gain Output Ratio and Performance Ratio of Solar Desalination System

For an active-type solar desalination system, solar energy basically is transferred into thermal energy and is then sent into the system. In this way, the evaluation index to desalination unit is GOR, which is defined as the ratio of total latent heat of evaporation of the distillate to the thermal energy input. It can be mathematically expressed as,

$$\text{GOR} = \frac{m_d \cdot h_{fg}}{Q_{in}} \tag{3.193}$$

where, Q_{in} is the heat sent into the desalination unit. Generally, an active-type solar desalination system may consume some electrical power. If the electrical power cannot be ignored, the evaluation index will change into PR. It is defined as follows,

$$\text{PR} = \frac{m_d \cdot h_{fg}}{Q_{in} + P} \tag{3.194}$$

where, P is the electricity power consumption. In this way, solar energy is only used as thermal energy.

GOR and PR are also used to evaluate the multiple-stage or multiple-effect system in which the thermal energy is usually used several times. So, GOR or PR is often larger than one. The more GOR or PR is, the higher the thermal energy utilized efficiency is.

3.13.3 Energy Recovery Ratio of Solar Desalination System

For improving the performance of the system, the thermal energy input into the system is often used multiple times. Especially, the latent heat of vapor condensation should be recovered. Therefore the recovery ratio (RR) is defined to express the recovery proportion as follows,

$$\mathrm{RR} = \frac{Q_{in,R}}{Q_{in}} \tag{3.195}$$

where, $Q_{in,R}$ is the recovered thermal energy from each effect. Q_{in} is the thermal energy supplied to this effect. For a multieffect system, the total RR is defined as follows,

$$\mathrm{RR} = \frac{\sum_{i}^{n} Q_{in,R,i}}{Q_{in}} \tag{3.196}$$

where, $Q_{in,R,i}$ is the recovered thermal energy from ith effect. Q_{in} is the thermal energy supplied to the whole system.

References

[1] Mink G, et al. Air-blown solar still with heat recycling. Sol Energy 1998;62(4):309−17.

[2] Fu Q, He Y. Thermodynamics and heat transfer basis. Xian Jiaotong University Press; 1995.

[3] DianHua X. Air conditioning. Beijing: Qing Hua University Press; 2001.

[4] Guosheng D. Heat transfer. Beijing: Higher Education Press; 1991.

[5] Dayou L, Guangjing Y, Gujun Z. Heat transfer basis. Beijing: Yuhang Publishing House; 1987.

[6] Weiquan X. Heat and mass transfer process and equipment. Beijing: Qing Hua University Press; 1987.

[7] Shi M, Gan Y, Ma C. Film condensation. Beijing: Higher Education Press; 1995.

[8] Mills AF. Laminar film condensation on a vertical wall, Heat Transfer. Homewood, IL, 60430, Boston, MA02116: IRWIN; 1992. p. 598−613.

[9] Sideman S. Some design aspects of thin film evaporator/condenser in water desalinization. In: Kakac S, editor. Heat exchangers. New York: Hemisphore Publishing Corporation; 1980. p. 681−703.

[10] Li X, Wang S, Wang Y. The heat transfer coefficient of film evaporation in horizontal tube. J Chem Eng 2003;54(3):299−304.

[11] Sernas V. Heat transfer correlation for subcooled water films on horizontal tubes. J Heat Transf 1979;101:176−8.

[12] Chun KR, Seban RA. J Heat Transf 1971;93:391−6.

[13] Ganic EN. On the heat transfer and fluid flow in falling film shell-and-tube evaporators. In: Kakac S, editor. Heat exchangers. New York: Hemisphere Publishing Corporation; 1980. p. 705−19.

[14] Berenson PJ. Int J Heat Mass Transf 1962;5:985.

[15] Webb RL. Heat Transf Eng 1981;2:46. Heat Transfer Eng, 4, 71, 1983.

[16] Bergles AE. Enhancement of heat transfer. Heat Transf 1978;6:89−108. Hemisphere Publishing, New York.

[17] Bergles AE. In: Hewitt GF, editor. Heat exchanger design handbook. New York: Begell House; 2002. Section 2.7.9.

[18] Griffith P. In: Hewitt GF, editor. Heat exchanger design handbook. New York: Hemisphere Publishing; 1990. Section 2.6.5.

[19] Incropera FP, Dewitt DP, Bergman TL, Lavine AS. Fundamentals of heat and mass transfer. U. S. A: John Wiley & Sons, Inc.; 2011.

[20] Standards of the tubular exchange manufacturers association. 6th ed. New York: Tubular Exchanger Manufacturers Association; 1978.

[21] Bemporad GA. Basic hydrodynamic aspects of a solar energy based desalination process. Sol Energy 1995;54(2):125−34.

CHAPTER 4

Traditional Solar Desalination Units

Chapter Outline

Solar Energy Desalination Technology. http://dx.doi.org/10.1016/B978-0-12-805411-6.00004-X

The conventional solar desalination device is also called a passive solar distiller or solar still, which refers to the device without any electricity-driven power elements, such as a water pump and fan, or any additional elements like the solar collector for active heating. The device operation is completed passively under sunlight.

Such a type of device is widely used for its simple design, convenient drawing materials, and extremely low operation cost. However, it also has some serious shortcomings, one of which is low water production per unit lighting area, so a large construction area is required.

The single-effect basin type solar still is the most typical and common type among passive devices. The first large-scale solar desalination device in the world, i.e., the 4700 m^2 device built in northern Chile Las Salinas in 1874, was made up of single-effect basin type solar stills. Such devices are still widely used because of the low overall cost.

Multiple forms have been developed from the passive system with the advance of technology, which will be introduced in detail in the following sections.

4.1 Operating Principle of Basin Type Solar Stills

The basin type solar still is one of the oldest devices among the solar desalination family, which has been investigated intensively and processes the best technology.

The basin type still is the simplest still, which is also known as the greenhouse type still. Though its performance is inferior to the complex structural and highly effective active solar stills, this type of still is better than other stills because of its simple structure, manufacture, operation, and maintenance, considered from the cost of producing the same quantity of fresh water. Therefore the basin type still has great market value.

The basin type solar still is similar to a greenhouse, as shown in Fig. 1.6 in Chapter 1. The seawater is spread thinly in a trough painted in black, and the whole trough is sealed by a glazing cover made of transparent plastics or glass. Most of the sunlight will be filtered through the glass cover, while a small portion of it will be reflected or absorbed. A part of the solar radiation filtered through the glass cover will be selected from the water, and the remaining will be absorbed by water through the black lining inside the trough, which will increase the seawater temperature and evaporate some water. As the solar energy absorbed by the head cover is limited and the cover dissipates heat directly into the air, the temperature of the glazing cover is lower than the water temperature. Therefore heat exchanges between the water surface and glass cover by radiation, convection, and evaporation. The vapor evaporated from water in the trough condenses on the lower surface of the glazing cover and releases latent heat of vaporization. As long as the cover has a suitable inclination, condensed water will flow down the glazing cover by gravity, gather in the water trough, and flow though the weep hole out of the still; then the fresh water product can be obtained.

This kind of solar still operates with relatively low energy consumption, and its running cost and maintenance fee are not high. Therefore the fresh water production cost depends largely on the equipment investment. To lower the water production cost, on the premise that the life span and efficiency of the still will not be overly reduced, it's better to use a simple structure and cheap materials to decrease the equipment cost or to collect rainwater by the outside surface of the cover to increase annual fresh water production of the still.

4.2 Performance Analysis of Basin Type Solar Stills

With the advancement of technology and increase of available materials, various forms of basin type solar stills have been developed and applied in some seawater desalination plants. Fig. 4.1 introduces some of the basic forms [1].

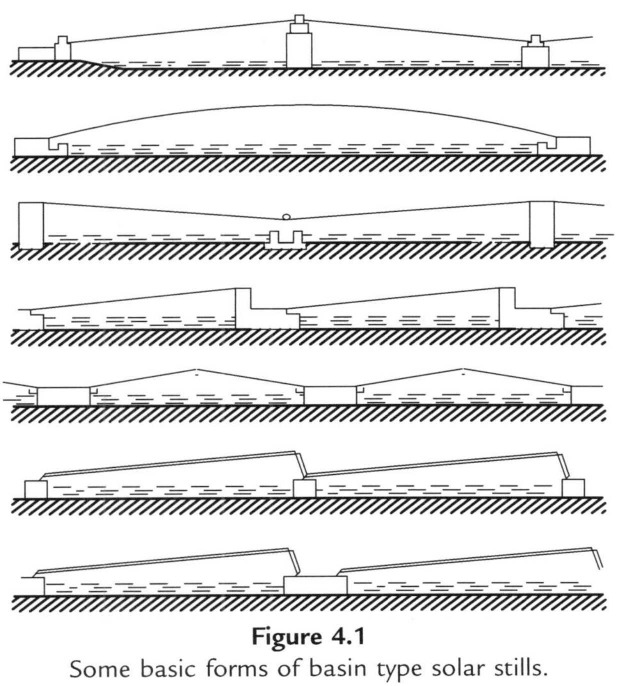

Figure 4.1
Some basic forms of basin type solar stills.

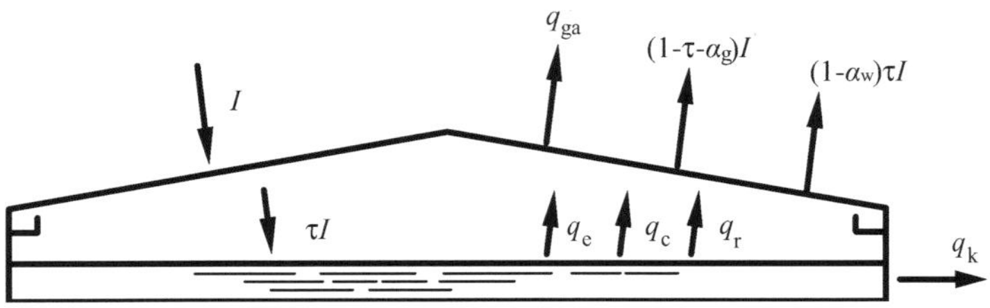

Figure 4.2
The energy transfer process inside the still.

Regardless of the diversified forms, an ideal basin type solar still can be simplified into the analysis model shown in Fig. 4.2 according to its installation method.

The most important performance index of the still is the fresh water productivity, which is generally expressed in the daily production of distilled water per unit area. To calculate fresh water production of the still and search for ways to improve its performance, the heat transfer process inside the still needs to be analyzed. The analysis was originally put forward by Dunkle [2], and was modified by other researchers later. Fig. 4.2 indicates the energy transfer process inside the still. The heat balance equation of the trough bottom is:

$$\alpha_w \cdot \tau \cdot H = q_e + q_r + q_c + q_k + C_{p,w} \cdot M_w \cdot \frac{dT_w}{dt} \tag{4.1}$$

where H is the intensity of solar radiation, τ is the sunlight transmittance of glass, α_w is the sunlight absorptivity of water in the trough and its bottom surface, $C_{p,w}$ is the specific heat of seawater, and M_w is the total mass of seawater. q_e,q_r,q_c refers to the evaporation, radiation, and convective heat transfer from the water surface to the glazing cover and q_k is the heat dissipation of the device wall.

This equation shows part of the radiation τH transmitted through the glazing cover, i.e., $a_w \cdot \tau \cdot H$ has been absorbed by water or the bottom of the trough. Some of the solar radiation energy absorbed by water and the bottom has been transferred to the glazing cover by evaporation, convection, and radiation (expressed in q_e,q_r,q_c), another part (q_k) has been lost in the ground below the still, and the remaining parts will increase the temperature of the bottom and the water inside. Heat capacity and temperature of the water are C_{wb} and T_w. In Eq. (4.1), the water and bottom of the trough are taken as a whole, and the temperature of the whole bottom will be replaced by the temperature of water. As for the glass cover, the heat balance equation can also be established.

$$q_e + q_c + q_r + a_g \cdot H = q_{ga} + C_{p,g} \cdot M_g \frac{dT_g}{dt} \tag{4.2}$$

where $C_{p,g}$ is the specific heat of the glass cover, M_g is the total mass of the cover, and $q_{g,a}$ is the transfer between the cover and the external. Strictly speaking, energy absorption of the glazing cover should also include $(1 - \alpha_w)\tau \cdot H \cdot a_g$.

As α_w is a number close to 1, $(1-\alpha_w)$ is very small, so $(1-\alpha_w)\tau \cdot H \cdot \alpha_g$ can be neglected. In other words, all of the energy reflected from the bottom $(1-\alpha_w)\tau \cdot I$ has been lost after passing through the glazing cover; secondary reflection and absorption of the cover will not be considered, which helps to simplify the whole analysis. Besides, total heat capacity of the cover M_g is pretty small compared with the heat capacity of the bottom M_w; if the change of heat storage caused by the change of cover temperature T_g is not considered, then the second term on the right side of Eq. (4.2) can be ignored. q_{ga} represents the heat of the cover dissipated outside, which consists of the radiation heat loss and the convective heat transfer in the air.

Add up Eqs. (4.1) and (4.2), then the heat balance equation can be established:

$$\alpha_g H + \alpha_w \cdot \tau \cdot H = q_{ga} + M_w \cdot C_{p,w} \frac{dT_w}{dt} + M_g \cdot C_{p,g} \frac{dT_g}{dt} \tag{4.3}$$

In fact, as it is influenced by many factors, it is difficult to get the accurate solution of Eqs. (4.1)−(4.3). The radiation heat transfer rate between the seawater in the trough and the glazing cover q_r can be solved by Eq. (4.2). But considering the small inclination of the cover, the water surface and the cover can be regarded as parallel plates that are very close. Let the angle factor be 1, and supposing the areas of both sides are equal, then Eq. (4.3) can be applied under the general condition. For easy calculation in engineering, Dunkle simplified the calculation by the following equation:

$$q_r = 0.9\sigma\left(T_w^4 - T_g^4\right) \tag{4.4}$$

It's the equivalent of taking the values $\varepsilon_w = 0.96$, $\varepsilon_g = 0.94$ approximately, which is reasonable in many cases.

The convective heat transfer rate between the water surface of the bottom and the glass cover is:

$$q_c = h_c(T_w - T_g) \tag{4.5}$$

The solution of the convective heat transfer coefficient h_c is complicated, which relates to nature convection between parallel plates with a hot surface underneath in a closed space. The heat transfer coefficient is calculated by empirical formulas obtained by an experiment, and the formulas given by different researchers are not the same. The specific calculation of h_c will be discussed in detail in the next section.

The heat transfer term q_e and convection term q_c occur simultaneously, and there is a certain relation between them. Generally, q_e is calculated by Eq. 2.13, i.e., $q_e = h_e(p_w - p_g)$. Dunkle proposed a relatively simple formula according to the relation between q_e and q_c under some conditions [2]:

$$q_e = 16.273 \times 10^{-3} q_c \frac{(p_w - p_g)}{(T_w - T_g)} \tag{4.6}$$

or

$$q_e = 16.273 \times 10^{-3} h_c (p_w - p_g) \tag{4.7}$$

Therefore the evaporation of water is:

$$m_e = \frac{q_e}{h_{fg}} \tag{4.8}$$

where h_{fg} is the vaporization latent heat of water.

The solution of the heat loss of the trough bottom and around the still is not easy, because the heat transfer coefficient varies with working conditions. What is more, the temperature of the soil below the still is unknown, and as a kind of approximate calculation method, it can be expressed as:

$$q_k = h_k (T_w - T_a) \tag{4.9}$$

Obviously, this formula is pretty simple formally, but actually it includes all the complicated factors in the heat transfer coefficient h_k. The value of h_k shall be decided by the structure and working condition of the still. For a well-designed still, especially the ones with a heat insulation layer at the bottom, the heat loss is little.

As discussed previously, heat transfer of the cover includes radiation and convection. The radiation occurs between the glazing cover and the air, which is:

$$q_{ga,r} = \varepsilon_g \sigma \left[T_g^4 - T_s^4 \right] \tag{4.10}$$

T_s is the temperature of the sky (K). The outward convective heat loss of the cover relates to wind velocity v and can be calculated by an approximate formula [3].

$$h_{ga,c} = 5.7 + 3.8v \tag{4.11}$$

To sum up,

$$q_{ga} = q_{ga,r} + q_{ga,c} = \varepsilon_g^4 \sigma \left[T_g^4 - (T_a - 12)^4 \right] + (5.7 + 3.8v)(T_g - T_a) \tag{4.12}$$

After the method to calculate each heat transfer term is educed, provided that the initial water temperature T_w and the time variations of solar radiation and air temperature $I(t)$ and $Ta(t)$ are known, fresh water output can be obtained by Eqs. (4.1) and (4.2). Obviously, the temperature of the transparent cover is obtained by iterative calculation.

If the time variation of I and T_a is not considered, and differential terms in Eqs. (4.1) and (4.2) are ignored, steady state characteristics of the still can be analyzed.

4.3 Performance Coefficient of Solar Still

The evaporation amount calculated by Eq. (4.8) is the fresh water output, which neglects leakage loss of the vapor in the still, reevaporation of the water condensed under the glazing cover, and loss of the fresh water that drops back to the trough. Instantaneous efficiency of the still is defined as:

$$\eta = q_e/H \tag{4.13}$$

As for the unsteady-state process, the average efficiency is

$$\overline{\eta} = \frac{\int q_e dt}{\int H dt} \tag{4.14}$$

The still efficiency decided by experimental method is

$$\eta = m \cdot h_{fg} / \int H dt \tag{4.15}$$

where m is the actual amount of the fresh water collected within a certain period; it is smaller than the integral value m_e. To make the still produce fresh water as much as possible, the solar radiation should be the furthest used to evaporate water in the trough, so that q_e can reach the maximum. This requires τ, i.e., transmittance of the material of the glazing cover, and α_w, i.e., absorptivity of the bottom trough, to be as large as possible. Meanwhile, q_c, q_r, and q_k should be further reduced. q_k can be reduced by the proper design and improvement of heat insulation performance of the bottom trough. Decrease of water temperature T_w in the trough can reduce q_c and q_r, but it should not be adopted because when T_w decreases, the reduction of q_e will be much faster than q_c and q_r, which will lead to efficiency decrease of the still. By applying the earlier relation, provided that $(T_w - T_g)$ is given, the respective share of q_c, q_r, and q_e in the total heat transfer amount can be calculated. Fig. 4.3 is the result of such calculation, originally given by Cooper [4]. As clearly shown in the picture, the higher the water temperature in the trough, the larger the share of the heat transferred by evaporation. Therefore to improve the efficiency of the still, the still should operate under the condition of high water temperature. Compared to water temperature in the trough, the difference value between the water temperature inside the trough and the top cover temperature has a less obvious effect on the efficiency of the still.

4.4 Determination of the Natural Convection Coefficient h_c in the Still

4.4.1 Natural Convective Heat Transfer in the Still and Correction of Gr

In the still, seawater evaporates after heating, and the vapor carrying heat flows into the upper air layer through natural convection and moisturizes the air. The moist air then keeps moving upward, condenses when meeting the glass cover with the lower

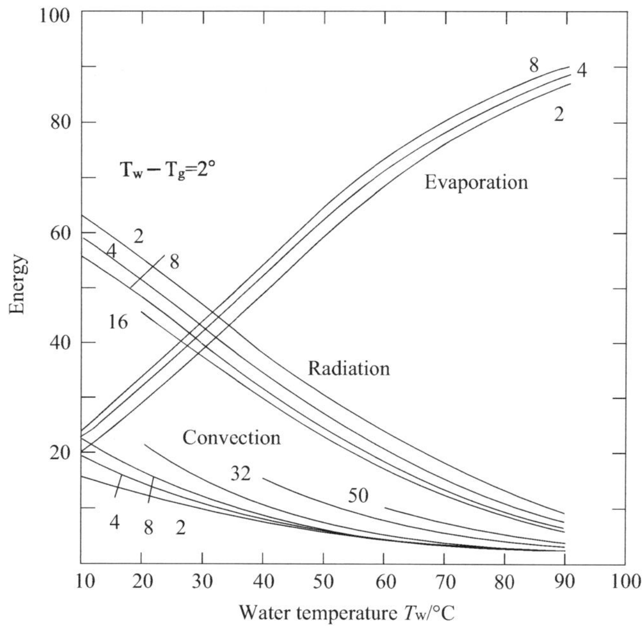

Figure 4.3
Energy share of evaporation, convection, and radiation varying with water temperature.

temperature, and releases its latent heat, which forms a complete cycle process of natural convection. Generally speaking, the natural convective heat transfer rate inside the still can be expressed as:

$$q_c = h_c \cdot \Delta T \tag{4.16}$$

where q_c is the natural convection heat passing through a unit area in unit time (W/m^2), h_c is the heat transfer coefficient of natural convection (W/m^2 °C), and ΔT is the temperature difference between the evaporation and condensation surface. Generally, it is easy to decide the temperature difference between the evaporation and condensation surface. However, the determination of heat transfer coefficient of natural convection h_c is complicated and is usually calculated by empirical formulas obtained by experiment.

Heat transfer relation of the natural convection heat transfer can be expressed by Nu, Gr and Pr, and Nu = f(Gr, Pr), where Nu, Gr, and Pr are the Nusull's number, Grashof number, and Prandtl number, respectively. However, it is noteworthy that besides the temperature difference, the vapor concentration difference also exists in the still. As the vapor is lighter than the air, the mass transfer process inside the still accelerates the transfer of heat. Therefore a modified Grashof number needs to be used for the solar still. Sharpley and Boelter define the modified Grashof number Gr′ by the following equation [1]:

$$Gr' = \frac{x_l^3 \cdot \rho^2 \cdot g}{\mu^2} \cdot \left[\frac{M_g' T_e}{M_e' T_g} - 1\right] \tag{4.17}$$

In the equation, x_l is the characteristic size of the cavity, ρ is the density of moist air, μ is the dynamic viscosity of the moist air, and the subscripts e and g denote, respectively, the state of evaporation and condensation surface. If M' denotes the molar mass of moist air, then the molar mass of moist air in the state of evaporation and condensation surface can be expressed as M'_e and M'_g; T_e and T_g are the temperatures of the evaporation and condensation surface, respectively. If there is no bid difference between T_e and T_g, and $M'_e \approx M'_g$, then the terms inside the bracket in the equation is changed to:

$$\frac{T_e}{T_g} - 1 = \frac{T_e - T_g}{T_g} = \frac{\Delta T}{T_g} \approx \beta \Delta T \tag{4.18}$$

That is, the expressions of Gr′ and Gr are basically the same. To further obtain the modified Grashof number expressed by physical parameters, to derive as follows [1,5,6]:

The subscripts w and a denote the vapor in moist air and dry air. Suppose the moist air inside the still is ideal gas, then:

$$p_{gw} + p_{ga} = p_{ew} + p_{ea} = p_T \tag{4.19}$$

where P_{gw} and P_{ga} are respectively the partial pressures of the vapor and dry air around the cover, P_{ez} and P_{ea} are the partial pressures of the vapor and dry air around the evaporation surface, and P_T is the total pressure of moist air, which shares the same value as the atmosphere pressure, i.e.,

$$P_T = 101.3 \times 10^3 \text{ Pa}$$

Hence the following can be obtained:

$$\begin{aligned}
\frac{M'_g \cdot T_e}{M'_e \cdot T_g} - 1 &= \frac{M_w p_{gw} T_e + M_a p_{ga} T_e}{M_w p_{ew} T_g + M_a p_{ea} T_g} - 1 \\
&= \frac{M_w \left(p_{gw} T_e - p_{ew} T_g\right) + M_a \left(p_{ga} T_e - p_{ea} T_g\right)}{T_g \left(M_w p_{ew} + M_a p_{ea}\right)} \\
&= \frac{\left[M_w p_{ew} + M_a (p_T - p_{ew})\right] \cdot (T_e - T_g) + (M_w - M_a)\left(p_{gw} - p_{ew}\right) T_e}{T_g \left[M_w p_{ew} + M_a (p_T - p_{ew})\right]} \\
&= \frac{(T_e - T_g)}{T_g} + \frac{(M_w - M_a)\left(p_{gw} - p_{ew}\right) T_e}{T_g \left[M_a p_T + (M_w - M_a) p_{ew}\right]} \\
&= \frac{1}{T_g} \left\{ (T_e - T_g) + \frac{(p_e - p_g) T_e}{\dfrac{M_a p_T}{M_a - M_w} - p_{ew}} \right\}
\end{aligned} \tag{4.20}$$

In the equation, $M_w = 18$ g/mol, which denotes the molar mass of vapor; $M_a = 28.96$ g/mol, which means the molar mass of dry air. By substitution of the value of M_w, M_a, and p_T into Eq. (4.20), obtaining:

$$\begin{aligned} \frac{M'_g \cdot T_e}{M'_w \cdot T_g} - 1 &= \frac{(T_e - T_g)}{T_g} + \frac{(p_{ew} - p_{gw})T_e}{(268.9 \times 10^3 - p_{ew})T_g} \\ &= \beta\left[(T_e - T_g) + \frac{(p_{ew} - p_{gw})T_e}{(268.9 \times 10^3 - p_{ew})}\right] \end{aligned} \tag{4.21}$$

Comparing Eq. (4.21) with the definition formula of Gr, it is clear that the term inside the bracket is actually an equivalent temperature difference $\Delta T'$ (as assumed in the equation that $\beta = \frac{1}{T_g}$). To keep in accord with the latter formulas and common symbols in the general literature, substitute p_{ew} and p_{gw} with p_w and p_g, which denote the partial pressures of the vapor around evaporation and condensation surfaces, and use T_w and T_g as the temperatures of the evaporation surface and the cover (when $T_w = T_e$), thereupon:

$$\begin{aligned} Gr' &= \frac{x_l^3 \cdot \rho^2 \cdot g \cdot \beta}{\mu^2}\left[(T_w - T_g) + \frac{(p_w - p_g)T_w}{(268.9 \times 10^3 - p_w)}\right] \\ &= \frac{x_l^3 \cdot \rho^2 \cdot g \cdot \beta}{\mu^2} \cdot \Delta T' \end{aligned} \tag{4.22}$$

where, $\Delta T'$ can be considered as a modified temperature difference.

4.4.2 Heat Transfer by Evaporation

The evaporation heat transfer term q_e(W/m^2) and convection heat transfer term q_c take place simultaneously. By analyzing the natural convection process inside the still, the mass of moist air transferred by natural convection per area per time can be obtained:

$$\dot{m}_{a-w} = \frac{q_c}{C_{pa-w} \cdot (T_w - T_g)} = \frac{h_c}{C_{pa-w}} \tag{4.23}$$

where C_{pa-w} is the specific heat capacity of the moist air. Assume that the air close to the evaporation and condensation surface is saturated inside the still (such assumption is rational). Then the mass of vapor is included in the unit mass of dry air, i.e., humidity ratio can be written as:

$$d = \frac{m_w}{m_a} = \frac{\frac{p_w V}{R_w T_w}}{\frac{p_a V}{R_a T_w}} = \frac{M_w}{M_a} \cdot \frac{p_w}{p_T - p_w} \tag{4.24}$$

In the equation, m_w and m_a are the masses of vapor and dry air when the air volume is V, R_w, and R_a are the air constants of vapor and dry air, and M_w and M_a are the molar masses of vapor and dry air, the values of which have been given earlier.

Therefore by making a simple derivation and considering that the mass of dry air plus the mass of vapor equals the total mass of the moist air, the mass of vapor transferred through the unit surface area in unit time can be obtained [7]:

$$\dot{m}_{ew} = \frac{d_e}{1+d_e} \cdot \dot{m}_{a-w}\big|_e = \frac{M_w p_w}{M_w p_w + M_a (p_T - p_w)} \cdot \frac{h_c}{C_{pa-w}} \tag{4.25}$$

In the same way, the mass of the vapor transferred in the unit area of the condensation surface in unit time is:

$$\dot{m}_{gw} = \frac{d_g}{1+d_g} \cdot \dot{m}_{a-w}\big|_g = \frac{M_w p_g}{M_w p_g + M_a (p_T - p_g)} \cdot \frac{h_c}{C_{pa-w}} \tag{4.26}$$

where d_e and d_g denote the humidity ratios of the moist air close to the evaporation and condensation surface. Hence the net mass transfer rate of vapor should be the difference between Eqs. (4.25) and (4.26), that is

$$\begin{aligned} \dot{m}_e &= \dot{m}_{ew} - \dot{m}_{gw} \\ &= \frac{M_w}{M_a} \cdot \frac{h_c}{C_{pa-w}} \left(\frac{p_w}{\frac{M_w}{M_a} p_w + (p_T - p_w)} - \frac{p_g}{\frac{M_w}{M_a} p_g + (p_T - p_g)} \right) \end{aligned} \tag{4.27}$$

Therefore the transfer ratio q_e (W/m^2) of the heat transferred from the water surface to the condensation surface caused by vapor mass transfer is:

$$\begin{aligned} q_e &= \dot{m}_e \cdot h_{fg} \\ &= \frac{M_w}{M_a} \cdot \frac{h_c}{C_{pa-w}} \cdot h_{fg} \cdot \left(\frac{p_w}{\frac{M_w}{M_a} p_w + (p_T - p_w)} - \frac{p_g}{\frac{M_w}{M_a} p_g + (p_T - p_g)} \right) \end{aligned} \tag{4.28}$$

where h_{fg} is the latent heat of vaporization (kJ/kg) of water. The equation can be further simplified as:

$$\begin{aligned} q_e &= \frac{M_w}{M_a} \cdot \frac{h_c}{C_{pa-w}} \cdot h_{fg} \cdot \frac{(p_w - p_g) \cdot p_T}{\left[\frac{M_w}{M_a} p_w + (p_T - p_w)\right] \cdot \left[\frac{M_w}{M_a} p_g + (p_T - p_g)\right]} \\ &= h_e \cdot (p_w - p_g) \end{aligned} \tag{4.29}$$

h_e in the equation can be defined as the evaporation heat transfer coefficient, the unit of which is W/m^2•Pa = m/s, also

$$h_e = \frac{M_w}{M_a}\cdot\frac{h_c}{C_{pa}}\cdot h_{fg}\cdot\frac{p_T}{\left[\frac{M_w}{M_a}p_w + (p_T - p_w)\right]\cdot\left[\frac{M_w}{M_a}p_g + (p_T - p_g)\right]} \tag{4.30}$$

If the operation temperature of the still is not too high, and the humidity ratios of the moist air of the evaporation and condensation surface are much smaller than 1, then $1 + d_e \approx 1 + d_g \approx 1$, a similar derivation as Eq. (4.27) can be obtained:

$$h_e = h_c\cdot\frac{M_w}{M_a}\cdot\frac{h_{fg}}{C_{pa-w}}\frac{p_T}{(p_T - p_w)(p_T - p_g)} \tag{4.31}$$

This is the relation between the evaporation and natural convection heat transfer coefficient derived theoretically by Malik (1982). Provided that $(p_T - p_w)(p_T - p_g) \approx p_T^2$ is considered, then Eq. (4.31) can be further simplified as:

$$h_e = h_c\cdot\frac{M_w}{M_a}\cdot\frac{h_{fg}}{C_{pa-w}}\cdot\frac{1}{p_T} \tag{4.32}$$

Substituting approximate values of relevant parameters into the equation, a simple relation between h_e and h_c can be obtained:

$$h_e = 0.013h_c \tag{4.33}$$

Attention should be paid that the coefficient 0.013 in the equation has a dimension (°C/Pa). There are certain differences between Eq. (4.33) and the relation between the evaporation and natural convection heat transfer coefficient obtained experimentally by Dunkle in 1961: $h_e = 0.0163h_c$. However, Malik believed that the reason for such difference is the neglect of affects from p_w and p_g in Eq. (4.31). It is noteworthy that when defining the relation, Dunkle assumed the still operation temperature is 50°C, and temperature difference of the evaporation and condensation surface is 17°C.

A high correlation has been found between the evaporation and the natural convection heat transfer inside the still. Therefore correct determination of their relation is the basis to calculate the production of the still. Although the solution of the convection heat transfer coefficient h_c is relatively complicated, generally it can be calculated by the empirical correlation obtained by experiment. Once the relation between h_e and h_c is determined, and the calculation expression of h_c is known, then the water production rate $\dot{m}_e$ (kg/m^2 s) can be calculated by Eq. (4.27) or (4.28), i.e.,

$$\dot{m}_e = \frac{q_e}{h_{fg}} = \frac{h_e}{h_{fg}}(p_w - p_g) \tag{4.34}$$

If the relation proposed by Dunkle is adopted (which is widely used at present), then

$$\dot{m}_e = 16.273 \times 10^{-3} \frac{h_c}{h_{fg}} \left(p_w - p_g\right) \tag{4.35}$$

where p_w, p_g, and h_{fg} can all be obtained by temperature measurement and table lookup. In the final analysis, the solution of h_c is essential to calculate the water production rate of the still.

4.4.3 Radiation Heat Transfer

The radiation heat transfer rate per unit area between the seawater and condensation surface in the solar still q_r can be calculated by the following formula:

$$q_r = \frac{\sigma \cdot \left(T_w^4 - T_g^4\right)}{A_w \left[\frac{1-\varepsilon_w}{A_w \varepsilon_w} + \frac{1}{A_w F_{wg}} + \frac{(1-\varepsilon_g)}{\varepsilon_g A_g}\right]} \tag{4.36}$$

where A_w and A_g are the areas of the evaporation and condensation surface, respectively, ε_w and ε_g are the emissivity of the seawater and condensation surface, σ is the Stefan–Boltzmann constant, and F_{wg} is the angle factor between the evaporation and condensation surface. If the two surfaces are roughly parallel to each other, and $A_w = A_g = A$, then the value of F_{wg} can be taken as 1. Then Eqs. (4.36) can be simplified as:

$$q_r = \frac{\sigma \cdot \left(T_w^4 - T_g^4\right)}{\frac{1}{\varepsilon_w} + \frac{1}{\varepsilon_g} - 1} \tag{4.37}$$

Three modes of heat transfer inside the still have been analyzed. To clearly see the percentage of the three modes in the whole heat transfer process, how the percentage changes with temperature is shown in Fig. 4.3. As indicated in the figure, evaporation heat transfer is the uppermost transfer mode, and it shows as an increasing function with the increase of temperature.

4.4.4 The Determination of Natural Convection Heat Transfer Coefficient

As discussed in the last section, the determination of natural convection heat transfer coefficient is the key to calculate the production of the still. In 1961, Dunkle proposed a set of complete natural convection relations by using a modified Grashof number:

$$\text{Nu} = C \cdot \left(\text{Gr}' \cdot \text{Pr}\right)^n \tag{4.38}$$

where Gr′ is the modified Grashof number and Pr is the Prandtl number:

$$\text{Gr}' = \frac{x_l^3 \cdot \rho^2 \cdot g \cdot \beta}{\mu^2} \cdot \Delta T' \tag{4.39}$$

$$\mathrm{Pr} = \frac{C_{pa-w} \cdot \mu}{k} \tag{4.40}$$

in,

$$\Delta T' = (T_w - T_g) + \left(\frac{p_w - p_g}{\frac{M_a \cdot p_T}{M_a - M_w} - p_w} \right) \cdot T_w \tag{4.41}$$

In the equation, k, ρ, β, μ, and C_{pa-w} are physical parameters of the moist air in the still, normally taken under the average temperature $T_{av} = (T_w + T_g)/2$. The coefficient C and n in Eq. (4.38) are decided according to the Grashof number Gr' under working conditions:

1. When $\mathrm{Gr}' < 10^4$, $C = 1$, $n = 0$.
2. When $10^4 < \mathrm{Gr}' < 3.2 \times 10^5$, $C = 0.21$, $n = \frac{1}{4}$.
3. When $3.2 \times 10^5 < \mathrm{Gr}' < 10^7$, $C = 0.075$, $n = \frac{1}{3}$.

According to the general size and working condition of the basin type still, Gr' normally falls in the third region. Supposing the average temperature of moist air is 50°C, and the temperature difference between evaporation and condensation surface is 17°C, the following well-known formula was given by Dunkle:

$$h_c = 8.84 \times 10^{-4} \left[(T_w - T_g) + \left(\frac{p_w - p_g}{268.9 \times 10^3 - p_w} \right) T_w \right]^{\frac{1}{3}}. \tag{4.42}$$

Dunkle further obtained the relation between the evaporation heat transfer coefficient h_e and natural convection heat transfer coefficient h_c by experiment:

$$h_e = 0.0163 h_c \tag{4.43}$$

It is worth noting that the dimension of the constant coefficient in the equation is °C/Pa. The Dunkle relation composed of Eqs. (4.42) and (4.43) have been acknowledged by most of the researchers and were adopted until now. However, a study shows that the Dunkle relation has a great deviation in practical application, especially when the solar still operates at a high- or low-temperature section. Obviously, this is because some factors were neglected. Also, the operation temperature of 50°C and temperature difference of 17°C were assumed when Dunkle established the earlier relations. For this reason, many researchers have proposed a lot of new relations according to experimental data. For example:

1. **Malik (1982)** [1]: the following relations were proposed according to theoretical derivation and relevant empirical formulas:

$$\mathrm{Nu} = 0.21 \cdot (\mathrm{Gr}' \cdot \mathrm{Pr})^{\frac{1}{4}} \quad \left(10^4 < \mathrm{Gr}' < 3.2 \times 10^5\right) \tag{4.44}$$

$$\mathrm{Nu} = 0.075 \cdot (\mathrm{Gr}' \cdot \mathrm{Pr})^{\frac{1}{3}} \quad (3.2 \times 10^5 < \mathrm{Gr}' < 10^7) \tag{4.45}$$

$$h_e = 0.013 h_c \tag{4.46}$$

where Gr' is the modified Grashof number. It is obvious by comparing Eqs. (4.46) and (4.43) that the coefficient 0.016 in Dunkle's expression is slightly bigger than the 0.013 in Eq. (4.46). Malik thought the reason would be the approximate assumption of $(p_T - p_w)(p_T - p_g) \approx p_T^2$ in the derivation process.

2. **Clark (1990)** [8]: by the steady-state test using solar simulation, the relation between the evaporation heat transfer coefficient h_e and the natural convection heat transfer coefficient h_c is concluded as:

$$h_e = 0.008 h_c \tag{4.47}$$

The water production indicated in the equation is only half of that in the Dunkle relation.

3. **Adhikari and Kumar (1990)** [9]: the following relation has been summarized by the steady-state test:

$$\mathrm{Nu} = 0.21 \cdot (\mathrm{Gr}' \cdot \mathrm{Pr})^{\frac{1}{4}} \quad (10^4 < \mathrm{Gr}' < 2.51 \times 10^5) \tag{4.48}$$

$$\mathrm{Nu} = 0.1255 \cdot (\mathrm{Gr}' \cdot \mathrm{Pr})^{\frac{1}{3}} \quad (2.51 \times 10^5 < \mathrm{Gr}' < 10^7) \tag{4.49}$$

where the relation between the evaporation heat transfer coefficient h_e and the natural convection heat transfer coefficient h_c is still taken as the Dunkle one.

4. **Tiwari (1991)** [10]: by applying Holland's natural convection formula in the tilt and closed cavity, the effect of the glass cover dip θ on the natural convection is considered in this equation:

$$\mathrm{Nu} = 1 + 1.44\left[1 - \frac{1708}{Ra' \cos\theta}\right]^{+} \cdot \left(1 - \frac{1708 \sin(1.8\theta)^{1.6}}{Ra' \cos\theta}\right) + \left[\left(\frac{Ra' \cos\theta}{5830}\right)^{\frac{1}{3}} - 1\right]^{+} \tag{4.50}$$

where the superscript "+" means only a positive number will be taken in the brackets; if it is negative, then the number will be 0. The Dunkle relation still applies to the relation between the evaporation heat transfer coefficient h_e and the natural convection heat transfer coefficient h_c.

5. **Shawaqfeh, Bulk, and Motion (1995)** [11]: the following relation has been evolved by experimental research and theoretical derivation:

$$h_c = 0.057 \mathrm{Ra}' \tag{4.51}$$

$$h_e = \frac{M_w \cdot h_{fg}}{2M_{a-w} \cdot C_{pa-w}}\left[\frac{1}{(p_a)_w} + \frac{1}{(p_a)_g}\right] \cdot h_c \tag{4.52}$$

In which M_{a-w} is the molar mass of moist air.

6. The literature [12] established the empirical formula of natural convection heat transfer of the air between parallel plates according to the literature [13], modified the Rayleigh number Ra, and proposed the following relation, which has been verified by steady-state test.

$$\mathrm{Nu} = 0.2\mathrm{Ra}'^{0.26} \quad \left(3.5 \times 10^3 < \mathrm{Ra}' < 10^6\right) \tag{4.53}$$

The Dunkle relation still applies to the relation between h_e and h_c.

The relations have been verified experimentally in a different extent, and the results are in good agreement with the experimental values under their respective conditions. However, all of them have some limitations, which are only applicable to their own experimental conditions. Therefore it is necessary to find the relation with a more extensive universality.

4.4.5 A New Group of Relations Forecasting Water Production

By comprehensive analysis of the literature, it is not hard to find that researchers have made approximate assumptions in different extents when establishing the heat and mass transfer relations, which is the important reason why their applications are limited. For example, the Dunkle relation made the assumption that the average temperature is 50°C, and temperature difference is 17°C, also the characteristic size x_l of the free water surface and the transparent cover is ignored when calculating the natural convection heat transfer coefficient h_c, which disables the reflection of the influence of the shape size on water production. The assumption $(p_T - p_w)(p_T - p_g) \approx p_T^2$, which Malik made, is only tenable when temperature is relatively low. Although the relations proposed by Adhikari consider the effect of the geometric size, it ignores the fact that the relation between the evaporation heat transfer coefficient and the natural convection heat transfer coefficient varies with temperature. Starting with the analysis on deficiencies of the earlier relations, the literature [14] has derived a set of relations, which have a relatively wide scope of application.

1. Empirical relation of natural convection heat transfer in limited space
   According to the experiment of the air transferring heat by natural convection between parallel plates, the literature [13] sorted out an empirical formula of natural convection heat transfer:

$$\mathrm{Nu} = 0.2 \times \mathrm{Ra}^{0.26} \quad \left(3.5 \times 10^3 < \mathrm{Ra} < 10^6\right) \tag{4.54}$$

   or

$$h_c = \frac{k}{x_l} \times 0.2 \times \mathrm{Ra}^{0.26} \tag{4.55}$$

   where k is thermal conductivity of the working medium (W/m°C) and x_l is the characteristic geometry size.

This relation is relatively simple, and the applicable Rayleigh number (Ra) has a pretty large range; more importantly, the exponential term of the Ra number is not $\frac{1}{3}$, so the characteristic geometry size x_l will be included in Eq. (4.55), which is very advantageous to the analysis on stills with different shapes. Therefore when deriving a new set of heat and mass transfer relations, the adoption of Eq. (4.55) as the basis of the relation between the convection heat transfer coefficient and Ra is beneficial. However, considering the abundant vapor existing in the still, Ra should be modified. According to the literature [1], the modified Ra′ is:

$$Ra' = \frac{x_l^3 \rho^2 g \beta}{\mu^2} \cdot \Delta T' \tag{4.56}$$

In,

$$\Delta T' = (T_w - T_g) + \left(\frac{p_w - p_g}{\frac{M_a \cdot p_T}{M_a - M_w} - p_w} \right) \cdot T_w \tag{4.57}$$

In the equation, take the average cavity height of the still as x_l; ρ and μ are the density and dynamic viscosity of the moist air under average temperature of the evaporation and condensation surface, and their values are set according to the average temperature of moist air. The subscripts w and g represent the physical states of the evaporation and condensation surface, respectively. M_a and M_w are the molar mass of the air and vapor; p_T is the atmosphere pressure.

2. Water production of the still

   Make an analogy of heat and mass transfer based on the boundary layer

$$\frac{\text{Nu}}{\text{Pr}^n} = \frac{\text{Sh}}{\text{Sc}^n} \tag{4.58}$$

where Sh is the Sherwood number and Sc is the Schmidt number. Substitute the numbers into the definition of Nu and Sh, obtaining

$$\frac{h_c x_l / k}{\text{Pr}^n} = \frac{h_m x_l / D_{wa}}{\text{Sc}^n} \tag{4.59}$$

Transpose the terms, obtaining

$$\frac{h_c}{h_m} = \frac{k}{D_{wa}} \cdot \left(\frac{\text{Pr}}{\text{Sc}} \right)^n \tag{4.60}$$

where k is the thermal conductivity of moist air, D_{wa} is the binary diffusion coefficient of water to air (m^2/s), and h_m is the convection mass transfer coefficient (m/s).

Considering $\frac{\text{Sc}}{\text{Pr}} = Le$, Le is the Lewis number, and $Le = \frac{a}{D_{wa}}$, a is the thermal diffusivity, so

$$\frac{h_c}{h_m} = \frac{k}{D_{wa}} \cdot \frac{1}{Le^n} = \frac{\rho C_{pa-w}}{D_{wa}} \cdot \frac{a}{Le^n} = \rho C_{pa-w} \cdot Le^{1-n} \tag{4.61}$$

C_{pa-w} is the specific heat capacity of moist air under average temperature. According to Eq. (4.54), $n = 0.26$, therefore

$$\frac{h_c}{h_m} = \rho C_{pa-w} Le^{0.74}. \tag{4.62}$$

Evaporation rate per unit area of the evaporation surface in the still is

$$\dot{m}_e = h_m \cdot \left(\rho_w - \rho_g\right) \tag{4.63}$$

Using Eq. (4.59), obtaining

$$\dot{m}_e = \frac{h_c}{\rho C_{pa-w} Le^{0.74}} \cdot \left(\rho_w - \rho_g\right). \tag{4.64}$$

where ρ_w and ρ_g are the densities of vapor corresponding to the states of evaporation and condensation surface. As for the basin type solar still, ρ_w and ρ_g can be solved by the state equation of the ideal gas, i.e., $\rho = \frac{M}{R_M} \cdot \frac{p}{T}$, substituting into the earlier equation:

$$\dot{m}_e = \frac{h_c}{\rho C_{pa-w} Le^{0.74}} \cdot \frac{M_w}{R_M} \cdot \left(\frac{p_w}{T_w} - \frac{p_g}{T_g}\right). \tag{4.65}$$

where M_w is the molar mass of vapor, R_M is the universal gas constant, and ρ is the density of moist air. Attention should be paid that the values of all the physical parameters in this section need to be taken under average temperature, except the ones with subscripts indicating the temperature range.

4.4.6 Experimental Verification of the Theoretical Relations

As discussed earlier, many researchers have done theoretical investigations on the water production of the basin type solar still and have given various sets of heat and mass transfer relations describing performance of the still, such as the relation Eqs. (4.44)–(4.46) by Malik, Eq. (4.47) by Clark, and Eqs. (4.48) and (4.49) by Adhikari, etc.

Among which, the most famous one is the Dunkle relation, i.e.,

$$\dot{m}_e = 1.63 \times 10^{-3} \frac{h_c}{h_{fg}} \left(p_w - p_g\right) \tag{4.66}$$

where

$$h_c = 8.84 \times 10^{-4} \left[(T_w - T_g) + \frac{(p_w - p_g) T_w}{268.9 \times 10^3 - p_w} \right]^{\frac{1}{3}} \quad (4.67)$$

which are Eqs. (4.35) and (4.42).

Experimental results show that the Dunkle relation can predict the water production of the basin type still within a relatively large temperature range, especially the low-temperature range ($T_w \leq 50°C$). However, in the high-temperature range ($T_w > 70°C$), the water production predicted by the Dunkle relation is 30% higher than the experimental result.

To find the heat and mass relation predicting water production of the basin type solar still in a more precise way, Adhikari (1990) designed the experimental device, as shown in Fig. 4.4, to do the steady-state simulation research. This device simulated solar heating by an electric heater, each basin is made of aluminum sheet, the bottom basin is square, with the area of 0.1681 m^2, and the average height of the cavity between the top and bottom basin is 0.07 m. The top basin is full of water and exposed to the environment to keep a lower temperature of the condensation surface.

Adhikari compared the experimental result with the prediction results of the heat and mass transfer relations proposed by many other researchers and found that their relations cannot predict the experimental values well. The prediction of the Dunkle relation is 30% higher, while the prediction result of the Clark relation is 30–50% lower. Therefore he summarized a set of relations according to the experimental result.

$$\dot{m} = \lambda (\Delta T')^n (p_w - p_g) \quad (4.68)$$

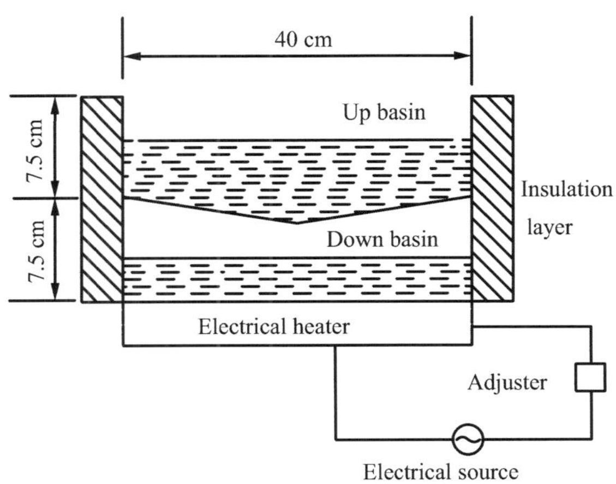

Figure 4.4
The steady-state simulation research device of Adhikari.

where $\Delta T'$ is the modified temperature difference and the value of n is taken according to the following condition:

$n = \frac{1}{4}$, when $10^4 \leq G_r^1 \leq 2.51 \times 10^5$

$n = \frac{1}{3}$, when $2.51 \times 10^5 G_r^1 \leq 10^7$, G_r^1 is the modified Grashof number, λ is a constant, and the values are taken as shown in Table 4.1:

The comparison between the water production predicted by Eq. (4.68) and the experimental result is shown in Table 4.2. The data in the table is the test result in 15 min. It can be seen from the table that the theoretical prediction and experimental result are well matched in terms of the overall trend, and the population deviation is within 5%, except some individual experimental points.

Table 4.1: The value of λ between different temperature ranges.

Temperature (°C)	$\lambda \times 10^9$	
	$G_r^1 < 2.51 \times 10^5$	$G_r^1 > 2.51 \times 10^5$
40	8.12	9.78
60	8.15	9.67
80	8.18	9.49

Table 4.2: The comparison between the water production predicted by equation and the experimental result [9].

The steady temperature (°C)		The water production of experimental result (g)	The water production predicted by equation (g)
T_w	T_g		
26.5	11.0	*6.0*	5.2
45.0	33.0	*11.0*	11.6
46.0	16.0	*26.0*	26.0
52.0	37.0	*16.5*	19.0
55.0	40.0	*18.0*	23.3
61.0	41.0	*35.0*	37.2
62.0	44.0	*26.0*	39.0
70.0	52.0	*48.0*	57.0
75.0	55.0	*76.0*	76.7
75.0	57.0	*68.0*	67.2
77.0	59.0	*79.0*	76.7
81.0	64.0	*122.0*	138.2
84.5	69.0	*141.0*	152.5
86.0	71.0	*154.0*	153.0
86.0	70.0	*155.0*	168.0
88.0	73.0	*165.0*	168.4
88.0	70.0	*206.0*	207.0
90.0	76.0	*208.0*	176.4
90.5	73.0	*228.0*	234.5
92.0	76.0	*236.0*	226.3

T_w, temperature of the evaporation; T_g, condensation surface; g, gram.

It is worth noting that, to verify the accuracy of the prediction of Eq. (4.65), the literature [14] also did a steady-simulation study by making a multistage-stacked tray still with three evaporation cavities. The experimental device is illustrated in Fig. 4.5, and the geometry sizes of each basin are shown in Fig. 4.6.

A comparison of the experimental test value and theoretical prediction value indicates that the relation given by Eq. (4.65) can predict the water production of the still in a relatively large Rayleigh number scope ($3.5 \times 103 < Ra < 2.26 \times 107$) and wide temperature range ($35°C < Tw < 86°C$). It is to be pointed out that, although Eq. (4.65) takes Eq. (4.55) as the basis, and its Ra range is $3.5 \times 103 < Ra < 106$, when applying to the Ra number above 106 and up to 2.26×107, it can still give a pretty satisfactory result.

Figs. 4.7–4.9 give the comparison between the measured values and the calculation values of the water production of each still, under the conditions of the device shown in Fig. 3.5

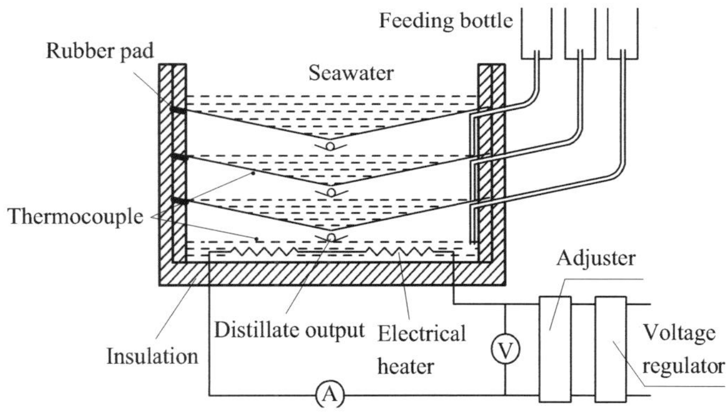

Figure 4.5
The sectional view of a multistage-stacked tray still.

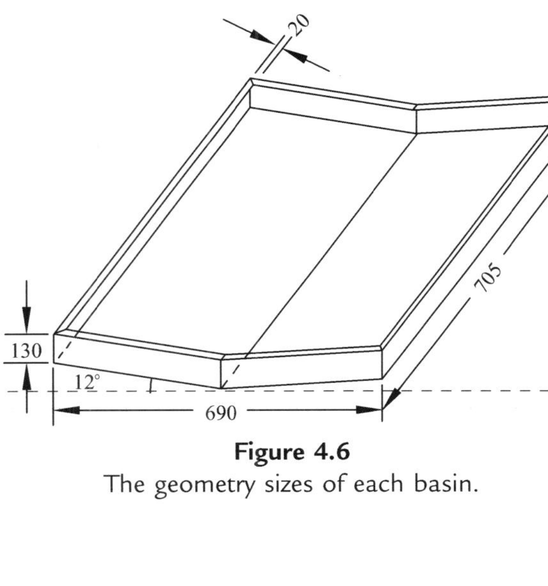

Figure 4.6
The geometry sizes of each basin.

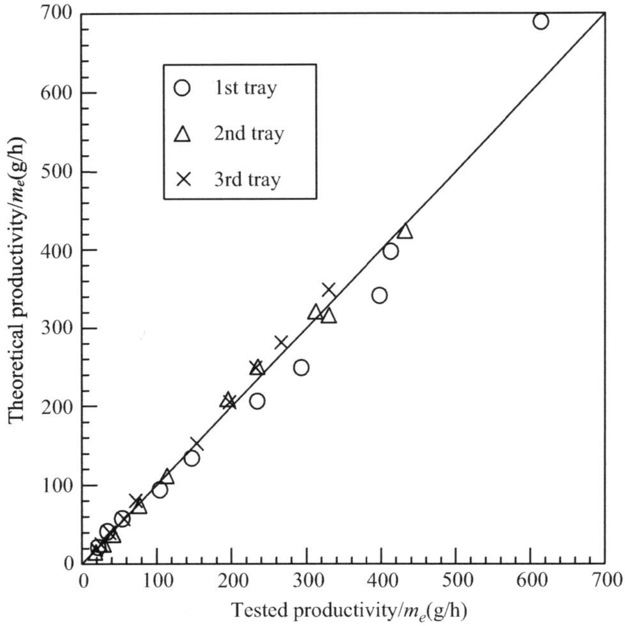

Figure 4.7
The comparison between the measured values and the calculation values of the water production of a three-stage distillation still.

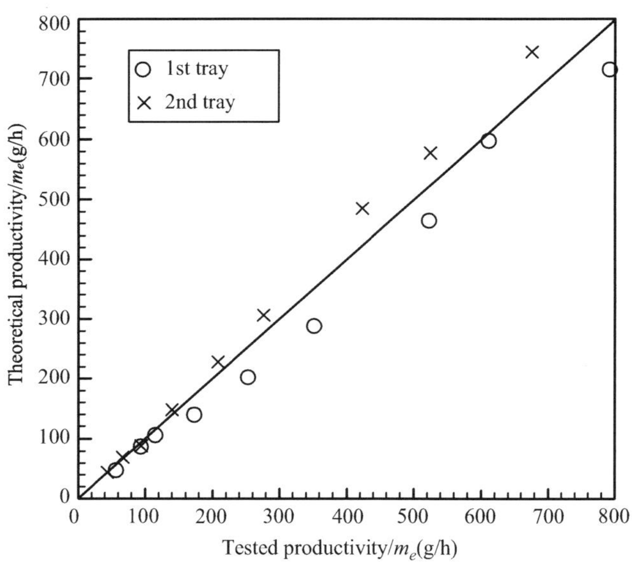

Figure 4.8
The comparison between the measured values and the calculation values of the water production of a two-stage distillation still.

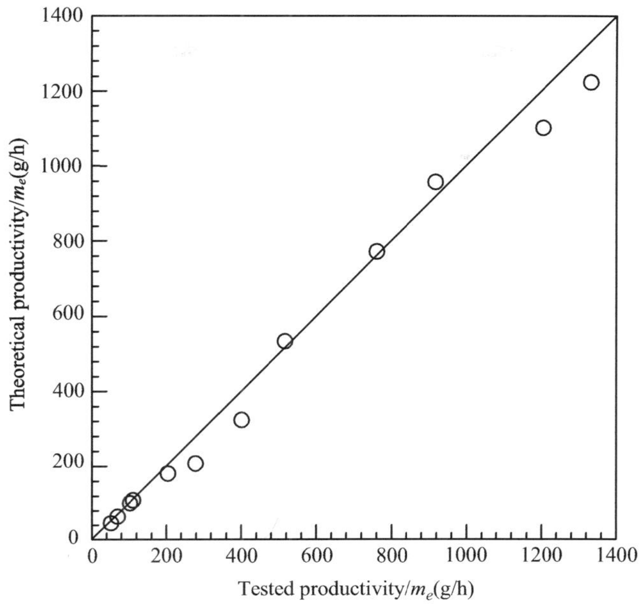

Figure 4.9
The comparison between the measured values and the calculation values of the water production of a single-stage distillation still.

taking three-stage, two-stage, and single-stage distillation. Related thermal physical parameters shall be calculated according to the following formula [15]. The calculation of all the physical parameters is done under the average temperature T_{av}.

$$C_{pa-w} = 0.9992 \times 10^3 + 1.4339 \times 10^{-1} \times T_{av} + 1.101 \times 10^{-4} \times T_{av}^2 - 6.7581 \times 10^{-8} T_{av}^3 \quad (4.69)$$

$$k = 0.0244 + 0.7673 \times 10^{-4} \cdot T_{av} \quad (4.70)$$

$$\mu = 1.718 \times 10^{-5} + 4.602 \times 10^{-8} \cdot T_{av} \quad (4.71)$$

$$\rho = \frac{353.44}{T_{av} + 273.16} \quad (4.72)$$

$$h_{fg} = 2324.6\left(1.0727 \times 10^3 - 1.0167 \cdot T_{av} + 1.4087 \times 10^{-4} \cdot T_{av}^2 - 5.1462 \times 10^{-6} \cdot T_{av}^3\right) \quad (4.73)$$

The saturation pressure of the vapor between the evaporation and condensation surface is calculated by the following formula [4]:

$$p = 6893.03 \exp\left(54.63 - \frac{12301.69}{T'} - 5.17 \ln T'\right) \quad (4.74)$$

In which, $T' = 1.8T_{av} + 491.69$.

In the formula, T_{av} is the average temperature in the still, i.e., $T_{av} = \frac{T_w + T_g}{2}$ (°C).

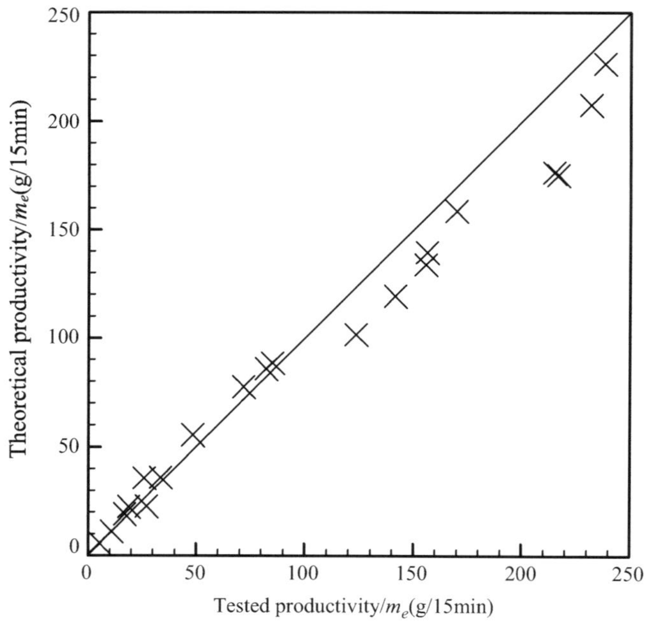

Figure 4.10
Experimental data of the literature [9] is compared with the calculation by Eq. (4.65).

Figs. 3.7–3.9 show that except for a few single points, the calculated values of the distilled water in each experimental stacked dish can agree well with the experimental values, and the average deviation is only 8.5%. So, it can be seen that the formula proposed in this article can predict the water productivity of the still pretty well. Furthermore, Eq. (4.65) can predict the experimental water yield calculated by Table 4.2, and the results are shown in Fig. 4.10. The large deviation between the calculated values and the experimental values only exists in a few single points, while the general trends are in good agreement, which further illustrates the applicability of Eq. (4.65). In addition, because the h_c in Eq. (4.65) is proportional to $Ra^{0.26}$, the h_c will contain x_l, the characteristic geometry size of the device, which overcomes the shortcoming of the Dunkle formula, where the h_c has no relationship with x_l.

4.5 Operation of Basin Type Solar Stills Under Actual Weather Conditions

4.5.1 Daily Operation Situation of Basin Type Solar Stills

As discussed before, the single-stage basin type solar still is the most widely used solar desalination device. Up to 1980, there were solar stills with a lighting area of around 60,000 m^2 operating all over the world except the Middle East [1]. The production practice shows that the performance coefficient of the single-stage basin type solar still is not very high, only approximately 30%. Under good weather conditions, the water production per unit lighting area is around 3.5 kg/m^2 per day, while for the whole year, total water production per unit lighting area is around 1000 kg/m^2. Therefore the improvement of water production rate per unit area is the key to improve the operation economy of the basin type stills.

To study the performance of basin type stills in a more accurate way, many researchers have done a great deal of experimental research and have given a lot of useful conclusions. Malik (1982) introduced the experimental study result of the device shown in Fig. 3.11 and compared it with theoretical analysis. The device possesses a lighting area of 0.56 m^2, made of galvanized sheet iron, using 3-mm white glass as cover plate, with an inclination of 10 degrees and vertical height of the inner cavity of 29 and 15 cm, respectively. Glass wool (5 cm) is used at the bottom and sidewall of the device for insulation. The comparison between experimental and theoretical results is shown in Fig. 4.12, and a comparison between the water temperature and cover plate temperature can be seen in Fig. 4.13, where the circle represents the experimental point, and the curve represents the theoretically calculated result. The theoretical calculation is to take the solar radiation value H and the value of T_a at ambient temperature in association with the expansion of the Fourier series, then substitute the first six terms and solve Eqs. (4.1)−(4.3).The result in the figure indicates that the theoretical result can predict the experimental result.

Tayeb (1992) introduced the operation situation of some types of basin type solar stills under actual weather conditions, as shown in Fig. 4.14 [16], and the design parameters of each still are listed in Table 4.3.

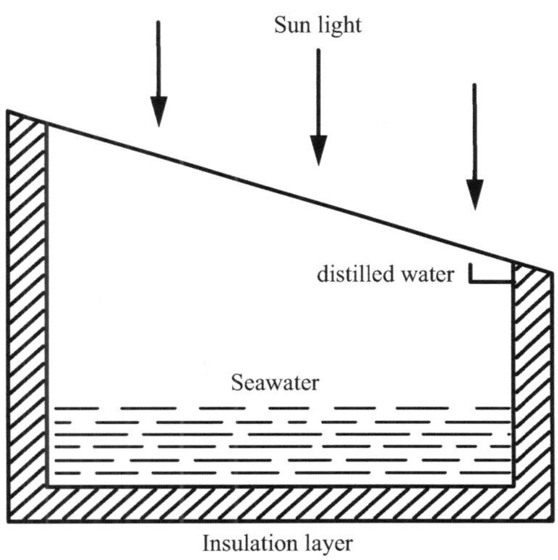

Figure 4.11
Solar still for the experiment through the equation.

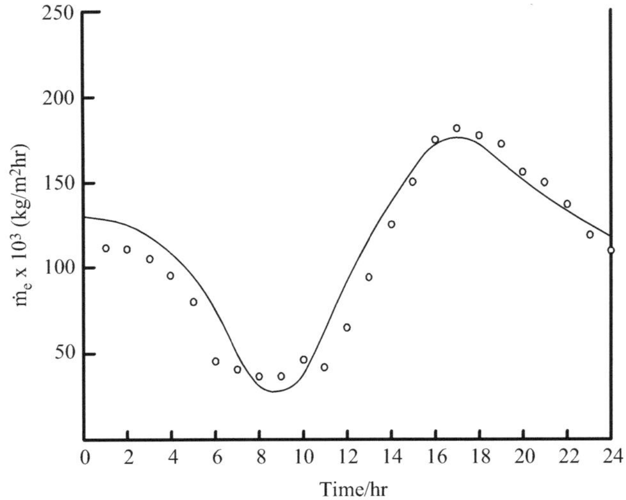

Figure 4.12
Variation of evaporation rate with time.

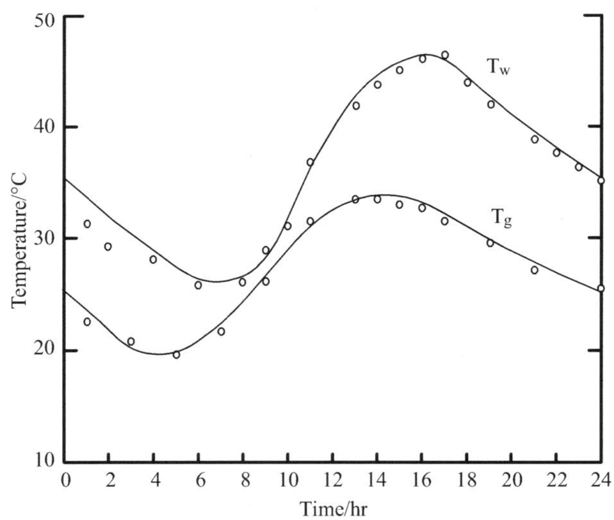

Figure 4.13
Temperature variation of water and glass cover with time.

The experimental result under a typical weather condition is shown in Table 4.4. Efficiency of the device is calculated by the equation:

$$\eta = \frac{m \cdot h_{fg}}{H} \tag{4.75}$$

where H is the average total solar radiation of the device with an area of 0.24 m^2, take $H = 14{,}630$ kJ/m^2 d, $h_{fg} = 2300$ kJ/kg.

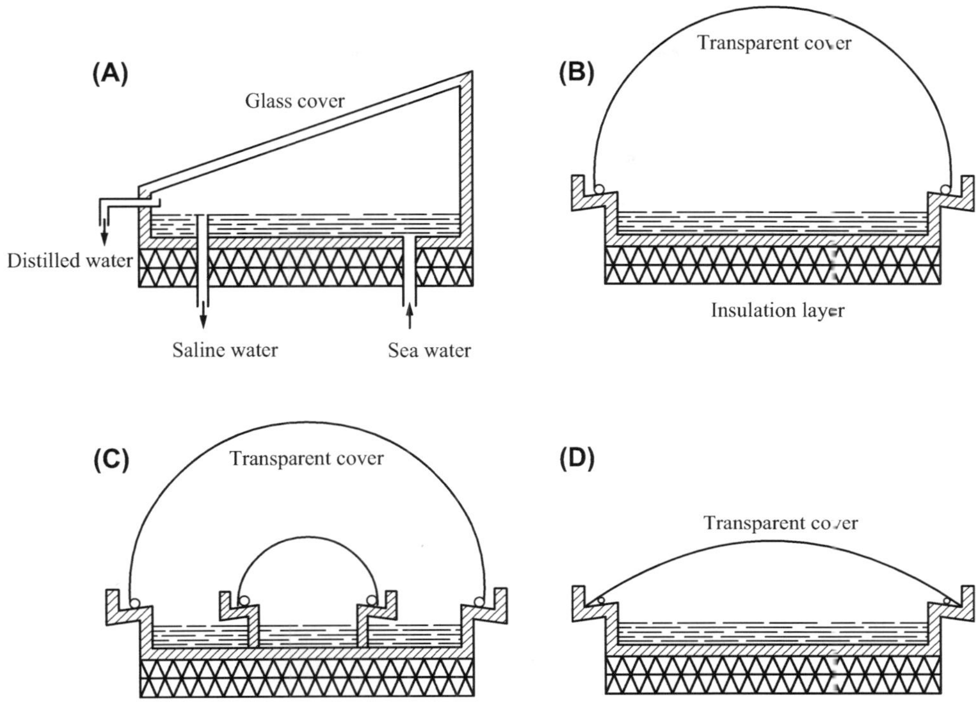

Figure 4.14
Basin type solar stills with a different type of cover.

Table 4.3: The experiment yields and plant efficiency under typical synoptic condition.

	Device number			
	1	2	3	4
Daily output (mL)	313	258	286	206
Total output (mL)	317	271	300	217
Computational efficiency (%)	21.8	18.6	20.6	14.9

As indicated from Table 4.3, in the situation of the same evaporating surface and seawater capacity, the increase of the condensation cover plate area not only cannot raise the water output, but also largely reduces the output, such as the case of device number 2 and 4. This might be because when the condensation surface (single layer) is added, the evaporating temperature of the seawater surface decreases with the increase of the lost heat.

For different types of solar stills with the same meteorological condition, how the water output per hour changes with the local time under a typical weather condition is shown in Fig. 4.15.

As shown in Fig. 4.15, fresh water output started at 8:00 in the morning, reached its peak from 11:00 to 15:00 (local time), and decreased after 15:00. But the still continued to

Table 4.4: Output and total solar radiation over each month.

Month	Fresh water yield (mL/m^2 day) Distill with insulation	Fresh water yield (mL/m^2 day) Distill without insulation	Total solar radiation (MJ/m^2 day)
1	1578	1392	16.58
2	2020	1758	19.24
3	3488	3163	21.30
4	4369	4124	21.18
5	5217	4981	27.04
6	4838	4645	23.53
7	4184	4059	18.68
8	1956	1871	15.23
9	2005	1655	17.59
10	1777	1323	18.53
11	1125	1109	15.05
12	1049	986	14.82
Average	2800	2588	19.06

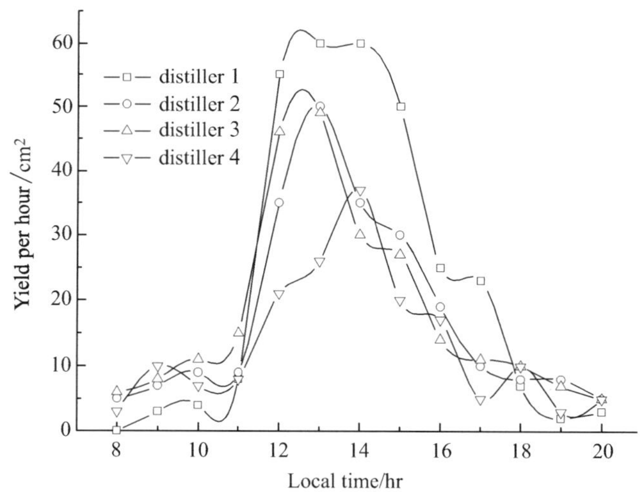

Figure 4.15
Variation of yield per hour with local time.

produce water until 20:00, which might be because the decrease of ambient temperature after sunset relatively increased the seawater temperature in the device.

As various stills have different cover plate areas, and the heat dissipation situations are not the same, temperatures of the cover plate and the water in the basin are different, and they will definitely affect the water output. Temperatures of the water in the basin and the cover plate during a day are illustrated in Fig. 4.16. How the cumulative water production amounts of different stills change over time in a day is shown in Fig. 4.17.

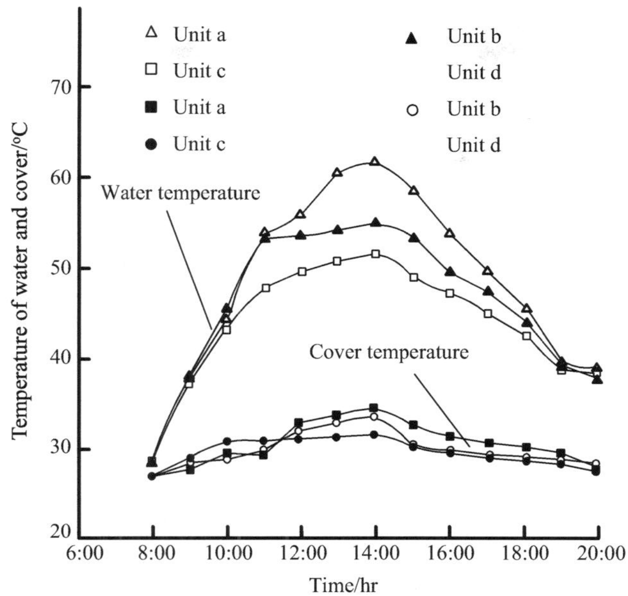

Figure 4.16
The change of water and cover temperature for different solar stills over time.

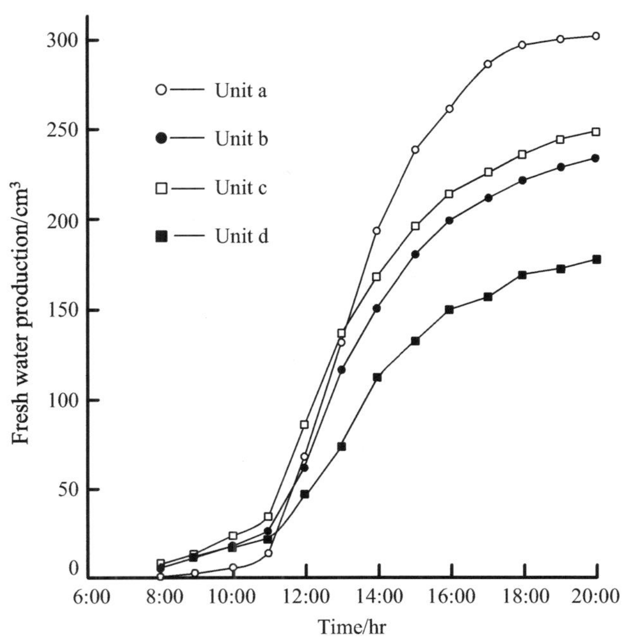

Figure 4.17
The water production by different solar stills over time.

4.5.2 The Whole-Year Operation of Basin Type Solar Stills

Yearly water yield of the basin type solar still is limited by many factors, such as the inclination and orientation of the cover plate, total solar radiation amount, ambient temperature, wind speed, dryness of the air, etc. Only when all of these factors are considered, can the maximum yearly total water yield be reached.

Lighting surface types of the conventional basin type solar still include a single inclined surface, double inclined surface, and V-type. To enable the fresh water condensed under the cover plate to drop into the collecting trough, inclination of the cover plate is normally above 15 degrees. But to reduce the average distance between the seawater surface and the cover plate, the inclination below 20 degrees would be better. It depends on specific circumstances, so the areas with low latitude could take smaller inclination, while the high latitude areas could make it bigger.

Orientation of the still also has a great influence on water production. Generally speaking, most of the solar stills with a single inclined surface should face the south (for the northern hemisphere). Considering the solar radiation can only reach the maximum after 12:00, installation position can be properly adjusted, i.e., 10—15 degrees west of south. But for the double inclined surface solar stills, the cases are not always the same. Some documents indicate that no matter whether it is a theoretical calculation or a measured result, or a south—north or east—west device orientation, there is no big difference among the total sunlight-collecting amounts. In some transition seasons or even for some cases with the east—west orientation, solar radiation could be larger. Therefore orientation or position of the solar still should be analyzed according to concrete situations, with the maximum solar radiation as the target.

In fact, water production of the still also relates to the ambient temperature, wind speed, and relative humidity. Garg (1976) introduced the whole-year operation of two small solar stills with single inclined surface illustrated in Fig. 4.11 [17]. One of them preserved heat with 25-mm insulation materials (sawing shoots) at the bottom of the still, while the other one took no insulation measures. Other parameters of the two devices are the same, with a basin area $A = 0.58\ \mathrm{m}^2$. Other indexes have no differences compared with typical solar stills.

Whole-year operation of the two stills is shown in Fig. 4.18, where the solar radiation is the measured value on the horizontal plane, and the abscissa takes a week as the step length, for 52 total weeks in a year.

The result of Garg's experiment indicates that the still with insulation materials at the bottom has higher water productivity, i.e., yearly water production can be increased by

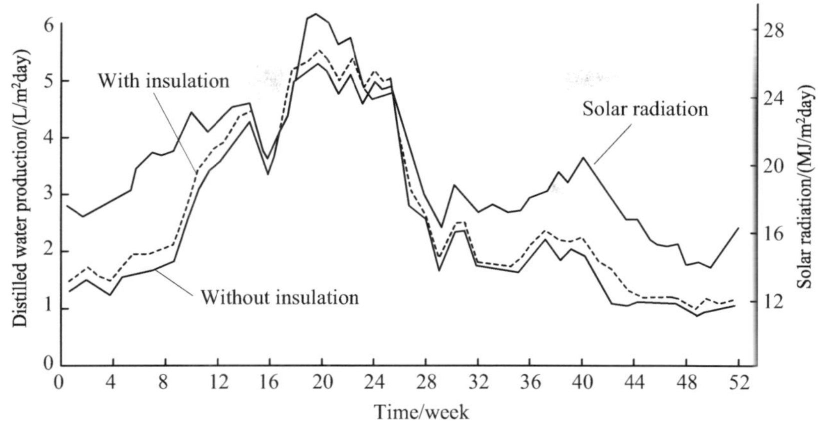

Figure 4.18
Distilled water production and solar radiation changes with time.

approximately 8%, and the device efficiency can be improved by 2–5%; energy utilization efficiency of the still changes in the range of 15.6–51.2%. For the still without insulation, the maximum daily water output is 5.27 l/m^2 per day (in the 19th week), while the minimum is 0.89 l/m^2 per day (in the 49th week).

By measuring the daily output and solar radiation through the whole year, and dealing with the data using least squares fit, Garg derived the approximate relation between distilled water output and daily solar radiation as follows:

$$Y = 0.00032X - 3.60 \tag{4.76}$$

where Y is the water production of the still (L/m^2 d) and X is the solar radiation (kJ/m^2 d). The relation can be used for estimation when the condition is not sufficient enough.

In fact, the ambient temperature and wind speed can also largely affect the water production, especially for the device without good heat preservation effect, which is worthy of special notice.

4.5.3 Influence of Other Parameters on the Performance of Basin Type Solar Stills

As mentioned earlier, many external factors, such as solar radiation, ambient temperature, relative humidity, and wind speed, can influence the water production of a still. However, such external factors are usually hard to control.

Inside the still, many structural parameters also have an effect on its water production rate, and some effects might be considerable. Therefore it is necessary to discuss these parameters respectively.

4.5.3.1 The Effect of Adding Dye on Water Production Performance

In conventional basin type solar stills, most of the solar energy is absorbed by the black substrate. After absorbing solar energy, the basin transfers heat to the seawater inside by convection, which increases the temperature of the seawater and makes it evaporate. There are two problems in collecting the solar energy by the material of the basin bottom. First, though a large part of the heat will be transferred to the seawater, some part of it may also be transferred to the earth or environment from the bottom (proportion of the lost heat depends on the heat preservation condition of the bottom). Secondly, the temperature increase of the seawater is slow because all the seawater in the basin is heated, which will postpone the time of water production. To overcome these two problems mentioned, it is ideal to make the seawater absorb heat by itself; one of the solutions is to add some deep-colored dye in the seawater.

The experimental result indicates that the sunlight absorbing capacity of the seawater has been enhanced because of the dye added, and almost all the devices with added dyes have a higher seawater temperature, which has improved the water production performance [18].

Rajvanshi and Hsieh (1979) did research on adding dyes in the basin type solar stills [1]. They made two identical solar stills, used normal seawater in one of stills, and the seawater with dye in the other. Meanwhile, a comparison was done among dyes with different colors.

In typical weather conditions, a comparison of the water production rates of the two devices is shown in Table 4.5.

The gain—output ratios and the changes of water production amount with seawater depth are shown in Fig. 4.19.

The change of hourly water output with local time during a day is shown in Fig. 4.20.

Table 4.5: Some pigment in the seawater influence on the yield.

Color	Daily output (seawater depth 0.1 m) With pigment (L/m^2)	With pigment (L/m^2)	Water yield ratio R'	Total solar irradiation $KJ/m^2/day$
Red	3.304	3.066	1.08	2.477×10^4
Purple	3.799	3.090	1.23	2.871×10^4
Black	3.862	3.066	1.26	2.477×10^4

Notice: R' = paint device with yield/paint device without yield, called water yield ratio.

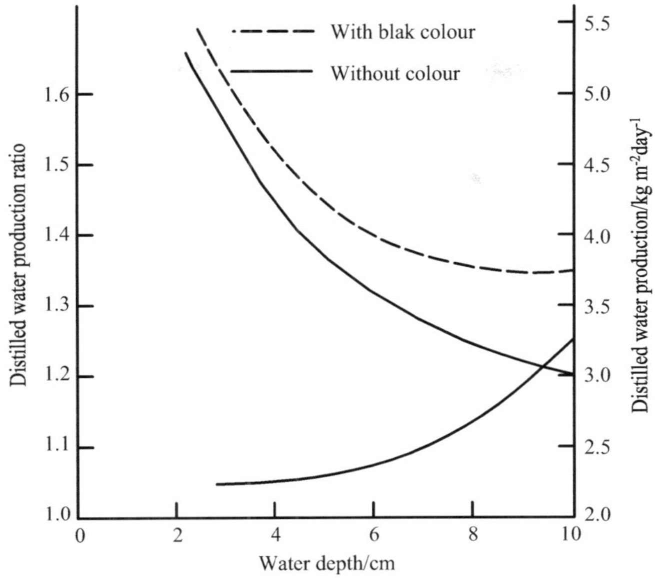

Figure 4.19

The relationship between the distilled water production ratio and solar stills with or without black pigment, as well as the relationship between distilled water production ratio and the water depth.

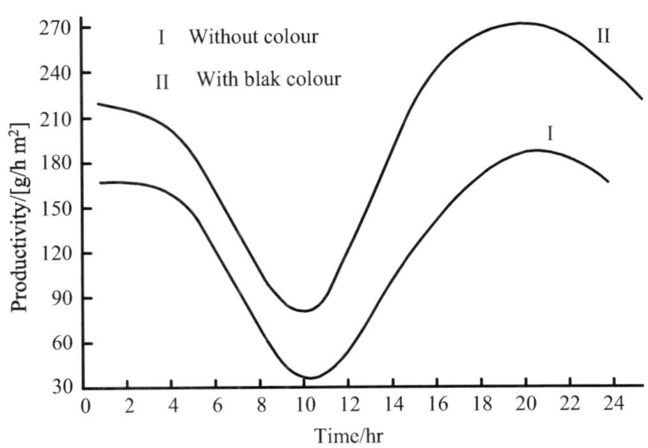

Figure 4.20

The change of water production per square meter with time.

The following conclusions can be drawn from Rajvanshi and Hsieh's experiment:

1. When the seawater depth is 10 cm, adding black dyes in the seawater can increase the water production rate by approximately 26%.
2. When the seawater depth is reduced, the effect of dye on the water production rate is thinner, and when the depth is increased, the water production rate will also be increased. Therefore adding black dyes to the solar stills with larger water depth makes more sense.
3. Adding the black and purple dyes is more effective because black and purple are better for sunlight absorption.

Besides adding dyes in the basin to increase the seawater temperature, some researchers also studied the effects of adding charcoal and a laying sponge on the water production rate and found that these measures could more or less improve the water production performance of the device. Particularly, laying the black sponge has great practical significance because the desalinated seawater can be discharged without special treatment, while discharge of the dyed desalinated seawater may lead to the waste of dyes.

4.5.3.2 Effect of the Seawater Depth in Dish on Water Productivity

The temperature difference between the sea surface and the transparent covering board is the driving force for the evaporation of seawater. Therefore increasing the temperature difference between them will be beneficial to increase the water productivity of the device.

Morse and Read (1968) [1] have studied the influence of seawater heat capacity on the water productivity in the still. They suggested that for the traditional solar still with a certain insulation layer on the ground, the depth of seawater should be 7.6 cm (equal to 1.41 kJ/m^2 heat capacity), but for the still without an insulation layer, 2.54-cm deep seawater is suitable. They also pointed out that when the daily average solar radiation intensity was 225 kJ/m^2 h, if the heat capacity decreased from 1.41 kJ/m^2 to zero (equal to 0.3 m deep water) the water productivity would increase by about 9%, and if the heat capacity increased from 1.41 kJ/m^2 to 5.64 kJ/m^2 (equivalent to 0.3 m depth), the water productivity would decrease by about 7%. Therefore in terms of the actual solar stills, the depth of seawater is not a crucial factor, but it should be as small as possible.

Cooper (1969) also studied the effect of the depth of seawater in stills on water productivity [1]. The results shown in Fig. 4.21 suggest that with the increase of the depth of water in a dish, the water productivity is slightly reduced. Especially for the stills whose bottoms did not take thermal insulation measures, this conclusion is more obvious. But for the stills whose bottoms took thermal insulation measures, the difference of the water productivity is more obvious when the depth of the water is small. The results indicate that when the depth of seawater is increased from 1.27 to 30.48 cm, the difference of the water productivity is about 30%.

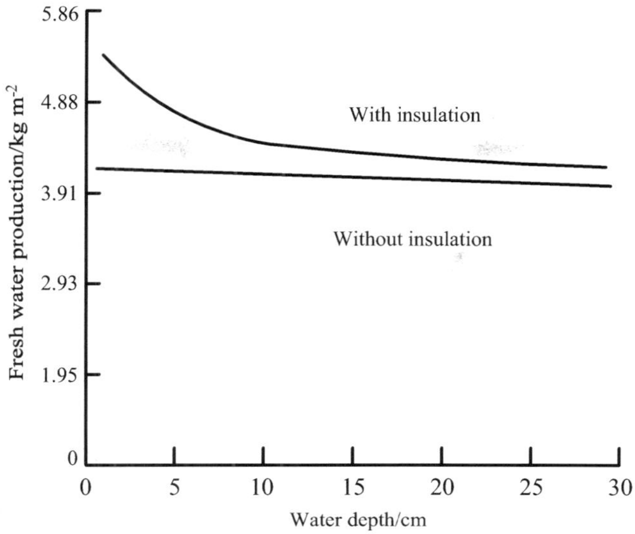

Figure 4.21
The relationship between fresh water production of solar stills with or without insulation layer and water depth.

It is worth pointing out that after entering the 1990s, Tiwari and Singh (1993) et al. had deeply studied the performance of the traditional dish solar still and especially calculated the energy budget in each component of the still [19]. The results show that for a double inclined cover board traditional solar still, energy gain and loss in the main parts of the device varying with the water depth in the dish can be seen in Tables 4.6 and 4.7 respectively. Note that the results of the following tables are obtained from the study of typical climatic conditions in Delhi, India, and there are some changes to the situation in China. So the data is only for reference.

Table 4.6: Variations of the percentage of solar radiation absorbed by different components in the double bevel plate solar still with the water depth in the plate.

The depth of the seawater (m)	Absorption of glass -cover board (%)	Absorption of the seawater (%)	Absorption of the dish bottom wall (%)
0.2	4.75	50.27	15.47
0.1	4.75	47.58	11.99
0.08	4.75	46.51	11.91
0.06	4.75	45.07	12.62
0.04	4.75	42.83	14.97
0.02	4.75	38.91	20.15
0.01	4.75	34.88	25.07
0.008	4.75	33.36	26.76
0.004	4.75	28.25	32.13

Table 4.7: Variations of the percentage of the heat loss in each unit with the depth of water.

Depth of water (m)	Convection of water in bottom (%)	Evaporation loss (%)	Radiation losses (%)	Convection loss (%)	Passing thought bottom loss (%)
0.2	15.47	12.08	14.11	4.85	19.51
0.1	11.99	21.99	18.97	6.83	25.68
0.08	11.91	25.10	19.72	7.15	26.83
0.06	12.62	28.65	20.00	7.26	27.56
0.04	14.97	32.58	19.40	7.05	27.45
0.02	20.15	36.49	17.60	6.51	26.19
0.01	25.07	38.05	16.54	6.25	25.30
0.008	26.76	38.30	16.35	6.22	25.13
0.004	32.13	38.69	16.03	6.18	24.87

4.5.3.3 The Influence of Environmental Wind Speed

Cooper (1969) [1] pointed out that when the wind speed increased from 0 to 2.15 m/s, the water productivity would increase about 11.5%, while from 2.15 m/s to 8.81 m/s, the water productivity would increase only by 1.5%. The conclusion shows that the breeze is good for the increase of water productivity, while the effect of strong winds on improving the water productivity is not obvious. The reason for this probably is that because the environmental wind speed contributes to heat dispersing of the cover board, the temperature difference between the cover board and the evaporation surface is relatively improved. However, when the wind speed is too strong, the cover board loses heat so fast that heat loss of the entire device increases rapidly, which is not conducive to the increase of the seawater temperature in the dish.

4.5.3.4 The Influence of the Environmental Temperature

Morse and Read pointed out that when the environmental temperature changed from 26.7°C to 37.8°C, the water productivity would increase by 11%, while from 26.7°C to 15.6°C, the water productivity would decrease by about 14% [1]. Cooper (1969) also summarized similar results.

4.5.3.5 Effect of Solar Radiation and Heat Loss Coefficient on the Water Production

The driving force for seawater evaporation in the solar still comes from solar energy. Therefore sunlight density certainly is a very important factor influencing the water production. Much research has been reported about the relation of solar radiation and the distilled water production.

Oder h_b represents the heat loss coefficient of the insulation layer of the basin bottom. For the unit shown in Fig. 4.11, there should be,

$$\frac{1}{h_b} = \frac{L}{\lambda} + \frac{1}{h'}$$

where h' is the convection coefficient of the outer boundary of insulation layer and the surroundings. L is the thickness of insulation layer. λ is the conductivity of the insulation layer. Obviously, h_b is one of the important factors to affect the distilled water production of solar stills.

Figs. 4.22 and 4.23 show the variation curve of water production with the solar radiation and heat loss coefficient, respectively. In the figures, T_a represents the ambient temperature and H_s represents the daily total solar radiation.

It can be found from the figures that solar radiation is the most important factor to affect distilled water yield rate. To some extent, the distilled water yield rate is related to the distribution of solar radiation during a day. The yield also has the relation with the ambient temperature. The higher the ambient temperature, the more the water yield. Fig. 4.23 also indicates that the heat loss coefficient has some effect on the water yield. The water yield will decrease when the heat loss coefficient increases, which emphasizes the importance of bottom insulation.

4.5.3.6 Effect of the Temperature of the Glazing Cover on the Performance of Solar Stills

The earlier discussion tells us the glazing cover absorbs about 4.5% solar energy. The condensation process will release latent heat to raise the temperature of the cover when the vapor condenses on the lower surface of the cover. As a result, the temperature

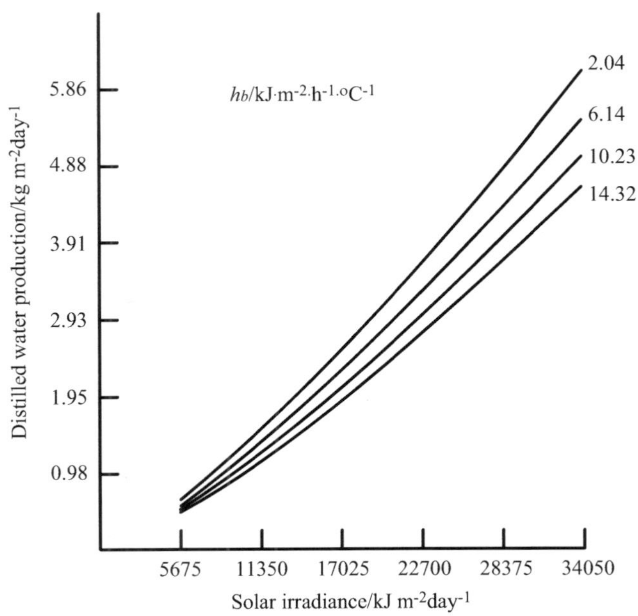

Figure 4.22
Variation of daily water production with solar irradiance.

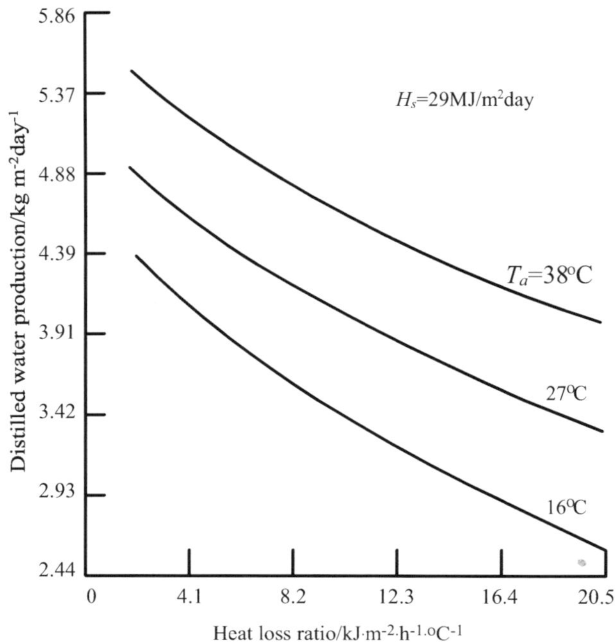

Figure 4.23
Variation of daily water production with heat loss ratio.

difference between evaporation surface and condensation surface will be reduced. That is to say, decreasing the temperature of the cover is good for elevating the distilled water yield.

For decreasing the temperature of the cover, besides increasing the wind speed on the cover (it increases about 11.5% in production, which is presented in the previous section), another effective method is to use the feeding water to cool the cover. To do so, there are two advantages. One is good for the condensation of vapor. Another is to recover part of condensation latent heat so as to raise the inlet temperature of feeding water.

Bassam (1996) [20] and Yadav (1991) [21] completed detailed research about the cover cooled by feeding water. The research results indicate that the solar still generally increases about 20% in production. When the relative humidity is 30%, the thickness of the water film is $x_f = 2 \times 10^{-4}$ m and volume flow rate of the feeding water is $V_f = 2 \times 10^{-6}$ m^3/s m, and the steady efficiency of the solar still changes with solar radiation, ambient temperature, and environment wind speed, as shown in Figs. 4.24–4.26.

The results from the figures indicate that the steady-state efficiency of the unit is improved after using the water film cooling. The effect is better, especially when the environmental temperature is high. It is notable from Fig. 4.26 that for traditional basin type solar stills,

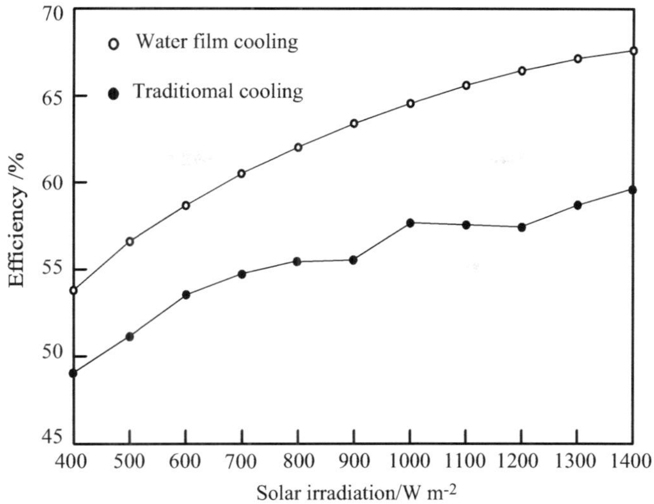

Figure 4.24
Variation of efficiency with solar irradiation.

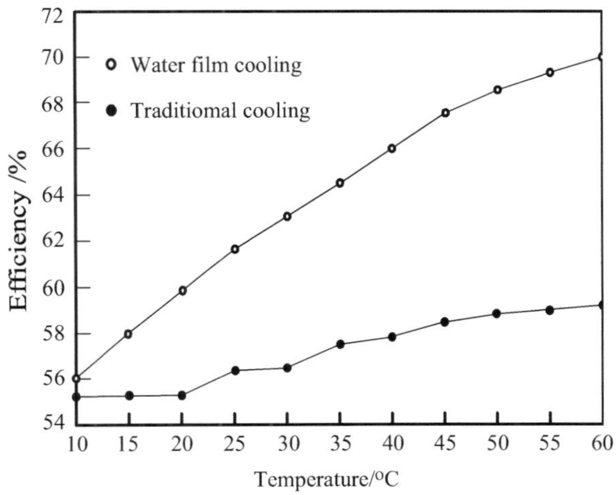

Figure 4.25
Variation of efficiency with temperature.

the steady-state efficiency of the unit increases with the wind speed when the wind speed is small. However, when the wind is big, on the contrary, the steady-state efficiency goes down with the wind speed. The relationship of efficiency and wind speed reaches the peak when wind speed is in range of 3–5 m/s, as shown in Fig. 4.26, which is similar to the results of Clark (1985). As a comparison, the efficiency of a unit with water film cooling does not change with the wind speed.

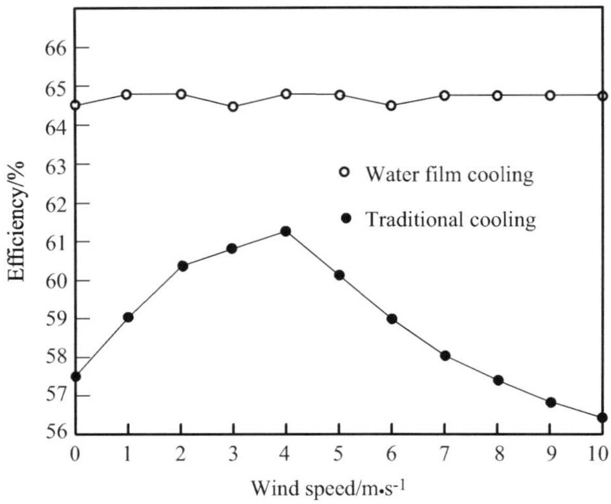

Figure 4.26
Variation of efficiency with wind speed.

4.5.3.7 Effect of Forced Convection of Internal Still on Water-Producing Performance

The heat and mass transfer processes of an internal traditional basin solar still generally are the natural convection processes. The seawater temperature is raised after absorbing the solar energy. At the same time, the evaporation process is formed. The water vapor first goes into the air near the water surface. In this process, the air is heated and humidified. Because the density of vapor is lower and the density of hot air becomes lower due to the expansion, under the action of buoyancy, the moisture air goes up to the near area of transparent cover. The moisture air occurs when the heat and moisture exchange with the glass cover. Part of the vapor is condensed on the lower surface of the cover. Air temperature decreases after the heat and moisture exchanges, and at the same time, the air density increases. Finally, the cold air goes back to the near area of the water surface under the weight action. The process is circulated repeatedly.

Generally, the heat and mass transfer coefficient of natural convection is lower than that of forced convection. Therefore it is propitious to increase the fresh water production when the forced convection process is used in the solar stills. Ali (1991) [22] had done the theoretical and experimental research on this aspect. Ali designed a traditional basin type solar still with 3 m^2 area for finding the difference of using natural convection and forced convection in a solar still. A fan was fixed in the solar still. Under the similar weather conditions, the solar still was tested when the fan operated or did not operate. As a result, the fresh water yield may improve 30–100% according to the forced convection situations, in which the fan forces the air circulation in the still. On typical days the productivity variation with the time and air circulation speed influence the improvement of the distilled water production, as shown in Figs. 4.27 and 4.28, respectively.

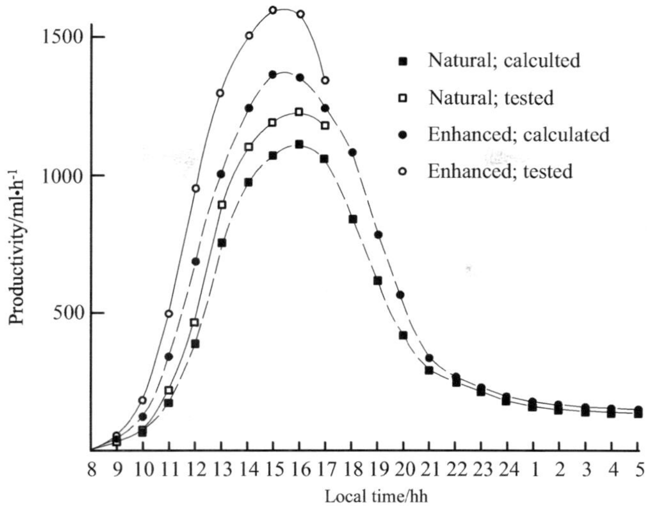

Figure 4.27
Productivity changes with time.

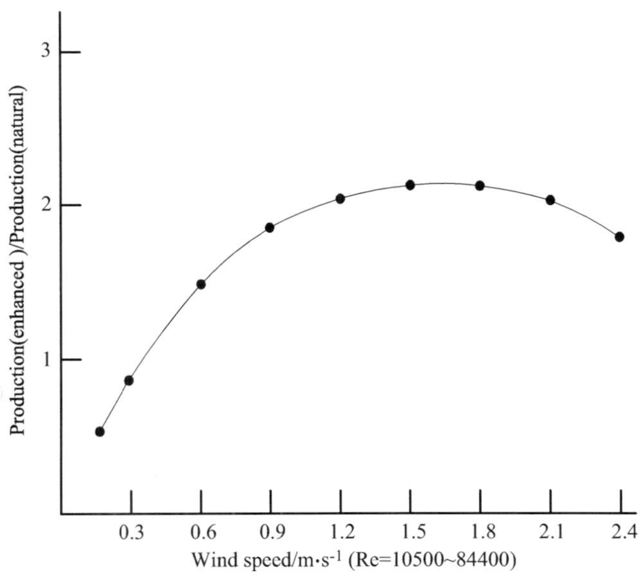

Figure 4.28
Influence of wind speed on productivity.

4.5.3.8 Effect of Other Enhanced Processes on the Distilled Water Production

To further improve the unit area production of basin type solar stills, many enhanced heat and mass transfer methods are proposed. Many works have been completed in the aspect of decreasing the heat capacity of saline water so as to shorten the heating time of saline

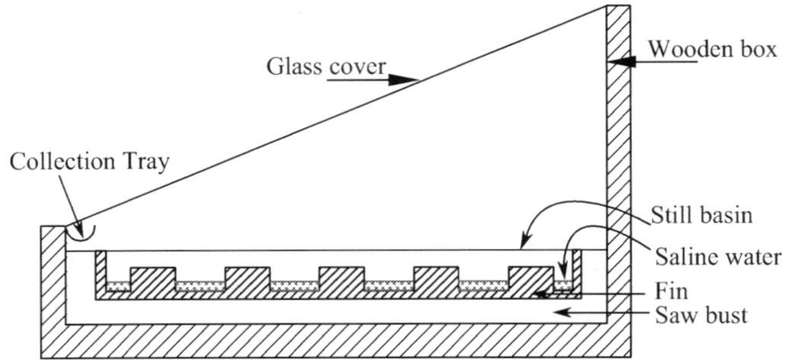

Figure 4.29
Solar still with convex and concave groove.

water. Besides adding pigment into the saline water (this way may cause pollution to the saline water and waste of pigment), concave groove plates usually are used in a water basin for decreasing the preheating time because the plate is able to absorb more solar light. In this way, the production generally is improved about 45% compared to the traditional solar stills.

The design of a solar still with concave groove plates is shown in Fig. 4.29. The common basin is replaced by a special basin with concave groove plates. The seawater is lessened due to the concave grooves, so that the thermal inertia of unit is decreased. Therefore the heating of seawater is much faster, and the temperature difference of seawater and glass cover is enlarged, which greatly enhances the evaporation and condensation processes. Also, the black concave groove reduces the reflectivity of the basin. More sunlight can be absorbed to elevate the utilization efficiency of solar energy.

According to the report of the literature [23], the production of the solar still with a concave groove plate in the basin is more than 45% than that of the traditional one. Under similar weather conditions, the production comparison of two kinds of solar stills with or without a concave groove plate is shown in Fig. 4.30. As a result, the still with a concave groove plate always has a higher production than the traditional one.

Certainly, for decreasing the thermal inertia of seawater in the basin, other methods are also available. The literature [23] gave other schemes, for example, floating porous plate or sponge on the seawater, laying black water absorbent cloth on an evaporating surface, and so on. The comparison experiments had been completed. The comparison results are shown in Table 4.8. It can be found from Table 4.8 that all enhanced schemes improve the production of the solar still, in which adding black concave groove plate and floating porous plate are the best.

To enhance the lighting and heat transfer of the inner still and reduce the seawater storage, the literature [24] introduced a spherical basin solar still with black water absorbent cloth,

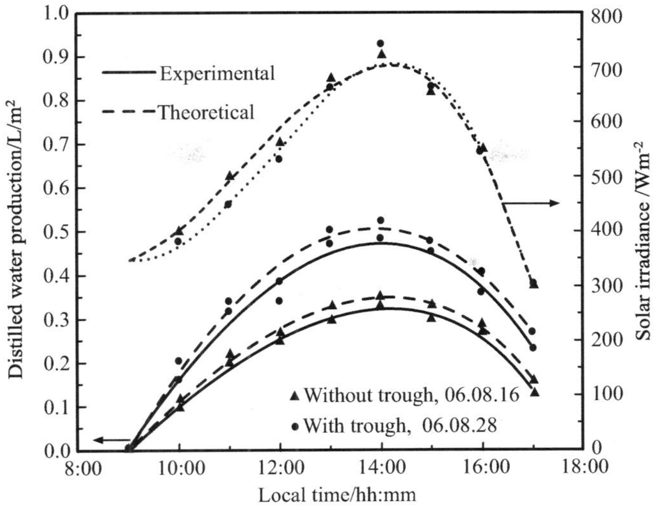

Figure 4.30
Comparison of solar distiller with or without convex and concave groove.

Table 4.8: The production improvement for different enhanced schemes.

No.	Enhanced scheme	Production improving/%
1	Floating sponge	18
2	With flat plate collector	52
3	Laying black rubber or rock smash	20
4	Floating porous plate	40
5	Floating baffling receiver	20
6	Adding black concave grooves	45

as shown in Fig. 4.31. The solar still includes several glass widows to receive the sunlight. Fig. 4.32 is the photo of the unit. According to the report, the solar still has good fresh water productivity. The experimental results on several typical days are shown in Fig. 4.33.

4.6 Multistage Basin Type Solar Stills

The single-stage basin type solar still has drawn much attention because of its simple structure, convenient drawing materials, and automatic operation. However, low water production efficiency and a large occupied area are its main defects. After analyzing the heat transfer process of the solar still, it is obvious that failing to make full use of the latent heat released from the vapor condensation on the cover plate is the most important cause of low water production per unit area.

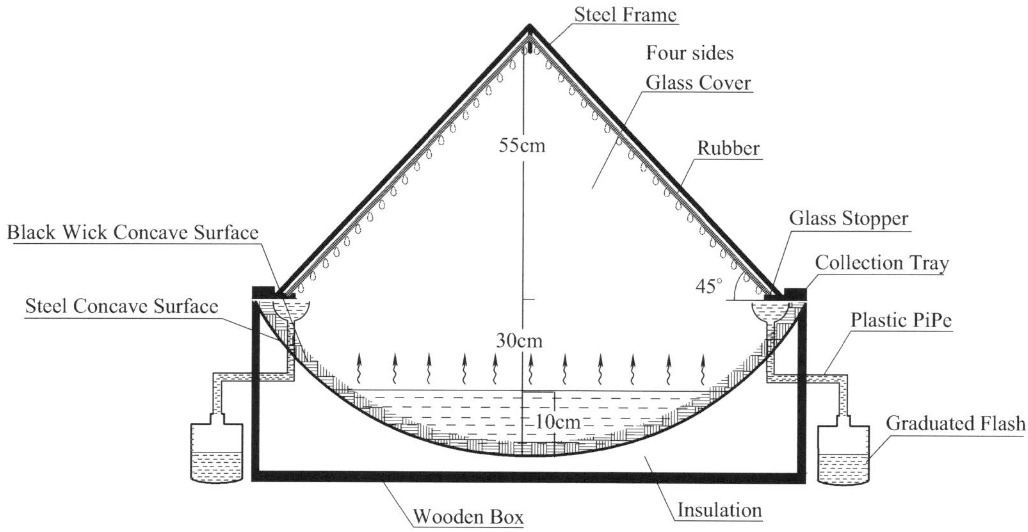

Figure 4.31
A spherical basin solar still with black water absorbent cloth.

Figure 4.32
The photo of the unit.

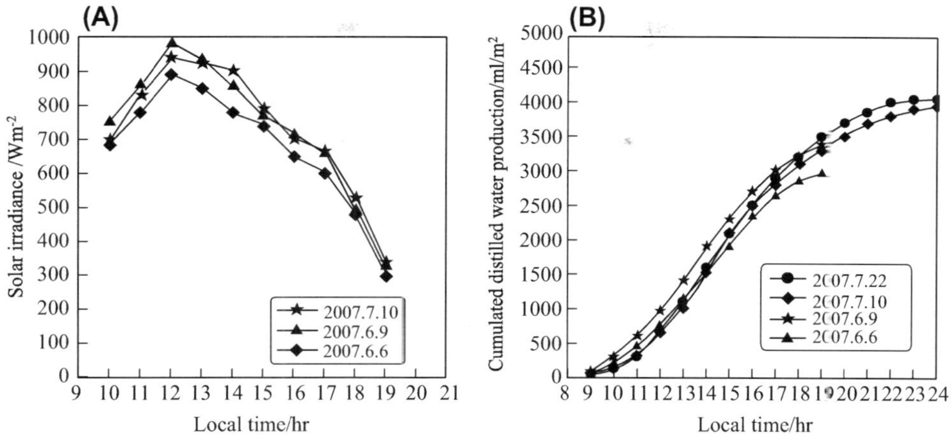

Figure 4.33
The experimental results of the spherical basin solar still: (A) solar radiation variation with local time and (B) cumulated production variation with local time.

In the conventional basin type solar still, when the vapor condenses on the surface of the glass cover plate, the latent heat of condensation transfers to the glass and increases its temperature, which may influence the effect of further condensation. Meanwhile, this part of heat dissipates because of convection and radiation, causing the waste of heat. To make the most of the latent heat of condensation, besides the method mentioned before to cool the cover plate with seawater and reuse it, researchers have also designed various multistage basin type solar stills. Some basic forms are shown in Figs. 4.34 and 4.35.

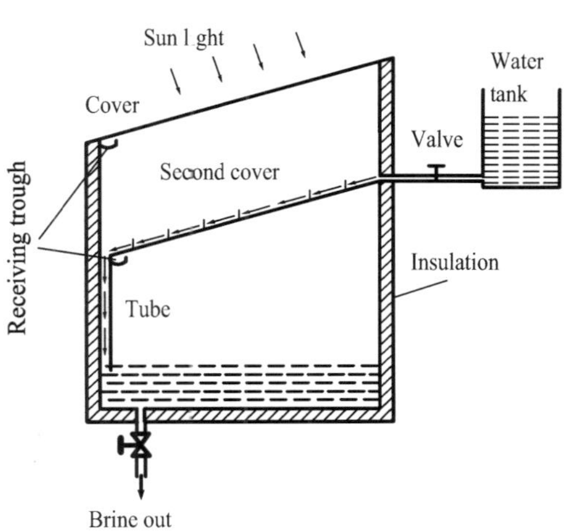

Figure 4.34
Two-stage basin still with single inclined surface [25].

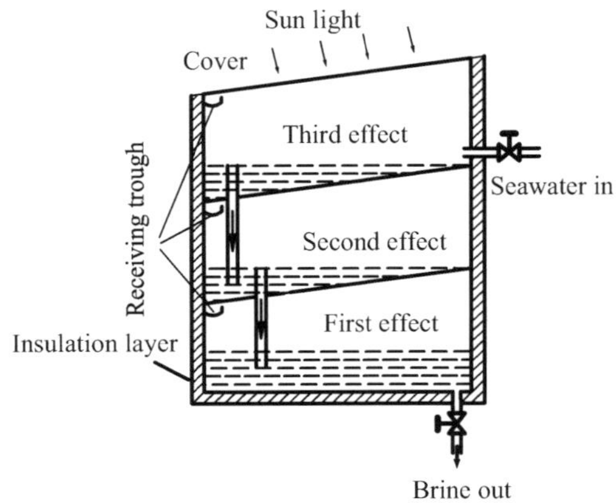

Figure 4.35
Three-stage basin still with single inclined surface [26].

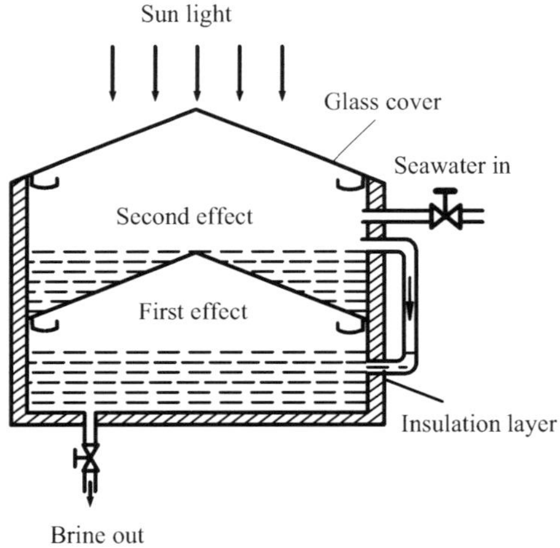

Figure 4.36
Two-stage basin solar still with two inclined surfaces [27].

Analysis and research finds that because of the reuse of the latent heat of vapor condensation, the multistage basin type solar stills shown in Fig. 4.34–4.36 can all obtain a higher water production rate per unit area. The theory of Mahdi (1990) [28] points out that under the condition of good weather, the relation of the stage of basin and the total water production per day can be approximately shown in Fig. 4.37.

As shown in the figure, the increase of the basin stage can help improve the water production per unit area. However, the result of theory and analysis also points out that when it increases to three stages and above, there is no practical significance, which is because the rise of total daily water output is limited with the increase of stages, due to

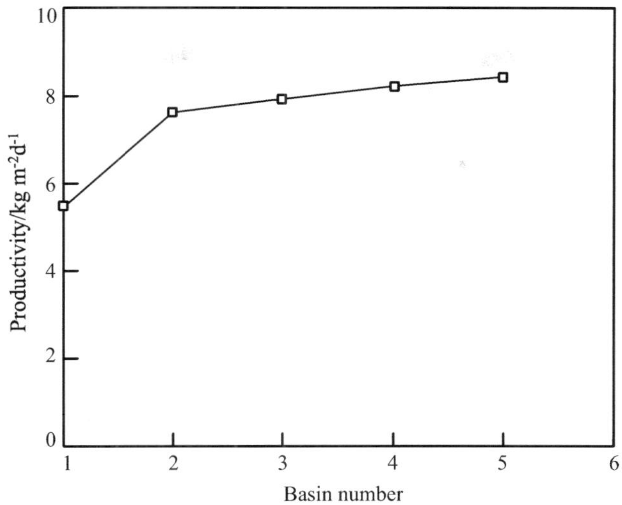

Figure 4.37
The relationship between the number of stages and daily production.

power reduction of the heat and mass transfer caused by the temperature decrease in the device. Generally speaking, such a device only takes two stages, three at most. The research of Kumar (1990) [27] also points out that, as for the double-sloped two-stage basin type still shown in Fig. 4.36, there is a chance that its water productivity can be increased by around one time of the single-stage basin type still.

4.7 Basin Type Solar Stills With Outer Condenser

Based on previous sections, evaporation of water in air has a relation to the difference of partial pressure between the vapor in the water surface and that already existing in air. If partial pressure of the original vapor in air is below partial pressure of saturated water vapor near the water surface, evaporation process can occur. If the difference of this partial pressure is larger, the more intense the evaporation can be. In the traditional basin type solar still, a transparent cover plate is used as the condenser. Though this method makes the system simpler, it has two disadvantages: (1) the latent heat released by water vapor condensation will increase the temperature of the cover plate. And it will result in the partial pressure increasing near the cover plate, making the difference of partial pressure between the evaporating surface and condensing surface become smaller. (2) The water film and droplet on the cover plate, formed by the condensation of water vapor, will make transmissivity of the cover plate decrease under some extent, causing the total solar energy received by salty water in the system lower. These will limit the improvement of the performance of the system. For this reason, Hassan (1992) et al. did research for the traditional basin type solar still with outer condenser. And the schematic diagram of the system they designed is shown in Fig. 4.38.

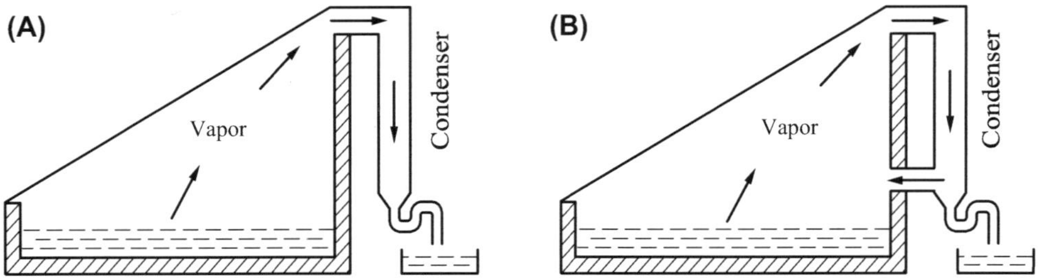

Figure 4.38
Basin type solar stills with outer condenser.

Hassan's theoretical and experimental researches show that, for the scenario in Fig 4.38A, when the condensation surface of the outer condenser is large enough (when it is close to the surface of collecting area of glass cover plate), the outer condenser can increase water production by 30%. For the scenario in Fig. 3.38B, the outer condenser can increase water production by nearly 50%. Thus in practical engineering, it is meaningful to add the outer condenser.

4.8 Basin Type Solar Stills With Porous Absorbent Material

In the traditional basin type solar still, because of the large heat capacity of seawater, the temperature of seawater increases slowly, resulting in the delay of the start time of the water production. To overcome this disadvantage, Sodha (1980) et al. developed solar stills with porous absorbent material (multilayer wicks) [30].

Fig. 4.39 gives the section view of this type of system. The operating principle is that seawater in the system is collected in a water sink rather than evenly distributed in the bottom basin. And the porous absorbent material that has a strong affinity interaction and capillarity to water, like hessian and cotton gauze, is formed into a thin fiber layer. One end is immerged in seawater, and the other end is placed on the top of an incline surface. The fiber layer extends to the bottom of the system. Water is drawn to the top of an inclined plane with the capillary action. Then, under the action of gravity, it flows along the fiber of the slope to the bottom, forming an evenly distributed seawater film. Because the heat capacity of seawater in the film is very small, seawater can easily evaporate under the sun. Thus the start time of water production is advanced. And this helps lower the heat loss through other parts of the system. To increase the absorptivity to sunlight, the fiber material can be dyed to black. The operating point of the system is focused on how to keep the water absorbent core wet, forming an evenly distributed water film on the inclined plane. In Sodha research, after adopting these methods, the water production per unit area of the solar still is higher than traditional basin type stills by 16–50%, and the efficiency is increased by 6.5–18.9%.

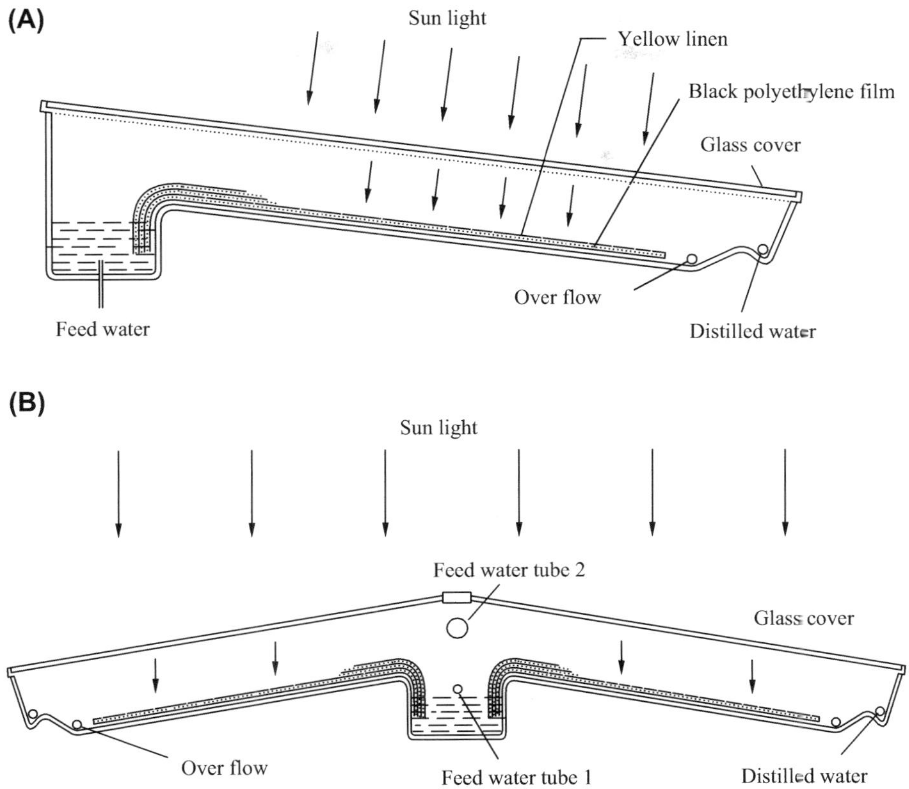

Figure 4.39
Section view of a solar still with multilayer wicks: (A) one inclined plane and (B) two inclined planes.

4.9 Concentrating Type Passive Solar Stills

Concentrating type passive solar stills generally are coupled with a parabolic dish, parabolic trough concentrator, or planar reflector. Because a single parabolic dish concentrator is limited by the dish area so that a single unit is not so big, the solar still coupled with a parabolic dish concentrator is not so popular. A large-scale planar reflection collector needs a center receiving tower. Therefore it is not suitable to be used together with passive solar stills. The left one, a parabolic trough solar concentrator, is the only one easy to be coupled with a passive solar still. Certainly, it not only can be combined with an active solar still, but also with a passive solar still. In an active system, a parabolic trough solar concentrator is used to collect the solar energy for following stills. That is the same as the active solar stills using a flat-plate solar collector or vacuum tube solar collector to gather solar energy for following stills.

Because the parabolic trough solar concentrator may do unlimited extension along the axial, its collection area is not limited by a single unit, which is usually convenient for application. Sometimes, this kind of trough solar concentrator usually consists of several curve surfaces or compound parabolic concentrators, which are so-called CPC. It is a bold attempt for CPC to be used in passive solar stills. A reverse receiving solar still is the typical example combined with CPC.

The glazing cover usually is fixed on the top of the unit in a traditional basin solar still. The solar light goes into the distilling chamber through the glazing cover to heat the seawater. To avoid the heat loss from the bottom of the still, the insulation layer is often installed under the bottom of the unit. This kind of passive solar still generally does not have a concentrating function, so the production per unit area is low. Also, the glazing cover only is made of transparent nonmetallic material. Therefore the cover is not suitable to take out the condensation latent heat rapidly, which is not conducive to rapid condensation of the vapor.

Hence the concept of reverse receiving solar still is provided, which is different than a conventional passive solar still. It uses the back of its bottom to receive the solar radiation. The vapor still condenses on the upper part, shown in Fig. 4.40. In this way, the condensation surfaces may not only be a transparent material, like glass, but may also be a common metallic plate. In Fig. 4.40, the solar light is reflected secondary after it is collected by CPC and sent to the back of the unit bottom where it is absorbed. By this

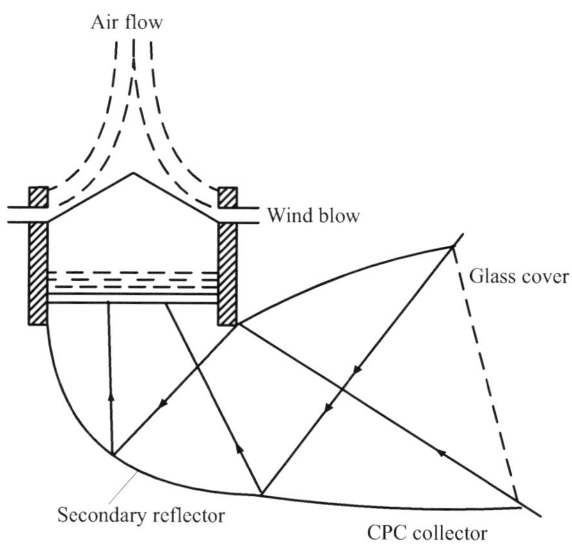

Figure 4.40
Sectional view of a solar still with asymmetric line compound parabolic concentrators [29].

way, the supplying heat density in the bottom is enhanced. Certainly, the enhanced extend is determined by the concentrating ratio. According to the prediction, it is suitable for the concentrating ratio to take 1.5−2. The condensation process in the still may be natural convection cooling, but it is better to be forced air convection cooling, which can decrease the temperature of the condensing surface more quickly.

The reverse receiving solar still is especially suitable to be used in a high-latitude region because the CPC is placed inclined. The solar zenith angle is small in a high-latitude region where the solar light is easy to be sent into CPC. Therefore the efficiency of CPC is high. However, in a high-latitude region the tilted angle of the glazing cover in a traditional basin solar still generally is relatively big for increasing the light collection ratio. As a result, the distance from evaporation surface to condensation surface will be increased, which is harmful to the unit efficiency. However, if using the reverse receiving method, the distance can be controlled effectively. Therefore the fresh water production per lighting area of the still will rise 20−30%, compared to a traditional passive basin type solar still.

Although the problem of heat transfer rate being low in the still is solved by using the reverse receiving method, the problem of heat capacity of seawater being too large in the basin still is not solved yet. It also affects the production.

For that, two improved reverse receiving solar stills were proposed, as shown in Fig. 4.41. The concentrator is still an asymmetrical CPC. In Fig. 4.41A, seawater is not stored in the still, but is flowing on the upper surface of the receiver. After that, the hot seawater flows on a vertical plate with the water film to produce the adiabatic evaporation. For forming the well-distributed water film on the vertical plate, the materials with high hydroscopicity may be used as the cover on the plate. The vapor is condensed on the condensing surface. The condensate flows to the output for the user. For elevating the condensation rate, the area of the condensation surface can be increased properly.

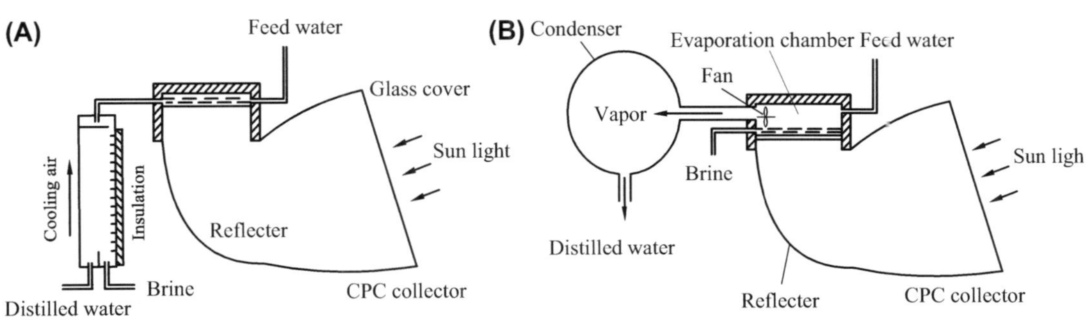

Figure 4.41
An improved reverse receiving solar still [31].

Fig. 4.41B shows the diagram of the reverse receiving solar still with a spherical condenser. The operation of seawater in a still is the same as in Fig. 4.41A. For increasing the condensation area, a spherical condenser is used in the system. The vapor produced in the evaporation chamber is sent into the spherical condenser by natural convection or microfan. The vapor is condensed rapidly on the inside of the spherical condenser. Due to the large area, the condensing process is enhanced greatly.

4.10 Inclined Type Solar Still

The simple disk solar still has some inevitable weaknesses, one of which is that as a collecting solar energy device, when the heat-absorbing surface (surface of water in still) is placed horizontally, the annual interception of total energy of solar radiation is always less than the heat-absorbing surface, which has the same size as the disk solar still and is properly tilted (except near the equator). According to this reason, the inclined solar still was designed. It is difficult to arrange an inclined surface, but there are some different ways to achieve this goal. Fig. 4.42 is a better scheme. The disk of this still is made into a step shape, and then the seawater will be evenly distributed in different steps. After the sun rises, the temperature of seawater will be increased rapidly. When the amount of sunlight reaches 2300 kJ/m^2 in one day, the daily water productivity can reach 4.8 kg/m^2, which is significantly higher than the common disc still.

The inclined still has the following two forms, as shown in Figs. 4.43 and 4.44, which also have high water productivity; they both raise the operating temperatures of seawater by reducing the amount of seawater in the disk.

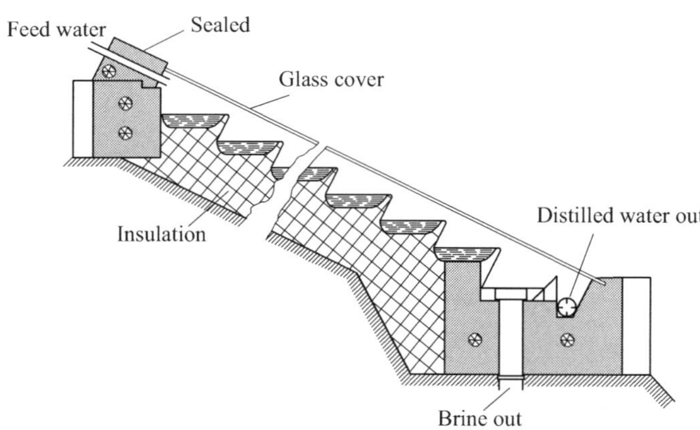

Figure 4.42
An inclined basin solar distiller [30].

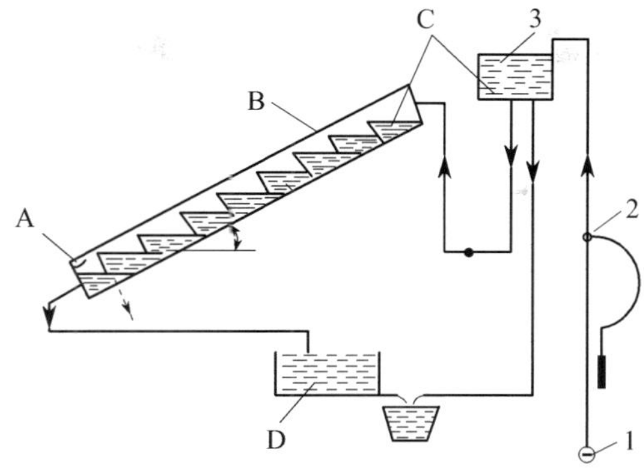

Figure 4.43
Cross section of multiple basin and tilted solar still: (A) distillate; (B) glass cover; (C) seawater; (D) saline water; (1) pump; (2) tree-pass valve; and (3) tank.

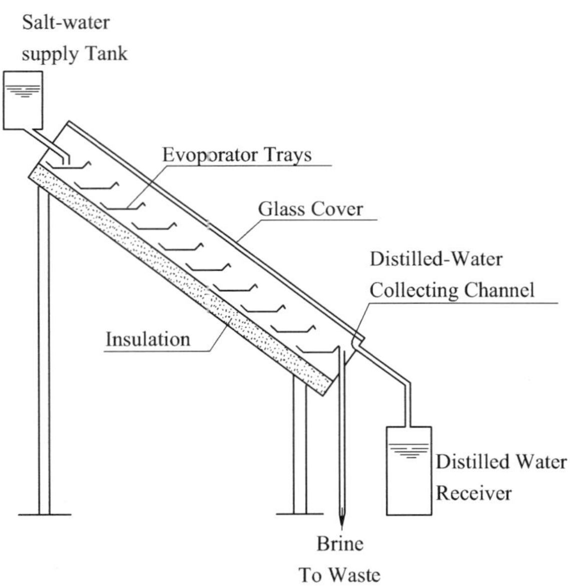

Figure 4.44
Schematic illustration of the natural flow multiple basin type solar still [1].

4.11 Other Shaped Passive Solar Stills

According to different requirements, many other forms of passive solar stills have also been proposed. The literature [4] did the review for these kinds of passive solar stills. Here, only several typical forms of passive solar stills are presented.

4.11.1 Simple Emergency Distiller

During the World War II, the United States produced hundreds of thousands of emergency distillers, as shown in Fig. 4.45A for Air Force and Navy use. After the pilot or sailor fell into the sea, he should blow the plastic cover of the distiller and hang the black towel with seawater in the center of the cover. After doing that, a simple solar still is completed. The simple solar still can be tied in the life raft and flows on the sea. Under the shining of solar light, the vapor will be formed from the wet towel and condensed on the inside of the cover. Finally, the condensate falls down to a bottle at the bottom of the cover. When the towel is dried, take it out, wash it clean, and fix it back again. The still works again. The simple solar still can be folded up when it is not used. Its weight is only 0.5 kg. During a sunny day, it can provide near 1 kg of drinking water.

Field workers will also encounter the problem of fresh water shortage. They need to find the drinking water locally. If the underground soil contains abundant water, they can construct a simple solar still to produce the fresh water. First, dig a hole in the ground and put a cup in the center of the hole. Secondly, cover the hole with a transparent plastic film and put one stone in the center of the film, as shown in Fig. 4.45B. Under the shining of solar light, the water in the soil will be evaporated. The vapor is condensed on the underside of the film, and the distilled water is collected into the cup. Finally, the drinking water is obtained. According to the report [33], if the water content of soil is 12%, a hole with diameter of 0.5 m can produce about 300 mL of drinking water a day.

4.11.2 Solar Still Combined With Green House

A still can develop a small-scale greenhouse cultivating at the places with only salt water or brackish water through the combination distiller and greenhouse. Certainly, the types of

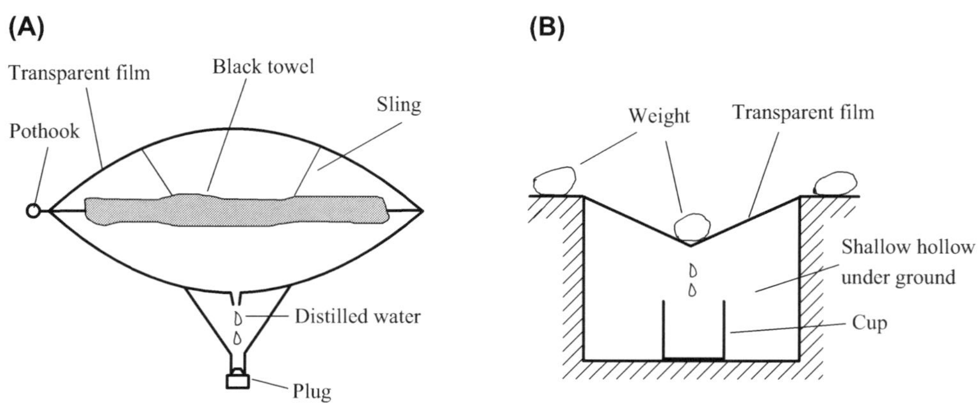

Figure 4.45
Simple emergency solar stills: (A) solar still used in sea survival and (B) distiller of shallow hollow under the ground.

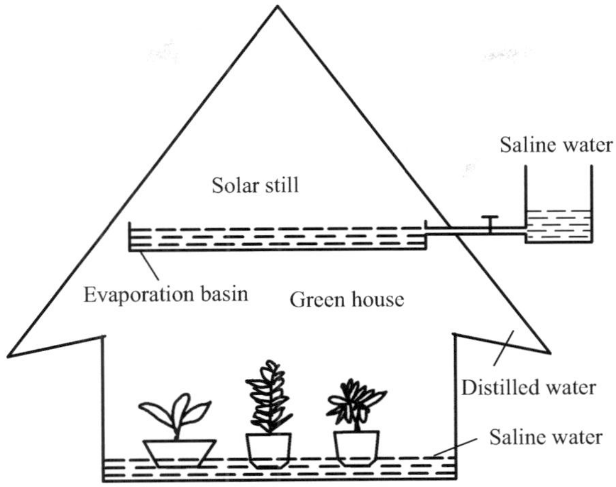

Figure 4.46
Greenhouse to be integrated with solar stills.

crops should be chosen properly, the design of greenhouse should be reasonable, and the environment should be suitable for the growth of crops. In this way, the fresh water that the crops need may be provided by the solar still. Fig. 4.46 shows a combination of a solar still with a greenhouse. It has a double structure in which the upper layer is the distiller and the lower part is the greenhouse. Maybe the bottom of the greenhouse still includes some troughs with brackish water to enhance the evaporating action. Inclined and transparent covers can let the sunlight go into the house and collect the fresh water down to the trough to be used.

If the solar stills are put on top of the flat-top house, they produce the distilled water, and at the same time, they also help to regulate the indoor temperature. The analysis results from researchers of India for the weather conditions of Delhi indicate that the solar distillation pond on the top of a house can reduce heat transfer into the home by 40% on a hot day in June [34]. In this case, the solar still also reflects part of the sunlight back to the sky. However, during the cold seasons, to enhance the light-receiving ability of the solar still is a requirement. For example, adding some dye material into the pond is available. There is an insulation layer between the bottom of the still basin and the flat top of the house, and the indoor temperature is higher than the ambient temperature, so the heat loss from the bottom of the still is quite small. Therefore its fresh water production can exceed that of a traditional solar still and reach 5 kg/m^2 per day. On a cold day, more heat energy is also transferred into a home compared to that without a solar still.

4.11.3 Distiller With Transparent Film as the Cover

A solar still with transparent film as the cover is shown in Fig. 4.47. The fresh water production of this kind of solar still may reach 4.2 kg/m^2 per day under the condition of

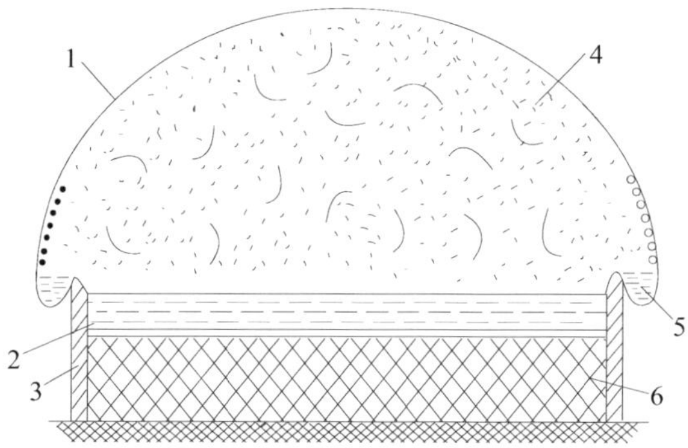

Figure 4.47
Cross section of a film-covered solar still [1]: 1, transparent film; 2, black fabric; 3, saline water; 4, vapor air mixture; 5, trough for distillate; and 6, thermal insulation. *After Achilov BM, Norov EZh. Solar heat-pumping still. Appl Sol Energy 1975;11(6):74–6.*

sunny weather. Under the same weather condition, its production may increase by 50% than that of traditional solar stills. Also, this kind of solar still has the advantage of low cost. The film can be replaced at any time. It is especially suitable to the temporary place.

4.11.4 Spherical Solar Still

The basic structure of the spherical solar still is shown in Fig. 4.48. Its best feature is that the sun in any direction can be very good to enter the device. Its seawater basin is put in

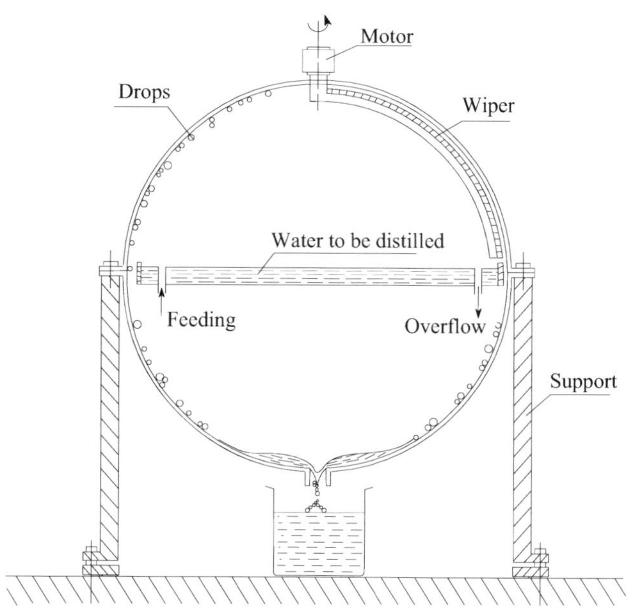

Figure 4.48
Spherical still with wiper (Menguy et al., 1980).

the center of the sphere. Its seawater capacity is easily adjusted. The vapor is condensed in the upper inside of the sphere after seawater has been evaporated. The fresh water flows along the inner wall to a tank under the sphere. Menguy (1980) developed this kind of solar still. A scraper is installed in the upper sphere so that the distilled water goes fast down to the tank to keep the transparence and decrease the thermal resistance. As a result, the scraper can improve production by 25%.

4.11.5 Integral Molding Plastic Solar Still

With the development of plastic integral molding technology, it becomes a reality to use plastic materials to make the solar still. Fig. 4.49 shows a diagram of a plastic solar still. The internal water plate of the still is separated in two parts. One part is used as a seawater container, and another part is a fresh water container. A black absorbent cloth is covered on the inside facing the sunlight and on the outside back to the sunlight, as shown in the figure. The absorbent cloth absorbs the seawater up to all inside facing the sunlight and down to the outside back. In this way, seawater is evaporated inside the wall, and the vapor goes to the back wall with outer evaporation cooling to condense. At the same time, the distilled water flows down to the fresh water pond for the user.

This kind of solar still has very good market value because it has the advantages of simple structure, low cost, and easy to handle, although its productivity is low. The integral plastic solar still will have better prospects if the transparency and aging resistance of plastics are improved further.

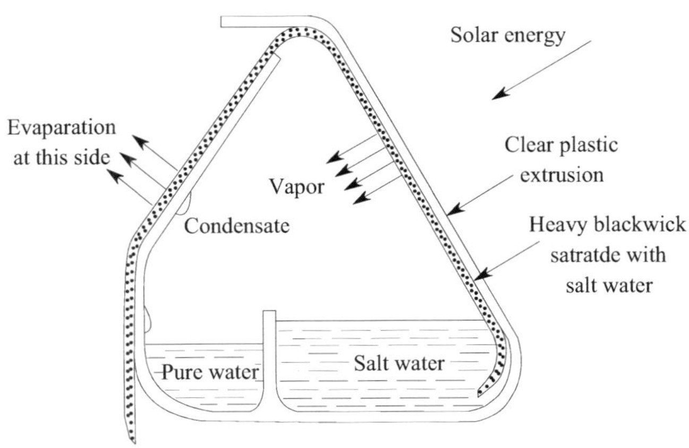

Figure 4.49
Integral molding plastic still with black wick for evaporation and cooling [1].

4.12 The Main Shortages and Improvement Direction of Basin Type Solar Stills

4.12.1 Three Main Shortages

A deep understanding of heat and mass transfer in basin type solar stills is very important to improve its performance of water production. As discussed in earlier sections, the solar energy entering stills is mainly absorbed by the water beside the loss from basin bottom, side wall, and vapor leaking. And then the heat exchange will happen between the water and the glazing cover through evaporation q_e, convection q_c, and radiation q_r. To make the device produce as much fresh water as possible, the solar radiation should be used to heat and evaporate the water in the basin by the maximum possible extent. That is to say, q_e is hoped to be the maximum. For this purpose, the transmission rate τ of the glazing cover and the absorptivity α_w of the basin should be as much as possible. At the same time, q_c and q_r should be as little as possible because they have no benefit for increasing the production.

From Fig. 4.3, we know that decreasing the seawater temperature T_w in the basin will let the terms of q_c and q_r decrease. However, the decreasing speed of q_e is faster than that of q_c and q_r when T_w is decreased. So, T_w is not suitable to be reduced. Also, it is easy to find from Fig 4.3 that the higher T_w is, the bigger the share of evaporation heat transfer is. Therefore for higher efficiency the solar still should operate at a relatively high water temperature. The temperature difference $(T_w - T_g)$ between the glazing cover and seawater also has quite an influence on the production, which is not more remarkable than that of T_w.

The distilled production of a solar still can be expressed as $m = q_e/h_{fg}$, where, h_{fg} is the vaporizing latent heat of water. Therefore increasing the share of q_e in the total energy transfer will have a decisive meaning for improving the distilled water production of solar stills. As we know, q_e is closely related to T_w. Finally, it converts to rising T_w. In the actual system, the seawater temperature in the basin is not only related to solar radiation, but also to the quantity of seawater and the capacity of the bottom and side wall. The total heat capacity of the device (also known as thermal inertia or thermal inertia) is often too large, reducing the increased rate of water temperature, and further delaying the starting time and fresh water production, which is one of the main shortages of solar stills.

The vapor produced from seawater rises to the glazing cover and condenses there. Finally, the latent heat is lost to the atmosphere through the cover. The process is related to the productivity. The bigger the productivity, the more heat lost to atmosphere, and also the faster the heat is lost from the seawater so that it limits the seawater temperature rise. Obviously, it is quite useful for elevating the productivity of a device if the heat loss rate from the cover can be controlled or the latent heat of condensation can be reused instead

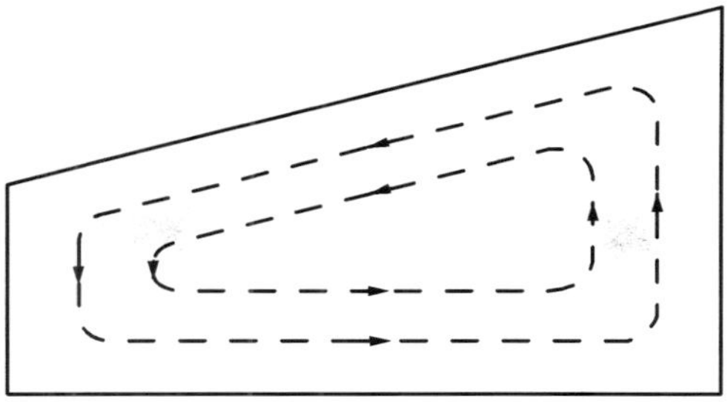

Figure 4.50
The basis circulation process in a solar still.

of letting it go to waste. That is the idea of multieffect distillation. Thus it is another shortage of single-stage basin solar stills because it cannot reuse the condensing latent heat of the vapor.

The movement process of the wet-air inner still is also a question to be studied. Baum (1964) did the theoretical and experimental research for that and validated the circulation inner still is similar to that as shown in Fig. 4.50 [27].

Vapor has a lower density compared to air. Therefore vapor has the natural upward trend, which is a benefit to improve the production. However, the vapor must first enter the air, which is with a lot of resistance. The air closed to the seawater surface does the heat exchange with the water surface and is humidified. The hot and wet air also have the upward trend to the cover. Finally, the hot and wet air goes to the vicinity of the cover and does the heat exchange. Portion vapor condenses on the back of the cover and another portion with air is cooled and goes back to the seawater surface vicinity. The next process is repeated.

Throughout the movement process of moisture in the interior of the still, it is completely a natural convection process when the operation temperature is not too high. Thus the heat transfer efficiency of the solar still is limited greatly because the natural convection process is not possible to have too high heat and mass transfer rate. That is the third shortage of the solar still. One of the ways to overcome this shortage is to add a fan in the still and let the wet air circulate forcedly [35]. However, this way will increase the cost of solar stills.

To improve the production of solar stills, it is possible to start with overcoming the three defects mentioned earlier. For example, decreasing the thickness of seawater is one of the effective measures to elevate the efficiency of a solar still. Because decreasing the thickness of seawater can effectively reduce the heat capacity of seawater in the basin so that the seawater can be raised the temperature rapidly when the still is radiated by sunlight. As a result, the daily average operation temperature will be elevated so as to obtain more daily average efficiency and production. Certainly, besides the heat capacity of seawater in the basin, the capacity of the basin bottom material and even the soil under the basin also should be taken into account, because their temperature is also increased when the operation temperature of the still increases. Although reducing the thickness of seawater can decrease the heat capacity of the device and improve daily efficiency, the heat loss from the bottom will become remarkable, so that the total effect may be negative if there is no insulation layer under the bottom of basin. Many research results have shown that the effect of the insulation layer is more obvious than decreasing the thickness of the seawater layer [4]. As long as the insulation layer is used, the soil temperature just does not rise dramatically with the seawater temperature. Only under this condition, decreasing the thickness of the seawater layer to reduce the heat capacity of seawater is significant. However, whether the insulation material is there or not, the thickness of seawater thinning must be with the limit. First, the thinner the seawater layer, the higher the flatness requirement of the basin bottom. Otherwise, part of bottom may be exposed from the seawater so that it may lead to decreased efficiency. Secondly, the seawater may precipitate salt when the seawater layer is too thin. Once the bottom is covered by the salt, the reflectivity of the bottom will increase greatly, which also led to the efficiency decrease.

4.12.2 The Maximum Efficiency of the Basin Solar Still

With the three main shortages discussed earlier, the efficiency of the basin solar still is limited significantly. Therefore overcoming the three shortages is the main direction to improve the productivity.

To determine the extent of this kind of still efficiency promoting, the maximum efficiency of an ideal still can be calculated. This so-called ideal still is without bottom and side wall heat loss. That is to say, $q_k = 0$. In the meantime, the so-called ideal still has such a small heating capacity of seawater that it is easy to reach the transient steady state under the conditions of any solar radiation density, ambient temperature, and wind speed. The calculation result of Cooper (1973) indicates that the maximum efficiency of a basin solar still is about 60% [4]. In fact, it is very difficult to reach 50% in actual systems. It is the daily average efficiency. The short period efficiency of a solar still may exceed the maximum efficiency. To give an extreme example, the solar still can also produce part of fresh water during no solar light using the heat storage. Even more, some solar stills are especially designed for producing fresh water during the night. Hence the transient efficiency of a solar still obtained by Eq. (4.75) will be infinite. Certainly, this evaluation is of no practical significance.

To further idealize the solar still, assuming that all of the solar light received by the still is used to evaporate the seawater, in this time, $q_c = q_r = 0$, the efficiency of the device is

100%. If the average solar radiation density is $\bar{I} = 800\ \mathrm{W/m^2}$ and the operation is 8 h a day, the total production of the ideal solar still is $\dot{m}_{e\max} = 9.6\ \mathrm{kg/d \cdot m^2}$(in calculation, the vaporization latent h_{fg} of water is taken 2400 kJ/kg), which is the maximum production of this kind of solar still. The actual production is much lower than this quantity. Even under a very good sunny day, a good single-stage solar still can only provide about 3.5–4.5 kg/d m^2 fresh water. The total production in one year is only about 1000 kg/m^2.

4.12.3 Energy Gain and Loss of Each Component of Solar Stills

The energy gain and loss of each component of solar stills are affected strongly by the depth of seawater in the basin and the material performance of the basin bottom and side walls. Singh (1993) analyzed the energy gain percentage of each component of the single-effect solar still with double inclined covers and discussed the variation of the energy gain percentage with the depth of the seawater, as shown in Table 4.6 [19]. Also, he introduced the variation of the energy loss percentage of each component with the depth of the seawater, as shown in Table 4.7.

Certainly, some less important but unavoidable losses under actual conditions are ignored when obtaining the earlier discussions. For example, the vapor leaking loss and condensed water drops reevaporated or dripped back to the basin and so on are not considered as the losses. And more, the sensible heat loss with the brine discharge is also not calculated.

The conclusions of Singh are too limited. Obviously, different units may have different energy loss percentages. To understand intuitively the distribution of solar radiation energy in a basin type solar still, the next energy loss percentages are with the probable quantity relationship [32]:

Assuming the solar radiation energy G on unit level surface is 100%, then
Energy loss on the glazing cover $= (1 - \tau - a) \approx 10\%$;
Absorption loss on the glazing cover $= \alpha_g G \approx 5\%$ (this part of the energy is not necessarily a direct loss);
Reflection loss from the bottom of the basin and the surface of seawater $= (1 - \alpha_w)\tau G \approx 5\%$;
Evaporation heat for seawater $q_e \approx 40\%$;
Radiation heat transfer between the basin and the top cover $q_r \approx 15\%$.
The convection heat transfer between the basin and the top cover $q_c \approx 15\%$;
The heat losses from bottom and side wall $q_k \approx 15\%$;
The leaking loss of the vapor $q_l \approx 15\%$;
Other losses are about $q_m \approx 15\%$.

References

[1] Malik MAS, Tiwari GN, Kumar A, Sodha MS. Solar distillation. UK: Pergamon Press Oxford; 1982. p. 33–6.

[2] Dunkle RV. Solar water distillation: the roof type still and a multiple effect diffusion still. In: International developments in heat transfer. University of Colorado; 1961. p. 895. Part V.

[3] Daffie JA, Beckman WA. Solar engineering of thermal progress. New York: John Wiley & Sons; 1980.

[4] Cooper PI. The maximum efficiency of single-effect solars stills. Sol Energy 1973;15:205–17.

[5] Sartori E. Solar still versus solar evaporator: a comparative study between their thermal behaviors. Sol Energy 1996;56(2):199–206.

[6] Sharma VB, Mullick SC. Estimation of heat-transfer coefficients, the upward heat flow, and evaporation in a solar still. J Sol Energy Eng 1991;113:37–42.

[7] Hongfei Z. Experimental studies on three kinds of solar stills: multi-stage stacked tray still, falling film evaporation still and air flow absorption still[D]. Hefei: University of Science and Technology of China; 1999.

[8] Clark JA. The steady-state performance of a solar still. Sol Energy 1990;44(1):43–9.

[9] Adhikari RS, Kumar A, Kumar A. Estimation of mass-transfer rates in solar stills. Int J Energy Res 1990;14:737–44.

[10] Tiwari GN, Lawrence SA. New heat and mass transfer relations for a solar still. Energ Convers Manage 1991;31(2):201–3.

[11] Shawaqfeh AT, Farid MM. New development in the theory of heat and mass transfer in solar stills. Sol Energy 1995;55(6):527–35.

[12] Liugang. Experimental study of heat and mass transfer correlations in solar stills[D]. Hefei: University of Science and Technology of China; 1992.

[13] Chen Z, Ge X, Sun X, Bar L, Miao Y. Natural convection heat transfer across air layers at various angles of inclination. J Eng Thermophys (suppl) 1984:211–20. Special Issue for the US-CHINA Binational Heat Transfer Workshop.

[14] Hongfei Z, Xiaoyan Z, Jing Z, Yuyuan W. A group of improved heat and mass transfer correlations in solar stills. Energ Convers Manage 2002;43(18):2469–78.

[15] Toyama S, et al. Simulation of a multi-effect solar still and its characteristics. J Chem Eng Jpn 1987;20(5):473–8.

[16] Tayeb AM. Performance study of some designs of solar stills. Energ Convers Manage 1992;33(9):889–98.

[17] Gang HP, Mann HS. Effect of climatic, operational and design parameters on the year round performance of single-sloped and double-sloped solar still under Indian arid zone conditions. Sol Energy 1976;18(2):159–64.

[18] Hailong M, Xinshi G, Yefa L. Experimental study of solar still with a piece of black plastic foam floating on the evaporative surface as a heat absorption layer. Acta Energiae Solaris Sin 1994;15(4):386–90.

[19] Singh AK, Tiwari GN. Long-term comparative study of solar distiller systems. Energy 1993;18(11):1161–9.

[20] Bassam AK, Hijleh A, Mousa HA. Water film cooling over the glass cover of a solar still including evaporation effects. Energy 1997;22(1):43–8.

[21] Yadav YP, et al. Parametric investigations on a basin type solar still. Energ Convers Manage 1991;31(1):7–16.

[22] Ali HM. Effect of forced convection inside the solar still on heat and mass transfer coefficients. Energ Convers Manage 1993;34(1):73–9.

[23] Velmurugan V, Gopalakrishnan M, Raghu R, Srithar K. Single basin solar still with fin for enhancing productivity. Energ Convers Manage 2008;49:2602–8.

[24] Kabeel AE. Performance of solar still with a concave wick evaporation surface. Energy 2009;34:1504–9.

[25] Dutt DK, Kumar A, Anand JD, Tiwari GN. Improved design of a double effect solar still. Energ Convers Manage 1993;34(6):507−17.
[26] Tiwari GN, Singh SK, Bhatnagar VP. Analytical thermal modelling of multi-basin solar still. Energ Convers Manage 1993;34(12):1261−6.
[27] Kumar A, Anand JD, Tiwari GN. Transient analysis of a double slope-double basin solar distiller. Energ Convers Manage 1990;21(1):129−39.
[28] Mahdi NAl. Performance prediction of a multi-basin solar still. Energy 1992;17(1):87−93. Solar Energy, 8, 78−82, 1964.
[29] Hassan ES, Elsherbiny SM. Effect of adding a passive condenser on solar still performance. Energ Convers Manage 1993;34(1):63−72.
[30] Sodha MS, Nayak JK, Tiwari GN, et al. Study on a double-basin multi-wick solar still. Energ Convers Manage 1980;20(1):20−5.
[31] Xu Q, Yang X. Development trend and performance analysis of solar still system design. New Energy 1995;17(9):38−41.
[32] Ge X, Gong bao, Lu W, Wang Y. Solar energy engineering—principle and application. Beijing: Academic Periodicals Publisher; 1998.
[33] Ahmadzaded J. Solar earth water stills. Sol Energy 1978;20:387−91.
[34] Sodha MS, Kumar A, Srivastava A, et al. Thermal performance of still on roof system. Energ Convers Manage 1980;20(20):181−90.
[35] Mink G, et al. Air-blown solar still with heat recycling. Sol Energy 1998;62(4):309−17.

CHAPTER 5

Active Solar Distiller

Chapter Outline

Solar Energy Desalination Technology. http://dx.doi.org/10.1016/B978-0-12-805411-6.00005-1

Passive solar distillers, especially basin solar distillers, have low efficiency, because the heat and mass transfer processes within the device mainly belong to natural convection. Even though some devices have taken measures to strengthen the heat and mass transfer, the operation temperature of the device is still hard to increase for the reason that thermal inertial of seawater within the basin and latent heat of condensation are underutilized. Thus, the water production rate per unit area of the device is not high, and the utilization of waste heat is also limited. Based on this reason, the development of this kind of solar distiller is cramped.

For this, Soliman et al. put forward the idea of an active solar distiller. For now, more than 10 types of active solar distiller designs have been conceived, and much research about these is under way [1,2]. Within stalled auxiliary equipment, the operation temperature of an active solar distiller is drastically increased, and the internal heat and mass transfer process is improved. Besides, most of the active solar distiller systems can actively reuse the latent heat that vapor releases during the condensation process; thus these systems can get water production more than double that of traditional solar distiller systems. This is the root cause of the current extensive attention given to active solar distillers.

5.1 The Basin-Type Solar Still Driven by Flat-Plate Solar Collectors

To overcome the defects of the traditional basin solar distiller that the operation temperature is too low and water production is not much, Rai and Tiwari [3] presented the basin active solar distiller combined with a flat-plate solar collector for the first time. The experimental results showed that the operational temperature of the distiller was drastically increased because of the adoption of an auxiliary collector. Thus, the water production rate per unit solar collecting area is much improved.

The structure schematic of basin active solar distiller with auxiliary flat-plate solar collector can be seen in Fig. 5.1.

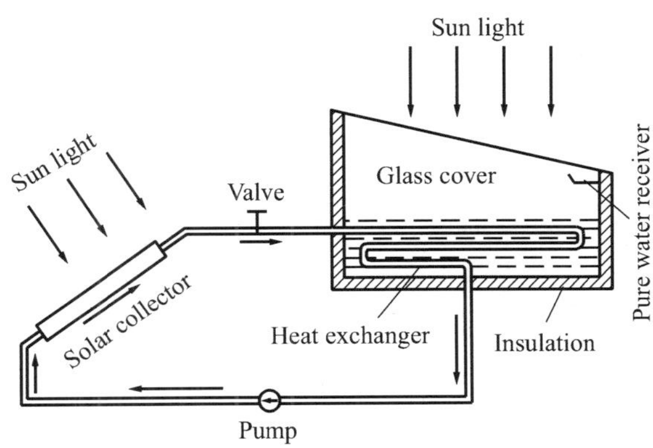

Figure 5.1
The schematic diagram of the system.

The distiller of the system is mainly used to evaporate and condensate, making heated seawater evaporate and condensate on a condensing plate to generate purified water. Technically speaking, there is no distinction between this process and the process of a traditional basin solar distiller. The distiller part can also receive solar power as the figure shows. The flat-plate solar collector is mainly used to harvest and store the solar energy. Because of its high efficiency, it can heat the water (or other working fluid) to a relatively high temperature. Through the pump and coil heat exchanger, the solar energy collected by the flat-plate solar collector can heat up the seawater in the distiller. In some types of distillers, seawater is used as working fluid in flat-plate solar collectors. However, because seawater is corrosive, to protect the collector, these kinds of devices are rarely used.

5.1.1 Performance Analyses of the Device

To theoretically predict the water production performance of the device, it is necessary to establish energy balance analysis for each part of the system. For this, the following predictions are made.

1. The heat capacity of the glass cover plate, heat exchanger, pipeline, structural material of the distiller body, and working fluid within the heat exchanger can be neglected.
2. There is no temperature gradient in the seawater and glass cover plate.
3. Loss caused by vapor leaking can be neglected.
4. The inclination angle of the glass cover plate is small; thus the surface area of the seawater can be considered the same.
5. The amount of purified water generated from the distiller can be neglected, because it is small compared to the amount of seawater. Thus the amount of water within the device can be considered the same.
6. The flat-plate solar collector is installed facing south, and its inclination angle is equal to local latitude.

5.1.1.1 Distiller Glass Cover Plate

$$\tau_1 H_s A_g + h_1(T_w - T_g)A_w = h_2(T_g - T_a)A_g \tag{5.1}$$

τ_1 is energy absorption ratio of the glass cover plate and can be calculated with Eq. (5.2).

$$\tau_1 = (1 - R_g)\alpha_g \tag{5.2}$$

5.1.1.2 The Basin of Distiller

Water in the distiller basin

$$\tau_2 H_s A_w + F(t) q_u = M_w c_w \frac{dT_W}{dt} + h_1 (T_w - T_g) A_w + (U_b A_w + U_s A_s)(T_w - T_a) + U_p A_p (T_w - T_a) F(t) \quad (5.3)$$

In Eq. (5.3), τ_2 is the energy absorption ratio of the water and can be calculated with Eq. (5.4).

$$\tau_2 = \tau_w + \frac{h_i \tau_b}{h_i + h_b} \quad (5.4)$$

Within Eq. (5.4), $\tau_w = (1 - R_g)(1 - \alpha_g)\alpha_w$, $\tau_b = (1 - R_g)(1 - \alpha_g)(1 - \alpha_w)\alpha_b$.

U_b is the total heat transfer coefficient and can be calculated as below.

$$\frac{1}{U_b} = \frac{1}{h_i} + \frac{1}{h_b} \quad (5.5)$$

Within Eq. (5.5) $h_b = \left(\frac{1}{h_0} + \frac{L_i}{\lambda_i}\right)^{-1}$

$F(t)$ is the time control function of the solar collector, and its value can be given as formula below.

$$F(t) = \begin{cases} 1 & \text{With solar radiation and connected to the collector} \\ 0 & \text{disconnected with the collector} \end{cases}$$

5.1.1.3 Solar Collector

$$\frac{T_{c0} - T_a - \frac{(\alpha\tau)H_c}{U_c}}{T_{ci} - T_a - \frac{(\alpha\tau)H_c}{U_c}} = \exp\left(\frac{F' U_c A_c}{\dot{m}_c C_w}\right) \quad (5.6)$$

5.1.1.4 Heat Exchanger

For the tube exchanger, if we pick Δx length to calculate, we can get:

$$\dot{m}_c c_w \frac{dT_f(x)}{dx} \Delta x = -\pi D U_h (T_f - T_w) \Delta x \quad (5.7)$$

Within the equation, $U_h = \left(\frac{1}{h_{fi}} + f\right)^{-1}$ (f: fouling coefficient),

h_{fi} is the convective heat transfer coefficient between the heat exchanger and seawater.

For fully developed fluid, it can be calculated as below.

$$\frac{h_{fi}L_f}{\lambda_f} = \begin{cases} 3.66 & \text{Laminar} \\ 0.027R_e^{0.8}p_{rf}^{1/3} & \text{Turbulence} \end{cases}$$

The fouling coefficient is chosen according to the actual condition. And the boundary condition of Eq. (5.7) is:

$$T_f(x) = T_{c0}$$

When $x = 0$, Eq. (5.7) can be solved as Eq. (5.8),

$$T_f(x) = T_w + (T_{c0} - T_w)\exp\left(-\frac{\pi DU_h x}{\dot{m}_c c_w}\right) \tag{5.8}$$

Thus the net heat that transfers from the heat exchanger to the seawater within the plate can be expressed as:

$$q_u = U_h\left[\frac{1}{L_h}\int_0^{L_h} T_f(x)dx - T_w\right] \tag{5.9}$$

After substituting Eq. (5.8), Eq. (5.9) can be integrated as:

$$q_u = \dot{m}_c C_w(T_{c0} - T_w)\left[1 - \exp\left(-\frac{\pi DU_h L_h}{\dot{m}_c C_w}\right)\right] \tag{5.10}$$

Using boundary condition, $T_f(x) = T_{ci}$ when $x = L_h$. Substitute this into Eq. (5.8), then we can get:

$$T_{ci} = T_w + (T_{c0} - T_w)\exp\left(-\frac{\pi DU_h L_h}{\dot{m}_c C_w}\right) \tag{5.11}$$

Solve Eq. (5.6) to get T_{c0} and T_{ci}, then put the results into Eq. (5.10) to get:

$$q_u = F'_R A_c[(\alpha\tau)H_c - (T_w - T_a)U_c] \tag{5.12}$$

Within Eq. (5.12), $F'_R = \frac{\dot{m}_c c_w}{U_c A_c}\left\{\frac{[1-\exp(-z_1)][1-\exp(-z_2)]}{1-\exp[-(z_1+z_2)]}\right\}$ and $z_1 = F'U_cA_c/\dot{m}_c c_w$, $z_2 = \pi DU_h L_h/\dot{m}_c c_w$.

Solve Eq. (5.1) to get T_g, then put the result into Eq. (5.3), and substitute to Eq. (5.12) to get q_u, the result can be seen in Eq. (5.13).

$$\frac{dT_w}{dt} + g(t)T_w = f(t) \tag{5.13}$$

Where,

$$g(t) = \frac{1}{M_w c_w}\left(\frac{h_1 A_w h_2 A_g}{h_1 A_w + h_2 A_g} + F(t)F'_R U_c A_c + F(t) U_p A_p + U_b A_w - U_s A_s\right)$$

$$f(t) = \frac{1}{M_w c_w}\left[\tau_2 H_s A_w + F(t) F''_R A_c (\alpha\tau) H_c + \frac{h_2 \tau_1 H_s A_w A_g}{h_1 A_w + h_2 A_g}\right]$$

Please notice that the U_s in the equation above can be calculated as in the equation shown below.

$$U_s = \left(\frac{1}{h_i} + \frac{1}{h_0} + \frac{L_i}{\lambda_i}\right)^{-1}$$

Because both the solar radiation intensity and environmental temperature are cyclical and vary for 24 h, the $f(t)$ and $g(t)$ can be calculated with Fourier expansion for time.

$$g(t) = \sum_{-\infty}^{+\infty} g_n \exp(in\omega t) \tag{5.14}$$

$$f(t) = \sum_{-\infty}^{+\infty} f_n \exp(in\omega t) \tag{5.15}$$

Generally speaking, the temperature of water can be described with Fourier expansion as below.

$$T_w(t) = \sum_{-\infty}^{+\infty} T_{wn} \exp(in\omega t) \tag{5.16}$$

Malik's research [4] pointed out that the relatively precise result can be given if we calculate the Fourier expansion above from $n = -6$ to $n = 6$. Thus T_w can be calculated after substituting combined Eqs. (5.14)–(5.16) into Eq. (5.13). Then, the instant water production and total water production can be given with the equations listed below.

1. Instant water production:

$$\dot{m}_e(t) = h_{ew}(T_w - T_g)A_b/h_{fg} \tag{5.17}$$

2. Water production per day:

$$M_e = \int_{day} \dot{m}_e(t)dt \tag{5.18}$$

3. Water production per year:

$$M_y = \sum_{year}\sum_{month} M_e \tag{5.19}$$

It is worth noticing that h_{ew} can be given by different relations. If Dunkle's relation [5] is used, the relation between h_1 and h_{ew} can be given as below:

$$h_1 = h_{cw} + h_{ew} + h_{rw} \tag{5.20}$$

$$h_{cw} = 0.884\left[T_w - T_g + \frac{(p_w - p_g)(T_w + 273)}{268.9 \times 10^3 - p_w}\right]^{\frac{1}{3}} \tag{5.21}$$

$$h_{ew} = 0.016273\frac{h_{cw}(p_w - p_g)}{T_w - T_g} \tag{5.22}$$

$$h_r = \frac{\varepsilon_e\sigma\left[(T_w + 273)^4 - (T_g + 273)^4\right]}{T_w - T_g} \tag{5.23}$$

The partial pressure of saturated vapor can be calculated with empirical relation:

$$p = 6893.03\exp\left(54.63 - \frac{12301.69}{T'} + 5.17\log T'\right) \tag{5.24}$$

Within Eq. (5.24), $T' = 1.8T_{av} + 491.69$, T_{av} is the average temperature within the distiller, and it can be calculated as $T_{av} = \frac{T_w + T_g}{2}$ (°C). h_2 can also be given with different relations. If Duffie and Beckman's relation is adopted, h_2 can be expressed as [6]:

$$h_2 = h_{cg} + h_{rg} \tag{5.25}$$

Within it, $h_{cg} = 5.7 + 3.8v$

$$h_{rg} = \frac{\varepsilon\sigma\left[(T_g + 273)^4 - (T_a + 261)^4\right]}{T_g - T_a}$$

From Eq. (5.1)–Eq. (5.25), the meanings of the parameters are as follows:

A, area, m^2;
c_w, specific heat capacity of water, J/kg °C;
D, diameter of heat exchanger pipeline, m;
F_R, heat removal factor of heat transfer;
F', efficiency factor of heat transfer;
h_b, heat transfer coefficient of the bottom of distiller, W/m^2 °C;
h_i, heat transfer coefficient from water plate of distiller to water, W/m^2 °C;
h_0, heat transfer coefficient from bottom of distiller to environment, W/m^2 °C;
h_{cg}, convection coefficient of glass cover plate to environment, W/m^2°C;
h_{cw}, convection coefficient from water surface to glass cover plate, W/m^2 °C;
h_{ew}, evaporate heat transfer coefficient from water surface to glass cover plate, W/m^2 °C;
h_{rg}, radiation coefficient from glass cover plate to environment, W/m^2 °C;

h_{rw}, radiation coefficient from water surface to glass cover plate, W/m^2 °C;
h_1, total heat transfer coefficient from water surface to glass cover plate, W/m^2 °C;
h_2, total heat transfer coefficient from glass cover plate to environment. W/m^2 °C;
h_{fg}, latent heat of vaporization of water, W/m^2 °C;
H_c, solar radiation intensity on plate collector surface, W/m^2;
H_s, solar radiation intensity on horizontal plate, W/m^2;
i, $\sqrt{-1}$, imaginary number;
λ_f and λ_i, heat conduction coefficient of working fluid within heat exchanger and thermal insulation material, W/m^2 °C;
L_h, length of heat exchanger, m;
L_i, thickness of thermal insulation material, m;
M_w, heat capacity per area of distiller, J/m^2 °C;
$\dot{m}_e$, mass flow rate of purified water, kg/s;
p_g and p_w, saturated partial pressure nearby the glass cover plate and water surface, Pa;
Pr, Prandtl number;
q_u, useful energy obtained from flat collector, W/m^2;
R_g, Reflectivity glass cover plate;
Re, Reynolds number;
T, Temperature, °C;
T_{ci}, T_{c0}, temperature of fluid at flat collector inlet and outlet, °C;
t, time, s;
U, total heat transfer coefficient, W/m^2 °C;
V, wind velocity, m/s;
X, length coordinate of heat exchanger pipeline, m;
α, absorptivity;
τ, transmissivity;
$(\tau\alpha)$, the product of absorptivity and transmissivity of glass cover plate;
ε_e, effective emissivity between glass cover plate and water;
ε_g, effective emissivity of glass cover plate;
σ, Stefan–Boltzmann constant, 5.67×10^{-8} W/m^2 K^4.

Subscripts:

a, environment;
b, plate of distiller;
c, heat collector;
f, working fluid in the heat collector;
g, glass cover plate;
h, heat exchanger;
s, distiller;
w, water body.

5.1.2 Experimental Operation of the Device

Tiweri and Dhiman [7] studied experimentally an active basin solar distiller with flat-plate collector equipment. The experimental devices are listed below. The distiller that has a $0.95 \times 0.95/m^2$ solar collecting area is made of galvanized iron. The glass and sidewall are sealed with a rubber mat to enclose the steam. The flat-plate collector is composed of two collector plates with 1.5 m^2 solar collecting area in series, a pipeline with 0.031 m diameter and a glass cover plate made of ordinary glass with 3 mm thickness. The installation angle of the flat-plate collector is 45°. The heat exchanger inside the distiller is made with 8 m long, 1.27 cm diameter copper pipe. The fluid inside the heat exchanger is ordinary water having a mass flow rate of $\dot{m} = 0.05$ kg/s. The specific heat volume per unit area of seawater inside the distiller is $M_w = 419$ kJ/m^2 °C.

The experiment of Tiwari et al. points out that, at the solar radiation intensity and the environmental temperature as Fig. 5.2 shows, the relation of water production per hour with or without active heating by flat-plate collector changed with time, as can be seen in Fig. 5.3. The curve in the diagram is a theoretical curve, and its parameters can be specified as below.

$$U_l = 8.0\ \mathrm{W/m^2\ °C}; \quad v = 2.0\ \mathrm{m/s}; \quad \tau_b = 0.7; \quad \tau_w = 0.0; \quad \tau_h = 0.7; \quad (\alpha\tau) = 0.81$$

As Fig. 5.3 shows, the active heating flat-plate collector can drastically increase water production of the distiller by about two to three times. The main reason is that with the flat-plate distiller actively increases the temperature; the operation temperature of the distiller can be significantly increased. It is worth pointing out that the total efficiency of the system is not able to increase for sure with the active heating flat-plate collector. Only with a reasonably matched solar collector area, can the total efficiency of the system be increased. Otherwise, its total efficiency might be lower than the efficiency of a traditional basin distiller. Like the total efficiency of Tiwari's experimental equipment, it is only around 20%.

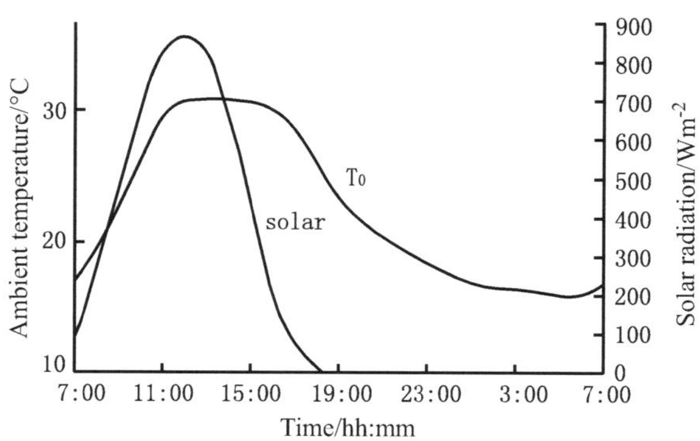

Figure 5.2
Variation of solar radiation and ambient temperature with time.

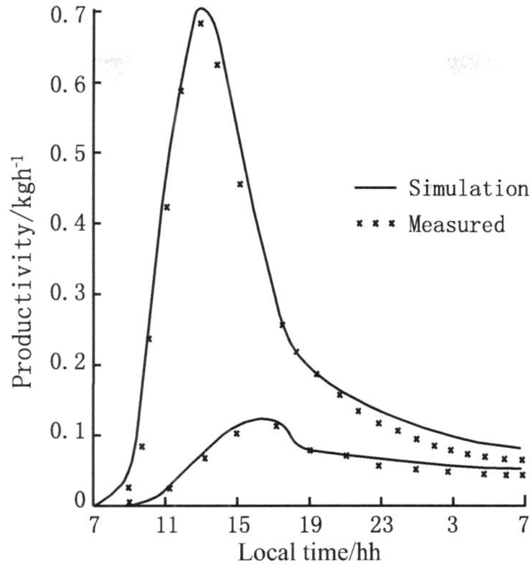

Figure 5.3
The variation of the production in a day.

G.M. Zaki [8] studied the basin solar distiller with the auxiliary heating flat-plate solar collector, and gave some theoretical results, and carried out the test under the actual weather. In Zaki's research system equipment lacks a heat exchanger, as shown in Fig. 5.1. The heated seawater is directed to flow into the flat-plate solar collector, then going into the double incline plate solar distiller to be evaporated.

The relative parameters, which are related to in Fig. 5.4, can be seen in Table 5.1. The theoretical and experimental results can be seen in Figs. 5.5 and 5.6. From the results, we can see that the water production of the system is largely increased with active reinforcement to the distiller.

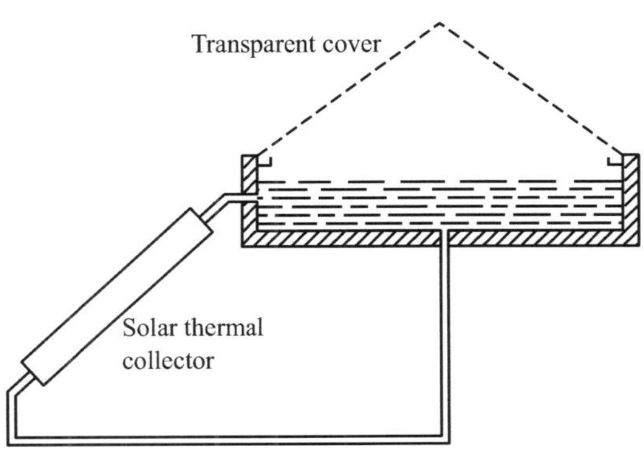

Figure 5.4
Solar still with auxiliary heated by flat-plate solar collector.

Table 5.1: Related parameter of system and experiment.

Matching collector	Plate type: facing south
Incline angle of collector	30°
Collector area	$A_c = 0.077–0.847\ m^2$
Area of water plate within distiller	$A_b = 1.68\ m^2$
Thickness of thermal insulation layer of basin, l_i	0.05 m
Thermal conductivity of thermal insulation layer, k_i	0.037 W/m K
Absorptivity of water within the basin	0.96
Transmissivity of cover plate, τ_g	0.8
Absorptivity of collector, α_c	0.96
Solar radiation intensivity, I	200–1000 W/m^2
Wind speed, v	3 m/s
Environment temperature, T_a	30°C
Area enhancement rate, A_c/A_b	0–2.3

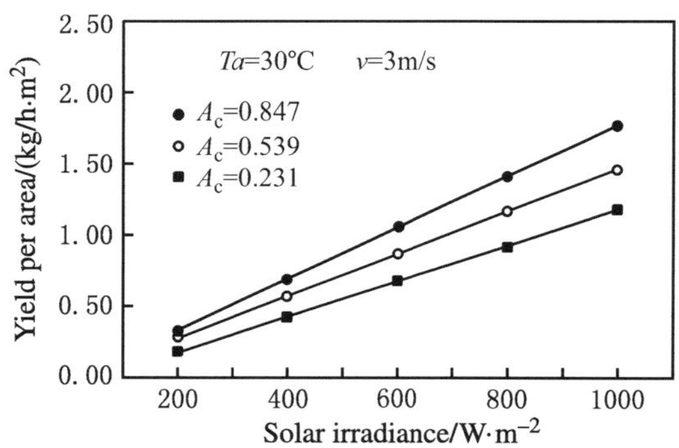

Figure 5.5
The variation of still yield with solar collector with solar irradiance.

5.2 The Active Basin-Type Solar Still With Glass Cover Cooling

Based on the research of the solar distiller with active heating flat-plate collector foundation, Singh and Tiwari [9] added an active cooling system on the cover plate of the distiller as Fig. 5.7 shows.

As Fig. 5.7 shows, the part of the solar collector and solar distiller are exactly the same as in the last section. The difference is an additional glass cover plate that is mounted on the original glass cover plate. Cooling water runs between these two glass cover plates, as the diagram shows, to cool down the glass cover plates.

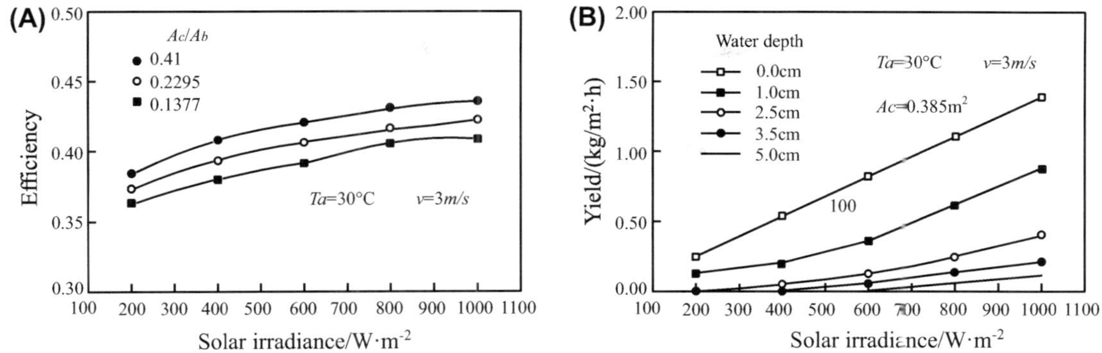

System efficiency variation with solar irradiance.

Yield variation with solar irradiance at different seawater depth

Figure 5.6
Variation of system efficiency and water production. (A) System efficiency variation with solar irradiance. (B) Yield variation with solar irradiance at different seawater depth.

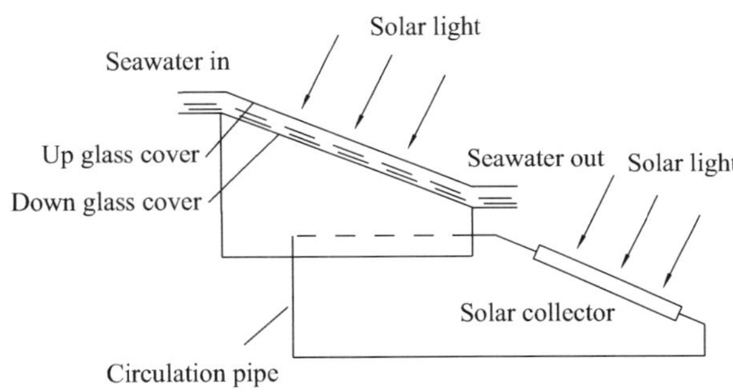

Figure 5.7
Active solar still with cover cooling.

The energy balance equations of water body within the distiller, heat exchanger, and flat-plate solar collector are the same as described in the last section. The energy balance equation of cooling water body within two glass cover plates can be written as:

$$m_{wu}C_W\frac{dT_{wn}}{dx}dx = bdx\left[h_1'(T_g - T_{wu}) - h_1''(T_{wu} - T_a)\right] \tag{5.26}$$

The energy balance equation of lower glass cover plate can be given as:

$$h_1(T_w - T_g) = h_1'(T_g - T_{wu}) \tag{5.27}$$

In the above two equations, m_{wu} is the mass flow rate of cooling water (kg/s); T_{wu} represents temperature of cooling water (°C); b is the width of the distiller cover plate that water flows through (m); h_1' is the total heat transfer coefficient from the glass cover plate to the upper cooling water (W/m^2°C); h_1'' is the total heat transfer coefficient from cooling water to upper glass; x represents the coordinate in cooling water flow direction.

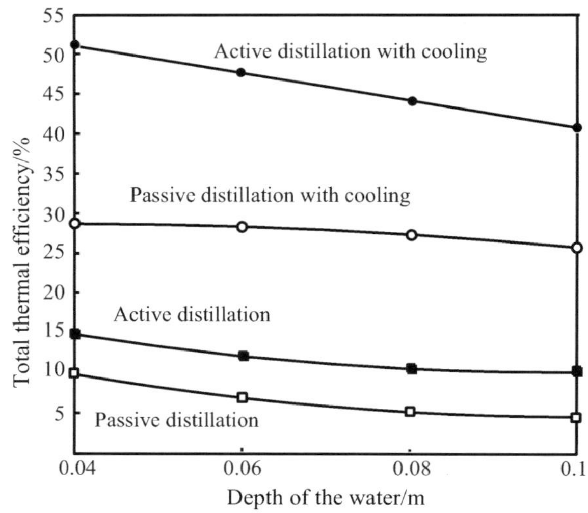

Figure 5.8
Total efficiency of active and passive still changing with the water depth.

The research carried out by Singh and Tiwari shows that, under the active heating operation mode, adding cooling to the glass cover plate can increase the water production of the system. Moreover, it can also drastically increase the total efficiency of system.

The study shows that, when the solar collecting area ratio of the distiller to the flat-plate collector is 1:3.5, the relation of total operating efficiency of the system with or without active heating changes with the depth of water in the plate can be plotted in Fig. 5.8. In the diagram, $b = 1$ m, and the value of m_{wu} is given.

Based on the plot, when the system, with a cooling cover plate is adopting passive operation mode (without flat-plate collector to heat), it can reach the highest efficiency, which is around 45–52%. An active operating system with cooling cover plate comes second. The total efficiency of the system can be around 30%. The worse case is the active operating system without the cooling cover plate. The total efficiency under this situation can only be 10%. It is not favorable to utilize energy. Thus when the system adopts an active heating method, it is important to recycle the latent heat of the cover plate. In this way, the system total efficiency and economy can be increased.

5.3 The Active Basin-Type Solar Still With Thermal Energy Storage Tank

The operation temperature of an active solar distiller with an added flat-plate collector is drastically increased, resulting in greater water production. However, when the operation temperature of this kind of system is highest, the environmental temperature is highest too. This results in the environmental temperature change not being able to be fully utilized.

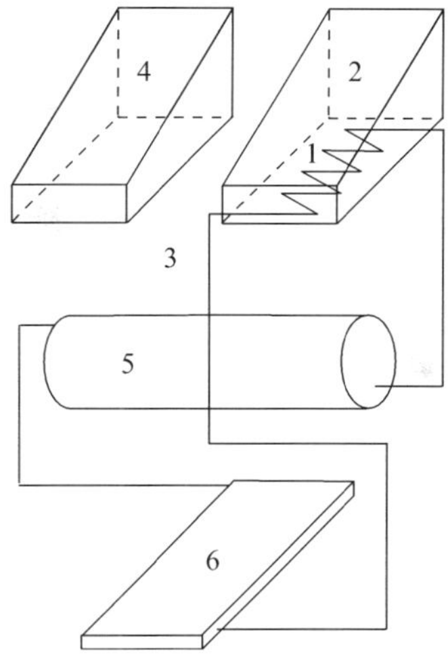

Figure 5.9
Schematic diagram of the active solar distiller with thermal energy storage tank: 1, heating coil; 2, active distiller; 3, circulation pipe; 4, passive distiller; 5, thermal storage tank; 6, solar collector.

To efficiently take advantage of a low-temperature environment, Varol and Yazar experimentally studied a system with a hot water storage tank mounted between a flat-plate collector and distiller [10], as shown in Fig. 5.9. In this system, the solar collecting area of the distiller is 1 m^2; the volume of the hot water storage tank is 150 L; the length of the copper pipe within the distiller is 15 m; the area of the flat-plate distiller is 2 m^2; the 4 mm thickness glass cover plate of the distiller made of low-iron glass is established at a 15° tilt angle. To evaluate this system, a traditional solar distiller that has same solar collecting area without any active heating auxiliary equipment is established near the system. Both experimental results are respectively recorded.

The experimental result shows that, because the water tank extends the distiller operating time in the high-temperature zone, and fully takes advantage of low temperature at night, the water production time of the system is extended for almost 24 h per day. Under typical clear day conditions, ambient temperature and solar irradiation changes as in Fig. 5.10A, the water production of the distiller with or without heating collector changes with time can be plotted as in Fig. 5.10B. According to the recorded results in the figure, during the day, when the environmental temperature is around 24–26.5°C, and solar radiation intensity is 17,820 kJ/m^2 day, water production of the system is 3.69 L/m^2 from 7 a.m. to 9 p.m. From 9 p.m. to 7 a.m. the next day, the water production of the system can still reach 3.05 L/m^2. This shows that the water production during the nighttime cannot be neglected.

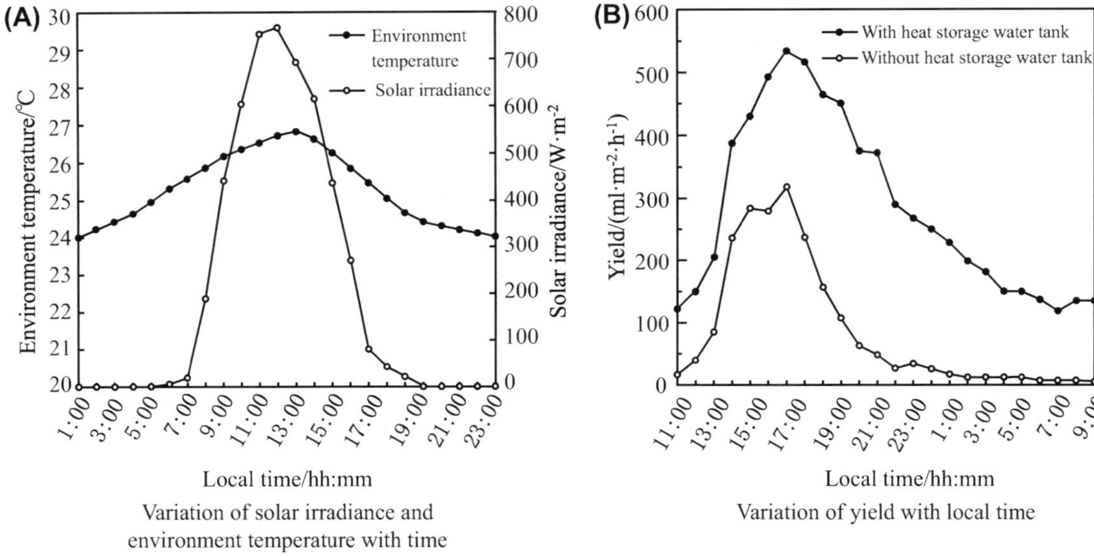

Figure 5.10
Variation of solar irradiance, environment temperature, and water yield with time. (A) Variation of solar irradiance and environment temperature with time. (B) Variation of yield with local time.

5.4 The Basin-Type Solar Still With Active Outer Condenser

For a traditional basin solar distiller, water vapor condensation on the glass cover plate is the heat and mass transfer of natural convection process, which limits the improvement of system performance. To enhance the process of water vapor evaporation and condensation, Mohamad et al. [11] came up with an active design with an external condenser as Fig. 5.11 shows.

Its operation principle is that the basin solar distiller collects heat to make seawater evaporate. However, water vapor that is generated by heated seawater cannot be completely condensed on a glass cover plate. Some of the water vapor is pumped into an external condenser by an electric fan. There is a cooling coil within the condenser. When hot water vapor sent from the distiller contacts the cooling coil, it will condense on the coil and generate distilled water as it cools down. The advantage of this design is that inside the distiller is in negative pressure because of the extraction of an electric fan, and this is favorable to evaporate water. The disadvantage is that the equipment is more complicated, making investment costs higher, and it needs an electric part to run.

Within this equipment, water vapor can be condensed on glass cover plate partly, or totally condensed on external condenser. ξ represents the ratio of condensation water on the

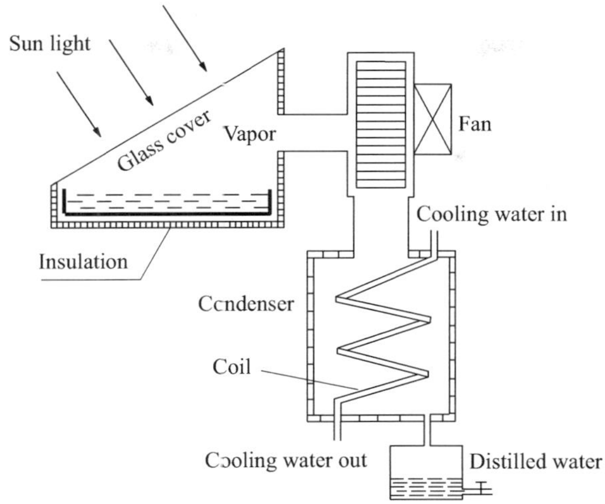

Figure 5.11
Schematic diagram of the basin-type solar still with an active outer condenser.

external condenser to total condensation water. The system total efficiency will change with ξ. Mohamad researched the effectiveness of some parameters from the system on system efficiency when ξ is different.

Figs. 5.12 and 5.13 give the relation of system efficiency changing with solar radiation intensity and environmental temperature when ξ is different. Based on the diagram, when $\xi = 1.0$, the system efficiency can be above 37% under general weather situations, and the efficiency can reach up to 45–47% on a sunny day. Thus, it can be seen that an external condenser structure can largely improve system efficiency.

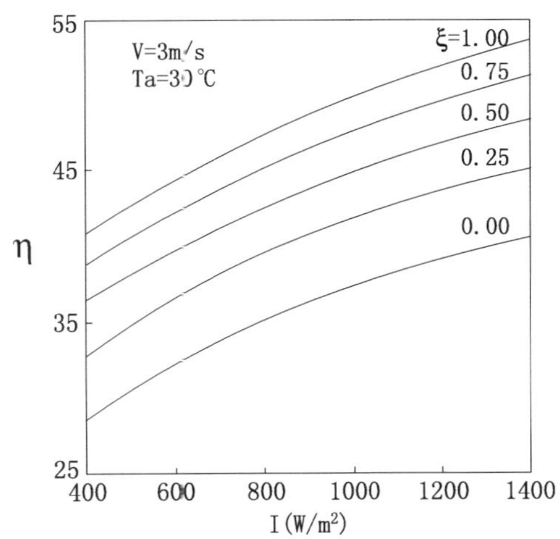

Figure 5.12
The system efficiency changes with solar radiation intensity.

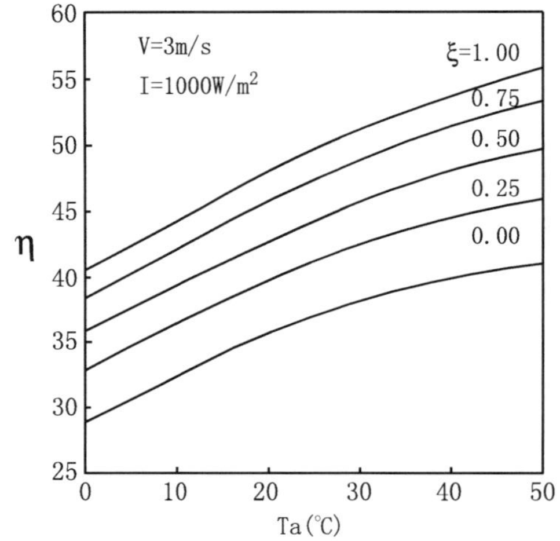

Figure 5.13
The system efficiency changes with temperature.

Because the system with an active external condenser requires consumption of a part of electric power, this kind of system is not suitable to be used in areas without electric power. Sinnocrot came up a solar distiller with an external condenser as Fig. 5.14 shows [12]. The experimental results show that this system can largely improve the water production of a traditional basin distiller, and its performance coefficient reaches up to 0.74.

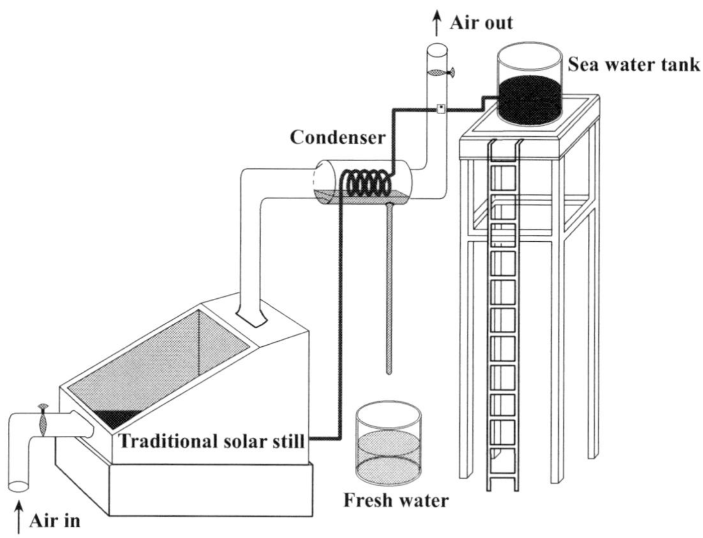

Figure 5.14
Solar still with passive outer condenser.

5.5 The Basin-Type Solar Still With Actively Recovering Latent Heat Outer Condenser

The biggest advantage of a basin solar distiller is that its process and equipment are simple, it is easy to maintain or repair, and its stable operation. However, its disadvantage is that its latent heat cannot be recycled as its heat and mass transfer process within the equipment is free convection. To overcome this shortage, Aboabboud et al. [13] designed a solar distiller that can actively recycle latent heat from an external condenser. Its operating principle can be seen in Fig. 5.15.

The distiller part of this kind of system has no difference from a traditional basin distiller. Both have the function to collect solar energy to evaporate seawater. After the seawater has evaporated, saturated air with high temperature is formed within the distiller. A fan pumps the saturated air out of the distiller to make saturated air condense in the external condenser. At the same time, the fan moves the same amount of air into the distiller to make ensure the air balance within the distiller. The significant difference of the system is that the air pumping process not only helps condense saturated air to produce purified water, but also recycles part to latent heat during the condensing process. The process is as follows: pumping seawater through a pipe passes the cavity submerged in hot saturated air, in which hot saturated air and a cold tube exchange heat, meaning that running up

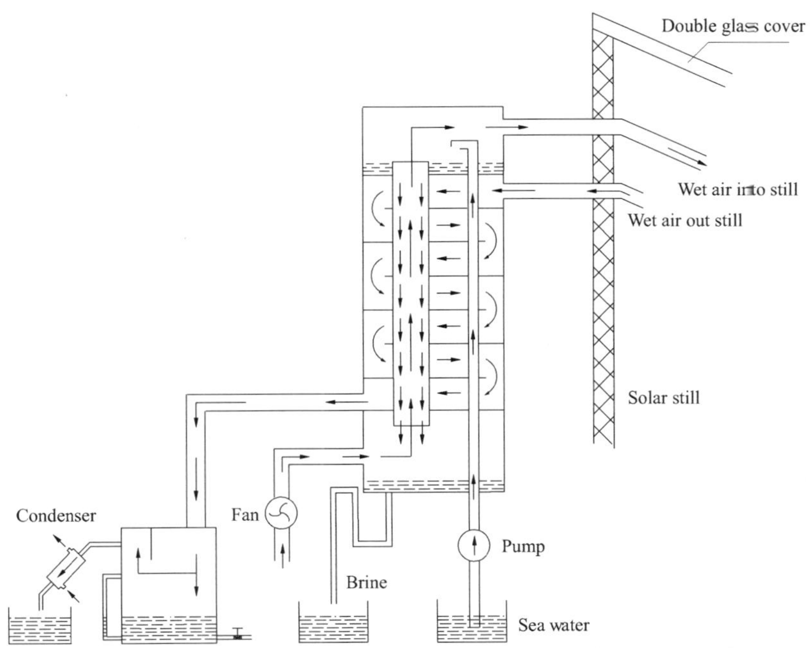

Figure 5.15
Schematic diagram of the solar still with active recycled latent heat of the external condenser.

seawater is heated and a part of the vapor is condensed. It can help heat transfer between hot saturated air and the cold pipe, and preheat the flow of seawater. When it rises to the top, seawater enters into an evaporating dish, then into the evaporating pipe through the distribution of the dish, in which seawater moistens the pipe wall and flows down by gravity. In this process, part of the seawater further evaporates by fully exchanging heat with upward-flowing air forced by the fan. Finally, seawater which had not been evaporated is excluded outside through a spill pipe. At this time, ambient air enters from the bottom of the device and flows upward along an evaporating pipe forced by the fan (as shown in Fig. 5.15). There is hot air outside the evaporating pipe and cool seawater inside the pipe, they exchange heat and hot air becomes saturated air by absorbing evaporated seawater generated in the upward process. This is an enthalpy-increasing process. Finally, saturated air enters the distiller through an evaporation tube. The hot saturated air which comes from the distiller enters into a cavity crossed with the evaporation tube driven by the fan, in which it exchanges heat with the evaporation tube and heats the seawater flow drawn upward by the pump. This is an enthalpy-reduction process. The air flows out from the bottom of the cavity and then runs into a moisture trap. The incompletely condensed water vapor is further condensed by a condenser of which the working fluid is external cooling water. This part of the produced purified water flows out of the equipment after this process.

Aboabboud et al. deeply studied the above equipment in laboratory using a steady light source ($\approx$700 W/m^2). The relative parameters of the equipment can be briefly listed as follows: 19 evaporating pipes, with inside diameter and outer diameter of 14 and 16 mm, respectively, and length 100 mm. Therefore, heat transfer areas are 0.083 and 0.102 m^2, respectively. The flow cavity of the vapor is a round cavity, for which the inside diameter is 110 mm and the outside is insulated. The lighting area of the basin distiller is 0.8 m $\times$ 1.25 m $=$ 1.0 m^2 and the inclination of cover is 20°. The cover is made of double glass with thickness 3 mm and the space between the two layers is 4 mm. The outside and bottom of the distiller are insulated. Ambient temperature ranges from 24°C to 28°C, supplied seawater temperature is 25 $\pm$ 2°C, the water depth in the distiller basin is 3–4 cm, and the average ambient wind speed is 2 m/s.

In order to compare the external condensation device with recovering the latent heat, the distiller is tested with irradiation of simulated solar light for 10 h, and then standing 14 h for measurement of total water production. The results are shown in Fig. 5.16.

The results showed that the distiller produces water at 3.05 kg/m^2 for light of 10 h. For the device with an external condenser, using the same simulation sunlamp irradiation for 10 h, but the pump and fan of external condensation open after 2 h of light, and stopping after 2 h of light, the total running time is also 10 h. In the case of 2.8 kg/h seawater pump flow rate and 10.5 m^3/h fan flow rate, the total water yield, the water

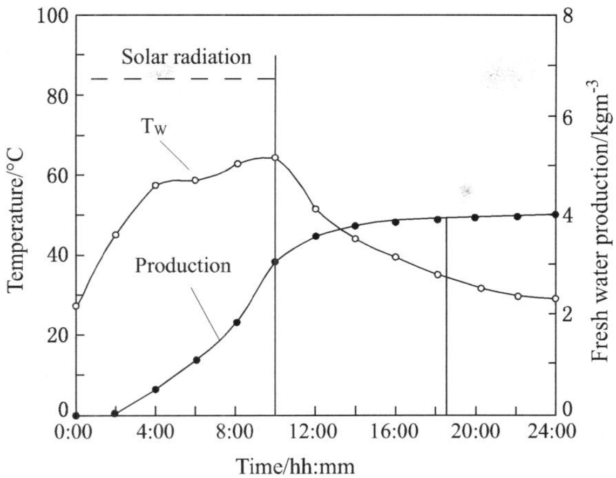

Figure 5.16

Variation of water temperature and total water yield with time in the case of using as traditional solar distiller.

yield of the gas-liquid separator (II) and the water yield of the cooler (I) change with the running time; the relations are shown in Fig. 5.17. The results showed that the water yield at 12 h is about 12 kg/m^2 excluding the failure productivity after 2 h of downtime, and the total water yield was about three times higher than that of the single basin distiller.

The key to improved water production is that the condenser has the ability to recover the latent heat. Therefore, it is significant to analyze and improve the energy condition

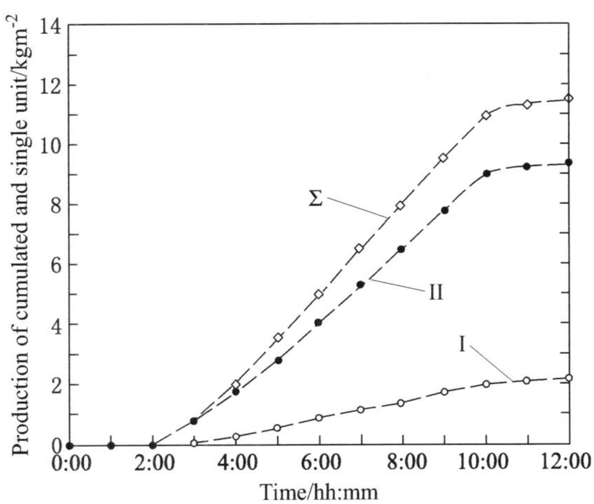

Figure 5.17

Variation of the total and each part's water yield with time for the solar distiller with active external condenser.

Table 5.2: Thermal equilibrium of an external condenser.

1. Input thermal energy
   - Seawater—26°C, 2.8 kg/h, 84.6 W
   - Air input—28°C, saturated, 10.5 m^3/h, 317.2 W
   - Entering wet air form distiller—64°C, 10.5 m^3/h (61°C saturated),1671.8 W
   - Total energy input—2073.6 W
2. Output thermal energy
   - Wet air entering the distiller from the top of the device—10.5 m^3/h, 56°C, saturated, 1290.0 W
   - Wet air from separator I—10.5 m^3/h, 44°C, saturated, 707.9 W
   - Freshwater condensed in separator I—1.2 kg/h, 44°C, 61.4 W
   - Output of concentrated brine—18 kg/h, 40°C, 83.7 W
   - Heat loss from the outer wall of the device—20.0 W
   - Total output 2163.0 W

of each part of the external condenser. For example, according to the steady-state experiments of Aboabboud et al., the energy input and output of an external condenser (or energy-recycling device) are analyzed and the results are shown in Table 5.2.

From Table 5.2, it is found that the energy output from the still is 1671.8 W, while the energy from the condenser to the distiller is 1290.0 W. The net energy loss of the still unit is only 381.8 W. The energy loss of the still is expected to be further reduced if the wet air outputting from the separator enters the circulation once again through the fan. Under the irradiation of 700 W/m^2 solar light, the energy per square meter is nearly 560 W/m^2. So it is greater than the energy output from the still. Although this part is greater than the energy loss of still cover, in the role of the double cover, the still can maintain a distiller running under high temperature. The experimental results show that the water temperature of the distilled water can reach 64°C under steady-state operation.

The system will have the following advantages with the active heat recovery of the external condenser:

1. The water production rate of the system is greatly improved, and the water yield per unit area is three times that of the traditional basin still.
2. The still cover is no longer the condensation surface of steam, so do not need to consider the wettability, it can be replaced by other transparent materials such as a plastic plate.
3. The heat loss rate of the cover plate to the environment can be reduced significantly because the double deck is available at this time.

This type of external condenser is also very suitable for other places with rich waste heat.

5.6 Multieffect Basin-Type Solar Still Combined With Solar Collectors

The most notable feature of an active heating solar basin distiller is that it can operate at relative high temperature to improve the intensity of evaporation. However, for a traditional disc distiller, when water vapor is condensed at the cover plate, latent heat of condensation will be lost to the environment through the cover plate in a very short time. This feature limits a further increase of system temperature, and it also hinders recycling latent heat. To solve this problem, the concept of an active multieffect solar distiller is presented. An experiment shows that the solution is feasible.

Baharna et al. [14] have studied the combination system of a flat-plate collector and triple-effect solar distiller. The system structure is shown in Fig. 5.18. Heated water that is generated by a solar collector flows into the bottom-most basin by natural convection. When the temperature of the water at the bottom disc is above the temperature of the water in the collector, the circulation will stop automatically. The solar collector is composed of a cuboid water sink with a double glass cover lid. Like Fig. 5.19 shows, thermal insulation material is covering all around the sink except the top surface, and the glass cover plate is insulated with a double-layer transparent honeycomb.

For a triple-effect solar distiller, energy balance equations of the glass cover plate and absorption bottom basin, respectively, are given below.

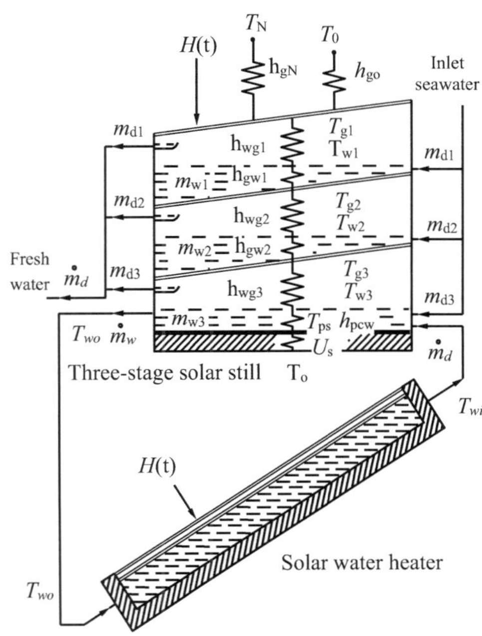

Figure 5.18

Schematic diagram of three-stage basin still with enhanced heating of solar water heater.

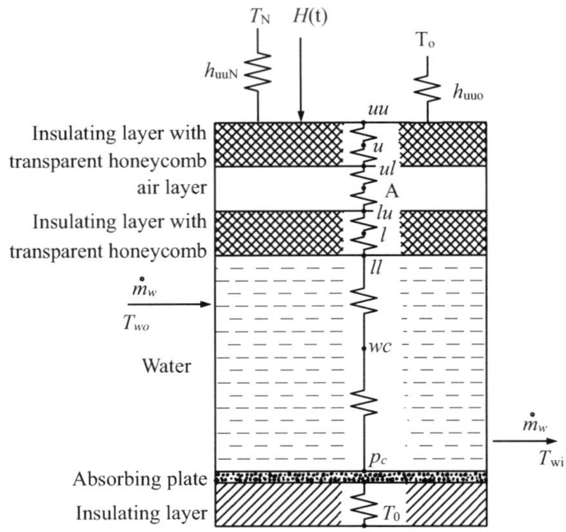

Figure 5.19
Cross-section of a solar water heater.

$$I_{g1} + h_{wg1}(T_{w1} - T_{g1}) = h_{g\infty}(T_{g1} - T_\infty) + h_{go}(T_{g1} - T_0) \tag{5.28}$$

$$I_{g2} + h_{wg2}(T_{w2} - T_{g2}) = h_{gw1}(T_{g2} - T_{w1}) \tag{5.29}$$

$$I_{g3} + h_{wg3}(T_{w3} - T_{g3}) = h_{gw2}(T_{g3} - T_{w2}) \tag{5.30}$$

$$I_{ps} = h_{psw}(T_{ps} - T_{w3}) + U_s(T_{ps} - T_0) \tag{5.31}$$

Here,

$$I_{g1} = \alpha_g H_s(t), \quad I_{g2} = \alpha_g \tau_g \tau_{w1} H_s(t), \quad I_{g3} = \alpha_g \tau_g^2 \tau_{w2} H_s(t) \quad I_{ps} = \alpha_{p1} \tau_g^3 \tau_{w1} \tau_{w2} \tau_{w3} H_s(t)$$

The transmissivity of water is changed with the depth of water x_i, and the relationship between them is represented by Eq. (5.32).

$$\tau_{wi} = (1 - r) \sum_{j=1}^{5} v_j \exp(-\mu_j x_i) \tag{5.32}$$

For this equation, μ_j and v_j, respectively, represent the absorption factor and related weight factor within the spectrum range of j. The values are shown in Table 5.3.

The energy balance equations for water body in the three basins are given below.

$$I_{w1} + h_{gw1}(T_{g2} - T_{w1}) + \frac{d(m_{d1} c_{w0} T_0 / A_s)}{dt} = h_{wg1}(T_{w1} - T_{g1}) + \frac{m_{w1} c_{w1}}{A_s} \frac{dT_{w1}}{dt} \tag{5.33}$$

$$I_{w2} + h_{gw2}(T_{g3} - T_{w2}) + \frac{d(m_{d2} c_{w0} T_0 / A_s)}{dt} = h_{wg2}(T_{w2} - T_{g2}) + \frac{m_{w2} c_{w2}}{A_s} \frac{dT_{w2}}{dt} \tag{5.34}$$

Table 5.3: Absorption factor and related weight factor of solar radiation spectrum for water.

j	Spectrum range	Weighting coefficient (ν_j)	Absorption factor (μ_j)
1	$0 \leqq \lambda \leqq 0.36$	0.237	0.032
2	$0.36 \leqq \lambda \leqq 1.06$	0.193	0.450
3	$1.06 \leqq \lambda \leqq 1.3$	0.167	3.000
4	$1.3 \leqq \lambda \leqq 1.6$	0.179	35.000
5	$1.6 \leqq \lambda \leqq \infty$	0.224	255.000

λ is the solar wavelength.

$$I_{w3} + h_{psw}(T_{ps} - T_{w3}) + \frac{d(m_{d3}c_{w0}T_0)}{dt}\frac{1}{A_s} + \dot{m}_w c_w (T_{w1} - T_{w0})\frac{1}{A_s} = h_{wg3}(T_{w3} - T_{g3}) + \frac{m_{w3}c_{w3}}{A_s}\frac{dT_{w3}}{dt} \tag{5.35}$$

For the above equations

$$I_{w1} = \alpha_{w1}\tau_g H_s(t), \quad I_{w2} = \alpha_{w2}\tau_g^2\tau_{w1}H_s(t), \quad I_{w3} = \alpha_{w3}\tau_g^3\tau_{w2}H_s(t)$$

For solar collector, the energy balance equations are listed below based on the thermal resistance circuit given in Fig. 5.19. And the meanings of subscripts including *uu*, *u*, *ul*, *l*, *ll*, *wc*, and *pc* are shown in the figure.

$$U_U(T_u - T_{uu}) = h_{uu\infty}(T_{uu} - T_\infty) + h_{uu0}(T_{uu} - T_0) \tag{5.36}$$

$$I_u + U_u(T_{ul} - T_u) = U_u(T_u - T_{uu}) + m_u c_u (dT_u/dt) \tag{5.37}$$

$$U_a(T_a - T_{ul}) + h_a(T_{lu} - T_{ul}) = U_u(T_{ul} - T_u) \tag{5.38}$$

$$U_a(T_{lu} - T_a) = U_a(T_a - T_{ul}) \tag{5.39}$$

$$U_l(T_l - T_{lu}) = U_a(T_{la} - T_a) + h_a(T_{lu} - T_{ul}) \tag{5.40}$$

$$I_l + U_l(T_{ll} - T_l) = U_l(T_{ll} - T_l) \tag{5.41}$$

$$h_{wu}(T_{wc} - T_{ll}) = U_l(T_{ll} - T_l) \tag{5.42}$$

$$I_{wc} + h_{wl}(T_{pc} - T_{wc}) = h_{wu}(T_{wc} - T_{ll}) + m_{wc}c_{wc}(dT_{wc}/dt)/A_c + m_p c_p (dT_{pc}/dt)A_c + \dot{m}_w c_w (T_{w0} - T_{wi})/A_c \tag{5.43}$$

$$I_{pc} = h_{wl}(T_{pc} - T_{wc}) + U_c(T_{pc} - T_0) \tag{5.44}$$

For the above equations

$$I_u = \alpha_u H_c(t), \quad I_l = \alpha_l \tau_u H_c(t), \quad I_{wc} = \alpha_{wc}\tau_u\tau_l H_c(t), \quad I_{pc} = \alpha_{pc}\tau_{wc}\tau_u\tau_l H_c(t) \tag{5.45}$$

For the given $H_c(t)$, $H_s(t)$ and T_a, the above Eqs. (5.36)–(5.45) can be solved to get the system water production and how temperature of the heated water within the system changed with time. From these, the optimum matching relationship of $A_S \sim A_C$ can be reasonably predicted.

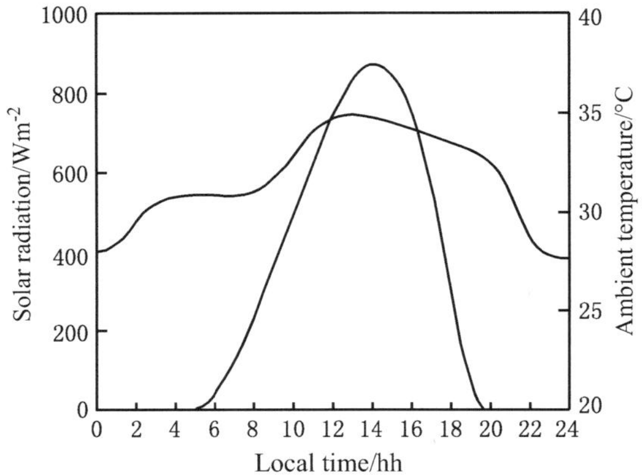

Figure 5.20
The change of solar irradiance and ambient temperature with local time.

Baharna et al.'s research result points out that because technology of active heating by collector is adopted for triple-effect distiller, water production could increase 2–2.5 times (when $A_S \approx A_C$). Thus the collector affects the system significantly. For given solar radiation intensity and environmental temperature change in one day, which is shown in Fig. 5.20, water production of the system and water temperature at the bottom-most of the system changing with time are represented by Figs. 5.21 and 5.22. The relationship between A_C/A_S and water production of the distiller can be seen in the curve of Fig. 5.23. The curve shows that, when $A_C \approx A_S$, the maximum increasing rate of water can be reached.

For the energy balance equations in this section, the meanings of relative physical signs are listed below.

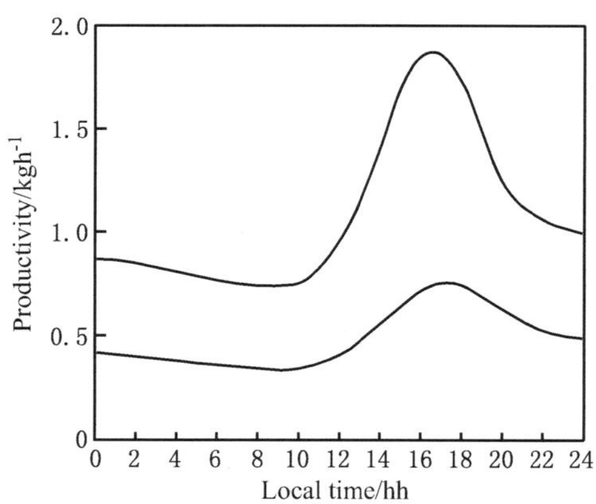

Figure 5.21
The change of water production with local time.

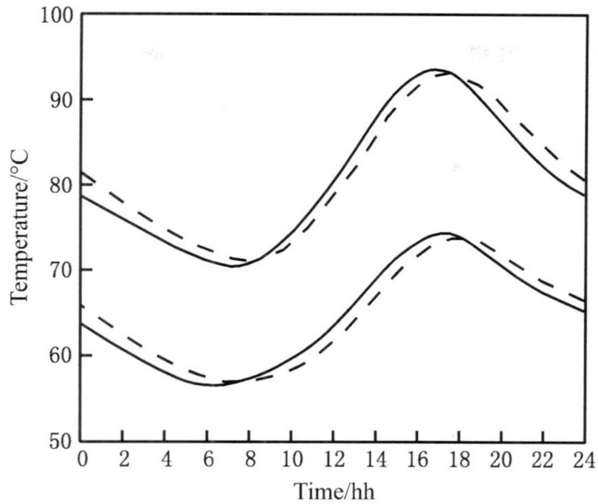

Figure 5.22
Variation of water temperature in the lowest basin with time.

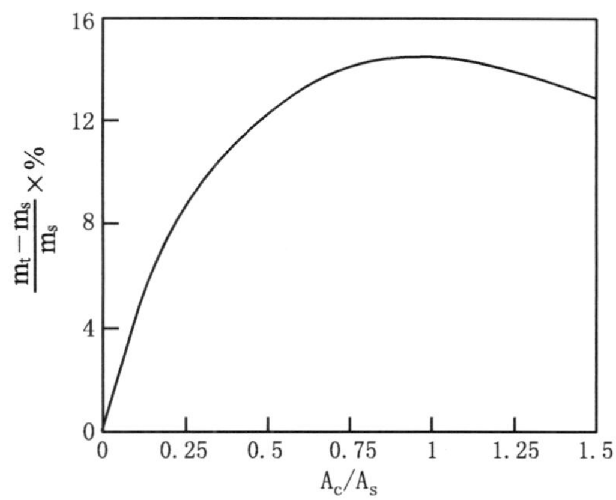

Figure 5.23
Productivity changes with A_c/A_s.

A, surface area, m^2;
c, specific heat, J/kg°C;
h, convective coefficient, W/m^2°C;
H, instantaneous solar radiation intensity, W/m^2;
m, mass of water in basin, kg;
$\dot{m}$, mass flow rate; kg/s;
r, reflectivity;
t, time;
T, temperature, °C;
U, total heat transfer coefficient, W/m^2;
x, thickness of water layer, m;

α, absoptivity;
τ, transmissivity.

Subscripts:

a, air;
c, collector;
g, glass cover plate;
i, surface, number of stage of basin or inlet condition;
o, environment or outlet condition;
p, absorption plate;
s, distiller;
w, water basin;
∞, sky;
m_t, mass of water of whole system;
m_s, freshwater production.

5.7 Multieffect Basin-Type Solar Still Heated Actively by Solar Collectors

There are a lot in common between a multistage basin still in this section and the multieffect basin solar still from the previous section. However, part of the multistage basin still does not receive solar energy directly, it only works as an evaporation and condensation chamber. Thus, this part can be designed so small that they can even be installed indoors to save investment of floor space and equipment.

Adhikari [15] studied a multistage basin solar still with active heating collector in the early days. The system that Adhikari gave is shown in Fig. 5.24. The shapes and dimensions of each basin are shown in Fig. 5.25. The number of stages of the system is

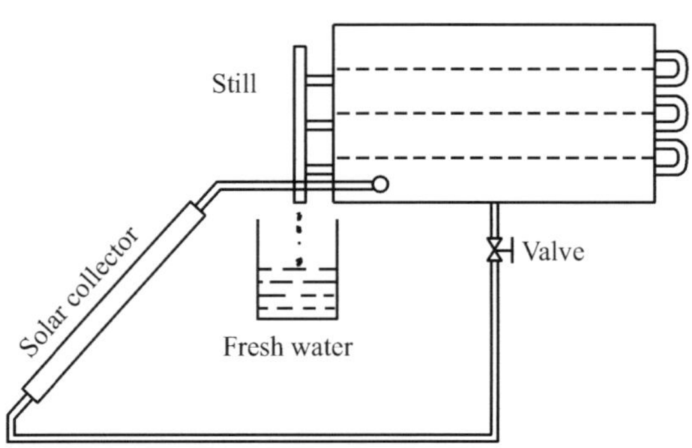

Figure 5.24
Multistage basin solar still.

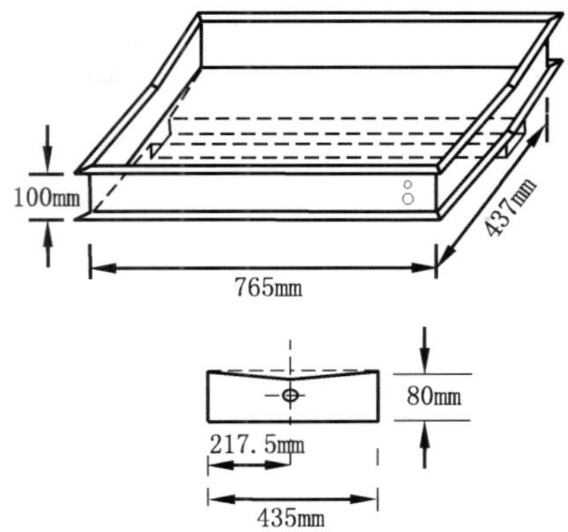

Figure 5.25
Dimensions of the basin.

adjustable from one to four. Each basin is made of aluminum plate. The opening area of the basin is 0.330 m^2, the average height of the basin is 0.09 m, 3 mm rubber sheeting is used to seal the gap between basins. During the experiment, an electrical heater is used to substitute for the solar collector. This test is carried out under the steady-state conditions. Computer simulation is also used to calculate the system.

The research of Adhikari points out that, under steady-state conditions, the water temperature of each basin will be changed with the power of the electrical heater that is provided at the bottom basin as shown in Fig. 5.26. The result shows that, as input of calorific power increases, the water temperature of each basin is increased notably. However, the temperature difference of water between the stages is almost the same.

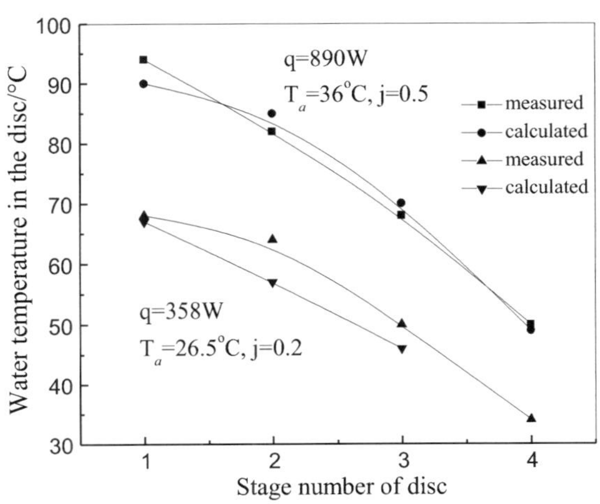

Figure 5.26
Variation of water temperature with electric heating power.

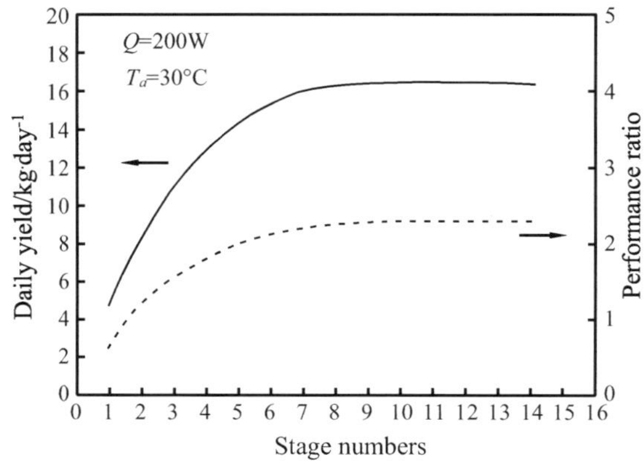

Figure 5.27
The variation of daily yield and PR with stage numbers.

How water production of the system and performance coefficient change with number of stages is shown in Fig. 5.27. By observing, when output power is not large, it is more appropriate to choose 4–5 stages. When there are more than six stages, the increase cannot improve performance sensibly, and even increases unnecessary investment and complexity.

The literature [16] also gave similar research about a multistage basin still. Steady-state experiment is simulated with the help of an electrical heater. The result is given under the actual weather condition, using an evacuated tube solar collector.

The simulation experimental results were completed under different stages. And the relationship between input power and total water production is shown in Fig. 5.28. It can be found from the figure that the three-stage unit is more of an advantage compared to the two-stage or single-stage unit. Under the same and steady heat input power situation,

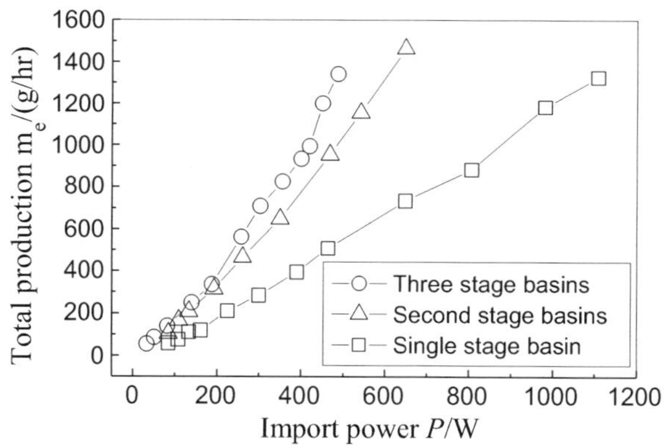

Figure 5.28
Curve of the relation between total yield and input power.

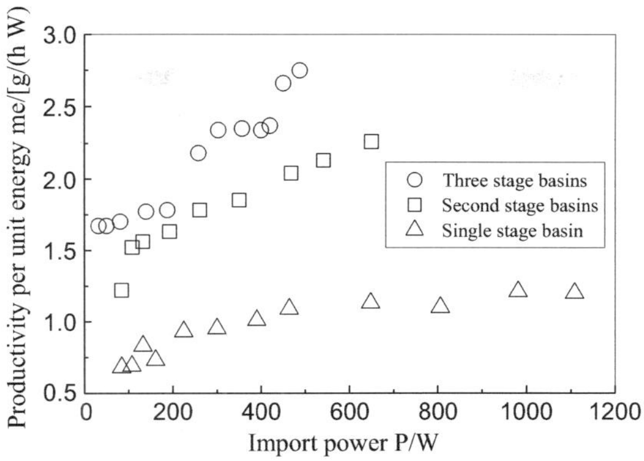

Figure 5.29
Curve of the relation between yield rate per unit energy consumption and input power.

a larger number of stages is favorable for distilled water production. It is also shown in the diagram that, for these three different numbers of stage situations, water production per hour shows a linear relationship with input power.

Distilled water total production per unit energy consumption is called water production rate per unit energy consumption of the system, or productivity. Water production rate is a significant index to evaluate system performance. Fig. 5.29 gives the relations between energy consumption and water production rate for three different basin numbers. The results show that, under the same heat input power, the three-stage system has better performance than the two-stage or single-stage. And the difference between them is more obvious when input power is increased.

The experimental results also show that to raise the water temperature in the bottom basin is good for elevating the distilled water production. Fig. 5.30 shows the

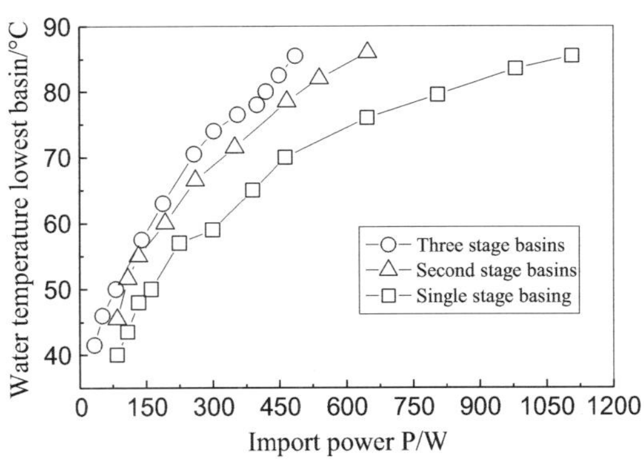

Figure 5.30
Curve of the relation between brine temperature of the bottom dish and input power.

relationship of the water temperature in the bottom basin and input power under the condition of three stacked basin numbers. The results indicate that increasing the number of stacked basins is good for elevating the water temperature in the bottom basin so as to increase production. It is noticeable that using high operation temperature needs more expensive solar collectors, which increases the investment of the equipment. At the same time, the high operation temperature may decrease the efficiency of the solar collection system. In the practical engineering system, comprehensive consideration is necessary.

To understand the actual operation cases of a multistage basin solar still, the literature [17] presented an actual operation system driven by a vacuum tube solar collector. The profile of the experimental system is shown in Fig. 5.31. The structure, sealing situation, and insulation of the evaporation chambers are the same as Adhikari's. The solar collector consists of 16 vacuum tubes installed in horizontal direction with the earth. The solar collector has a stainless steel plate as the subreflector under the vacuum tubes to increase the solar heat. The hot water produced by the solar collector is sent into the lowest evaporation chamber to produce the evaporation process. The total area of the solar collector is 1.32 m^2. The environment wind speed is in the range of 0.0−2.5 m/s.

The experiment was carried out mainly for the three-stage unit. During the test period, the variations of solar irradiation, the temperature in each basin and total distilled water production on a typical sunny day with the local time are shown in Figs. 5.32−5.34. In the results, it can be seen that the solar still with three-stage basins can raise the temperature rapidly and operate in relatively high temperature range, which is favorable to increase freshwater yield rate. In the figure, T_{w1}, T_{w2}, T_{w3}, and T_a are the water temperature in the first, second, and third basins and ambient temperature, respectively. m_1, m_2, m_3, and m are the distilled water productivity in first, second and third evaporation chambers and total productivity.

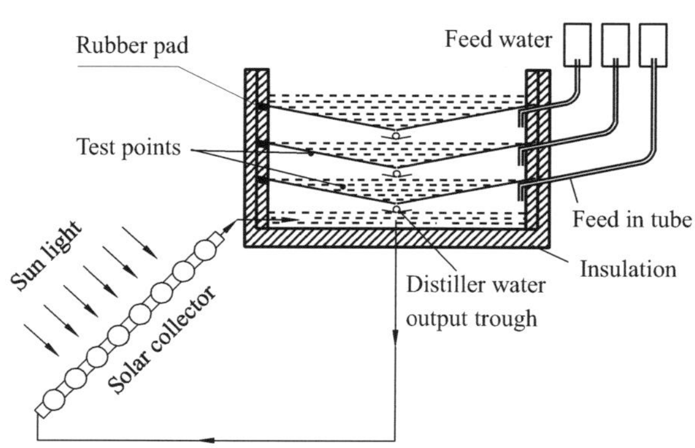

Figure 5.31
Sectional view of the experimental apparatus.

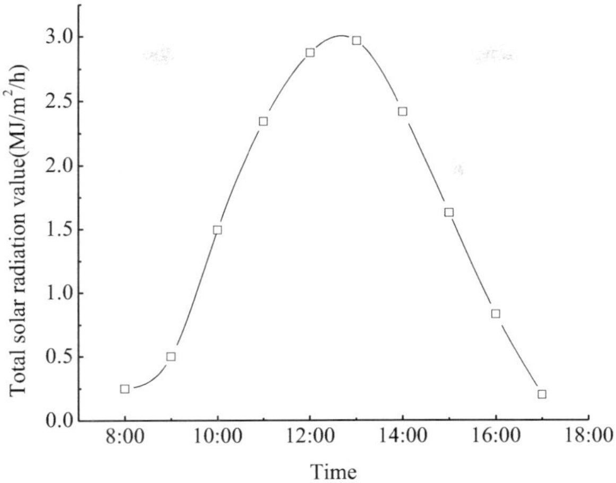

Figure 5.32
Variation of the total solar radiation value with local Beijing time.

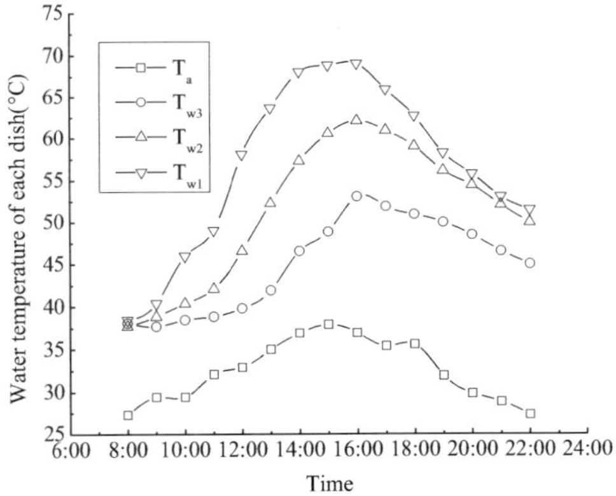

Figure 5.33
Variation of the water temperature of each dish with local Beijing time.

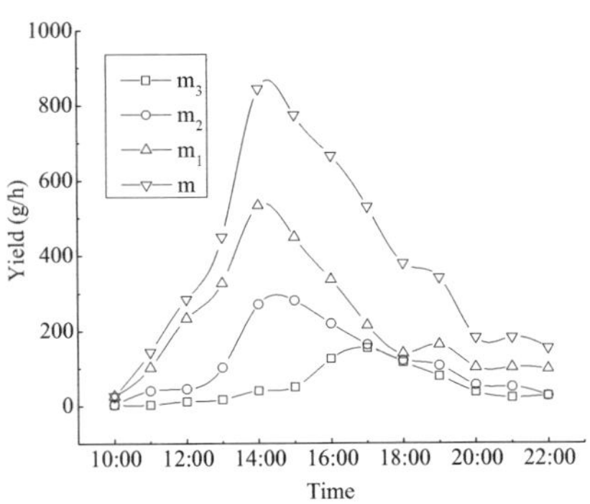

Figure 5.34
Variation of the yield per hour with local Beijing time.

According to the formula to calculate efficiency of traditional solar still

$$\eta = \frac{m \cdot h_{fg}}{H} \tag{5.46}$$

where H and h_{fg} are the total solar radiation and vaporization latent of vapor, respectively. The total efficiency was calculated by tested data to be 50–60%. Although, the efficiency of this system is still low, it still has some practical value because it has the characteristics of no moving parts, material easy gotten, and low operating cost.

Based on the experimental results, some conclusions are obtained as follows.

1. Multistage basin solar still has more efficiency because the latency of the vapor is reused many times. Generally, the device is taken to be suitable for three stage, four stage, and five stage when the operation temperature is 70°C, 80°C, and 90°C, respectively.
2. The better the sunshine, the higher the operation temperature and the more productivity. Therefore, the vacuum tube solar collector is available.
3. The yield water time of the device can extend to night or even to the following morning because the insulation is good. This way, it can utilize the temperature difference between day and night to increase the distilled water production.

5.8 Multistage Stacked Tray Solar Still With Enhanced Condensing Faces

The bottom crease multieffect basin solar still is an improved system from that discussed in the previous section. The multieffect basin solar still presented in the previous section is still with several shortcomings as follows. (1) The basin only constitutes of one angle trough, for the distillate water flowing to the bottom of the trough with falling film, the trough angle is not greater than 150°, otherwise the distilled water drops may fall down to the saline water basin again because of gravity. As a result, the seawater mass in one basin becomes too great and even more when the device is larger so that the thermal mass is also big which is not conducive to enhancing the temperature of the device. (2) The bottom of the basin is only two plate surfaces, the condensation surface is limited. (3) The single-trough structure of the basin lets the distance between basins become larger, which increases the heat and mass transfer resistance so as to reducing the performance of the device.

In order to overcome the above defects, the literature [18] designed a bottom crease multieffect basin solar still as shown in Fig. 5.35 and tested its performance with the electric heater as a simulation source. That is to say, the solar collector (the dotted line part of Fig. 5.35) is replaced by the electric heater.

The basins are made of galvanized iron sheet with 0.8 mm thickness. The entrance area is $0.45 \times 0.45\ m^2$. The outside of the device is insulated by a polystyrene foam boa with

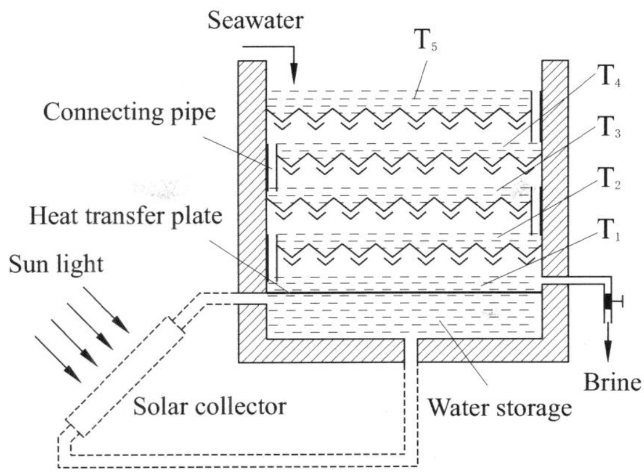

Figure 5.35
Schematic diagram of the bottom crease multieffect basin solar still.

thickness 5 cm. The edges of the basins are separated by rubber gaskets with 3 mm thickness to avoid the vapor leaking out and causing direct conduction heat transfer. The bottom crease is formed by a regular triangle with 2 cm length on a side, as shown in Fig. 5.35. The distilled water on the back of the basin flows down to the bottom of each regular triangle and drops to the V trough just below the regular triangle. Finally, the distilled water is sent outside from the V trough. The average height from the free water surface to the back of the above basin is 4 cm.

In the experiment, the seawater is sent into the device from the top basin, and then sent into the next basin and so on until the lowest stage. The seawater is heated step by step in every basin. Also, the seawater is evaporated in every stage and at the same time receives latent condensation. In this way, the latent heat of vapor and sensible heat of seawater are used sufficiently, which is beneficial to improve the productivity per unit energy consumption.

It can be found from the experiment that the multieffect basin solar still has higher efficiency because it can operate under higher temperature. Therefore, the actual solar still, especially the multieffect solar still, should work under high temperature as far as possible. For example, the vacuum tube solar collector or concentrating solar collector should be chosen to provide the heat energy for the stills. However, it is noticeable that the solar collection system will have lower efficiency when it operates under a higher temperature section. Thus, the suitable operation temperature should be considered for the total efficiency maximization of the solar collector and distiller. The experimental results indicate that only for the distiller, three stages are suitable when the operation temperature is 70°C and four stages are available when the operation temperature is 80°C. If the operation temperature reaches 90°C, a five stages basin still may be chosen. Generally, the more the stage numbers, the less the temperature difference between the stages is. Therefore, the stage numbers should not be too many.

5.8.1 The Transient Performance of the Device

Because the multiple channel enhanced heat and mass transfer measures have been used in every basin, the heat transfer rate is enhanced. On the other hand, the seawater layer in every basin is thin at only about several kilograms. Therefore, the device is easy to be heated. Depending on the measurements, the still can produce freshwater after only 20 min operation. However, it needs about 180−200 min to reach the steady state. After starting, the water temperature variation in the basins with time is shown in Fig. 5.36. In this time, the power supply is 2.2 A × 220 V and the device has four stages. The distilled water productivity variation per 20-min interval in two power supplies with time is shown in Fig. 5.37. From these figures, it is easy found that to reach the steady state, including steady productivity and steady temperature, needs about 200 min.

5.8.2 The Relationship of Productivity and Operation Temperature

The temperature of the saline water in the lowest basin is defined as the operation temperature. The total freshwater production in unit time is called productivity. Under the conditions of steady-state operation temperature, the variation of total productivity of the device with different stacked basin numbers with the operation temperature is shown in Fig. 5.38. When the stacked basin number is four, Fig. 5.39 shows the change in production in each basin with the operation temperature.

Fig. 5.38 indicates that the operation temperature has a strong influence on production. The production increases rapidly when the operation temperature and the temperature difference between evaporation and condensation surfaces increase. And the higher the operating temperature, the faster the water yield is increased. That is to say, the device has an advantage when it operates in higher temperature. Fig. 5.39 shows that the operation

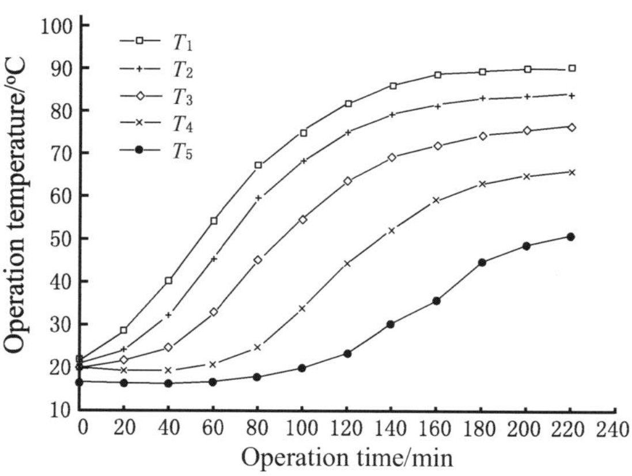

Figure 5.36
Variation of water temperature in different basins with time.

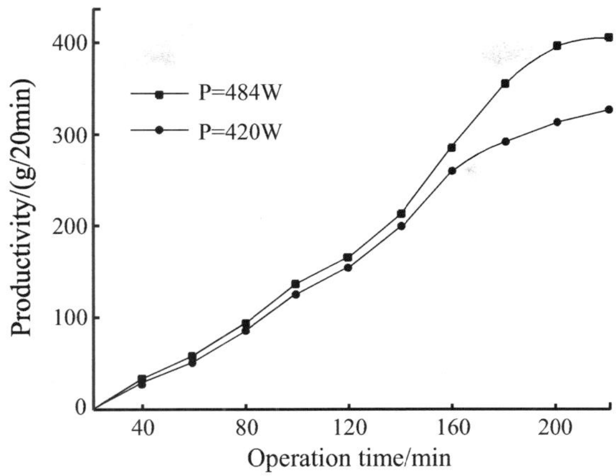

Figure 5.37
Variation of productivity with time.

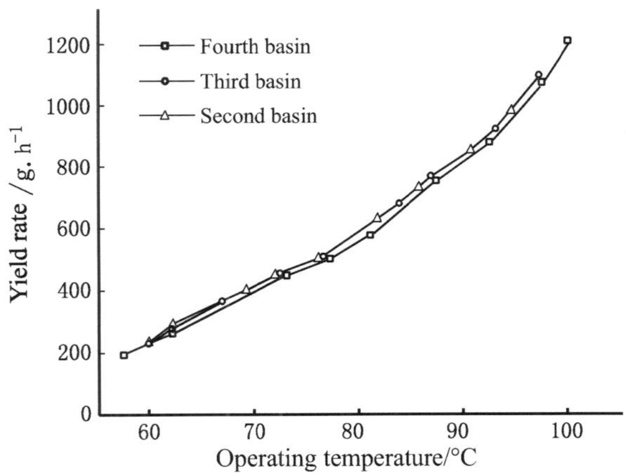

Figure 5.38
The variation of yield rate with operating temperature.

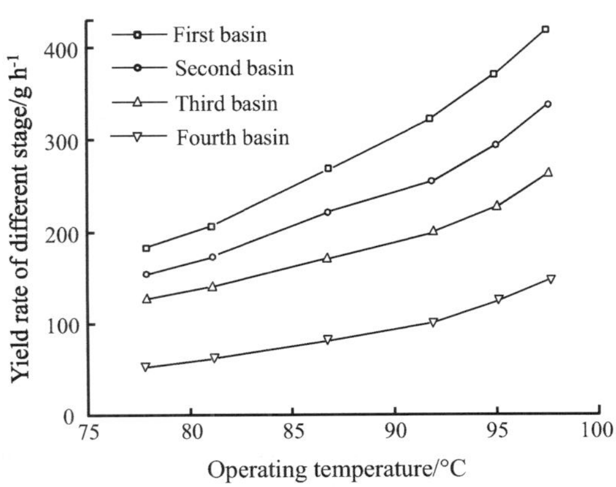

Figure 5.39
The variation of yield rate of different stages with operating temperature.

temperature also has a strong influence on each basin. However, the top basin or above basin only has low production when the operation temperature is relatively low. Only in high operation temperatures does the freshwater yield rate in the above basin obviously increase. Therefore, it is significant to add the basin number only when the operation temperature is high.

The experimental results also indicate that the freshwater yield rate of the device with different basin numbers is basically the same when the operation temperature is the same. It indicates that when the difference between the operation temperature and the ambient temperature is determined, the basin numbers only have a small influence on total productivity. However, their power consumption is different. When the basin number is small, it needs more power to maintain the same operation temperature.

5.8.3 Steady-State Performance of the Device

The performance coefficient is a very important index to evaluate the solar still. The bigger the performance coefficient is, the better the device performance. According to the energy import relationship, the steady-state performance coefficient of this kind of device is given by

$$\eta = \frac{m_e h_{fg}}{Q} = \frac{m_e h_{fg}}{P\Delta t} \tag{5.47}$$

where m_e is the distilled water productivity; h_{fg} is latent vaporization of water; Q is total heat energy to the system; P is the total power consumption; and Δt is the operation time. Because the water temperature in different basins is different, the latent vaporization is also different. Therefore, the performance coefficient should be changed to

$$\eta = \sum_i \frac{m_{e,i} h_{fg,i}}{P\Delta t} \tag{5.48}$$

where, the subscript i is the stage number from bottom to top, $m_{e,i}$ is production in the ith basin.

As a result, it can be found that the device has a relatively high performance coefficient, generally more than 100%, which is a guarantee for the device to achieve more freshwater. Fig. 5.40 shows the variation of the performance coefficient of the device with different basin numbers with the steady operation temperature. The results show that the higher the operation temperature, the more performance coefficient. They indicate that the system will have an advantage when it operates in the higher temperature range. Generally, when the operation temperature is the same, a four-stage unit is better than a three-stage unit and better than a two-stage unit. Experimental results also illustrate that many factors can affect the performance coefficient, especially the feeding flow rate having a relatively large

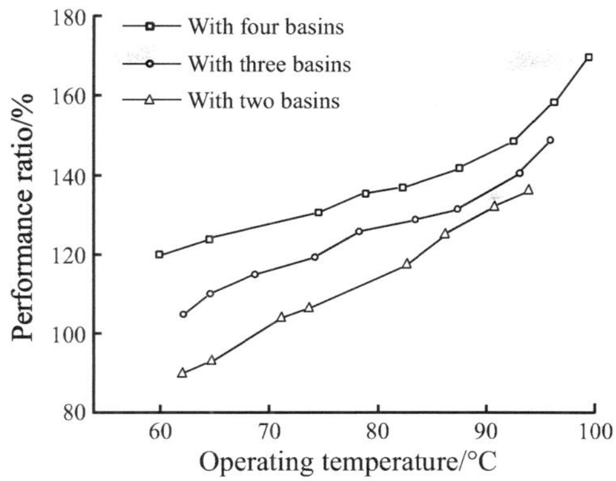

Figure 5.40
The variation of performance ratio with operating temperature.

effect on the performance coefficient because the brine will carry some of the sensible heat when it is removed. The performance coefficient basically rises when the feeding flow rate decreases. However, the feeding flow rate has the lowest requirement for preventing brine crystallization. It is suitable for the feeding flow rate to be 1.5 times that of freshwater production.

5.8.4 The Relationship of the Operation Temperature and Input Power

Obviously, to raise the operation temperature (the saline water temperature in the lowest basin) is beneficial to improving the production of distilled water. Therefore, a higher operation temperature is hoped for in the actual system. Fig. 5.41 shows the relationship of the operation temperature and input power in three-unit structures. As a result, adding the basin numbers is beneficial to elevating the operation temperature.

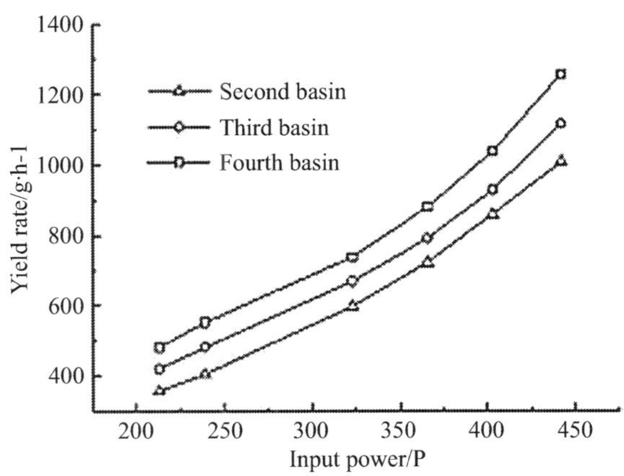

Figure 5.41
Variation of the yield rate with input power.

5.8.5 Conclusions

1. A multistage basin solar still can utilize sufficiently the latent heat of vapor. The device will have better performance coefficient when it works in a high operation temperature range (>70°C). And then, the higher the operation temperature, the more stacked basin numbers are available and the better the performance coefficient.
2. This device replaced the flat-bottomed single trough in the literature [3] to bottom crease multiple troughs, which reduces the distance between basins. As a result, this device has higher productivity per unit energy consumption. For example, the four-stage device in this work increases by 30% the yield rate compared with that in the literature [3] when they all operate at 90°C. It indicates that this device has better performance.
3. The total production of the device has a linear relationship with the input energy. The more the stacked basin numbers (more stages), the bigger the gradient of the linear relationship.

5.9 Multistage Stacked Tray Solar Still Driven by Double Heat Sources

The earlier sections introduced the multieffect stacked tray solar distiller, to which the solar collector and distiller are detached. This is not of benefit for management and commercialization, meanwhile, the water vapor on the top basin is wasted and it cannot collect solar energy from the top disc, which reduces the performance of the system. For this reason, the literature [19] further proposed a multistage stacked tray solar still driven by double heat sources based on the former foundation. The literature [19,20] has reported its steady state and performance under real weather.

5.9.1 Characteristics and Operating Principle of the Solar Still

1. Structure parameters and characteristics
   The schematic of the multieffect solar still with enhanced condensation surface is shown in Fig. 5.42. The seawater in a solar still is heated by two kinds of heat sources. One is the vacuum tube collector installed on the bottom with dimensions of 0.1 m (diameter) × 1.88 m (length) (16 pieces), and the other is the black titanium alloy coatings on the upper surface of the top stacked tray which can adsorb the sunlight and heat the seawater, with absorptivity of 0.6. The dimension of the glass cover on the top is 2 m × 0.6 m × 0.005 m, with transmissivity of 0.9. The inclinations of the vacuum tube collector and glass cover are all 20 degrees, which is applied for a low-latitude zone. The stacked trays are made of stainless steel with thickness of 1 mm, and are folded into a corrugated shape (isosceles right-angled triangle) with right side length of 3 cm. The total number of stacked trays is three for this solar still. The V-shape troughs downside the stacked trays are used to collect the freshwater condensed on the lower

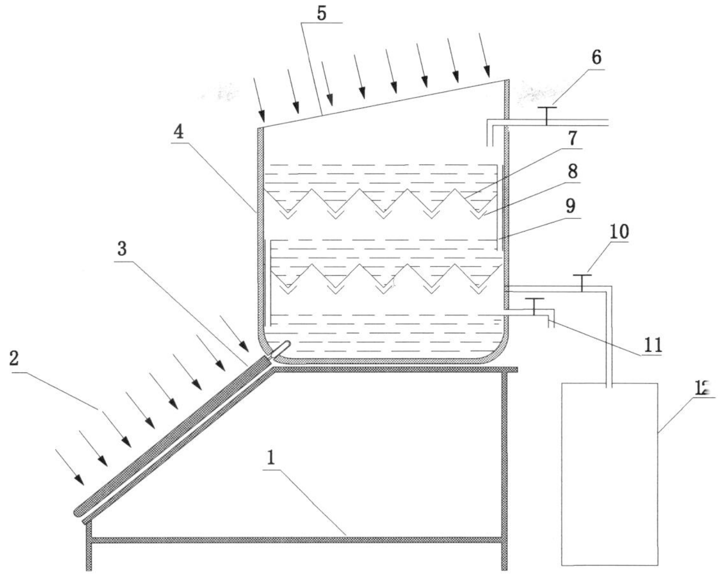

Figure 5.42
Schematic of the multieffect solar still: 1, holder; 2, sunlight; 3, vacuum tube collector; 4, water tank; 5, glass cover; 6, seawater valve; 7, stacked tray; 8, V-shape troughs; 9, connecting tube; 10, freshwater valve; 11, overflow valve; 12, freshwater tank.

surface of the stacked trays. The surroundings of the solar still are insulated with polyurethane foams (8 cm thick) to reduce heat loss. The average height between the water surface and the lower surface of the stacked trays is 6 cm, and the condensation area of the corrugated shape stacked tray is 1.7 m^2. For the bottom-stacked tray, the seawater storage capacity is about 140 kg, while for other stacked trays the capacity is about 30 kg. Therefore, the total seawater capacity of the solar still is about 200 kg. For the designed multieffect solar still, the main characteristic is the application of the corrugated shape stacked trays. Compared to the traditional solar still, it has the following merits: (1) the generated freshwater can quickly drip because the flow distance of the condensed water on the condensation surface is short and the inclination of the condensation surface is large; (2) the condensation resistance of the water vapor is reduced due to the thin liquid film; (3) the area of the condensation surface is increased, which can enhance the heat transfer efficiency of the water vapor. Therefore, the designed structure (corrugated shape stacked trays) is called an enhanced condensation surface. In addition, for the present solar still, the simultaneous heating from the top and bottom efficiently uses the solar energy and will increase the freshwater yield.

2. Operating principle

   When the solar still works, the vacuum tube collector first absorbs the solar energy and heats the seawater in the bottom stacked tray. The generated water vapor due to heating flows upward and then condensates on the corrugated shape enhanced condensation surface, while the latent heat of condensation continuously heats the seawater in the superior stacked tray. The freshwater due to condensation flows along the corrugated shape condensation surface and drips into the V-shape troughs. The heat transfers from the bottom to the top, with a small proportion dissipating to the surroundings through the polyurethane foam insulation. In the meantime, the upper sunlight penetrates the glass cover and heats the seawater in the top stacked tray, making it evaporate into water vapor and then condensate on the downward surface of the glass cover. All the generated freshwater finally flows into the freshwater tank.

5.9.2 Experiment Results With Steady Heating

In steady-state experiments, an electric heater is used to replace the vacuum tube solar collector, which shared power of 2 kW with eight electric heaters and without sunlight on the top tray. The temperature of each measuring point inside the device is used to extract the signal through the calibration of the K type thermocouple and read by the XMD-16P type temperature indicator. A photograph in steady-state experiment is shown in Fig. 5.43. The the experiment under the actual weather is shown in Fig. 5.44.

The electric power consumed by the electric heater is measured by the voltmeter and the current meter. In order to make the input power stable, the voltage regulator is adopted in the circuit. In the experiment, tap water is used instead of seawater, the temperature is determined every 10 min and the amount of distilled water is measured every half an hour by commercial use electronic measurement with an error margin of 5%. This experiment is carried out in the room, the change scope of the environment temperature and the relative humidity is small (range 9.0−11.0°C and 15−30%, respectively), basically this is a stable environment.

Figure 5.43
Photograph of the apparatus of experiments with constant power heating.

Figure 5.44
Photograph of the apparatus worked on actual weather.

Because of the multichannel enhanced heat and mass transfer measures adopted in each iteration stack of the device, the seawater layer of each stack is very thin. In addition, the amount of water stored in the two lap trays is only about 30 kg, and it is easy to be heated by the latent heat. However, due to the heat transfer between plates mainly being based on the seawater steam, the heat transfer rate is limited and the system is difficult to reach a stable state. In the experiment, when the device is completely cooled, the seawater in the lower layer of the device is applied to the thermal energy of 2 kW and the temperature rise rate of the system is measured. After the power is switched on, the variation of water temperature in each effect with the running time is shown in Fig. 5.45. The variation of the total water yield and the water yield per 30 min is shown in Fig. 5.46.

It can be seen from Fig. 5.45 that, after the heating system, the temperature of all effect basins are exhibited increased linearly and do not obvious slow until increased to more

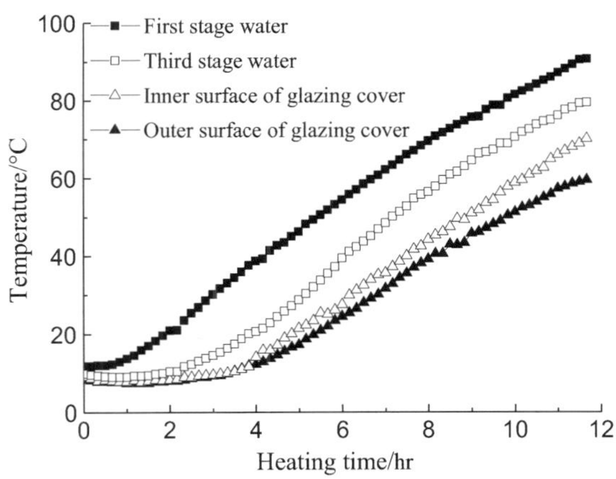

Figure 5.45
Variation of water temperature of each measuring point with the running time.

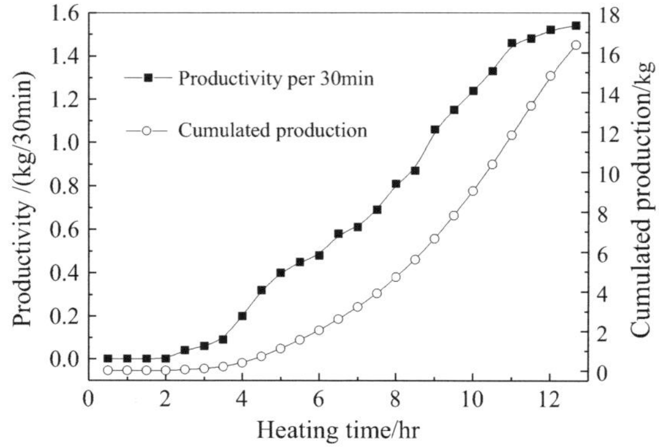

Figure 5.46

Variation of the total water yield and the water yield per 30 min with running time.

than 90°C. This phenomenon indicates that system power dissipation is very low, sound insulation and heat transfer outward mainly through the glass cover plate. Fig. 5.46 shows that, under the conditions of 2 kW constant power heating, the water production rate reaches a stable state after the system has run for 11 h and the lowest temperature of the device reaches more than 90°C at this time. As the starting temperature of the system is very low, distilled water starts to be produced about 2 h after starting running. When the water temperature reached 65°C, the yield was significantly improved.

As the raw water in the device and each basin has a large heat capacity, so there will be some freshwater output in the night. In order to evaluate the performance of the device at night, the power down cooling and water yield performance test experiments were done, and the results are shown in Figs. 5.47 and 5.48. The results of Fig. 5.47 show that, when the

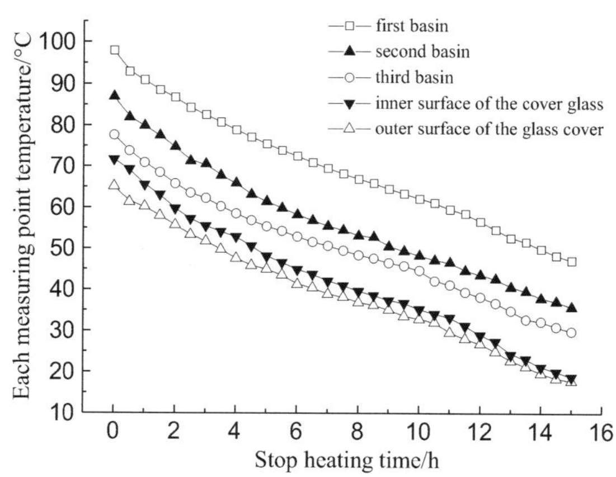

Figure 5.47

The temperature changes with time without thermal input.

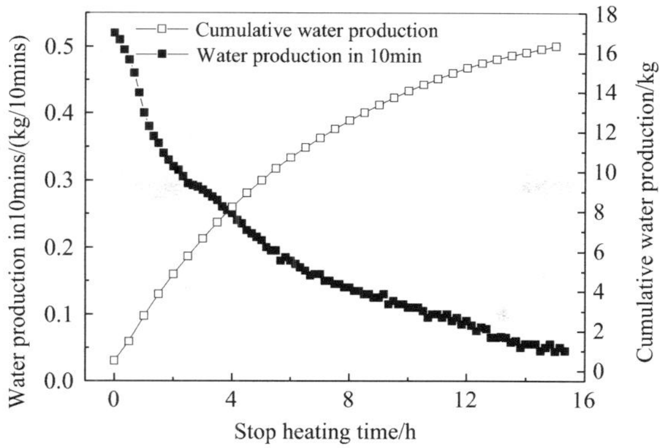

Figure 5.48
The water production changes with time without thermal input.

device in the environment of 10°C, the lowest level of raw water temperature can still be maintained at about 50°C after continuous cooling for 15 h. Therefore, when the device is running the second time, the starting temperature is higher and the device can quickly enter the stage of efficient water production. Fig. 5.48 shows that when the cool down test was performed with the chassis temperature of the device from 90°C, the total water production is about 16 kg after 15 h of running, which shows that the water production at night cannot be ignored. Night water production has also made full use of the low temperature of the night environment, if the device is running at a certain wind speed, the heat dissipation of the glass cover will be more intense, and water production may be higher.

5.9.3 The Relationship Between Water Production Rate and Temperature

Fig. 5.49 shows the relationship between the temperature of the measuring point and the water temperature of the base basin in a stable operating condition. As the heat capacity of the device is relatively large, it is difficult to ensure that the whole device is operating in a stable condition. Fig. 5.49 given the results after 3 h of stable conditions, the temperature of each measuring point may also have a small amount of deviation. Fig. 5 50 shows the variation of the water yield with the operating temperature.

Fig. 5.50 shows that the operating temperature of the system has a considerable impact on the production of water. The water yield increases rapidly when the operating temperature is increased and the difference between the evaporation temperature and the condensing temperature increases. The higher the operating temperature is, the faster the production of the system. When the operating temperature of the device is greater than 65°C, the device enters the stage of efficient water production, and the water yield will reach even more than 1 kg/h. It can clearly be seen that the system operating at high temperature will have a greater advantage, which is similar to the results presented in other literature.

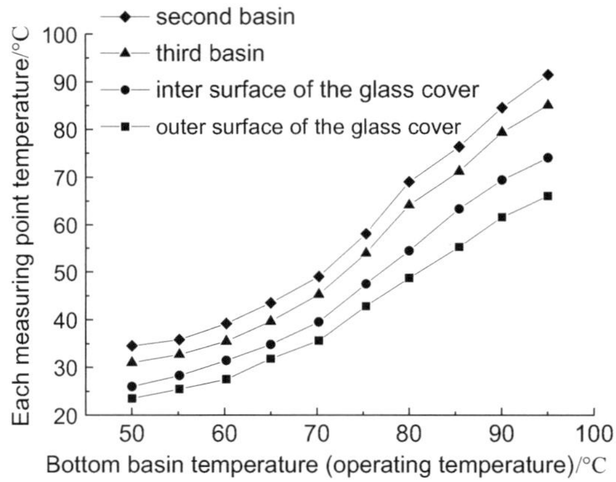

Figure 5.49
The temperature of the measuring point with the temperature of the bottom basin.

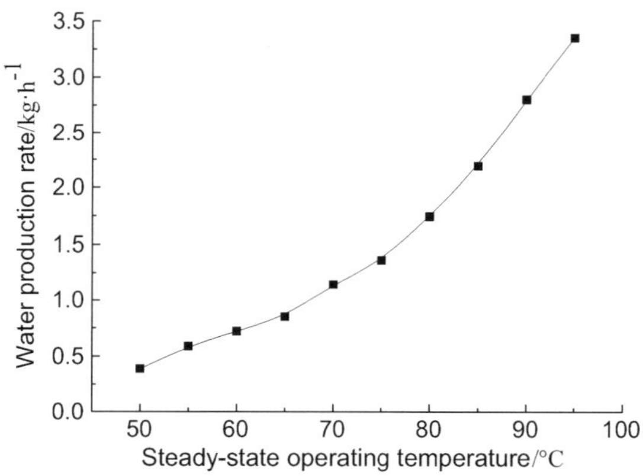

Figure 5.50
The variation of water yield with constant power heating.

5.9.4 Performance Coefficient of the Device

Performance ratio (PR) is an important indicator to characterize system performance. The system performance is superior when the PR is larger. According to the input and output relations, the PR of the device can be given by the following formula:

$$PR = \frac{\sum m_e h_{fg}}{Q} = \frac{\sum m_e h_{fg}}{P\Delta t} \tag{5.49}$$

where, $\sum m_e$, total freshwater production in a working day, kg; h_{fg}, latent heat of vaporization of, approximate to 2.3×10^6 J/kg; Q, the total heat energy of heater supplied to the system, J; P, power consumption of the system, W; Δt, running time of the electric heater, s.

Taking into account the normal circumstances, especially those in South China and even in the southern island (the inclination of the receiver is designed as 20°), ambient temperature is high and the initial temperature of the device on the second day will not be less than ambient temperature. In spite of the device changing some of the water every day, it can be assumed that the starting temperature of the device is 30°C. Based on the temperature, the device is heated to 95°C, and then stops heating to cool and produce water. For the estimation of PR, the all-day measurement results in Figs. 5.46 and 5.48 were calculated. The total water yield of the device can be calculated as 32.6 kg. When heated by an electric heater of 2 kW, the basis basin temperature of the device is heated from 10°C to 96°C for 12.5 h and the water temperature of the device is heated from 30°C to 96°C within 9.5 h. According to the calculation of the above, when the initial water temperature of the device is 50°C, 30°C, 10°C, the PR is 1.66, 1.23, 0.92, respectively. Obviously, the starting temperature of the device has a great impact on its performance, so as far as possible that the system should be ensured to have a higher starting temperature.

5.9.5 Experiment Under Actual Weather Conditions

5.9.5.1 Temperature Characteristics Under Actual Weather Conditions

To accurately test the performance of the designed solar still, experiments under real weather conditions were performed. During the experimental period, the solar radiation meter was installed with an inclination angle of 20°, the same as the angle of the solar collector. Seawater was supplied from the top stacked tray at 8 a.m. every day during the operation period.

Fig. 5.51 shows the measured solar irradiance and accumulated solar irradiance on the solar still on three typical days. Fig. 5.52 shows the temperature change during the test.

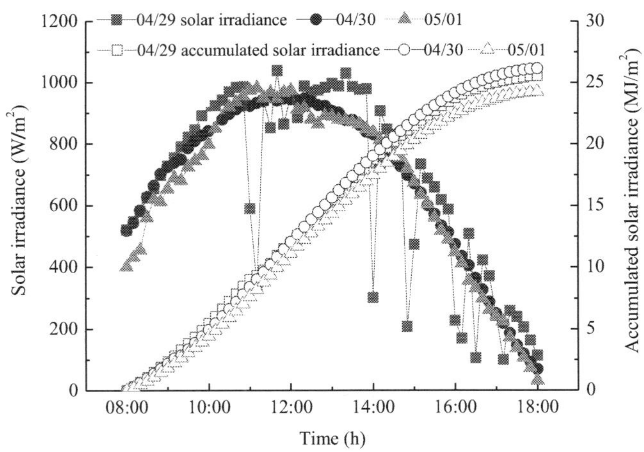

Figure 5.51
Solar irradiance and accumulated solar irradiance of the solar still under real weather.

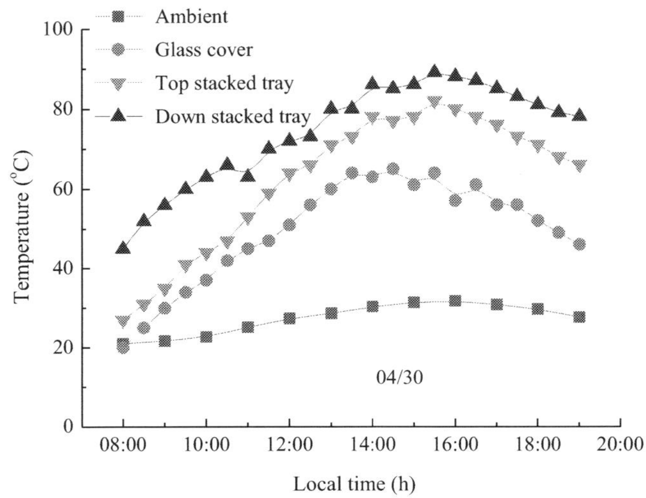

Figure 5.52
Temperature change of the solar still under actual weather on a typical day.

For the test field, the electricity is supplied to 7 p.m., thus the temperature change from 7 p.m. to 8 a.m. of the following day is not given in Fig. 5.52. These two figures indicate that the solar irradiance decreases after 12 a.m., while the operation temperature of the solar still decreases after 4 p.m. This means that the temperature change lags behind the solar irradiance change for about 4 h. For the 3 test days, the accumulated solar irradiance is 25.6, 26.1, and 24.3 MJ/m^2, respectively, and the corresponding mean temperature of seawater in the down stacked tray is 69.8°C, 74.2°C, and 73.0°C, respectively. From these two figures, we can also see that the operation temperature is not only related to the solar irradiance but is also related to the starting temperature (seawater temperature in the down stacked tray). For the first day, due to the relatively low starting temperature (43°C), the mean operation temperature of the seawater in the down stacked tray is lower than that of the next 2 days, although the solar irradiance of this day is not weak. In addition, the starting temperature increases with the increase of operation time of the solar still, changing from 43°C to 49°C. Therefore, for the third test day (01/May), even though the solar irradiance is lower than the other 2 days, the operation temperature still keeps high.

5.9.5.2 Freshwater Yield Characteristics Under Actual Weather Conditions

The freshwater yield is measured every 30 min and the results are shown in Fig. 5.53. The change of accumulated freshwater yield is also included in this figure. By comparing this figure with Figs. 5.51 and 5.52, we can see that the freshwater yield of the solar still under real weather conditions is closely related to the solar irradiance, operation temperature,

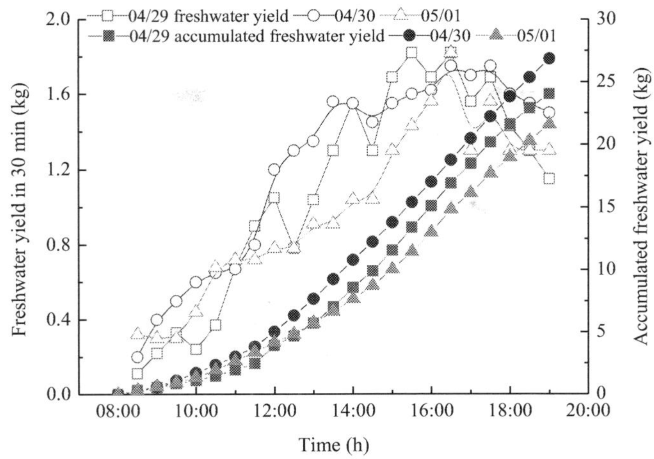

Figure 5.53
Relationship of water production rate and freshwater yield with operation time.

with very similar change tendency among them. However, the change of freshwater yield lags behind the change of solar irradiance for about 5 h.

For the first test day (2010/04/29), since the starting temperature was relatively low, the freshwater was generated after running for 2 h. At 1 p.m., the freshwater yield started to increase significantly, and at this time the operation temperature of the down stacked tray was about 80°C. For the second test day (2010/04/30), the freshwater was generated at the beginning of the experiment due to the high starting temperature. And the freshwater yield enhanced greatly at 12 a.m., with the operation temperature of the down stacked tray also keeping on 80°C. For the last test day (2010/05/01), although the starting temperature was the highest among the 3 days, the solar irradiance was the lowest, so the freshwater yield started to keep a high level at 1 p.m., which was later than that of the second test day. At this time, the operation temperature of the down stacked tray also stayed at 80°C, which was similar to the previous two test days.

For the present solar still, the seawater storage capacity is comparatively large (200 kg), so a large proportion of heat originated from the solar irradiance is stored in the seawater in the form of sensible heat. After sunset, the heat and mass transfer process inside the solar still also carries on, which makes the sensible heat of the seawater continuously transferred into the latent heat of water vapor. Therefore, the freshwater yield of the solar still in the night is also abundant. In addition, considering the ambient temperature in the night is very low, this factor also increases the freshwater yield indirectly. Table 5.4 lists the measured freshwater yield in the daytime, in the night, and the total amount. It indicates that the proportion of

Table 5.4: Accumulated freshwater yield and performance ratio of the solar still during three test days.

Time (m/d)	Accumulated freshwater yield				PR
	Daytime (kg)	Night (kg)	Total (kg)	Ratio (night/total)	
04/29	24.0	15.5	39.5	39.2%	1.65
04/30	26.9	16.5	43.4	38.0%	1.76
05/01	24.1	16.5	40.6	40.6%	1.76

freshwater yield in the night amounts to about 40% of the total freshwater yield, which is incredible.

5.9.5.3 Performance Analysis of the Solar Still

Based on the accumulated freshwater yield and solar irradiance, the performance of the solar still can be analyzed. For the solar still, the overall efficiency, η, is expressed as:

$$\eta = \eta_0 \cdot \mathrm{PR} = \frac{m_0 h_{fg}}{H_1 A_1 + H_2 A_2} \tag{5.50}$$

where PR is the performance ratio, which standards for the ratio of freshwater yield to energy consumption, i.e., the heat usage efficiency of the solar still; η_0, the average efficiency of the collector; m_0, the accumulated freshwater yield in 1 day, kg; h_{fg}, the latent heat of vaporization of water, kJ/kg; H_1, the accumulated solar irradiance of the collector, MJ/m^2; A_1, the total receiving area of the collector, about 3 m^2; H_2, the accumulated solar irradiance of the absorb coatings on the top stacked tray, MJ/m^2; A_2, the area of the absorb coatings on the top stacked tray, about 1.2 m^2.

By substituting the measured parameters into Eq. (5.1), the η for the three test days can be calculated and the result is 0.81, 0.86, and 0.86, respectively. For the average efficiency of the collector, η_0, detailed analysis indicates that this parameter can be taken as 0.49 for every test day. Therefore, the performance ratio of the solar still, PR, represented as η/η_0 according to Eq. (5.50), can be obtained. For the three test days, PR is 1.65, 1.75, and 1.75, respectively, which is also listed in Table 5.3. The determined PR is relatively high, demonstrating the designed multieffect solar still with enhanced condensation surface is very efficient, thus will be very effective and convenient for engineering application.

5.9.6 Conclusions

This multieffect solar still with enhanced condensation surface can generate freshwater not only in the daytime but also in the night, with maximum freshwater yield amounting to 43 kg. When the starting temperature is relatively high, the performance ratio of the equipment can reach 1.75, which is also relatively high.

5.10 Tubular Solar Desalination Device

Tubular desalination devices including horizontal and vertical tubular distillers are the most common components in desalination systems. However, it is an innovation when tubular desalination devices are used in solar desalination fields.

The concept of tube-type solar still was proposed by Tiwari in 1988. From that time, the basic heat and mass transfer relations of tube-type solar stills have been discussed by many researchers [21].The results indicate that the tube-type solar still has quite a good yield performance. For example, Murase gave a numerical simulation to the heat and mass flow process of a single-effect tube-type solar still [22]. Islam and Ahsan analyzed the heat and mass transfer to tubular solar still under steady condition. Ahsan discussed the condensation mass transfer in unsaturated humid air inside tubular solar still. Oslam contributed an effort to a special experiment to observe the evaporation, condensation, and production instantaneous performance in a new tubular solar still. In their study, the cover material of the tubular solar still was changed to a polythene film instead of stainless steel [23,24]. All of the researches indicate that there are some advantages of a tube-type solar still, such as its structure is compact, it is ease of fabrication, it can work with a combination of multieffect, it can utilize other waste heat, and the cover can be made of nonmetallic material, and so on.

5.10.1 Single-Effect Tubular Distiller

The structure of the single-effect tubular distiller device is shown in Fig. 5.54. Raw water from the water inlet (3) goes into the storage tank (4) and the water in the storage tank outflows through the overflow pipe (1) when it reaches a certain height. The heating

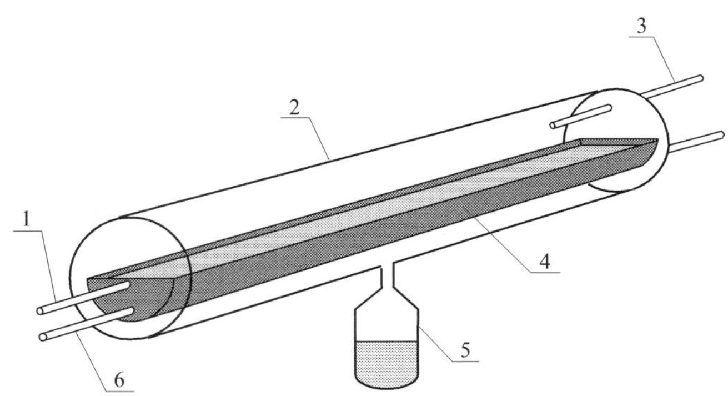

Figure 5.54

Schematic diagram of the single-effect tubular distiller. 1, Over flow tube; 2, tubular shell as condenser; 3, feed water input; 4, semicircular seawater storage trough; 5, freshwater tank; 6, heating tube.

working fluid heats seawater in the water tank through the heat exchange tube (6), the seawater is heated and evaporated to produce steam and natural convection can be formed on the surface of the hot water and the cold surface of the tube. The heated water vapor rises to the condenser tube (2) with lower temperature, then condensed into freshwater and flows along the wall of the pipe to the bottom of the condenser tube, eventually ending in the freshwater tank (5).

The energy transfer process of the device is shown in Fig. 5.55. Heat working fluid heats seawater through the heat transfer tube and transfers energy to the seawater in the tank. As the seawater is heated, there is a certain temperature difference and water vapor density difference between the sea surface and the condenser tube. The temperature difference brings about the heat transfer process, including heat conduction, convection, and radiation. The density difference generated mass transfer, which is accompanied by evaporation and condensation of water vapor and also accompanied by a phase change heat transfer process. Since there is a smaller thermal conductivity of wet air, the heat transfer through the wet air can be neglected. With heat transfer to the condenser tube, the condensation tube is heated, so that the condensation tube and the external environment will form a certain temperature difference. Therefore, the condensation tube mainly emitted energy into the environment by convection and radiation, so as to complete the energy transfer process of the whole device.

It is not difficult to get the following energy balance equation from the view of energy balance.

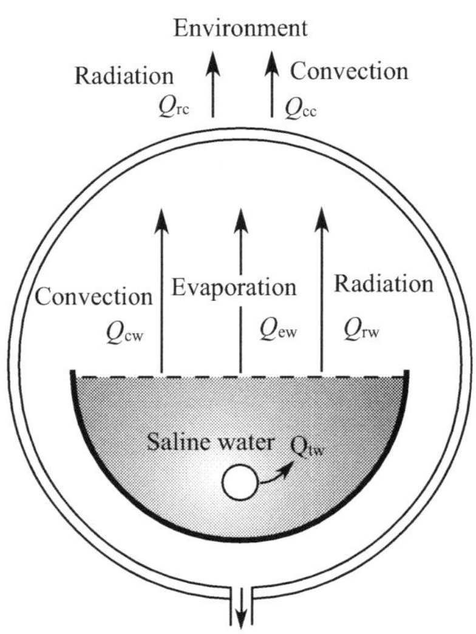

Figure 5.55
Energy transfer process of single tube-type desalination system.

Energy balance equation of seawater:

$$Q_{tw} = Q_{ew} + Q_{cw} + Q_{rw} + (\rho \cdot C \cdot V)_w \cdot \frac{dT_w}{dt} \tag{5.51}$$

Energy balance equation of condensation outer tube:

$$Q_{ew} + Q_{cw} + Q_{rw} = Q_{cc} + Q_{rc} + (\rho \cdot C \cdot V)_c \cdot \frac{dT_c}{dt} \tag{5.52}$$

where:

$Q_{tw} = P_t \cdot t$
$Q_{ew} = h_{fg} \cdot h_{ew} (\rho_{vw} - \rho_{vc}) \cdot A_w$
$Q_{cw} = h_{cw} (T_w - T_c) \cdot A_w$
$Q_{rw} = h_{rw} (T_w - T_c) \cdot (A_w + A_t)$
$Q_{cc} = h_{cc} (T_c - T_a) \cdot A_c$
$Q_{rw} = h_{rw} (T_c - T_a) \cdot A_c$
P_t, electric heating power, W;
t, heating time, s;
h_{fg}, evaporation latency of water, J/kg
h_{ew}, the mass transfer coefficient from water surface to condensation tube, m/s;
h_{cw}, the convection heat transfer coefficient between water surface and condensation tube, W/(m^2 K);
h_{rw}, the radiation heat transfer coefficient between water surface and condensation tube, W/(m^2 K);
h_{cc}, the convection heat transfer coefficient between condensation tube and environment, W/(m^2 K);
h_{rc}, the radiation heat transfer coefficient between condensation tube and environment, W/(m^2 K);
A_w, evaporation area of water, m^2;
A_t, surface area of the trough, m^2;
A_c, surface area of the condensation tube, m^2;
Q_{ew}, the phase change heat transfer between water surface and condensation tube, J/s;
Q_{cw}, the convection heat transfer between water surface and condensation tube, J/s;
Q_{rw}, the radiation heat transfer between water surface, trough, and condensation tube, J/s;
Q_{cc}, the convection heat transfer between condensation tube and environment, J/s;
Q_{rc}, the radiation heat transfer between condensation tube and environment, J/s;
ρ_{vw}, the density of vapor on the water surface, kg/m^3;
ρ_{vc}, the density of vapor on the inner surface of the condensation tube, kg/m^3;
T, temperature, °C;
ρ, density, kg/m^3;

C, specific heat capacity, J/(kg K);
V, volume, m^3.

Subscripts: *w*, *c*, *t*, *a* represent water, condensing tube, water trough, and environment, respectively.

According to the equations above, the energy processes of the tubular distiller can be solved; and the freshwater productivity can be predicted. The solving solutions have been reported in a great deal of literature.

5.10.1.1 Performance Test Under Natural Sunning Conditions

Amimul Ahsan reported the experimental results of a natural boring tube evaporator with a transparent outer tube, for which the operating principle is shown in Fig. 5.56 and the structure is different to Fig. 5.54 in that the heating energy is directly provided by sunlight [23]. The sunlight through the outer tube directly projected to the black water collection tank to heat seawater. The seawater is evaporated and condensed into freshwater inside the outer tube. The experimental device structure size is: the transparent shell is made of polyethylene film (0.15 mm thick), thin film is supported by spiral steel wire (diameter 2 mm), transparent shell diameter 0.18 m, length 1.2 m, internal square water tank 1.09 m, width 0.13 m.

Amimul Ahsan and others measured in the actual weather and Fig. 5.57 given the results of the experiment in the different typical weather. The figure results show that, in fine weather, the highest temperature of wet air in the device can reach about 70°C, a day of water yield can reach 5 $kg/m^2/d$, which is better than general basin solar stills.

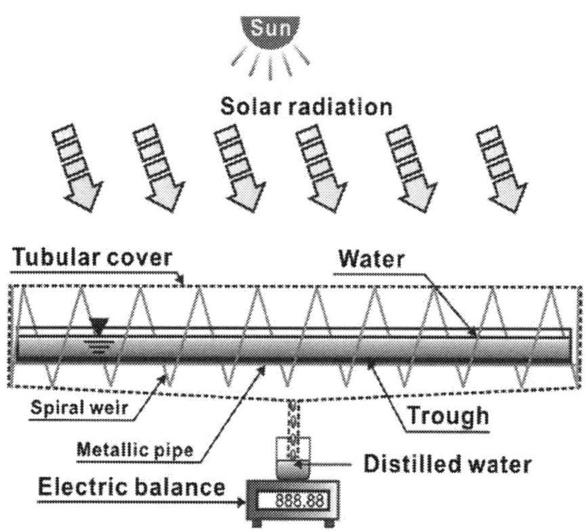

Figure 5.56
Natural sunning tubular distiller covered by transparency film.

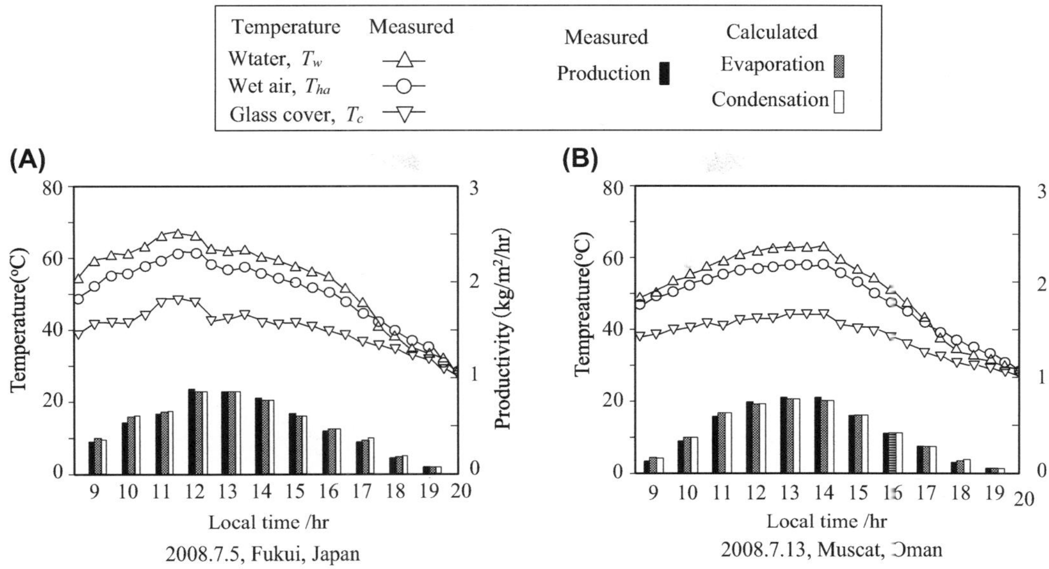

Figure 5.57

The operation situations of a tubular distiller covered by transparency film under actual weather conditions: (A) 2008.7.5, Fukui, Japan; (B) 2008.7.13, Muscat, Oman.

5.10.1.2 Performance Test Under Steady Conditions

For validating the steady performance of this kind of tubular distiller with single effect, the literature [25] introduced a single-effect tubular distiller. A half-cut tube was installed as a basin. The width and length of the basin are 110 and 1900 mm, respectively. The tubular shell is 2000 mm in length and 133 mm in diameter. The evaporation area of the basin is 0.209 m^2.The condensation area of the device in the shell is 0.835 m^2. The diameter of the heating tube in the basin is 10 mm. The material of the distiller is 304 stainless steel. In order to predict accurately productive performance of the device, a steady-state experiment was carried out indoors. The experiment was implemented in two situations: one is that the system was heated by an electric heater under the condition of fixed power. Another is that the system was heated by a fixed-temperature working medium. And then, distillate productivity was measured under the atmospheric pressure. The experimental results are presented as follows.

1. Fixed temperature heating experiments
   In the experiment, the heating working medium temperature was regarded as the operating temperature. The experiment was carried out indoors, the ambient temperature was about 25°C. In the experiment, the steady condition must be kept to exceed 1 h. The temperature, the yield, and the other correlative parameters had been measured. The productivity of the single-effect tubular distiller along with heating temperature is shown as in Fig. 5.58.

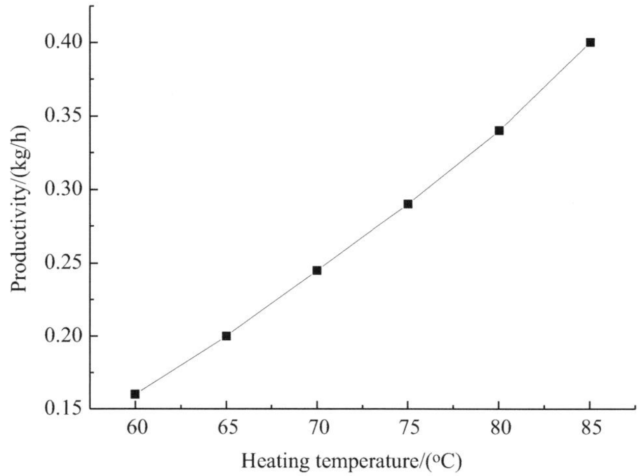

Figure 5.58
Variation of the freshwater productivity of the single-effect tubular distiller with heating temperature.

It can be found easily from Fig. 5.58 that the freshwater productivity will increase with the heating temperature. That is to say, the distiller will have higher efficiency when it works under higher operation temperature.

2. Fixed heating power experiment
   In this experiment, different heating powers were selected. Fig. 5.59A and B show the experimental results at heating power of 300 and 400 W, respectively.
   It can be found from Fig. 5.59 that the temperature profile increases quickly at the beginning of the experiment, 3 h later the degree of increase slows. It reaches a

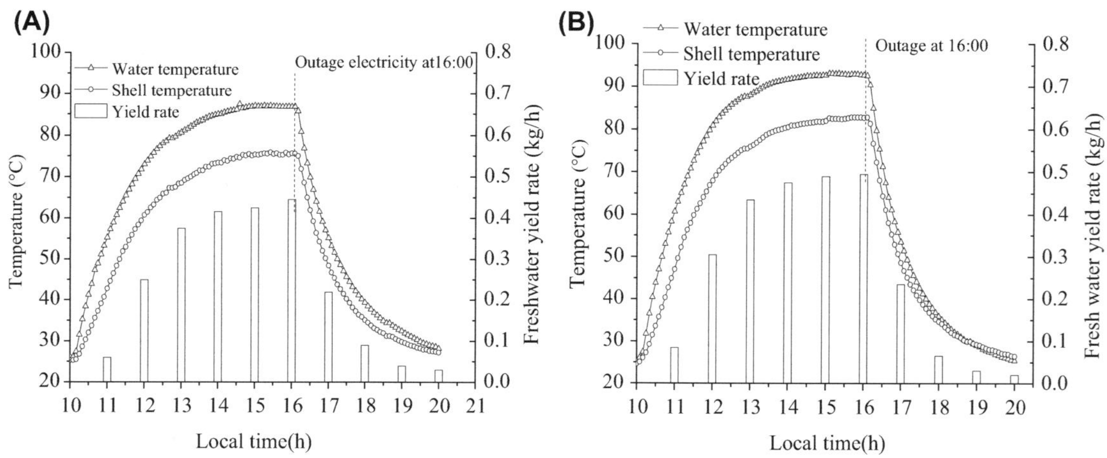

Figure 5.59
Variation of temperature and distillate productivity in single-effect device with the operation time: (A) heating power 300 W; (B) heating power 400 W.

relatively steady state from 14:00 to 16:00. During this time the changes in temperature are small and the temperature profiles reaches the maximum. The biggest steady-state temperature at heating power of 300 and 400 W is about 86°C and 93°C, respectively. It also can be found that the water and shell temperature fall sharply when the power was switched off at 16:00 p.m. The distillate productivity a the similar tendency with that of the temperature. The maximum distillate productivity under steady-state conditions is 0.445 kg/h and 0.495 kg/h, respectively, for heating power of 300 W and 400 W.

The experiment results illustrate that seawater evaporation temperature, shell temperature (representing the condensation temperature), and evaporation area have an important influence on the distillate productivity. The higher the seawater evaporation temperature is, the more the proportion of evaporation heat transfer and the greater the water production rate is. The bigger the temperature difference between the seawater and condensation shell is, the bigger the water vapor density difference between the seawater and condensation shells, and then the evaporation mass transfer rate is bigger, and accordingly the bigger the distillate productivity may be.

It is also found that, after power being interrupted at 16:00 p.m., seawater and shell temperatures fall sharply. This was because all the evaporating energy only comes from the seawater heat capacity, and no outer energy is entered. In this time, the temperature difference between the seawater and condensation shell becomes smaller, which leads to the distillate productivity falling sharply. After switching of the power, the total cumulated fresh production under different heating power is shown in Fig. 5.60.

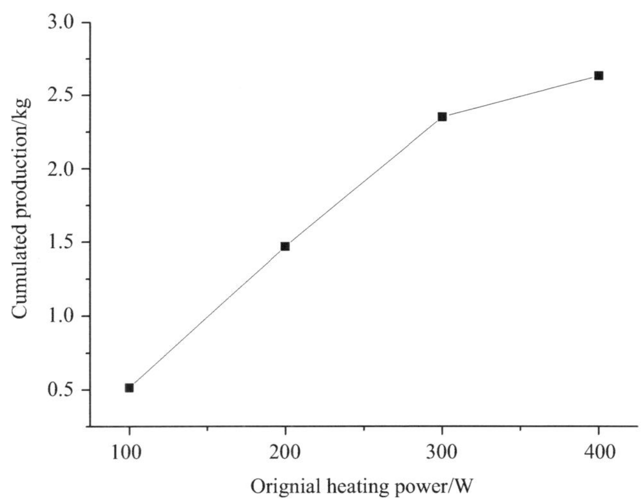

Figure 5.60
The cumulated freshwater production under different heating powers.

5.10.2 Two-Effect Tubular Distiller

A single-effect tubular distiller has the advantage of structure simple. However, the vapor latent is not re-utilized, so the GOR is only 0.7–0.8. The literature [25] also describes a designed two-effect tubular distiller (Fig. 5.61).

The operation principle of the experimental device can be presented as follows: the two-effect tubular solar still consists of two circular tubular shells and two semicircular troughs which fix inside the tube shells, respectively. Seawater enters into the second-effect semicircular trough 8 and the first-effect semicircular trough 1 through input tubes 3 and 4, respectively. The seawater in the first-effect semicircular trough is heated to produce evaporation by the heating tube 10. The water vapor density in the humid air will increase and some of it will be condensed on the inner surface of the tubular shell 7 to release its latent heat of vaporization. The condensed water naturally trickles down toward the bottom of the tubular shell due to gravity and enters into a distillate collector 6 through the first-effect freshwater outlet tube 9 equipped at the underside of the tubular shell. At the same time, the latent heat released by the first-effect vapor on the inner surface of the tubular shell 7 will again heat the seawater in the second-effect semicircular. And then the seawater in the second-effect semicircular also produces the evaporation process. Because the temperature of tubular shell 2 is lower than that of water vapor, the water vapor is condensed on the inner surface of tubular shell 2. The condensed water naturally trickles down toward the bottom of the tubular shell and then is stored in a freshwater collector 6 through the second-effect freshwater outlet tube 5.

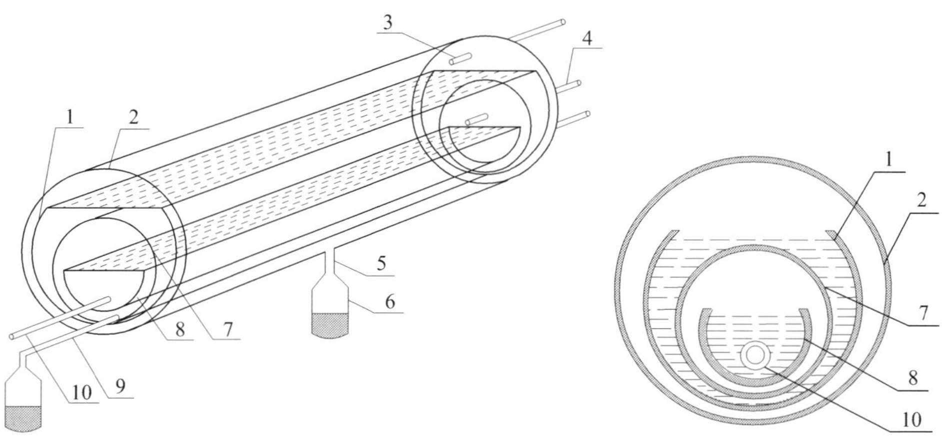

Figure 5.61

Structure diagram and cross-sectional view of two-effect tubular distiller: 1, the second-effect semicircular trough; 2, tubular shell; 3, the second-effect input; 4, the first-effect input; 5, the second-effect freshwater outlet; 6, freshwater collector; 7, the first-effect tubular shell; 8, the first-effect semicircular trough; 9, the second-effect freshwater outlet; 10, heating tube.

A two-effect tubular solar still consists of four stainless steel tubes ordered from small to large. The width and length of the first-effect basin is 100 and 1900 mm, the first-effect tubular shell was 1950 mm in length, 114 mm in diameter. The width and length of the second-effect basin is 124 and 1950 mm, the second tubular shell was 2000 mm in length, 168 mm in diameter. The evaporation area of the two-effect tubular still is 0.19 and 0.242 m^2,respectively, and its condensation area is 0.698 and 1.055 m^2, respectively.

5.10.2.1 Fixed-Temperature Heating Experiments

The productivity characteristics of the two-effect tubular distiller are shown in Fig. 5.62. It is easily found from Fig. 5.62 that the first effect and second effect have the same freshwater yield rate tendency. However, the production in first effect is bigger than that in the second effect because the working temperature in first effect is higher.

5.10.2.2 Fixed Heating Power Experiment

In the two-effect device experiment, a different heating power value was chosen. Fig. 5.63 shows the seawater temperature, the shell temperature, and the distillate production rate variation profiles with the operation time for heating power of 300 W and 400 W. The water temperature and shell temperature change with the operating time had a similar trend to that of the single-effect device. The temperature increased quickly at the beginning of the experiment, and reached the relatively steady state at 16:00, and meanwhile the variation of the temperature and distillate productivity was relatively small. After the electricity power was interrupted at 18:00 p.m., the seawater and shell temperatures fell sharply.

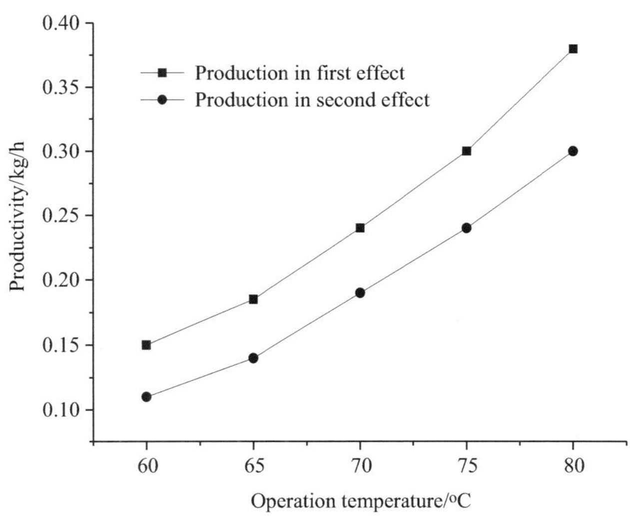

Figure 5.62
Variation of the freshwater productivity in first and second effect with operation temperature.

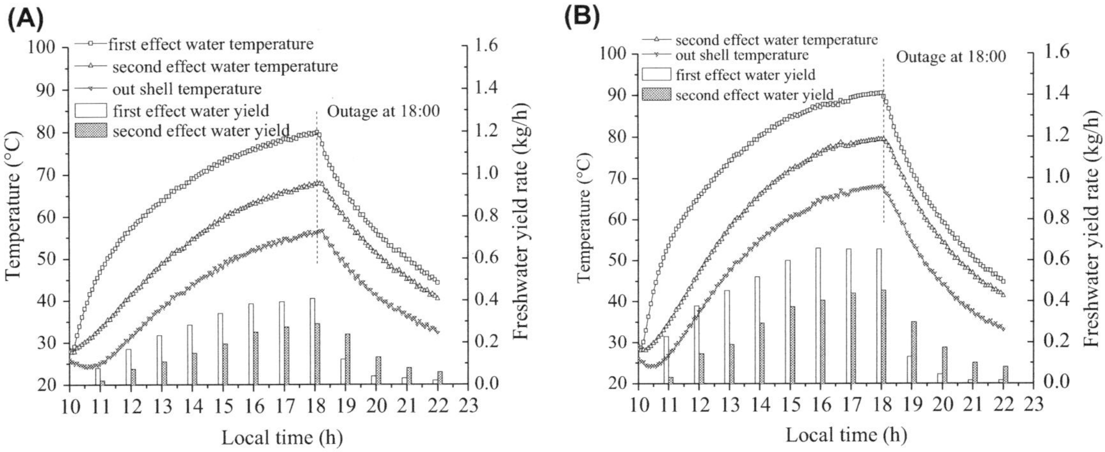

Figure 5.63
Variation of temperature and distillate productivity in a two-effect device with the operation time: (A) heating power 300 W; (B) heating power 400 W.

It can be seen from Fig. 5.63 that distillate productivity of the first-effect is bigger than that of the second-effect. However, after the power is cut off, the distillate productivity of the first-effect fell much faster, and was less than that of the second-effect. This may be explained as follows: firstly, heating was carried out in the first-effect, so the seawater temperature of the first-effect was higher than that of the second-effect. In this time, the temperature difference between the seawater surface in the first basin and the first-effect shell was the same as that of the second-effect. Then the first effect had a bigger distillate productivity than the second-effect. However, when the electricity power was cut off, although the temperature in almost measuring points fell quickly, it could still be found that the temperature difference between seawater and the condensation surface in the first-effect is less than that in the second-effect. In this situation, even if seawater temperature in first-effect is higher, the productivity of first-effect is still less than that of second-effect because the temperature difference between the evaporation and condensation surface is the main factor influencing the mass and heat transfer.

5.10.3 Experiment With a Three-Effect Device

For re-utilizing further the condensation latent, the literature [20] also described the design of a three-effect tubular distiller, as shown in Fig. 5.64.

The three-effect tubular distiller consists of six stainless steel tubes ordered from small to large. From inside to outside, three semicircular troughs and their corresponding tubular shells form the first, second, and third water production cavities, respectively. The water storage capacity of the three semicircular troughs is 4.66, 5.05, and 5.82 kg, respectively. The evaporation area is 0.08, 0.14, and 0.13 m^2, respectively. The condensation area is

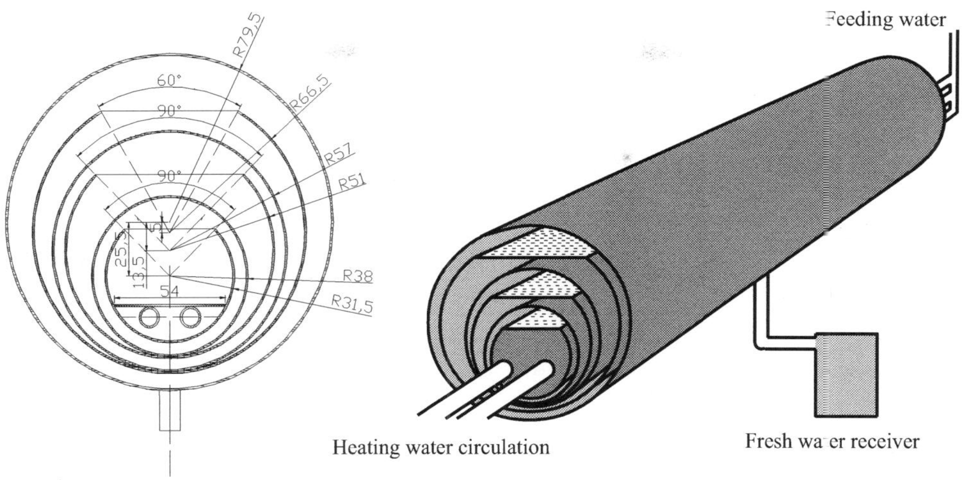

Figure 5.64
Structure diagram and cross-sectional view of three-effect tubular distiller.

0.11, 0.17, and 0.16 m^2, respectively. The operation method is similar to the two-effect device.

Heat transfer rate in the three-effect tubular solar still is relatively slow. If too-large a power heater is used, the seawater temperature of the first-effect may reach 100°C, which is not good for the experiment. Therefore, the experiment was only carried out on two models with heating power of 300 and 350 W (400 W would be too large). The variation curves of the seawater temperature in each effect with operating time are shown in Fig. 5.65, in which the heating time is 8 h and the cooling time is 16 h.

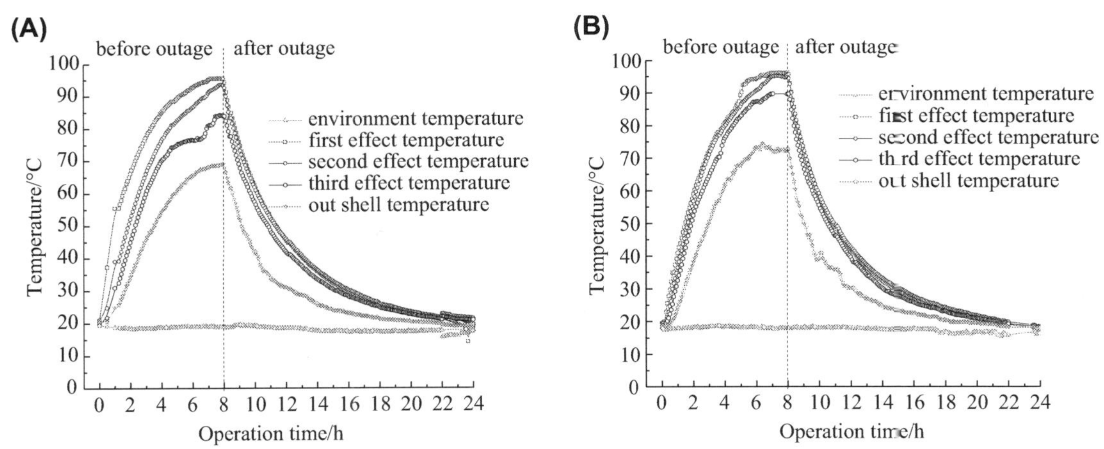

Figure 5.65
Variation of temperature of the three-effect device with the operation time: (A) 300 W; (B) 400 W

It can be observed from Fig. 5.65 that, at the beginning, the temperature variation curve is very steep. However, the temperature increases with the heating process and becomes gradually slower until the power is interrupted, and then the temperature goes down sharply. When the heating power is 350 W, the temperature of the first-effect evaporation trough exceeds 95°C after it is heated for some time. Afterward, the temperature at each measuring point reaches a steady state and stops rising until the power is interrupted. Hence, there appears a "platform" of the temperature variation curves throughout a whole day.

Fig. 5.66 shows the distillate productivity variation curves at the heating power of 300 and 350 W. It reveals that the greater the heating power, the sooner the distillate productivity can reach its peak. When the heating power is reduced, the peak of the distillate productivity is gradually delayed. From the distillate productivity point of view, greater heating power leads to greater distillate productivity.

5.10.4 Performance Comparison of Three-Effect Devices

Experimental results show that with the increase of input power, hourly distillate yield will increase for all three devices. It is seen that the hourly distillate yield of a three-effect device is more than that of a two-effect device, similarly, the hourly distillate yield of a two-effect device is more than that of a single-effect device under the same input power condition. Apparently, it is because the latent heat of vapor along with sensible heat of brine in multieffect devices is successfully recycled, also the condensation area and the evaporation area in the multieffect devices are larger.

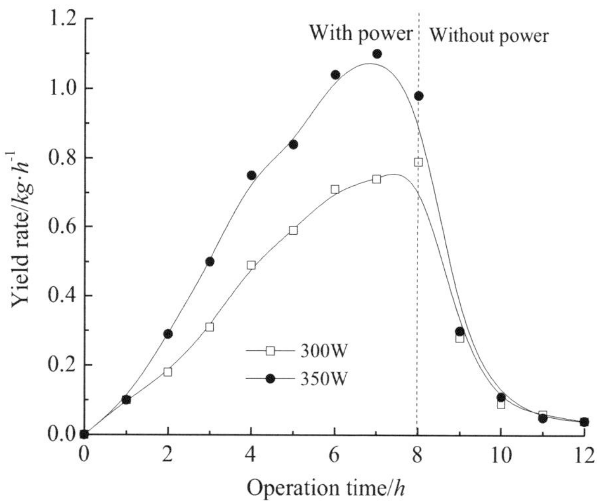

Figure 5.66

Variation of distillate productivity of the three-effect device with the operation time.

Table 5.5: The total distillate water yield and performance ratio at different heating powers.

	Single-effect device		Two-effect device		Three-effect device	
Heating power/W	Productivity (kg)	GOR	Yield (kg)	GOR	Yield (kg)	GOR
300	0.440	0.93	0.640	1.36	0.790	1.68
350	—	—	—	—	1.05	1.91
400	0.495	0.79	1.05	1.67	—	—

The energy utilization efficiency of the device is represented by the PR. This is an important indicator reflecting the device quality. A larger PR of a desalination system means that the energy efficiency rate is higher, and the distillate production performance is better. The steady-state PR is given as follows:

$$PR = \frac{M_e h_{fg}}{Q} = \frac{m_e h_{fg}}{P \Delta t}$$

where, m_e is the total distillate yield, h_{fg} is the vapor latent heat, P is the heating power of the heater, and Δt is the heating time. In the experiment of the single-effect device, the experimental period is from 10:00 a.m. to 20:00 p.m. every day, in which the electricity power is turned on at 10:00 a.m. and turned off at 16:00. In the experiment of the two-effect device and the three-effect device, the experimental period is from 10:00 a.m. to 22:00 p.m. every day, in which the electricity power is turned on at 10:00 a.m. and turned off at 18:00. The heat loss of the device is ignored and h_{fg} is considered to be 2300 kJ/kg in the calculation. The total yield and the PR are listed in Table 5.5.

As shown in Table 5.5, the total yield of the three devices increased with the input heating power. Experimental results also indicate that the incremental rate of yield will decrease with the increase of the input power under the same conditions. It may be that the evaporative area in the device is not constant, and what is more, it depends on the water level, especially for a multieffect device, after running for some time, the declining water surface led to the evaporative area decreasing.

The multieffect device has a larger PR than the single-effect device, as a result of successful recycling of the latent heat of vapor condensation as well as the sensible heat of brine in the multieffect device. This led to a higher energy utilization ratio and greater yield.

5.10.5 The Performance of a Tubular Still Under Negative Pressure

A very obvious advantage of the tubular still is its good bearing pressure characteristics. It can even be made of nonmetal material, such as glass or plastic, which can also withstand high vacuum degree (the vacuum degree means atmosphere pressure minus working pressure). Other still devices do not have this characteristic. In fact, only the

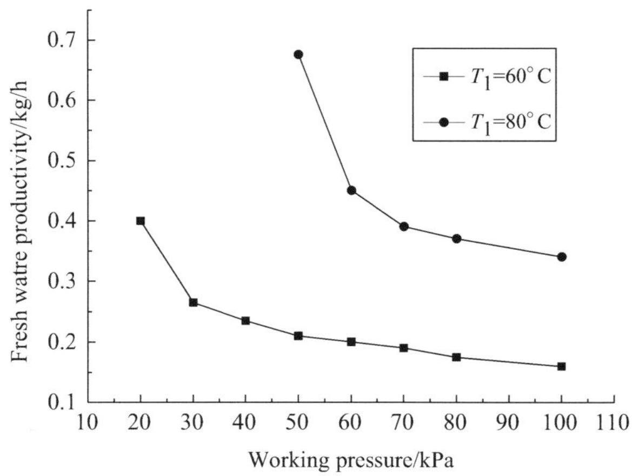

Figure 5.67

Distillate productivity of a single-effect device under different pressures and temperatures.

outer tubular shell has to bear the external pressure when the device works in vacuum condition, thus the bearing pressure performance of the inner material does not need to be considered.

From the view of heat transfer, under the negative pressure conditions, noncondensable air in the evaporation chamber will decrease, which is beneficial to the evaporation transfer process. In this situation, the share of the convective heat transfer from effect to effect will be reduced, which will lead to an increase of the share of the evaporation heat transfer. Therefore, the tubular still has more advantages if it works under negative pressure. Fig. 5.67 shows the distillate variation curve of a single-effect device under different pressures and different temperatures.

The experimental method in Fig. 5.67 was the same as the abovementioned, but the working pressure in the experiment was different. Fig. 5.67 shows that the yield rate tended to decrease with the increase in working pressure. And low vacuum degree has little influence on the yield rate. When the internal pressure in the device achieves the saturated steam pressure for the heating temperature, the yield rate obviously increases. On the one hand, low working pressure can decrease the boiling temperature and enhance the evaporation and condensation process. At the same time, it reduces the effect of the noncondensable gas to reduce the share of convection heat transfer. On the other hand, when the internal pressure of the device reaches the saturated steam pressure, the seawater may boil and give off a large amount of vapor, which leads to a rapid increase in yield rate.

The same results can also be obtained from the two-effect device experiments under negative pressure. Fig. 5.68 shows the distillate productivity of two-effect device variation with different working pressures and different temperatures.

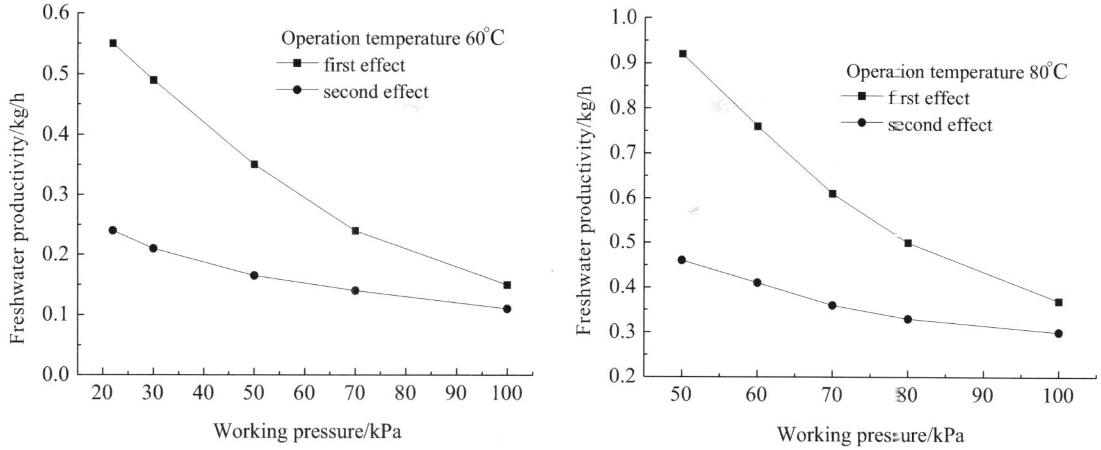

Figure 5.68
Variation of distillate productivity of the two-effect device with the pressure and operation temperature.

It can be seen from Fig. 5.68 that the yield rate of the first-effect decreases obviously with the increase in working pressure, which reflected the trend of the single-effect device. However, in the second effect the yield rate decrease is relatively slow. This is because under the same working pressure, the first-effect has higher water temperature than the second-effect. Besides, it is closer to the boiling point. The first-effect will thus have the bigger evaporation rate.

5.10.6 Experiments With Different Gas Media

In the experiment, the different gas media were inflated into the three-effect tubular still, and the total pressure in the device was kept the same as the ambient pressure, which was obtained by the pressure buffer balloon. The vacuum pump was used to withdraw the air in the tube, then inflate the chosen gas and repeat this action four times. The air in the device was replaced by helium, oxygen, and carbon dioxide, respectively. Some densities of the gas medium are greater than the air and some are less than the air. The seawater was prepared by dissolving salt into tap water.

The thermostatic water bath was used to provide heat to the energy for the device. Under the conditions of the stable temperature, the performance of the device was tested. For each gas medium, the freshwater yield was measured for six heating temperatures (60°C, 65°C, 70°C, 75°C, 80°C, and 85°C). The water yield rate under each temperature was obtained and is shown in Fig. 5.69.

As shown in Fig. 5.69, when the operating temperature increases from 60°C to 85°C, the performance of the device with the oxygen medium is always the best among the four

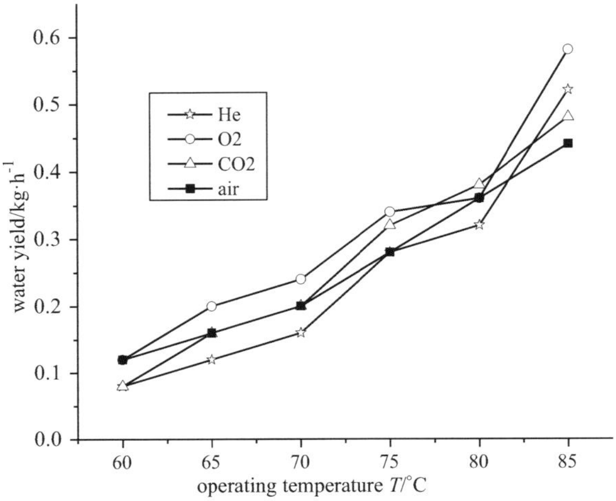

Figure 5.69
The variation of water yield rate under different gas media, with operating temperature.

chosen media, although the freshwater yield at 80°C has a slightly strange decrease. This may be caused by experimental error and other factors. The second suitable medium is carbon dioxide. When the gas medium is helium, the performance of the device is worst. However, when the operating temperature is 85°C, the performance of helium medium was improved greatly. That is to say, within high temperature range, the gas helium may be a perfect medium, while air is worst.

As shown in Fig. 5.69, in the condition of different gas media, the water yield per hour increased with the increase in operating temperature. Especially in high operating temperature range, water yield will have a large increase. When the gas medium is air, water yield grows linearly as the operating temperature increases. But when the gas medium is oxygen or helium, the water yield has a sharp rise when the operating temperature is from 80°C to 85°C. If the operating temperature is higher, the performance may be better. In general, oxygen is found to be the best gas medium to achieve more desalinated water. When the operating temperature is 85°C, using oxygen as gas medium, the water yield per hour of the device can reach 0.58 kg/h. The water yield per hour of the device is only 0.44 kg/h when the gas medium is air.

5.10.7 Conclusions

The experimental investigations of a series of tubular solar stills with multieffect were carried out indoors. The experimental results indicate that tubular solar stills, especially for the multieffect still device, have good productivity performance. In the multieffect device, the latent heat of vapor can be used multiple times so that it has a high PR. Because the condensation area of the device is always larger than the evaporation area, condensation processes are strengthened, which is of benefit in improving the evaporation heat transfer.

In the fixed power heating experiments, the PR of the single-effect device can reach 0.7–0.8, the PR of the two-effect device can reach about 1.4, and the PR of the three-effect device can reach about 1.7 at heating power of 300–400 W. In the fixed-temperature heating experiments, when the operating temperature is about 80°C, the yield of the two-effect device can be 1.711 kg/(m^2 h). Under the conditions of the same input power, the yield of the two-effect device is about one time more than that of the single-effect device, the yield of the three-effect device is about 23% more than that of the two-effect device. It indicates that the condensation latent heat of the vapor along with some part of sensible heat of brine in a multieffect device can be successfully recycled.

Furthermore, this kind of device can be operated under negative pressure, which can enhance the productivity performance. The experimental results show that the distillate productivity will have an obvious increase when the tubular solar still works in negative-pressure conditions. Especially when the internal pressure of the first-effect achieves the saturated steam pressure, a tubular solar still can get 100% more yield increase than that of working at a common pressure.

It has been found that the heavier gas as a medium in the tubular distiller could increase the water yield rate. For example, when the gas medium in the still is oxygen, the water production rate can reach 0.58 kg/h at the operating temperature of 85°C, which is about 31.82% higher than that for air as the gas medium. At the same operating temperature, the water production rate of the device with carbon dioxide is also higher than that of air. If the gas medium is suitable, the water production may be improved further. Among the four gas media studied, oxygen seems a good gas medium for a tubular solar desalination device.

In a word, a tubular solar desalination device has its unique advantage. It can even be made by nonmetal material, for example glass and plastic. It creates a good condition to reduce the construction cost of a solar desalination device.

5.11 Multieffect Concentric Vertical Tube Solar Still

As discussed above, the tubular solar still has many advantages, such as its compact structure, pressure-bearing, no thermal insulation requirement, ease of fabrication, the nonmetallic material for manufacture, and so on. At the same time, its condensing area is always bigger than its evaporating area so that the evaporation thermal resistance is matched with the condensation thermal resistance.

However, the horizontal tubular solar distiller has some drawbacks, such as water vapor is directly condensed at the top of the tubular shell because the mouth of the storage water trough is always upward, thus leading to low usage rate of condensation area; the evaporation area in the device is not a constant after running for some time, especially for the multieffect device, which affects the stability of the unit; the seawater containment in every effect is relatively more, that is the thermal inertia of the device is larger. In order to strengthen the merits of a multieffect tubular solar desalination system, it is necessary to overcome its limitations mentioned above. Therefore, the concept of a multieffect vertical concentric tubular solar distiller is provided. For that, Chang [26] designed a group of triple-effect vertical concentric tubular solar desalination devices, with three operational modes, i.e., single-effect, double-effect, and triple-effect. The experimental investigation on the performance of the device was carried out under the conditions of fixed heating power and controlled heating temperature to compare these modes.

5.11.1 The Structure and Operational Principle of the Device

5.11.1.1 Structure Parameters and Characteristics

The device has three operational modes so as to test the performance of single-effect, double-effect, and triple-effect unit, respectively. Fig. 5.70 illustrates a schematic diagram of the triple-effect vertical concentric tubular solar still.

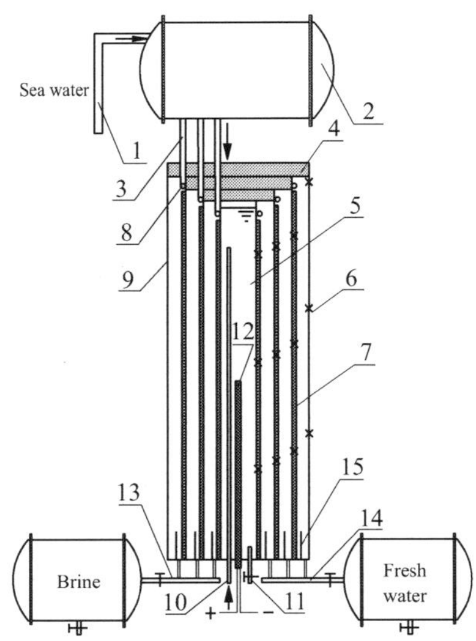

Figure 5.70
Structure diagram of a triple-effect tubular solar still: 1, the seawater input tube; 2, the water storage tank; 3, the seawater drainage pipe; 4, insulation layer; 5, heating water tank; 6, the positions of thermocouples; 7, water absorption materials; 8, perforated tube; 9, tubular shell; 10, hot water tube; 11, cold water tube; 12, electric heater; 13, the brine pipe; 14, freshwater tube; 15, water linerboard.

The triple-effect vertical tubular still is composed of four concentric circular stainless steel pipes, which form three spaces that are used as the first-effect, second-effect, and third-effect distillation chambers, respectively, while the innermost pipe also serves as a hot water container. For the convenience of description, three pipes within the outer shell may be numbered as the first, second, and third, from the innermost to the outer. Water absorption material (hydrophilic wick) is adhered to the outside surfaces of the first, second, and third pipes to hold seawater flow from water tank 2 on the top of the desalination device. Seawater flows into the water absorption materials of every effect via a perforated tube 8 acting as a water distributor.

The seawater in the water absorption materials on the outside surface of the first pipe is heated up directly by the hot water inside the pipe, i.e., the hot water tank which is heated with the electric heater 12. In the solar system, the hot water should come from a solar collector. Upon being heated up, evaporation of the seawater will increase the vapor density in the humid air within the space between the first and second pipes, so the vapor will be condensed on the inside surface of the second pipe to release its latent heat of vaporization. This is the first-effect distillation. The condensed water then naturally trickles down toward the bottom of the tubular shell due to gravity and enters a distillate collector through the first-effect freshwater outlet tube 14 fixed on the underside of the tubular shell.

At the same time, the latent heat released by the vapor in the first-effect distillation will heat the seawater in the water absorption materials on the outside surface of the second pipe. The water vapor evaporated from the seawater will condense on the inside surface of the third pipe due to the temperature difference to achieve the second-effect distillation. Similarly, the evaporation and condensation process occurs in the space between the third pipe and the outer shell of the distillation device, where the third-effect distillation is obtained. At the bottom of the tubular shell there are three brine pipes, through which brine is rejected into the brine tank, and then is discharged to the outside. However, the condensed water is collected independently for each effect in order to have a comparison between them.

The several advantages of the designed distillation device can be summarized as follows: (1) For each effect distillation, the area of the condensation is always larger than that of the evaporation, which is beneficial to the condensation of vapor. (2) The surface of the condensation is vertical circular, which is beneficial to the collection of distillate. (3) A relatively large percentage of the latent and sensible heat of condensation along with brine is successfully directly recycled and reused so that the thermal performance of the system is greatly improved. (4) The water mass in the water absorption materials of every effect is relatively smaller, which means the thermal inertia of the device is smaller. (5) The utilization efficiency of condensation area is increased because the condensing process occurs around the circumference.

Table 5.6: Technical specification of three operational modes (unit: mm).

	Heating water tank		The first pipe		The second pipe		The third pipe	
Number of effects	Diameter	Length	Diameter	Length	Diameter	Length	Diameter	Length
1	100	1470	130	1480	–	–	–	–
2	100	1470	130	1475	150	1485	–	–
3	100	1470	130	1477	150	1490	210	1500

5.11.1.2 Specifications of Operational Modes

The main specifications of the experimental device for three operational modes are summarized in Table 5.6. The evaporation area of the single-effect operational mode is 0.446 m^2 and its condensation area is 0.584 m^2. The evaporation area of the double-effect operational mode is 0.446 and 0.582 m^2, respectively, and its condensation area is 0.582 and 0.721 m^2, respectively. The evaporation area of the triple-effect operational mode is 0.446, 0.584, and 0.723 m^2, respectively, and its condensation area is 0.584, 0.723, and 0.983 m^2, respectively.

5.11.1.3 Experimental Setup and Procedure

A series of indoor experiments was carried out to investigate the performance of the device. The first was when the system was heated by an electric heater under the condition of fixed power. The heating process started when the electric heater was turned on and finished when the distillate yield reached steady-state conditions. Then, the electric heater was turned off and the cooling period began. The cooling experiment finished when the temperature of the heating water tank was below 40°C. In this case, distillate productivity and temperatures were measured for different modes. The other is when the system was heated by fixed heating temperature. The distillate productivity was measured at each effect under different operation temperatures. The steady-state productivity and transient-state productivity were also studied. The experiment was carried out in a temperature- and relative humidity-controlled laboratory to keep the external environmental conditions surrounding the device constant.

5.11.2 Results and Discussion

The productivity of the multieffect tubular still under the designed conditions was researched experimentally. There is one electric heater installed in the heating water tank to represent a solar water heater. Before each heating process, the tank was fully filled with water at ambient temperature. In addition, the heating power can be adjusted so as to meet the experimental requirement. The tests were performed to assess the influence of the following parameters on the performance of the unit.

1. The feed seawater mass flow rate;
2. Temperature of the water in the heating water tank;
3. Heating power of electric heater.

5.11.2.1 Fixed Heating Power Experiment

In this experiment, the yield and feed water temperature of the single-effect mode, the double-effect mode, and the triple-effect mode under heating and cooling processes were studied at heating power of 305 W. The total yield of the single-effect mode, the double-effect mode, and the triple-effect mode under heating processes were compared at heating power of 400 W. In addition, the feed water mass flow rate was 30 g/min. The 20-min yield of every effect and total yield of the unit were recorded. Under the same conditions, the temperature of every measured point inside the unit was recorded. Fig. 5.71A–C show the experimental results of the single-effect mode, the double-effect mode, and the triple-effect mode, respectively.

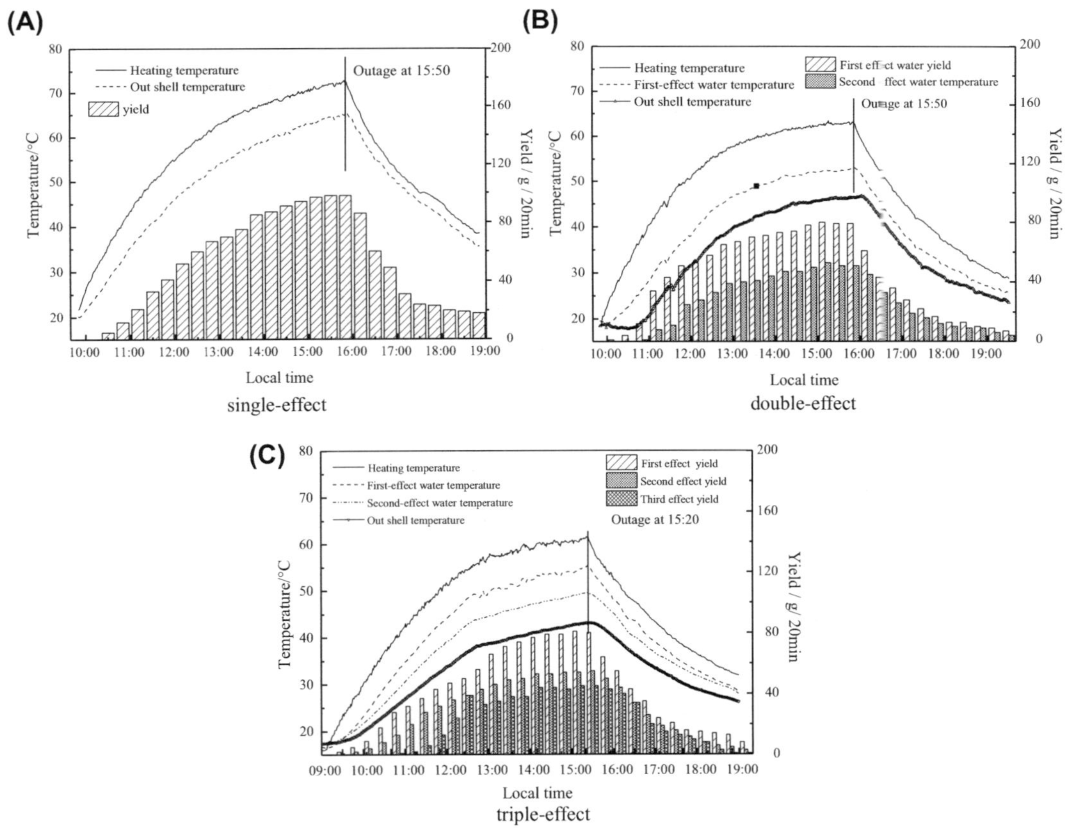

Figure 5.71

Variation of temperature and distillate productivity of three operational modes with the operation time: (A) single-effect; (B) double-effect; (C) triple-effect.

Fig. 5.71A demonstrates that the temperature of measuring points increases rapidly at the beginning of the experiment, and reaches a relatively steady state after 3.5 h. At this time the change of the water yield rate is small, the temperature differences keep stable. The highest steady-state temperature at heating power of 305 W is about 73°C. It can also be discovered that the feed water and shell temperatures fall sharply when the power is turned off at 3:50 p.m. Meanwhile, the distillate productivity has a similar tendency with the temperature. The maximum distillate productivity under steady-state conditions is 0.294 kg/h.

The experimental results illustrate that heating temperature (representing feed water evaporation temperature) and outer shell temperature (representing the condensation temperature) have a great effect on the productivity of the device. The freshwater yield rate increases with the feed water evaporation temperature, which means more heat energy was used to evaporate the seawater. The larger the temperature difference between the feed water and condensation shell, the larger the water vapor density difference between the feed water and outer shell will be, and so is the evaporation mass transfer rate and the higher the distillate yield accordingly.

Fig. 5.71A also illustrates that, after the power being cut off at 3:50 p.m., the feed water and the outer shell temperature falls sharply. The reason is that all the evaporating energy only comes from the water heat capacity of the heating water tank. Meanwhile, the temperature difference between the feed water and outer shell becomes smaller, thus, the productivity of the operational mode falls sharply.

Fig. 5.71B shows the change of the feed water temperature, the outer shell temperature, and the yield when operation is in the heating mode. The trend of the feed water temperature and outer shell temperature change with the operation time is similar to that of the single-effect mode. The temperature increases quickly at the beginning of the heating phase, and reaches relatively steady state from 3:00 p.m. to 3:50 p.m. During this period the variation of the temperature is small. After electric heater was turned off at 3:50 p.m., the feed water and shell temperature fell sharply.

It can be seen from Fig. 5.71B that the productivity of the first-effect is larger than that of the second-effect during heating and cooling processes. The reason behind this behavior can be explained as follows: at first, heating was carried out in the innermost water tank, so the feed water temperature of the first-effect was higher than that of the second-effect. Meanwhile, the temperature difference between the feed water in the water absorption materials of the heating water tank and the first-effect shell was larger than that of the second-effect, which leads to more yield of the first-effect. Furthermore, when the electricity power was cut off, the temperature at every measured point fell quickly, it was found that the temperature difference between feed water and the condensation surface in the first-effect was also larger than that in second-effect, which may be due to heat transferred to feed water flow. As a result, the temperature difference between the evaporation and condensation surface and the heating temperature have a significant influence on the mass and heat transfer process.

Thermal resistance in the triple-effect vertical tubular still is largest among three operational modes, which leads to slower heat transfer rate. The variation curves of the feed water temperature and yield of each effect with operating time are shown in Fig. 5.71C, in which the heating time is 5.3 h, which is longer than that of the double-effect device. In addition, in Fig. 5.71C it can be observed that at the beginning, the temperature variation curve is very steep. However, the temperature increases with the heating process gradually becoming slower until the power is cut off, and then the temperature drops sharply.

Fig. 5.71C also shows that for the heating process, the yield of the first-effect was much larger than that of the second-effect by 49.54%, because the temperature difference between the feed water in the water absorption materials of the heating water tank and the first-effect shell was larger than that of the second-effect, that is, the effect of convection heat transfer and evaporation mass transfer are strengthened.

The comparison profiles of the total yield of three operational modes with the heating time under the same operation conditions are shown in Fig. 5.72. As shown in Fig. 5.72, the 20-min total yield increases quickly at the beginning of the experiment, and 4 h later the yield increases slowly. Among them, distillate productivity of the triple-effect device is larger than that of the single-effect and the double-effect. This is due to the successful utilization of the latent heat of vapor condensation in the multieffect mode, and results in an increasing total yield.

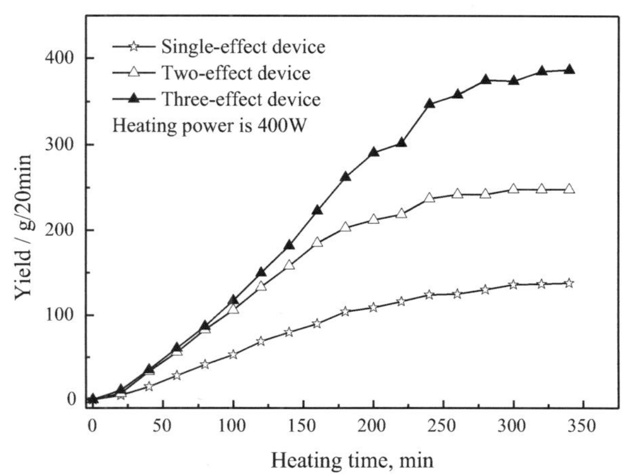

Figure 5.72
Variation of the yield in the system with the operation time.

It is noteworthy that there exists a time lag among total yield of the single-effect mode, the double-effect mode, and the triple-effect mode, which reach relatively steady state. The 20-min total yield of the single-effect mode reaches relatively steady state after 3 h, and meanwhile the variation of the yield is relatively small. The 20-min total yield of the double-effect and triple-effect operational mode reaches relatively steady state after 4 and 5.2 h, respectively. This is caused by the bigger thermal resistance in the multieffect operational modes than that in the single-effect mode, thus the heat transfer rate in the triple-effect mode is relatively slow.

5.11.2.2 Fixed-Temperature Heating Experiments

Several tests were performed in steady-state conditions within the range of 50−80°C, with 5°C temperature steps, for the experimental procedure validation. The effect of feed water mass flow rate and heating temperature on the device productivity were studied. The experiment took into account the water temperature of the heating water tank as the operation temperature. In each experiment, the steady condition must be kept for more than 2 h. The experiment was performed indoors at the ambient temperature of about 18°C. The experimental results in the three operational modes are shown in Fig. 5.73A−C.

Fig. 5.73 illustrates the effect of the feed water mass flow rate on the unit yield. It can be observed in this figure that the hourly yield of the three operational modes increases with the operating temperature, and the yield rate of triple-effect mode is larger than that of the single-effect and double-effect. When the operation temperature is 80°C, the hourly yield of the triple-effect mode is more than that of the double-effect and single-effect by 47% and 148%, respectively, at the feed water mass flow rate of 30 g/min. This may be explained as follows: Firstly, the triple-effect mode has larger total evaporation area than the single-effect and the double-effect mode. Secondly, the latent heat released by the vapor of the first-effect and the second-effect in the condensation process of the triple-effect mode is re-utilized. Finally, the temperature of the outer tubular shell of the single-effect mode was higher because vapor was directly condensed on the inner surface of the tubular shell. This leads to the decrease in the temperature difference between the condensation and the evaporation surface.

In Fig. 5.73 it can also be seen that the hourly yield of the three operational modes increases with a decrease in the feed water mass flow rate. This behavior can be explained by Fig. 5.74, which shows the temperature variation of every effect of a double-effect device. Fig. 5.74 depicts the reduction of the water temperature difference naturally leading to a decrease in the heat transfer driving force between the feed water and the tube shell, and the reduction of the freshwater yield and the performance of the device.

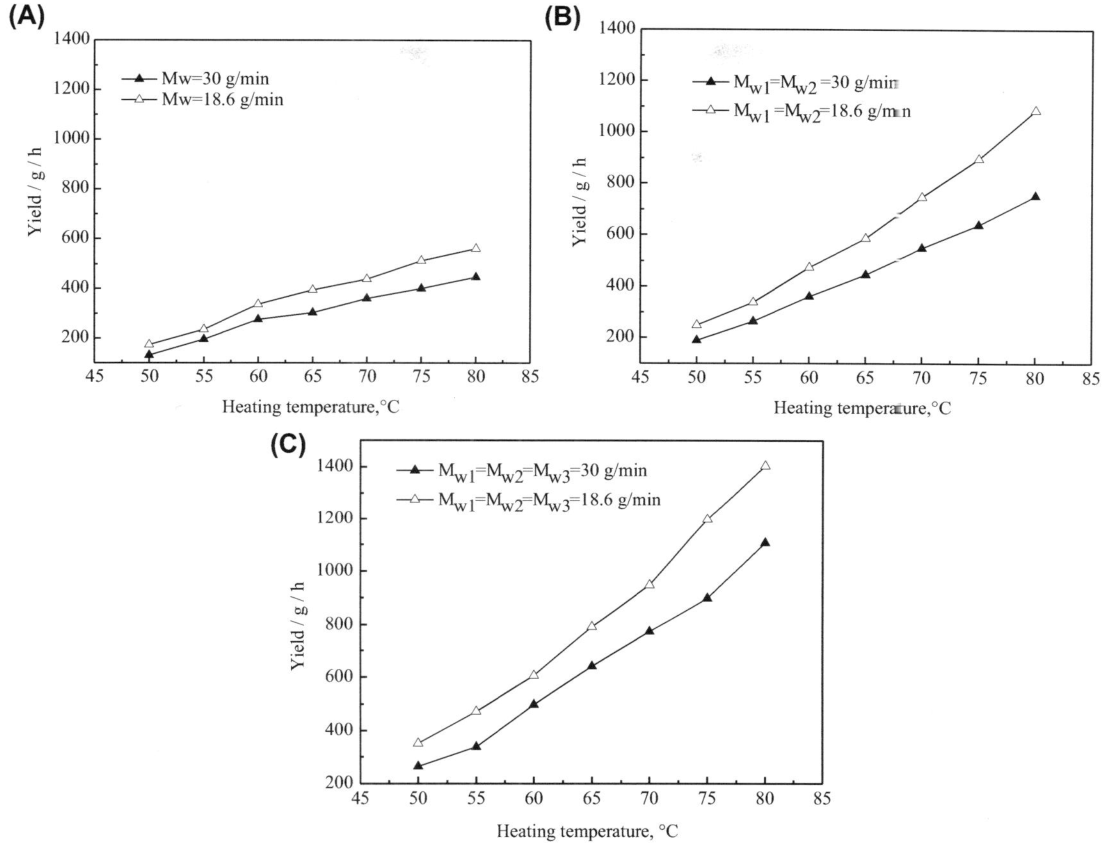

Figure 5.73

Variation of the yield with heating temperature in the mode at different feed water mass flow rates.

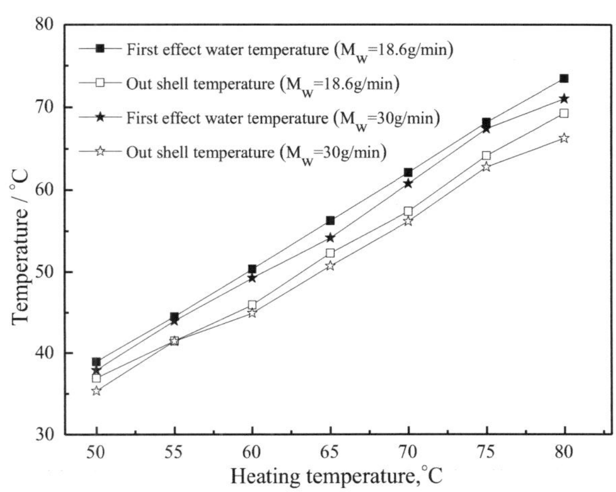

Figure 5.74

Variation of the water temperature of double-effect mode with heating temperature.

5.11.2.3 Performance Comparison of Three Operational Modes

Experimental results indicate that the productivity rate of all three operational modes will increase with the decrease of feed water mass flow rate. It is observed that the latent heat of the vapor can be successful re-utilized in the multieffect operational modes, and the multieffect modes have a larger condensation area and evaporation area. For this reason, the triple-effect mode produces more hourly distillate production than that of the double-effect mode. Similarly, the double-effect mode has more hourly distillate production than that of the single-effect mode under the same operation condition.

Solar desalination system performance basically depends on the solar radiation and environmental conditions. The performance or energy efficiency of vertical tubular desalination system is defined as the gained output ratio (GOR). A higher value of GOR in a desalination system means the performance of the still is better and the energy reuse ratio is higher. For comparison, the total yield and GOR of the system are shown in Table 5.7 when it works under two operating models.

Table 5.7 shows the hourly productivity of three operational modes increases with input heating power increase. However, the incremental rate of the yield of the single-effect mode is relatively smaller than that of multieffect cases. Experimental results indicate that the yield of the triple-effect mode increases by 115% than that of single case. The multieffect mode also has larger GOR than the single-effect mode, as a result of successful recycling of the latent heat of vapor condensation in multieffect mode. This led to a higher energy utilization ratio and more yields.

5.11.3 Conclusions

The experimental results indicated that this unit, especially for the multieffect operational mode, has a good productivity performance. In the multieffect mode, the latent heat of vapor can be used many times so that it has a high GOR. The condensation processes are strengthened because the condensation area of the device is always larger than the evaporation area, which has a beneficial effect in improving the evaporation heat transfer. In the steady-state experiment, heating temperature, the feed water mass flow rate, and the temperature difference between the evaporation and condensation have shown a great effect on the yield of the unit.

Table 5.7: The water yield and the GOR of the operational modes under steady state.

	Single-effect mode		Double-effect mode		Triple-effect mode	
Heating power/W	Yield (g/h)	GOR	Yield (g/h)	GOR	Yield (g/h)	GOR
305	306	0.65	390	0.84	540	1.17
400	414	0.67	744	1.21	1162	1.89

5.12 The Solar Distiller With Single-Stage Falling Film Evaporation and Condensation

Falling film evaporation and falling film condensation were used in many fields, such as refrigeration, air separation, and traditional desalination industry and other fields. Its advantage is minimal refrigerant heat capacity, it can quickly reach the temperature close to the heating body. And in the process of falling film, the area of film expanded is quite large. It has an advantage for evaporation of liquid or condensation of steam, and it also has an advantage for the heat exchange between steam and heating body. By these advantages, the falling film evaporation or condensation were used in a solar distillation unit. A solar desalination apparatus having vertical plate or standpipe falling film evaporation and condensation chamber is the more common form.

Fig. 5.75 shows a single-effect vertical plate falling film evaporator and condensed vertical plate falling film type solar still working principle [27]. It consists of solar collectors, vertical plate falling film evaporator and condenser, and the water storage tank and other components. Where the riser evaporator and condenser are assembled in the same chamber, this is an evaporation—condensation chamber. Vertical plate evaporator and condenser have many forms; in order to make a film on the water evaporator flow evenly, usually vertical panel auxiliary absorbent pads of some porous material, such as cotton cloth or glass fiber are used. They can also be installed in a grid in the vertical plate or pressed in to strengthen the board with a variety of channels, and do hydrophilic treatment. The condenser also take many forms, some are made of flat cassette, intercom cold seawater; some are composed of a finned pipe, intercom cold water; some absorbing film is simply made with freshwater, allowed to flow directly in the vertical version, absorb steam directly, and so on. Running of such devices can be explained as follows: seawater from the storage tank 3, pumped through 5, to the lower end of the condenser 11, then flows upward along the flow path in the condenser, in the process it takes away the latent heat of steam condensation at the condenser surface, so that the temperature is increased. On the rear of the condenser, seawater passes through a pipe into the solar collector 1, where the water absorbs sunlight energy, the temperature is further increased to the highest temperature of the design, and then comes to a falling film evaporator via line 8 entrance into the evaporator at the top of the solution distribution tube (as shown in Fig. 5.75), after the allocator, seawater under the action of gravity along the surface of the evaporator is evenly landed, forming a layer of film-like water. At this point in the process of the water constantly evaporated falling in, while the temperature is decreasing, it reaches the bottom of the evaporator, and the temperature becomes the lowest, the final concentrated solution outlet 10 is excluded from the device. Freshwater condensing in the condenser 11 is generated on the outer surface, also under the influence of gravity, landing at the bottom of the condenser, and finally discharged through the outlet 12 of the

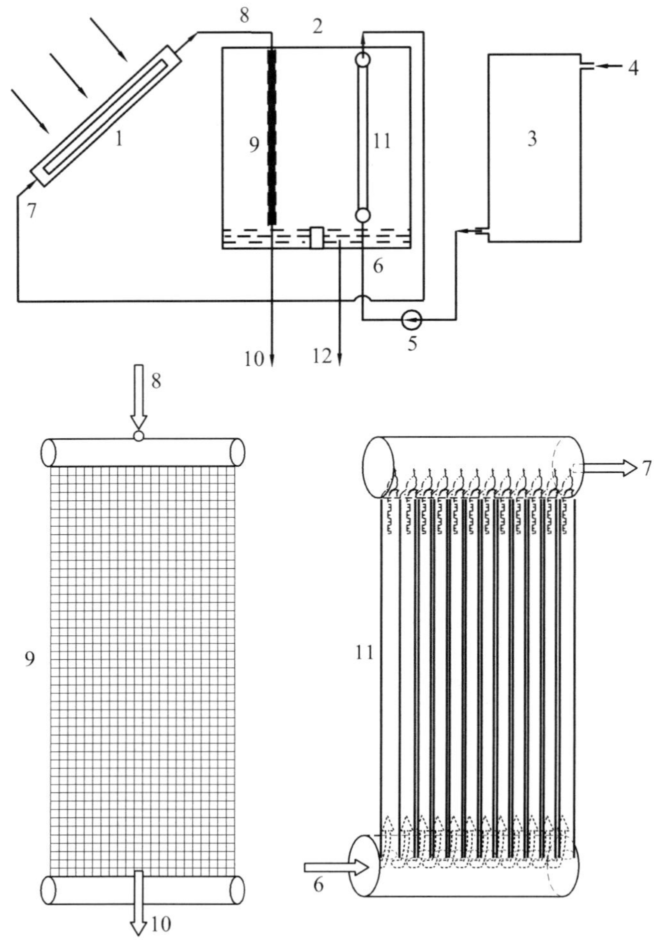

Figure 5.75

The falling film evaporation solar distiller: 1, solar collector; 2, evaporation/condensation chamber; 3, storage tank; 4, cold water inlet; 5, pump; 6, cooling water inlet; 7, preheat water inlet; 8, Hot water inlet; 9, evaporator; 10, waste water outlet; 11, condenser; 12, freshwater.

evaporation−condensation chamber, and the finished product is collected to give freshwater.

Kudish et al. studied the theoretical and experimental aspects of the device shown in Fig. 5.75 [27], where the timber size and design of the device are as follows: The total width of the evaporator and condenser are 0.3 m, namely $b_n = b_t = 0.3$ m; evaporator height is $L_e = 0.085$ m, made by polypropylene with 5 mm × 5 mm mesh in the outer surface, the thickness is 1 mm. Condenser height is $L_c = 0.9$ m; condenser flow passage is a circular tube, wall thickness is $\delta_t = 1.5$ mm, and the inner diameter of the tube is 3 mm, thermal conductivity of the material is 0.3 W/m K, and the number of tubes is 90.

The solar collectors' lighting area is 2 m^2. Note that all the above materials are constructed using an organic polymer, including solar collectors. The results show that in fine weather conditions, when the seawater flow rate is in range of 40–50 kg/m^2 per day and the evaporator and condenser interval is 2 cm, the production of freshwater will reach 11 kg/m^2 per day per square meter of collector area in which the freshwater productivity has been improved greatly than conventional solar distillation unit's 3–5 kg.

It is worth noting that, except for the amount of solar radiation, the decisive influence of factors on plant capacity of water are pitch of evaporator and condenser, water flow, seawater into the condenser temperature and hot water temperature input at the top of evaporator also have a greater impact on water production rates. An established theoretical model, at the given actual weather conditions (a given amount of solar radiation and ambient temperature versus time relationships), the influence of the device's water production by evaporator and condenser's pitch can be theoretically calculated, as shown in Fig. 5.76. The figure shows that when the pitch increases, the device's water production tends to be reduced. Thus, the pitch between the two should take as small a value as possible. In fact, this is well understood, when the pitch increases, the distance of steam's mass transfer is increased, thereby increasing the mass transfer resistance, it is unfavorable for steam rapid coagulation, thereby reducing the production of distilled water.

Similarly, given the weather conditions, according to the theoretical model, the influence of water production by the flow rate of seawater, the temperature of water into the condenser, and the temperature of hot seawater input at the top of evaporator, can also be calculated. The computer simulation results given by Kudish et al. are shown, respectively, in Figs. 5.77 and 5.78.

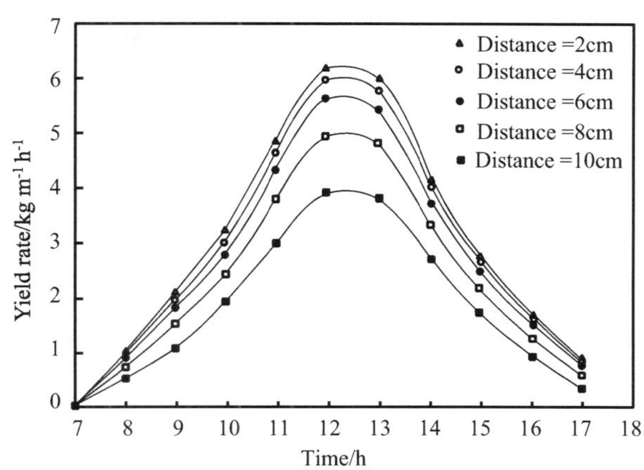

Figure 5.76
The effect of spacing of the evaporator and condenser on the water yield.

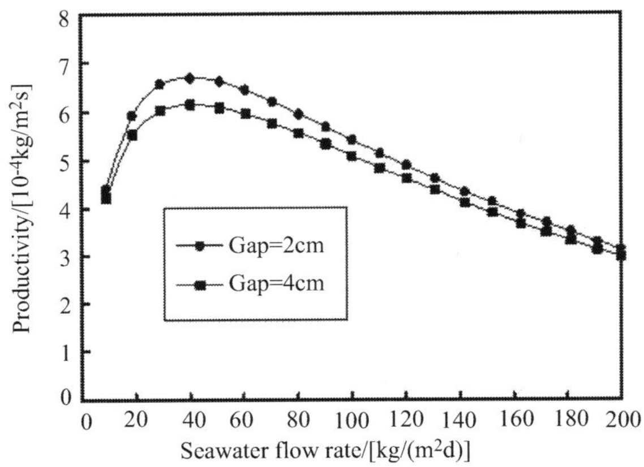

Figure 5.77
Effect of seawater flow rate on productivity.

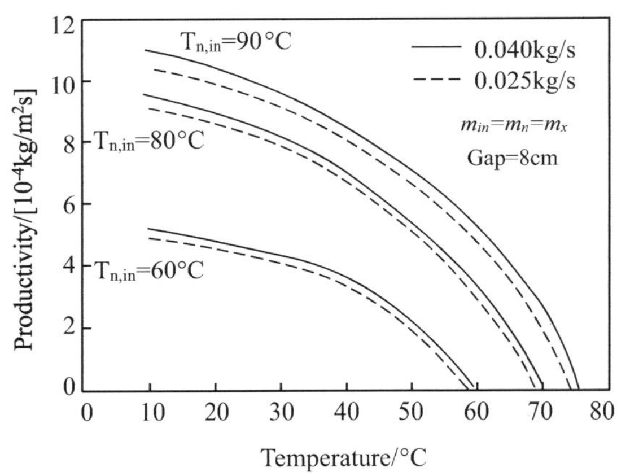

Figure 5.78
Effect of temperature on productivity.

Chinese scholars also studied a similar device, more representative of the literature [28], a solar distillation unit capable of recovering the latent heat of the presentation, as shown in Fig. 5.79. The device, with 3 m of vacuum tube collectors, collects solar energy for heating the evaporator, between the evaporator and the condenser to achieve heat and mass transfer by natural convection mode. Experimental results show that when the temperature of the seawater in the condenser is 25°C, the daily amount of solar radiation is 25.2 MJ/(m^2 d), the system average yield per square meter of collector area of freshwater is about 9 kg/d. When the water flow is obtained at 45 L/h, the evaporator inlet water temperature is 75°C, and the system has large water production efficiency results.

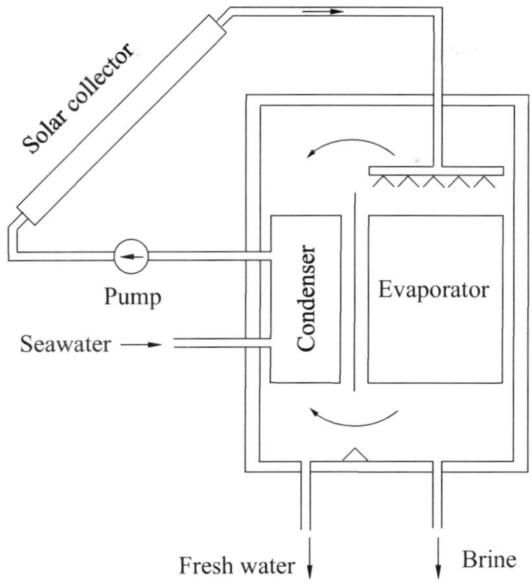

Figure 5.79
The solar distiller with recovering of steam latent heat.

5.13 The Solar Distiller With Multiple-Stages Falling Film Evaporation and Condensation

The most significant advantage of the falling film solar distiller is that the amount of water stored is small, the heat capacity of the device is small, the water is heated quickly, and temperature rises rapidly. It is conducive to the operation of the device at a higher temperature, but also conducive to the production of freshwater as soon as possible. The surface area of unit mass seawater is far bigger than the expanded surface area of the seawater which is stored at the transmission type solar energy distillation device. It is conducive to the evaporation of seawater. Therefore, the falling film distillation has gained more attention.

However, in single-stage (or single-effect) falling film solar distiller discharge the brine temperature is higher. Heat loss is large, which is unfavorable for the efficient use of solar energy. It is also not fully recovered in the latent heat of condensation of steam. To overcome these shortcomings, many scholars put forward a multistage falling film design.

The design of a more traditional multistage falling film solar still is shown in Fig. 5.80 [29]. It consists of a collector and water storage tank of the solar water heater and a distillation unit. There is a set of metal plates which are placed vertically in the distillation unit. Hot water, which is stored at the storage tank, flows down from the pipe, fixed at the left side of the leftmost metal plate. After reheating the metal plate, the temperature of hot

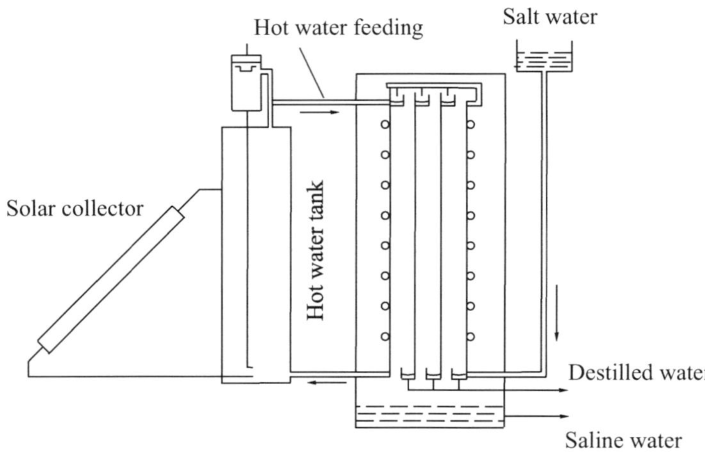

Figure 5.80
The solar distiller with three-stage falling film evaporation.

water decreases and it flows into the storage tank. At the same time, salt water flows up from the tube of the metal plate on the right where the distillation temperature is low. This causes sheet metal cooling. This water enters into the sink at the top of the each piece of metal plate as the supply water of the distillation (except the first one on the right panel), and it slowly drips from the right side of each plate. Since the left of the metal plate is heated by the hot water which comes from the tank, a portion of salt water is evaporated along the right side of the plate. This then condenses on the left surface of the second plate. After latent heat is absorbed by the metal plates, it is used to make salt water evaporate along the right side of the second plate. This process continues until it reaches the rightmost piece of the board. On the rightmost panel, cold sea water flows up along the fixed coil on the right which comes from the seawater supply device. It makes the temperature of the riser decrease, and it promotes condensation of steam on its left side; the steam comes from the previous level riser. In this case, the steam condenses on the left side, where the released heat heats the supply water. Condensate water enters into the collection tank along the left side of each plate and leads to producing freshwater. However, the concentrated brine is collected in the bottom of the distillation, then it is discharged.

In order to further reduce intermediate links of solar collector, reduce heat loss, saving the cost of investment, Goff [30] proposed a design, shown in Fig. 5.81.

The distiller's part of Goff's system is basically the same as the structure in Fig. 5.80. But the collection of solar energy is very different. It does not use conventional solar collectors, which use solar panels to absorb solar energy. The left side of the leftmost solar absorber plate is coated with a selective solar absorber coating, stamped heel glass cover to avoid heat transferring to the outside. The temperature of the plate rose rapidly under

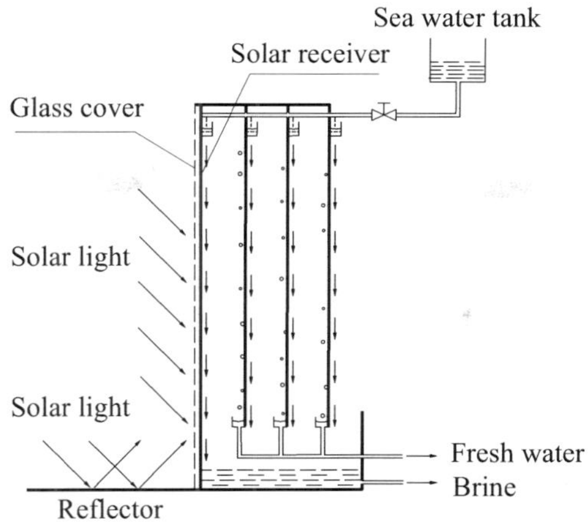

Figure 5.81
The multistages falling film distiller with direct solar collection.

the irradiation of sunlight. At this time, on the right side of the plate the seawater flowing down with falling film, sunlight is absorbed and the temperature of the seawater rises rapidly to achieve a falling film evaporation. The rising steam condenses on the left side of the second metal plate. It releases latent heat of vaporization which is absorbed by the metal plate, and then it allows the downstream heated water to evaporate on the right. This is repeated, until it arrives at the extreme right side of the piece of metal vertical plate. At this point, the right side of the right vertical metal plate is cooling. There are a variety of cooling forms, it may be a cooling coil as shown in Fig. 5.80, it also can be the same as the vertical plate in front of the right side of the falling film cooling with seawater, as shown in Fig. 5.83.

Some useful results are given in the literature [31] of reinforcement plate as shown in Fig. 5.82, with a multistage falling film type solar still, in a simulated experiment.

The geometry size of the various components is shown in Fig. 5.83. The vertical plate is made of 0.5-mm-thick copper plate through a folding suppression. It improved the performance of the heat and mass transfer, and it reduces the size of the device. The spacing of plates is 4 cm. The whole distillation chamber system is installed in a 0.5-mm-thick galvanized iron plate made of a rectangular box. One side of the box is open and brine water tries to cool the final stage of the vertical plate at ambient temperature. Five-sides outside the long box use 5-cm-thick polystyrene foam board as insulation. All stages of plate seams are sealed with a sealing material to prevent the leakage of steam between the stages. The top of the riser is provided with a water dispenser in order to allow the saline to distribute evenly over the riser. In spite of this, the water cannot ensure uniformity everywhere on the whole piece of vertical plate, some places will appear dry. Therefore, the evaporation surface of the

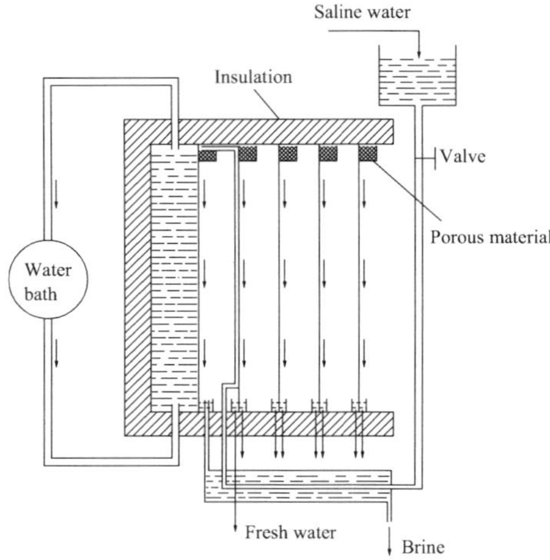

Figure 5.82
Multistage falling film solar distiller with enhanced board.

vertical version is evenly bedded with a layer of absorbent gauze. Water spray on the gauze which comes from a seawater distributor, then slowly flows evenly onto the vertical plate.

At the simulation experiment, the hot water generated by the super constant temperature water bath is pumped into the first stage of the flat container for heating the first-stage risers. The heating and coagulation mode of the rest of the board is the same as the introduction of device.

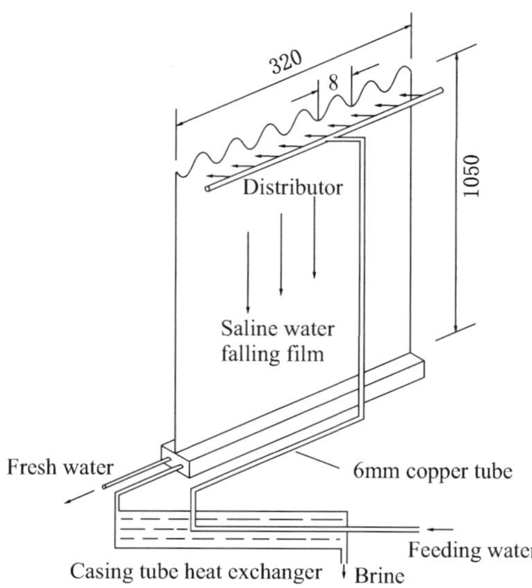

Figure 5.83
The vertical plate and dimensions of the dispenser (unit: mm).

The temperature of strong brine is generally high, which comes from the bottom of the vertical plate. In order to take use of this part of the waste heat, firstly make it flow through a pipe heat exchanger, making preheated brine enter the device. The process is shown in Fig. 5.83. Try to insulate the outer casing of the heat exchanger. For simplicity, the figure only depicts a tube heat exchanger. In fact, one to four strong brine exports have a similar tube heat exchanger. Experiments are carried out in the indoor stable environment. The variation range of environment temperature is about 26–28°C.

The experimental results and preliminary analysis are as follows:

1. Comparison of the evaporation surface with or without gauze
   Analysis shows that the vertical plate folded into the vertical groove will be good for heat transfer enhancement of a condensation surface, because it can make the condensate quickly gather dripping, thus reducing the liquid film thermal resistance. But the vertical groove wall of falling film evaporation is likely to have a bad effect, because it is difficult to make the whereabouts of vertical tank brine evenly distributed on the shaft wall. It easily produced a "dry" area on the wall "backbone." If bedding yarn on the surface of the evaporation network, it will help to wet the surface, and it can strengthen the effect of the evaporation. This is one reason why we need bedding yarn. Although bedding gauze can increase the thermal resistance, gauze can cling to the shaft wall after pumping, and gauze is very thin. Therefore, the negative effect of the increase in thermal resistance is much smaller than when it reduces the "dry" area of interest. It can be said that even in the case of flat surfaces, bedding gauze can effectively improve water production. For this reason, the comparative experiment with or without gauze had been done.
   The experimental results are shown in Fig. 5.84. The curve shows that the effect of laying or not laying gauze on the water yield is obvious. In many circumstances, the water production can nearly double, because laying gauze reduces the evaporation surface area of the drying zone.
2. Influence of operating temperature on the water yield
   Import and export with an average temperature of hot water is as the operating temperature of the device. The pumping flow rate of the water bath is 406.8 kg/h Fig. 5.85 shows the relationship between freshwater rate and operating temperature when the device is one and four stages. The curves corresponding to the brine flow stages are marked in the figure. The experimental results show that the water rate is increasing quickly when the operating temperature increases, both for single stage and stage 4 of the device. The higher the temperature, the faster the rise. Multistage distillation is described as being more meaningful at a high temperature. It can also be found in Fig. 5.85 that the water production of a 4 stage distiller is much higher than single stage. This is the result of using the water vapor condensation latent heat many times.

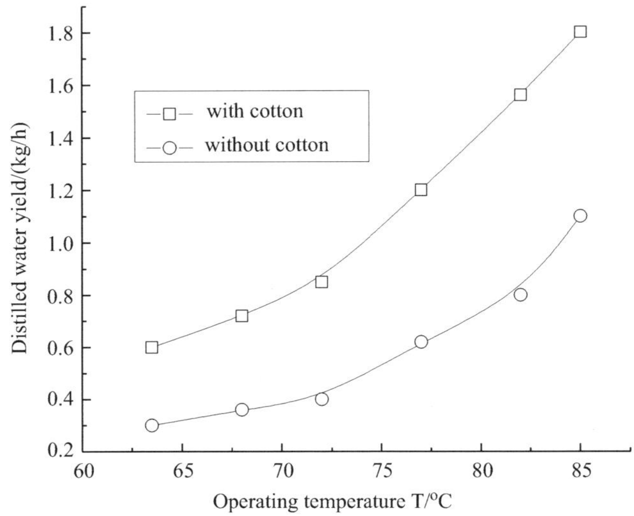

Figure 5.84

Comparison of water production with or without cotton for a four-stage device: 1, $M_1 = 147$, $M_2 = M_3 = 110$, $M_4 = 120$; 2, $M_1 = 133$, $M_2 = 136$, $M_3 = M_4 = 126$. (M_i is the brine flow rate in different stages [ml/min].)

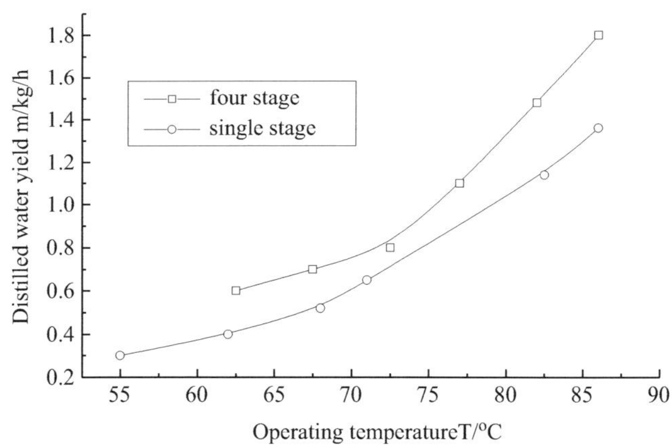

Figure 5.85

Relationship of water yield with operating time for four-stage and single-stage devices (brine flow rate of each stage is about 120–147 mL/min).

When the device is of four stages, the relationship of water production and operating temperature is shown in Fig. 5.86. The graphic shows that the fourth stage (sometimes even the third stage) may have no water output in the low-temperature working segments. Therefore, the operating temperature should be taken to set the number of device stages. In general, it takes stage 4, when the temperature is greater than 80°C. Otherwise, it is more reasonable to use stage 3.

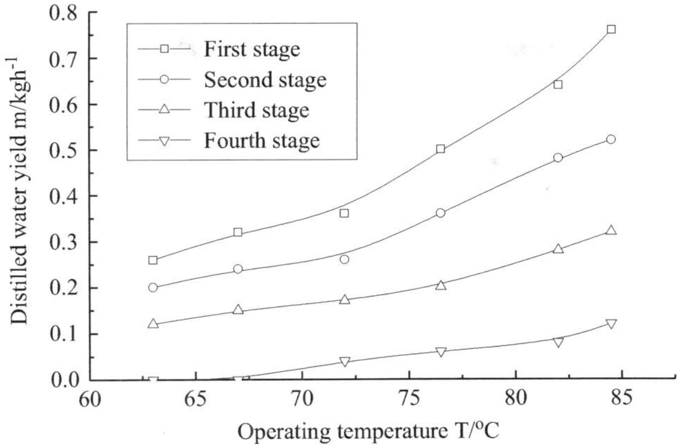

Figure 5.86
Relationship of water production of each stage with operating temperature for a four-stage device (brine flow rate of each stage is about 120—147 mL/min).

3. The influence of evaporation surface tilt on water yield
   It is easy to understand that the ramp will affect the evaporation and condensation of the falling film, because water vapor density is lighter than air. When the distillery tilts (evaporation surface up, condensation surface down) it will be good for the flow of steam. Therefore it is beneficial to the evaporation and condensation process, but its effects require further validation. In the experiment, a four-stage distillation unit was tested at a different angle of deviation from the plumb line, respectively.
   The results are shown in Fig. 5.87. The figure indicates that the deviation angle of the device has a strong influence on the water yield, increasing the deviation angle will be useful to increase water production. The reason can be explained as follows: in the case

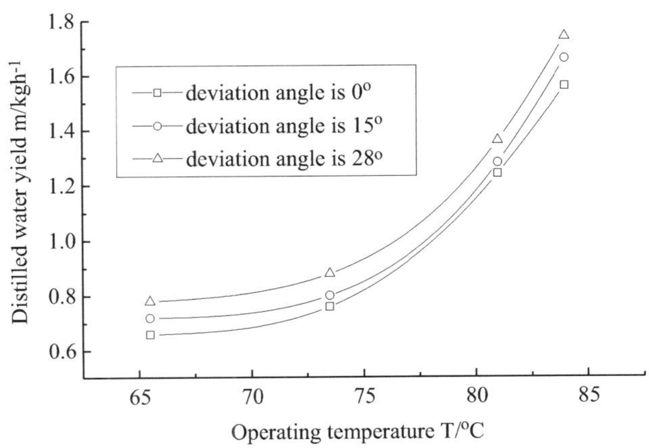

Figure 5.87
Variation of water yield with different deviation angle (brine flow rate of each stage is 150 mL/min).

of the vertical plate, when a mixture of air and vapor contact the condensation surface, once the steam condenses, it will form a thin layer of air near the condensation surface. The thin layer will continue to hinder the steam contact with the condensing surface, thereby affecting water production. Under the condition of the rake face, because the steam is lighter than air and the temperature of steam is higher generally, subsequent steam is easier to break through the air layer block and contact the condensation surface under the action of gravity. The water production rate is improved naturally. Whether this is the best angle of deviation of plumb line needs further study.

4. The influence of brine flow rate of the water production

Obviously, the flow rate of brine will affect the output of distilled water. If the flow rate is too large, the salt water film is too thick, it can increase the thermal resistance of water film, and it is not conducive to the evaporation of the brine. At the same time, if the flow rate is too large, salt water is too quickly dropped to the bottom of the riser, the salt water does not evaporate then be ruled out, it will also reduce the production of distilled water. On the other hand, if the brine flow rate is too small, it is also not conducive to the improvement of the water production rate. This is because if flow rate is too small, it will be conducive to the uniform distribution of vertical plate brine, it may cause some "dry zone" to appear on the vertical plate, thereby reducing the water rate. In addition, if the flow rate is too small, it can make the concentration of brine increase too fast in the walls. It is not conducive to evaporation, it may even cause precipitation of crystalline salt in medium yarn, and make the performance of the system reduced.

Thus, the flow rate of the brine should have a reasonable quantity. Under the condition of a stage 3 device with yarn nets, operation temperature 82.5°C, and brine flow rate take different values at different stages, respectively, Fig. 5.88 shows the change of all stages and quantity of output. The results show that the water rate of primary distillation chamber increases with the increase in brine flow rate, but the second stage does not have an increasing trend, and even has some experimental point decreases, and the third level of the water rate is gradually reduced. However, the total water production is essentially unchanged. The conclusion has a guiding significance to the steady-state operation of the device. It is easy to be understood that the sensible heat loss taken by the brine will increase when the brine flow rate increases. This subsequently reduces the evaporation temperature, thus reducing all stages of water rate. However, as a result of sensible heat being increased, this makes the temperature of the first-stage still's condensation decreased, which is beneficial to the first-level stills to increase water production. The total water production is essentially unchanged, because the evaporation surface of the first level keeps the same temperature under the steady-state operation, ensuring enough energy for the sake of brine evaporation. However, in the actual solar installation, we should choose low brine flow rate, so as to avoid the loss of sensible heat.

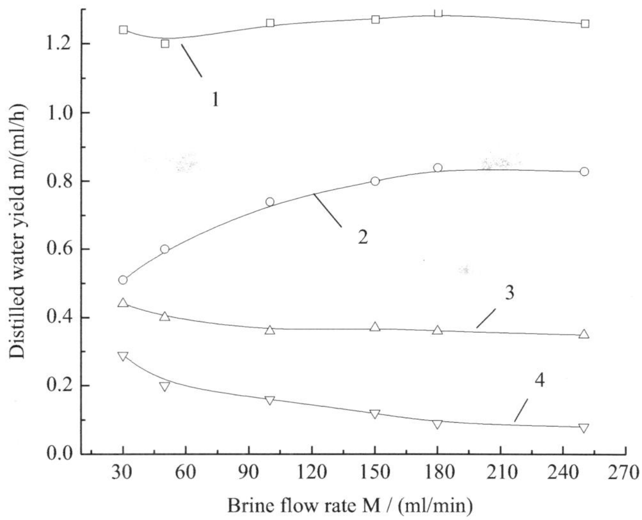

Figure 5.88
Relationship of water production with brine flow rate: 1, total water yield; 2, first-stage yield; 3, second-stage yield; 4, third-stage yield; 5, unit energy consumption of water production rate.

Per unit energy could produce an amount of distilled water, known as the system of water production rate per unit energy consumption, also known as the water production rate. The water production rate is an important indicator of system performance. In our experiment, the calculation formula is:

$$\dot{m} = \frac{m_e}{c_p \cdot M_p (T_{in} - T_{out})}$$

where, m_e is distilled water production per unit time; c_p is the specific heat capacity of hot water; M_p is the pumping flow rate of constant temperature water bath; T_{in} and T_{out} are the temperature of the hot water when it enters or goes out the flat box before the distiller, respectively.

Fig. 5.89 gives the relationship of water production rate and operating temperature in cases of the single stage and stage 4. The results show that in the single stage or stage 4, unit energy consumption of the water production rate increased with the increase of operating temperature. It shows that the higher the temperature, the better performance of device. The results showed that the stage 4 situation is better than single stage under the condition of equal operating temperature. The higher the temperature, the more the former is advantageous.

5. Conclusion
   a. Multistage strengthening falling film solar distiller is an ideal solar distillation unit which can run at the higher temperature range (>70°C). The higher the operating temperature is, the better the result. Due to the system of solar collector and

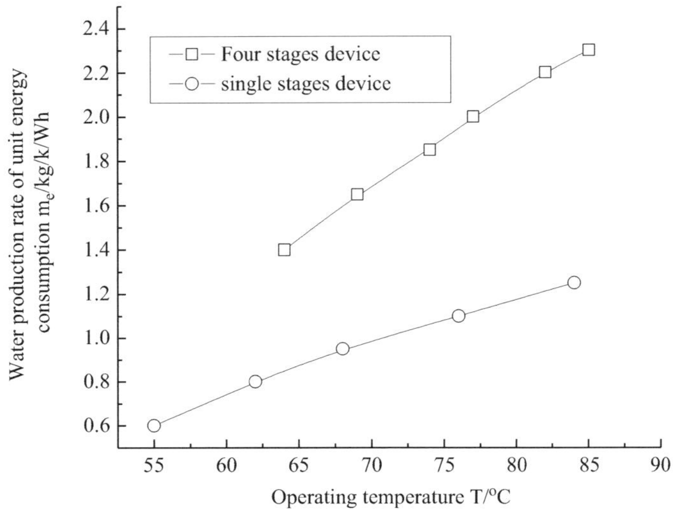

Figure 5.89
Relationship between operating temperature and specific water production.

distillery being separated, it is easy to use industrial waste heat or other renewable energy as a heat source.

b. In general, the higher the operating temperature, the more series can increase. But stage 3 has been able to get a high water production rate and save investment at the heating temperature of the solar collector (60°C, 90°C).
c. It can effectively improve water production, bedding absorbing porous material at the evaporation surface, and strengthening surface moisture.
d. Enhancing the heat transfer surface, reducing the condensation surface liquid film thermal resistance, not only can it reduce the size of the device, it can also effectively increase the yield.
e. Tilting device deviation from the plumb line, evaporation face up, condensation face down, can obviously increase the water rate of the plant.

5.14 The Solar Distiller Enhanced by Power

5.14.1 Basin Solar Distiller With a Water-Cooled Condenser Outside

A taditional basin solar still's steam condenses on the cover, the latent heat of condensation of water vapor increases the temperature of the glass cover, which reduces the temperature between evaporation water surface and condensation surface, diminishing the water vapor generation rate, thereby reducing the efficiency of the distiller energy and water production.

Fig. 5.90 shows a schematic view of a basin solar distiller having a water-cooled condenser outside. Its main idea is to pump saturated steam outside from the disc distiller, and let it condense in the cooling coil outside the distiller, while taking advantage of the

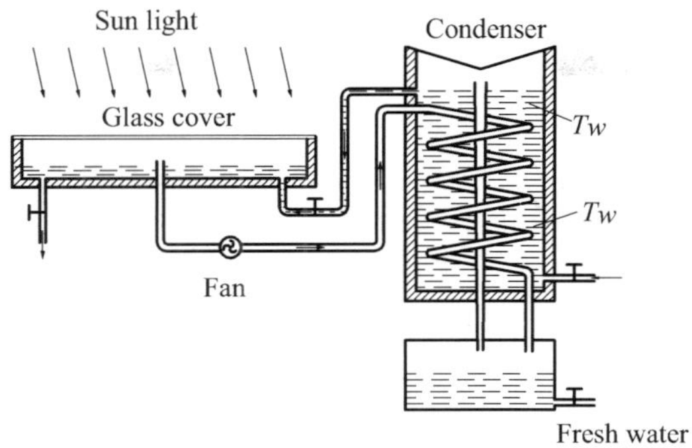

Figure 5.90
Basin solar distiller with a water-cooled condenser outside.

latent heat of condensation of steam to preheat the distiller's incoming water, so that latent heat of condensation of water vapor is repeatedly used [32].

The main device consists of several parts of the solar still, a small electricity power wind turbine, a water-cooled condenser coil type, a freshwater collector, and coupling pipes. The distiller lid is placed level at this time, because it does not require a condenser. The dish absorbs solar energy to evaporate seawater, steam passes through a water-cooled cooling coil by a low-power fan, and with per square meter of area lighting with a 100-W electric fan. Condensed freshwater in the condenser coil re-enters the freshwater collector storage for users. When the cooling water outside the coil is gradually heated, there will be water evaporated in the condenser, so that in this part of the steam condensate on the cooler upper end of the ramp, freshwater is produced by the other pipe dripping into the freshwater, as shown in Fig. 5.90. The amount of this part of the additional freshwater is dependent on the temperature difference between the cooler water and environmental temperature. The greater the difference, the greater the amount of additional freshwater. The water is preheated to provide additional water for the distiller in the condenser, so that the temperature is high when seawater enters into the distiller, and is conducive to rapid evaporation in the distiller.

Such a device has the following three advantages:

1. Latent heat of condensation of water vapor is used again to improve energy efficiency.
2. The cover glass is not a condenser, it is conducive to increasing the temperature difference between the water surface and the cover, and strengthening the evaporation process.
3. As the role of the fan, the gas pressure inside the distiller is slightly lower than ambient atmospheric pressure, which also strengthens the evaporation of water, it is advantageous to improve the water production rate of the apparatus.

Nikolai results show that such water production apparatus unit area is increased by 2.5 times compared to traditional disk distiller water production. It is worth noting that without considering the consumption of electric power, device efficiency can be increased by 2.5 times, but if you want to take into account the consumption of certain electric power, solar energy utilization efficiency of the apparatus is only increased ;by about 1.25 times higher than the efficiency of a conventional distillation basin.

5.14.2 Focus Solar Still of Small Vacuum Strengthened Evaporation

Nikolai [33] has introduced an experimental system as shown in Fig. 5.91, the system consists of a focusing solar collector, absorbing tube, condenser, freshwater, freshwater storage tanks, and water storage tanks for seawater and other parts. Absorbent tube is mounted in the focal line of a parabolic trough's focuser. Seawater enters the solar absorber pipe through the reservoir tank, where the water absorbs solar energy and the temperature rises to around 80°C. Due to the vacuum pump, a pressure lower than the pressure inside the absorption tube environment, the seawater boils at about 80°C. Steam passes along the pipeline into the water-cooled condenser, exchanging heat with the condensing coil, condense, steam latent heat is taken away by the cooling water. The condensed freshwater flows into the freshwater tank, and finally to the user.

Experimental results show that due to the vacuum pump, the absorption tube pressure is lower than the ambient atmospheric pressure, reducing the water boiling temperature,

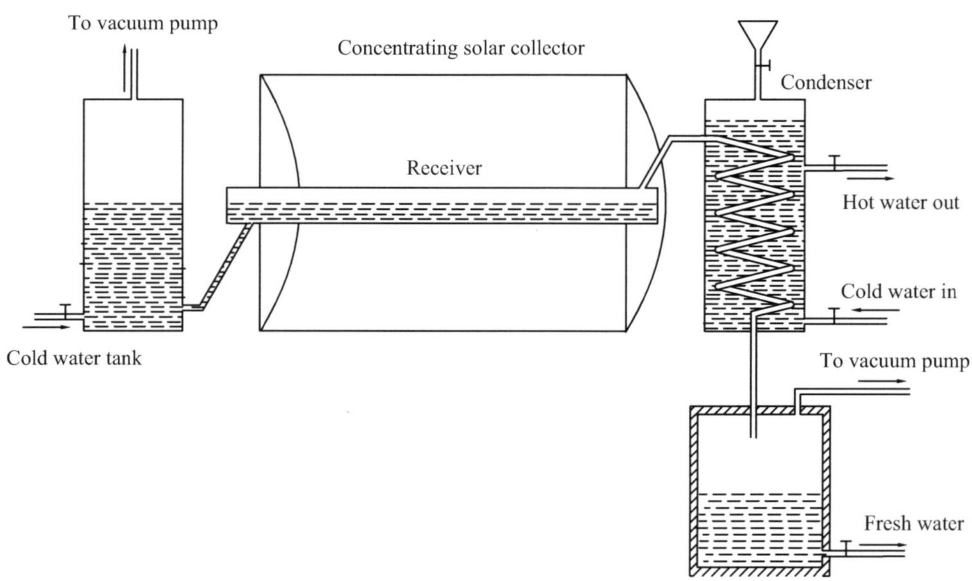

Figure 5.91
Focus solar still of small vacuum strengthened evaporation.

and strengthening the evaporation of seawater, but also reducing the absorption tube's convective and radiative heat loss to the environment, which is conducive to improving freshwater productivity. The results show that the freshwater production rate per unit area concentrator in the same solar radiation and ambient temperature, if the water in the absorption tube boiled in the vicinity of 80°C, is higher 100% than that of traditional basin distiller. Since the degree of vacuum required is not high, a smaller vacuum pump can be used to reduce the power consumption of the system.

Also worth noting is that the system's degree of vacuum is not good as high as possible; results showed that the system's degree of vacuum is optimal when the water boils at 80°C−85°C. If using a high-power vacuum pump, seawater is indeed possible to boil at a lower temperature and can be further reduced from the absorber tube's heat losses by convection and radiation. However, lower pressure will not help these vapors condense in the condensing coil, especially when the temperature difference between steam and the condensing coil is small. The lower the degree of vacuum, the more the condensed vapor cannot be taken away by vacuum pump. Therefore, we must choose a reasonable degree of vacuum range.

For further energy saving, it can use the heated cooling water as the seawater supplied to the apparatus, in order to increase the temperature of seawater into the device.

5.15 A Multieffect Thermal Regeneration Solar Desalination Unit With Horizontal Tube Falling Film Evaporation and Closed Circulation

5.15.1 Operational Principle of the Experimental Unit

This kind of device was introduced by Chen [34] and Zhang [35]. The device is a distinctively active solar desalination distiller in which many new technologies were used, such as horizontal tube falling film evaporation, multiple heat regeneration, active enhanced heat and mass transfer, and so on. However, it still keeps the characteristics of solar distillers such as required operation temperature not so high, without vacuum, and easy to be small scale. The literature [34−36] has introduced steady-state experimental results for the unit.

The schematic diagram of the unit is shown in Fig. 5.92. It is mainly composed of an evaporation chamber (a rectangular box with many horizontal tubes) with a seawater sprayer, a seawater distributor, a circulation water pump, a hot water pump, a circulation air-fan, a stacked tray still, a gas−liquid separator, and a solar collecting system (in steady-state research with a simulation heat source). The operational principle of the unit is described as follows [36].

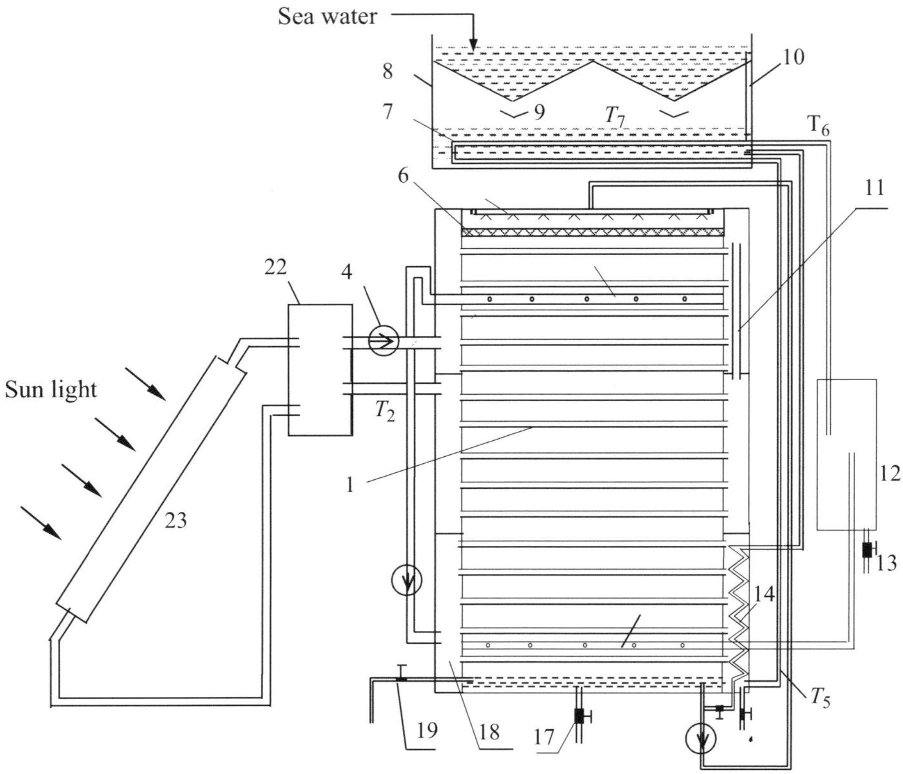

Figure 5.92

Schematic diagram of the unit. 1, horizontal tubes; 2, distribution chamber; 3, drawing vapor tube; 4, hot water pump; 5, sprayer; 6, holed plate distributor; 7, heat exchanger; 8, stacked tray still; 9, V-shape troughs; 10, link tuber; 11, link tube; 12, gas—liquid separator; 13, freshwater valve; 14, heat regenerator; 15, freshwater valve; 16, circulation water pump; 17, brine out valve; 18, distribution chamber; 19, overflow valve; 20, delivering air tube; 21, air-fan; 22, hot water storage tank; 23, solar collector.

The hot water heated by solar collectors or waste thermal energy enters the unit from the hot water pump 4 and passes through many horizontal tubes 1 in a shell-tube type of distribution chamber 2, where the seawater flowing outside the horizontal tubes is heated and evaporated. The hot water flows into the first stage of the distribution chamber, then into the second stage via a link tube 11, and finally leaves the unit to go back to the hot water storage tank 22.

The feedstock enters the unit from the first stage of the stacked tray still 8, and enters continuously the second stage of tray still by the link tube 10. It passes through a serpentine copper pipe 14 in the right-hand side of condensation cavity to reach the bottom of the unit, where it is pumped up to the top of the evaporation chamber by water pump 16 and circulated repeatedly, and finally reaches the bottom of the evaporation chamber. At the top of the evaporation chamber, the mixture of feedstock and brine is sprayed to a holed plate distributor 6 through a sprayer 5 so as to be equally sprinkled on the horizontal tubes. The brine forms falling film on the surfaces outside of the horizontal tubes, flows downwards due to gravity from the upper to lower tubes in the droplet shapes, and finally reaches the bottom of the evaporation chamber. The brine is discharged from the overflow valve 19.

When the brine flows from one horizontal tube to another, it absorbs large amounts of heat from the hot water inside the tubes and evaporates into water vapor. Especially under negative-pressure conditions, this process is enhanced. The vapor is sent into the condensation cavity, where it passes through and continuously condenses into distillates inside the horizontal tubes under positive-pressure conditions and releases most of its latent heat, by the circulation air-fan 21 from the distribution chamber via the drawing vapor tube 3. During this process, the latent heat is transferred to the falling film of brine outside the horizontal tubes and makes the brine evaporate. In the right-hand side of the condensation cavity, some of the vapor also exchanges heat with the feedstock flowing inside the serpentine copper pipe 14 and becomes freshwater. Then the mixture of condensed water and steam is sent to heat exchanger 7 in still 8 and cooled again to produce more freshwater. Finally, the vapor—water mixture with lower temperature is sent into gas—liquid separator 12, where the air-stream is condensed further and more freshwater is distilled. The rest of the saturated air-stream is recycled into the evaporation chamber via the delivering air tube 20. Then a closed circulation is completed. The freshwater is drained from valves 13 and 15, respectively.

In the stacked tray still, the seawater is heated through exchanger 7 and then evaporated. The distillate condensed on the underside of the first stage of the tray still is collected with V-shape troughs 9. The waste heat is discharged into the air through the first stage of the tray still.

5.15.2 Description of the Experimental Unit

1. *The evaporation chamber*: The evaporation chamber is the main component of the unit. It is a rectangular box of 1000 × 550 × 280 mm^3, which is made of stainless steel plate with thickness of 1.5 mm. In the box, the evaporation chamber is traversed with 117 horizontal copper tubes in which the hot water or steam flows through, as shown in Fig. 5.92.
   In order to assure all of the tubes are equally wetted, a sprayer and a seawater distributor are put on the top of the chamber (see Fig. 5.92.). The seawater distributor is a holed plank, on which the holes are arranged in 20 mm × 20 mm and their diameter is 2 mm. The horizontal tubes are composed of copper tubes with the thickness of 1.0 mm and outer diameter of 20 mm. They are crosswise arranged with the width of 40 mm in

column and 50 mm in line. The length of the tubes is 550 mm. Therefore, the total inner and outer heat exchanger areas of the tubes in the evaporation chamber are 3.839 and 4.041 m^2, respectively.

2. *External energy source*: The external energy inputted into the unit is hot water (about 65–90°C), which is heated by three electric heaters when it operates. The temperature of hot water can be adjusted so as to research the influences of the temperature on the yield of freshwater.
3. *The heat exchangers*: The heat exchanger 2 is made of a copper pipe with outer diameter of 8 mm, thickness of 0.5 mm and length of 4000 mm, which is used to preheat the feedstock. The heat exchanger 9 is made of a copper pipe with outer diameter of 12 mm, thickness of 0.5 mm, and length of 3500 mm.
4. *Drawing vapor tubes and delivering air tube*: The drawing vapor tubes 13 are composed of two tubes, which are connected in parallel, and the outer diameter and length of these tubes are 25 and 550 mm, respectively. There are many holes on the two opposite sides of the tubes so as to draw in the air stream. The delivering air tube 17 is constructed of a copper pipe with outer diameter of 20 mm and length of 540 mm. Also there are many holes on the tube wall to deliver the air stream.
5. *The hot and circulation water pumps as well as air-fan*: The power, lift head, and flow rate of the hot water pump 14 are 40 W, 1.6 m^3/h and 3.6 m, respectively. Those of circulation water pump 21 are 40 W, 12 L/min and 8 m. The power of the air-fan 16 is 180 W. The maxima of pressure difference and air flow rate produced by the air-fan are 15 kPa and 25 m^3/h, respectively. Considering the vapor corrosion problem, the blades of this air-fan are made of plastic.
6. *The stacked tray still*: This is constructed of aluminum sheets with thickness of 0.8 mm and tray area of $0.75 \times 0.40\ m^2$. The average height between the bottom of the second stage and the base surface of the first stage of the tray still is 50 mm.
7. *The gas–liquid separator*: The gas–liquid separator 4 is constructed of aluminum sheets with thickness of 0.8 mm. It is a rectangular box with length of 200 mm, height of 280 mm and width of 80 mm. The distillate yield is measured once every 30 min.
8. *Insulation*: The outer walls of the main body and the still are insulated with thickness of 50 mm polystyrene foam.

5.15.3 The Performance of the Unit

5.15.3.1 The Transient-State Performances of the System

Due to the relative independence of the operation, the main body of the unit and the stacked tray still are taken into account separately.

The main body of the unit began to operate in steady state after approximately half an hour of start-up and its temperatures and productivity remained essentially constant. The variation curves of relevant temperatures with the operation time are shown in Fig. 5.93

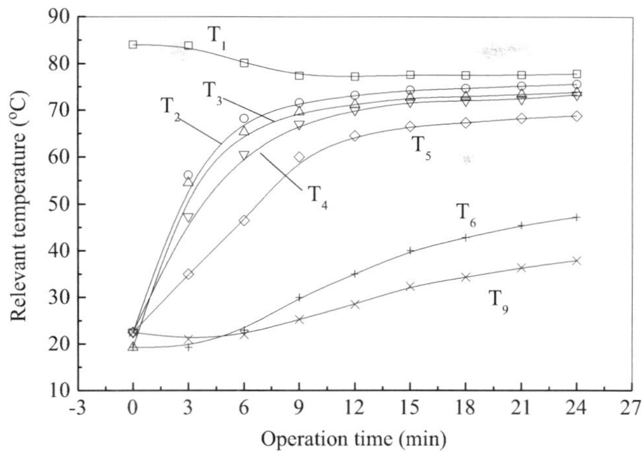

Figure 5.93
Variation curves of the temperatures with operation time under steady-state temperature of 85°C.

(The temperature of hot water input is about 85°C. The flow rates of feedstock, air, and hot water are 8.6, 6.5, and 930 kg/h, respectively.). It can be seen from Fig. 5.93 that quasisteady temperature is achieved after 20 min operation. The remaining operation conditions are unchangeable, the relation curves of distillate yield with operation time under different T_1 are shown in Fig. 5.94. It is shown that the system can provide freshwater and reach the steady state after 5 and 30 min operation, respectively.

The variation curves of the relevant temperatures and the yields in the still with the operation time are shown in Fig. 5.95. It can be seen that the quasisteady temperature and yield in the still are achieved after 90 min operation.

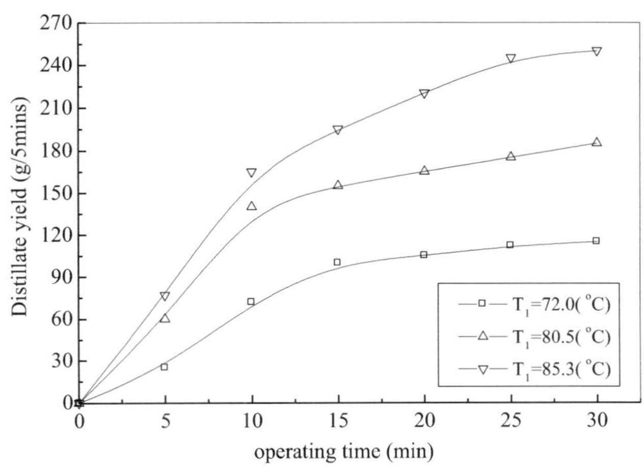

Figure 5.94
Relation curves of distillate yield with operation time under different T_1.

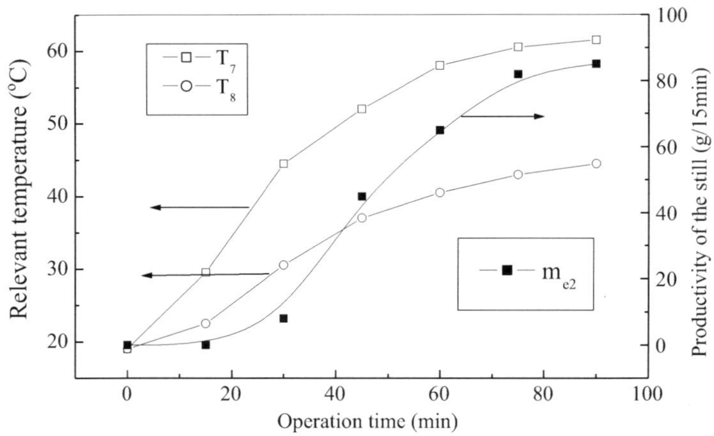

Figure 5.95
Variations of the testing temperatures and the yield with the operation time.

5.15.3.2 Steady Performance of the Unit

Under the steady-state conditions, the variation curves of total productivity with operation temperature are shown in Fig. 5.96. For a definite operation temperature, the system will have a corresponding steady state. The parameters and relevant temperatures under a certain steady state are shown in Table 5.8.

The experimental results illustrate (see Fig. 5.96) that the operation temperature of unit has a fair influence on the yield. The higher the operation temperature, the more the yield of the system. Therefore, the unit has more advantages when it operates under high-temperature conditions.

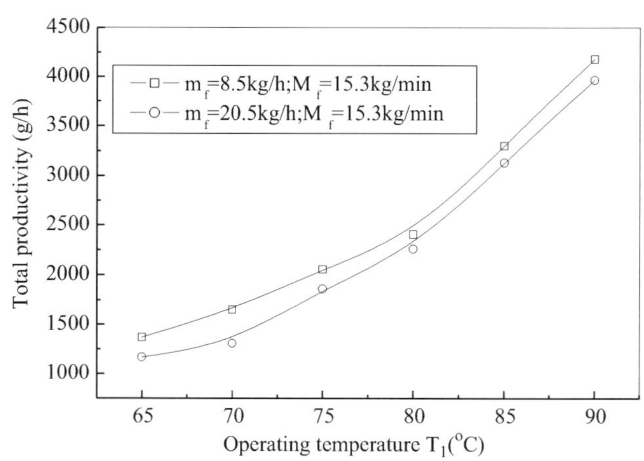

Figure 5.96
Variation curves of total productivity with operation temperature T_1 under different operation conditions.

Table 5.8: Productivity and relevant temperatures under steady state when T_1 is about 85°C.

T_1	T_2	T_3	T_4 (°C)	T_5	T_6	T_9	T_7	m_f (kg/h)	M_f (kg/h)	V_{am} (m^3/h)	m_{e1} (kg/h)	m_{e2} (kg/h)	PR (%)
84.9	83.0	80.7	81.1	75.5	59.7	47.5	66.5	20.0	830	6.50	2.85	0.41	103.9

The performance ratio of this system is given by

$$PR = \frac{m_e h_{fg}}{Q} = \frac{(m_{e1} + m_{e2})h_{fg}}{M_f C_p (T_1 - T_2) + W_p}. \tag{5.53}$$

Experimental results illustrate that the performance ratio of this system is very good and generally reaches 80%. From experiment results it is found that many factors influence PR, especially the hot water flow rate. The PR of the system will increase when the hot water flow rate decreases (but actually the freshwater yield may decrease). For example, under 70°C operation condition, the PR of the system is 87.5% when the hot water flow rate is 760 kg/h and the feedstock flow rate is 14.0 kg/h. But it will get to 101.3% when the hot water flow rate decreases to 300 kg/h and the feedstock flow rate remains unchangeable. When the hot water flow rate continuously decreases to 150 kg/h, the PR will be greatly more than 100%. It may be explained as follows: when the hot water flow rate decreases, the temperature of the hot water output decreases, which makes the temperature of the brine droplet falling to the bottom of the evaporation chamber decrease too. It is more helpful to the condensation of vapor in the tubes. Therefore, more vapor latent heat can be recovered.

Experimental results also illustrate that the contribution of the yield of the still to PR of the system is about 10–15% (when $T_1 \geq 70$°C). The variation of PR with the operation temperature is shown in Fig. 5.97. It is seen that PR increases with the increase in operation temperature under the same conditions. It illustrates that the system has more advantages in higher operation temperature.

5.15.4 The Unit Operating Under Practical Weather Conditions

Zhang [35] carried out the test for the desalination unit under practical weather conditions which was coupled with a solar collector with 2 m^2 collecting area including 18 vacuum tubes. The installation angle of the solar collector is about 30°. The capacity of the water storage tank is 50 kg. Solar radiance is measured with a solar radiation meter laid on a slope-plane with the same slope-angle as the collector. The error of the meter is 5%. The outdoor experiments were carried out continuously for 2 months. The solar irradiation intensity and the ambient temperature were also measured at the same time.

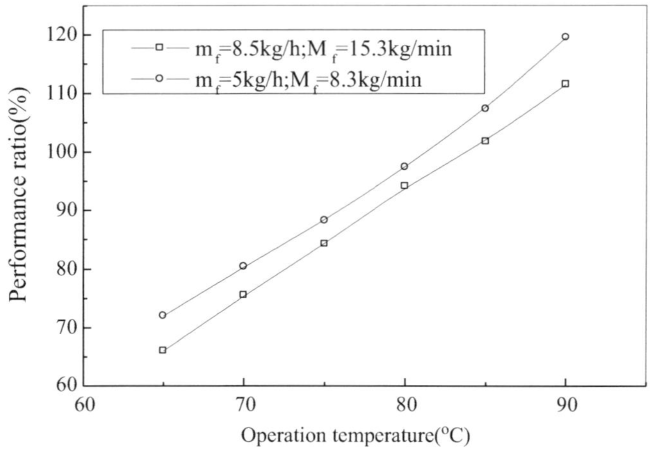

Figure 5.97
Variation curves of PR with the operation temperature.

Table 5.9: The experimental results on several typical sunny days.

Test day	Solar radiation, H (MJ/m^2 day)	Start temperature, T (°C)	Yield, m_e (kg/day)	Feedstock rate, m_f (kg/h)	Daily efficiency, η (%)
3 June	19.86	65	13.570	20.0	69.5
5 June	21.92	70	15.505	15.0	73.5
15 June	22.28	75	17.035	13.6	79.7
17 June	21.52	75	17.410	12.0	83.9
28 June	21.78	78	17.828	10.0	85
4 July	22.30	80	17.905	8.5	83.7
7 July	23.47	85	18.308	8.5	81.8
15 July	22.10	85	17.875	10.5	84.2
16 July	23.50	85	18.010	8.0	80.4
18 July	23.09	84	18.225	6.5	82.7

The experimental results for several typical sunny days are listed in Table 5.9. The variations of the solar radiation intensity throughout the day (Beijing time) on four typical days are shown in Fig. 5.98. The variations of the yield per hour during the same day are shown in Fig. 5.99.

The experimental results illustrate that on sunny days this system has a fairly high efficiency. It can be observed in experiments that the starting temperature exerts an influence on the yield. The higher the operation temperature is, the more the yield of the system is. For example, on 7 July and 16 July, the starting temperature is higher than that on 5 and 15 June, with greater productivity on 7 and 16 July being obtained. At the same time, the feedstock flow rate has some influences on the yield. But, in the 5–12 kg/h range, the influences of the feedstock flow rate to the yield can be omitted. The system

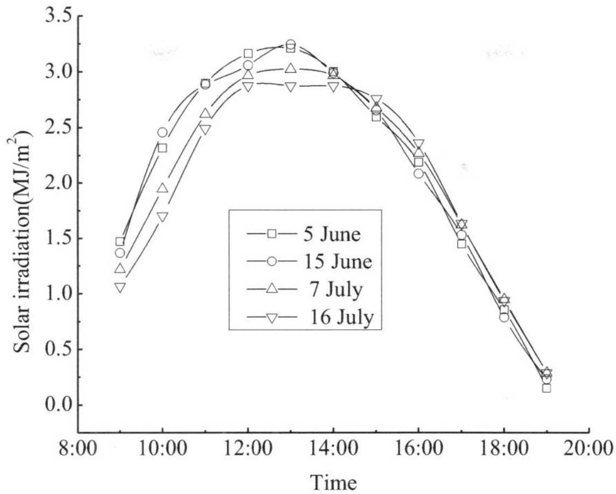

Figure 5.98
Variation of solar radiation with time.

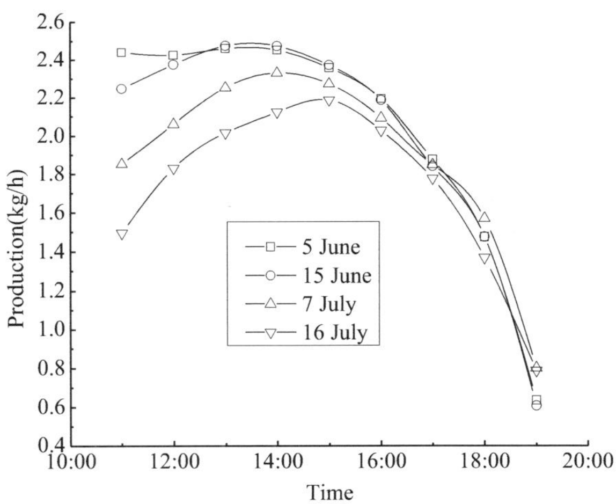

Figure 5.99
Variation of distillate yield per hour with time.

operates best under high-temperature conditions. So, the starting temperature should be above 70°C.

The ultimate purpose for using the falling film evaporation technique is to overcome the shortcomings of the conventional basin-type solar still with an overly large thermal mass so that the temperature of the system increases more rapidly and the distillate water is generated earlier. Fig. 5.100 shows the progress of the most characteristic temperature of the system over time on 16 July. It is seen from Fig. 5.100 that at the start time, the operating temperature (T_1) of the system is almost the highest on the day. The starting

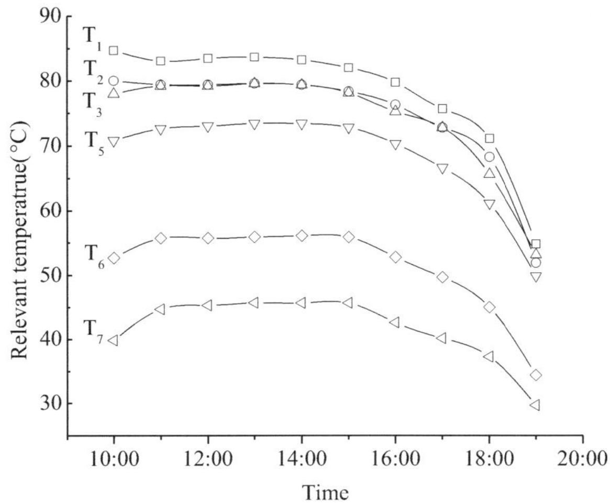

Figure 5.100
Process of the most characteristic temperature with time on 16 July.

time is 10:00 a.m. and the starting temperature is about 85°C. The productivity of the system is relatively higher, which can be seen from Table 5.8.

Considering that some electric power W_p is required, the daily efficiency of the system is given by:

$$\eta = \frac{\sum m_e h_{fg}}{H + W_p \Delta t}$$

where, H is solar radiation during the day, $\sum m_e$ is the daily freshwater production, and Δt is the operation time of the unit. Using the above equation, the daily efficiency of the system on several typical days is listed in Table 5.2. From the experimental results in Table 5.8 it can be found that the unit can produce two or three times more freshwater than traditional solar stills.

It is worth pointing out that in Zhang's system the basin still on the top of Fig. 5.92 was not used. This basin still can produce about 10–15% freshwater.

5.15.5 Conclusions

1. In this system, a considerable fraction of latent and sensible heat is successfully recycled and utilized to preheat the brine and recycle air. The total reused heat is up to 100%.
2. Because of the falling film evaporation and multieffect thermal regenerative technology being used and the flash distillation and vapor compression characteristics, the yield rate of the specific energy consumption of the system is enhanced greatly. Its yield rate is more than twice as much as that of a conventional single basin type of solar still.

3. The system exhibits a wide range of flexibility with regard to the variations of the feedstock flow rate and has more advantages in higher operation temperature range.
4. Due to the stacked tray still being added to the system, the yield of the system is increased by 10−15%.
5. The system can be directly operated with a solar system or supplied with waste heat. For a solar energy system, the starting temperature has an influence on the yield of the system. Generally, the starting temperature should be above 70°C and the feedstock rate should be at the range of 5−12 kg/h.

5.16 A Self-Storing Water Vertical Plate Solar Distiller

A vertical multieffect diffusion solar still (VMEDS) has the advantages of rapid speed of the condensate, leading to the decreased heat resistance of the condensation film over the vertical plate. In Section 5.14, a vertical multieffect diffusion solar still was described, and its excellent performance has been proved. However, the wick material may gradually peel off from the vertical plate after operating for a certain time, probably leading to the blockage of the gap and the contamination of the freshwater. In addition, controlling the evaporation temperature and the feeding rate of the system have to be much more careful to prevent the deposit and bubble formation of the wick material in that design. Thus, the cost of the conventional VMEDS has to be increased for precise control of the system.

In order to overcome the shortfalles of the VMEDS described above, a novel VMEDS has been designed in Ref. [20], and its structure is shown in Fig. 5.101. The proposed VMEDS is actually formed by welding four cylindrical barrels (partitions) concentrically on a top and a bottom seal plate. The four barrels are made of stainless steel, with diameters of 215, 317, 430, and 540 mm, respectively, and with a wall thickness of 1 mm. The outside barrel was rolled by a flat plate, while the other three were rolled by corrugated plates with transverse ridges, forming annular water troughs on the outside of the barrel walls for seawater storage combined with a metal strip welded on each ridge. Except the outside barrel, each of the others has eight ridges, total of 24 troughs are then set inside the VMEDS. These ridges increase the heat transfer area between the vapor and seawater in the troughs. The top and bottom seal plates are covered by a 10-mm-thick rubber and a 5-mm-thick engineering plastic plate, along with four inlet tubes (one for the circulation water and the other three for seawater) and seven outlet tubes (one for circulation water, three for excess seawater, and three for freshwater).

The four barrels and two seal plates form four cavities inside the VMEDS, of which the central one is used as a water tank for the solar collector, and the others as distillation compartments. The other three cavities are referred to as the first, second, and third stage distillation cavity from inside to outside. Two pipelines connected the solar collector and the central water tank to form a loop combined with a pump to circulate water between

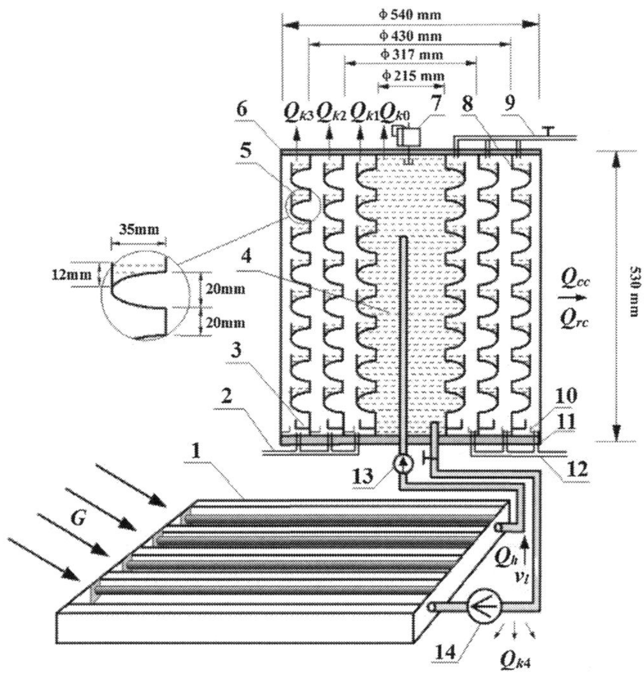

Figure 5.101
Schematic of the solar still with vertical ripple surface: 1, solar collector; 2, excess seawater outlet tube; 3, excess seawater collection trough; 4, central water tank; 5, evaporation trough; 6, top cover plate; 7, overflow valve; 8, vertical ripple condensation surface; 9, seawater inlet tube; 10, freshwater collection trough; 11, bottom cover plate; 12, freshwater outlet tube; 13, check valve; 14, pump.

the two places continuously. Meanwhile the seawater in evaporation troughs in the first-stage distillation cavity are heated by high-temperature water in the water tank to vaporize, after which it is condensed by the lower temperature in the second stage, thus, the heat from the solar collector was transported to the VMEDS continuously.

Fig. 5.101 also shows the flow path of the freshwater and seawater. The seawater is supplied to the evaporation troughs from the top. The freshwater and excess seawater flow down from different sides of the barrels, and are respectively collected in the troughs welded on the lower surface of the barrels. Besides, several valves are set on the top and bottom of the still for controlling the feeding rate of water. The design specification of the still is shown in Table 5.10.

Compared with other VMEDSs, the proposed still owns the following characteristics: (1) the saline-soaked wicks (porous) are replaced by the water troughs on the vertical ripple plates in order to give better reliability and antiscale property, while the gaps among partitions become wider; (2) the compact structure decreases the heat loss and facilitates

Table 5.10: Design specifications of the proposed solar still.

Shape	Cylindrical outside, corrugated barrel inside		
Overall dimensions	D 540 mm × H 530 mm; effective area of the solar collector = 1.48 m^2; length of the connection pipeline = 2.5 m		
No. of multiple effects	Three		
Solar still material	304 stainless steel		
Detail parameters	First-stage distillation cavity (inmost)	Second-stage distillation cavity	Third-stage distillation cavity (outmost)
Diffusion gap	50 mm	50 mm	38 mm
Seawater capacity	3.1 kg	5.0 kg	8.0 kg
Evaporation area	0.27 m^2	0.39 m^2	0.51 m^2
Condensation area	1.15 m^2	1.51 m^2	0.85 m^2

the operation of the solar still, for the cavities surrounding the central water tank layer by layer.

In Ref. [20], the steady heating power experiments were performed with different electrical heating power with each power lasting for 6.5 h during the experiment. When the heating power was 300 and 600 W, the temperature variations of the device are shown in Fig. 5.102. The maximum value of the water temperature in different heating power is also shown in Fig. 5.103.

The humid air inside the distillation cavity also moves along in the vertical direction, due to the temperature difference between the humid air at the bottom and at the top.

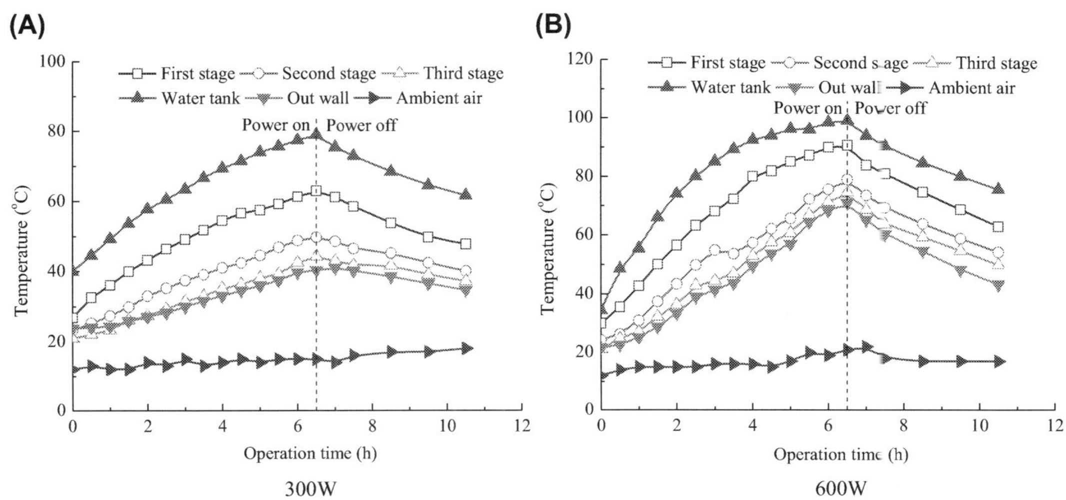

Figure 5.102
Variation in temperature of the solar still at different heating powers: (A) 300 W; (B) 600 W.

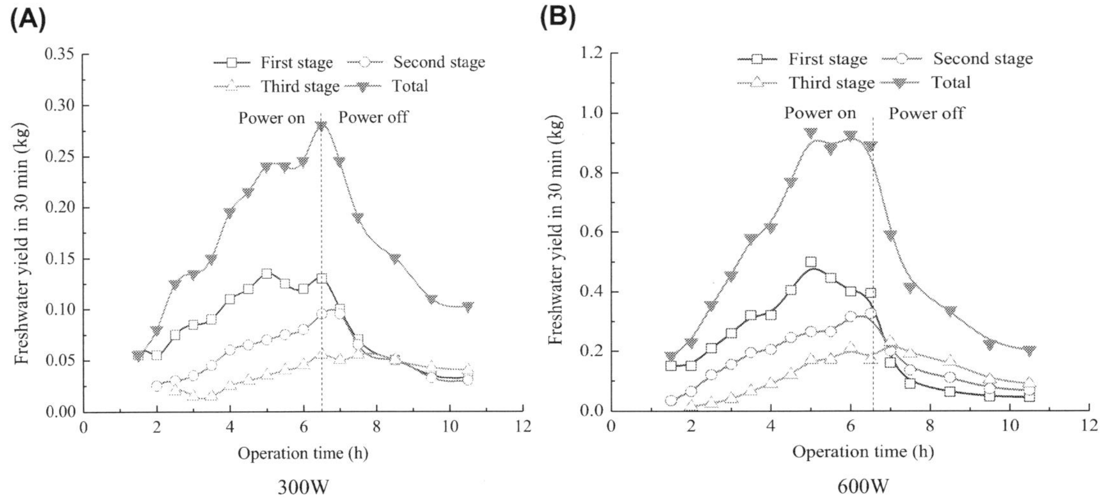

Figure 5.103
Freshwater yield of three-stage evaporation troughs at different heating powers: (A) 300 W; (B) 600 W.

Therefore, the temperature distribution of the vertical plate is also not uniform, leading to the nonuniform evaporation and condensation process on the plate. The temperature gradient exists both in the horizontal and vertical directions in the solar still, resulting in the more complex temperature distribution of the distillation cavity, compared to the basin-type solar stills. The temperature in Fig. 5.102 is the average value of the measuring point on the vertical plate, where the temperature is not uniform, rather than the uniform temperature distribution on the wall of the basin-type solar still. In addition, the degree of temperature distribution on the wall also changes with the heating power.

During the test, it was found that the temperature was different in the upper, intermediate, and lower parts of the barrels, thus for a certain distillation cavity the averaged temperature was taken as the qualitative temperature in the calculations. Fig. 5.102 shows the variations in temperature with time at heating powers of 300 and 600 W. The highest temperature in the water tank reached nearly 100°C when the heating power was 600 W. After power-off, the temperature decreased sharply and then gradually presented a tardy tendency.

Fig. 5.103 shows the change tendency of freshwater yield every 30 min of the three-stage evaporation troughs of the solar still at heating powers of 300 and 600 W. The freshwater yield (total three stages) in 30 min increased as the heating power increased. However, when the heating power was larger than 600 W, the increase of heating power couldn't contribute to the yieldobviously. Also note that the yield started to increase once the system was actuated and after the power was turned off it instead turned to decrease in heating range of 300–500 W. This tendency would be changed when heating power was larger than 600 W, wherein the yield almost kept constant after reaching a maximum till the power was turned off. After power-off, the freshwater was generated by the sensible heat in all stage evaporation troughs and its yield at any heating power decreased sharply in the first 1−2 h and then dropped slowly thereafter.

In more detail, the freshwater yield of the first stage increased at first and then decreased after its temperature reached 100°C (before power-off) at 600 W heating power. The unexpected decrease in the yield before power-off is due to insufficient cooling of the vapor in the first stage. The vapor which could not be condensed overflowed from the bottom tube with the freshwater, leading to the waste of the heat and freshwater. This phenomenon can also be observed at heating power of 700 and 800 W.

Fig. 5.104 shows the accumulated freshwater yield on the whole day of all stage evaporation troughs, with an increasing trend from the first stage to the third stage. The differences among the yields of the three stages gradually became smaller as the heating power increased. For example, the accumulated freshwater yield of the first stage cavity in "on + off" process was 70% higher than the third stage at the heating power of 300 W, while the excess came to 8.4% at the heating power of 800 W.

According to the experiment data, the efficiency of every stage and the performance of the solar still can be calculated as:

$$\eta_i = \frac{m_i h_{fg}}{Q} = \frac{m_i h_{fg}}{P\Delta t} \tag{5.54}$$

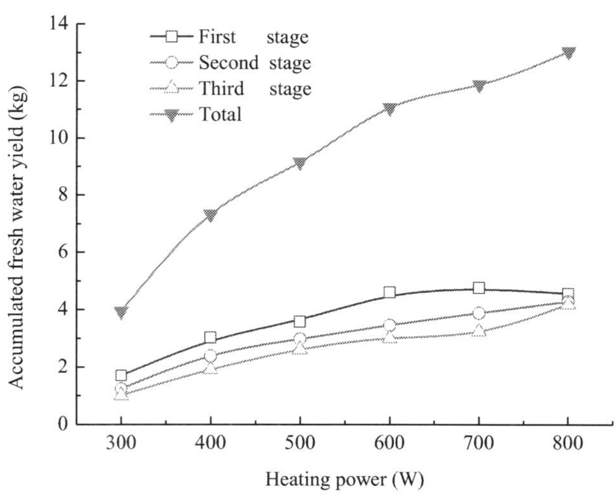

Figure 5.104
Accumulated freshwater yield of all stage evaporation troughs under different heating powers.

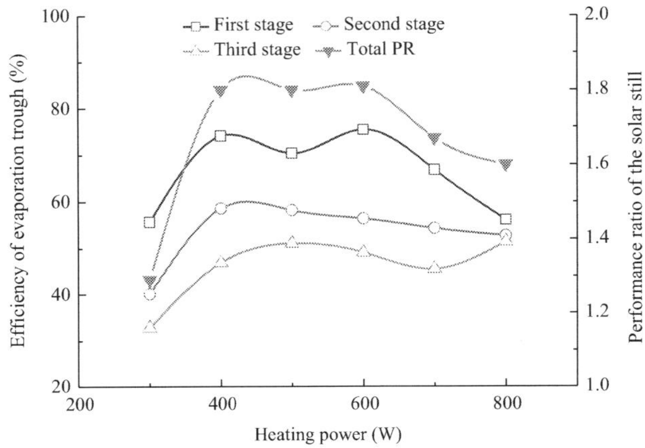

Figure 5.105
Efficiency of all stage evaporation troughs and the performance ratio of the solar still.

where m_i, accumulated freshwater yield of every stage, kg/s, $i = 1,2,3$ indicates the first, second and third distillation cavity; h_{fg}, latent heat of water vaporization, 2.3×10^6 J/kg; Q, heating power of the device, W; P, energy consumption of the system; Δt, duration of the electrical heating.

From Eq. (5.54), η_i and PR can be calculated and the results are shown in Fig. 5.105. The calculated efficiency η_i is highest in the first stage and lowest in the third stage. The total efficiency reaches its maximum of 1.81 at the heating power of 600 W, which is relatively high compared with the other triple-effect solar stills.

Ref. [20] also involved the performance of the proposed solar still under environment conditions. The environment test of the solar still was performed on June 13, 2011. The measured solar irradiance and the accumulated values are shown in Fig. 5.106. The measured maximum solar irradiance and accumulated value on the test day were 908 W/m^2 and 22.77 MJ/m^2, respectively.

Fig. 5.107 shows the temperature variation of seawater in all stage evaporation troughs and outer stainless steel wall under environment conditions. The tendencies of the variation of the temperature at different positions in the solar still are similar. The highest temperature in the water tank was near to 90°C at about 14:00 p.m. After 17:00 p.m., the temperature decreased sharply and then gradually presented a tardy tendency.

The simulated freshwater yield in 1 h in the experiment is shown in Fig. 5.108. The highest freshwater yield is 1.26 kg/h at 14:00 p.m., 2 h later than the maximum of the solar irradiance. Comparing that with a general diffusion solar still, increasing the gap size and water storage inside the proposed still leads to a more obvious time lag of water production and vapor generation, which is influenced significantly by the diffusion process inside the cavities.

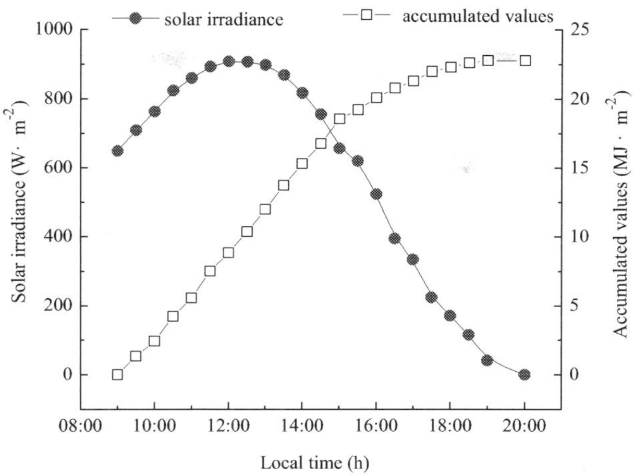

Figure 5.106

The measured solar irradiance and the accumulated value on June 13, 2011.

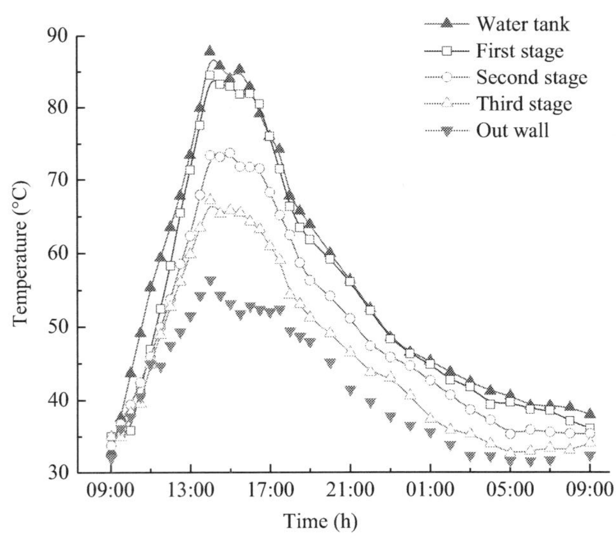

Figure 5.107

The temperature variation of seawater in all stage evaporation troughs and outer stainless steel wall under environment conditions.

By further observation, there is an intersection of the three freshwater curves. Before 18:30, the first-stage cavity generated the most freshwater, while the third generated the least. After this time, the tendency was reversed as the third-stage cavity generated the most. The explanation is that the sun set after this time (no solar energy input), and the freshwater yield is just arisen from the sensible heat of the water in the troughs. The third stage stores the most seawater, resulting in the most sensible heat storage and

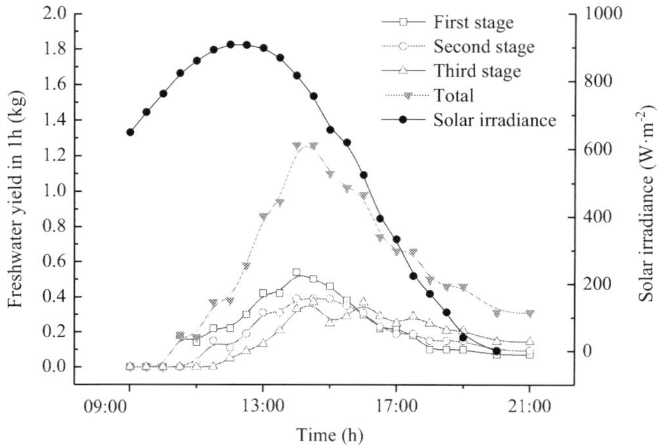

Figure 5.108
Simulated results of freshwater yield and comparison with experimental data.

freshwater yield after sunset. This is also consistent with the results from steady heating power tests in which the third stage prevails during the power-off period.

Detailed calculation indicates that the accumulated freshwater yield per effective collecting area of solar collector for the first, second, and third stage was 2.09, 1.76, and 1.89 kg/m^2. The average performance ratio of the proposed solar still in the experiment under environment condition was 1.35.

The experimental results show that the proposed solar still has a relatively high performance ratio of 1.35 under actual weather conditions, which is about 400% of the current basin-type solar stills. Moreover, the proposed solar still is highlighted by its better reliability and operability in long-term operation, and its characteristic still needs deeper investigation.

5.17 Solar Water Desalination Using an Air Bubble Column Humidifier

5.17.1 Introduction

Air bubble columns are suitable for humidification and dehumidification. Its great interface area allows a great value of heat and mass transfer rate consequently improves productivity and efficiency.

Vlachogiannis et al. [37] studied experimentally and theoretically a novel desalination concept, combining between the principles of air humidification–dehumidification and mechanical vapor compression. Air was injected into the evaporation chamber through a porous bottom wall and dispersed as small-diameter bubbles.

Inaba et al. [38] studied experimentally heat and mass transfer of air bubbles in a hot water layer. The results showed that the mean diameter of generating air bubbles increased with an increase in the superficial air flow velocity.

Zhang et al. [39] studied experimentally operating factors that affect bubbling humidification by using a single sieve plate. The result showed that air relative humidity reached 100%. Moreover, humidification capacity was increased by about 80% when water temperature was increased by 10°C. Zhang et al. [40] studied experimentally the influence of working design parameters of a solar bubbling humidification desalination unit on its gained output ratio, electric power consumption, and freshwater production cost. They found that gained output ratio was increased as the humidification temperature was increased, while electrical power consumption decreased.

Emily and John [41] experimentally studied the heat flux and effectiveness of a bubble column dehumidifier. It was found that the effectiveness was decreased while the heat flux was increased with decreasing coil area and increasing the temperature and air flow rate. Ghazal et al. [42] examined air bubble regeneration on the performance of a solar still using a humidification–dehumidification desalination (HDD) process. The results showed that the exit air from the humidification process was fully saturated. The effectiveness of the solar humidifier and the mass and heat transfer were upgraded by air bubble regeneration.

5.17.2 Experimental Setup and Efficiency Calculation

5.17.2.1 Experimental Setup

Fig. 5.109 presents a schematic diagram of the experiment setup which was proposed by A. Khalil et al. [43]. The saline water, which is working in a closed loop, is heated by an evacuated tube solar collector as an energy source in order to save electrical energy. Air working in an open loop is injected into the evaporation chamber from the bottom through sieve plates with 1200 holes of 1, 3, and 5 mm diameter. The characteristics of generating bubbles are changed by using different hole diameters of sieve plates. The study includes the effect of water temperature and level, air flow rate, and orifice diameter of the sieve plate on the unit performance.

The unit main components are HUM (humidifier), DEH (dehumidifier), and SWS (solar water collector). The system consists of two loops, a closed water loop and open-air loop. In the closed water loop, the hot water at (1) is pumped to the HUM inlet (2) through CV (control value) and FM (flow meter) that heats in the evacuated tube solar collector. At HUM exit (3) the hot water is pumped through CV and FM to TST (thermal storage tank) at (5). The hot salt water is circulated in a closed loop between HUM and TST. High salt concentration saline water is rejected from CV at (6). HUM is made of an acrylic plastic sheet of 10 mm thickness with 580 mm × 580 mm cross-section, and 900 mm height.

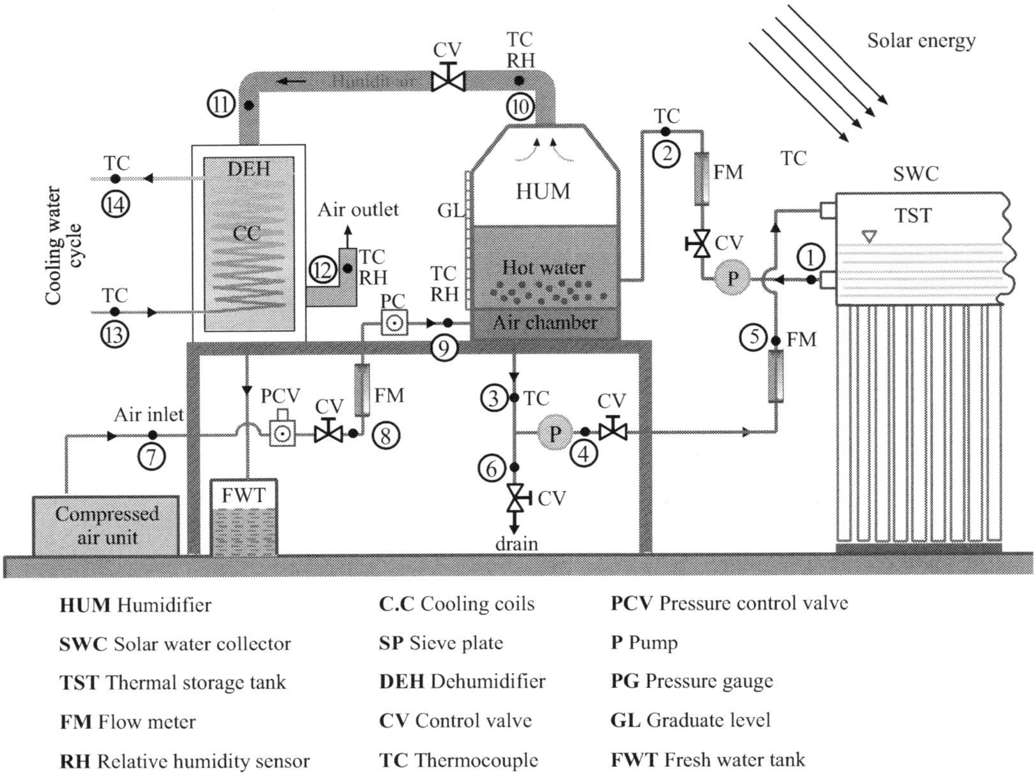

Figure 5.109

Experimental setup schematic diagram [43]. *C.C*, cooling coils; *CV*, control valve; *DEH*, dehumidifier; *FM*, flow meter; *FWT*, freshwater tank.; *GL*, graduate level; *HUM*, humidifier; *P*, pump; *PCV*, pressure control valve; *PG*, pressure gauge; *RH*, relative humidity sensor; *SP*, sieve plate; *SWC*, solar water collector; *TC*, thermocouple; *TST*, thermal storage tank.

In the open-air loop, air at (7) is supplied from the reservoir of a compressor containing 220 L at a pressure of 10 bar. The desired pressure and flow rate of air at (8) could be adjusted by a PCV (pressure control valve) and CV (control valve) then it is passed through the FM and PG (pressure gauge) that indicate the pressure before the HUM. The air at (9) enters the HUM through the sieve plate which is located at the bottom of the HUM. Three different sieves are used, with 1200 holes each, with 1, 3, and 5 mm hole diameter. The air flow is humidified by passing through the water level in HUM and carries vapor water to DEH. Then, the produce water through a tube is desalinated to the FWT (freshwater tank). DEH is a shell and tube heat exchanger. The shell is made of 0.7-mm-thick steel with dimensions of 400 mm × 400 mm, and 900 mm height. Inside the shell, the cooling water tube is fixed.

The type of SWC is an evacuated tube solar collector. It consists of 25 evacuated glass tubes with 58 mm diameter and 1.8 m length. The capacity of the TST is 250 L. The tilt angle of the solar collector is 30° all day.

5.17.2.2 Efficiency and Gained Output Ratio of the System

It is important to calculate the hourly and daily efficiency of the desalination unit to know the performance achieved in real systems, which is calculated as:

$$\eta_h = 100 \times \frac{(M_w \times LH_{w\cdot av})/3600}{A_{SWS}\cdot I_S + P_p + W_c} \tag{5.55}$$

$$\eta_d = \frac{1}{n}\sum_{1}^{n} \eta_h \tag{5.56}$$

$$LH_{w\cdot av} = 10^3 \times \left(2501.9 - 2.40706\cdot T_w + 1.192217 \times 10^{-3}\cdot T_w^2 - 1.5863 \times 10^{-5}\cdot T_w^3\right) \tag{5.57}$$

where η_h and η_d are the hourly and daily efficiency, M_w is the hourly productivity, $LH_{w\cdot av}$ is the average of the latent heat of vaporization of water [44], A_{SWS} is the solar water collector area, I_S is the incident solar radiation on the collector surface, $P_p = I \times V$ is the power of pumps that are measured with different flow rates of cold and hot water, W_c is a power of compressor, and n is the duration time of solar radiation.

The GOR is calculated as [45]:

$$\text{GOR} = \frac{M_w \times LH_w}{\left[(\text{m} \times h)_{w,i} - (\text{m} \times h)_{w,o}\right]} \tag{5.58}$$

where LH_w is the latent heat of vaporization at the ambient condition, and $[(m \times h)_{w,i} - (m \times h)_{w,o}]$ is the total enthalpy difference between entering and leaving the humidifier for hot water. The hot water flow rate, leaving the system is calculated as:

$$m_{w,o} = m_{w,i} - M_w \tag{5.59}$$

5.17.3 Results and Discussion

Test rig external ambient conditions are very important. Average relative humidity of the ambient air is 15–40%. Wind speed during the experiment is 1.5–3 m/s in that period of time. The inlet water flow rate to HUM is 10 L/min, while the exit water flow rate is measured by FM and checked by using Eq. (5.59). Four different values of air flow rates are used, 11, 13.2, 15.4, and 17.7 kg/h, each value has been used during a whole working day. In addition, four different values of water height in humidifier are used, 5, 10, 15, and 20 cm.

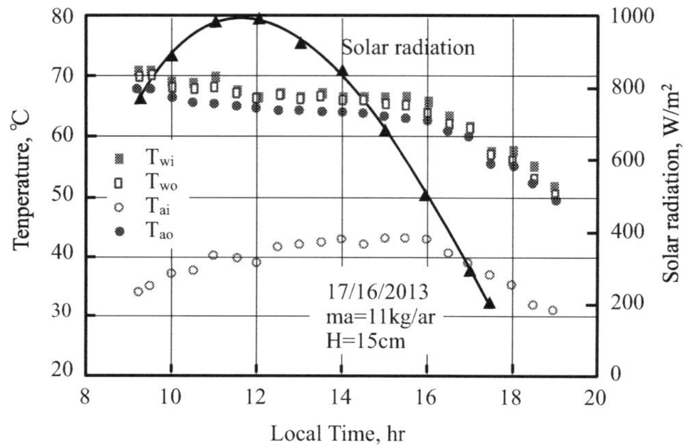

Figure 5.110
Solar radiation, inlet, outlet water, and air temperatures at the humidifier at an air flow rate of 11 kg/h [43].

5.17.3.1 Effect of 1-mm Hole Diameter

Depending upon weather conditions, which are different during the daytime, including the wind speed, atmospheric temperature, and solar intensity, Fig. 5.110 shows the general trend of solar radiation, inlet and outlet water temperature, and air temperature at the humidifier inlet and outer for 11 kg/h air mass flow rate. It can be seen that the solar radiation and air temperature at the humidifier inlet increases in the morning hours, reaching its maximum values around midday, and then decreases in the afternoon. In the case of water, water temperature started from high temperature and then decreases in the morning hours, almost remains constant around midday, and then decreases in the afternoon. The high temperature of the water at the beginning is because the water in the solar collector tank heating from sunrise to the desalination unit starts at 9.50 a.m. In the morning hours, the water temperature decreases because the energy lost from water in the humidifier is larger than the energy gain by solar radiation to the solar collector. In the midday, the water temperature is nearly constant. This is because the energy lost from water in the humidifier is nearly equal to the energy gain by solar radiation to the solar collector in this period.

Fig. 5.111 shows the effect of average hot water temperature on daily productivity, daily efficiency, and GOR at different inlet hot water flow rates. It indicates that both the average daily water temperature and water flow rate promote the daily productivity, daily efficiency, and GOR increases. The results indicate that the water flow rate and temperature play an important role in the system performance.

5.17.4 Conclusion

Experimentally a small-scale solar air bubbling humidification–dehumidification unit was studied. The effects of the water temperature, water height, air flow rate, and hole

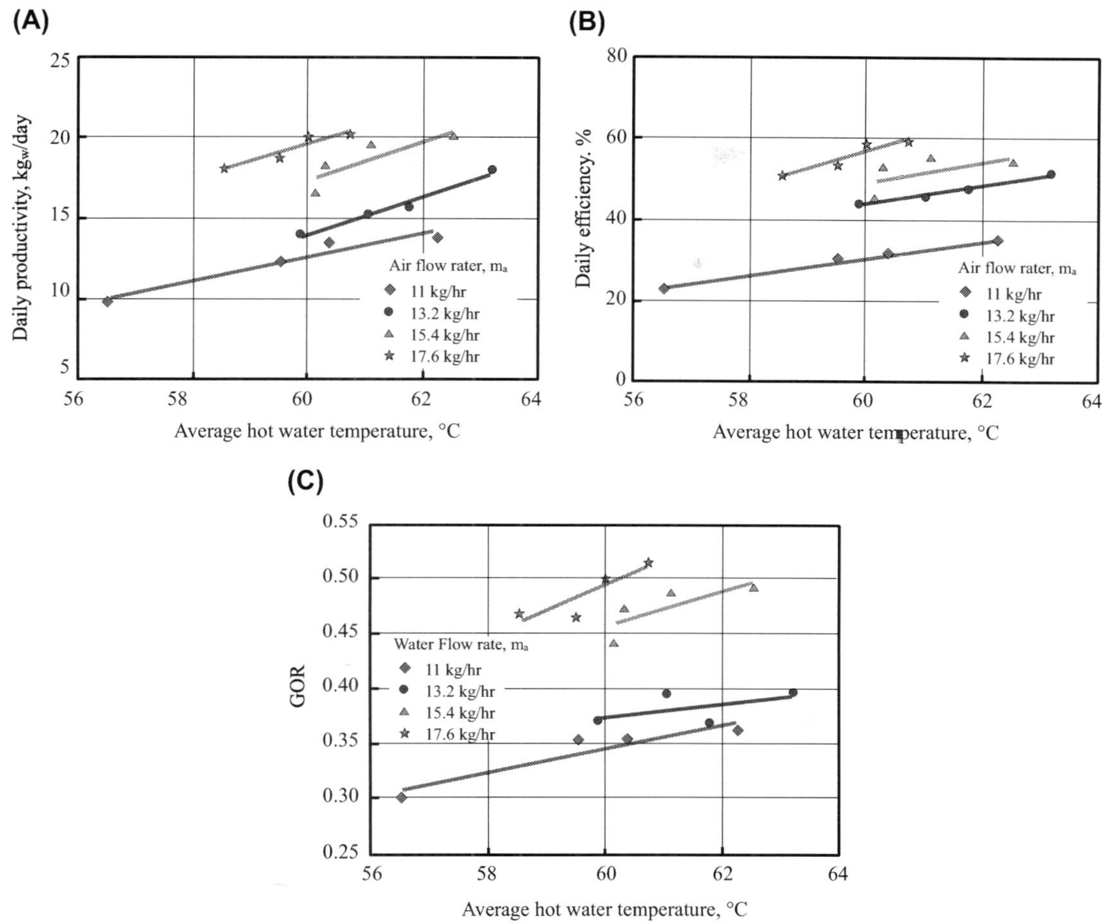

Figure 5.111
Effect of average hot water temperature on [43]: (A) daily productivity; (B) daily efficiency; (C) GOR.

diameters on the performances of the system were observed during the present experiments and the following points are concluded.

1. Air bubbling humidification is an efficient technique which can be used in HDH systems and the outlet air from the bubble columns is always saturated.
2. The productivity, efficiency, and GOR are affected strongly by inlet water temperature and air flow rate to the humidifier.
3. The value of the amount of extracting water is increased slightly with the increasing water height.
4. The temperature difference of water is constant for all the water height and air flow rate conditions. The change in the temperature difference along the column is less than 2.5°C for all measurements.

5. The present system achieved daily productivity, daily efficiency, and GOR of 21 kg, 63%, and 0.53, respectively, at inlet water temperature of 62°C.
6. The best performance is observed from a sieve with 1-mm hole diameter.
7. The air bubble column achieves higher performance than that for the conventional humidifier.

5.18 Theoretical Analysis of a Vertical Multiple-Effect Diffusion Solar Still Coupled With a Tilted Wick Still

5.18.1 Vertical Multiple-Effect Diffusion Solar Still Coupled With a Tilted Wick Still

A schematic diagram of a proposed still is shown in Fig. 5.112 [46]. The proposed still consists of a vertical multiple-effect still (called the multiple-effect section) and a tilted wick still (called the tilted wick section). The tilted wick section consists of a double glass cover and wick and the bottom of the wick is insulated. The multiple-effect section consists of a double glass cover and a number of vertical and parallel partitions. The wicks are attached to the rear surfaces of all of the partitions except for the outermost one. Saline water is constantly fed to the wicks. The wick of the first partition and the wick of the tilted wick section are connected, and the concentrated saline water flowing through

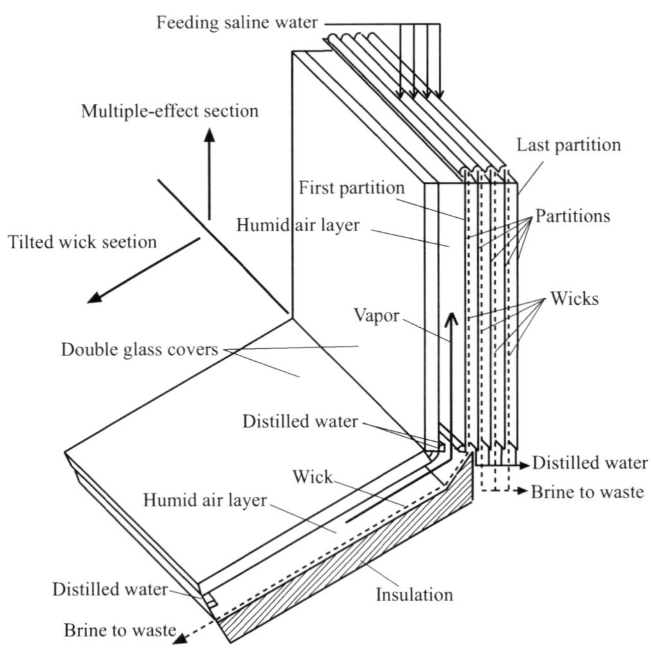

Figure 5.112
Schematic diagram of the proposed still [46].

the wick of the first partition is fed to the wick of the tilted wick section. Concentrated saline water from the wicks of the partitions except for the first partition is discharged from the bottom of the wicks. The humid air layers of both the tilted wick section and multiple-effect section are connected and sealed so that the vapor evaporated from the wick of the tilted wick section can be transported to the humid air layer of the multiple-effect section by natural convection.

Solar radiation transmits through the double glass covers and is absorbed on the wick of the tilted wick section and the front surface of the first partition of the multiple-effect section. Water vapor from the wick of the tilted wick section is transferred by natural convection through the humid air layers and condenses on the surfaces of the inner glass cover of both the tilted wick section and the multiple-effect section. The remainder of the vapor condenses on the front surface of the first partition, and the latent heat of condensation and solar radiation directly absorbed on the front surface of the first partition are utilized as heat sources for the multiple-effect section. The vapor evaporated from the wick of the first partition diffuses through the humid air layer between partitions and condenses on the front surface of the second partition. The latent heat of condensation causes further evaporation from the wick of the second partition. In this manner, the energy absorbed on the first partition (both the latent heat of condensation of vapor evaporated from the wick of the tilted wick section and solar radiation directly absorbed on the first partition) is recovered and recycled in the multiple-effect section to increase distillate productivity. Condensates on each glass cover and each partition are gathered by channels attached to the bottom of each condensing surface. Here, in order to increase the amount of condensation on the first partition (or decrease the amount of condensation on both the inner surfaces of the glass covers) of vapor evaporated from the wick of the tilted wick section, double glass covers are used for both sections to insulate the glass covers. The inclination angle of the tilted wick section is assumed to be adjustable to increase solar radiation incident on the tilted wick section.

5.18.2 Results and Discussion

The weather, design, and operating conditions and physical properties employed in this calculation are listed in Table 5.11.

Theoretical predictions of hourly variations of global solar radiation on a horizontal surface (Global) and the distillate production rates on the three enclosed inner condensing surfaces of the humid air layer, e.g., the inner glass covers of the tilted wick section (gti) and the multiple-effect section (gmi) and the first partition of the multiple-effect section (p1), on a spring equinox day are shown in Fig. 5.113. The overall production rate (Total), which is the sum of all of the distillate production rates in the still, is also shown in 1/10 scale. Here, the distillate production rate as well as the daily

Table 5.11: Weather, design, and operation conditions and physical properties.

Weather conditions	Latitude and longitude: 33° and 130° Transmittance of atmosphere: 0.7 $T_a = 25°C$ on the spring equinox and 33°C and 20°C on the summer and winter solstices $P_{total} = 101.3$ kPa
Design conditions	$u_a = 1$ m/s $w_s = l_t = l_m = 1$ m $\theta_t = 30°$ $\delta_{gp} = 5$ mm Thickness of each glass plate, $\delta_g = 4$ mm Thickness of humid air layer: 60 mm Distance between glass plates of double glass covers: 6 mm Thickness of the first partition: 1 mm Thickness of the partitions from second to the last partitions: 0.5 mm
Operation condition	$m_{f,pi}$: Twice as large as sum of the steady-state evaporation rates from the wick of the first partition and the wick of the tilted wick section for $m_{f,pi}$, and twice as large as the steady-state evaporation rate from the wick of *i*th partition for $m_{f,pi}$ ($i = 2$ to n), which have been calculated on the assumption that the global solar radiation on a horizontal surface will be constant and equal to the maximum of the day
Physical properties	$\alpha_w = \alpha_{p1} = 0.9$ $\alpha_g = 0.085$ $\varepsilon_w = \varepsilon_g = \varepsilon_{pi} = 0.9$ $\tau_g(\beta) = 0.697 \times \cos^4\beta - 0.310 \times \cos^3\beta - 2.096 \times \cos^2\beta - 2.560 \times \cos\beta$ [44] $\lambda_g = 1.03$ W/mK Thickness and thermal conductivity of insulation: 50 mm and 0.04 W/mK Density and specific heat of glass: 2520 kg/m^3 and 0.80 kJ/kg K Density and specific heat of partition plate (stainless steel): 7900 kg/m^3 and 0.499 kJ/kg K

amount of distillate is shown as those per effective glass cover area of the tilted wick section in this chapter. The distillate production rate on each inner surface of humid air layer as well as the overall production rate is roughly parallel to the global solar radiation. The distillate production rates on gmi and p1 are almost equivalent, and production on gti is about a half of that on gmi and p1. The peak value of the overall production rate is about 1.0 g/m^{2}s.

Temperature distribution in the still at peak value of the overall production rate is shown in Fig. 5.114, and the distribution of the daily amount of distillates on the same day is shown in Fig. 5.115. Here, the temperature drop within each glass plate was taken into account in the calculation as mentioned above, and the temperature drop of about 1K

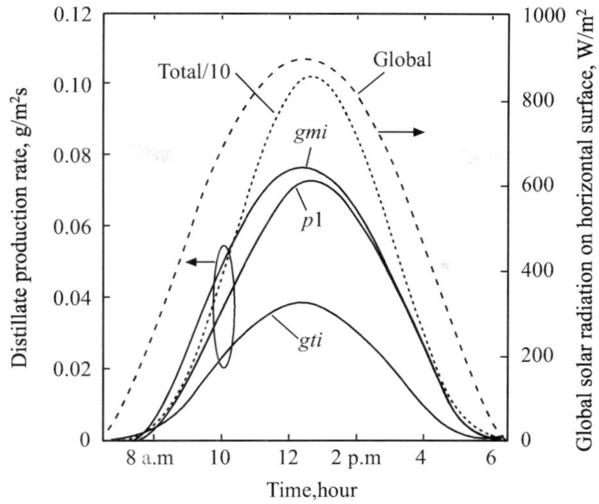

Figure 5.113
Hourly variations of global solar radiation on a horizontal surface (Global) and distillate production rates of three inner condensing surfaces (gti, gmi and p1) of the humid air layer as well as overall production rate (Total) on the spring equinox

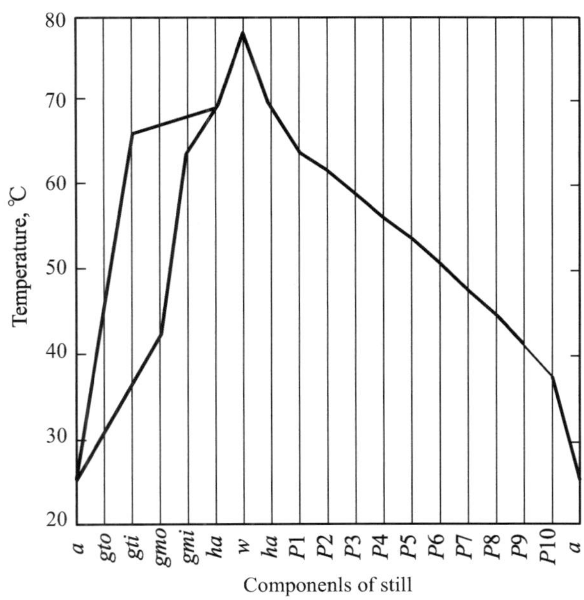

Figure 5.114
Temperature distribution in the still at the peak value of the overall production rate on the spring equinox.

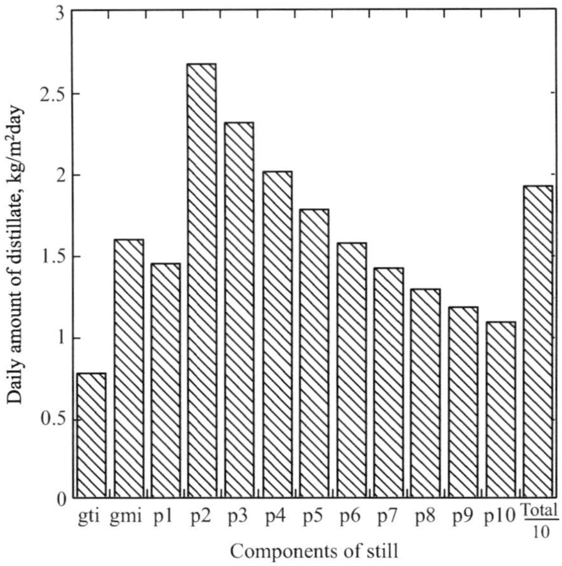

Figure 5.115
Daily amount of distillates in the still on the spring equinox.

was observed within each glass plate. Overall daily amount of distillate, i.e., the sum of all of the daily amounts of distillate in the still is shown as Total in the 1/10 scale. Temperature is highest at the wick of the tilted wick section (w) and decreases on a trajectory going through the double glass covers and the multiple-effect section. There is a considerable temperature drop between each double glass cover, indicating good thermal insulation.

Three enclosed inner condensing surfaces of the humid air layer (gti, gmi, and p1) share the vapor in the humid air layer and the ratio of condensation on each inner surface would be basically determined by a temperature difference between the humid air layer and each inner surface. The temperature differences between the humid air layer and the inner surfaces of the multiple-effect still (gmi and p1), $T_{ha} - T_{gmi}$ and $T_{ha} - T_{p1}$, are almost the same (about 6.3 and 5.4K), and the difference between the humid air layer and the inner glass cover of the tilted wick section (gti), $T_{ha} - T_{gmi}$, is about 3.3K, since the inner glass cover of the tilted wick section absorbs more solar radiation than the multiple-effect section for the day. This causes the daily amount of distillate for gti to be the smallest of the three inner surfaces of the humid air layer.

The second partition p2 produces a greater daily amount of distillate than the first partition p1. This is because the first partition p1 can absorb solar radiation directly, and latent heat of condensation as well as solar radiation directly absorbed on the first partition p1 is utilized to evaporate the saline water in the wick of the first partition p1.

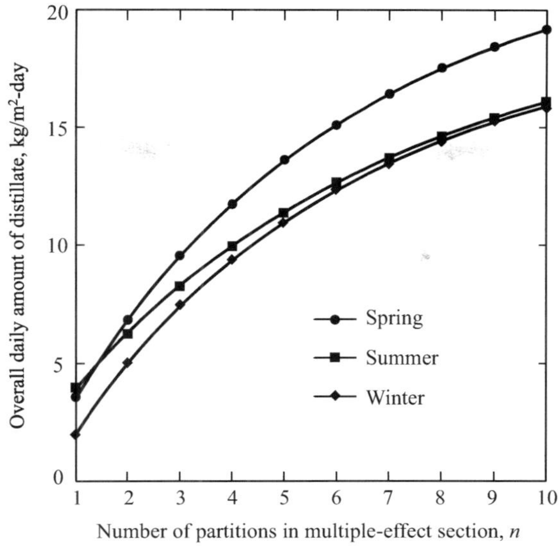

Figure 5.116

Variation of overall daily amount of distillate with a number of partitions n on three days (spring equinox and summer and winter solstices).

The daily amount of distillate decreases progressively from the second partition p2 to the last partition p10. The reason is that the mean temperature of the evaporating and condensing surfaces in each distilling cell decreases and the thermal energy which goes through each distilling cell decreases due to energy consumption used to increase enthalpy of the saline water fed to the wicks.

5.18.3 Conclusions

A vertical multiple-effect diffusion solar still coupled with a tilted wick still is proposed and numerically analyzed on three typical days (the spring equinox and summer and winter solstices) at 33°N latitude. The results of this work are summarized as follows:

1. A sufficient amount of solar energy can be transferred to the multiple-effect section on three typical days.
2. The major energy source for the multiple-effect section is solar radiation directly absorbed on the first partition on the winter solstice and the latent heat from condensation of vapor evaporated from the wick of the tilted wick section on the summer solstice.
3. Both solar radiation directly absorbed on the first partition and the latent heat of condensation can be utilized for the multiple-effect section on the spring equinox.
4. The ratio of condensation on the enclosing three inner condensing surfaces of the humid air layer is basically determined by the temperature difference between the humid air layer and each condensing surface.

5. The overall daily amount of distillate produced by the still is predicted to be about 19.2, 16.0, and 15.9 $kg/m^2/day$ on the spring equinox and summer and winter solstices, respectively, when the diffusion gaps between partitions are 5 mm and the number of partitions is 10. The overall daily amount of distillate varied with the number of partitions n in the multiple-effect section on the spring equinox and summer and winter solstices are shown in Fig. 5.116.

References

[1] Soliman HS. Solar still coupled with a solar water heater. Mosul (Iraq): Mosul University; 1976. p. 34.
[2] Prasad B, Tiwari GN. Effect of glass cover inclination and parametric studies of concentrator-assisted solar distillation system. Inter J Energy Res 1996;20:495–505.
[3] Rai SN, Tiwari GN. Single basin solar still coupled with flat plate collector. Energy Convers Manage 1983;23(3):145–9.
[4] Malik MAS, Tiwari GN, Kumar A, Sodha MS. Solar distillation. UK: Pergamon Press Oxford; 1982.
[5] Dunkle RV. Solar water distillation: the roof type still and a multiple effect diffusion still, International developments in heat transfer. University of Colorado; 1961. p. 895. Part V.
[6] Duffle JA, Beckman WA. Solar engineering of thermal processes. New York (USA): Wiley; 1978.
[7] Tiwari GN, Dhiman NK. Performance study of a high temperature distillation system. Energy Convers Manage 1990;32(3):283–91.
[8] Zaki GM, Radhwan AM, Balbeid AO. Analysis of assisted coupled solar stills. Sol Energy 1993;51(4):277–88.
[9] Singh AK, Tiwari GN. Thermal evaluation of regenerative active solar distillation under thermosyphon mode. Energy Convers Manage 1993;34(8):697–706.
[10] Varol HS, Yazar A. A hybrid high efficiency single-basin solar still. Int J Energy Res 1996;20:541–6.
[11] Mohamad AQ, Bassam AK, Olhman ON. Experimental Study and numerical simulation of a solar still using an external condenser. Energy 1996;21(10):851–5.
[12] Sinnokrot RO, AbuArabi M, Tamimi A. Modeling of a conventional solar still with draft and humidification-dehumidification process. In: Proceedings of the 1st International Nuclear & renewable energy Conference (INREC10). Amman, Jordan; 2010.
[13] Aboabboud MM, Horvath L, Szepvolgyi J, Mink G, Radhika E, Kudish AI. The use of a thermal energy recycle unit in conjunction with a basin-type solar still for enhanced productivity. Energy 1997;22(1):83–91.
[14] Al Bahana NS, Al Mahdi, Zaky F. Performance analysis of a triple-basin still integrated with a natural circulation solar heater. Energy Convers Manage 1993;34(7):545–56.
[15] Adhikari RS, Kumar A, Sootha GD. Simulation studies on a multi-stage stacked tray solar still. Sol Energy 1995;54(3):317–25.
[16] Hongfei Z, Xinshi G. A group of improved heat and mass relationship in solar still. Acta Energae Solaris Sin Chin 2000;20(1):72–5.
[17] Zhang X, Zheng H, Zhang lianying. Experimental research of multiple stage tray solar still. Water Treat Technol Chin 2003;29(4):233–5. 8.
[18] Zheng H, Yang Y. Simulation experimental research of multiple stage tray solar still with wrinkling bottom surface. Acta Energae Solaris Sin Chin 2004;25(2):200–3.
[19] Cheng J, Zheng H, Xie G, He K. Steady state research of passive multiple stage tray solar still with enhanced condensation surface. Acta Energae Solaris Sin Chin 2012;02:308–12.
[20] Xie G, Xiong J, Liu H, Xu B, Zheng H, Yang Y. Experimental and numerical investigation on a novel solar still with vertical ripple surface. Energy Convers Manage 2015;98:151–60.

[21] Tiwari GN, Kumar A. Nocturnal water production by tubular solar stills using waste heat to preheat brine. Desalination 1988;69:309−12.

[22] Murase K, Yamagishi Y, Iwashita Y, Sugino K. Development of a tube-type solar still equipped with heat accumulation for irrigation. Energy 2008;33:1711−8.

[23] Ahsan A, Fukuhara T. Mass and heat transfer model of tubular solar still. Sol Energy 2010;84:1147−56.

[24] Islam Kh Md S, Fukuhara T. Production analysis of a tubular solar still. Dob Gakkai Ronbunshuu B 2007;632.

[25] Wang H, Zheng H. The performance research of multiple effect tube solar distiller. J Beijing Inst Technol 2012;8:147−51.

[26] Chang Z, Zheng Y, Chen Z, Zheng H, Zhao M, Su Y, et al. Performance analysis and experimental comparison of three operational modes of a triple-effect vertical concentric tubular solar desalination device. Desalination 2015;375:10−20.

[27] Kudish AI, Evseev EG, Walter G, Priebe T. Simulation Study on a Solar desalination system utilizing an evaporator/condenser chamber. Energy Convers Manage 2003;44:1653−70.

[28] Zhao Jun, Chengchao Fang, Qi Xiling. A recoverable latent heat solar distiller. Acta Energae Solaris Sin Chin 1998;19(2):195−7.

[29] Xinshi G, et al. Solar energy engineering−application and principle (in Chinese). Beijing: Academic Journal Publishing House; 1988.

[30] Le Goff P, Le Goff J, Jeday MR. Development of a rugged design of high efficiency multi-stage solar still. Desalination 1991;82:153−63.

[31] Zheng H, Xinshi G. Experimental research of multi-stage falling film solar still. Acta Energae Solaris Sin Chin 1999;4:45−50.

[32] Mink G, et al. Air-blown solar still with heat recycling. Sol Energy 1998;62(4):309−17.

[33] Kalogirou S. Survey of solar desalination systems and system selection. Energy 1997;22(1):69−81.

[34] Ziqian C, Hongfei Z, Kaiyan H, Chaochen M. Steady-state experimental studies on a multi-effect thermal regeneration solar desalination unit with horizontal tube falling film evaporation. Desalination 2007;207:59−70.

[35] Zhang L, Zheng H, Wu Y. Experimental study of a horizontal tube falling film evaporation and closed circulation solar desalination system. Renew Energy 2003;28(8):1187−99.

[36] Zheng H, Wu Y. Experimental research on the horizontal tube falling film evaporation and inner regeneration solar desalination system. Acta Energae Solaris Sin Chin 2002;23(3):265−9.

[37] Vlachogiannis M, Bontozoglou V, Georgalas C, Litinas G. Desalination by mechanical compression of humid air. Desalination 1999;122:35−42.

[38] Inaba H, Aoyama S, Haruki N, Horibe A, Nagayoshi K. Heat and mass transfer characteristics of air bubbles and hot water by direct contact. Heat Mass Transf 2002;38:449−57.

[39] Zhang L, Gao S, Zhang H. Experimental researches of factors affecting bubbling humidification. In: Jin D, Lin S, editors. CSISE, vol. 106. AISC; 2011. p. 359−64.

[40] Zhang L, Gao S, Chen W. Performance comparison of different solar desalination processes by bubbling humidification. Adv Mater. Res 2011;282−283:584−8.

[41] Emily WT, John HLV. Experiments and modeling of bubble column dehumidifier performance. Int J Therm Sci 2014;80:65−75.

[42] Ghazal MT, Atikol U, Egelioglu F. An experimental study of a solar humidifier for HDD systems. Energy Convers Manage 2014;82:250−8.

[43] Khalil A, El-Agouz SA, El-Samadony YAF, Abdo A. Solar water desalination using an air bubble column humidifier. Desalination 2015;372:7−16.

[44] El-Dessouky HT, Ettouney HM. Fundamentals of salt water desalination. Elsevier Science BV; 2002.

[45] McGovern RK, Thiel GP, Narayan GP, Zubair SM, Lienhard JH. Performance limits of zero and single extraction humidification-dehumidification desalination systems. Appl Energy 2013;102:1081−90.

[46] Tanaka H. Theoretical analysis of a vertical multiple-effect diffusion solar still coupled with a tilted wick still. Desalination 2016;377:65−72.

CHAPTER 6

Humidification—Dehumidification Solar Desalination Systems

Chapter Outline

Solar Energy Desalination Technology. http://dx.doi.org/10.1016/B978-0-12-805411-6.00006-3

6.1 The Principle and Category of Humidification–Dehumidification Solar Desalination Systems

A noticeable interest in Humidification–Dehumidification Desalination (HDD) technology has emerged since 2006. The HDD process is a new desalination technology different than traditional distillation or membrane processes. HDD has several advantages, such as operating at low temperature ($<80°C$) and near-ambient pressures, required minimum technical features, low-cost materials, and integration with renewable energy, i.e., solar or wind energy.

Saturated and unsaturated moist air touches hot seawater with temperature T_2. Its temperature will increase. Its moisture content also will change into d_2. And then if the moist air contacts a cool plate or cool water with temperature T_1, its temperature and moisture content will decrease and some freshwater will be released. Assuming that the final state moisture content of the air is d_1, then the production of freshwater in one cycle is $d = d_2 - d_1$. For example,

$$T_2 = 60^\circ\text{C}, \quad d_2 = 0.177\ \text{kg(v)/kg(a)}$$
$$T_1 = 30^\circ\text{C}, \quad d_1 = 0.032\ \text{kg(v)/kg(a)}$$

The freshwater production in one cycle is

$$d = d_2 - d_1 = 0.177 - 0.032 = 0.145\ \text{kg(v)/kg(a)}$$

With this way, there is no need to vacuum, there is a low operation temperature, hot seawater is conveniently produced by the solar collector, and so on. However, HDD also has some shortcomings: it needs a relatively big power fan, which will consume more electricity. Also, it needs more heat exchange area due to relatively low operation temperature.

In fact, using the humidification and dehumidification processes to gain pure water is not a new concept and method. These processes are universal in nature. For instance, the seawater is heated to produce the vapor when the sun is on the sea and the vapor goes into the air, which is just a humidification process. On the contrary, when the hot, wet air is cooled to form the rain and falls down to the land, that is a dehumidification process. It is the important source that lands to obtain the freshwater in nature. These processes maintain the ecological system of the Earth's surface to progress and multiply.

Humidification and dehumidification processes made by humankind to acquire pure water are not a new concept, because the heat and mass transfer processes in a man-made basin solar still are going through the typical humidification and dehumidification processes. The seawater in the basin is heated by sunshine to produce the vapor. The vapor goes into the air, which is a humidification process. The hot, wet air rises up to cool the cover plate and releases freshwater, which is a dehumidification process. Although using humidification and dehumidification processes to produce freshwater has been around for a long time in nature, the natural process or conventional method by HDD to produce freshwater is not efficient. As we know, the efficiency of the traditional basin solar still is under 45%. The reason is that the evaporation and condensation processes happen in one space, and the condensing latent heat is almost lost into the surroundings by passing through the glass covers. To overcome the shortcomings of traditional solar stills, a new HDD process uses flowing air as the carrier of water vapor. The evaporator and condenser are separated. At the same time, the temperature of the evaporator and condenser can be controlled respectively. In this way, air is humidified in the evaporation chamber and dehumidified in the condensation chamber. The freshwater is collected in the condenser. The condensing latent heat generally is recovered by the feed water.

As we know, only when the temperature of the moist air rises to over 70°C, 1 m^3 air can include more than 1 kg vapor. In a natural process, it's hard to reach such a high temperature, even in the basin solar still. Therefore it is very difficult to obtain the high productivity by using the natural convection process. For improving performance, enhancing flow or the heat transfer process should be used. From 1961, the University of Arizona started to pay close attention to the HDD process and technology and provided the "humidification and dehumidification circulation desalination" process. In this process, solar energy was used to preheat seawater or the feed water. The heat energy mainly is utilized to evaporate the seawater. The condensation process was completed in another chamber. So it was not the traditional solar distillation process.

Generally, one solar HDD system includes four processes: solar collection, heating seawater, releasing the freshwater, and air circulation. The four processes are completed in a solar collector, humidifier (evaporator), dehumidifier (condenser), and air circulation system, respectively. Certainly, it should be coupled with water pumps, pipes, a freshwater storage tank, and so on.

According to different ways of a solar collection system supplying heat to a dehumidification and humidification system, the solar HDD systems generally are divided into three categories: (1) heating seawater type, (2) heating air type, and (3) heating seawater and air compound type. Also, the heating seawater type may be separated into two forms: direct heating and indirect heating modes. In direct heating, solar collectors directly heat the seawater, and the seawater will go into the solar collectors. In this way, the collection effect is optimum and the middle heat exchanger can be omitted. However, the solar collectors will be corroded more heavily because seawater goes into the collectors. The indirect heating mode will use another working medium; for example, water is heated firstly and then the hot working medium is used to heat the seawater and let it evaporate. In this way, an extra heat exchanger is added, which will decrease heat transfer efficiency, but the benefit is to avoid the corrosion of solar collectors by seawater.

At present, the direct heating seawater mode is the main type of solar HDD systems. The heat energy obtained by solar collectors is first used to heat seawater and is stored in seawater as sensible heat energy. The sensible heat makes seawater evaporate, which brings about the air humidification. In this way, the HDD system generally includes an evaporation chamber, condensation chamber, heater, seawater pumps, and fans. The heating water type solar collection system is getting more and more mature, and it is easy to get 60−80°C hot water in high efficiency. Therefore more and more heating water type solar collection systems have been used in HDD systems.

The heating air type solar HDD system also has other characteristics. (1) The seawater is centralized in a small range, so that the corroded case by seawater can be reduced to the lowest. (2) Only air is circulated in the solar collection system and pipes, so that the requirement for the structural material of system becomes lower and there are no scaling problems. (3) Even if there is a little leakage in the solar collection system or in the pipes, the system still can operate. However, its disadvantage is that fan power consumption will increase considerably.

In a word, the HDD technology is quite different from traditional distillation and membrane desalination technology. (1) The scale of device is flexible and its investment is small. Besides, the device is easy to dispersedly use. Especially, for a solar desalination system, the scale may be big or small in order to adapt the different users. (2) The operation pressure is atmospheric pressure and the operation temperature is in the range of 70 – 90°C, which is easily obtained by traditional solar collectors and is also combined with other low quality heat energy. The device also can be made with low-cost materials because there is no vacuum in the system, which is good for reducing the cost of the device. (3) The vaporization process only happens in the interface of gas and liquids other than in heating surface. As a result, the scaling tendency of equipment is small so that the pretreatment requirement for feed water is relatively low. (4) The vaporization process is not too strong so that the small seawater drops carried by the air is not heavy. Thus the freshwater quality is higher.

6.2 Process Description of Humidification–Dehumidification (HDH) Desalination

6.2.1 The Operation and Evaluation Methods of HDH Desalination System

The desalination systems based on the humidification–dehumidification (HDH) principle are considered as the most viable among solar desalination systems. Moist air is used as a carrier gas of vapor. In the humidifier, moist air is humidified by the vaporized seawater. Then it flows into the dehumidifier, where it is condensed to produce the freshwater. The condensation latent heat can be recycled to preheat seawater. Compared with other desalination systems, it is an innovative technology with a promising application due to its flexibility, simplifying design, low-maintenance cost, extended lifetime of over 20 years, low capital cost, utilization of the low-grade energy (such as solar energy, geothermal energy, and industrial waste heat), and so on.

The desalination systems based on the HDH principle have undergone a substantial development owing to the pioneering work. As the HDD system developed from the one-effect to the multieffect, the performance of the HDD system was improved because of the recovery of the latent heat of condensation. Now, the research works about the HDD system are mainly focused on how to use the thermal energy multiply, simplify the system structure, optimize parameters, and reduce the cost.

A traditional HDD unit is shown in Fig. 6.1. The unit consists of two vertical ducts connected by the top and the bottom to form a closed loop for air circulation. The unit is operated in a forced draft mode by using a fan. A large surface condenser is fixed in one

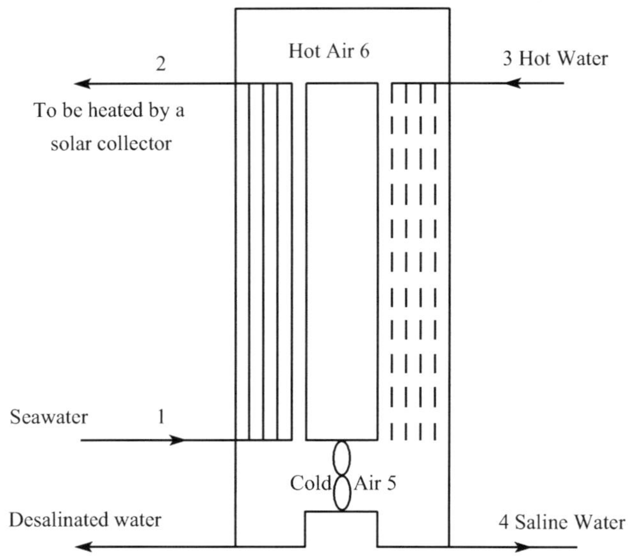

Figure 6.1
Sketch of the desalination process.

of the ducts, while wooden packing, plastic ball bed, or other hydrophilic material is used in another duct for efficient humidification of the air.

Firstly, the saline water at 1 is fed to the condenser to partially condense the water vapor from the air at 6 to produce freshwater. The latent heat of condensation is reused to preheat the seawater to 2. At the same time, the freshwater is sent to the user from the output at the bottom of dehumidifier. The saline water at 2 is further heated in the solar collector. Lastly, the seawater at 3 is sprayed over the wooden packing in the humidifier. The saline water is finally removed to the outside at 4. In air circulation, the air from the bottom of the condenser at 5 is blown upward in the humidifier to get to state 6, in which the air is continuously heated and humidified. The warm brine at 4 is rejected from the bottom of the humidifier.

Before proceeding with our discussion of approximations, several key performance parameters for the HDD processes should be defined. The first parameter of interest concerns the energetic performance of the system, measured by considering the mass (or moles) of water produced per unit of heat input to the cycle. Since this quantity is dimensional, it is common to multiply by the latent heat of vaporization to obtain the gained output ratio (GOR), a dimensionless quantity.

$$\mathrm{GOR} = \frac{m_e h_{fg}}{Q_h} \tag{6.1}$$

where h_{fg} is the latent heat of vaporization, which is evaluated at the ambient air temperature, Q_h is the heat rate that is needed to heat water with a constant flow rate, and m_e is the freshwater production, given as,

$$m_e = m_a\left[\omega\left(T_{a,t}\right) - \omega\left(T_{a,b}\right)\right] \tag{6.2}$$

where, m_a is dry airflow rate in the system, ω is the content of the moist air, and $T_{a,t}$ and $T_{a,b}$ is the top and bottom air temperature of the system, respectively.

The second measure of interest concerns the fraction of feed water that the system is capable of recovering as a product. The recovery rate (RR) captures this fraction and is defined as the ratio of product water, m_e, to feed water, m_f,

$$\mathrm{RR} = \frac{m_e}{m_f} \tag{6.3}$$

The third evaluation index is the thermal energy recovery ratio, which is defined as,

$$\gamma = \frac{Q_{rec}}{Q_h} \tag{6.4}$$

where, γ is the thermal energy recovery ratio of the system, and Q_{rec} is the recovered thermal energy in one cycle. In one cycle, the moist air changes from state 6 to state 5 in a dehumidifier in which its enthalpy change is $\Delta H = H_6 - H_5$. If the heat loss can be ignored, then the thermal energy recovery should be

$$Q_{rec} = H_6 - H_5 - Q_{h,min} \tag{6.5}$$

$Q_{h,min}$ is the minimum heat energy requirement for the system.

6.2.2 Energy Analysis of HDH Desalination System

For a traditional HDD system, the heat and mass transfer mechanisms in the humidifier and the dehumidifier can be simplified as shown in Fig. 6.2.

For the convenience of calculations, the mass flow of the spray water and the circulating air is normalized by the mass flow of the feed saline water. Corresponding to the unit mass flow of the feed water, the mass flow of the air and the spray water are $\dot{m}_a$ and $\dot{m}_h$, respectively. The temperature and enthalpy-changing processes of the spray water, the circulating air, and the feed saline water in different state points is shown in Fig. 6.3. Note how, due to the enthalpy difference required, we can think the input enthalpy of the feed water at state 1 and the output enthalpy of spraying water at state 4 is the same as the enthalpy of saturated air at state 5. At the same thing, the output enthalpy of the feed water at state 2 and the input enthalpy of spraying water at state 3 are same as the enthalpy of saturated air at state 6.

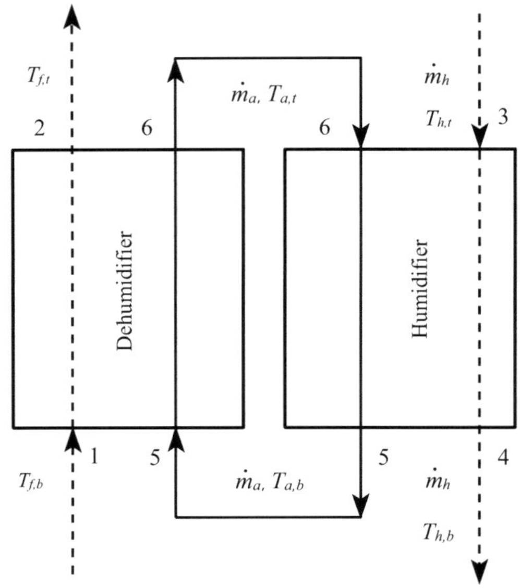

Figure 6.2
Control volumes of the humidifier and dehumidifier.

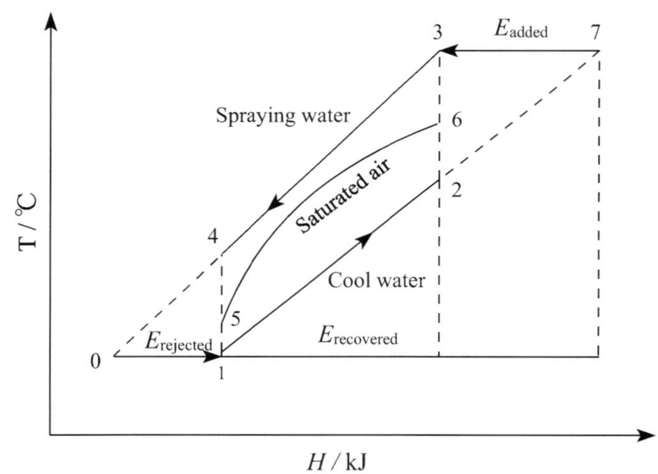

Figure 6.3
The *T-H* relation curves of both the humidification and dehumidification process.

It is noticeable that the gradient of spraying water and feed water is different, when the mass flow rate of spraying water and feed water is different. Generally, they are same so that the line 1−2 and line 3−4 are parallel.

It can be found from Fig. 6.3 that the temperature of feed saline water will increase eventually from input to output in a dehumidifier, which undergoes line 1−2.

Oppositely, the temperature of the moist air will decrease from the top to the bottom of the dehumidifier, which undergoes curve line 6–5. Under perfect conditions, the moist air can be considered to be saturation anywhere in the dehumidifier. Therefore the moist air will change downward only on the saturation line, which is a curve line. However, due to the mass flow rate, which is much more than that of airflow rate, the rising temperature curve of the feed saline water is near to a sloped straight line in the *T-H* diagram, as shown in Fig. 6.3. A bit different in the humidifier, the temperature of the moist air will increase from bottom to top. In the meantime, the moist air still will change only on the saturation line, but upward is the curve line 5–6. On the other hand, the temperature of spray water will be decreased in the humidifier in which some water will be evaporated. So the mass flow rate of the spray water at input and output is different. However, due to evaporated water being so small (only about 4% in each stage), it may still be considered as a constant flow rate. Therefore the process curve of the spray water in the *T-H* diagram still can be thought to be a straight line, as shown in Fig. 6.3.

Since the dehumidifier is externally adiabatic, the enthalpy change in the feed water exactly matches that in the moist air. That is to say, the thermal energy obtained by the feed water is just as same as the lost energy from moist air. Based on the energy balance, someone is easy to get,

$$c_{pw}\left(T_{f,t} - T_{f,b}\right) = \dot{m}_a\left[h\left(T_{a,t}\right) - h\left(T_{a,b}\right)\right] \tag{6.6}$$

where c_{pw} is the specific heat of the feed water, $T_{f,t}$ and $T_{f,b}$ are the temperature of feed water in the top and bottom, respectively, $\dot{m}_a$ is the mass flow rate of air in the system, and $h(T_{a,t})$ and $h(T_{a,b})$ are the enthalpy of the moist air in the top and bottom, respectively. Since the cycle is closed for the air, the top air temperature in the humidifier and dehumidifier is also identical. Similarly, the bottom air temperature is also identical. Finally, an energy balance equation is easy to obtain because the thermal energy given out by spray water just equals the thermal energy obtained by moist air for the external adiabatic humidifier:

$$\dot{m}_h c_{pw}\left(T_{h,t} - T_{h,b}\right) = \dot{m}_a\left[h\left(T_{a,t}\right) - h\left(T_{a,b}\right)\right] \tag{6.7}$$

where $\dot{m}_h$ is the mass flow rate of the spray water, and $T_{h,t}$ and $T_{h,b}$ are the temperature of spray water in the top and bottom, respectively. Note how, due to the shape of the saturation curve, there are pinch points within the inlet and outlet of the humidifier at which there is the minimum temperature difference ΔT_{min} between spraying water and the moist air. The process path of the spraying water is close to, but not exactly, a straight line due to the evaporation of water. Since the mass of water evaporated is small, the curvature is negligible. The spraying water line must be parallel to the tangent of the saturation curve at the pinch point in the humidifier, and the pinch point temperature difference is ΔT_{min}. In other words,

$$h(T_{a,p}) = \frac{\dot{m}_h c_{pw}}{\dot{m}_a} \tag{6.8}$$

$$\dot{m}_h c_{pw}\left[(T_{a,p} + \Delta T_{min}) - T_{h,b}\right] = \dot{m}_a\left[h(T_{a,p}) - h(T_{a,b})\right] \tag{6.9}$$

where $T_{a,p}$ is the temperature of the moist air at pinch point, $\dot{m}_a$ is the mass flow rate of dry air in the system, and $h(T_{a,t})$ and $h(T_{a,p})$ are the enthalpy of the moist air at pinch point.

6.3 Performance Optimization by Pinch Technology

To reduce production cost and utilize resources properly, Professor Bodo Linnhoff and his colleagues at the University of Manchester presented the optimization system for heat exchanger networks, which gradually developed into the pinch technology. The concepts and methodology of pinch technology are well explained in the works of Linhoff et al. [1], Eastop and Croft [2], Linnhoff [3], and Mubarak Ebrahim [4]. Compared with the traditional design method, the pinch technology can save 30–50% energy in the new device. For its low modification cost and good energy-saving effect, this technology is gradually adopted in the modification of an old device.

In 2005, Hou [5,6] first used the pinch technology in the field of desalination and proposed the air-extraction multistage HDD system based on the pinch technology. McGovern et al. [7,8] further analyzed and improved the air-extraction multistage HDD desalination system based on the pinch technology. And then, Zamen et al. [9,10] verified the two-stage air-extraction desalination in the experiment.

6.3.1 Single Effect System Optimization

As we know, there is a pinch point within the inlet and outlet of the humidifier at which there is the minimum temperature difference ΔT_{min} between spraying water and the moist air due to the shape of the saturation curve. It is easy to find the pinch point when we describe the process paths of the moist air, spraying water, and feed saline water in the *T-H* diagram. We can also find from Fig. 6.3 that the supplying thermal energy to the system in one cycle is E_{added} and the rejection heat from system is E_{rejected}. E_{added} is the enthalpy difference of feed water or cooling water from state 2 to state 7. The state 7 has the same temperature as the state 3. Certainly, E_{added} can be changed by adjusting the flow rate of spraying water and feed water. When E_{added} takes the minimum value, the GOR of the system should be best. If the input temperature of feed saline water and output temperature of the spraying water are fixed, obviously we can adjust the mass flow rate to change the slop of the process lines so that E_{added} (just the length of line 7 to 3) takes the minimum. For example, if the state point 1 and state point 3 are fixed, we can reduce the mass flow rate of the feed saline water so that the output temperature of the feed saline water can increase. By this way, the length of line 7 to 3 is shorter. That means E_{added} decreases. That is the reason why we can do some optimization for HDH processes. However, the minimum pinch temperature difference ΔT_{min} is definitely not zero. Generally, $\Delta T_{min} = 3 - 5°C$ is available.

6.3.2 Multieffect System Optimization

Fig. 6.3 gives the temperature and enthalpy change of the spray water, the circulating air, and the feed saline water in a single-effect system. Although some measures can be taken, for example, adjusting the mass flow rate of feed water or spray water to let ΔT take the minimum at pinch point and optimize the GOR of system, there still is a quite large temperature difference between saturated moist air and feed saline water or spraying water in other places. Obviously, that will produce quite large thermodynamic loss. For further improving the performance, many researchers proposed the multieffect HDD system, in which the heat transfer temperature difference can be decreased further. In this way, there still is a pinch point in each effect. Based on this pinch point, each effect can be optimized as in a single-effect system. Finally, the optimized performance of the total system can be obtained. In the next sections, two kinds of multieffect systems are discussed as examples to explain the optimization methods by pinch technology.

6.3.2.1 Regenerative Multistage Desalination

Fig. 6.4 shows the regenerative multistage HDD system presented by Zheng et al. [11,12]. In the regenerative multistage HDD, the highest stage (*N*th stage) is heated directly by the solar energy, while the other stages (*i*th stage) are heated by latent heat of condensation that was released by the moist air in the dehumidifier of the higher stage. In this way, it realizes the cascade utilization of the thermal energy.

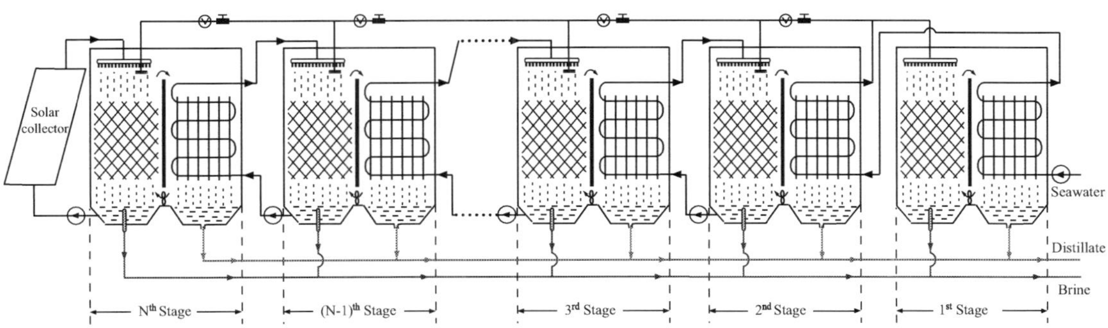

Figure 6.4
The schematic drawing of the regenerative multistage HDD system.

In the first stage, the saline water is fed to the dehumidifier and then cools the water vapor in the saturated air from the humidifier.

$$T_{1,f,b} = T_{feed} \tag{6.10}$$

Then the saline water enters the dehumidifier of the second stage to be further heated, and eventually is divided into many parts. The main part enters the humidifier of the first stage to spray.

$$t_{2,d,b} = t_{1,d,t} \tag{6.11}$$

$$t_{2,d,t} = t_{1,h,t} \tag{6.12}$$

In the ith stage, the cooling water of the dehumidifier is pumped from the bottom of the humidifier of the $(i-1)$th stage and is heated by the latent heat of condensation. Then the saline water is mixed with the additional water from the dehumidifier of the second stage, which can ensure the amount of hot water supply and the exhaust brine. After that, the mixture water is injected into the humidifier of the $(i-1)$th stage. Thus,

$$t_{i,d,b} = t_{(i-1),h,b} \tag{6.13}$$

$$t_{i,d,t} = \dot{m}_{(i-1),h} t_{(i-1),h,t} - \left(\dot{m}_{(i-1),h} - 1\right) t_{2,d,t} \tag{6.14}$$

and

$$\dot{m}_{i,h} = \begin{cases} 1 - \dot{m}_{i,0}(N-1) & i = 1 \\ 1 + \dot{m}_{i,0} & i = 2 \sim N \end{cases} \tag{6.15}$$

where $\dot{m}_{i,0}$ is the additional water from the dehumidifier of the second stage.

In the Nth stage, the saline water is heated by the solar collectors. Then the hot water is mixed with the additional water, and the mixture water is injected into the humidifier again. Thus,

$$t_s = \dot{m}_{N,h} t_{N,h,t} - \left(\dot{m}_{N,h} - 1\right) t_{2,d,t} \tag{6.16}$$

$$Q_{in} = c_{pw}\left(t_s - t_{N,h,b}\right) \tag{6.17}$$

Now, the equations of the regenerative multistage desalination are completed.

Combining the equations of the regenerative multistage desalination and the pinch analysis of Section 2.2, the temperature–enthalpy profile of the regenerative multistage desalination system can be obtained, as shown in Fig. 6.5.

As shown in Fig. 6.6A, the GOR increases as the number of stage increases. But it decreases as Δt_{min} increases. Furthermore, with the increase of Δt_{min}, the GOR decreases more and more slowly. For example, when $\Delta t_{min} = 3°C$, the GORs of stages 1–6 are 1.94, 2.87, 3.72, 4.48, 5.12, and 5.62, respectively. When $\Delta t_{min} = 5°C$, the GORs of stages 1–6 are 1.57, 2.34, 2.91, 3.27, 3.42, and 3.42, respectively.

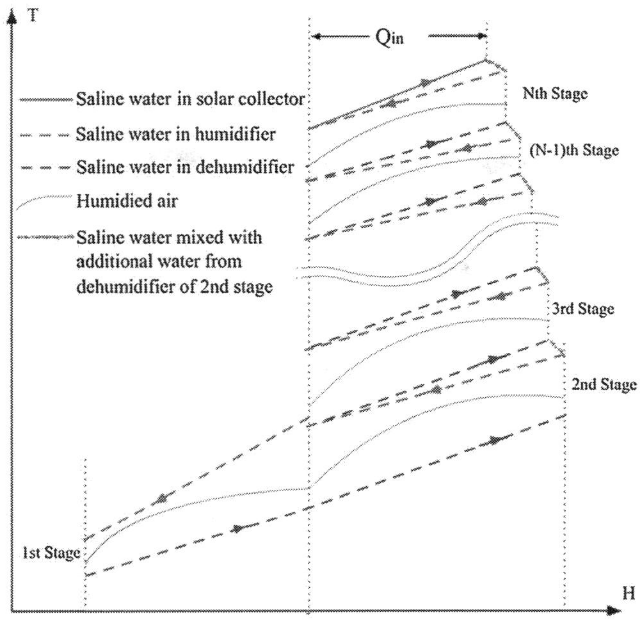

Figure 6.5
Temperature–enthalpy profile of both humidification and dehumidification in the regenerative multistage HDD system.

As shown in Fig. 6.6B, the limit RR (when $\Delta t_{\min} = 0$) increases as the number of stages increase. But as $\Delta t_{\min}$ increases, there is an optimal stage number. For example, when $\Delta t_{\min} = 3°C$, the RR of the four-stage system reaches the maximum (0.078). When $\Delta t_{\min} = 5°C$, the RR of the three-stage system reaches the maximum (0.064).

As shown in Fig. 6.7, when the single-stage system is upgraded to a two-stage system, the performance of the system is improved significantly. But as the number of stages increases further, this enhancement effect is gradually weakened, and even the performance becomes worse. For example, when $\Delta t_{\min} = 5°C$, upgrading a single-stage system to a two-stage system can increase GOR by 49.0% and RR by 17.0%; upgrading a two-stage system to a three-stage system can increase GOR by 24.4% and RR by 3.23%; upgrading a three-stage system to a four-stage system can increase GOR by 12.4% but reduce RR by 3.13%.

6.3.2.2 Air-Extraction Multistage Desalination

Fig. 6.7 shows the air-extraction multistage HDD system present by Hou et al. [5,6] and McGovern et al. [7,8]. As shown in Fig. 6.7, the saline water is fed into the first stage, and after passing through the first dehumidifier, it enters to the second stage and so on. The saline water temperature increases after each stage. When the saline water leaves from the dehumidifier of the *N*th stage, it is heated by the solar collectors. Then hot water enters

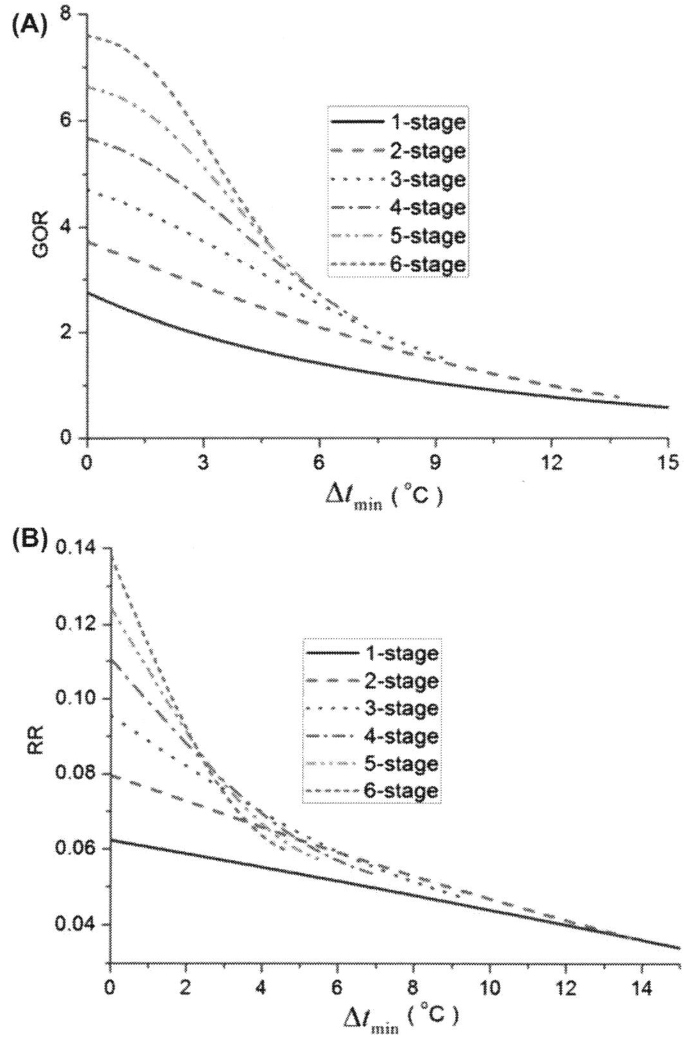

Figure 6.6
Gained output ratio (GOR) and recovery rate (RR) for varying Δt_{min} with a different number of stages in the regenerative multistage desalination system [13]: (A) GOR; (B) RR.

the humidifier of the highest stage (*N*th stage), cascades in the other lower stages, and exits from first stage at last. Thus,

$$t_{1,d,b} = t_{feed} \tag{6.18}$$

$$t_{i,d,b} = t_{(i-1),d,t} \tag{6.19}$$

$$t_{i,h,d} = t_{(i-1),h,t} \tag{6.20}$$

$$t_s = t_{N,h,t} \tag{6.21}$$

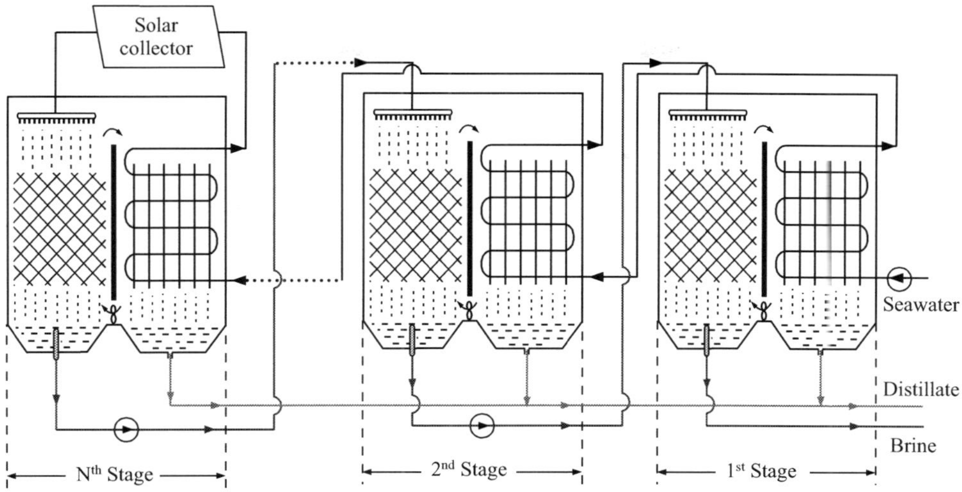

Figure 6.7
The schematic drawing of the air-extraction multistage HDD system.

and

$$Q_{in} = c_{pw}(t_s - t_{N,d,t}) \tag{6.22}$$

Now, the equations of the air-extraction multistage desalination are completed.

Combining the equations of the air-extraction multistage desalination and the pinch analysis of Section 2.2, the temperature–enthalpy profile of the air-extraction multistage desalination system can be obtained, as shown in Fig. 6.8.

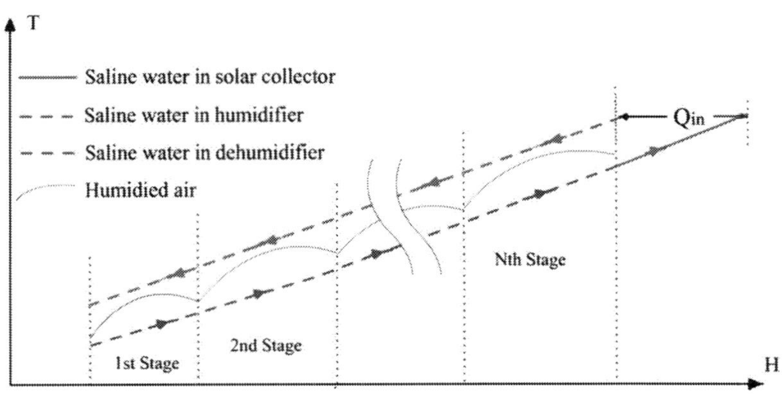

Figure 6.8
Temperature–enthalpy profile of both humidification and dehumidification in the air-extraction multistage HDD system.

As shown in Fig. 6.9A, the GOR increases as the number of stages increase. But it decreases as Δt_{min} increases. Furthermore, with the increase of Δt_{min}, the GOR decreases more and more slowly. For example, when $\Delta t_{min} = 3°C$, the GORs of stages 1−6 are1.94, 3.67, 4.75, 5.38, 5.72, and 5.97, respectively. When $\Delta t_{min} = 5°C$, the GORs of stages 1−6 are1.57, 2.55, 3.04, 3.28, 3.41, and 3.49, respectively.

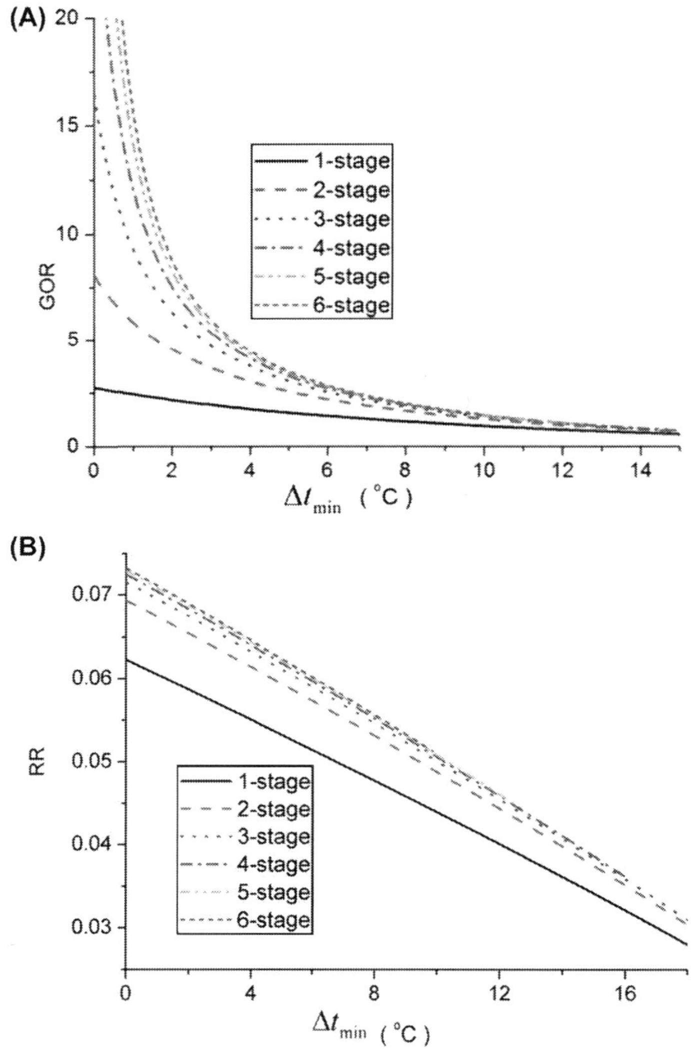

Figure 6.9
Gained output ratio (GOR) and recovery rate (RR) for varying Δt_{min} with a different number of stages in the regenerative multistage desalination system [13]: (A) GOR; (B) RR.

As shown in Fig. 6.9A, the limit RR (when $\Delta t_{min} = 0$) increases as the number of stages increase. But as Δt_{min} increases, the enhancement effect of RR is gradually weakened. For example, when $\Delta t_{min} = 3°C$, the RRs of stages 1–6 are 0.057, 0.063, 0.065, 0.066, 0.067, and 0.067, respectively. When $\Delta t_{min} = 5°C$, the RRs of stages 1–6 are 0.053, 0.059, 0.061, 0.062, 0.062, and 0.062, respectively.

As shown in Fig. 6.9B, when the single-stage system is upgraded to a two-stage system, the performance of the system is improved significantly. But as the number of stages increases further, this enhancement effect is gradually weakened. For example, when $\Delta t_{min} = 3°C$, upgrading a single-stage system to a two-stage system can increase GOR by 89.2% and RR by 10.5%; upgrading a two-stage system to a three-stage system can increase GOR by 29.4% and RR by 3.17%; and upgrading a three-stage system to a four-stage system can increase GOR by 13.3% and RR by 1.54%. Therefore when the single-stage system upgrades to a two-stage system, the performance is significantly improved. But as the number of stages increases further, this enhancement effect is gradually weakened.

6.3.3 Comparisons

According to the earlier analysis, the limit RRs (when $\Delta t_{min} = 0$) of both regenerative and air-extraction multistage desalination increases as the number of stages increases. However, in the practical situation, it is impossible to transfer heat when $\Delta t_{min} = 0$. Additionally, too small Δt_{min} will increase the system investment cost. Thus in engineering applications, there is an optimal Δt_{min}, which is partly dependent on the economic condition. The increment of GOR with the increase of the stage number will become smaller or even negligible. In addition, the RR does not monotonically increase with the increase of stage number. Thus as shown in Figs. 6.10 and 6.11, the two-, three-, and four-stage regenerative and air-extraction multistage desalinations are compared at the condition of $\Delta t_{min} = 3°C$ and $\Delta t_{min} = 5°C$.

As shown in Figs. 6.10 and 6.11, when $\Delta t_{min} = 3°C$ and $\Delta t_{min} = 5°C$, the GOR of the two-, three-, and four-stage regenerative multistage desalination is less than that of the air-extraction multistage desalination, while the RR of the regenerative multistage desalination is larger than that of the air-extraction multistage desalination. As Δt_{min} and the number of stages increase, the performance difference between the two multistage systems becomes small. It is even negligible in a four-stage system when the pinch point temperature difference is 5°C.

The research content will help to select the right device in the design of the HDD system. When the GOR is the main design goal, it is better to select air-extraction multistage desalination, and when the RR is the main design goal, it is better to select regenerative multistage desalination.

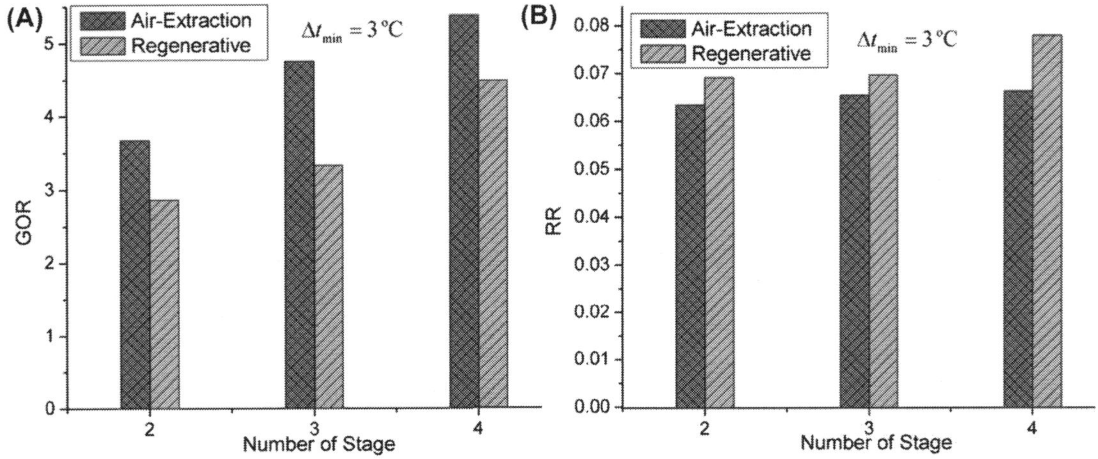

Figure 6.10

Comparison of the regenerative and air-extraction multistage desalination with $\Delta t_{min} = 3°C$ [13]: (A) Gained output ratio (GOR); (B) recovery rate (RR).

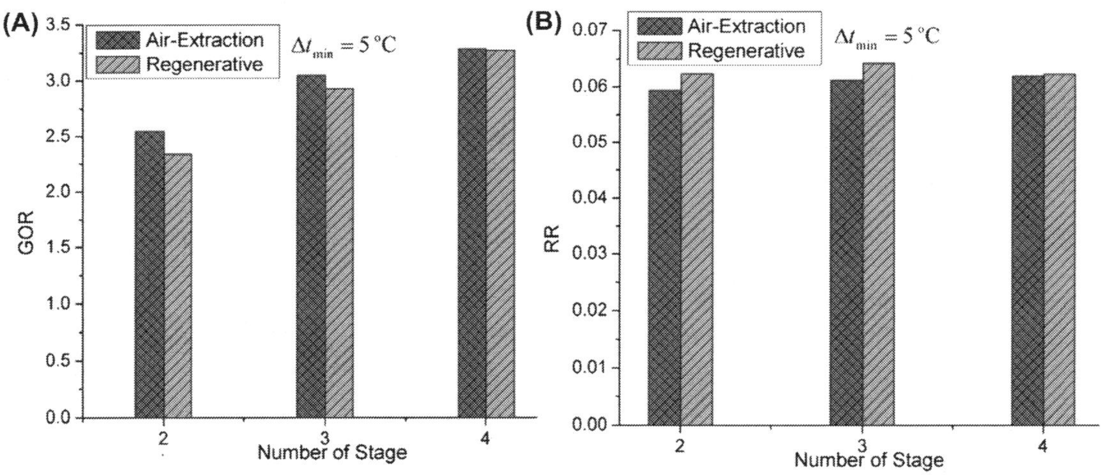

Figure 6.11

Comparison of the regenerative and air-extraction multistage desalination with $\Delta t_{min} = 5°C$ [13]: (A) Gained output ratio (GOR); (B) recovery rate (RR).

6.4 The Minimum Work Required for a Solar HDD Process

From the perspective of thermodynamics, the humidification process is actually the mixing process of air and water vapor, while the dehumidification process is the separation process of air and water vapor [14–16]. The former produces exergy losses, while the latter demands minimum work required. Three steps are needed to complete the whole process of HDD. First, evaporate water vapor out of seawater, which needs energy that can be recovered from the condensation of the system. Then mix vapor with air, which costs no energy but leads to work loss. Finally, separate water vapor from air, namely the dehumidification process, which requires energy and demands minimum work required. Therefore the sum of work of the two steps to finish an HDH desalination cycle is needed. If each part gets the minimum work, the work of the desalination process required reaches its minimum value. There are, actually, a lot of scholars who have studied the minimum work required of the desalination. For example, Yunus Cerci [17,18] deduced the minimum work required of the seawater evaporation process and gave the minimum work required equation of seawater evaporation process and also the minimum work required for separating freshwater of unit mass when the saline is at different concentrations. He has proven that the minimum power consumption increases when the feed salinity increases. Qian [19] worked out the theoretical equation for the minimum work required of the desalination process based on the chemical potential change of seawater composition. Soteris [20] analyzed the desalting process and constructed exergy steady balance equation by adopting the classical theory of thermodynamics.

There are also many scholars who have studied the minimum work required for the dehumidification process. Majed [21] discussed and calculated the minimum work required for the dehumidification process. Camacho [22] and Cengle [23,24] studied the work required of the isothermal process for condensing water from the atmosphere and gave a theoretical derivation. These studies all made a foundation to further explore the minimum work required of the solar-powered desalination based on air HDD.

However, an idealized analysis of a complete solar-powered HDD system is yet to be made. Conducting research on this issue is beneficial for the actual system designers to improve their design ideas and conducive to evaluate the actual project as an evaluation indicator of the system, which makes this research meaningful.

6.4.1 An Ideal Humidification–Dehumidification Desalination (HDD)

Fig. 6.12 shows the basic operation process of an ideal single-effect humidification–dehumidification (HDH) desalination system. In this system, cold seawater first enters the dehumidifier, cooling humid air circulated from the fans and precipitating freshwater. At the same time, the seawater is heated and then returns to the humidifier to evaporate. (In an actual solar desalination system, seawater should enter the solar thermal system to be further heated.) Finally, the seawater becomes concentrated brine and is discharged out of the humidifier. Humid air in the humidifier was transferred into the dehumidifier by a fan to complete the whole cycle. In the dehumidifier, the temperature of humid air changes

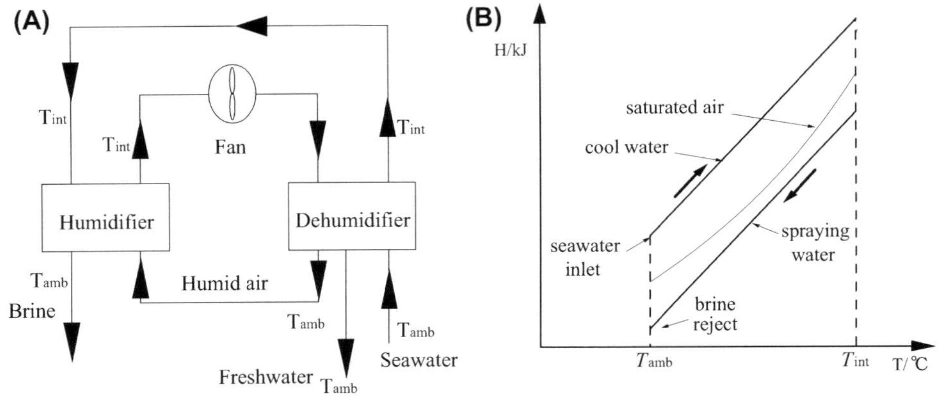

Figure 6.12
Ideal single-action HDH desalination system (A) and the *T*–*H* relation curves (B).

from T_{int} to T_{amb}. The humid air is always saturated and changes according to the saturation line. Dehumidification for unit air circulation is $\omega_{int} - \omega_{amb}$, ω with indicating the moisture content of humid air, kg_w/kg_a.

In the humidifier, when humid air temperature changes from T_{amb} to T_{int}, it still can be considered that the humid air is completely saturated. Its process returns from environment state to internal state along the saturation line. And the unit circulating air humidification remains $\omega_{int} - \omega_{amb}$. For an ideal process, it can be considered that humid air and seawater exchange heat sufficiently in the humidifier and dehumidifier. We can also ignore the fan power consumption and the heat capacity of seawater and freshwater. At the same time, presume that the amount of circulating water is very limited. In this case, the total power consumption of the system only contains power consumption of seawater evaporation and power needed for humid air dehumidification process.

6.4.2 The Minimum Work Required for the Dehumidification Process

In an ideal HDD, it can be assumed that the humid air is always saturated. Consequently, for calculating the minimum work required for the dehumidification process, we only need to work out the minimum work required of humid air from internal state to environment state in Fig. 6.13. For the ideal desalination process, we can make the following assumptions about the wet air: (1) the humidification process is simplified as the mixture of two kinds of ideal gases of air and water vapor. (2) The liquid water is always saturated at the local temperature, and the specific heat of the air and water vapor is constant. (3) The dead state is assumed to be the dry ambient while the final state 3 for complete dehumidification is taken to be saturated air at 273.16K. (4) In the process of cooling, it is considered that the air is always carried along the constant humidity until the air is saturated.

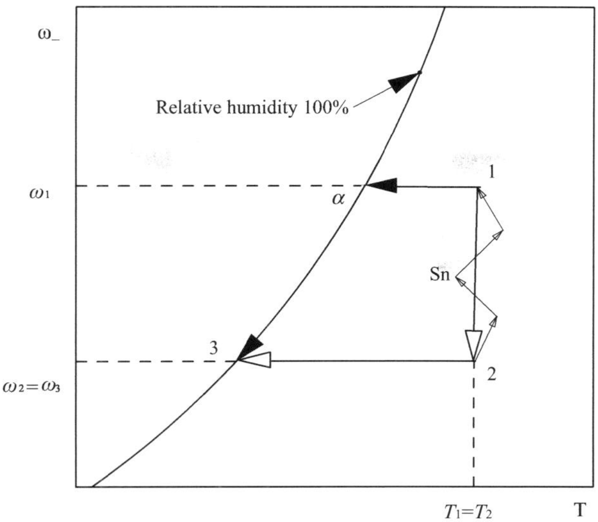

Figure 6.13
Representation of the reversible process of hypothetic isothermal dehumidification on the psychometric chart.

In fact, the process of dehumidification is the separation of water vapor and air, and its inverse process is the mixing process. The mixing process is a typical irreversible process, so there will be power losses or exergy losses. This is, actually, the minimum work required for the separation process. As long as we can calculate the exergy losses of the mixing process, we can know the minimum work required for separation. Therefore the inverse process of the isothermal dehumidification process 1 → 2 is the isothermal mixing process 2 → 1 shown in Fig. 6.13. The isothermal humidifying process can be realized by making the air slowly flow through a well-designed humidification channel as shown in Fig. 6.14. There are many different steam spray and cold spray stages in the channel, with steam spraying being humidification and the warming process and the spraying of cold

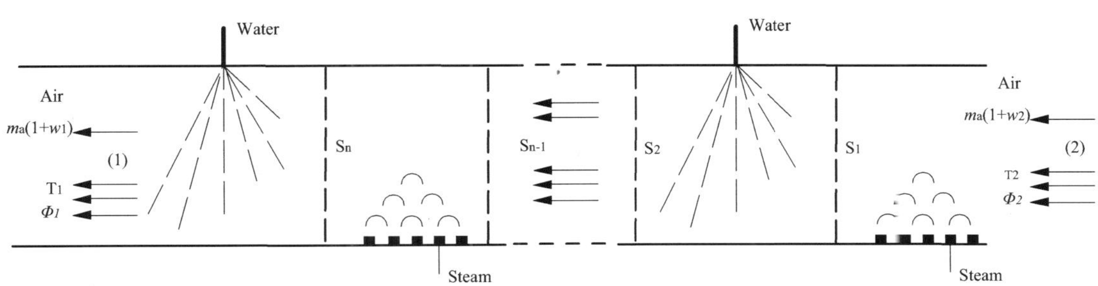

Figure 6.14
An approximate isothermal humidification process.

water being humidification and the cooling process. If such a process undergoes slowly, it can be considered as an isothermal humidification process, just as line 2 → 1 is shown in Fig. 6.13.

Based on the earlier assumptions, Majed [15] did the full derivation and analysis for the minimum work required for the air dehumidification process starting from the unsaturated state 1 in Fig. 6.13 and developed the theoretical expressions. In the process of derivation, Majed made a further hypothesis: the dehumidification process is a combination of two independent processes, cooling processes of equal moisture rate and dehumidification, and considered that state 3 is the dead state of the system. Majed thought that process 1-a-3 can be equivalent to the process of 1-2-3 and assumed that sum of exergy increasing in processed 1 → 2 and 2 → 3 is the minimum work required for a total dehumidification process. The minimum work required is:

$$\mathrm{w}_{\min} = \mathrm{w}_{12} + \mathrm{w}_{23} \tag{6.23}$$

where $\mathrm{w}_{\min}$ is the minimum work required for dehumidification on the basis of a unit mass of dry air and w_{12} and w_{23} are represented respectively by the following equations [15]:

$$\mathrm{w}_{12} = \psi_2 - \psi_1 + (\omega_1 - \omega_2)\psi_{l2} + \left(1 - \frac{T_0}{T_1}\right)[\mathrm{s}_2 - \mathrm{s}_1 + (\omega_1 - \omega_2)\mathrm{s}_l] \tag{6.24}$$

$$\mathrm{w}_{23} = -(c_{pa} + \omega_2 c_{pv})\left[(T_2 - T_3) - T_0 \ln\left(\frac{T_2}{T_3}\right)\right] \tag{6.25}$$

where ψ indicates specific exergy of humid air, kJ/kg, ω represents moisture content of humid air, $\mathrm{kg_w/kg_a}$, and c_{pa} and c_{pv} indicate specific heat of dry air and water vapor, respectively. T is the temperature of each state point, s is the specific entropy of humid air, and subscript 1 and 2 indicate the state of 1 and 2, respectively. Subscript l indicates the state of liquid water. ψ_{l2} is the specific exergy of liquid water in state 2. The specific exergy of each state can be written as [25]:

$$\begin{aligned}\psi_\mathrm{i} &= (c_{pa} + \omega_\mathrm{i} c_{pv})T_0\left[\frac{T_\mathrm{i}}{T_0} - 1 - \ln\left(\frac{T_\mathrm{i}}{T_0}\right)\right] + (1 + 1.608\omega_\mathrm{i})R_a T_0 \ln\left(\frac{\mathrm{p}}{\mathrm{p}_0}\right)\\ &+ R_a T_0\left[(1 + 1.608\omega_\mathrm{i})\ln\left(\frac{1 + 1.608\omega_0}{1 + 1.608\omega_\mathrm{i}}\right) + 1.608\omega_\mathrm{i} \ln\left(\frac{\omega_\mathrm{i}}{\omega_0}\right)\right]\end{aligned} \tag{6.26}$$

$$\psi_{li} \cong [h_f(T_i) - h_g(T_0)] - T_0[s_f(T_i) - s_g(T_0)] + [p - p_g(T_i)]v_f(T_i) - R_v T_0 \ln \phi_0 \tag{6.27}$$

R_a and R_v are gas constants of air and water vapor, h represents specific enthalpy, ϕ_0 indicates the relative humidity of the environment, and $v_f(T_i)$ is the specific volume of saturated liquid water at the temperature T_i. The subscript f and g respectively represent the state of saturated water and saturated vapor, p represents total gas pressure, and $p_g(T_i)$ is saturated steam pressure at temperature T_i. The relationship between the moisture content of humid air, the relative humidity of the air, and the partial pressure of saturated water vapor is as follows:

$$\omega_i = 0.622\frac{\phi_i p_g(T_i)}{p - \phi_i p_g(T_i)} \tag{6.28}$$

$$p = p_{ai} + \phi_i p_g(T_i) \tag{6.29}$$

Note that the derivation is based on state 1 to the dead state ($T_3 = T_d = 273.16\text{K}$), and the final state of the air is the saturated air at 273.16K.

By the Eq. (6.23) we get the minimum work required per unit mass of dry air in the unsaturated humid air cooling dehumidification process. In this process, the moisture content of humid air changes from ω_1 to $\omega_2 = \omega_3$, and the amount of freshwater it produces happens to be $\omega_1 - \omega_3$. Therefore the minimum work required for the production of unit mass freshwater can be expressed as:

$$w_{w,min} = \frac{w_{12} + w_{23}}{\omega_1 - \omega_3} \tag{6.30}$$

For an ideal HDH desalination system, when it operates in a steady state, the humid air can be considered to be saturated at local temperature. That is, the initial point of the wet air is saturated and the end of the dehumidification process is not dead state 3, but saturation state 2, as shown in Fig. 6.15. The dehumidification process goes directly along the saturation line from the state 1 to the state 2. It can be found from Fig. 6.15 that the process from the state 1 to the dead state 3 can be equivalent to 1-a-b-c-3, while that from

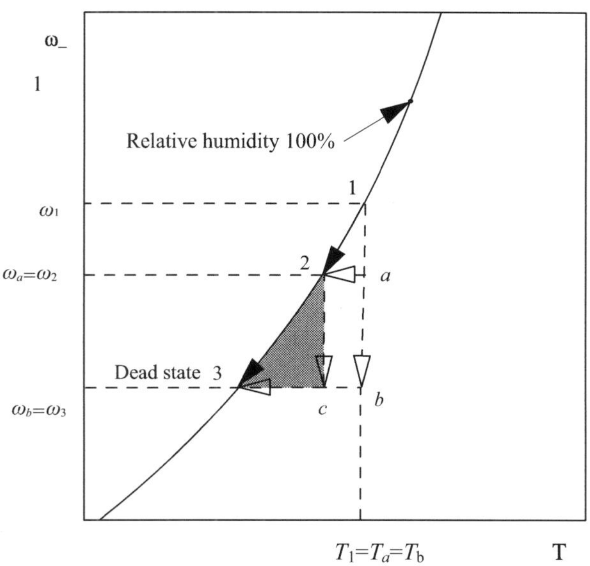

Figure 6.15
Dehumidification process in the HDH desalination system.

the state 2 to dead state 3 can be equivalent to 2-c-3. So the minimum work required from state 1 to state 2 can be derived from the following formula:

$$w_{12,\min} = w_{13,\min} - w_{a3,\min} + w_{a2,\min} \tag{6.31}$$

If the values of the T_1 and T_2 are given, then $w_{13,min}$ and $w_{a3,min}$ can be directly calculated according to Eqs. (6.23)–(6.29) and $w_{a2,min}$ can be obtained by Eq. (6.25). In order to give a quantitative analysis of the minimum work required for a certain dehumidification stage, we could set $T_2 = 293$K. Then, we get effects of initial temperature T_1 on $w_{13,min}$ and $w_{a3,min}$, as shown in Figs. 6.16 and 6.17. Among them, the relevant parameters are obtained through the look-up table and calculation.

$$\omega_2 = 0.0147 \text{ kg/kg}_\text{a}$$
$$c_{pa} = 1.007 \text{ kJ/kg}_\text{a}\text{K}$$
$$c_{pv} = 1.881 \text{ kJ/kg}_\text{v}\text{K}$$

The minimum work required from state a to state 2 can be calculated with the following formula:

$$w_{a2,min} = -1.03465\left[(T_1 - 293) - 293 \ln\left(\frac{T_1}{293}\right)\right] \tag{6.32}$$

Fig. 6.16 shows that the higher the temperature of the initial state 1 is, the greater the minimum work that will be for the same final state. And with the increase of the initial temperature of T_1, the minimum work increases more quickly, for the curve is concave upward. It is not difficult to understand that the reason is that all points on the curve are corresponding to the same final state, namely, the earlier assumptive state T_3. It can be obtained from Fig. 6.16 that $w_{23,min} = 2.65$ kJ/kg$_\text{a}$, while $T_1 = T_2 = 293$ K. Fig. 6.17 plots

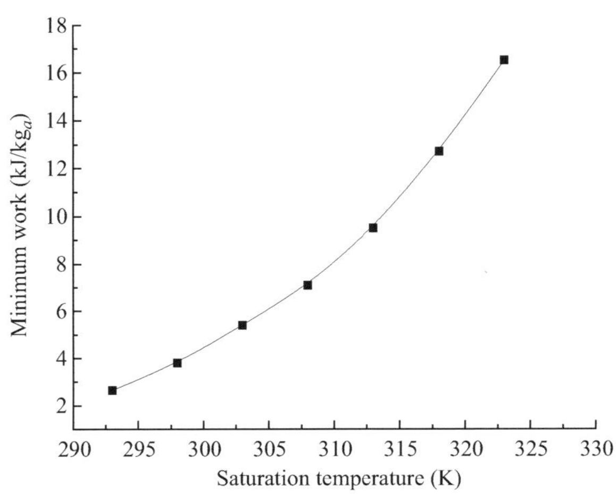

Figure 6.16
Effect of T_1 on the minimum work for 1–3 process.

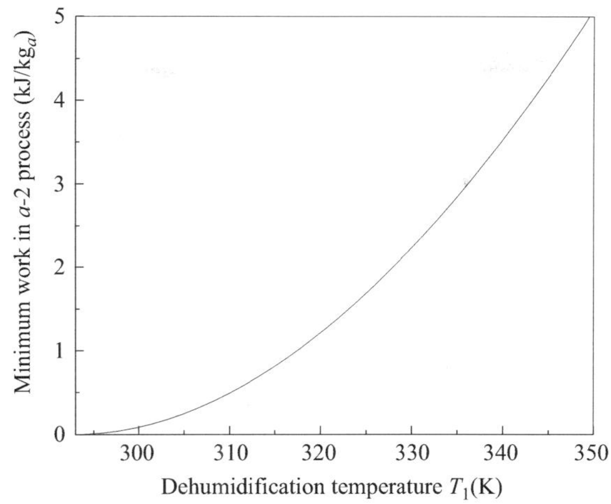

Figure 6.17
Effect of initial temperature T_1 on minimum work for a-2 process.

the effect of temperature T_1 on the minimum work required for a-2 process. For the same final state, when the temperature of initial state 2 is higher, the minimum work required is greater. And with the increase of the initial temperature T_2, the minimum work required increases faster for the curve and is also concave upward.

According to the parameters given earlier, we can calculate the minimum work required $w_{12,min}$ of the dehumidification process from state 1 to state 2, and this minimum work required is concerned with the quality of the unit dry air. The calculation results are shown in Fig. 6.18. The figure illustrates that when the initial temperature dehumidifier T_1 is higher, the minimum work required is greater due to the larger gap between T_1 and T_2, and the growth trend is rapider for the concave upward curve. In fact, more attention is paid on the minimum work required to produce the unit mass of freshwater. As shown in Fig. 6.19, the minimum work required for the production of freshwater of unit mass changes with T_1 under state 2. It can be seen that the minimum work required increases along with the increase of T_1. But the curve is concave downward. With the rising of temperature, the condensation latent heat of water vapor becomes smaller, so the energy required to condensate the unit mass of freshwater is relatively reduced. The point in the graph has some deviation, which is caused by the choice of the decimal point in the calculation process. The larger temperature difference between the T_1 and T_2 is, the greater the minimum work required is, indicating greater process irreversibility. If temperature difference $T_1 - T_2$ remains constant, the minimum work required to produce unit quality freshwater will be reduced with the increase of temperature.

The earlier discussion is based on the given environment temperature and a given state point 2. If the state point 2 changes or the air is not saturated in the HDH desalination

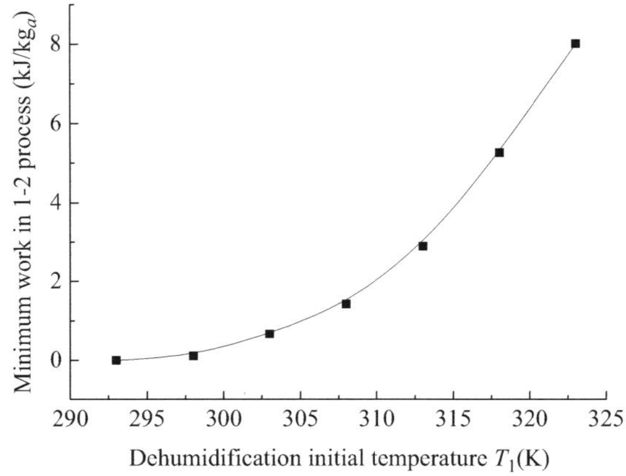

Figure 6.18
Effect of initial temperature T_1 on minimum work for 1−2 process.

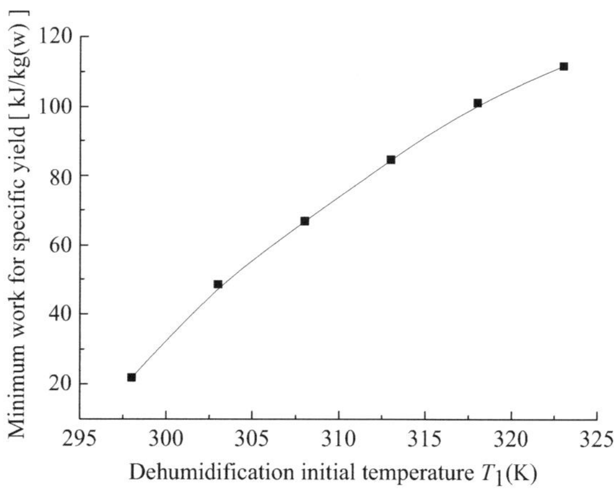

Figure 6.19
The minimum work required for the freshwater per unit mass.

system, the earlier statistics and derivations need to be modified. According to the results shown in Fig. 6.19, if the dehumidification initial temperature is 303K, it will take about 12.72 kWh to produce a ton of freshwater. This value is considerable, 10 times the minimum work required of the flash process. Therefore the energy recovery process is badly needed for the HDH desalination system to improve system efficiency.

6.4.3 The Minimum Work Required for Evaporation of Seawater

The discussion in the previous section gives the minimum work of the dehumidification process. Except for the dehumidification process, there are still the evaporation process and the humidification process in the HDH desalination system. The humidifying process is a natural mixing process, which does not need to work but has exergy losses. On the contrary, the evaporation process needs work.

The heat required of the evaporation process can be obtained from the heat recovery for the steam condensation process. However, it should be noted that there is a difference between the evaporation on the surface of seawater and condensation on the surface of freshwater. It is mainly because vapor pressure at the evaporation surface and condensation surface are different. The saturated vapor pressure at the freshwater surface is generally calculated by the Antoine formula.

$$lgp_w = 7.07406 - \frac{1657.46}{T - 46.13} \tag{6.33}$$

where p_w is the saturated vapor pressure of the freshwater surface with kPa as its unit and K as temperature T's unit. This formula is more accurate in $283\text{K} \le T \le 440\text{K}$. According to this formula, the temperature changes when the pressure changes dp_w can be calculated as:

$$dT/dp_w = \frac{(T - 46.13)^2}{3816.44}\frac{1}{p_w} \tag{6.34}$$

Assume that seawater is just a mixture of sodium chloride and water and make that the mole percentage of sodium chloride in seawater is x, then we know the saturation vapor pressure at sea surface according to the Henry's law.

$$p_{w,s} = (1 - x)p_w \tag{6.35}$$

Thus at a certain temperature, the vapor pressure difference between the surface of the seawater and freshwater is:

$$\Delta p_w = xp_w \tag{6.36}$$

Substitute Eq. (6.36) into Eq. (6.34), and we obtain the temperature difference caused by pressure difference:

$$\Delta T = \frac{(T - 46.13)^2}{3816.44}x \tag{6.37}$$

For the general seawater, it can be considered that the mass concentration of sodium chloride is 3.5%, thus $x = 0.01105$. For example, if the seawater temperature is 323K, then $\Delta T|_{T=323} = 273.222\text{K}$.

The temperature difference will produce exergy loss in the heat transfer process. Assuming heat transfer is Q, the exergy loss will be:

$$\delta W_i = Q\left(1 - \frac{T - \Delta T}{T}\right) = Q\frac{\Delta T}{T} \tag{6.38}$$

We should note that this is internal loss of the system, which will eventually produce the external loss by chain reaction. According to the relationship between internal loss and external loss ($\delta W_g = \frac{T_0}{T}\delta W_i$, T_0 is the ambient temperature), we can get the external loss.

$$\delta W_g = \frac{T_0}{T}\delta W_i = \frac{T_0}{T^2}\Delta TQ \tag{6.39}$$

Assuming that the condensation latent heat of pure water is h_{fg}, according to $Q = m_e h_{fg}$, the minimum work required about evaporating per unit mass of freshwater from the seawater can be calculated by:

$$w_{e,min} = \frac{\delta W_g}{m_e} = \frac{T_0}{T^2}\Delta T h_{fg} \tag{6.40}$$

When $T_0 = 293\text{K}$ and $T = 303\text{K}$, the minimum work required of the evaporation process is $w_{e,min} = 1.4634$ kJ/kg, the minimum work required of the dehumidification process is 47.89 kJ/kg, and the total minimum work required for the evaporation and dehumidification process is:

$$w_{min} = w_{e,min} + w_{12,min} = 49.353\ \text{kJ/kg} \tag{6.41}$$

6.4.4 The Maximum GOR of the Solar HDD System

A solar heat collection system is included in an ideal solar HDH desalination system, as shown in Fig. 6.20. Seawater enters into the dehumidifier, which causes temperature to rise, then into the solar collectors for further heating until the temperature reaches T_1 and finally returns the humidifier humidifying the air by spraying.

GOR of the seawater desalination system is a performance index to evaluate the system. Generally, it can be expressed as the energy required for producing a unit mass of freshwater, namely:

$$\text{GOR} = \frac{m_e h_{fg}}{Q_{in}} \tag{6.42}$$

where m_e is the production of freshwater, kg, h_{fg} indicates the latent heat of condensation, kJ/kg, and Q_{in} is the heat supply into the system, kJ.

On the other hand, for an ideal solar collector, if its conduction and convection heat loss can be ignored and only radioactive heat transfer is considered, a unit area of the collector can produce the maximum useful heat [26]:

$$q_{u,max} = Cf\sigma T_s^4 + (1 - Cf)\sigma T_0^4 - \sigma T_{abs}^4 \tag{6.43}$$

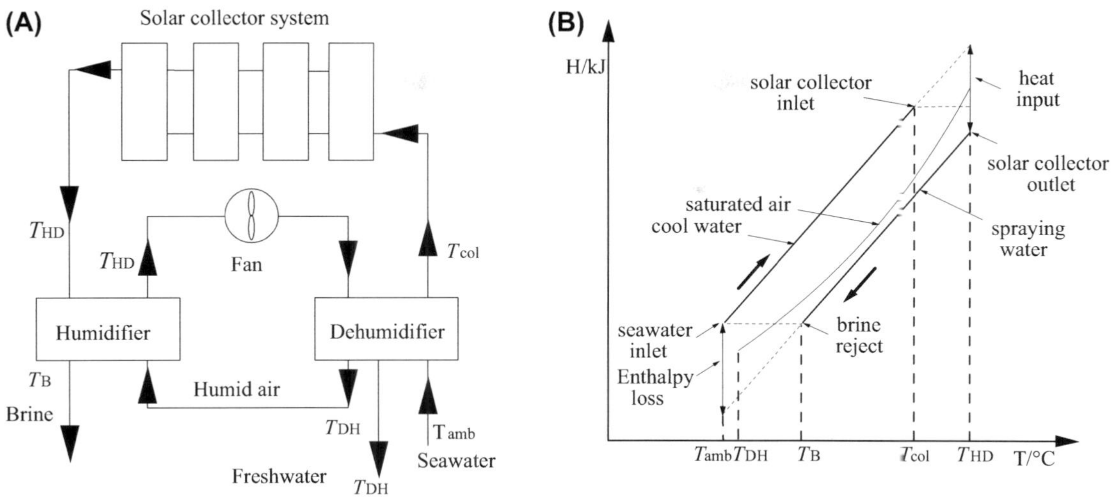

Figure 6.20
An ideal solar HDH desalination system (A) and the *T*–*H* relation curves (B).

where C is concentration ratio of solar heat collection relative to the concentrator aperture, f is the view factor of solar concentrator, σ is the Stephen Boltzmann constant, and T_s and T_{abs} are temperature of the sun surface and absorber surface, respectively. In the ideal case, heat can be transported unhindered to a reversible Carnot heat engine. Reversible Carnot heat engine achieves the largest amount of work by working at temperatures between T_{abs} and T_0. This work is the maximum exergy output of the collector under environmental conditions. In this case, the amount of work that can be output through the Carnot engine is:

$$w_{c,max} = q_{u,max}\left(1 - \frac{T_0}{T_{abs}}\right) \tag{6.44}$$

where $w_{c,max}$ is the maximum mechanical power provided by per unit area of the receiver. So the maximum production of freshwater per unit area of the receiver can be generated as follows:

$$m_{e,max} = \frac{w_{c,max},\ \mathrm{kW/m^2}}{w_{min},\ \mathrm{kJ/kg}} = \frac{q_{u,max}}{w_{min}}\left(1 - \frac{T_0}{T_{abs}}\right),\ \mathrm{kg/s} \tag{6.45}$$

Accordingly, the maximum GOR of the solar system can be expressed as follows:

$$\mathrm{GOR_{max}} = \frac{m_{e,max}h_{fg}}{q_1} = \frac{q_{u,max}}{q_1 w_{min}}\left(1 - \frac{T_0}{T_{abs}}\right)h_{fg} \tag{6.46}$$

q_1 is the solar energy projected onto a unit area of receiver surface, generally:

$$q_1 = Cf\sigma T_s^4 \tag{6.47}$$

Substituting Eqs. (6.43) and (6.47) into Eq. (6.46), we get the maximum GOR of the system:

$$\mathrm{GOR_{max}} = \frac{Cf\sigma T_s^4 + (1 - Cf)\sigma T_0^4 - \sigma T_{abs}^4}{Cf\sigma T_s^4 w_{min}}\left(1 - \frac{T_0}{T_{abs}}\right)h_{fg} \tag{6.48}$$

Here we can do a simplified analysis. Assuming that the receiver operating temperature is very low and equals to the temperature T of the humidifier, while ignoring the radiative heat transfer between the surrounding and the collector, then Eq. (6.48) becomes:

$$\mathrm{GOR_{max}} = \frac{1}{w_{min}}\left(1 - \frac{T_0}{T}\right)h_{fg} \tag{6.49}$$

Taking $h_{fg} = 2400$ kJ/kg, $T = 303$K, and $T_0 = 293.15$K, it can be calculated that $\mathrm{GOR_{max}} = 1.60$. It can be seen clearly that if a solar HDH desalination system operates at a lower solar collection temperature, as usually seen in a single-effect system, its GOR is low. GOR of a single-stage is low.

In order to further analyze $\mathrm{GOR_{max}}$ as a function of temperature T, we can fit the relationship of minimum work $w_{12,\mathrm{min}}$ with the change of temperature, as shown in Fig. 6.18, and get:

$$w_{12,min} = -7096 + 42.641T - 0.063T^2 \tag{6.50}$$

Eq. (6.48) is obtained under completely ideal conditions.

Assume $T_3 = T_0 = 293.15$K and then substitute it into Eq. (6.47). However, considering that the concentrating collector only collects direct light and environment radiation is mostly scattered radiation, thus the contribution on the receiver is inversely proportional to concentration ratio C. Meanwhile, in order to facilitate the discussion, suppose that the irradiation of solar light on the receiver is equivalent to $f\sigma T_s^4 = I$, where I is the solar irradiance on the collector surface. So Eq. (6.48) can be rewritten as:

$$\mathrm{GOR_{max}} = \frac{\left(C\cdot I + \frac{\sigma T_0^4}{C} - \sigma T^4\right)\left(1 - \frac{T_0}{T}\right)h_{fg}}{C\cdot I\cdot\left[\frac{T_0}{T^2}\Delta T h_{fg} - 7096 + 42.641T - 0.063T^2\right]} \tag{6.51}$$

In the formula, the unit of T is K and I can represent 1000 W/m^2 or 600 W/m^2. According to the formula, we can get the relationship between $\mathrm{GOR_{max}}$ and T, as shown in Fig. 6.21.

When studying Fig. 6.21, you can see a very interesting phenomenon. From 293 to 294K, there is a slump, and the maximum GOR of the system nearly falls to 1.5. With the increase of temperature, GOR rises slowly and increases quickly after 350K. Finally, GOR reaches 6.5 at 373K (100°C), and the highest value reaches 12 under the condition of

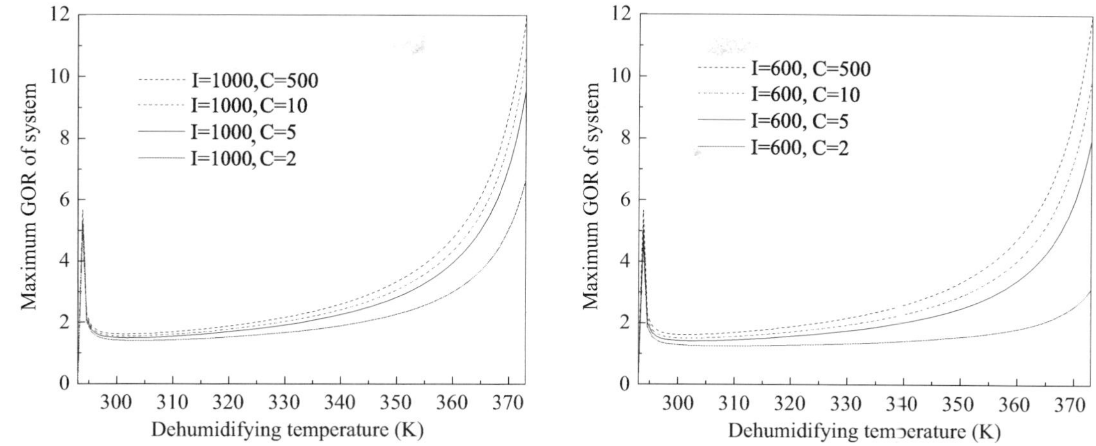

Figure 6.21
The relationship between the maximum gained output ratio (GOR) and the operating temperature of different concentration ratio.

1000 W/m^2 solar radiation in different concentration ratio. Obviously, it is the highest performance efficient value of a single-action HDH desalination system driven by the solar heat collector. From the analysis, we can see that, for the solar HDH desalination system, system operation is more dominant at higher temperatures. Fig. 6.21 also proposes that the higher the concentration ratio is, the greater the GOR becomes, given the condition that other conditions remain unchanged.

At the same time, the solar radiation has effects on the GOR. Fig. 6.22 shows the influence of the solar radiation on the GOR under the environment temperature 293K.

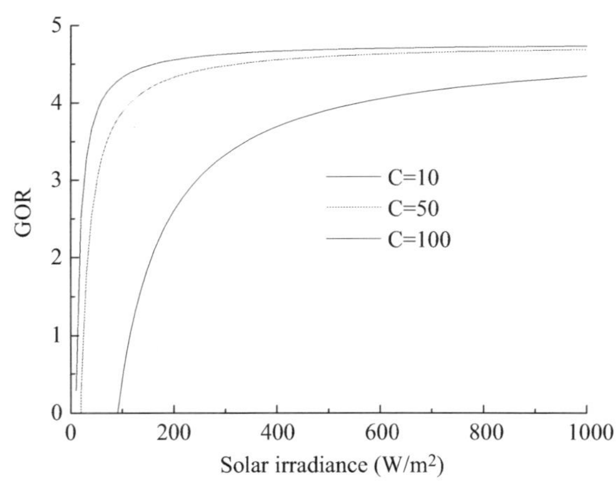

Figure 6.22
Effect of solar irradiance on the gained output ratio (GOR) with different concentration ratio.

Generally speaking, the bigger the solar irradiance is, the higher GOR is. But the influence weakens when the solar irradiance reaches a certain value. Particularly, when the concentrating ratio is relatively large, solar irradiance almost has no effect on GOR. From an overall point of view, the GOR of a single-action solar HDH desalination system is not high, generally 5 or less. This is a value in an ideal condition and that of the actual condition is much smaller than this. But in the actual system, through multiaction design, the repeating use of water vapor condensation latent heat can also increase the coefficient of performance of the system. It is worth pointing out that all the earlier derivations are based on the Majed's theory. His calculations are correct in relatively low temperature; however, when the temperature is larger than 60°C, the error will become larger and the results will be more and more inaccurate. Therefore it can only be used as a reference.

6.4.5 Conclusions

For an ideal solar HDD desalination system, the minimum work required and the maximum GOR are analyzed. It is pointed out that the minimum work required only in the HDH process is 47.89 kJ for the production of 1 kg freshwater when the ideal HDH desalination system operates at 303K. And with the increasing of the operating temperature, power consumption increases. The maximum GOR for a solar HDH desalination system is generally in the range of 1.5 and 12.0 according to the different concentration ratio. This result seems lower than that of the GOR of flash technology. For further improvement of its performance coefficient, the multistage regenerative technology can be adopted.

6.5 Heating Water Type Solar HDD System

Solar water-heated technology has made great development since water has a large heat capacity and easy-flowing characteristics; especially in low temperatures, the efficiency of the solar water-heated system is the highest. Thus for a solar humidification–dehumidification system, the first use of solar collectors for heating water to provide energy for the system is the most reasonable choice.

One of the most typical solar humidification–dehumidification desalination systems is shown in Fig. 6.23. According to experimental data, such a system's water production rate is over three times higher than the solar distiller's, generally.

When seawater entering the solar system directly is heated, and the solar collector efficiency is relatively high, the evaporation temperature of seawater entering the device can be controlled by adjusting the water flow rate. Generally speaking, the temperature of water entering the humidification chamber is higher, and the effect of the humidification–dehumidification process is better. But if so, the time on water remaining in the collector

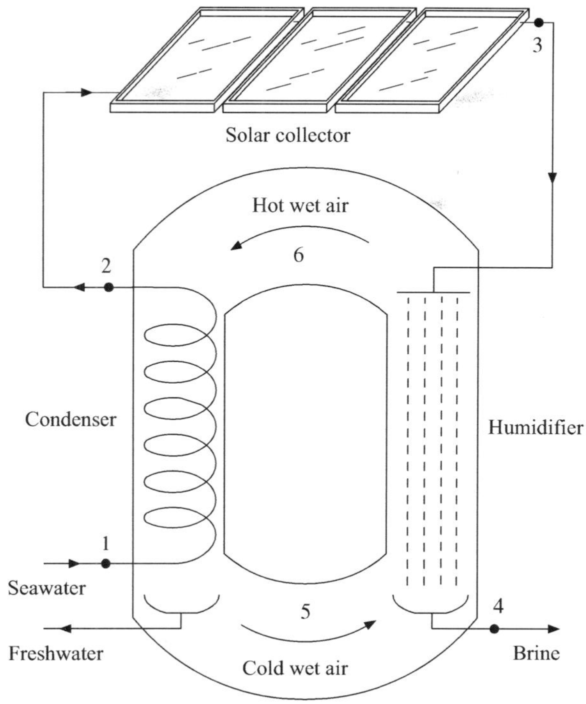

Figure 6.23
Sketch of a natural draft air circulation multi-effect humidification desalination unit.

will be longer, the water flow rate will become smaller, and when solar collectors are at a higher operating temperature, the efficiency will drop. Therefore it's necessary to choose the reasonable temperature of water entering the humidification chamber. When the humidification–dehumidification process takes place, the air is to achieve a closed cycle in the cavity, which can be a forced or a natural cycle. Thermal energy obtained from solar collectors heats the seawater and humidifies air, and the latent heat of condensation of steam is recovered in the condenser and preheats the incoming seawater. Experimental results show that such a device also has a good effect at a lower temperature.

In the all-closed air cycle of the humidification–dehumidification desalination system, the air states are experiencing two significant processes. In the humidifier, it's time when the air gets nearly saturated (generally not full saturation) through the heated humidification process at a higher temperature than when the air contains a lot of water vapor. As the air reaches the condensation chamber, it also experiences a cooled dehumidification process in which temperature is cooled to the dew point first and is then cooled to the outlet temperature of the condensation chamber along the saturated line. In the process, the moisture content of air decreases and separates out a lot of water vapor, which became low-temperature saturated steam, and then is sent to the humidifier to get on the humidification–dehumidification process, and the process cycles repeatedly.

Therefore during the air cycle process, it contains three balance equations: the quality balance of dry air, the quality balance of wet vapor, and the energy balance. They are respectively:

$$m_{a1} = m_{a2} = m_a \tag{6.52}$$

where, m_a is mass circulation rate of dry air, kg/s and subscript1、2 represent the air states of the humidification side and condensation side, respectively. The mass balance of water and water vapor is:

$$\omega_{a1} m_a = m_w = \omega_{a2} m_a \tag{6.53}$$

where, m_w is the flow rate of condensation water and w_a is the moisture content of the air that is kg (water)/kg (dry air). The energy balance in the cycle process is:

$$q + m_a h_{a1} = m_a h_{a2} + m_w h_\omega \tag{6.54}$$

$$q = m_a[C_{pm}(t_{a2} - t_{a1}) + h_{fg}(\omega_{a2} - \omega_{a1})] + m_w h_w \tag{6.55}$$

where, C_{pm} is average specific heat capacity of air and water vapor, $C_{pm} = C_{pa} + \omega C_{pv}$, C_{pa} is specific heat capacity of dry air, C_{pv} is specific heat capacity of water vapor, q is heat transfer rate, h_a is the enthalpy of air, h_ω is enthalpy of condensed water, t_a is dry bulb temperature of air, and h_{fg} is latent heat of vaporization from water evaporation. The literature [27] gives the specific enthalpy of saturated air, as expressed by the empirical formula:

$$h = 0.00585T^3 - 0.497T^2 + 19.87T - 207.61$$

The system is given by Fig. 6.23, of which the structure is simple, but heat transfer is not necessarily sufficient; it means that part of the latent heat of steam was recovered in the process. But hot saltwater, the result of being heated by solar, is discharged directly and wastes the thermal energy at the same time. To this end, Mehrgoo (2011) [28] gives a very similar device, shown in Fig. 6.24.

This system uses an array of vertical tubes instead of the condensing coil in Fig. 6.23 and porous material instead of the direct dropping of seawater in Fig. 6.23, which strengthens the process of heat and mass transfer overall in the system. The calculation shows that the effect is significantly improved after the system is improved.

Mehrgoo does simulation analysis according to design parameters from Table 6.1, giving the relationship between circulating airflow and heated water temperature and the condensed water temperature, shown in Fig. 6.25, and giving the relationship between water production and heated water temperature and the condensed water temperature, as shown in Fig. 6.26.

Results in Figs. 6.25 and 6.26 show that when the water temperature in the humidification chamber is higher, the water production rate will be higher, the same as the greater volume

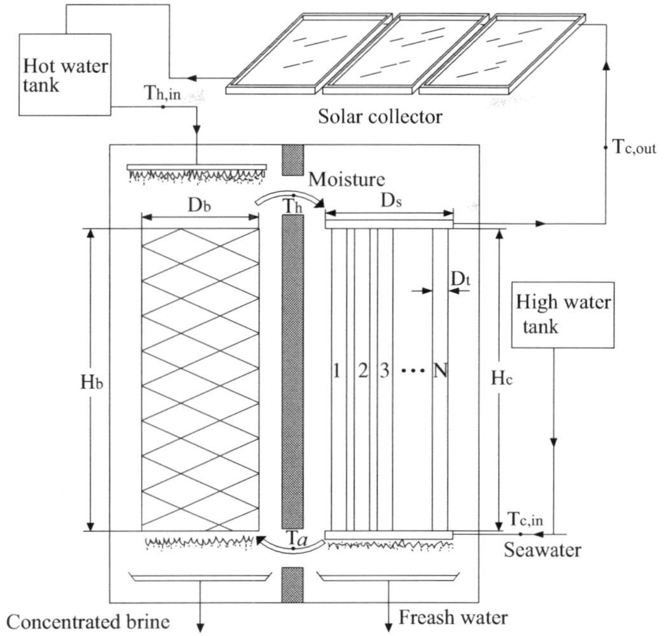

Figure 6.24
Sketch of advanced solar humidification–dehumidification desalination system.

Table 6.1: Values of design and operating parameters.

Project name	Design parameters
Inner diameter condenser tube/D_i	0.35 m
Humidifier height/H_b	3.5 m
Condenser height/H_c	2.8 m
Total volume/V	3 m^3
Number of condenser tubes/N	100 ind.
Air relative humidity in condenser and humidifier/φ	100%
The thermal conductivity of the condenser tube/K_w	0.04 kW/(mK)
Temperature of water entering the humidifier	50–90°C
Inlet temperature of cooling water	5–25°C
Temperature of air entering and leaving the humidifier	19–23°C

of circulating wind. Additionally, they are basically a linear relationship. It also instructs that when the water temperature is lower, the water production rate and volume of circulating wind required are also higher. Of course, it is ensured that water temperature into the humidification chamber, which is called operating temperature, is constant. Generally, the seawater temperature entering the device is lower, which leads to the lower temperature into the solar collectors. The water temperature from the solar collectors will be lower at this time, and the hot water temperature into the humidification chamber will

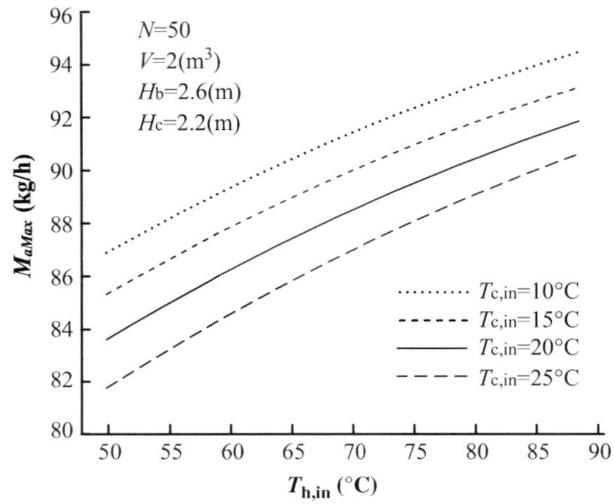

Figure 6.25
Variations of maximum air mass flow rate versus inlet cold and hot water temperatures for $N = 50$, $V = 2$ m, $H_b = 2.6$ m, and $H_c = 2.2$ m.

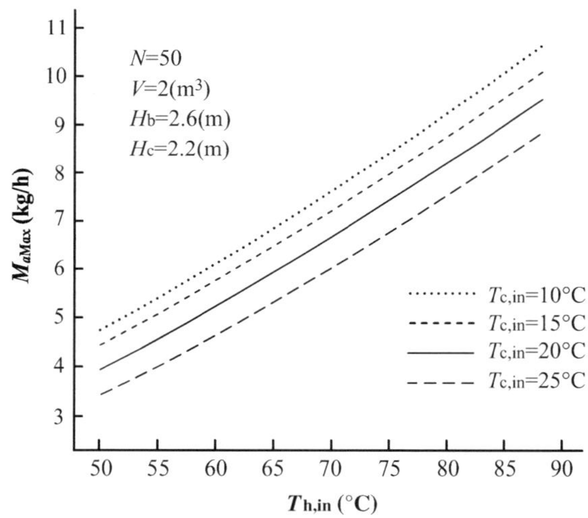

Figure 6.26
Variations of maximum water production flow rate versus inlet cold and hot water temperatures for $N = 50$, $V = 2$ m, $H_b = 2.6$ m, and $H_c = 2.2$ m.

be reduced correspondingly; thus we cannot guarantee desalination at the same operating temperature. Therefore the cooling water temperature and flow rate of air circulation require being optimized in order to obtain the best benefit. At the same time, we must note that the normal collector efficiency of the solar collector will decrease as the collector temperature increases. Especially the efficiency of the flat-plate solar collectors with enhancing operational temperature decreases very fast, so the system should not set to operate at a toohigh temperature, generally choosing to work in 70°C annex. For all-glass vacuum tube collectors or heat pipe vacuum tube collectors, we can select the higher operating temperature, which is also particularly suitable for the collector system of heated water.

Although the structure shown in Fig. 6.23 is compact, it's difficult to be used in a large device. So the humidification chamber and the dehumidification chamber are separated into two independent components generally, which connect to each other with pipes, shown in Fig. 6.27. Fig. 6.27 also adds a water tank of preheating or heat storage for regulating the unevenness of solar irradiance. In order to enhance the condensation process in the condensation chamber, the heat exchangers may have a variety of forms, such as the type of coil shown in Fig. 6.27, horizontal row tube, shell tube, and horizontal tube with fin. All in all, the purpose of it is increasing the contact area with humid air. The structure in the humidification chamber also has various forms, which are cream type and a type of vertical tube, and it is filled with a porous material, for example, porous paper core, plastic pellets, and gravel, etc. Only one purpose is to increase the heat transfer area between water and air and contact sufficiently between the incoming air and water to achieve the heated humidification process of air. The circulation of air is forced to carry on by outer-connecting pipes and fans, and how to reduce wind resistance and save fan power become very important.

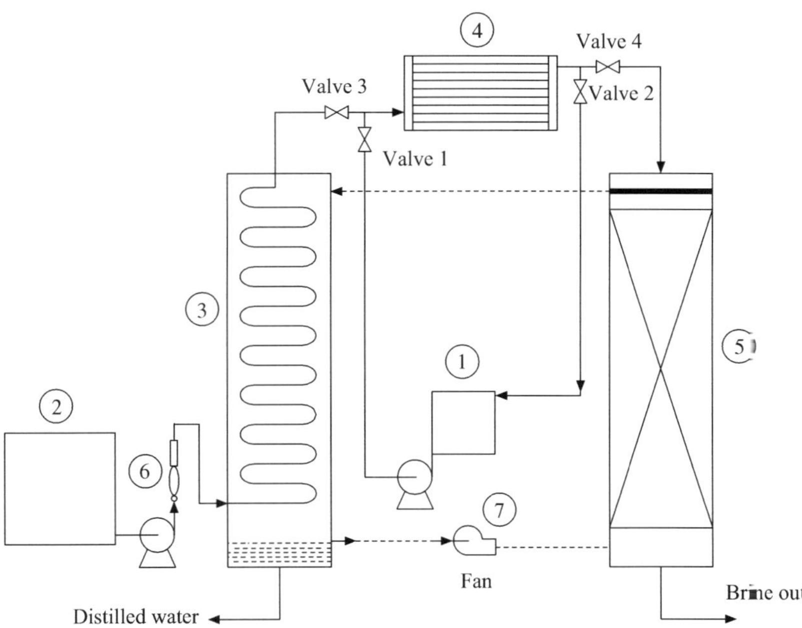

Figure 6.27
Sketch of humidifier–dehumidifier separated desalination system device.

Seawater entering the solar collectors will cause collators to be corroded. Most large-scale systems use the technology of heat transfer indirectly, which first heats in the solar system with a noncorrosive working medium such as water and antifreeze and then uses the hotworking medium to heat seawater. Thus seawater will be confined to a small area, reducing the corrosivity of seawater to the system.

The number of hot brine discharge of this type of system is relatively large, and the temperature of brine excluded is also high. How to recover this sensible heat is also worthy of attention; someone has put forward a program that uses the heat pump to recover this part to reenergize their vitality, but it needs to be considered from an economic standpoint.

In order to verify the long-term performance of such a solar desalination system, Adel M. (2009) actually built a solar desalination system with an isolated humidifier and dehumidifier, taking a long-term test under three different operating modes. Adel M.'s systems and three models are shown in Fig. 6.28 [29].

In Fig. 6.28, System I takes an operation mode with indirect heat transfer and adds an auxiliary heater to reduce the heat exchangers, System II uses the mode of indirect heat transfer, and System III takes the direct heating mode. For devices in Fig. 6.28, the structure parameters are as follows: area of the flat-plate solar collectors is 8.9 m^2, placed south, tilted 30°. The solar collector is connected to the water tank with heat preservation, which has a volume of 200 L. In System I and System II, a connecting solar water tank with desalination systems is an immersion coil heat exchanger. The power of the auxiliary system of the electric heater is 6 kW and the capacity of the electric heating tank is 200 L; it immediately stops heating until it is heated to 70°C. Sea spray and the desalination chamber is $186 \times 118 \times 160$ cm^3. A condenser made of galvanized iron has a 5-cm thickness and an area of a 1 m^2 box. The storage tank joining seawater is placed at a position where it's higher than the condensation chamber by 2 m. When the device operates, the seawater entering the device will be firstly preheated. Then it will mix with some brine and be sent to a solar heating system which is good for recycling a part of waste heat from salt water.

In such three systems, since the heat exchange process is different, there is a large difference in heat transfer thermal resistance, which leads to water production efficiency of the different system. Figs. 6.29 and 6.30 give the changes of operating temperature and water production under the similar weather with time respectively. Experimental results show water production characteristics under these three operating modes. Due to the electric auxiliary heating, temperature from the first operation mode is stable and the water production process is relatively steady, the degree of declining is not big (although it falls in the afternoon), and water production remains between 0.7 and 0.5 kg/h basically. The second operation mode with increasing heat exchangers and without

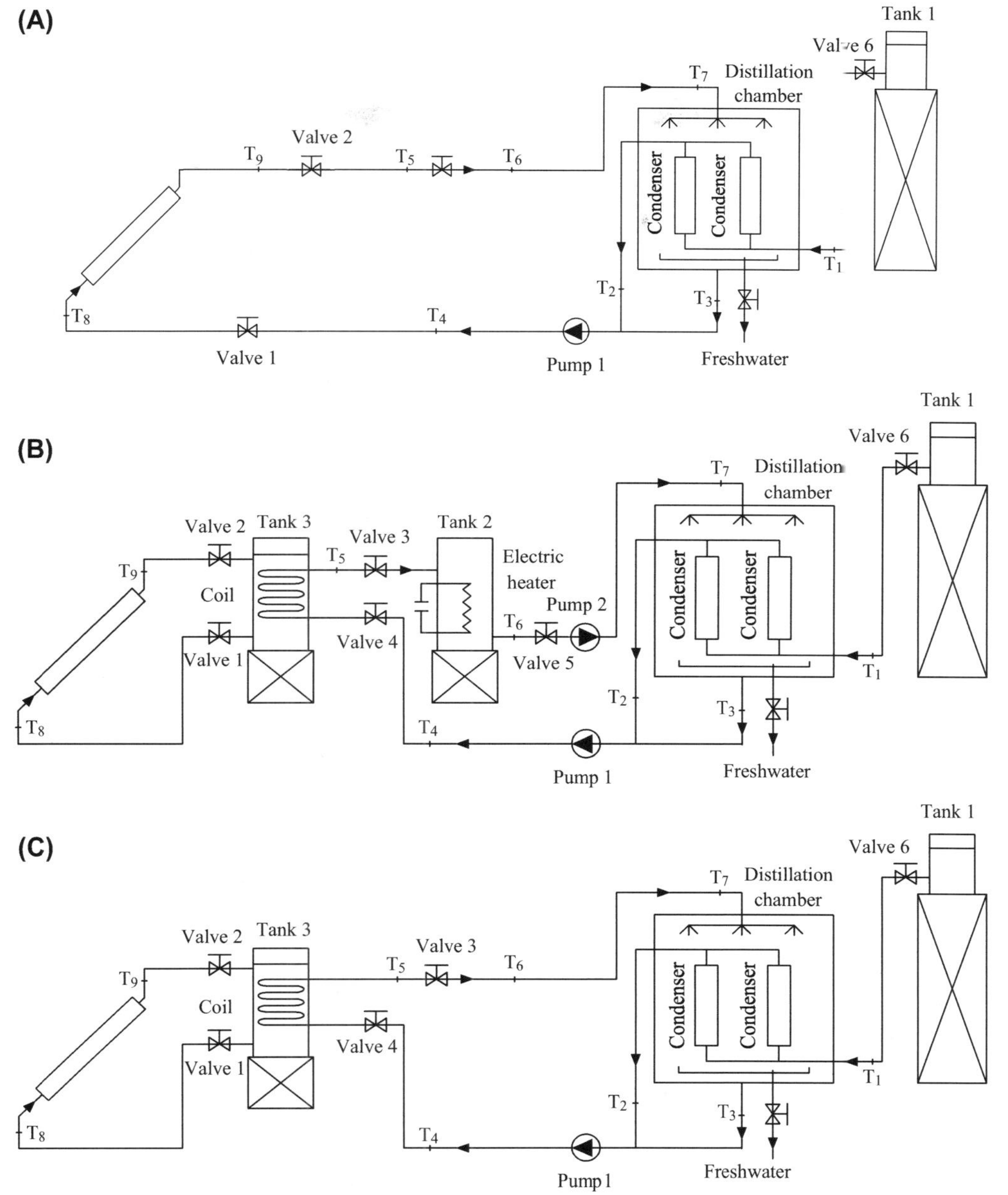

Figure 6.28
The operation route of a solar desalination system: (A) System I; (B) System II; (C) System III.

auxiliary heating, which the temperature changes as solar irradiance changing and variation range, is larger. When the test starts at 10:00, water production increases basically from zero and decreases rapidly after reaching the maximum, reducing to zero substantially at 18:00 to the end. The third model without the heat exchanger uses the direct heat transfer mode and increases the heat transfer efficiency of the system.

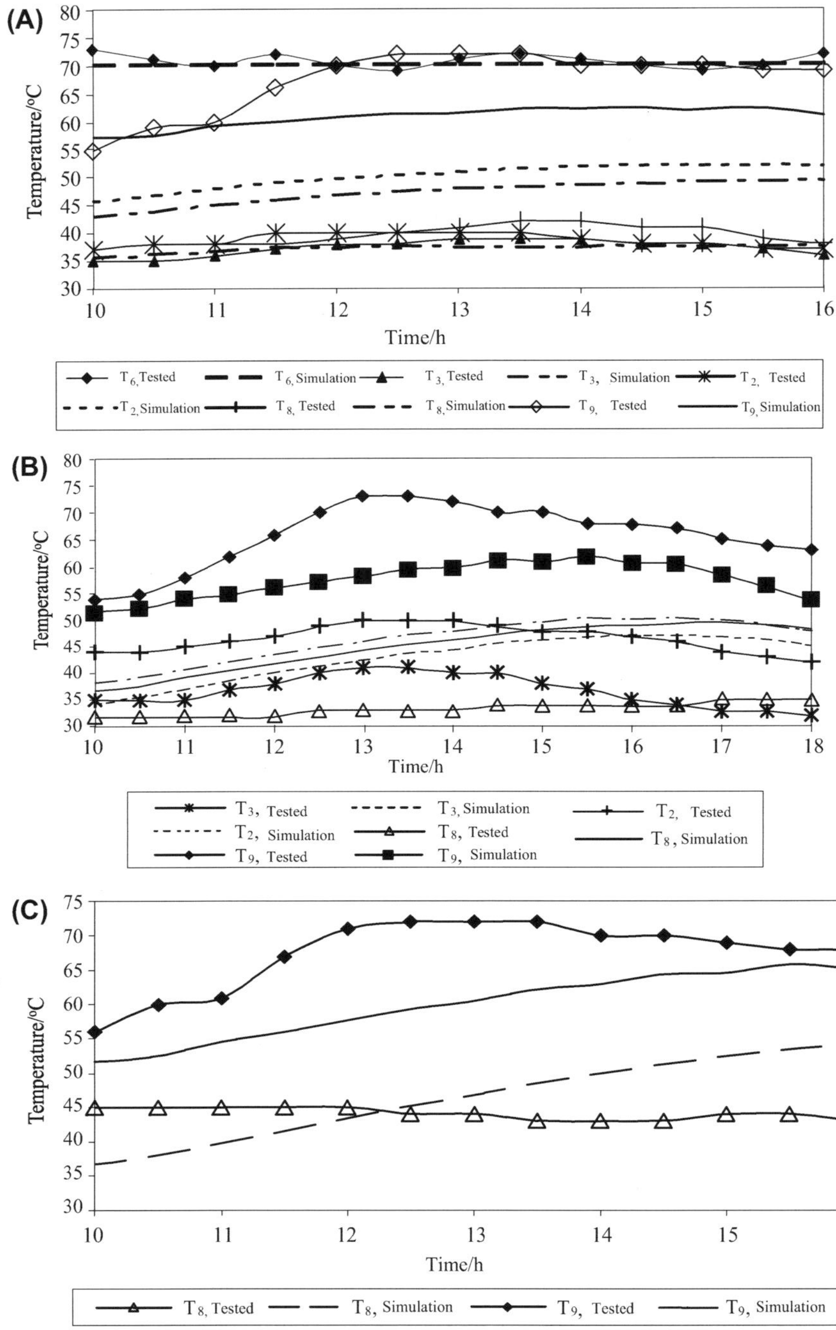

Figure 6.29

Operation temperature comparison of three kinds of operation modes: (A) experimental and simulation results of the System I; (B) experimental and simulation results of the System II; and (C) experimental and simulation results of the System III.

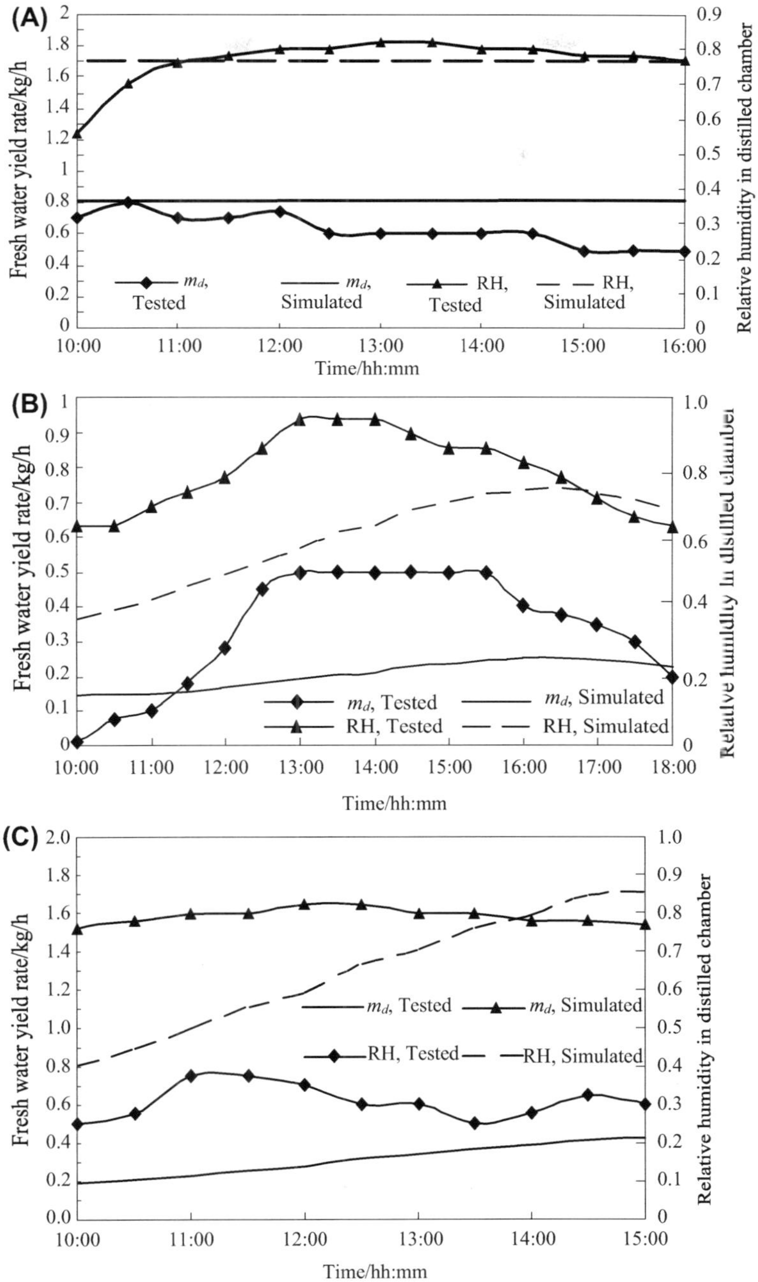

Figure 6.30

Water productivity comparison of three kinds of operation modes. (A) System 1; (B) System 2; and (C) System 3.

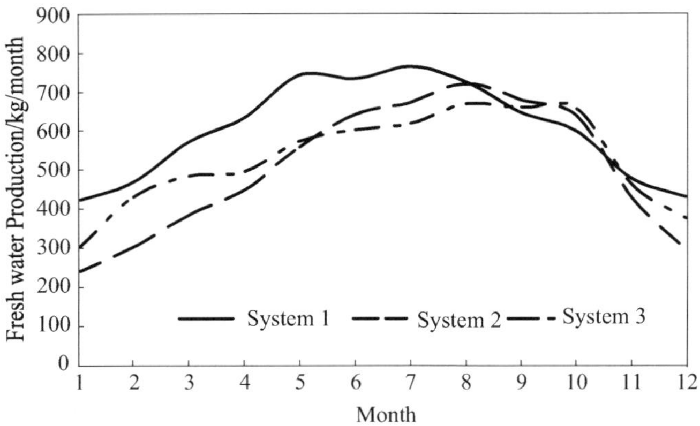

Figure 6.31
Variation of water productivity in three models *r* with the month.

Therefore there is freshwater output shortly after the experiment beginning at 10:00, then it rises. But the rate of increase is not fast, and the speed of decrease after reaching the maximum is not fast either. It still has the high water production until the end of the experiment at 18:00. It shows that the way of direct heat transfer has the highest efficiency than those two modes.

Changes in annual water production with the month are shown in Fig. 6.31. From the figure, the first model has the largest water production, especially in winter with the insufficient sun, along with more advantages. For the second way, water production is greater than the third mode in the summer when the sun is very adequate for a short time, but is not stronger than the third mode in the season of weaker sunlight. Hence, as long as conditions permit, we can add the auxiliary heating system or else use the way of heating seawater directly.

6.6 Heating Air Type Solar HDD Systems

Solar air heating technology has made considerable progress. With the characteristics of high collection efficiency, no need to prevent frosting, insignificant leaks without affecting the efficiency, and low corrosion problem, the solar air collector is more appropriate to be applied to the humidification–dehumidification desalination system. However, there are still some flaws in this kind of system. First, the small heat capacity of air and the limited energy carried by air requires a large amount of circulated air, which increases electric power requirements of the fan. Second, without energy storage in solar air collectors, the system is greatly affected by weather conditions. Furthermore, the outlet temperature of the solar air collector is not so high that it limits the operation

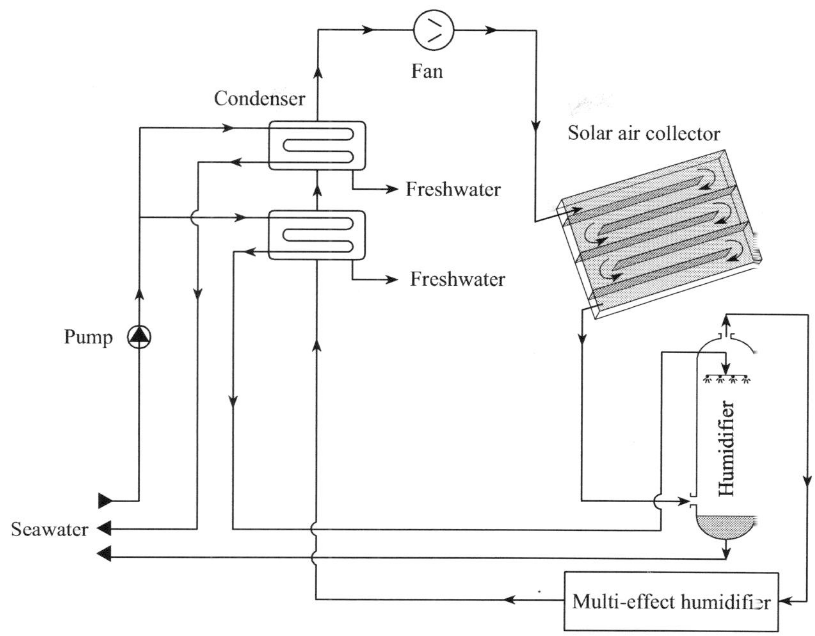

Figure 6.32
Solar desalination system with multistage humidification and dehumidification processes [30].

of the entire system in a high temperature range. The concentrating air collector emergence makes it possible to raise the outlet temperature of the air up to about 140°C, greatly improving the operating temperature of the heating air type humidification–dehumidification desalination system.

The most common solar heating air humidification–dehumidification desalination system is shown in Fig. 6.32. The system still contains four major parts: solar air collector, humidifiers, dehumidifiers, and fan and piping systems.

Usually, the heating air type solar desalination system uses the closed air circulation. In the whole cycle, the total amount of dry air remains constant, while the state of the air changes continuously.

In the solar collector, the air undergoes a pure heating process. After entering the humidifier, it undergoes an adiabatic humidification process and its temperature then decreases, but the humidity or moisture content will increase, nearly reaching saturation. At this time the air is sent to the dehumidifier to be cooled and dehumidified. After the dehumidification process the air is supplied to the solar collector to be heated to repeat the same process. For a humidification–dehumidification solar desalination system in which the air goes through single process in each component, the state change process of the air

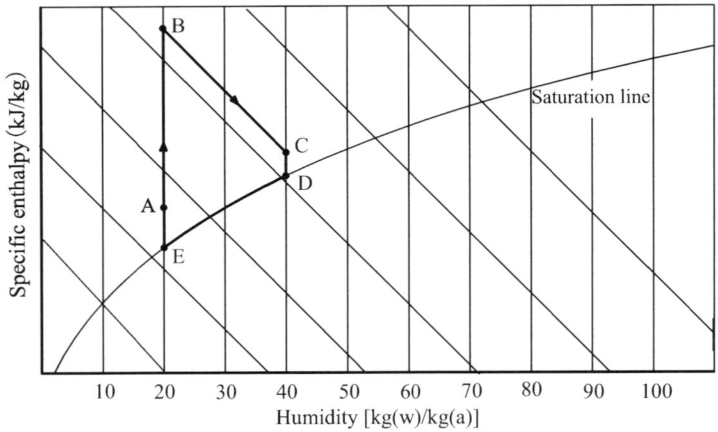

Figure 6.33
Variation of air state in a single humidification–dehumidification process.

in a loop process is shown in the lines of Fig. 6.33. The low temperature and low moisture content in the air starts from point A, then enters the solar air collector to be simply heated to point B, and then is transferred to the humidifier for humidification, which is an adiabatic humidification process, that is, the enthalpy of moist air results from the decreased air temperature to point C. C is close to the saturation point, but it is not the complete saturation state. At this moment the air is fed into the dehumidifier to be cooled and dehumidified. Having gone through C to D and then D to E, in which the dehumidification process is completed, the air is sent to the solar collector again to be heated to start another cycle.

However, for a single humidifying process, humidity or moisture content often increases for a limited range, and the amount of water carried by the air is not too much. Therefore in order to increase the moisture content per unit mass of air and reduce the amount of circulating air, the multistage humidification mode is often adopted in the humidification process. The case is the same in the dehumidification process. For a single condenser, the condensation process may not be fully completed. So the multistage structure condenser is usually used to reduce the circulating airflow rate and to improve system efficiency. Fig. 6.33 shows a solar heating air type seawater desalination system with a multistage process of humidification. The variation of enthalpy with moisture of the air in the psychrometric chart is shown in Fig. 6.34 (including an eight-stage humidification process).

The heating and humidification process can be explained as follows: the low temperature and low moisture content air starts from point A and enters the solar air collector to be simply heated to point B, and then it is transferred to the humidifier for humidification. It

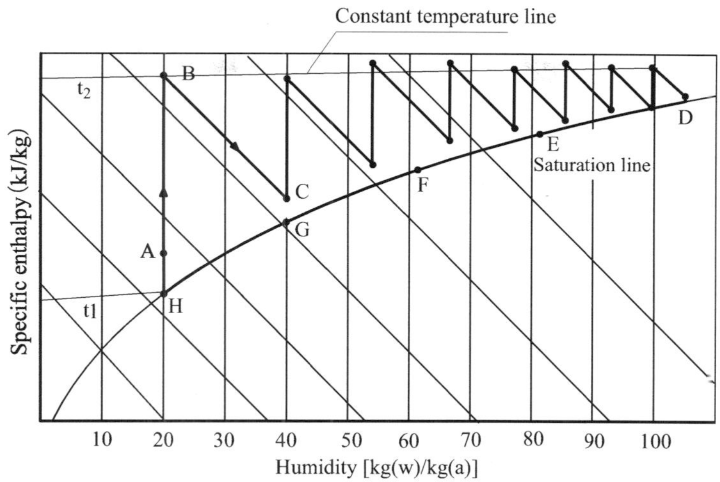

Figure 6.34
Variation of the air state in a solar system with eight stages of the humidification–dehumidification processes.

is an adiabatic humidification process, that is, the enthalpy of moist air is resulted from the decreased air temperature to point C. C is close to the saturation point but is not completely saturation state. At this time, the air is fed into the solar air collector of the next stage for reheating. After elevating temperature, it then enters the humidifier for humidification to repeat the process of A-B-C until it reaches the last stage. It is fed into the dehumidifier for dehumidification. The dehumidification process is started from the point D and is finished when it returns to H, in which the process goes through E, F, and G and finally completes the whole cycle.

Fig. 6.35 shows the flow diagram of a four-stage humidified solar heating air type humidification–dehumidification desalination system. Fig. 6.36 is a photograph of its actual system. In this system, the solar air collector is divided into five sections, with each section heating the air, so that temperature is rising continuously until it goes into the humidifier. The area of the first and second stage solar air collector is 44.1 m^2 respectively. The total area of the third and fourth stage collector is 45 m^2 and the fifth stage area is 27 m^2. When the air is heated, it has also been humidified to raise its moisture content. In this way the temperature and moisture content of the air are elevated at the same time. Taking into account the recovery of sensible heat and considering that it is difficult to extract water exactly from the moist air by only passing through one dehumidifier and only undergoing one time condensing process, four dehumidifying chambers were designed for dehumidifying at different cooling temperatures. It can be said that such a system is a relatively comprehensive planed one,

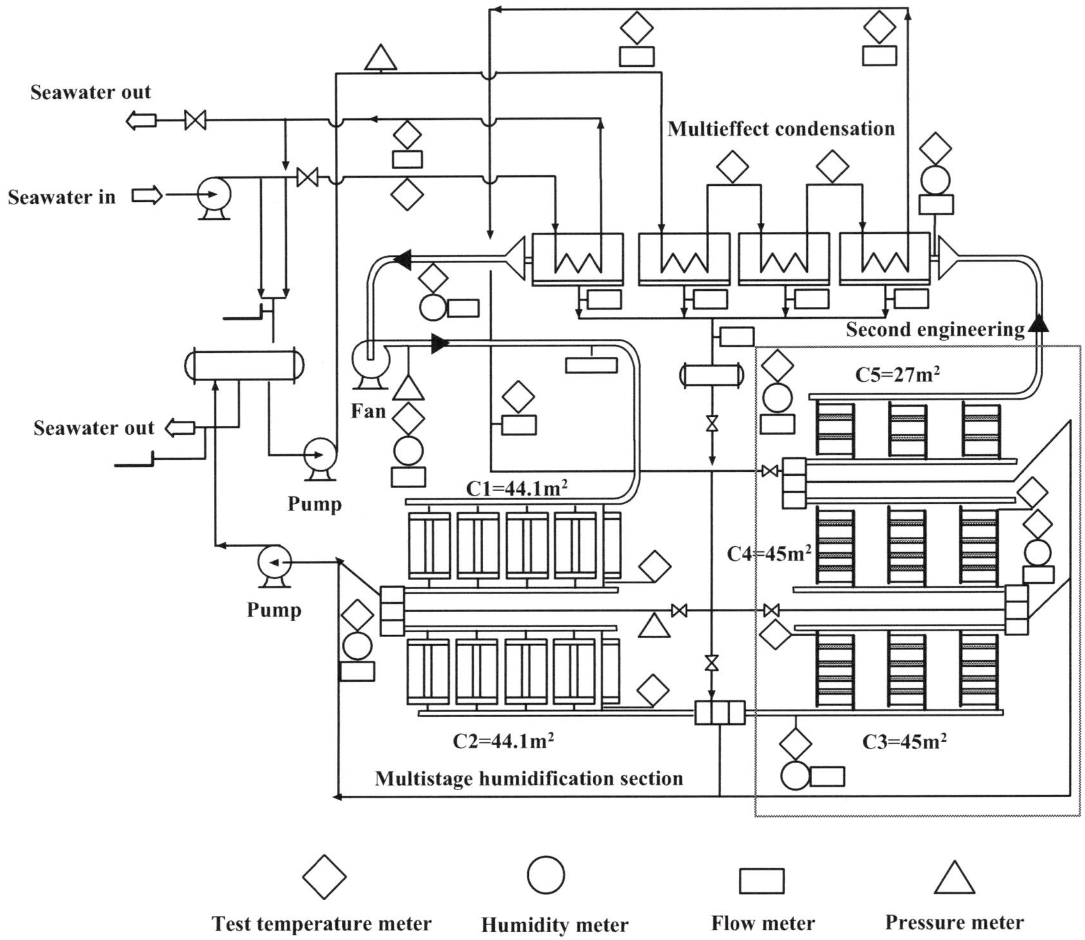

Figure 6.35
Flow chart of the solar heating air type desalination system with four stages of humidification and multieffect condensation processes.

and its heat recovery rate is relatively high, therefore the water production efficiency of the system is high when it is compared with others. Fig. 6.37 shows the total amount of water production per day and the cumulative value of solar radiation in June in 2003. (Note: The statistic provided is only the heating result of the first and second stage solar air collector, while the third, fourth, and fifth stage collectors had not been installed.) As it can be seen, the water production per day of the system is between 250 and 400 L, and the maximum water production is about 430 L. As a result, the water production rate per square meter solar collector reaches 4.25 L/m^2, and the maximum water production rate is about 4.9 L/m^2. Related operation parameters are as follows: circulating dry air mass flow rate is 1500 kg/h; seawater flow rate into the humidifier is 1000 kg/h; flow

Figure 6.36
Solar heating air type desalination system with four stages of humidification processes [30].

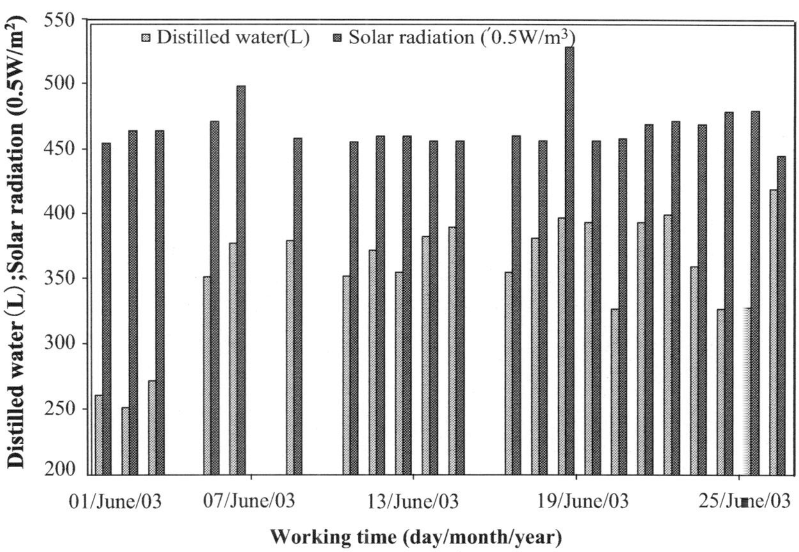

Figure 6.37
Distilled water production and solar radiation in June [30].

rate of cooling water through the condenser for heat recovery is 1000 kg/h; flow rate of seawater entering the condenser for cooling is 3000 kg/h; and ambient temperature is between 30 and 37°C.

Under the similar operating parameter conditions, average daily water production and cumulative value of solar irradiation from March to August are shown in Fig. 6.38. The result indicates that solar heating air type humidification–dehumidification desalination system relates closely to the solar irradiance values. When the system operates in sunny weather, the system performance will be better.

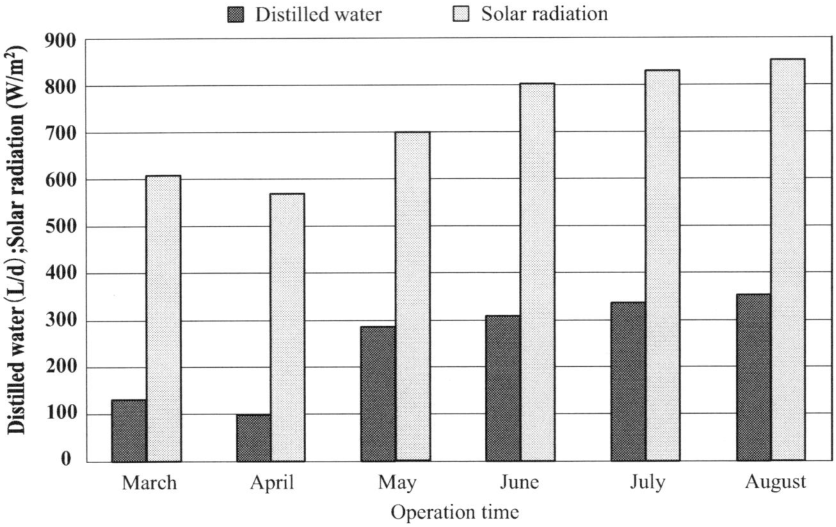

Figure 6.38
The average daily yield and average solar irradiance per month from March to April [30].

6.7 Solar Water and Air Heating Compound HDD System

The earlier discussion tells that the solar water heating HDD system has the characteristic of high heating efficiency. However, the solar air heating HDD system has the advantage of high humidifying efficiency. Therefore the combination of the two can play their common characteristics.

The simplest and oldest solar water and air heating compound HDD system is shown in Fig. 6.39. It is similar to an integral solar water heater and also to a solar basin still. Its characteristic is that the air can realize the natural circulation in the interior. Furthermore, the condensing latent of the vapor can be recovered partially because partial freshwater will be condensed on the back of the basins, in which a portion of condensing latent is sent back to seawater in the basins.

The theoretical analysis of Hassan E.S indicates that the efficiency of this system reaches 56%. Under the conditions of a sunny day, the freshwater production reaches 5.2 kg/m^2, in which the production on glass cover is 2.2 kg/m^2 and on the back of the basins is 3.05 kg/m^2. It indicates that this kind of system is worthy of further study.

K. Zhani (2010) provided a heating water and air compound type HDD system, as shown in Fig. 6.40. The humidifier and dehumidifier are set apart, and the dehumidifier can recover part of the vapor latent. The area of solar air collector is 16 m^2 and the area of solar water collector is 16 m^2. The volume of the humidifier and dehumidifier is $0.6 \times 0.8 \times 0.56$ m^3 and $0.88 \times 1.5 \times 2.07$ m^3, respectively. K. Zhani tested the system

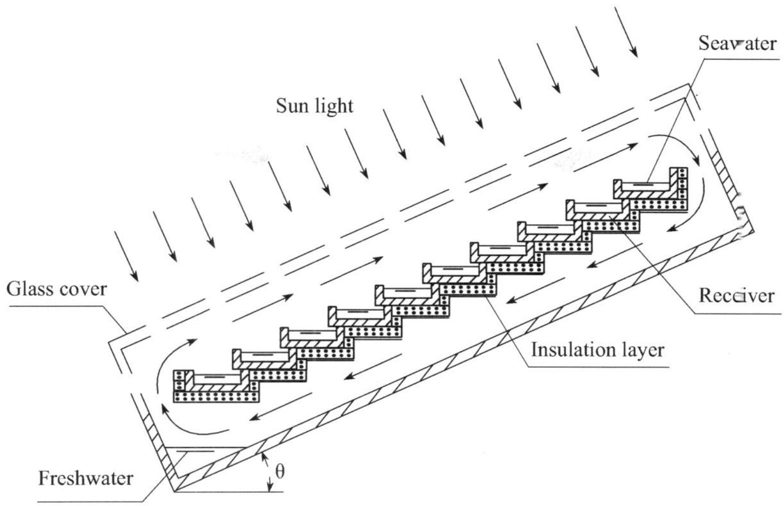

Figure 6.39
Slop basin solar still with recovering latent function [31].

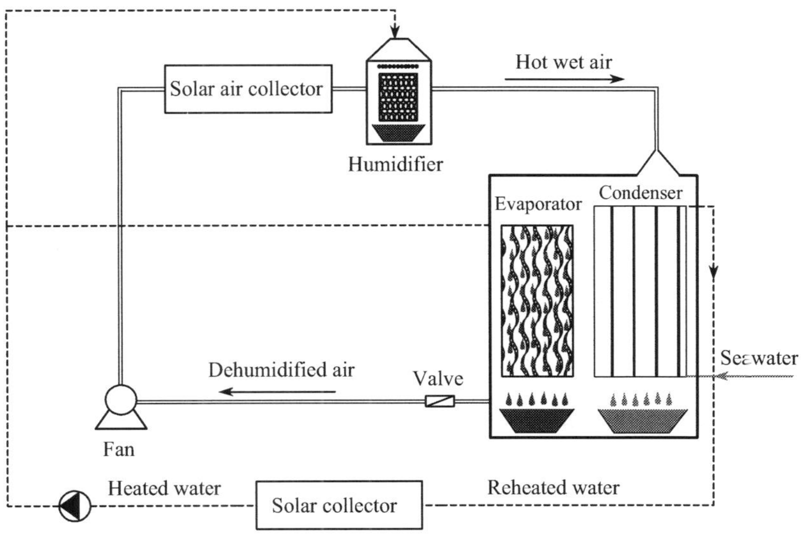

Figure 6.40
Heating water and air compound type HDD system [32].

and reported the results. Somehow, the production of this system is only about 20 kg per day.

The literature [33] introduced a scale compound system, as shown in Fig. 6.41. Fig. 6.42 is its practical operation diagram. Figs. 6.43 and 6.44 are the photos of the solar air collection system and the HDD system, respectively. The main parameters of the system are listed in Table 6.2.

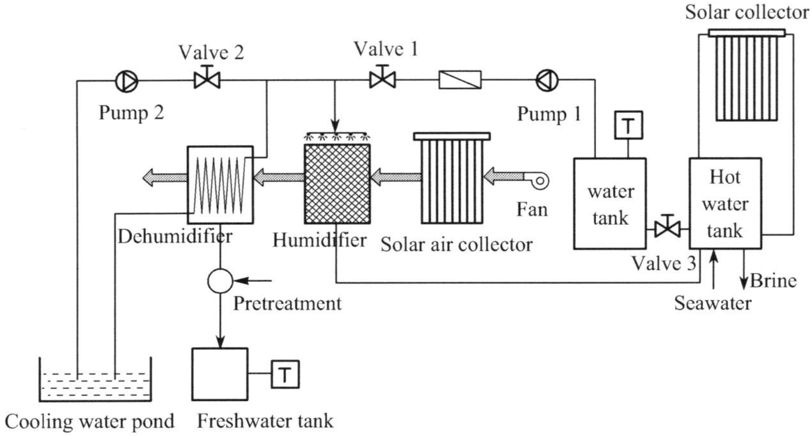

Figure 6.41
Hot water and air compound HDD solar system [33].

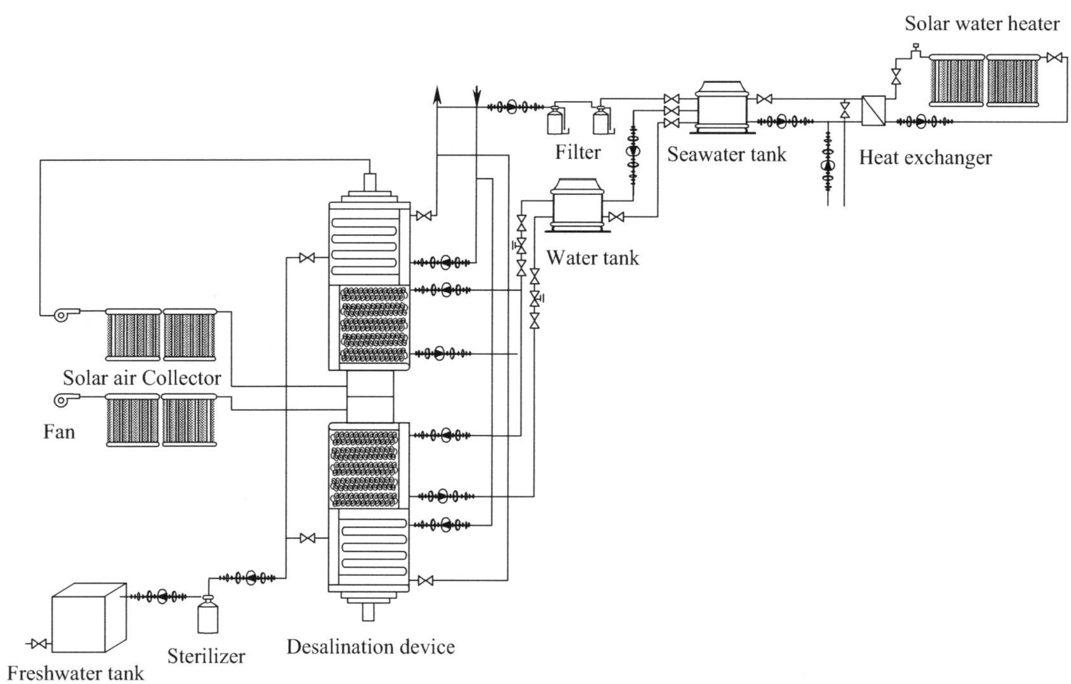

Figure 6.42
Diagram of the actual system.

Figure 6.43
The photo of the actual solar air collection system.

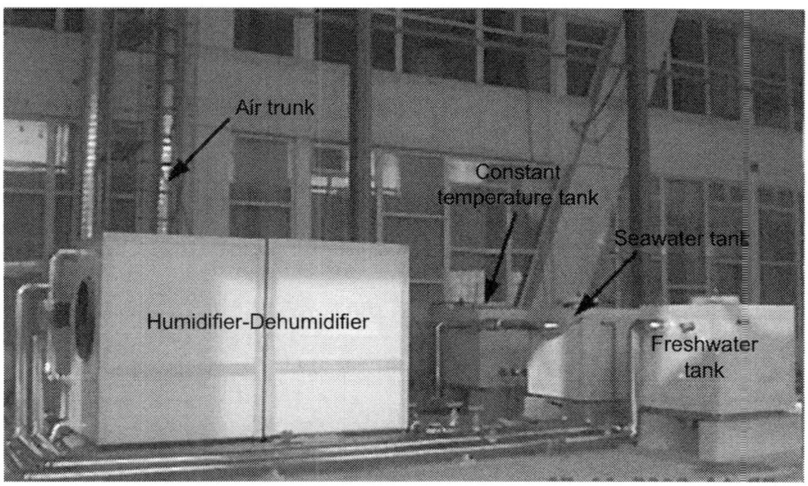

Figure 6.44
The photo of the dehumidifier of the actual system.

Table 6.2: Detail and design parameters of main equipment.

Equipment name	Design parameters
Solar air collector	100 m^2
Solar water collector	14 m^2
Humidifier–dehumidifier integration	Designed productivity 125 kg/h
Hot water tank	Designed capacity 1 ton
Constant temperature water tank	Designed capacity 1 ton
Freshwater tank	Designed capacity 1 ton
Quartz sand filter	Processing capacity 0.5 m^3/h
Activated carbon filter	Processing capacity 0.5 m^3/h
Ozone sterilizing device	Processing capacity 0.2 m^3/h

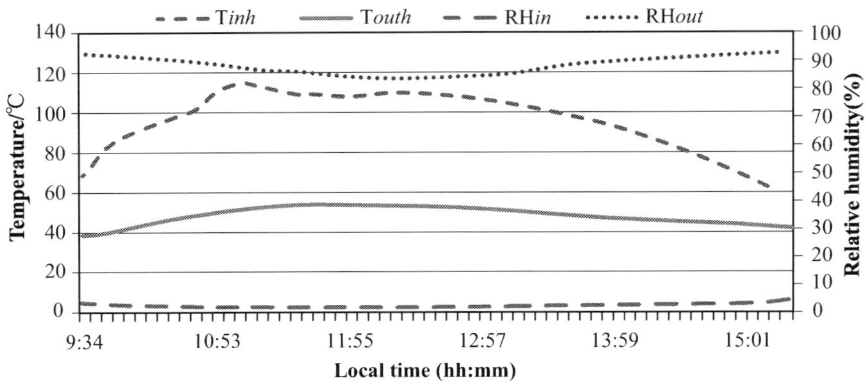

Figure 6.45
Variations of the temperature and relative humidity at input and output of the humidifier.

The measured results indicate that the output temperature of the solar air collector reaches 118°C when solar radiation is 760 W/m^2. However, solar radiation has quite the influence on the output temperature. Sometimes it is only 65°C, which will directly affect the production of the compound HDD system. The temperature change trend at the output is the same as that of solar radiation, maybe a little less.

Fig. 6.45 shows the air temperature and relative humidity at output of the humidifier on a typical day. It is easy to find that after heating by the solar air collector, the relative humidity of the air at the input of the humidifier is generally low, only about 1%. However, the relative humidity of the air at the output of the humidifier will reach 80–90%. The output temperature is in the range of 40–55°C.

The freshwater production of the system was measured continuously under the conditions of a sunny day or a basic sunny day. Fig. 6.46 shows the situations of the freshwater production and solar radiation on several typical days. It can be found from the experimental results that solar radiation has a very important influence on the production. The production is only 300 kg/day when the average solar radiation is about 300 W/m^2. However, the production reaches more than 500 kg/day when the average solar radiation is more than 500 W/m^2. The production even reaches more than 600 kg/day under sunny days. The production decreases rapidly under the cloudy days.

6.8 Multieffect Humidification–Dehumidification Solar Desalination System

Because the solar system has two weaknesses, namely, low water-making ratio and small scale, it has been segregated by the traditional seawater desalination system. Therefore how to improve the water-making ratio and water production per unit area for the solar

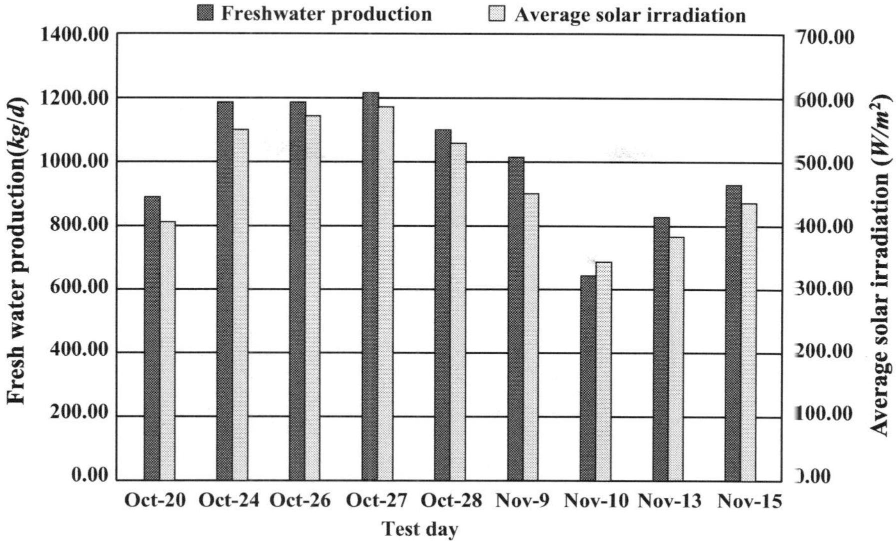

Figure 6.46
The production and average solar radiation on several days.

energy system has always been an important issue. Based on the thermodynamic analysis of the humidification–dehumidification solar desalination system, it is not difficult to find that when the system works, there is thermodynamic irreversible loss in the system, especially under large temperature difference. For reducing the irreversible thermodynamic loss, some researchers have put forward a multieffect humidification and multieffect heat recovery system.

For example, a four-stage multieffect humidification–dehumidification system given by ImedHoucine is a typical multieffect process [30]. For making humidification fuller and dehumidification more thorough and reducing the irreversible thermodynamic loss, four pad humidifiers shown in Fig. 6.47 have been set up in different parts of the system. Each humidifier has a different humidification capacity and different temperature of the seawater for humidification, which can avoid the large temperature difference of heat transfer and the thermodynamic loss.

Similarly, in the dehumidification process, ImedHoucine also gives multieffect process arrangement that the dehumidification chamber is divided into four parts, and the temperature for cooling water in every dehumidification chamber is also different. One of the operation processes can be shown in Fig. 6.48. It shows that the temperature of the moist air gradually decreases with the moist air passing through different parts in the system, namely, the first-level temperature is reduced by 25°C, the second by 4°C, the third by 4°C, and the fourth by about 22°C. In contrast, the temperature of the cooling water gradually increases with the increasing level, namely, the temperature of cooling

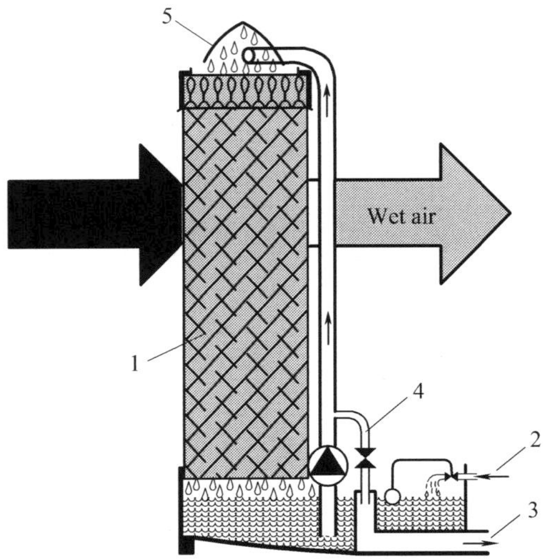

Figure 6.47
Schematic sketch of a pad humidifier: 1, humidifying pad GLASdek or CELdek; 2, water inlet; 3, drain; 4, bleed-off; 5, water distributor.

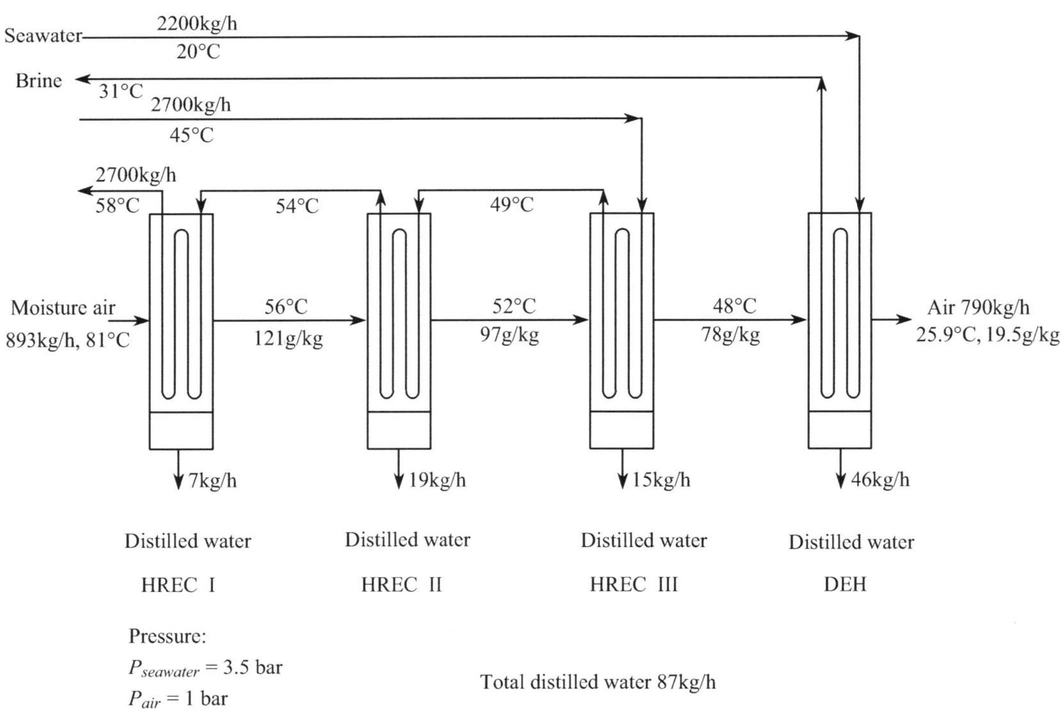

Figure 6.48
Heat recovery and dehumidification system.

water entering the fourth level is 20°C, the third 45°C, the second 49°C, and the first 54°C. With each level being separated, it is good for reducing the system heat transfer temperature difference. However, it will also bring a disadvantage that fan power and pump power will increase, and at the same time, because the system structure become more complex and parts increase, the manufacturing cost of the system will increase. Considering the profit of the enterprise, it is worth controlling the system economy when the increase in the cost of the device must be offset by the increase in water production.

The humidification–dehumidification solar heating system can also adopt multieffect humidification and multieffect heat recovery, which can improve efficiency and the water production ratio of the system. For example, the literature [6] has put forward a two-stage multieffect humidification–dehumidification desalination system, shown in Fig. 6.49 and given the theoretical analysis of the system. The principle of this system is as follows: the humidification–dehumidification chamber is divided into two stages. In the high stage, the hot air with a slightly higher temperature exchanges heat with the slightly higher temperature cooling water. At the same time, in the low stage the hot air with a slightly lower temperature exchanges heat with the slightly higher temperature cooling water. Based on this principle, the heat transfer under the large temperature difference can be avoided, which is good for reducing heat transfer irreversible loss.

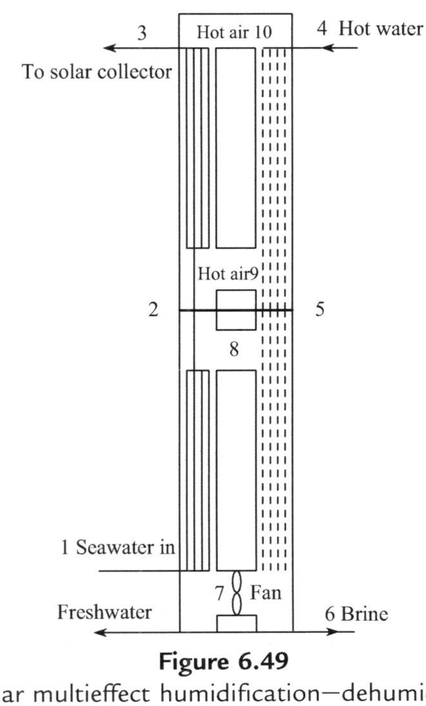

Figure 6.49
A sketch of the two-stage solar multieffect humidification–dehumidification desalination process.

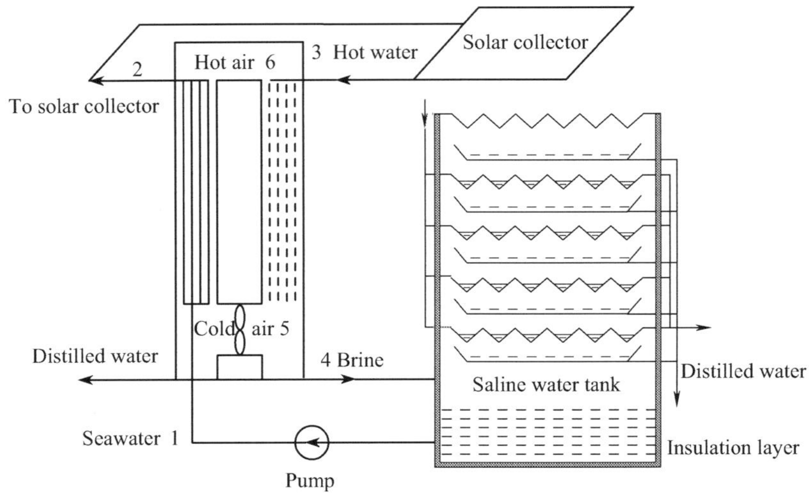

Figure 6.50
Sketch of a hybrid solar desalination process of the multieffect humidification–dehumidification and basin-type unit [34].

Because of the high temperature of the concentrated brine discharged from the traditional humidification–dehumidification desalination system, it contains a certain amount of sensible heat. How to recycle this part of sensible heat and improve the system efficiency are the matters of interest to people. In the literature [34], the idea of the combination of the humidification–desalination system and the traditional multistage tray solar distillation system is proposed, which is shown in Fig. 6.50. The left part in Fig. 6.50 is a traditional humidification–dehumidification desalination system, and the right part is a stacked tray solar still composed of many stacked trays. According to the analysis, the system can improve the total efficiency of the system by using the residual heat of the multistage stacked tray distillation system to produce some freshwater.

With the solar system mainly operating in the day, the temperature of the brine excluded from the system is higher in high environmental temperatures. In the evening, the temperature of the environment will become low, which will enlarge the temperature difference between the inside and the outside of the stacked tray distillation system. Under the high temperature, the stacked tray distillation still can use the surplus heat to produce freshwater, which is one of advantages of the stacked tray distillation still.

In short, the purpose of taking the multistage technology mainly is to reduce the thermodynamic loss in the heat transfer process and improve the quality and quantity of heat recovery. In the present advanced large-scale solar desalination system, the heat recovery and multieffect technology must be considered. The reason for this is that the cost is huge and the ability of collecting heat is limited in a solar heat collection process. How to make full use of the heat energy that has been obtained and to improve the system economy has become an important problem that every designer has to consider.

6.9 A Closed Circulation Solar Still With Enhanced Falling Film Evaporation and AirFlow Absorption

Just like the discussion in the earlier chapters, there are three main directions to improve the productivity of the stills: (1) reutilizing the latent of the vapor condensation; (2) enhancing the heat and mass transfer processes in the stills; and (3) decreasing the heat capacity of the seawater in the stills. In the modern desalination industry, many methods based on this idea have been investigated. Some of them have been applied in actual engineering. However, more research should be done for these technologies to be applied in the solar field.

The solar still system presented in this section not only reutilizes the vapor latent many times, but also decreases the capacity of seawater in the system. At the same time, it can also actively enhance the evaporation and condensation processes in the still.

As we know, although the passive enhanced technology also can improve the evaporation process from 50–300%, such as adding channels, convex edges, or cold fingers and so on, on the vertical plate, vertical tube, and horizontal tube, the effect is not so remarkable as in the condensation process. The reason is that the heat resistance from the evaporation process and condensation process is different. The heat resistance of the condensation process comes mainly from the conduction resistance of film thickness of the condensed liquid. However, the heat resistance of the evaporation process not only comes from the film thickness, but also comes mainly from the saturation vapor pressure on the boundary between gas and liquid. If the vapor cumulates continuously on the boundary, the vapor pressure on the boundary will increase step-by-step so that the vapor driving force becomes weak. Finally, the evaporation process will stop when the cumulated vapor reaches to some degree. Therefore for enhancing the evaporation process, it is very important to draw out the evaporated vapor in a timely manner, besides decreasing the thickness of liquid film. For that, active enhancement technology should be used. For example, to use mechanical vibrance, drawing out the gas, ultrasonic vibration, and so on, the evaporation rate can be improved more than one time. Especially, the drawing method has a very good effect and is used widely in the modern seawater desalination industry. In vapor compress and multistage boiling processes, the vapor should be drawn out in a timely manner. So it is very significant to use this kind of technology in the solar desalination field.

6.9.1 The Energy and Mass Balance in the Closed Circulation Solar Still

When the unsaturated wet air or low temperature air is blown over the surface of the hot water with higher temperature, the vapor produced from the water will diffuse to the air. The wet air will become more and more moist. This system was provided out just based on this principle.

In a channel with water, as shown in Fig. 6.51, the air will be wetted and heated when the unsaturated air with temperature T_1 passes over the water surface with temperature $T_2(T_2 > T_1)$. If the channel is long enough, the final temperature of the wet air will become a saturated air with temperature T_2.

Under the conditions of the steady state and flow, if one ignores the change of the kinetic and potential energy of the air and supposes the total pressure in input and output to be similar, the following equations can be obtained based on the energy conservation.

$$\dot{m}_{a1} = \dot{m}_{a2} = \dot{m}_a \tag{6.56}$$

$$\dot{m}_{v1} + \dot{m}_w = \dot{m}_{v2} \tag{6.57}$$

where, $\dot{m}_a$ is the mass flow rate of dry air, $\dot{m}_v$ is the mass flow rate of the vapor, $\dot{m}_w$ is the evaporating rate from the water surface, and Subscript 1and 2 describes the inlet and outlet state of the air, respectively. Using the definition of the content of the moist air $d = \frac{m_v}{m_a}$ and Eq. (6.55), Eq. (6.56) will be rewritten as

$$\dot{m}_a d_1 + \dot{m}_w = \dot{m}_a d_2 \tag{6.58}$$

Finally, the evaporating rate from water surface is obtained as follows:

$$\dot{m}_w = \dot{m}_a (d_2 - d_1) \tag{6.59}$$

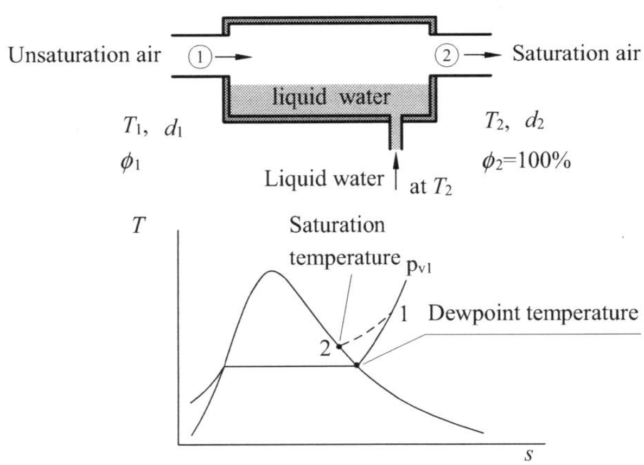

Figure 6.51
Air humidification process and its description in *T-s* diagrams.

If the wet air can be considered as an ideal gas, the content of the moist air can be calculated as follows:

$$d = \frac{0.622\varphi \cdot p_s}{p_T - \varphi \cdot p_s} \tag{6.60}$$

where p_T is the total pressure, p_s is the saturation partial of the vapor under the saturation temperature, and φ is the relative humidity. Eq. (5.54) can be rewritten as

$$\dot{m}_w = 0.622 \cdot m_a \left[\frac{\varphi_2 p_{s2}}{p_T - \varphi_2 p_{s2}} - \frac{\phi_1 p_{s1}}{p_T - \phi_1 p_{s1}} \right] \tag{6.61}$$

where, p_s is obtained by checking the water vapor table through testing the temperature. p_T and φ can be obtained by measuring directly. Therefore the evaporating rate finally can be calculated out. For a simple situation, if the wet air is saturated at input and output at the local temperature, the vapor content $m_{v,1}$ and $m_{v,2}$ $\left(\text{kg}/\text{m}^3_{\text{dry air}}\right)$ in unit volume dry air may be obtained by checking the table through T_1 and T_2. When the volume flow rate of the dry air is V (m^3/h), the water vapor mass carried out by the dry air is calculated as

$$\dot{m}_w = V \cdot \left(m_{v,2} - m_{v,1}\right) \tag{6.62}$$

This equation is very convenient to analyze the performance of the still.

6.9.2 Operation Principle of the Closed Circulation Solar Still

The schematic diagram of the presented solar still is shown in Fig. 6.52. It is mainly composed of the following parts: a still body, a tilted rectangular box that is separated into two chambers by a central partition; a vertical tube falling film evaporator-condenser; and a gas–liquid separator with a heat exchanger and an air fan. The operation mode of the system can be explained as follows [35].

The feedstock enters the system at the bottom of the vertical tube condenser ③ via a serpentine copper pipe ④ and finally reaches the overflow weir ⑦ at the top of the central partition of the still body through ③ and the lower chamber ② of the rectangular box. The feedstock overflows the weir and uniformly wets the evaporation film ③ where it is heated by solar irradiation to evaporate. The brine, which leaves the evaporation chamber, ① enters the vertical tubes ⑥ of the falling film evaporator where it wets the inner walls of the tubes and flows downward by gravity. In the meantime, the cooler air coming from the air fan is blown into ⑥ from its bottom and flows upward where it forces the falling film brine to evaporate. Then, the air vapor mixture formed in ⑥ is sent back to the upper chamber where it absorbs the higher temperature vapor and changes into high temperature humid air. The humid air is blown directly into the lower chamber ② in which a part of it condenses and preheats the entering feedstock. Afterward the mixture of the distillate and

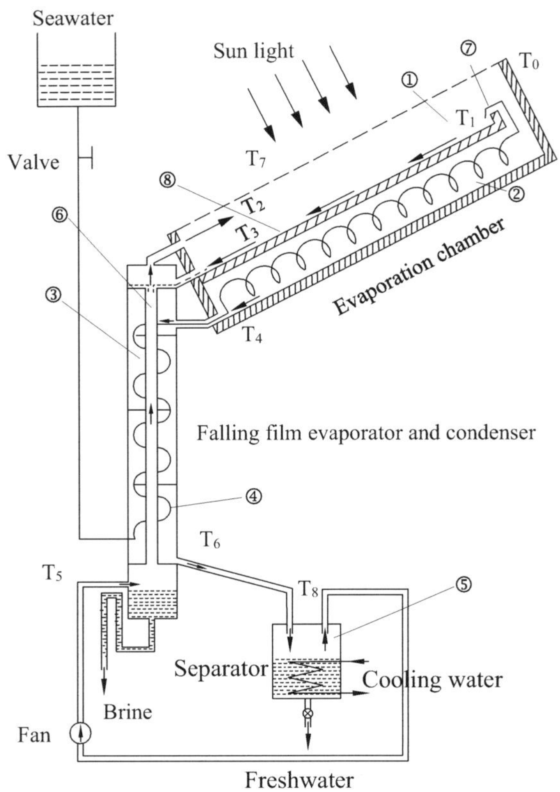

Figure 6.52
The schematic diagram of the solar still with heat recovery [35].

the relatively low temperature humid air is sent into the condenser ③. In ③, the latter condenses continuously on the outer walls of the vertical tubes ⑥ and gives out most of its latent heat. During this process, the latent heat is conducted into the falling film brine on the inner walls of the vertical tubes and further drives the brine to evaporate. Finally, the mixture of the condensed water and the lower temperature air steam enters the gas–liquid separator ⑤ with a heat exchanger, where it is further cooled down close to ambient and releases most of its water content. The surplus mixture air is recycled through the air fan.

6.9.3 Description of the Experimental Unit

6.9.3.1 The Still Body

The still body is a relatively thin rectangular box ($1960 \times 550 \times 220$ mm^3), constructed from 0.8 mm-thick galvanized iron sheets and insulated externally with 50 mm-thick polystyrene foam. The still window is double glazed with 15 mm spacing (also with a single glass cover and sun plate, respectively, for contrast). The body is separated into two chambers by a central partition, which is covered with a black wick. The central partition is constructed from 5 mm common glass plate and connected on three sides to the still walls by a glass and metal sealant. A 15 mm-wide slot between the central partition and the fourth wall is left to connect the two chambers to each other. The tilted angle of the still body is 20°. The serpentine conduit of the feedstock entering is 6 mm diameter × 0.5 mm thickness × 4000 mm length copper pipe, serving for preheating the feedstock.

6.9.3.2 The Falling Film Evaporator–Condenser

The outer wall of the unit is constructed from 0.8 mm-thick galvanized iron sheets and is insulated externally with 50 mm-thick polystyrene foam. The inside diameter of the unit casing is 120 mm. The falling film evaporator–condenser contained three evaporation–condensation vertical tubes in a triangulate arrangement. Its diameter and length are 17/18 and 1000 mm, respectively. Thus their total inner and outer heat exchange surface areas are 0.151 and 0.160 m^2, respectively. To distribute the brine coming from the still body in three tubes uniformly, a distributor is installed at the top of the tubes. All of the other connecting pipes are the plastic tubes of 25 mm diameter.

6.9.3.3 The Gas–Liquid Separator

The gas–liquid separator includes a heat exchanger in it. It is constructed from 0.8 mm-thick galvanized iron sheets, and its dimensions are 150 mm diameter × 250 mm height. The heat exchanger is made from 6 mm diameter × 0.5 mm thickness × 3500 mm length copper pipe. Its heat exchange area is about 0.066 m^2. The yield is measured at 15 min intervals.

6.9.3.4 Air Fan

The maximum power of the air fan is 80 W.The maximum pressure difference produced by the air fan is 35 kPa. The practical flow rate of air is measured in experiments.

6.9.3.5 The Solar Simulator

The solar simulator is made from four halogen lamps of 1000 W, which are installed in a frame. The simulator provides different incident energy to the solar still with the different distance between the frame and the still window. The practical incident intensity is measured in experiments.

6.9.3.6 Temperature and Yield Measurement

The relevant temperature $T_0 \sim T_8$ shown in Fig. 6.52 is measured by an electric potential difference meter through the signal provided by the thermocouples. The yield of freshwater collected in separator is measured at defined time intervals through a cylinder.

6.9.3.7 The Other Conditions

The testing of the solar still was performed in a laboratory and outdoors, respectively, and continued for five months. The ambient temperature and relative humidity is in the range of 15–22°C and 70 – 80%, respectively, during the period of the indoor experiments. In the outdoor experiments, the real solar irradiation intensity, the ambient wind velocity, and the ambient temperature are measured practically. In these experiments, city mains water (16 – 20°C) served as the feedstock.

6.9.4 Results and Discussion

6.9.4.1 The Transient-State Performance of the Still

As a result of the low thermal mass of the system, steady-state operating conditions are achieved after approximately 2.5 h of system start-up, after which temperature and productivity rate remained essentially constant. The progress of the most characteristic temperature with the time of the irradiation is shown in Fig. 6.53 (irradiation intensity: 582 W/m^2, airflow rate: 3.10 m^3/h, feedstock flow rate: 2.4 kg/h, double glazing).

Under the same conditions, a comparison of the progress of the yield per 15-min interval with the time for different covers is given in Fig. 6.54. It can be seen that quasi-steady productivity rates are achieved after about 150 min. Fig 6.54 also shows that the double glazing case gives more slow productivity increasing rate but the highest quasi-steady yield. The single glazing cover has no advantage to the steady-state yield. Using a sun plate as the cover, the steady productivity is within the average.

Because of the good thermal insulation, the operation temperature of the system does not decrease fast despite its heat capacity being litter when the solar irradiation simulator is switched off. For validating the performance, the transient-state decreasing temperature

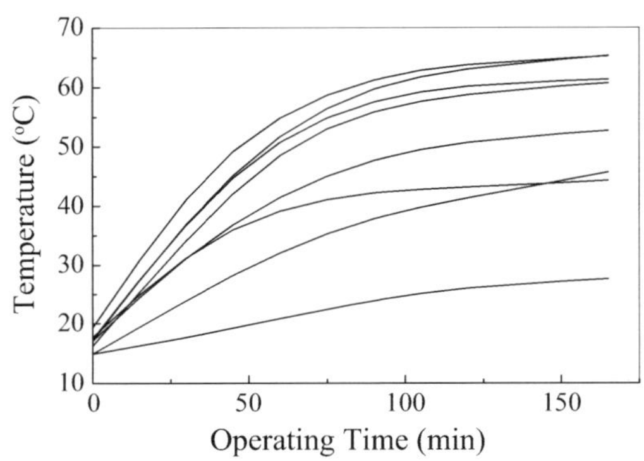

Figure 6.53
Attainment of the steady-state temperature for the double glazed unit.

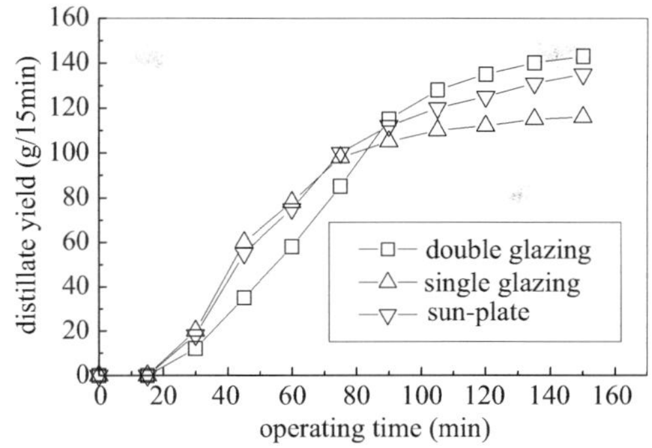

Figure 6.54
The progress of the distillate yield versus time.

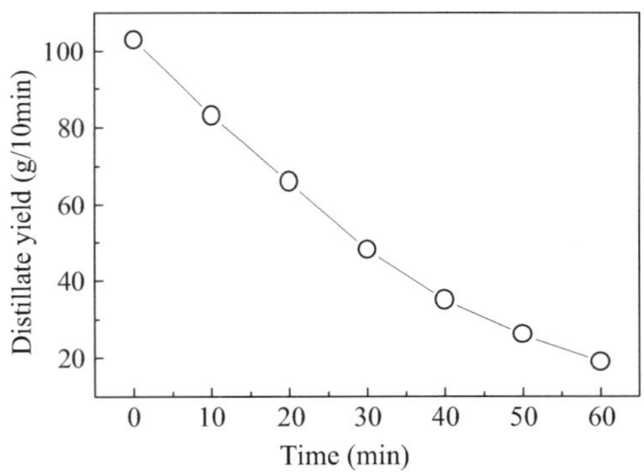

Figure 6.55
The decreasing process of the temperature and the yield of the system after stopping irradiation.

experiments are carried out for the mode of double glazing, and the results are shown as Fig. 6.55. In the experiments, the airflow rate and feedstock flow rate is kept to be 3.10 m^3/h and 2.4 kg/h, respectively.

6.9.4.2 The Steady-State Performance of the Still

The window of the solar still gains the different irradiation intensity when the distance between the solar simulator and the window is different. For definite irradiation intensity, the system will obtain a definite steady state. Table 6.3 shows some steady-state parameters and relevant temperature when the irradiation intensity is 555 W/m^2. It can be seen from Table 6.3 that the operation mode of double glazing achieves the highest efficiency and the highest operating temperature (T_0). It indicates that it is very important to prevent the thermal diffusion from the cover of the solar still.

Table 6.3: Different cover influences on the steady-state performance of the system.

	T_0	T_1	T_2	T_3	T_4	T_5	T_6	T_7	T_8	T_b	T_a	m_e (g/h)	η (%)
Sun plate	65.7	65.3	53.8	63.5	62.6	42.1	46.8	63.5	31.3	34.5	15.0	621	66.7
Single glazed	61.2	61.0	49.2	58.4	56.6	38.5	40.2	59.1	26.4	31.5	13.0	543	58.4
Double glazed	66.6	65.8	53.3	62.3	61.8	44.6	44.5	75.4	30.1	33.5	13.5	662	71.1

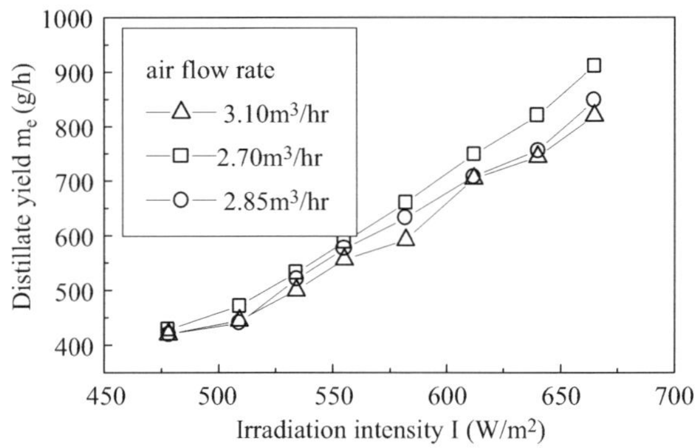

Figure 6.56
Variation of the yield per hour versus the incident energy. Feedstock flow rate: 35 – 40 mL/min.

Under steady-state conditions, the variation of the total yield with the incident energy is shown in Fig. 6.56 for double glazing. It can be seen from the figure that the total yield increases smoothly with the irradiation intensity beside few points. The main causes producing these peculiar points are the variation of the ambient temperature. It is observed that the ambient temperature has a fair influence on the yield. The higher ambient temperature is an advantage to the yield. As shown in Fig. 6.56, the airflow rate has a litter influence on the steady-state yield, about 5–10%. The practical airflow rate is measured in the connecting pipe before air fan.

6.9.4.3 Feedstock Flow Rate Influencing the Yield

The system exhibits a wide range of flexibility with regard to variations in the feedstock flow rates, as shown in Fig. 6.57. For the double glazing cover, when the feedstock is changed from 1.8 to 4.0 kg/h, the biggest difference of the steady-state yield is only about 15%. At higher feedstock flow rates, the decrease of the yield is also slow. For example, only about 20% decrease is found at a feedstock flow rate of 6 kg/h with regard to the rate of 2.4 kg/h. On the other hand, the temperature of brine in outlet is only on the range of 30–51°C for different feedstock flow rate. Thus these cases provide many convenient conditions to the practical operation. It can be concluded that the major fraction of the

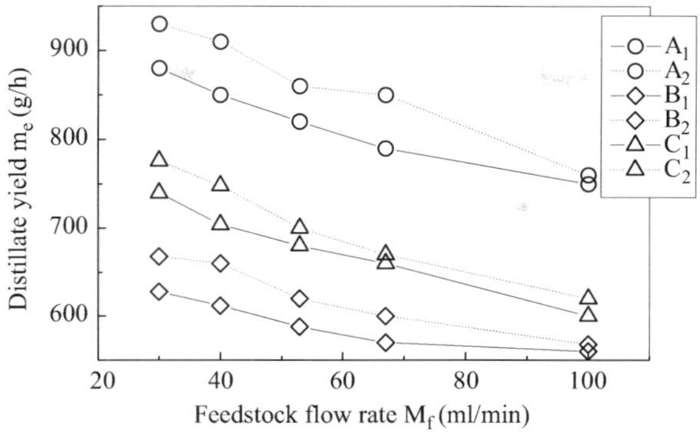

Figure 6.57
Effect feedstock flow rates on the yield.

enthalpy of the hot saturated air and the surplus heat of the brine is successfully recycled and used for evaporation.

Radiation:	$A_1,A_2 = 665\ W/m^2$	$A_1,A_2 = 582\ W/m^2$
	$B_1,B_2 = 582\ W/m^2$	$B_1,B_2 = 555\ W/m^2$
	$C_1,C_2 = 612\ W/m^2$	
Airflow rate:	$A_1,B_1,C_1 = 2.70\ m^3/h$	$A_1,B_1 = 3.10\ m^3/h$
	$A_2,B_2,C_2 = 3.10\ m^3/h$	$A_2,B_2 = 2.70\ m^3/h$

6.9.4.4 The Steady-State Efficiency of the System

The efficiency of the conventional solar stills is usually defined as follows:

$$\eta = \frac{m_e \cdot h_{fg}}{I \cdot A \cdot \Delta t} \tag{6.63}$$

where η is the efficiency of the still, m_e is the yield of the distillate, h_{fg} is the evaporation latent heat of the distillate, I is the intensity of solar irradiation, A is the area of the solar still, and Δt is the operation time. Considering some electric power W_p required, the efficiency of this system is given by

$$\eta = \frac{m_e \cdot h_{fg}}{(I \cdot A + W_p) \cdot \Delta t} \tag{6.64}$$

Fig. 6.58 shows the variation of the efficiency with the irradiation intensity on the window in which the feedstock flow rate is 40 mL/min. It is seen that the efficiency increases for the double glazing cover with the intensity of the irradiation. It illustrates that the still has more advantages in a higher temperature range.

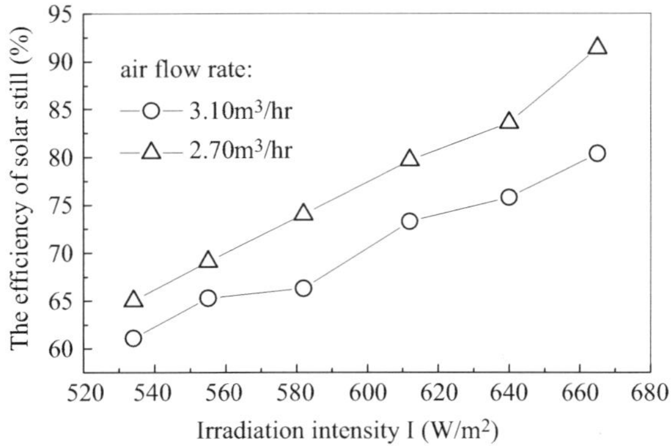

Figure 6.58
The steady-state efficiencies of the still versus the irradiation intensity.

6.9.5 System Operating Under the Practical Weather

The illuminant temperature of the solar simulator in this work is only 2800K, which is quite different from the temperature of the sun's surface. In addition, the wind velocity over the cover and the ambient temperature under practical conditions will change rapidly, which exerts a great influence on the performance of the still. So it is necessary to know the situation of the system operating under sunlight. For this reason, the outdoor experiments were carried out (for double glazing). The practical solar irradiation intensity and the ambient temperature were measured in experiments at the same time.

Fig. 6.59 shows the variation of the total amount of solar radiation with the time (Beijing time) on three typical days. The variation of the yield per hour on the same day with the time is shown in Fig. 6.60 when air circulation rate was 3.10 m^3/h and seawater flow rate was 2400 mL/h, 3500 mL/h, and 2800 mL/h, respectively.

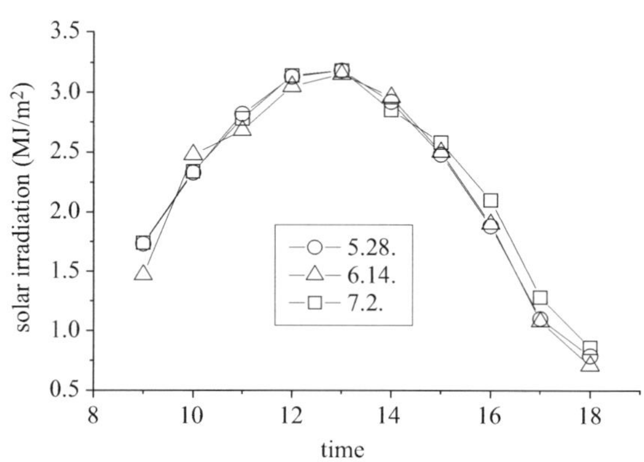

Figure 6.59
The variation of the solar radiation with the time (Beijing time).

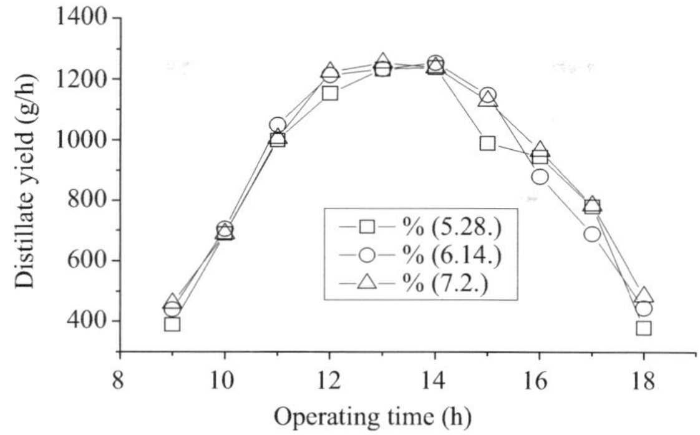

Figure 6.60
The variation of the distillate yield per hour with the time.

6.9.6 Heat and Mass Transfer Analysis of the System

In this operation mode, the recycling air in the system undergoes significant changes in temperature and water vapor content and, as a consequence, its flow rate and enthalpy also changes. Thus it is very difficult to calculate accurately the heat and mass transfer balance in the system.

In the following calculations, only the changes of the water vapor content and enthalpy of 1 m^3 recycling air in the system are considered. To simplify the heat and mass transfer calculations, the recycling air is thought to be saturated at anyplace at local temperature in the device (except the range of the outlet of the air fan to the bottom of the evaporation tubes, see Fig. 6.52). The heat losses through the bottom and the side walls of the still body can be ignored. At the same time, to facilitate our analysis, only one operation mode was taken to calculate, in which the cover is double glazing, the irradiation intensity is 665 W/m^2, the feedstock flow rate is 2.4 kg/h, and airflow rate is 3.10 m^3/h (before entering the air fan), respectively. The relevant temperature and other conditions are listed in Table 6.4. Thermodynamic properties of water and saturated air are based on the data of the literature [35].

Table 6.4: Steady-state productivity rates and relevant temperatures under mentioned conditions.

T_0	T_1	T_2	T_3	T_4	T_5	T_6	T_7	T_8	T_a	T_b	T_f	m_e(g/h)
76.7	72.0	59.6	68.7	69.5	49.2	52.0	91.9	39.0	17.0	51.0	16.0	820

The mass flow rate G_v of the water vapor and the total heat flow rate $Q_{a,s}$ of the fluid can be given as

$$G_v = V_{a,298} \cdot m_v \quad \text{and} \quad Q_{a,s} = V_{a,298} \cdot H_{a,s} \tag{6.65}$$

where $V_{a,298}$ is the flow rate of air at a reference state of $T = 298\text{K}$ and $p = 101.3$ kPa, m_v is the amount of water vapor in saturated air, and $H_{a,s}$ is the enthalpy of saturated air. $H_{a,s}$ and m_v in different temperatures are given in the literature [35].

As in this work, the temperature of the air that enters the air fan is close to ambient. For draft estimations, $V_{a,298}$ can be substituted by the measured flow rate $V_{a,m}$. To obtain more precise data, the following correction is used,

$$V_{a,298} = V_{a,m} \times \frac{298}{T_8} \times \frac{101.3 - p_w}{101.3} \tag{6.66}$$

where p_w is the vapor pressure and T_8 is the absolute temperature of the saturated air entering the air-fan (see Fig. 6.52).

The approximate heat and mass transfer balances of the major components are listed in Table 6.5. From Table 6.5, we can find that the distillate produced in the lower chamber, the falling film condenser, and the separator account for 38.1%, 48.3%, and 13.6% of the total distillate yield, respectively. It illustrates that the effect of the falling film condenser is very important. The total distillate yield is theoretically 1.18 kg/h. In fact, this

Table 6.5: Heat and mass balances of the system ($T_{reference} = 0°C$).

	Chamber					
	Upper	Lower	Evaporator	Condenser	Separator	Air fan
1. Mass in (kg/h)						
Feedstock	2.40	2.40	1.55	2.40	0	0
Water vapor	0.48	1.33	~0.15	0.88	0.31	0.15
Distillate	0	0	0	0.45	1.02	0
2. Mass out (kg/h)						
Feedstock	1.55	2.40	1.22	2.40	0	0
Water vapor	1.33	0.88	0.48	0.31	0.15	0.15
Distillate	0	0.45	0	1.02	1.18	0
3. Enthalpy in (kJ/h)						
Feedstock	723	440	445	161	0	0
Airstream	1488	3941	~560	2574	998	523
Distillate	0	0	0	131	222	0
4. Enthalpy out (kJ/h)						
Feedstock	445	723	260	~440	0	0
Air-stream	3941	2574	1488	998	523	560
Distillate	0	131	0	222	192	0
5. Heat loss (kJ/h)	−2175	953	−743	1045	505	−37

productivity is only 0.82 kg/h and the difference is 43%. But the vapor mass produced in the upper chamber is 0.85 kg/h, which is much closer to the measured yield 0.82 kg/h. It can be also seen from Table 6.5 that there is a little heat loss through the bottom and the side walls of the still body and the connecting pipes. The heat loss from the lower chamber, evaporator–condenser, and separator is 953, 302, and 505 kJ/h, respectively. As shown in Table 6.5, due to the existence of the lower chamber and the falling film evaporator–condenser a considerable amount of condensation heat is reused for the evaporation processes.

6.9.7 Conclusions and Discussions

In this still a considerable fraction of the latent and sensible heat is successfully recycled and utilized for preheating the feedstock and the recycling air via a falling film evaporator-condenser. The forced thin layer evaporation and thin film condensation, which are the efficient enhanced processes for heat transfer, are applied in this unit. The performance of the still is improved greatly. As a result, the yield is about two to three times greater than that of a conventional basin-type solar still. It is found from the experiments that the external electric power is not critical. If the scale of the solar still is increased, the specific electric power consumption (consumed electric power per unit yield) can be decreased further. It makes this kind of solar still able to be used in the remote area where the electricity is comparatively poor.

Since the inner surface of the glazing does not function as a condensing surface, it is possible to utilize nonwetting glazing as the window cover. Because of the utilization of double glazing, the rate of thermal energy loss from the window can be reduced significantly. As a result, the solar still can be operated in a higher temperature. If the waste heat is abundant, the system can be modified simply to adapt to the hybrid operation.

The analysis indicates that the productivity in the lower chamber is about 38.1% of the total productivity. It illustrates that the lower chamber of the still is very important. However, it is easy to find that the main function of the lower chamber is to preheat the feeding water entering the upper chamber. Therefore the lower chamber could be replaced by a double-pipe heat interchanger. For enhancing the function of the lower chamber, an improved system was designed, as shown in Fig. 6.61. The remarkable feature of the new still is that the hot wet air not only preheats the feed water entering the upper chamber, but also preheats the feed water between the upper chamber and the lower chamber when it goes into the lower chamber from the upper chamber. In this way, the latent heat of the vapor in the lower chamber can be recovered more sufficiently, so that the temperature of the wet air entering the falling film condenser is much lower. However, this process still can preheat the air entering the upper chamber. According to the theoretic calculation, the new still can produce 1.6–1.7 times more freshwater than that of the old one.

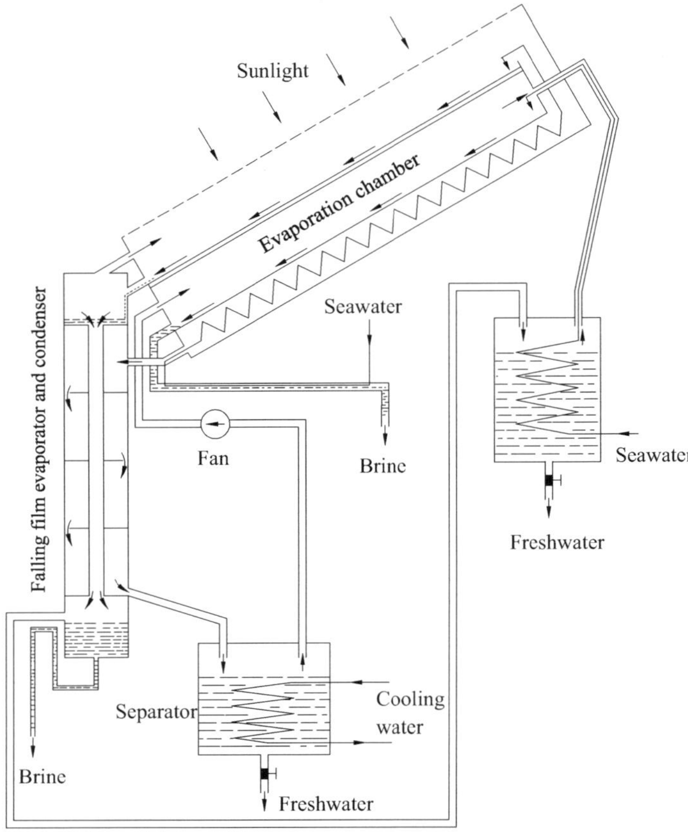

Figure 6.61
Schematic diagram of an improved system.

6.10 Tandem Multieffect Isothermal Heating Solar Desalination System Based on the Humidification–Dehumidification Processes

As we know, many researchers have contributed their efforts to solar desalination systems by HD processes, especially in improving the system energy usage efficiency. However, they do not pay enough attention to the production rate per unit volume of the system, which is a shortcoming of HDD systems relatively compared to other desalination methods. In traditional solar HDD systems, only parts of stages work in high temperatures. The operation temperature in other stages will decrease step-by-step with the increasing stage number. On the other hand, all of the seawater entering in the system will be heated to the highest temperature, which will waste too much sensible heat energy. To further improve the performance of solar HDD, Gang Wu (2016) [36] presented a novel tandem isothermal heating HDD method and a three-stage multieffect HDD system. Compared to the previous multieffect HDD systems, this system cannot only reuse the latent heat of condensation and recycle the residual heat in the brine, but also most stages can operate in the highest temperature except for the lowest stage, which can further improve specific volume yield. Moreover, it only needs to heat part of the feed water, resulting in saving energy.

6.10.1 Description of the Experimental Setup and the Working Principle

6.10.1.1 Structure Parameters and Working Principle

The humidification–dehumidification process is based on the fact that the moisture content of air increases progressively with elevated temperature. For instance, 1 kg of dry air can carry almost 1.25 kg vapor or more when its temperature increases from 60 to 90°C.

The system presented in this paper consists of four closed loops: a solar loop, which can produce hot water, and three water desalination loops with the function of producing desalinated water. The solar loop is a thermosyphon water heater connected with a heat exchanger to deliver heat to the seawater.

The desalination loops include three evaporation towers with hot seawater sprayers and three condensation towers, which are configured to form an upper loop and a lower loop. Air is circulated by fans within each loop. Fig. 6.62 illustrates a schematic diagram of the experimental setup and Fig. 6.63 shows its photograph.

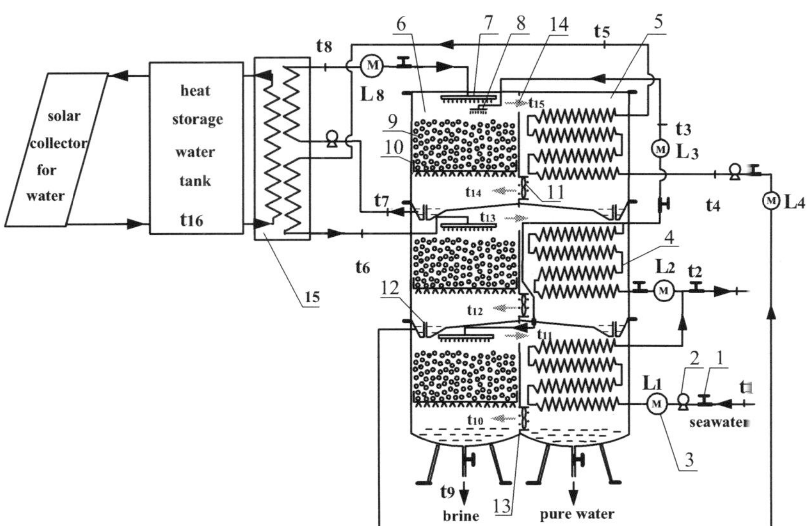

Figure 6.62

Schematic drawing of the experimental setup and the thermocouple locations [36]. 1, value; 2, pump; 3, water flow meter; 4, condenser; 5, dehumidifier; 6, evaporator chamber; 7, evaporating sprayer; 8, moisturizing sprayer; 9, plastic ball; 10, basket; 11, fan; 12, water limited pipe; 13, battle; 14, humid air; 15, heat exchanger; $t_1 \sim t_9$, Temperature of water at different locations; $t_{10} \sim t_{15}$, Temperature of air at air channel; t_{16}, Temperature of water tank.

Figure 6.63
A photograph of the experimental setup. (A) The inner shell; (B) external shape.

In the condensation sections, three heat exchangers, which are made of copper tubes with corrugated aluminum fins, are connected to each other in series. The porous plastic balls are chosen to establish the packed bed. To increase the contact surface between air and seawater, the porous plastic balls were packed in the evaporation sections to achieve a larger wetted surface and an efficient humidification of the air. The shell of evaporators and condensers were constructed of 3 mm stainless steels by welding and were insulated by Styrofoam layers with a thickness of 25 mm. The operation principle of this desalination process is presented as follows.

First, the seawater is fed to the low condensation tower (LCT) to condense the moist air coming from the low evaporation tower (LET). The latent heat of condensation is used to preheat the feed water. At the exit of the LCT, a small part of seawater is expelled from the system, and the remainder enters the middle condensation tower to condense the water vapor in the moist air flowing from the middle evaporation tower (MET). After that, the heated seawater is divided into two parts. One enters the high evaporation tower (HET) through the moisturizing sprayer in HET and the other flows into the LET by the evaporating sprayer. The brine accumulated at the bottom of the HET is circulated through the heat exchanger, where the brine is heated up and returned to the HET through the evaporating sprayer. In the HET, due to heat and mass transfers between the hot brine and the air stream, the moisture content of air increases. At the bottom of the HET there is a liquid collector where brine is collected as it drains down to the bottom. In the liquid collector, a vertical overflow pipe is set to ensure the seawater level to a desired degree.

In the middle stage, the brine accumulated at the base of the MET is circulated through the condenser in the high condensation tower (HCT) where the brine is heated up and then reheated in heat exchanger. Finally, the brine is returned to the MET through evaporating sprayer in MET.

The novel system operates at atmospheric pressure using air as a carrier for vapor. A forced convection heat and mass transfer processes between air and hot brine will occur in the evaporation tower. The forced convection process is assured by an axial flow fan fixed at the bottom of the evaporation tower. The resulted saturated moist air is transported to the condensation tower, where it contacts the condensers and is cooled by the seawater to its dew point. A part of the water vapor in the moist air therefore turns into freshwater, which is collected at the bottom of the condensation tower.

Compared with the previous multieffect HDD system, the main characteristics of the three-stage multieffect HDD system are outlined as follows. (1) It cannot only reuse the latent heat of the condensation but also successfully recycles the residual heat in the saline water, so the thermal efficiency of the system could be greatly improved. (2) It has a few merits, namely, compact structure, stable running, easy control, and small floor space occupation. (3) For units operating under atmospheric pressure with no requirement for pressure-bearing material, it has a low fee. (4) The unit can operate within a very wide temperature range from 60°C to 95°C.

6.10.1.2 Size of the Device

The condensation towers have the same structure as the evaporation towers of the unit. The dimension of the single-stage distillation chamber is 2.0 × 1.2 × 1.15 m. The air channel, through which the humid air circulates from the humidifier to the dehumidifier, is with the section of dimensions 0.7 m × 0.26 m. Each chamber has one observation window in the wall. The diameter of the porous plastic ball is 26 mm, and the thickness of the packed bed is 0.2 m. The capacity of the heating water tank is 250 L. The total area of heat exchanger is 5 m^2. The maximum seawater flow rate entering the LCT by the pump is 3 m^3/h. In addition, feed seawater mass flow rate can be adjusted to a desired value by a valve.

6.10.2 Results and Discussion

In the experiment the tap water (sometimes using the mixture of tap water and NaCl salt) was used instead of seawater. Some electric heaters are installed in the heating water tank to simulate a solar water heater. In addition, the heating power can be adjusted to meet the experimental requirement. The tests were performed to assess the influence of the following parameters on the yield of the system:

1. water temperature in the heating water tank;
2. mass flow rate of the circulate seawater;
3. flow rate of the humid air; and
4. circulate seawater temperature.

6.10.2.1 System Performance Under Transient Conditions

Transient characteristic is an important indicator for observing system stabilization. Reaching a steady state in a short time is necessary to increase the desalination efficiency of the unit, especially for the solar desalination system.

In order to measure the transient characteristic, when the temperature of heating water t_8 reached 60°C, the unit started to operate, and other parameters mentioned earlier were measured at 5-min intervals. Continuous measurements were carried out, lasting 90 min in one time test. In addition, the flow speed of the humid air was 2.2 m/s. The 5-min yield and accumulated yield of the unit were recorded. Under the same conditions, the temperature of every measured point inside the tower was recorded. The variation curves of the yield with the operation times are shown in Fig. 6.64, and the temperature variations within 90 min are shown in Fig. 6.65.

Fig. 6.64 shows profiles of the 5-min yield and variation of the accumulated yield with the operation time. The 5-min yield increased quickly at the beginning of the experiment and reached a quasi-steady state at 11:10 am. Meanwhile, the accumulated yield increases with the operation time all the time. The reason of this reaction can be explained by Fig. 6.65. It can be seen from Fig. 6.65 that the trend of the variation of the temperature with the

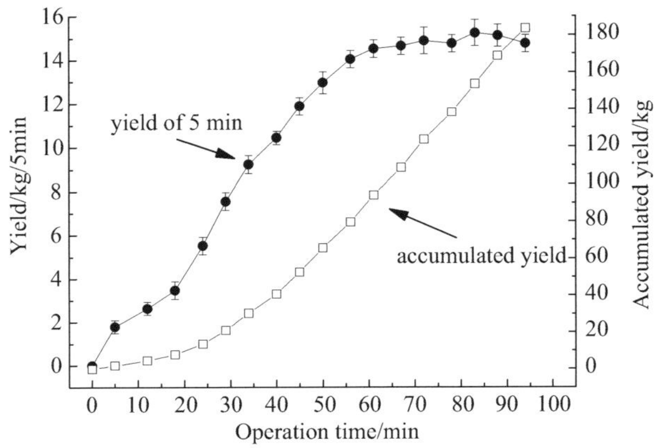

Figure 6.64
The variation of the yield in the system with the operation time.

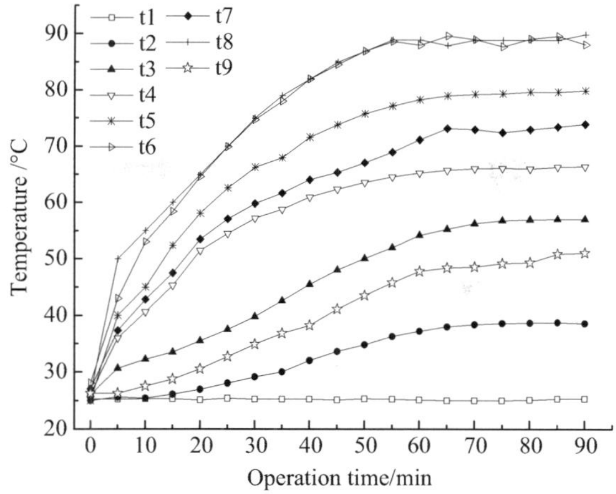

Figure 6.65
The variation of the temperature in the system with the operation time.

operation time at every measured point is the same as that of the 5-min yield. This is due to the increase of the heating temperature causing an increase of the temperature of the moist air entering the dehumidifier. It reveals that the productivity of the system will increase with the increasing relative humidity of the air.

However, the increase of the contact time between the spray seawater and the humid air induces a quickly elevated temperature of the feed seawater, which leads to a less condensation effect and slows the yield growth. After 90 min the quasi-steady state of the system was obtained.

6.10.2.2 System Performance Under Steady-State Conditions

In the process, the recycled air in the unit undergoes significant changes in heat and mass transfer. Generally, it is difficult to make accurate calculations of the heat and mass transfer in the tower and the yield of the system. In order to simplify the yield calculations, the recycled air is considered as saturated at the set temperature. The heat loss from the system is ignored. It is supposed that the pressure inside the unit is 101.3 kPa. The humidity d and enthalpy H of the saturated air were calculated using the following empirical correlation developed from reported data [37].

$$d = 2.19 \times 10^{-6}t^3 - 1.85 \times 10^{-4}t^2 + 7.06 \times 10^{-3}t - 0.077 \tag{6.67}$$

$$H = 0.00585t^3 - 0.497t^2 + 19.87t - 207.61 \tag{6.68}$$

where t is the temperature of the saturated air, °C, d is the humidity of saturation air, kg_{water}/kg_{dryair}, and H is the enthalpy of moist air, J/kg.

Then the freshwater yield of the system produced in the humidifier is

$$G_v = (d_{in} - d_{out}) \times v_a \times A \times 3600 \tag{6.69}$$

where v_a is the humid airflow speed (m/s), A is defined as the effective area of the air channel, and A is calculated to 0.18 m^2.

On the basis of ensuring the smooth operation of the process, the test of feed water mass flow rate, heating temperature, and humid airflow rate on the unit productivity were studied. Every 10 min, the desired parameters were measured, and the setting temperatures in the heating water tank were selected as 60°C, 65°C, 70°C, 75°C, 80°C, 85°C, and 90°C separately for half an hour.

Fig. 6.66 illustrates the productivity comparison between the calculated and the measured with the heating temperature under different feed seawater mass flow rates. M_w represents the water mass flow rate, kg/s. Due to the limited conditions, this experiment only did the test at the maximum water flow rate of 2 t/h. Theoretical calculation indicates that the feed water flow rate should be 3 t/h so as to achieve higher productivity. Unfortunately, the system is hard to achieve the flow rate due to the limitations of the pipe diameter of the condenser. For the case studied, the relative deviation of the experimental data is less than 5%. Computer simulation methods become necessary to support the insufficient amount of feed water flow rates. Based on the mathematical model, a steady-state simulation of yield rate at different water mass flow rates has been carried out. It shows that the yield rate increases with an increase of feed water mass flow rate and heating temperature, in which the flow speed of the humid air is assumed to be 2.2 m/s. However, the increase of yield in

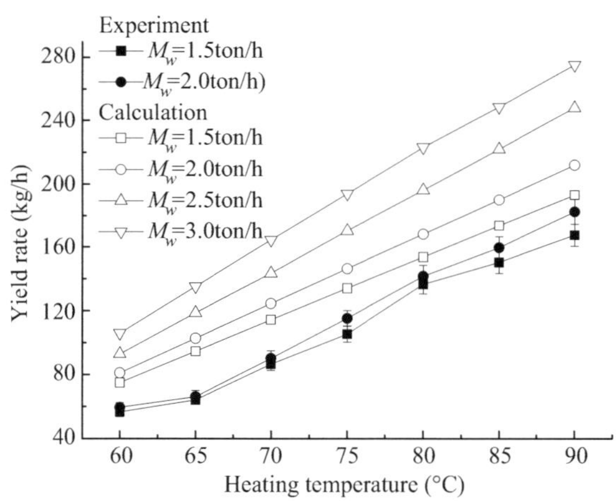

Figure 6.66
The variation of the yield with heating temperature in the unit at different water mass flow rates.

the heating process varies slowly until the heating temperature reaches 70°C, after which the trend grows sharply. It can be explained as follows:

1. Under the same feed water rate, less saline water will be evaporated in a packed bed when the system operates under a low temperature range. The excess saline water through the vertical overflow pipe flows to the lower humidifier stage. Eventually, more sensible heat along with the overflowed saline water is evacuated to the outside.
2. The system has some instability. Maybe it did not reach complete steady when we did the test because it will spend more time to reach the steady state when its operation temperature is in low range.

The calculated values of the yield rate agreed well with the experimental results. This also can be explained by Fig. 6.67. The temperature of the humid air leaving the evaporation tower increases as the heating temperature increased and the temperature difference of the air between the inlet and the outlet of the condensation tower becomes larger along with the increase of relative humidity of the air, which leads to more production. Since the condensate temperature is too high in HCT, this will result in a low temperature difference between inlet and outlet air ports of the high stage.

Increasing the feed seawater mass flow rate causes a decrease of the temperature of the water, leaving the condensers in the LET and HET. The heat transfer between condenser and humid air is strengthened, for the fact that the temperature difference between the feed seawater and the humid air in the condensation tower becomes larger and then induces more yield of the unit.

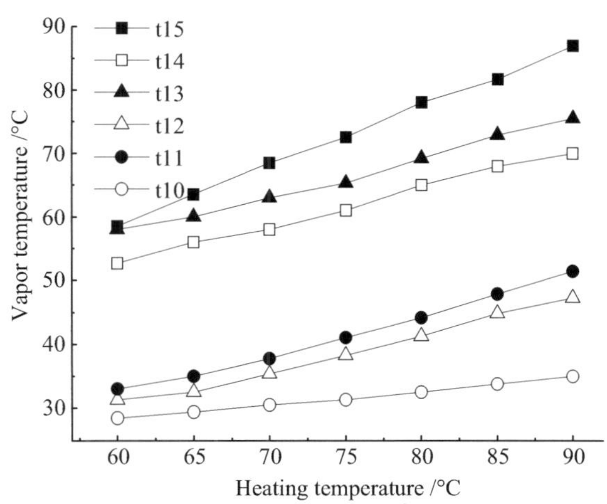

Figure 6.67
The variation of the inlet and outlet humid air temperature with heating temperature.

Fig. 6.68 shows the effect of the humid airflow rate on the unit productivity. It can be observed from the figure that the yield at the humid airflow speed of 2.2 m/s is more than that at the humid airflow speed of 1 m/s. It reveals that the productivity of the unit will increase with the increasing humid airflow rate because air would carry more water vapor to the condensation tower than before.

This behavior can also be explained by Fig. 6.69, which shows that when the humid airflow rate increases, more water vapor condenses. Since the temperature difference of the humid air at the flow speed of 2.2 m/s is larger than that at the flow speed of 1 m/s, the moisture content of the former leaving the dehumidifier is more. As a result, an increased amount of freshwater can be obtained from the unit.

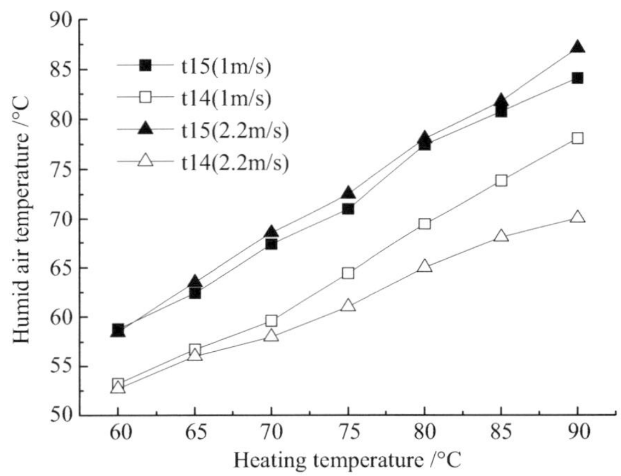

Figure 6.68
Effect of air mass flow rate on the system productivity.

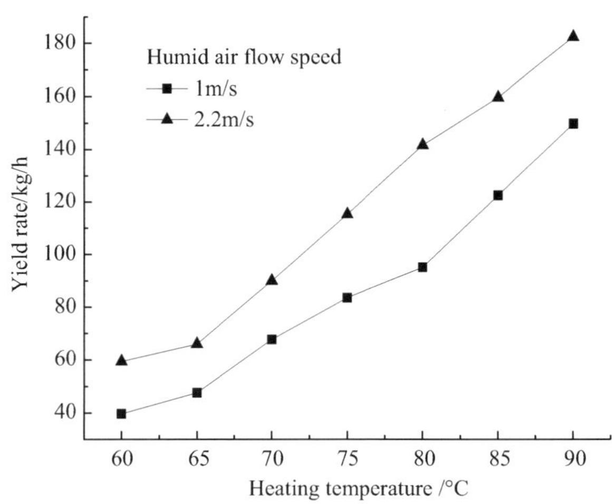

Figure 6.69
Effect of air mass flow rate on the humid air temperature.

6.10.2.3 Performance Evaluation of the System in Steady-State

Solar desalination system performance basically depends on solar radiation and operation conditions. The performance or energy efficiency of the humidification–dehumidification desalination system is defined as the performance ratio (PR). A larger value of PR in a desalination system means high energy efficiency and better distillate production performance. Based on the relationship of the input and output energy, the steady-state PR is given as follows:

$$\mathrm{PR} = \frac{M_e h_{fg}}{Q_{in} + P \cdot t} \tag{6.70}$$

where M_e is the freshwater production, h_{fg} is the latent heat of vaporization at the ambient condition, $P \cdot t$ is consumption work of circulating water pumps and fans, and Q_{in} is the heat energy input to the system. In this study the heat loss of the heaters in the water tank is ignored, and the latent heat is calculated at the average partial pressure of water vapor. The total yield and the PR under steady-state are listed in Table 6.6.

It can be seen from Table 6.6 that the heating temperature and the feed seawater mass flow rate have significant influence on the unit yield. This result can be explained as follows: the increase of heating temperature increases the temperature of water spraying to the porous plastic balls. Therefore higher moisture content of the humid air leaves the evaporation tower. Consequently, the amount of freshwater obtained from the system is increased. And increasing the feed seawater mass flow rate enlarges the heat transfer temperature difference of the heat exchanger, that is, the effect of condensation heat transfer is strengthened. Increasing the feed seawater mass flow rate also causes a reduction in the moisture content of the air leaving the condensation tower. In the experiment, the PR of the unit can reach a value between 1.8 and 2.65, indicating that the latent heat of condensation of water vapor can be successfully utilized in the system. Actually, the PR of the unit should increase with the increase of the heating temperature, since air would carry more water vapor to the

Table 6.6: The freshwater yield and the performance ratio of the equipment under steady-state.

	M_w = 1.5 t/h		M_w = 2 t/h	
Heating temperature/°C	Yield/kg	PR	Yield/kg	PR
60	56.50	1.80	59.41	1.87
65	64.11	1.83	66.04	1.92
70	86.58	1.72	90.15	2.12
75	105.41	1.80	115.24	2.20
80	136.51	2.06	141.61	2.33
85	150.34	2.18	159.63	2.65
90	167.62	2.28	182.47	2.58

condensation tower. However, the PR of the unit does not always increase. In fact, increasing the heating temperature causes an increase in the heat loss. It can be observed from Table 6.6 that thermal insulation has a significant influence on the PR of the unit, especially when the operation temperature is above 85°C.

6.10.3 Conclusions

This experimental system presents a tandem multieffect isothermal heating desalination system based on the humidification–dehumidification process and its working principle. Using porous ball humidifiers enhances the condensation and evaporation process by enlarging the contact surface area. Experimental results show that the system has good transient and steady-state natures. The test results indicate that the system parameters (heating temperature, feed seawater flow rate, and humid airflow rate) will influence the process of freshwater production.

Compared to the previous multieffect humidification–dehumidification desalination system, it not only reuses the latent heat of condensation but also successfully recycles the residual heat in the saline water between the desalination loops. When the heating temperature increases from 60 to 90°C and the feed water flow rate is 2 t/h, the freshwater yield increases from 59.41 to 182.47 kg/h. The yield per unit volume is calculated as 22 kg/m^3 h and the PR can reach about 2.65.

6.11 Other Typical Air Humidification–Dehumidification Solar Desalination Systems

6.11.1 Tower Type Air Humidification–Dehumidification Solar Distiller

As we know, the humidity ratio of saturation air changes remarkably with its temperature. A saturation air humidification–dehumidification type solar distiller is designed just by using this performance of saturation air. It forces the air through the desalination system to undergo the humidification and dehumidification processes, respectively, thereby to achieve freshwater from seawater.

The saturation air humidification–dehumidification type solar distiller consists of three parts: solar collector, evaporator, and condenser. For improving the efficiency, some systems add an air preheater and the heat exchanger between cold and hot seawater. Figs. 6.70 and 6.71 show two tower solar distillers according to air humidification–dehumidification processes.

The solar collector generally is made of vacuum tube solar receivers or flat plates, which provide the heat energy for the solar distiller. The evaporator usually is filled up within corruptible solid porous material. Seawater flows through the heap material from up to

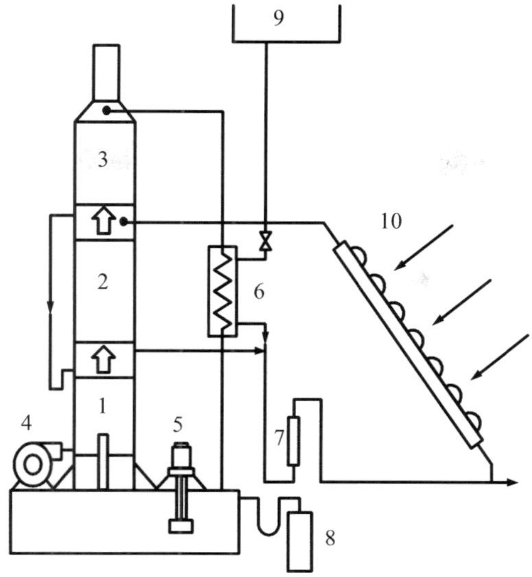

Figure 6.70
Distillation system with direct contact condensation: 1, Preheating air section; 2, Evaporation section; 3, Condensation section; 4, Fan; 5, Cooling water circulating pump; 6, Freshwater cooler; 7, Rotor flowmeter; 8, Cylinder for freshwater; 9, High seawater tank; 10, Evacuated tube solar collector.

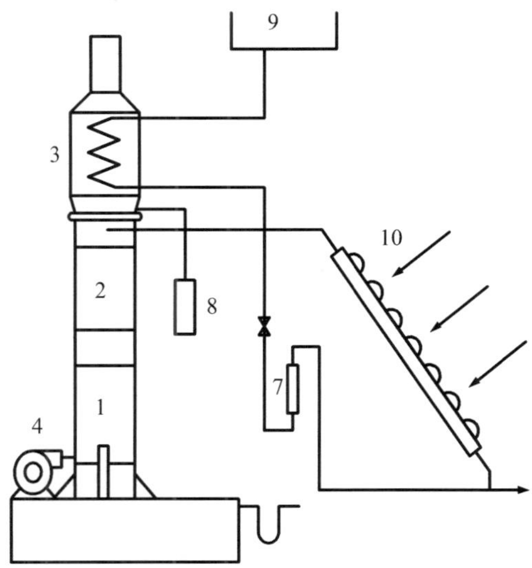

Figure 6.71
Distillation system with fin condenser: 1, Preheating air section; 2, Evaporation section; 3, Condensation section; 4, Fan; 5, Cooling water circulating pump; 6, Freshwater cooler; 7, Rotor flowmeter; 8, Cylinder for freshwater; 9, High seawater tank; 10, Evacuated tube solar collector.

down and the air passes through the heap material from down to up to form the counter flow heat exchange with seawater. The condenser general consists of a gas–liquid heat exchanger with fins. Sometimes cold pure water also can be used as the medium, which is circulated by a pump, to contact with hot saturation air so as to form a directly contacting type condenser.

During the operation, the hot seawater heated by the solar collector is sprayed downward at the evaporation section of the tower. Sprayed seawater inversely contacts the air to produce heat and mass exchanges in which air is heated and at the same time is humidified continuously to reach a saturation state. Finally, the saturation hot air is sent into the condenser, where it is cooled to release the freshwater. In the heating air process, the hot seawater decreases its temperature and goes out of the tower with a little higher temperature than ambient temperature. The hot saturation air is cooled in the condenser to produce freshwater. And at the same time, the hot saturation air also heats the entering seawater. The waste gas is removed from the top of the tower.

Experimental results indicate that the strongest factors influencing the evaporation rate are the spraying rate W of seawater, mass flow rate G of circulation air, and spraying temperature T_h of seawater.

The experimental results of the literature [38] show that when $W = 50$ kg/h and $W/G = 1.67 \sim 3.0$, the evaporation rate had the relationship with spraying temperature T_h of seawater as follows,

$$lgm_e = 7.062 + 3.446T_h \tag{6.71}$$

where, m_e is the freshwater production, kg/m^2 h. In the experiment, the condenser is a directly contact type heat exchanger. Under the condition of sunny days (solar irradiance is about 19 MJ/m^2 day) with vacuum tube solar receivers as the solar collector, the freshwater productivity was about 5–7 kg/m^2 day, which is two times the production as that of a traditional solar still.

Theoretical and experimental results all show that the spraying temperature of seawater has an important influence on the production. The higher the temperature is, the more fresh production will be. But from the energy utilization efficiency, it is not the higher spraying temperature the better. According to W/G, it has an optimum quantity. In the humidification process, only when the circulation air temperature T_S is similar to seawater spraying temperature T_h, the evaporation rate can reach the maximum in this temperature. For satisfying this condition, the temperature of discharge brine will increase so that the heat energy utilization efficiency will decrease. Therefore the suitable spraying temperature and circulation air rate have to be chosen. Based on theoretical calculation, selecting W/G = 1.6–3.0, $T_h = 60°C$, and W/G = 11, $T_h = 90°C$ is available.

Fig. 6.72 shows the effect of gas–liquid ratio W/G to the energy consumption per unit water production when spraying temperature $T_h = 90°C$. As the result, the bigger W/G is, the less the energy consumption rate is when the spraying temperature is fixed. Fig. 6.73 also shows the effect of the spraying temperature to the energy consumption per unit water production when W/G = 11. It is seen from the curve that the spraying temperature is with an optimum value other than the higher the better or the less the better. In practical engineering, it should be noticed to find the optimum point to achieve the maximum efficiency.

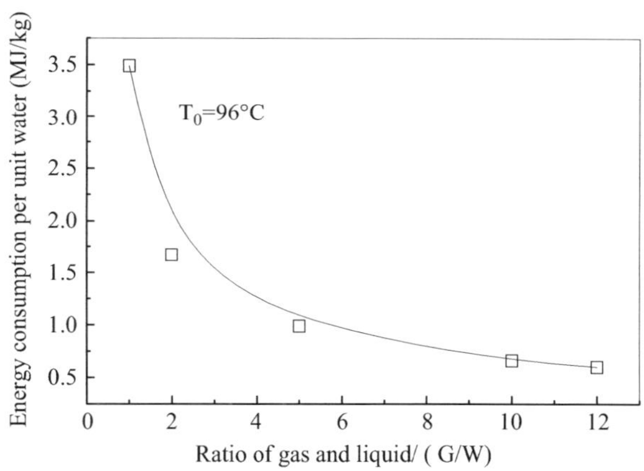

Figure 6.72
Variation of energy consumption rate with ratio of gas–liquic.

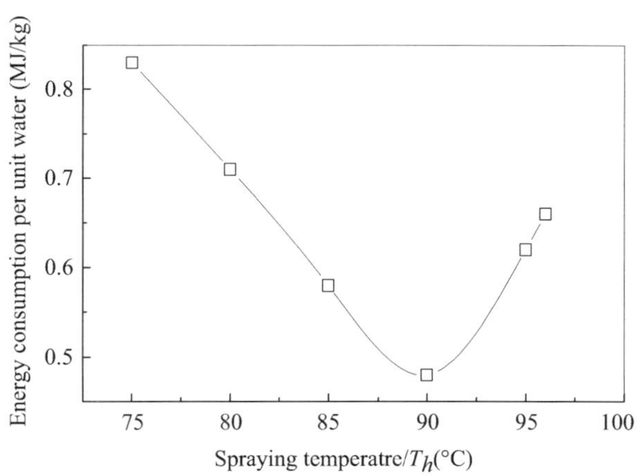

Figure 6.73
Variation of energy consumption rate with ratio of gas–liquid rate with the spraying temperature.

6.11.2 Humidification–Dehumidification Solar Desalination System With Falling Film Evaporation

6.11.2.1 The Structure and Principle of the System

The literature [39] introduced a saturation air humidification–dehumidification type desalination system with falling film evaporation. It can be driven by solar energy or other waste heat. Its operation principle is shown in Fig. 6.74. The system used falling film evaporation technology, which usually is used in industry fields, so that the unit scale is reduced largely and the efficiency of evaporator and condenser is elevated. The basis processes in the system are to use the solar heat or waste heat to elevate the temperature of seawater in a seawater storage tank and use the height difference to distribute the seawater with flow rate G_A and temperature T_A to the evaporator to form falling film evaporation. Fig. 6.75 shows the filler in the evaporator with a honeycomb structure, which is made of paper. The air is forced to pass through the evaporator and lets the hot seawater evaporate to produce the vapor. After the evaporator, the air is closed to saturation state. Its relative humidity is more than 95%. When the air is cooled, the vapor is condensed on the surface of the condenser to produce the freshwater. The brine goes down to the bottom of the evaporator and is recycled to the hot water tank to be heated again. In this way, the freshwater is continuously produced. The system will have better performance if the air in the output can be recycled to the input to be used again.

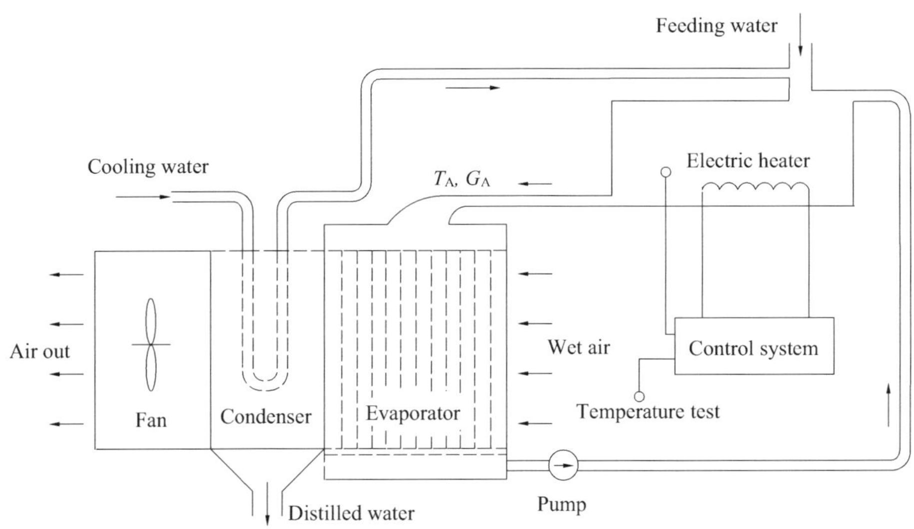

Figure 6.74
Schematic diagram of humidification–dehumidification solar desalination system.

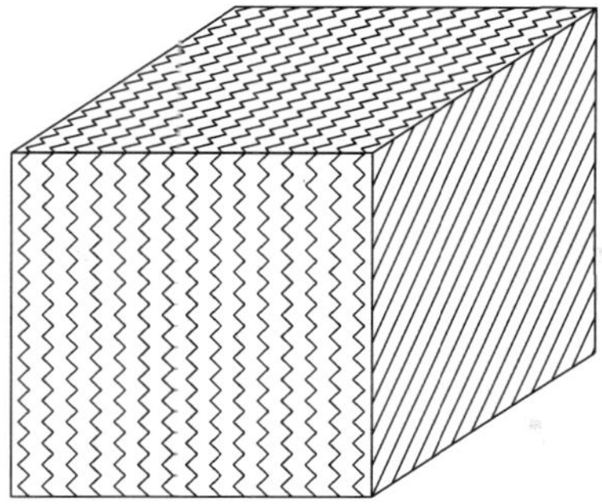

Figure 6.75
Filler in evaporator with paper honeycomb structure.

Compared with other desalination systems with similar working principles, this device has the following characteristics.

1. The heat and mass transfer is improved greatly by using enhanced convection process. Evaporation efficiency is higher.
2. The wet air flows in straight channels so that the pressure loss of air flowing is small. Evaporation and condensation effects are better.
3. The contradiction between rising evaporation temperature and decreasing condensation temperature in a conventional solar still is overcome.
4. The warm saline water can be reused through the circulation, which is good for reducing the heat loss.
5. The vaporization latent is reused sufficiently to preheat the entering seawater.

6.11.2.2 Performance Index

Two indexes are used to evaluate the performance of the system. One is freshwater productivity, M_e, and another is system efficiency, η. Productivity is defined as the freshwater production in unit time per unit solar collector area. The efficiency is defined as useful heat, which is used to produce freshwater over the total input heat energy as follows,

$$\eta = M_p h_{fg} / Q_T \tag{6.72}$$

where, h_{fg} is the vaporization latent of water, Q_T is the total heat energy input, and M_e is total freshwater production.

6.11.3 Experiment of the System

6.11.3.1 Description of the Experimental System

System size is 1 m × 1 m × 1.5 m, in which the humidifying section is 0.6 m long. Measured parameters include water temperature at the input of the humidifier, flow rate, and temperature of hot seawater and cooling water, fan speed, temperature, and humidity of air at different input and output. In the experiment, the hot water is produced by a boiler to replace the solar collector.

6.11.3.2 Experimental Results and Analysis

The influence of hot seawater flow rate passing through the humidifier on system performance is shown in Fig. 6.76. The figure indicates that productivity and efficiency all increase linearly with the seawater flow rate. Air is heated and humidified to near saturation in the honeycomb structure filler in the evaporator. The bigger the seawater flow rate is, the more temperature and humidity of the air at the output of the humidifier. It is beneficial to increase the production of freshwater.

The influence of airflow rate on thermal efficiency is shown in Fig. 6.77. It can be seen from Fig. 6.77 that for 85°C and 65°C seawater temperature at input of humidifier, there are the optimum airflow rates for two conditions. If the seawater temperature is lower, then the optimum airflow rate is bigger. The thermal efficiency reaches 0.85 under the condition of optimum airflow rate. In the actual operation of the device, the airflow rate can be adjusted by changing the rotation speed of the fan. Too big or too small airflow rate is harmful to improving efficiency and production.

The variations of thermal efficiency and freshwater production with heating water temperature under two airflow rates are shown in Figs. 6.78 and 6.79, respectively. When other conditions are fixed, the hot seawater temperature at the input of the humidifier has a great influence on the efficiency and production. The higher seawater temperature can provide more heat energy for wet air, so more vapor enters into the air. In conventional solar stills or solar desalination systems, the evaporation and condensation processes are

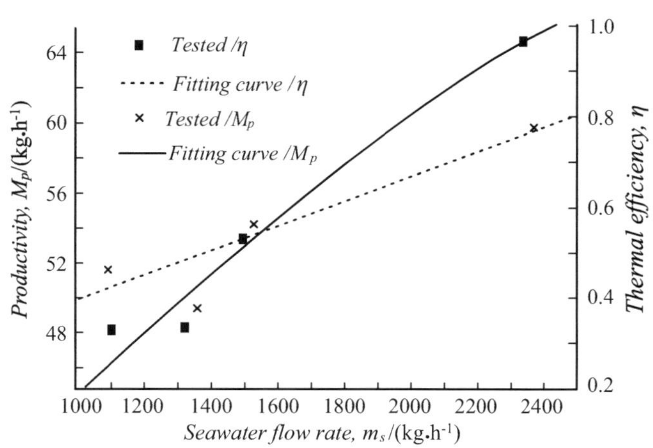

Figure 6.76
Effect of seawater flow on water productivity and thermal efficiency.

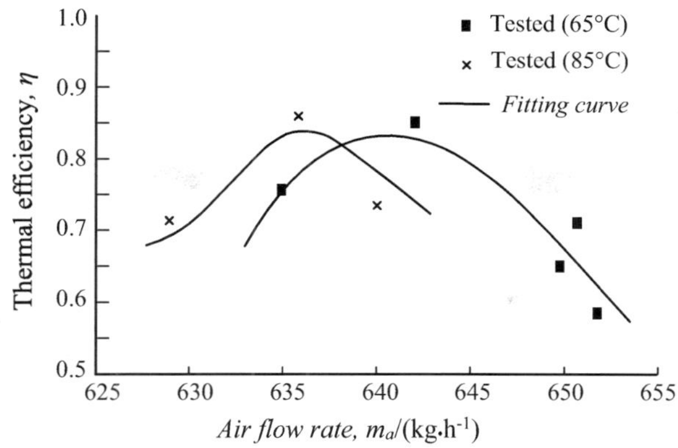

Figure 6.77
Effect of airflow on system thermal efficiency.

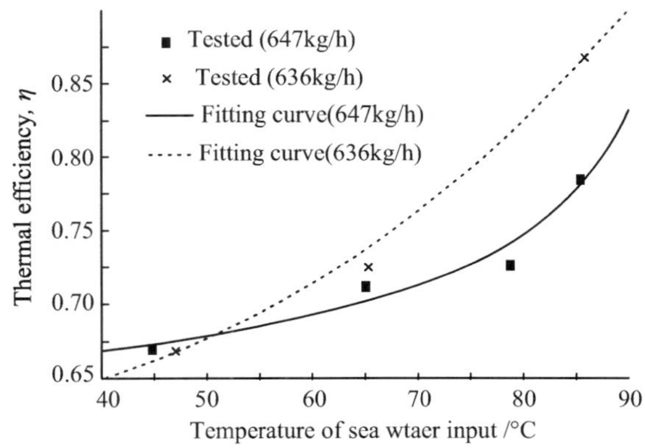

Figure 6.78
Effect of temperature of seawater input on thermal efficiency.

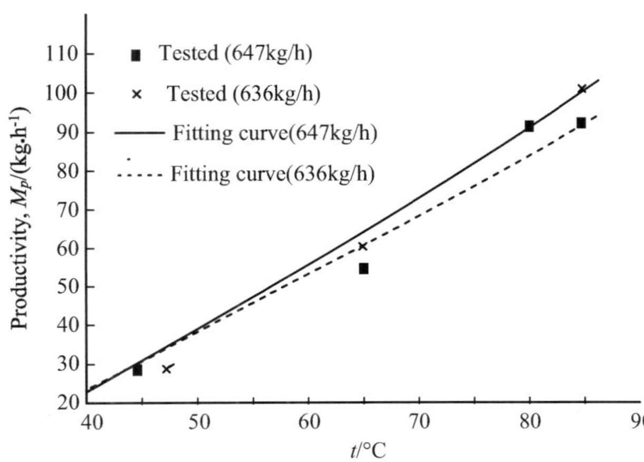

Figure 6.79
Effect of temperature of seawater input on water productivity.

collected in one space. And then the condensation temperature is also elevated when the evaporation temperature is elevated, so that it is difficult to obviously improve the production. In HDD systems, the evaporation process and condensation process is separated. It is convenient to increase the evaporation temperature while keeping the condensation temperature constant, which can improve the performance largely.

References

[1] Linnhoff B, et al. User guide on process integration for efficient use of energy. UK: AIChE; 1982.

[2] Eastop TD, Croft DR. Energy efficiency for engineers and technologists. UK: Longman Scientific and Technical, John Wiley & Sons; 1990.

[3] Linnhoff B. Pinch analysis — a state-of-the-art overview. Trans Icheme Part A September 1993;71:503–22.

[4] Mubarak Ebrahim A. Pinch technology: an efficient tool for chemical plant energy and capital cost saving. Appl Energy 2000;65(1–4):45–9.

[5] Hou S, Ye S, Zhang H. Performance optimization of solar humidification dehumidification desalination process using pinch technology. Desalination 2005;183:143–9.

[6] Hou SB. Two-stage solar multi-effect humidification dehumidification desalination process plotted from pinch analysis. Desalination 2008;222:572–8.

[7] McGovern RK, Thiel GP, Narayan GP, Zubair SM. Performance limits of zero and single extraction humidification–dehumidification desalination systems. Appl Energy 2013;102:1081–90.

[8] Chehayeb KM, Narayan GP, Zubair SM, Lienhard JH. Use of multiple extractions and injections to thermodynamically balance the humidification dehumidification. Int J Heat Mass Transf 2014;68:422–34.

[9] Zamen M, Soufari SM, Amidpour M. Improvement of solar humidification–dehumidification desalination using multi-stage process. In: 14th international conference on process integration, modelling and optimisation for energy saving and pollution reduction, Florence, Italy, May 2011.

[10] Zamen M, Soufari SM, Abbasian Vahdat S, Amidpour M, Zeinali MA, Izanloo H, et al. Experimental investigation of a two-stage solar humidification–dehumidification desalination process. Desalination 2014;332:1–6.

[11] Chang ZH, Zheng HF, Yang YJ, Su YH, Duan ZC. Experimental investigation of a novel multi-effect solar desalination system based on humidification–dehumidification process. Renew Energy 2014;69:253–9.

[12] Kang HF, Yang YJ, Chang ZH, Zheng HF. Performance of a two-stage multi-effect desalination system based on humidification–dehumidification process. Desalination 2014;344:339–49.

[13] Kang H, Wang T, Zheng H. Comparative analysis of regenerative and air-extraction multi-stage humidification–dehumidification desalination system using pinch technology. Desalination 2016;385:158–66.

[14] Fath HES, Ghazy A. Solar desalination using humidification-dehumidification technology. Desalination 2002;142(2):119–33.

[15] Goosen MFA, Sablani SS, Shayya WH, Panton C, Al-Hinai H. Thermodynamic and economic consideration in solar desalination. Desalination 2000;129(1):63–89.

[16] Goosen MFA, Sablani SS, Paton C, Perret J, Al-Nuaimi A, Haffar I, et al. Solar energy for arid coastal regions: development of a humidification-dehumidification seawater green-house. Sol Energy 2003;75(5):413–9.

[17] Cerci Y. The minimum work requirement for distillation processes. Exergy Int J 2002;2(1):15–23.

[18] Cerci Y, Cengel Y, Wood B. The minimum separation work for desalination processes. Proc ASME Adv Energy Syst Div AES 1999;39:545–52.

[19] Qian H, Zhou Z, Zhang L, Wu F, Zhang Q, Zhang Z. Thermodynamic analysis on the theoretical energy consumption of seawater desalination. Desalination 2013;320(1):73–9.

[20] Kalogirou SA. Seawater desalination using renewable energy sources. Prog Energy Combust Sci 2005;31(3):242–81.

[21] Alhazmy MM. Minimum work requirement for water production in humidification–dehumidification desalination cycle. Desalination 2007;214(1–3):102–11.

[22] Camacho CG. Exergo-environmental analysis of processing condensed water from atmospheric air. Desalination 1999;121(2):103–16.

[23] Cengle Y, Cerci Y, Wood B. Second law analysis of separation processes of mixtures. Proc ASME 1999;39:537–43.

[24] Cengle Y, Boles M. Thermodynamics: an engineering approach. 4th ed. New York: McGraw-Hill; 2002.

[25] Mistry KH, Lienhard JH, Zubair SM. Effect of entropy generation on the performance of humidification-dehumidification desalination cycles. Int J Therm Sci 2010;49(9):1837–47.

[26] De Vos A. Thermodynamics of solar energy conversion. Germany: Wiley-VCH, Verlag GmbH & Co. KGaA; 2008.

[27] Stoecker WF, Jones JW. In: Refrigeration and air conditioning. 2nd ed. Singapore: McGraw Hill; 1982. p. 418–9.

[28] Mehrgoo M, Amidpour M. Derivation of optimal geometry of a multi-effect humidification–dehumidification desalination unit: a constructal design. Desalination 2011;281:234–42.

[29] Dayem AA, Fatouh M. Experimental and numerical investigation of humidification/dehumidification solar water desalination systems. Desalination 2009;247:594–609.

[30] Houcine I, BenAmara M, Guizani A, Maâlej M. Pilot plant testing of a new solar desalination process by a multiple-effect-humidification technique. Desalination 2006;196:105–24.

[31] Fath HES, E1-Sherbiny SM, Ghazy A. Transient analysis of a new humidification-dehumidification solar still. Desalination 2003;155:187–203.

[32] Zhani K, BenBacha H. Experimental investigation of a new solar desalination prototype using the humidification dehumidification principle. Renew Energy 2010;35:2610–7.

[33] Yuan G, Wang Z, Li H, Li X. Experimental study of a solar desalination system based on humidification–dehumidification process. Desalination 2011;277:92–8.

[34] Hou S, Zhang H. A hybrid solar desalination process of the multi-effect humidification dehumidification and basin-type unit. Desalination 2008;220:552–7.

[35] Hongfei Z, Xinshi G. Steady-state experimental study of a closed recycle solar still with enhanced falling film evaporation and regeneration. Renew Energy 2002;26(2):295–308.

[36] Wu G, Zheng H, Kang H, Yang Y, Cheng P, Chang Z. Desalination 2016;378:100–7.

[37] Nawayseh NK, Farid MM, Omar AA, Sabirin A. Solar desalination based on humidification process: II—computer simulation. Energy Convers Manag 1999;40(13):1441–61.

[38] Liang F-S, Huang J-M. Experimental study on a saturated air solar still. Acta Energiae Solaris Sin 1982;3(4):360–7.

[39] Dai YJ, Wang RZ, Zhang HF. Parametric analysis to improve the performance of a solar desalination unit with humidification and dehumidification. Desalination 2002;142:107–18.

CHAPTER 7

Solar Desalination System Combined With Conventional Technologies

Chapter Outline

Solar Energy Desalination Technology. http://dx.doi.org/10.1016/B978-0-12-805411-6.00007-5

The development direction of solar desalination technology is the combination of traditional desalination technologies. With the help of sophisticated schemes, measures, material, and administrative experience from traditional desalination technologies, the development of solar desalination technology will be greatly promoted.

For existing solar desalination technology, neither its energy utilization efficiency, nor its water productivity per solar collector area are very high. We might even say that they are far below traditional desalination systems. And existing solar desalination technology has to face extraordinary obstacles to the aspects of scale and industrialization. Thus, many experts point out that it is critical for development of solar desalination technology to combine with the traditional desalination industry.

The running principles of traditional desalination technologies fall broadly into two categories: (1) phase transition process. This includes multistage flash (MSF), multiple-effect boiling (ME), and vapor compression (VC). (2) Dialysis process. This mainly includes reverse osmosis (RO) and electrodialysis (ED). The specific energy consumption and specific water production of these running methods are shown in Table 7.1. The most significant advantage of phase transition process is the ability to reuse the latent heat of evaporation and condensation processes. This advantage makes it possible to preheat the seawater that enters the device, and condensate some vapor into freshwater production at the same time. In this case, the coefficient to describe system performance is the PR. PR is a dimensional coefficient, meaning that the mass of produced freshwater per energy consumption or per unit mass vapor consumption, and it could be kg/2326 kJ, because 2326 kJ is the latent heat for 73°C water. The dialysis process represents the method that is able to directly produce freshwater with electricity. Its prominent strengths are device simplicity, easy for realizing miniaturization, and high specific water production rate.

Using a combination of solar energy and regular desalination devices, we need to consider many economic and technical factors, which are: (1) the device must be applicable to solar

Table 7.1: Energy consumption of each type desalination method.

Desalination method	Type of feed water	Specific volume of energy consumption (kWh/m^3)	Rate of energy recovery (%)	Cost of water production (dollar/m^3)
MSF	Seawater, saline water	81–144 (thermal)	0.6–6	1–5
MED	Seawater, saline water	50–194 (thermal)	6–38	2–9
MD	Seawater, saline water	100–60 (thermal)	3–5	13–18
VC	Seawater, saline water	11–15 (electric)	—	5–20
RO	Seawater, saline water	1.2–19 (electric)	10–51	3–27
ED	Saline water	0.6–1 (electric)	25–50	3–16
Solar distillation	Seawater, saline water	100–650 (thermal)	—	15–30

energy technology, like nowadays, solar energy application can only be run at mid–low-temperature working condition; (2) about the given solar heating device, the desalination system should have relative high efficiency; (3) based on the freshwater we need, the desalination process should be decided; (4) the process of seawater treatment needs to be considered; (5) issue of equipment investment; (6) device-mounting area, etc. The solar heating device can be solar collector, like a flat-plate solar collector, and evacuated tubular solar collector; solar pond, and it could also be a solar photovoltaic (PV) system. The energy consumptions of MSF, ME, VC, RO, and ED are listed in Table 7.1 [1].

So far, there are lots of running systems combined with solar energy and traditional desalination technology around the world; Fig. 7.1 shows a solar desalination system in China [2]. However, among these systems, solar energy and desalination are considered as basically two separated parts, because the solar collector nowadays just plays an energy-supply role for the desalination system (like the desalination device that drives the ED process by utilizing electric power, which is converted from solar irradiation). Thus, these systems are not yet complete solar desalination devices. Solar desalination in the future will be an integration and interpenetration of solar technology and up-to-date desalination technology.

7.1 Multistage Flash Solar Desalination System

7.1.1 The Flash Principle and Technology [3]

7.1.1.1 The Principle of Multistage Flash

Multistage-flash (or MSF) is short for multistage flash distillation.

Figure 7.1
Solar desalination system in China.

The principle of the multistage flash (MSF) process is as follows: pumped seawater is heated to a certain temperature into the flash room, where the pressure is controlled at lower than the heated seawater saturated vapor pressure in the condition of the temperature. Thus, the heated seawater becomes superheated water after flowing in to the flash room and it will vaporize fast so that the temperature of the seawater will decrease. The water vapor will condense on the wall of the exchanger pipe and become pure water. Meanwhile, the incoming seawater is preheated in the exchanger. The MSF is based on this principle, making seawater flow into several flash rooms where pressure declines one by one. The temperature of the seawater decreases and the concentration of seawater is increased, until the temperature is approaching the natural seawater temperature (actually, it is a little higher).

Fig. 7.2 is a schematic of the two-stage flash. The heated seawater flows into flash rooms successively, where the pressures are declining gradually, and vaporizes and condenses.

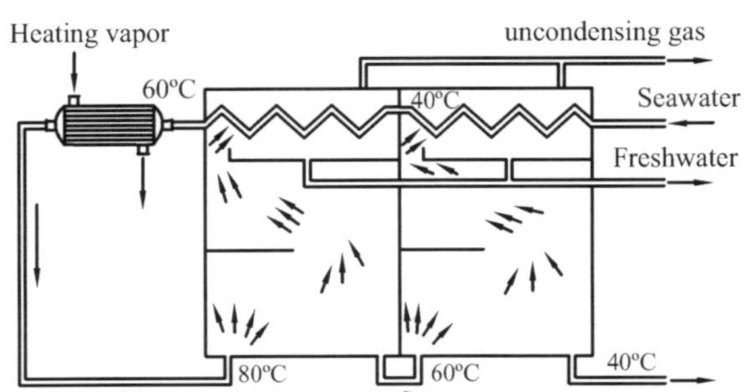

Figure 7.2
Multistage flash principle.

The temperature of each stage of seawater is decreasing. As for normal equipment, the minimum decrement is 1.6–2°C. In order to ensure the necessary condensing rate, a proper temperature difference should be maintained, it is 4–6°C generally. During this process, the latent heat which is needed for vaporization is provided by the remaining heat of the condensate seawater. For example, when the temperature of seawater declines from 100°C to 60°C, the percentage of vaporized mass is about 7.1%.

MSF is developed to improve the demerit of scaling of multiefficiency desalination, which has several advantages such as simple and reliable devices, uneasy scaling, easy for large-scale realization, operating flexibility and importantly, it can utilize low-quality heat and wasted heat. So it is applied and developed fast once it comes out. Nowadays, MSF is not only used for desalination, but extensively for thermal power plants, boiler water supply of petrochemical works, management and recovery of industrial wastewaters and brackish waters of mines, etc.

7.1.1.2 Process of MSF

The operation of MSF can be classified as tubular type and circular type. In the tubular type the seawater is heated to the highest temperature and then goes through the heat exchanger. Then stage by stage process evaporation takes place until the last stage, when the concentrated seawater is wasted out. Its working process and temperature curve are shown in Fig. 7.3. The circular type is where the concentrated seawater in the last stage is not poured out totally, instead, a large portion is recycled with the new incoming seawater and processing flash evaporation. Its working process and temperature curve are showing in Fig. 7.4.

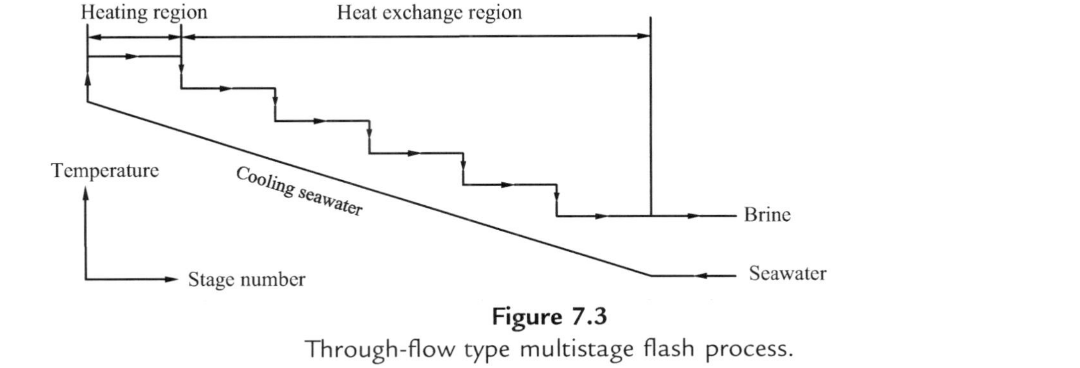

Figure 7.3
Through-flow type multistage flash process.

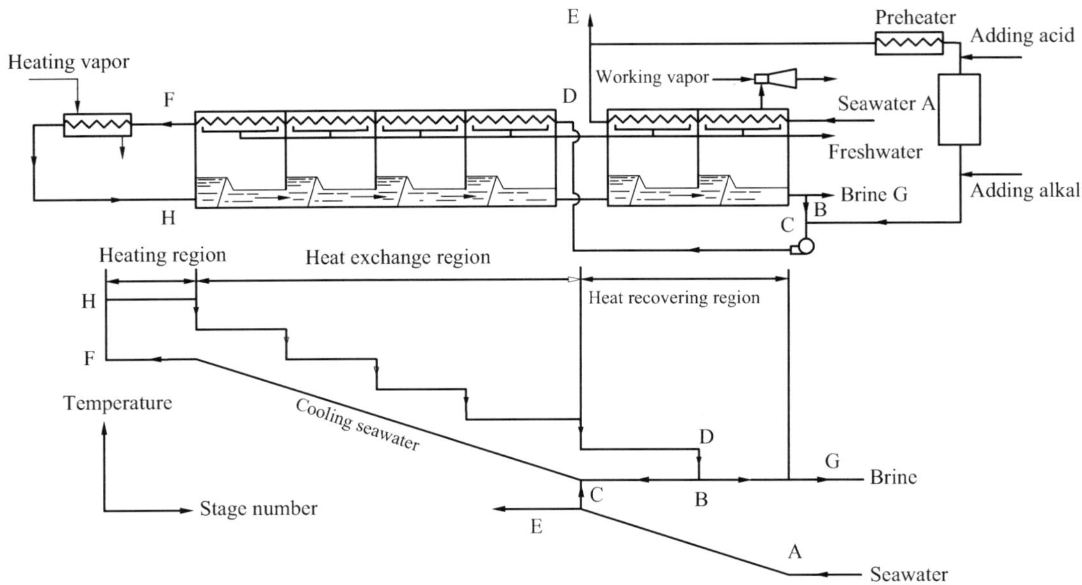

Figure 7.4
The process of recycling multistage flash.

As in the above description, MSF utilizes the remaining heat of concentrated seawater in each stage, which illustrates that the pure water production is directly relative to the flow and flash evaporation temperature.

Set parameters are as follows:

$C_{p,f}$, Specific heat capacity of seawater (kJ/kg °C)
M_f, Flow rate of seawater (kg/h)
T_h, The highest temperature of seawater (°C)
$T_{b,N}$, The lowest temperature of concentrated seawater (°C).

The seawater released heat q is:

$$q = C_{p,f} \cdot M_f \left(T_h - T_{b,N}\right) \tag{7.1}$$

Then presume h_{fg} is the latent heat of vaporization (kJ/kg).

Thus the pure water production M_d is (kg/h):

$$M_d = \frac{C_{p,f}}{\overline{h}_{fg}} \cdot M_f \left(T_h - T_{b,N}\right) \tag{7.2}$$

where $\overline{h}_{fg}$ is the average latent heat of evaporation.

Eq. (7.3) shows that the pure water production M_d can be improved by increasing the flow rate of seawater M_f and the temperature range of flash evaporation $(T_h - T_{b,N})$. $(T_h - T_{b,N})$

is usually constant, because T_h is constrained by scaling and $T_{b,N}$ is effected by temperature of cooling water, in order to increase pure water production, the flow rate should be increased. The tubular type has a large wasted volume of concentrated seawater so that much more seawater needs to be added. In this way, the managed volume of seawater is too large to realize high efficiency and economy. Therefore using the circular type is more popular than the tubular type.

7.1.1.3 Main Procedure Parameters

1. Coefficient of evaporation and concentrated ratio

   Coefficient of evaporation is defined as the division of pure water production and the flow rate of seawater. Combining with Eq. (7.2), the total coefficient of evaporation of the system is:

$$R = \frac{\dot{M}_d}{\dot{M}_f} = \frac{C_{p,f}}{h_{fg}}\left(T_h - T_{b,N}\right) \tag{7.3}$$

   As for the i stage, the temperature of the seawater is $T_{b,i}$ and the outlet temperature is $T_{b,i+1}$. Considering the seawater's specific heat capacity, latent heat, and flow rate as constant in the small temperature difference, the pure water production is:

$$\dot{M}_{di} = \frac{C_{p,f}}{h_{fg,i}}\dot{M}_{f,i}\left(T_{b,i} - T_{b,i+1}\right) \tag{7.4}$$

   So the coefficient of evaporation in i stage is:

$$R_i = \frac{C_{p,f}}{h_{fg,i}}\left(T_{b,i} - T_{b,i+1}\right) \tag{7.5}$$

   Although Eq. (7.5) is calculated about the specific case, it can also apply in other series for actual advices, nevertheless total evaporation temperature difference reaches 50–90°C. The main reasons are that, for actual advices, seawater circulated is greater than water production, and seawater saltiness increases a little.

   The concentrated ratio is defined as the ratio of the concentration of seawater in the last stage (total dissolutive solid, TDS) and the supplied seawater. When the feed water is seawater, the concentrated ratio is generally no more than two for the scaling. This aims at normal seawater concentration. The concentration of normal seawater is 3.4483% (mass) and the concentration of exhausted seawater (TDS) is approaching 70,000 mg/L. If the feed water is river water or brackish water with different salinity, the concentrated ratio could be even improved. On the contrary, the concentration of seawater in some inland bays is higher than the normal value; the concentrated ratio might be lower than 2.0. Practical experience points out that the salinity should be no more than 70,000 mg/L no matter what kind of feed water is.

2. Heat equilibrant equation
 The heat equilibrant equation of the system is shown below (for the circular type):
 Input heat:
 a. The net heat input by heater, Q (kJ)
 b. Sensible heat carried by the cooling seawater, $M_{f,a} \cdot C_{p,a} \cdot T_{f,a}$. Here, $M_{f,a}$ is flow of cooling seawater, $C_{p,a}$ is the specific heat at constant pressure of seawater, $T_{f,a}$ is the temperature of seawater. Note that the replenishment heat supplied for seawater is ignored here.

 Output heat:
 a. The heat of the poured cooling seawater, $(M_{f,a} - M_{f,e}) \cdot C_{P,e} \cdot T_{f,e}$; Here $M_{f,e}$ is the seawater's flow into the circulation, $T_{f,e}$ is the temperature, and $C_{P,e}$is the specific heat at constant pressure.
 b. The heat of the output pure water, $M_d \cdot C_{p,d} \cdot T_d$; Here M_d is the production of pure water, $C_{p,d}$ is the heat of the poured cooling seawater, and T_d is the temperature.
 c. The heat of the poured concentrated seawater, $M_{f,b} \cdot C_{p,b} \cdot T_{b,N}$; Here $M_{f,b}$ is the flow of the concentrated seawater in the last stage, $C_{p,b}$ is the specific heat at constant pressure and $T_{b,N}$ is the temperature.

 Here the heat loss is ignored and the equation is:

$$Q + M_{f,a} \cdot C_{p,a} \cdot T_{f,a} = \left(M_{f,a} - M_{f,e}\right) \cdot C_{p,e} \cdot T_{f,e} + M_d \cdot C_{p,d} \cdot T_d + M_{f,b} \cdot C_{p,b} \cdot T_{b,N} \tag{7.6}$$

 or:

$$Q = \left(M_{f,a} - M_{f,e}\right) \cdot C_{p,e} \cdot T_{f,e} + M_d \cdot C_{p,d} \cdot T_d + M_{f,b} \cdot C_{p,b} \cdot T_{b,N} - M_{f,a} \cdot C_{p,a} \cdot T_{f,a} \tag{7.7}$$

 On the other hand, heat Q is provided by heater. Thus:

$$Q = M_{f,h} \cdot C_{p,f}(T_h - T_f) \tag{7.8}$$

 In the above equation, M is the mass flow rate of fluid water (kg/s), C_p is the constant-pressure specific heat of salty water or seawater (kJ/kg °C), T is pure water and a,b,e are A,B,E points in Fig. 7.4, respectively, which represent the status of salty water or seawater.
3. Temperature difference between two adjacent stages and evaporation capacity
 Assume that the equipment has N stages and the temperature differences are equivalent between each other. The temperature relation of the circulation from the output of heater to the output of the last stage is:

$$T_h - T_{f,1} = T_{f,1} - T_{f,2} = \cdots = T_{f,N-1} - T_{f,N} = \frac{T_h - T_{b,N}}{N} \tag{7.9}$$

 where, T_f is the temperature of salty water (°C). For the gas phase, the relation is:

$$T_{V,1} - T_{V,2} = T_{V,2} - T_{V,3} = \cdots = T_{V,N-1} - T_{V,N} = \frac{T_h - T_{V,N}}{N} \tag{7.10}$$

where T_v is the temperature of the vapor (°C). Generally, $T_{f,i}$ is not equal to $T_{V,i}$, instead, $T_{f,i} > T_{V,i}$, which is caused by the increment of the salty water boiling point, the static pressure of deep salty water and the resistance loss of water flowing into the condensing room, etc. Set $\Delta T_{fVi} = T_{f,i} - T_{V,i}$, which is the temperature difference loss of i stage. The evaporation capacity from the first stage to N stage is [4]:

$$M_d = M_f\left[1 - (1 - R)^N\right] \quad (7.11)$$

where, R is the average evaporation coefficiency of each stage.
From Eq. (7.11), we can obtain that:

$$\frac{M_d}{M_f} = 1 - (1 - R)^N \quad (7.12)$$

The total evaporation coefficient is:

$$\beta = 1 - \left[1 - \frac{C_{pf}\left(T_h - T_{b,N}\right)}{Nh_{fg}}\right]^N \quad (7.13)$$

Thus:

$$M_d = M_f\left[1 - (1 - R)^N\right] = M_f\beta \quad (7.13a)$$

Eq. (7.13a) illustrates the relation between water production, circular water, and quantity of stages, which is of benefit in designing and estimating relative parameters.

4. Gain output ratio
The so-called gain output ratio is the ratio between evaporation capacity of the evaporator and saturated steam consumption of heat. It is usually replaced by energy consumption for convenience. r can be defined as follows:

$$r = \frac{M_d \cdot h_{fg}}{Q} \quad (7.14)$$

As the approximate calculation, considering the temperature difference is equal between two adjacent stages and ignoring the heat loss during the circulation, r is equal to the quantities ratio of flash stages and heat dissipation stages. In practice, the r of mid–small-type MSF equipment is about 3–6. It is larger than 10 for large equipment.

5. Heat transfer area
Three parts are included when calculating heat transfer areas, which are areas of heater, heat recovery part, and heat dissipation part. The formulas are given as follows:
① Heat transfer area of heater A_i

$$A_i = \frac{Q}{K_i \Delta Q_i} = \frac{M_d \cdot h_{fg}}{rK_i \Delta Q_i} \quad (7.15)$$

where K_i is the coefficient of heat transfer (kW/m^2 °C), ΔQ_i is the logarithmic mean temperature difference.

② Heat transfer area of heat recovery part A_r

$$A_r = \frac{(N-j)M_d h_{fg}}{NK_r\Delta Q_r} = \frac{N-j}{N}\cdot\frac{1}{K_r\Delta Q_r}\cdot r\cdot M_f C_{p,r}\left(T_h - T_{b,N}\right)\frac{j}{N} \tag{7.16}$$

where K_r is the coefficient of heat transfer (kW/m^2 °C), ΔQ_r is the average logarithmic mean temperature difference of total flash stages, $C_{p,r}$ is the average constant-pressure specific heat capacity (kJ/kg °C), and M_f is the mass flow rate (kg/s).

③ Heat transfer area of heater A_p

The cooling water flow rate $M_{f,a}$ is approximately equal to circular salty water flow rate M_f and the specific heat capacity is very close, which illustrates that the changes of temperature of the cooling water in heat recovery part and the circular salty water in heat dissipation part are equivalent. So the heat dissipation area in each stage is equal to the heat recovery area and the total heat transfer area is:

$$A_p = \frac{j}{N-j}\cdot A_r \tag{7.17}$$

It is worth pointing out that the result calculated by Eq. (7.17) is not accurate enough, so that it can only be used for estimation. If the exact result is needed, it should be done considering the actual heat dissipation condition to calculate the average logarithmic mean temperature difference of each stage and then figuring out the total heat transfer area.

6. The highest temperature of salty water

The so-called highest temperature of salty water, T_h, illustrates the water heated by heater and flow into the first flash stage. According to Eq. (7.3), it is of benefit for improving evaporation coefficient R and total evaporation coefficient β if increasing T_h. Actually, it is also of benefit for increasing the pure water production ratio r and declining the heat transfer area. However, the highest temperature of salty water is limited by many factors. Firstly, the thermal dynamical analysis shows that the increment of temperature makes no contribution to heat-power efficiency when T_h is about 160−170°C.

Secondly, the higher it is, the more harmful are scaling and corrosion. The temperature of salty water is limited at 120°C for the acid method of a scale-prevention system. As for the medicament method, the highest temperature depends on the anticrustator. For example, using phosphoric acid as an anticrustator, the highest temperature is limited at 90°C, while adopting a pH-adjusting method, the temperature approaches 125°C. T_h should be controlled within 104−121°C and the temperature of steam should be no higher than 127°C in actual use. When in summer, offshore seawater is diluted by river water, T_h can be increased up to 128−130°C.

7. The relations among gain output ratio with flash stages number and temperature difference between stages
   Pure water production ratio r is a basic factor for designing MSF equipment. It is also an indication of the economic efficiency of distillation plants. When designing MSF equipment, r should be determined based on production scale and requirement of technology and economy. Generally, small equipment should be designed with small r and vice versa. At present, the equipment that produces pure water is lower than 5000 t, r is about 5–8. As for production capacity lower than 10,000 t, r is about 10 or larger, and if the production capacity is about 30,000–40,000 t, r is larger than 10.
   On the other hand, when T_h and $T_{b,N}$ are defined, the temperature difference is larger if flash stages become lower so that the process of evaporation is more deviating from the reversible process. Thus, the quality of heat that is recovered by cool feed water in an inner condenser becomes lower, which will make the pure water production ratio become lower. Therefore increasing flash stages properly and decreasing the temperature difference is better.
   In actual designing, the temperature difference between stages is variable, which is related to the scale and price of energy etc. As for the large equipment with production of 1000–10,000 ton levels, the temperature difference between stages is 2–3°C and thus the stages are about 30–40. With small-scale equipment with production of hundreds of tons or tens of tons, the temperature difference between stages is 5–6°C. As for some small equipment using wasted heat, the temperature difference between stages is a little higher considering the cost of the device.
8. The relations of evaporation ratio, flash stages, and heat transfer
   The cost of the heat transfer pipe is a large proportion of the cost of MSF equipment. According to statistics, the cost of pipe materials is 20% of the total investment for an MSF device with production 5 ~ 50 kt per day. The fabrication and installation fees of pipe materials share 50% of the total fabrication and installation cost. Therefore decreasing the heat transfer area is necessary to save copper pipe and alloy when designing an MSF device.

When analyzing Eq. (7.17), it can be obtained that the heat transfer area is declining with the increment of flash stages when other parameters are constant. However, this tendency becomes slow when flash stage numbers are large enough, which is shown in Fig. 7.5 [4].

Fig. 7.5 illustrates how to choose flash stages and evaporation ratio. From the view of reducing the investment fee for the heat transfer pipe, when r is given, the design points should choose between two dashed lines. If the flash stages are too little, although the equipment structure is simple, the heat transfer area requirement is too much. If flash stages are too much, not only is the effect of reducing the heat transfer area not evident, but also the complexity of the equipment is increased and the manufacturing cost rises rapidly.

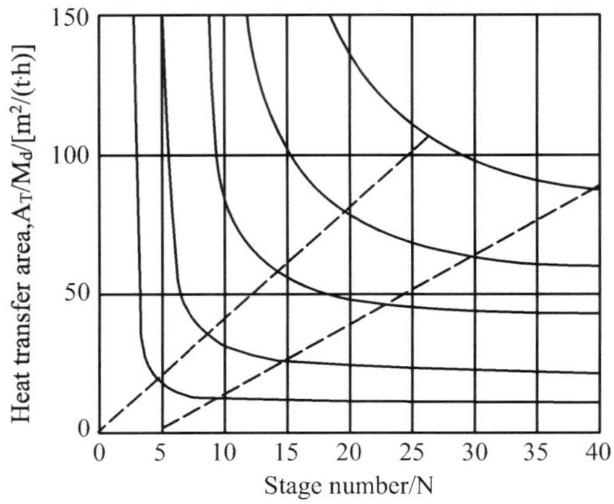

Figure 7.5
Relationship between stage number and heat transfer area.

7.1.2 Solar Desalination With Multistage Flash Processes

7.1.2.1 Theoretical Estimation of Solar MSF Desalination System

Fig. 7.6 shows a schematic of a typical solar MSF desalination system. It can be divided into three parts according to energy flow, including heat input part, heat recovery part, and heat dissipation part. The heat input part includes solar collector, water tank, auxiliary heater, and heat exchanger 1, etc. It is mainly used for providing heat for flash stages or heat recovery part. The heat is provided by a solar collector directly. There are several kinds of solar collectors that can be chosen, including flat plate collector, vacuum tube collector, and focusing type collector, etc.

The temperature of seawater will rise to T_h after flowing through the heat exchanger 1, and then flow into the heat recovery part. It will evaporate and condense several times and the

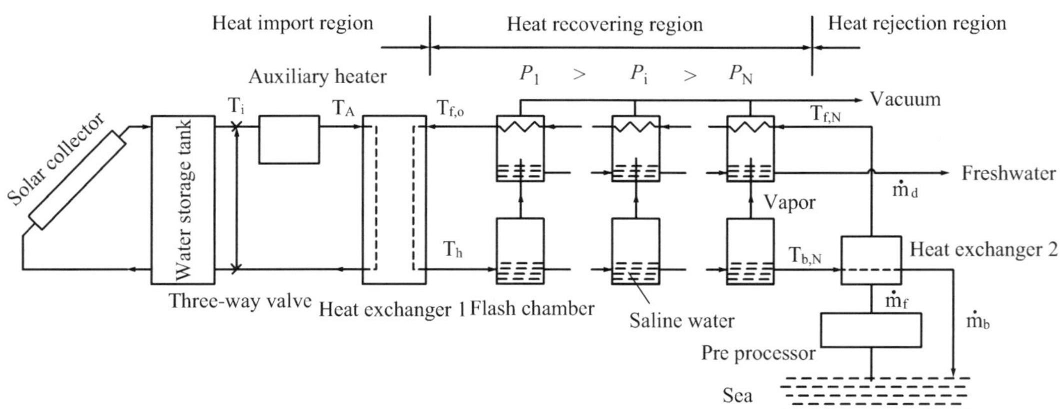

Figure 7.6
Schematic of typical solar MSF desalination system.

heat will transfer to the new incoming cool seawater that enters the exchanger pipe. The concentration of the hot seawater increases continuously and then flows out the last flash stage with temperature $T_{b,N}$. The concentrated seawater flows into the heat dissipation part, where it will change heat with cool water via heat exchanger 2. Finally, the concentrated seawater is poured out. The new incoming seawater will flow into the water treatment equipment to process conventional treatment. Then it will flow into the heat dissipation part and get heat from heat exchanger 2. The temperature of it will rise to $T_{b,N}$, and then it will flow into the heat recovery part. The temperature will rise continuously by absorbing the vapor's latent heat stage by stage. When flowing out the heat recovery part, the temperature will be $T_{f,o}$ and follow on, it flows into heat exchanger 1 and gets heat from the solar heater. Finally, the temperature will reach to T_h.

The MSF system has two critical parameters, which are pure water production ratio r and performance coefficient R. For solar MSF systems, they can be described as follows:

$$r = (T_h - T_{b,N})/(T_h - T_{f,o}) \tag{7.18}$$

$$R = \dot{m}_d/\dot{m}_f \tag{7.19}$$

where $\dot{m}_d$ and $\dot{m}_f$ are mass flow rate of pure water and seawater, respectively. The temperature range of flash evaporation is the heat recovery measure of total production of pure water and heat recovery.

For i stage, the ratio of pure water production and salty water flow rate can be calculated by a heat equivalent equation:

$$r_i = \Delta T_{b,i} C_i / h_{fg,i} \tag{7.20}$$

where $\Delta T_{b,i} = T_{b,i-1} - T_{b,i}$. T_{bi} is the temperature of salty water of i stage, which is defined as 1.4–2.8°C in actual devices. It is limited by the boiling point and pressure. C_i is the specific heat capacity, $h_{fg,i}$ is the latent heat of vaporization, and $T_{v,i}$ is the temperature of vapor. C_i and $h_{fg,i}$ are the function of $T_{v,i}$.

According to mass balance relation, the pure water production of i stage and the total system production can be written as:

$$\dot{m}_{d,i} = r_i \dot{m}_{b,i-1} \tag{7.21}$$

$$\dot{m}_d = \sum_{i=1}^{N} \dot{m}_{d,i} \tag{7.22}$$

where

$$\dot{m}_{b,i} = \dot{m}_{b,i-1}(1 - r_i), \quad \dot{m}_{b,o} = \dot{m}_f \tag{7.23}$$

$$N = (T_h - T_{b,N})/\Delta T_{b,i} \tag{7.24}$$

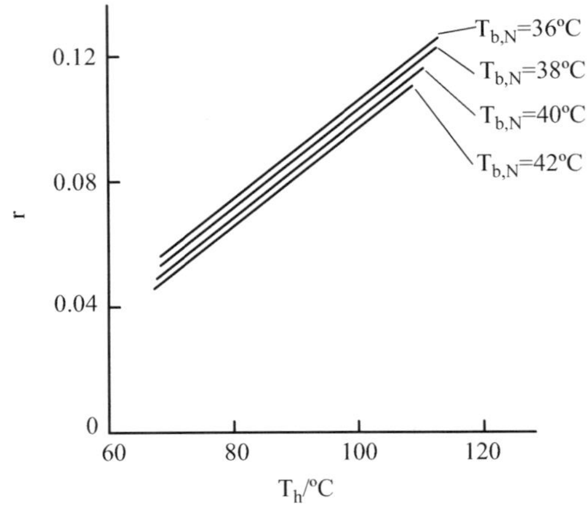

Figure 7.7
Change of GOR with T_h under different $T_{b,N}$.

It can be also obtained that $\Delta T_{V,i} = \Delta T_{b,i}$ from the heat balance relation of the i stage and $\Delta T_{V,i} = T_{V,i-1} - T_{V,i}$. Generally, $T_{b,N} = 36–43°C$. In this case, set $\Delta T_{b,i} = 2°C$, the relations among r, T_h, and $T_{b,N}$ are shown in Fig. 7.7. According to Eq. (7.24), r is the function of N and the change of T_h is related to $\Delta T_{b,i}$.

The ratio of salinity of poured concentrated seawater X_b (the total dissolved solids mass, parts per million (ppm)) and incoming seawater X_f can be written as follows:

$$X_b/X_f = \dot{m}_f/\dot{m}_b \tag{7.25}$$

where $\dot{m}_f$ and $\dot{m}_b$ are mass flow rates of seawater flowing in and flowing out.

In the heat dissipation part, the mass flow rate of poured concentrated seawater can be written by the mass balance relation:

$$\dot{m}_b = \dot{m}_f - \dot{m}_d = \dot{m}_f(1 - r) \tag{7.26}$$

The temperature of incoming seawater (flowing out of heat exchanger 2) $T_{f,N}$ can be written as:

$$T_{f,N} = T_{tr} + (1 - r)E(T_{b,N} - T_{tr}) \tag{7.27}$$

T_{tr} is the temperature of incoming seawater that flows out of primary treatment equipment and E is the effective factor of heat exchanger.

The total heat input is:

$$\dot{Q}_L = \dot{m}_f C_{p,f}(T_H - T_{f,o}) \tag{7.28}$$

where $C_{p,f}$ is the constant-pressure heat capacity. For a solar desalination system, the heat input is provided by a solar collector. The total heat that the solar collector can supply is related to several factors including solar irradiance, lighting area of collector, temperature of surroundings, and the heat loss of the collector, etc. Thus, different solar collectors have different heat-supplying capacities of unit area. As for the flat plate collector and compound parabolic concentrators (CPC) collector, the heat-supplying capacities of unit area are given as follows:

$$\text{Flat plate collector:} \quad \dot{Q}_u = A_c F_R [Q - U_L (T_{in} - T_a)] \tag{7.29}$$

$$\text{CPC collector:} \quad \dot{Q}_u = A_a F_R [Q - (U_L / C)(T_{in} - T_a)] \tag{7.30}$$

where A_c and A_a are lighting areas, respectively, F_R is the factor of heat transfer, Q is the total received solar energy, U_L is the coefficient of heat loss, C is the concentration ratio of CPC, T_{in} is temperature of the fluid flowing into the collector, and T_a is the temperature of surroundings. Generally, the total energy consumed by the desalination system is provided by a solar collector. When the solar energy cannot match the energy consumption, that is $\dot{Q}_u < \dot{Q}_L$, the energy lacking, $(\dot{Q}_L - \dot{Q}_u)$, is supplied by an assistant heat source. Thus, a scale factor can be defined to illustrate the percentage of solar energy supply, which is $f = \dot{Q}_u / \dot{Q}_L$.

Farwati (1997) employed the above equations and analyzed the solar MSF desalination system shown in Fig. 7.6 numerically with several assumed conditions as follows [5]:

1. In the heat recovery part, set $\Delta T_{b,i} = 2°C$, $T_{b,N} = 36°C$ and assumed that every stage of the system is in the heat equilibrium state.
2. The coefficient of heat exchanger is $E = 0.7$.
3. As for flat plate solar collector, $A_c = 1.0\ m^2$. The collector has a single glass cover plane. The inner of the collector is set as a receiver with selective coating. $U_L = 4.5\ W/m^2°C$, $F_R = 0.9$. The collector is aslant assembled and the tilt angle is equal to local latitude.
4. As for the CPC collector, $A_a = 1\ m^2$. The collector has a single glass cover plane. The inner of the collector is set as a receiver with selective coating. The tilt angle is 16° and the reflectivity of parabolic reflectors is 0.92. $U_L = 6\ W/m^2°C$, $F_R = 0.87$. The aperture of the collector is changed every day so that the energy collection efficiency at noon-time is highest.
5. The average values of climate parameters in a month are chosen and the system works in the daytime.
6. The mass flow rate of seawater is properly chosen so that the energy $\dot{Q}_u$ received by collector with 1 m^2 can be consumed and the pure water production is maximum. So the average daily pure water production in a month is $\dot{M}_d = f\dot{M}_f r$. $\dot{M}_d$ and $\dot{M}_f$ are the mass flow rate of pure water and seawater every day. $M_f = Q_L / C_{pf}(T_h - T_{f,o})$, $\dot{M}_d = Q_u r / C_{p,f}(T_h - T_{f,o})$. In this case, $f = \dot{M}_d / \dot{M}_{d,\max}$.

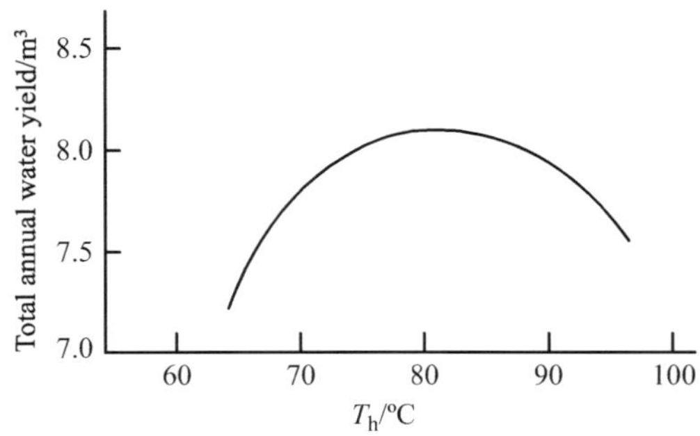

Figure 7.8
Total annual pure water yield changes with T_h.

Farwati found that the total pure water production of a year $\sum \dot{M}_d$ is the function of T_h and $\dot{M}_d$ is also related to T_h. That is, considering the performance of the solar collector, an optimal T_h can be found to obtain the maximum pure water production.

Fig. 7.8 shows the annual pure water production changes with T_h. When $T_h = 122°C$, the maximum annual pure water production is 8.2 m^3 per square meter.

For the flat plate solar collector with 1 m^2 area, $T_h = 80°C$, the daily pure water production and *f* in each month are shown in Fig. 7.9. The maximum daily pure water production is 43.7 kg, which was obtained in August.

For CPC collector with 1 m^2 area, $T_h = 122°C$, the daily pure water production and *f* in each month are shown in Fig. 7.10. The maximum daily pure water production is 64.9 kg, which was also obtained in August.

It can be seen that the solar MSF desalination system has high water production in theory and it is worthy of further study.

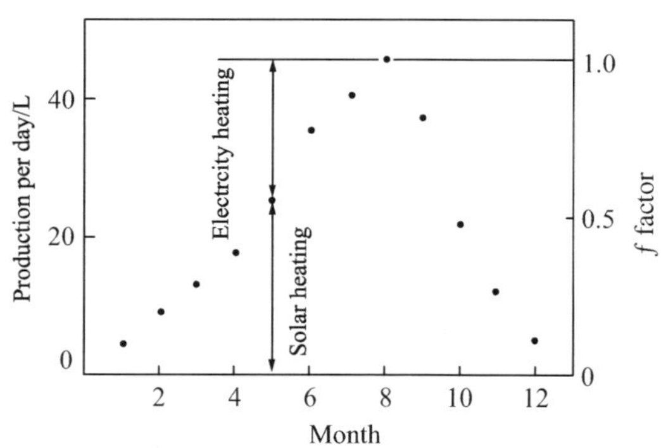

Figure 7.9
Monthly average daily production changes with time when $T_h = 80°C$.

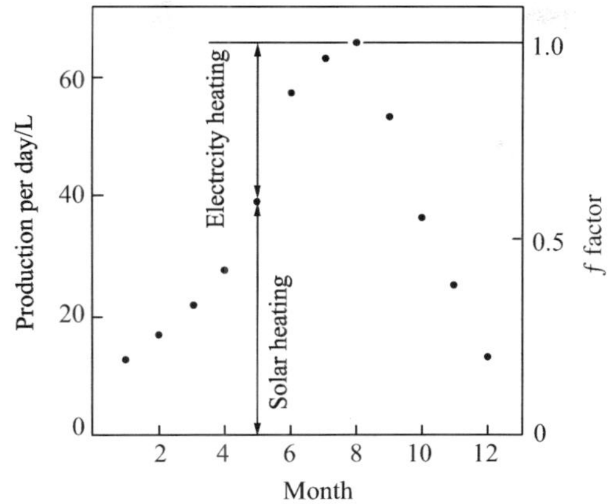

Figure 7.10
Monthly average daily production changes with time when $T_h = 122°C$.

7.1.2.2 The Function of the Solar MSF Desalination System

The MSF desalination system is the most popular desalination device which is widely applied in industry at present. The current level of technology needs heat 221.9–276.3 MJ and electricity 3–4 kW h [4] for a product of 1 m^3 pure water. Therefore from the point of energy usage, thermal energy is the main consumed energy of the MSF system, which creates a good condition for the use of solar energy. Besides, the operation temperature of MSF is not very high and the highest temperature is about 120°C. If the stages decrease, MSF can even act with T_h lower than 90°C. The present solar collectors can easily provide heat energy with this temperature. At present, the mid–high-temperature solar collection technology is maturing gradually. It can easily provide heat energy with temperature higher than 120°C. So it could become possible to develop out the high-efficient desalination devices by using a solar energy technology and modern desalination technology combination.

The MSF process needs precise pressure control and the pressure difference should be maintained between two adjacent stages, which makes it hard to control the parameters in the transient state. Meanwhile, it creates a challenge for the solar heat system as the irradiance of the sun is changing all the time, sunny and cloudy, day and night. To solve this difficulty, a heat-reserving device or assistant heater can be added.

In 1985, Maustafa et al. reported on a solar MSF desalination system built in Kuwait [6]. The structure of it is similar to Fig. 7.1. The system produces water at 10 m^3 per day. The solar collector is a parabolic trough with a 220 m^2 aperture area. The water storage tank is 7 m^3. This system has 12 stages. The stored heat can drive the system acting continuously in cloudy weather and night. According to the report, this system could produce 10 times more water than a solar distiller with the same lighting area.

With the development of MSF technology, there are several solar MSF desalination systems that have been established. The main area is the Middle East, including Kuwait, Saudi Arabia, etc. The stages of the system are designed as 10–30 (with temperature difference of stages at 2°C). The daily water production is 60–100 kg/m^2.

The literature [7] also describes a combination of MSF evaporator with a fluidized bed evaporation (FBE) system, as shown in Fig. 7.11.

7.2 Multieffect Solar Distillation System

Heat energy is the main consumption of the traditional multieffect evaporation desalination system, so solar energy can be used to take the place of other energy. In recent years, due to the development of technology, a variety of falling film evaporation processes and various kinds of strengthening heat transfer processes have been recognized gradually, which promote the development of multieffect evaporation desalination systems greatly. It is also combining solar energy much more close than ever before. Inventions of various kinds of multieffect evaporation desalination systems with solar energy have become a major highlight in the desalination field.

7.2.1 The Principle of Multieffect Distillation

Fig. 7.12 shows the working process of multieffect evaporation desalination using solar energy. The solar collector can take place by industrial heat if the heat source is not solar energy [8].

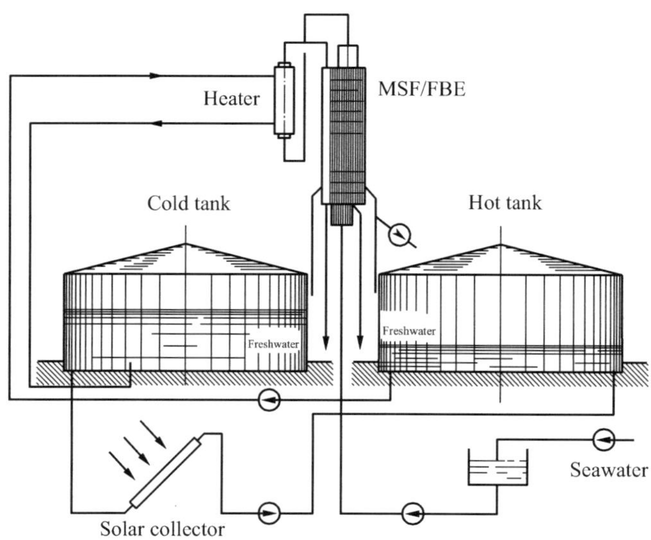

Figure 7.11
Solar energy MSF/FBE system.

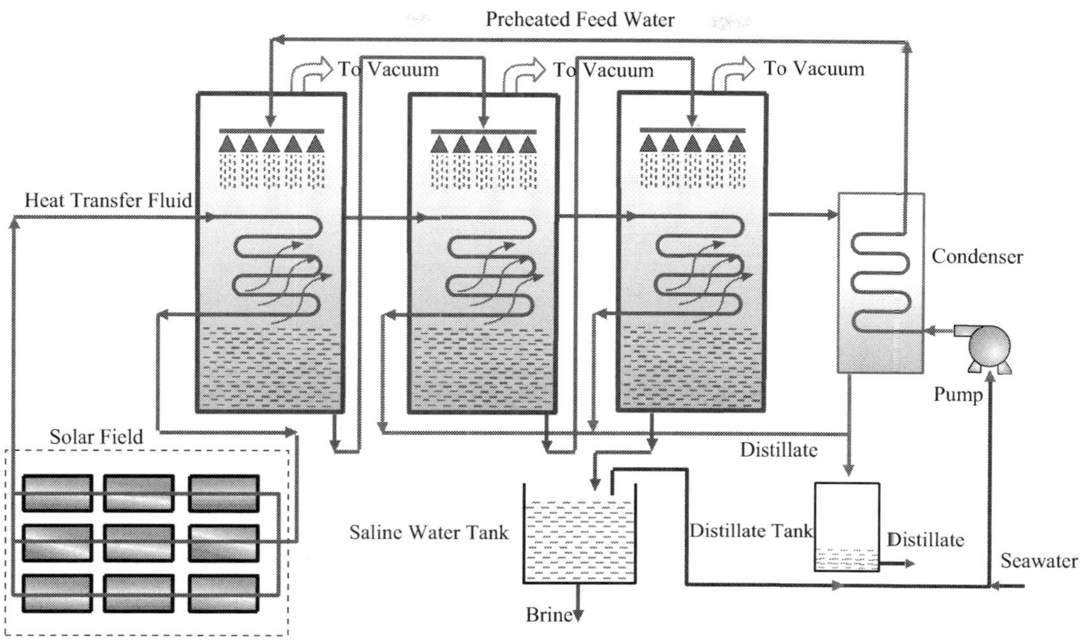

Figure 7.12
Schematic diagram of solar multistage evaporation seawater desalination.

This is similar to the MSF system, multieffect evaporation desalination also consists of several working units and these units are usual called "effects."

The multieffect evaporation desalination system is an integrated system that includes several single evaporators. The vapor flow from the last evaporator will go into the next evaporator as heated vapor and condense and the not liquefied vapor will flow out and go to the next evaporator. The vapor will be liquefied in each evaporator successively and the pressure and temperature from left to right decline. Each process in the evaporator is one effect. Thus, there will be a double effect, triple effect, and multieffect. The feed water has several ways to flow into the system include reverse flow, advection (flow into each effect, respectively), cocurrent flow (flow into the system from the first effect) and reverse flow to preheat then cocurrent flow, etc. In the large-scale desalination device, the last way is usually used and the others are usually used in chemical evaporation. Multieffect evaporation has been applied in desalination and large–mid-sized thermal power plants for boiler water supplement.

According to the energy balance relationship and ignoring the heat loss, 1 kg of heated steam condensation can produce about 1 kg of secondary steam in an evaporator. If this 1 kg secondary steam is flowing into another evaporator as the heated steam, the condensation can vaporize out again at about 1 kg of secondary steam. So it will produce

2 kg of pure water by using 1 kg heated steam if combined in two evaporators in series. Similarly, with the number of *N* evaporators in series, each 1 kg of heating steam consumption can vaporize *N* kg pure water. It can be seen that this is the root cause of the high water rate of multieffect evaporation. Also, it is the reason that these devices are paid more attention to.

In order to realize the multieffect evaporation, the water's boiling point in the second evaporator must be lower than the water in the first evaporator. It can also be comprehended that the pressure in the second evaporator must be lower than the water in the first evaporator. Otherwise there is no heat transfer temperature difference, and evaporation could not occur. The effect of the second evaporator is to cool down the secondary steam that is flowing out the first evaporator. In the secondary evaporator, when the water's boiling point is decreasing due to decompression, the secondary steam coming from the first evaporator will heat the water in the second evaporator. In a similar way, in the triple effect, the decompression will work on the last effect, the last effect will decompress the middle effect and the middle effect will decompress the first effect. Thus, for multieffect, only one decompression device is needed to ensure the normal operation of the desalination device.

7.2.2 The Classification of the Multieffect Distillation Technological Process

The multieffect distillation has three kinds of technological processes including normal flow, contra flow, and parallel flow [4].

1. Normal flow
   Normal flow indicates that the feed water going through the desalination system from the first effect to the last effect is by sequence. It has several advantages. First, because the vacuum degree from the first effect to the last effect is increase gradually, the pressure of each effect is decreased successively. Thus, there is no need to use a pump to pump the feed water from the last effect to the next. Instead, the pressure difference between two adjacent effects can drive the feed water to flow automatically. Second, because of the temperature difference that always exists between the effects and it is decreasing by sequence, so when the feed water flows from the last effect to the next effect, the water is turning to superheated water for the relatively low pressure and thus it will evaporate a little part of extra steam and produce more pure water. It is similar to flash evaporation. Third, for the high-concentrated feed water with high viscosity in the latter effects, the heat transfer coefficient is relatively low and the boiling point of the water is increasing, which makes it hard to maintain the temperature difference between two adjacent effects. But for the seawater desalination, this question is not very serious because of the concentration among every effect is not high enough.

2. Contra flow

   Contra flow indicates that the flowing direction of feed water is reversed to the flowing direction of heated steam. The feed water flows into the system from the last effect that has the highest vacuum degree. The concentration of feed water from the last effect to the first effect increases orderly. Because the pressure of the front effect is higher that the latter effect, there needs to be a pump to drive the feed water to flow. And as the temperature is turning high with the feed flowing into the front effects one by one, there is no flash evaporation when feed water goes from one effect to the last and even it is needed to preheat to the boiling point. So it can be seen that the demerits and advantages are opposite. Contra flow is appropriate for high-concentration and high-viscosity feed water as the last evaporation is in the first effect, which has the highest temperature. So even if the feed water has a high concentration, the viscosity can be decreased. It is commonly used for chemical processes.
3. Parallel flow

   Parallel flow indicates that each effect has an independent feed water inlet and they are feeding parallel. The heated steam flow into the first effect is the original steam and then flows into the rest of the effects. Parallel flow is appropriate for the easily frozen feed such as salt manufacturing. Once the water is heated and evaporated, the liquid water reaches saturated status and the salt will crystallize. Thus, there is no need to pump the concentrated water to the other effects.

   As for the water treatment process, the main objective is to obtain pure water. So reverse flow and parallel flow are not available. Besides, the heat utility efficiency of normal flow is higher than the two other ways.

7.2.3 The Classification of Multieffect Distillation Equipment

Multieffect distillation equipment has many various types. As for different feed water with different concentration, the evaporator can be different.

1. According to the arranging direction of the evaporating tubes, it can be classified as vertical tube evaporator (VTE) and horizontal tube evaporator (HTE).
2. According to the flowing type of feed water, it can be defined as a forced convection evaporator and film type evaporator. In the membrane type evaporator, the feed water is beginning to flow naturally after flowing out the water distributor.
3. The film type evaporator can also be classified as a climbing film evaporator and falling film evaporator according to the flowing direction of the feed water.
4. The falling film evaporator can be classified as a vertical tube falling film and horizontal tube falling film.
5. According to the structure of the combination of all the effects, it can be classified as horizontally combined evaporator and tower type evaporator. VTE and HTE can both be combined to be a tower type evaporator.

There are several types of evaporator that can consist of multieffect evaporation system. The commonly used are three kinds as follows:

1. Sink tube
   Sink tube evaporator is the large class of evaporation device in which the heating tube is immersed by feed water. Generally, the sink tube has many types, such as line tube, snake-like tube, U-type tube, stand tube, horizontal tube, etc. The feed water has two kinds of circulation, including natural convection circulation and forced circulation. The sink tube evaporator is invented relatively earlier and it is an easy operation. It has some demerits such as easily scaling, feed water has relatively deeper immersion and the large temperature difference loss. So the number of effects is usually lower than six. In recent years, forced circulation has been applied in seawater desalination, which makes the number of effects reach 10. A desalination system with a 10-effect forced circulation evaporator can produce about 1500 m^3/d. The heat transfer area of each evaporator is about 1600 m^2 and the diameter of the separator between two adjacent effects is about 5–8 m. The other kinds of sink tube evaporators are widely used for chemical evaporation. Some desalination and purification systems in the power plants and ships are also using sink tube evaporators. However, this kind of evaporator is not very popular at present on the whole. On one hand, it is hard to realize scaling prevention and descaling, which makes the system more complex and hard to operate. On the other hand, the heat transfer coefficient is relatively smaller and the structure looks giant. Fig. 7.13 shows the structural and flowing schematic of one kind of sink tube seawater desalination system.

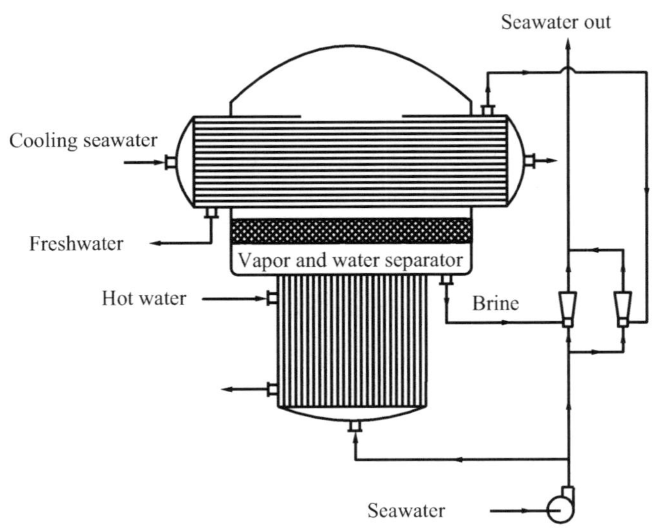

Figure 7.13
Immersion tube type single-effect distiller process.

2. Vertical tube evaporation
   Here we mainly illustrate the inner tube falling film vertical tube evaporator, which is shown in Fig. 7.13. This kind of evaporator has two advantages. One is that the evaporation is inner tube film-like evaporation. Two sides of the tube wall have phase change. Therefore the heat transfer coefficient is much higher. Besides, the temperature difference loss caused by the feed water's static liquid column is eliminated. This kind of system has a high pure water production ratio. Generally, the number of effects is about 11–13, and the pure water production ratio is about 9–10. Another merit is that the feed water goes through the system only once. The feed water consumption is relatively lower, so that the cost of water pretreatment is low and water pumping power dissipation is small. Thus, the cost of operation is low. However, the scaling cannot be ignored for this kind of evaporation, especially when the feed water is distributed unevenly or there is a lack of feed water. In this case, the tube wall will appear as a dry area and thus scaling appears. So it has a high requirement for scaling prevention and descaling. Generally speaking, this kind of evaporator is not suitable for seeding method. The main descaling method is a chemical method and scaling prevention can be realized via the rational design of temperature and concentration.
   Fig. 7.14 shows the structure of a VTE. A feed water distributor is set in the top of the evaporator. The function of it is to distribute the feed water from the bottom evenly on each vertical tube. The feed water will flow along the inner wall of the tube slowly and become liquid film and then realize film evaporation. A separation chamber used for separating water and vapor is underneath the evaporator.
   Fig. 7.15 shows the flowing process of a modernized multieffect falling film using VTE [3]. The temperature and pressure of each effect are declining in order from left to right. The feed water that separated from the condenser will be pretreated and then pumped into preheaters E_n, E_{n-1}, …, E_3, E_2, E_1 by pump G1 to preheat. Then it is

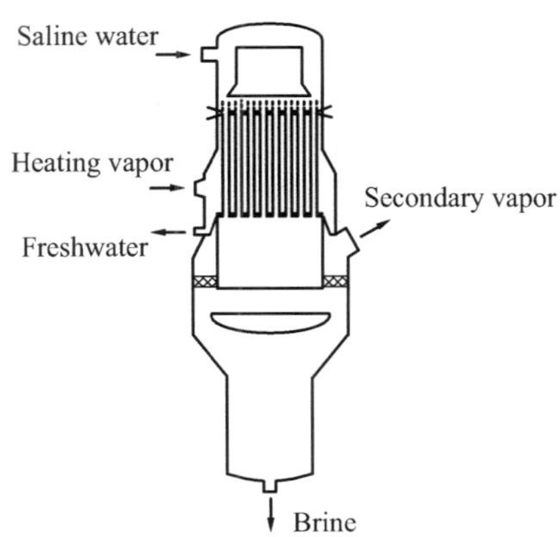

Figure 7.14
Schematic diagram of a vertical tube evaporator.

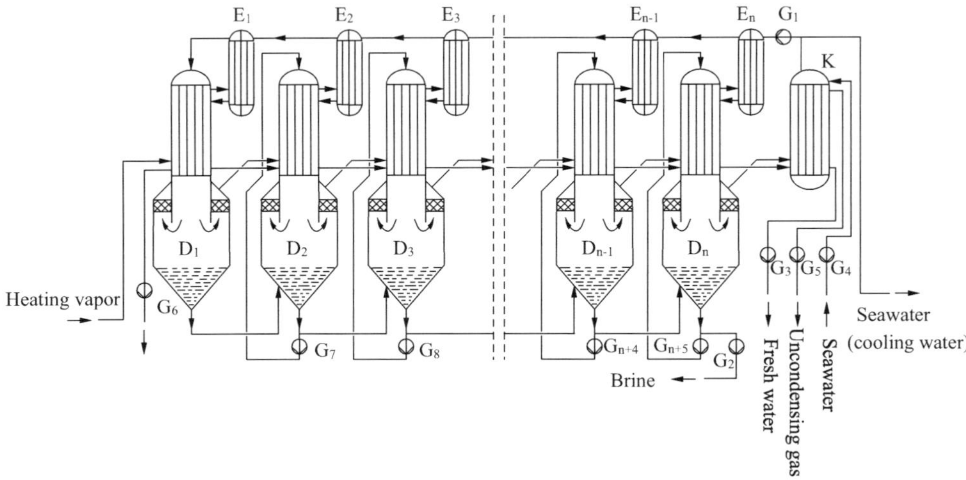

Figure 7.15
The process of multieffect falling film with vertical tube evaporation. D, evaporator; E, preheater; K, condenser.

pumped into the top of the first effect D1 and distributed onto the transfer tubes' inner wall according to the requirement. The outside of the tube holds the heated steam. The new evaporate steam from the inner tube wall will reach these parathion chambers together with the falling feed water and they will be separated here. The secondary steam will be pump in the next effect after passing through the demister. The rest of the feed water will be driven into the neat effect evaporator by the pressure difference. Starting from the second effect, each effect has its own circulation pump G_7, G_8, ..., G_{n+4}, G_{n+5}, which will pump the feed water onto the top of evaporator and distribute the water on the vertical tubes evenly. It will circulate this operation until the last effect Dn. The pure water produced by every effect would also flow through every effect along with the pressure dropping. Sensible heat will be reused and, finally, it will flow into the condenser K. The accumulated pure water will be pumped out. The concentrated feed water will be poured out from the bottom of Dn.

Compared with single-effect evaporation, the multieffect evaporation can reuse the heat energy several times and the pure water production ratio will multiply with the number of effects. However, the cost of the device would also double with the number of effects. Thus, the number of effects should not be increased unnecessarily.

3. Horizontal tube evaporation

The horizontal tube evaporation makes the feed water spray onto the outer walls of the horizontal tubes with a spray thrower. The water will became a water film attached to the wall of the tubes. The steam (or the secondary steam from the last effect) will condense in the inner tube. It has the same merits and demerits as the VTE, while the height of the device is much lower than the VTE. The device is compact because of all the tubes, spray pipes, and water-vapor separator being integrated into one cylinder. It has low heat transfer resistance and energy consumption. Recently, the newly developed aluminum-alloy tube HTE has been investigated by many countries. It has relatively lower temperature so that the scaling and erosion are reduced greatly, which makes sure of the high heat transfer coefficient. Besides, because the resistance of the gaseous phase is lower than liquid, the static pressure heat loss is eliminated. The heat transfer temperature difference can be much smaller so that the low-quality heat can still be used. In some designed devices, the evaporation temperature of the first effect is only about 75−55°C. Therefore it is very beneficial for connecting the low-temperature air exhaust of the power plant. At present, the aluminum-alloy tube HTE can be used for the assembling seawater desalination system with single pure water production of 20,000 m^3/d. The energy consumption of unit pure water production compares favorably with RO, but the water quality is better than RO. The development of falling film horizontal tube evaporation or low-temperature multieffect is better than VTE in recent years. The schematic of HTE is showed in Fig. 7.16.

Fig. 7.17 shows the structural schematic of one kind of HTE low-temperature multieffect desalination (MED) system. This device was also called a (MED) system,

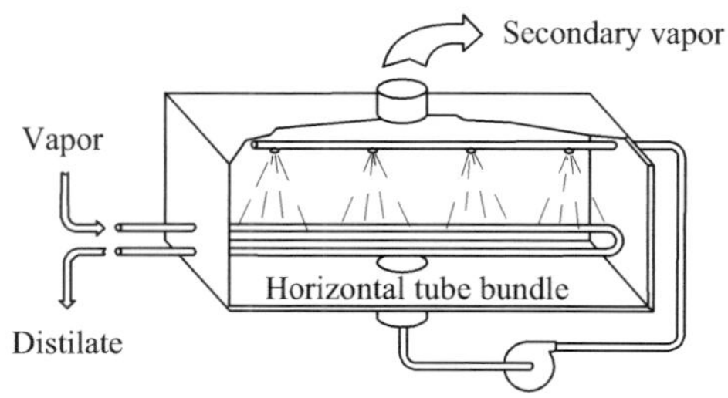

Figure 7.16
Horizontal tube bundle falling film evaporator.

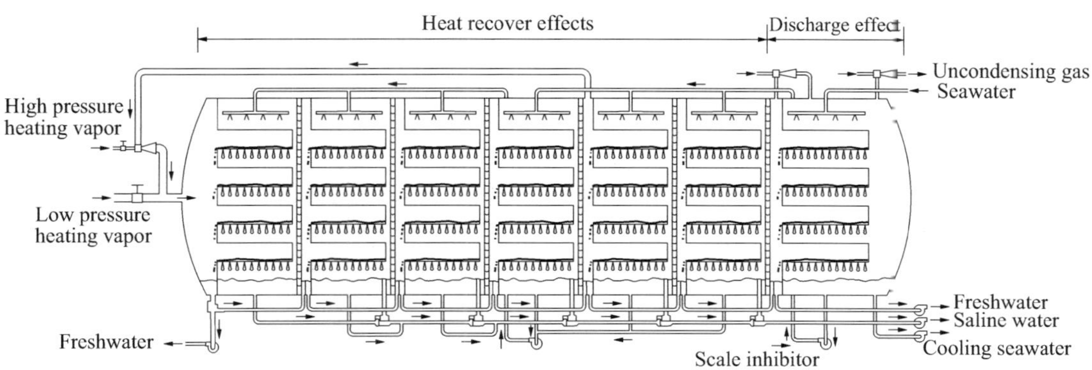

Figure 7.17
Low-temperature multieffect distillate process.

which was developed by America and Israel [4]. The system has seven effects and it was divided into two parts for circulation. The front six effects are heat recovery effects and the last effect is the heat dissipation effect. The large part of concentrated water is poured out while the remaining part flows into the fourth, fifth, and sixth effects to evaporate again. And then the concentrated salty water would be pump into the first, second, and third effects to continue evaporation. Finally, the concentrated salty water is poured out of the system. The pure water flows through every effect and is finally pumped out. The low-pressure steam pumped out of the steam turbine of the power plant flows into the first effect as the heat source. The steam is pumped back to the thermal equipment of power plant after condensing in the tube.

7.2.4 The Temperature Difference Between the Effects and the Factor Influencing Effect Number

7.2.4.1 The Temperature Difference Distribution Between Two Effects

When the working state is approaching the steady state of the multieffect evaporation system, if there is no thinking of temperature loss, the heat transfer rates should be equivalent to each other. For example, a three-effect evaporation system has the following relationship:

$$\begin{aligned} q_1 &= q_2 = q_3 \\ K_1A_1(t_0 - t_1) &= K_2A_2(t_1 - t_2) = K_3A_3(t_2 - t_3) \\ \Delta t_1 &= t_0 - t_1, \quad \Delta t_2 = t_1 - t_2, \quad \Delta t_3 = t_2 - t_3 \\ R_1 &= \frac{1}{K_1A_1}, \quad R_2 = \frac{1}{K_2A_2}, \quad R_3 = \frac{1}{K_3A_3} \end{aligned} \tag{7.31}$$

where Δt_1, Δt_2, Δt_3 are the heat transfer differences of each effect, respectively. R_1, R_2, R_3 are the heat resistance.

Thus

$$\frac{\Delta t_1}{R_1} = \frac{\Delta t_2}{R_2} = \frac{\Delta t_3}{R_3} \tag{7.32}$$

Eq. (7.32) illustrates that for the multieffect evaporation, the heat transfer difference of each effect has direct ratio relations to the heat resistance. When the initial temperature of steam in the first effect and the boiling point of the water in the last effect are confirmed, the condensing temperature of the secondary steam in each effect is determined by the heat resistance in the next effect. So it can be seen that during the multieffect evaporation, if the heat resistance of one arbitrary effect is increased, such as the wall of the device is scaling, bad circulation, the condensed water accumulated too much and the water level declined, causing the heat transfer area to decrease, etc. All these would lead to the heat transfer rate slowing down and the heat transfer rate of other effects will slow down to the same extent. Meanwhile, the temperature difference in this effect will increase and the temperature difference in other effects will decrease. As a result, the steam condensing temperature in effects before the broken effect are rising, while the effects that are behind the broken effect have falling temperature.

7.2.4.2 Factors Influence on Number of Effect

The application of multieffect evaporation is to deduce the consumption of steam and cut the operation cost. However, it is limited by temperature loss and the extent of energy utility, the number of effects is limited.

The effects that cause temperature loss are as follows:

1. The increase of boiling point will lead to the declination of temperature. The temperature difference of evaporation indicates that the temperature of the heated steam compares to the boiling point of seawater. Because the vapor pressure of seawater is lower than pure water, under the condition of some intensity of pressure, the boiling point of seawater is higher than that of pure water. This phenomenon is defined as the elevation of boiling point. The higher the concentration is, the higher the boiling point is.
   In the evaporator, the temperature of steam produced by boiling seawater is higher than the steam produced by pure water when the intensity of pressure is equal. In other words, the water surface of boiling seawater will produce steam and the steam is the superheated steam related to the pressure of the evaporator. So when the steam moves to the space of the evaporator, the temperature will decline to the temperature of the space which is conforming to the pressure. This phenomenon is defined as temperature declination. The degree centigrade of temperature declination is equal to the elevation of boiling point. Temperature declination directly influences the utility extent of secondary steam.
2. The elevation of boiling point caused by static pressure head. The seawater filled in the evaporator has a depth, which leads to the pressure in the deep area being larger than the surface. Actually, it is the addition of the water column and air pressure. Thus, the boiling point of the deeper water is higher than at the surface. The elevation of boiling point caused by static pressure head makes the actual heat transfer temperature difference lower than the temperature difference between the boiling point of seawater and condensing temperature of steam.
3. Mechanical resistance loss and heat loss. On the walls of the pipes between two effects, the frictional resistance and heat loss are unavoidable, which will lead to the decrement of pressure when the steam moves from one effect to the next. The condense temperature would also decrease.

Because of the temperature loss caused by the above three reasons, the effective temperature difference of multieffect will decrease. The effective temperature difference is distributed to every effect so that the increase in the number of effects will makes the temperature difference between two effects decrease. According to these experiences, the effective temperature difference between two effects should not be lower than 5−7°C. So the temperature loss not only slows down the heat transfer rate, but also the effect on the production capacity of multieffect device. Besides, it limits the number of effects.

The number of effects is not only limited by temperature loss, but also the extent of heat utility.

Because the seawater's pressure in the evaporator is lower than that of the secondary steam, the latent heat of evaporation of seawater is higher than the latent heat of condensing steam (shown in Fig. 7.18). When producing 1 kg pure water, the actual consumption of heated steam is larger than 1 kg. The experience shows that to produce 1 kg pure water, the consumption of heated steam is larger than the following:

Single-effect	1.10
Double-effect	0.57
Triple-effect	0.40
Four-effect	0.30
Five-effect	0.27

It can be seen that when turning single-effect to double-effect, the vapor mass can be saved by about 50%, and turning four-effect to five-effect, the vapor mass can be saved only by about 10%. When turning 10-effect to 11-effect, the vapor mass can be saved by only about 1%. If the cost of additional effect is equal to the saved cost of vapor, the number of the effects is the limit value.

7.2.5 Process Conditions

1. The highest concentration of seawater and concentration ratio. This is limited by the scaling prevention method, the highest temperature of seawater of VTE is not higher

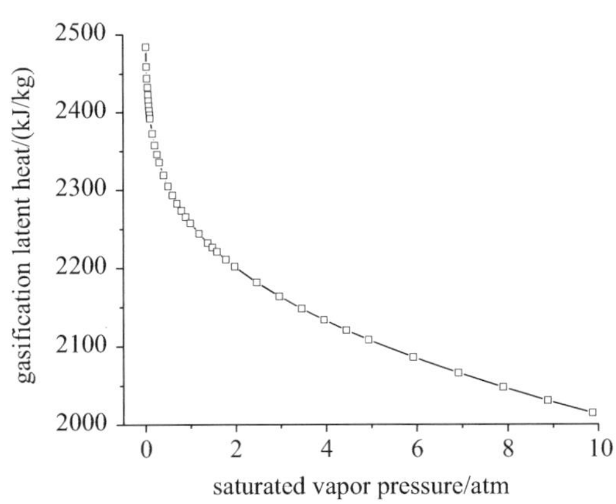

Figure 7.18
Relationship between the latent heat and the saturated vapor pressure.

than 125°C (the temperature of the first effect). From the first effect to the last effect, the temperature of the concentrated seawater declines gradually. The concentration ratio is approaching 3.

2. Pure water production ratio. This is defined as the pure water production per kilogram vapor (= pure water production/consumption of vapor). It is also called the operating ratio, which points at the evaporation ratio of the operation in the chemical unit. Its value is determined by the optimal design of the costs of device and operation.
3. The number of effects. This is mainly influenced by temperature loss. If the temperature of heated steam is 121°C and the largest temperature difference is 96°C, the number of effects is no more than 19 when set the temperature difference between two adjacent effects is 5°C. Because of the temperature loss, the actual number of effects is only 12. The commonly used number is 4−6.

7.2.6 The Advantages and Disadvantages of Multieffect Distillation

Multieffect evaporation, especially VTE, is the earliest large-scale seawater desalination device. In 1961, a large desalination plant with 12-effect can produce 3900 m^3 pure water per day. This type can also be used for producing pure water in a power plant.

The multieffect evaporation has many advantages. However, it also has some demerits. When talking about the advantages and demerits, it is usually compared to the MSF system.

The advantages of multieffect evaporation are as follows: (1) The heat transfer processes of multieffect evaporation are boiling and condensing, which are phase changes. Therefore the heat transfer coefficient is very large. For example, the total heat transfer coefficient of a brass smooth tube is about 3000−3850 W/(m^2 K). Use of the double-trough tube is shown in Fig. 7.19, and the enhanced heat transfer coefficient is about 7200−10,300 W/(m^2 K). The above data are measured from seawater. If using freshwater as the feed, the values will be a little lower. Of course, if employing material with low heat conductive coefficient, the total heat transfer coefficient would decrease. In total, the heat transfer area applied in multieffect is less than in MSF. (2) Multieffect evaporation is a once-only

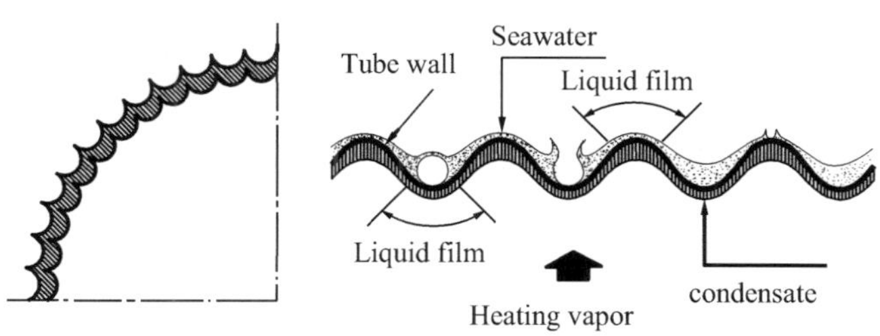

Figure 7.19
Evaporation and condensation process of double-trough tube inner and outside surfaces.

process, which is not like the MSF system where the large amount of liquid is circulating in the device. Therefore the driving power consumption is lower. For example, when producing 1 ton pure water, the VTE falling film will expend 2.5 Wh electricity, the HTE tower-type falling film is 1 Wh, while with MSF it is 4 Wh. (3) The multieffect evaporation can make the concentration ratio of seawater much higher than MSF, so the feed water consumption is relatively lower. (4) The allowable range of multieffect evaporation is very wide. As the load range changes from 100% to 25%, the device can act imperviously. And most importantly, the pure water production ratio does not decline.

Multieffect evaporation has several demerits compared to the MSF system. First, the structure of a multieffect evaporation device is too complex. Second, because the feed liquid is boiling on the wall of the vessel, it is easily scaled. Therefore it needs scaling prevention and clearing frequently. For the falling film process, the serious scaling can be considered as the deadly demerit. However, because the falling film process has high concentration ratio and high heat utility ratio, it is widely used at present.

As multieffect evaporation has many advantages and the structure and working process are flexible, it has been quickly developed in the past several years, and many new highly effective processes and devices have been developed.

7.2.7 The Experimental Operation of a Multieffect Solar Distillation System

Under the conditions of the present technological level, the MSF system and multieffect evaporation system have nearly the same heat and electricity consumption. In various occasions, because of multieffect evaporation using falling film evaporation and condensing process, the multieffect evaporation can operate in relatively lower temperature when the pure water production is the same. Thus, to some extent, multieffect evaporation is easier to connect with a solar heating system.

Zarza et al. [8] produced a 14-effect evaporation system using solar energy. This system was built in the south of Spain. The solar collector is a parabolic trough with 2672 m^2 lighting area. The water production is 3 m^3 per hour. A heat preservation tank was appended on the system which has a volume of 155 m^3. The working fluid of the solar collector is a kind of synthetic oil (3M Santo therm55). The PR of the system (the energy consumption of unit water production) is about 9.3−10.7. Kalogirou [9] gave a further analysis to illustrate that if the wasted heat of the system could be recovered better, such as by adding a heat pump, the PR could be developed significantly.

Wang [4] also introduced two examples of an HTE falling film multieffect solar desalination system. They were set in Takami Island in Japan and Abu Dhabi in the United Arab Emirates, respectively. The former is 15-effect and the latter is 18-effect. The performances are shown in Table 7.2.

Table 7.2: Performances of two sets of multieffect system.

Project	(Takami Island) Japan		Abu Dhabi (United Arab Emirates)	
Water temperature of collector (°C)	100	82.0	99	78.5
The evaporation temperature of the first effect (°C)	82	63.6	82	67.5
The evaporation temperature of the last effect (°C)	40	41.6	42.5	40.8
The average temperature difference of each effect (°C)	2.8	1.47	2.25	1.48
Water production (annual average value) (t/d)	16.4	120	118	
Total solar irradiance (kJ/(m^2 d))	14,721		23,488	
Annual average temperature of surroundings (°C)	18		27.4	
The largest water production (t/d)	23.3		143	
The average water production (t/d)	10		117	
Heat consumption (kJ/kg)	223		175	
Electricity consumption (kW h/t)	14.2		4.8	

The long-term action shows that the declination of the performance of solar collector due to the dust accumulation, corrosions accumulation, and the decrease in the degree of vacuum of some vacuum tube receivers.

The tower type HTE of a VTE multieffect evaporation system using solar energy also has several types shown in Figs. 7.20–7.22.

7.3 Solar Vapor Compression Desalination System

7.3.1 Compression Distillation Principle

In the last two sections the descriptions of the two MSF and multieffect evaporations have both emphasized heat recovery and have shown that heat recovery is the key to improving the energy utilization efficiency of the whole system. However either the MSF system or multieffect evaporation system unavoidably has temperature difference and pressure difference between each stage. Besides, the temperature and pressure decrease from the first stage to the last stage, which is harmful to the energy utilization efficiency.

Based on the second law of thermodynamics, the minimum power required for the distillation process is 19.512 J/kg (freshwater) [10], but the energy required in actual

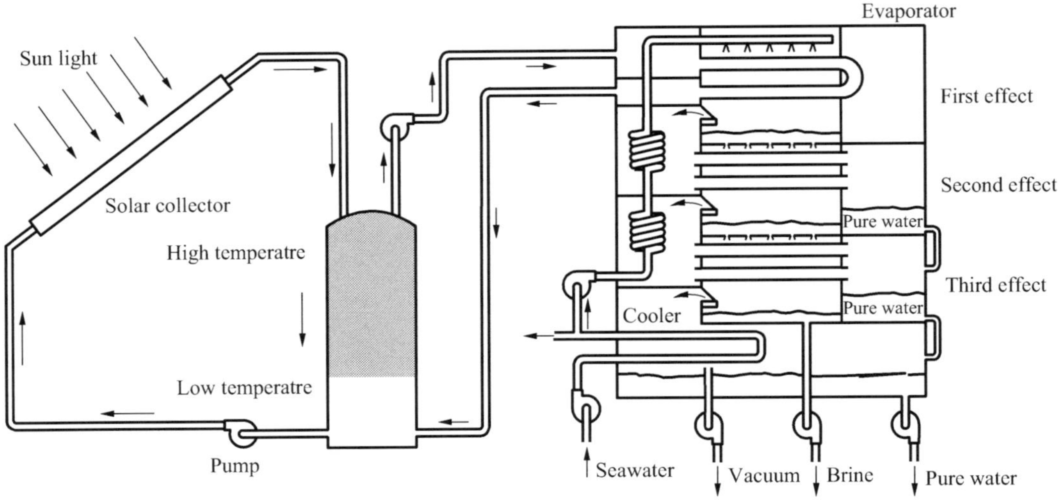

Figure 7.20
Solar energy HTE system.

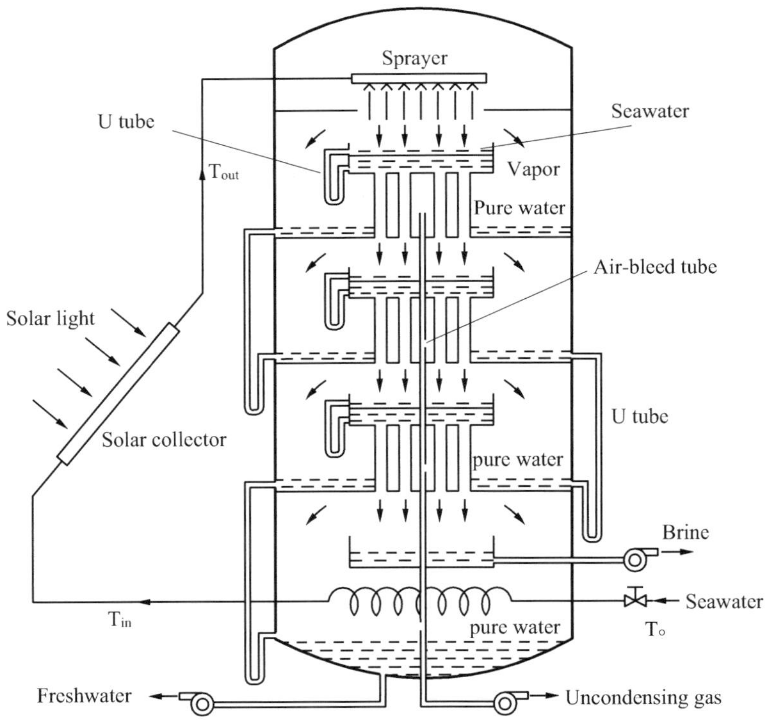

Figure 7.21
Solar tower vertical tube falling film distillation system.

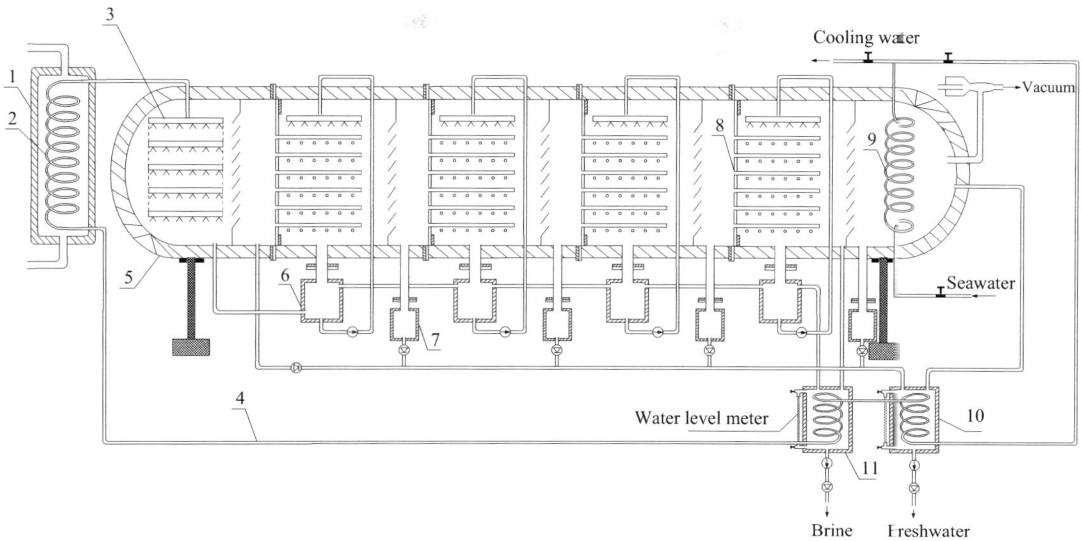

Figure 7.22
A new low-temperature MED system driven by a solar collecting system. 1, hot water tank of solar collector; 2, coiling heat exchanger; 3, water distributor; 4, seawater inlet; 5, device housing; 6, salty water circulation can; 7, pure water collection can; 8, condensing-evaporating equipment; 9, coiling condenser; 10, pure water tank; 11, concentrated seawater tank.

process is much greater than this value. Take multieffect evaporation for example, the thermal-work efficiency in this process is discussed.

Providing that this system works in an ideal situation, the steam consumption is 1:1 in the single-effect evaporation process, the steam consumption in the process of N-effect evaporation is $1/N$ in single-effect evaporation process. Meanwhile, supposing the temperature of heating steam is 104°C and the pressure 0.12 MPa, the latent heat of water vaporization is 2240.48 kJ/kg by checking Appendix Table 1A. With all these assumptions, heat to power efficiency of the single-effect evaporation is $\eta \approx 0.86\%$. Even for 10-effect evaporation process, the efficiency is only 8.6%. Thus, even repeatedly using heat energy in the multieffect evaporation process, the heat to power efficiency is still low, as is the MSF process. The reason for this situation is that the vast majority of heats (60–90%) are used to increase the sensible enthalpy of the seawater and make the seawater vaporize. However, the secondary steam containing high heat enthalpy cannot be fully used. Even in the multieffect evaporation, the final high heat enthalpy in the secondary steam is abandoned.

In fact, the abandoned enthalpy in the secondary steam is similar to the enthalpy in heating steam. They can be calculated as follows.

When the seawater under 0.1 MPa is heated by the heating steam 0.12 MPa (saturation temperature is 104°C), the secondary steam temperature, without considering the water boiling point elevation, is 100°C, and two kinds of steam enthalpy can be obtained by checking Appendix Table 1A as follows

$$i\big|_{t=104^\circ C} = 2675.2\ \text{kJ/kg}$$
$$i\big|_{t=100^\circ C} = 2671.0\ \text{kJ/kg}$$

The enthalpy difference between the heating steam and secondary is only $\Delta i = 4.2$ kJ/kg, the temperature difference is 4°C, and the pressure difference is 0.02 MPa. Supposing that there is a machine which can increase the secondary steam pressure from 0.1 to 0.12 MPa, and the temperature from 100°C to 104°C, the secondary steam can also be used as the heating steam. Based on the cycle, a small amount of mechanical work is needed. However the enthalpy included in the steam can be reused indefinitely. The VC is based on this idea in the seawater desalination system.

The easiest way to improve the steam temperature and enthalpy is adiabatic compression. The compression work converts to heat so that the secondary steam temperature is increased. Then the secondary steam, whose temperature, pressure, and heat enthalpy are increased, is reused to heat steam.

7.3.2 Compression Distillation Processes

The process of single-effect compression distillation is shown in Fig. 7.23. The process generally is as follows: Firstly, the seawater under environment pressure P_1 and temperature T_a is introduced into the evaporation chamber and heated to saturation state (p_1, T_1) by the cooling coil. Then the saturated steam (p_1, T_1) compressed by the compressor compression is turn into superheated steam (p_2, T_3), which will release heat and turn into the new saturation steam (P_2, T_2) and a part of saturated water in

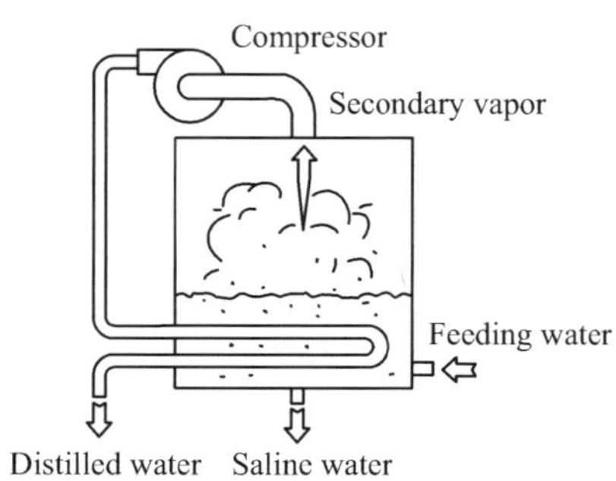

Figure 7.23
Compression distillation (VC) process.

the cooling coil (also called the heating coil) of the evaporation chamber. This part of saturated water formed in the last stage will exchange heat with new seawater, and then continue to release a part of sensible heat to the new state (p_1, T_1). Finally, the saturated steam (p_1, T_1) is discharged from the system in the form of freshwater.

In many cases, the vapor compression distillation process does not require an external heat source. However only a small amount of heating steam will be introduced to start the device at the beginning of the operation and after that, the secondary steam evaporation can be used to automatically vaporize the seawater in the future. Because the steam enthalpy is reused in the compression distillation process, the input of mechanical work is not much. For example, when the steam pressure increases from 1 to 1.2 atm the adiabatic work produced by the compressor is only 24.48 kJ/kg. Therefore, the theoretical efficiency can reach 80%. Although much lower than the theoretical efficiency, the actual efficiency in large steam compression distillation process can reach 40%. It can be seen that the seawater desalination system of vapor compression distillation has some advantages over other systems of seawater desalination.

Based on the analysis of the line thermodynamic graph, the steam compression process can be better understood. Because the pressure–enthalpy diagrams (p–i) are widely used in the analysis and calculation of steam process, the corresponding changes of VC process in the pressure–enthalpy diagrams can be shown to analyze its technical advantages.

Fig. 7.23 shows the compression distillation process which also can be presented in the pressure–enthalpy diagrams as shown Fig. 7.24. The process is as follows.

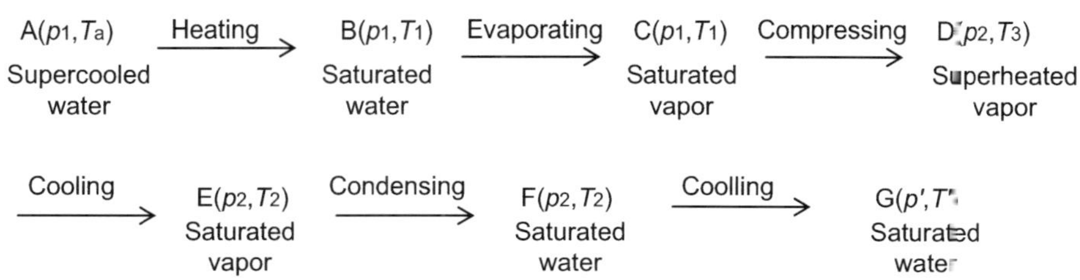

As the adiabatic compression is an isentropic process, the pressure goes along the isentropic line. In order to make the secondary steam pressure increase from 1 atm to 1.2 atm, the temperature T_3 at adiabatic compression end must be the temperature of superheated steam under $p_2 = 1.2$ atm. T_3, which can be estimated by ideal gas equation in the adiabatic process, is about 116°C and there is a large temperature difference between T_3 and the seawater temperature ($T_1 \approx 100$°C) which needed to be heated, which guarantees the rapid heat transfer of condensation tube.

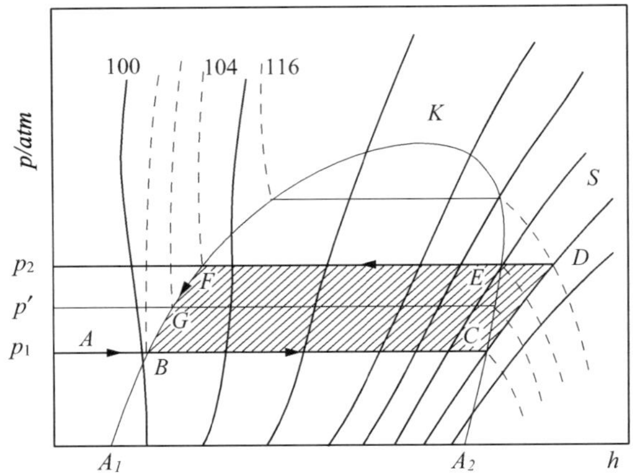

Figure 7.24
Vapor compression distillation process p–h diagram.

7.3.3 Main Operation Modes of Compression Distillation

Although the compression distillation principle was recognized early (before the 1970s), due to the limitation of compressor technology, the compression distillation in the field of seawater desalination slowly developed. After 1970, with the highly efficient centrifugal compressor appearing, some difficult problems of the roots compressor, such as overweight, slow-speed, large-scale, were overcome, especially the compressor shaft seal technology is guaranteed, which promotes rapid development of the compression distillation technology. Today, the compression distillation technology is already quite mature and has become the preferred technology combined with other desalination technology. In view of the use of devices and energy in the compression process, the compression distillation process is divided into mechanical compression and thermal compression.

1. Mechanical compression
   This utilizes the compressor to absorb the secondary steam, which generally is applied on a small to medium scale (a few hundred tons production of freshwater per day). The compressor contains centrifugation, screw, roots etc. The principle of mechanical vapor compression distillation can be seen in Fig. 7.25. In normal operation, the energy

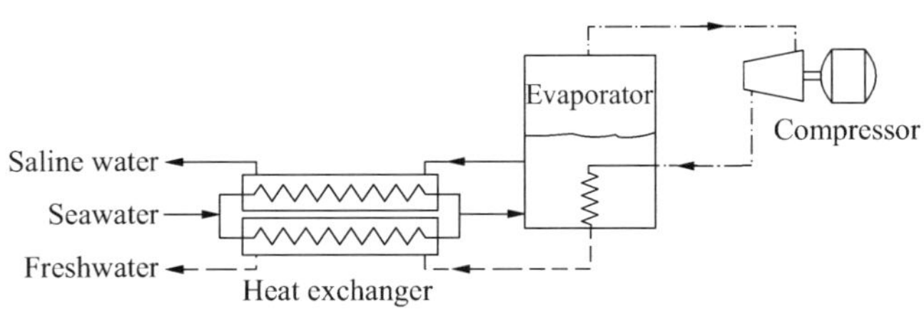

Figure 7.25
Mechanical compression type distillation process.

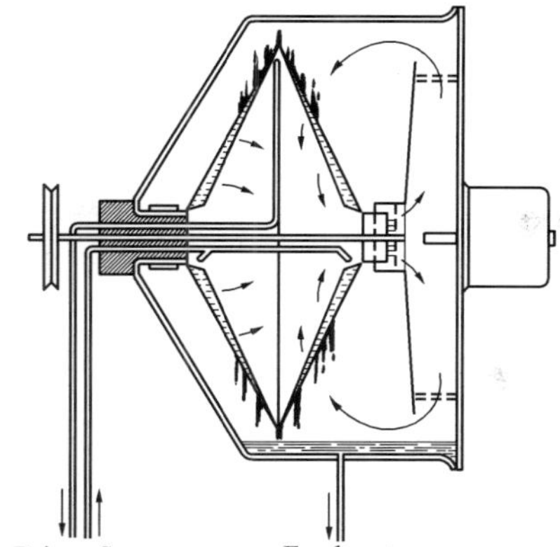

Figure 7.26
Compression distillation flowchart.

required by the mechanical compression distillation unit is basically obtained from the compression work and a little from heat.

A simple centrifugal compression distillation process is shown in Fig. 7.26. There is a space which is limited in two cones and partly evacuated. The seawater enters the inner surface of the space from the rotating two entrances of two cones, diffuses under the influence of centrifugal effect and is partially converted into steam which is guided to another space and compressed by the compressor. When condensed on the outer surface of the cone the compression steam will become freshwater, meanwhile the concentrated seawater will be discharged constantly from the surrounding outlet.

The main part of this equipment is a high-speed rotating heat exchange wall. With the seawater diffusing into the film, the high heat transfer coefficient also can be obtained in lower temperature. Besides, because of low operation temperature (50–60°C), the scaling is no longer a serious problem and the heat transfer temperature difference between the two surfaces is small, which causes the corresponding pressure difference to also be small, for example, if $\Delta t = 53 - 50°C = 3°C$, then $\Delta p = 0.019$ atm. At last, such a small pressure difference has low requirements for a compressor, and can even be supplied by these machine-like ventilators.

Fig. 7.27 is a typical single-effect mechanical vapor compression (MVC) which is used by Sasakura, a Japanese company in the United Arab Emirates. The gas turbine will generate power to drive the MVC and the exhaust gas from the turbine will be provided to the MVC as an auxiliary heat, which has perfect energy-saving effect, whose field data compared with MSF and RO [5] is shown in Table 7.3. Some authors have carried out a detailed analysis of various desalination methods, and think the lowest cost of producing freshwater is MVC [6,7].

The mechanical compression steam distillation technology can also be used for a multieffect evaporation system as shown in Fig. 7.28.

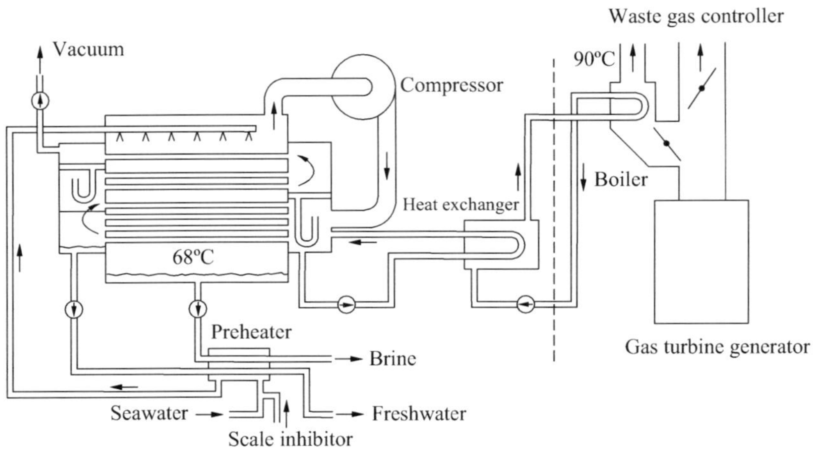

Figure 7.27
Combination of low-temperature compression distillation and gas turbine generator.

Table 7.3: Comparison of small and medium-sized MVC with MSF and RO.

Energy consumption	Low-temperature MVC	MSF	RO
Capacity (m^3/d)	206	1200	300
Steam consumption (kg/m^3)		125	
Waste heat consumption (kJ/m^3)	22,860		
Electricity consumption ($kW\ h/m^3$)	88.81	3.8	12.6
Total energy consumption (measure with electricity) ($kW\ h/m^3$)	11.43	37.34	12.6
Total energy consumption (measure with heat) ($kW\ h/m^3$)	102,920	369,600	113,420

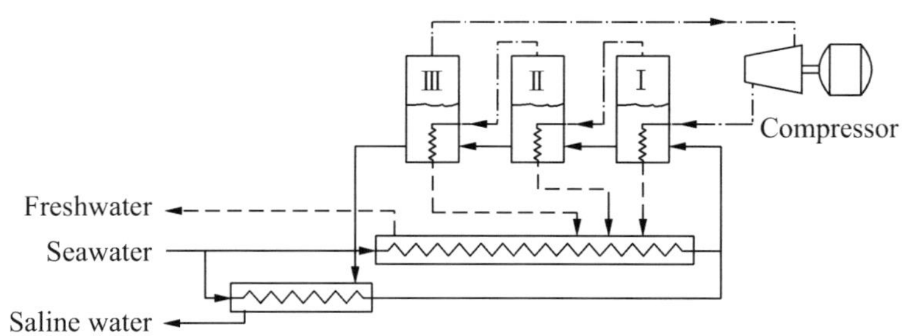

Figure 7.28
Multieffect compressing distillation system [4].

2. Thermal compression
   This takes advantage of high-pressure steam in injectors to attract the secondary steam so that it can reuse low-pressure steam. This system is usually for large and medium-sized scale and daily output of freshwater can reach kilotons. Because its structure is simple and easy to operate, the system has developed rapidly in recent years.
   The system of thermal compression distillation needs to provide more high-parameter steam to drive the steam jet pump. In addition to maintaining the required amount of evaporation, the excess heat energy in the system will be supplied to other desalination devices. The principle of this system can be seen in Fig. 7.29.
   The thermal compression distillation technology can also be used to double (Fig. 7.30) or multieffect evaporation system. In 1993, Adil proposed that the multieffect distillation can be combined with the thermal compression [4], that is to say, the steam is compressed to increase temperature and is introduced to the next effect as the heating steam, which can improve the heat utilization rate. His research showed that the water-making ratio of a typical seven or eight effect horizontal tube falling film evaporator is less than 8, but by adding a thermal compression unit (each effect compression ratio is 1.25), the water ratio can reach 11; the performance of five-effect distillation compression is similar to the nine-effect without compression.

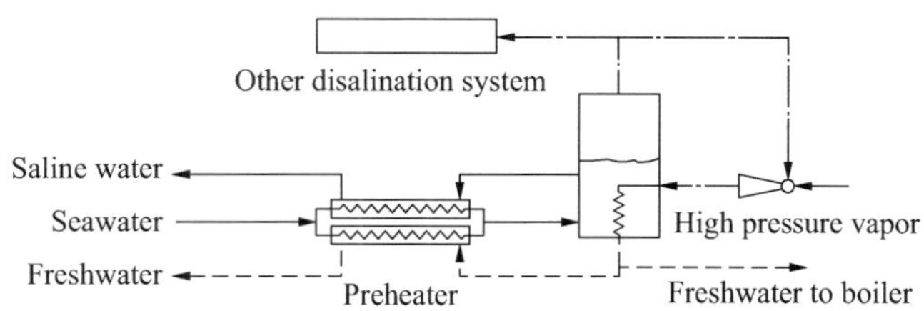

Figure 7.29
Thermal compressing distillation system [4].

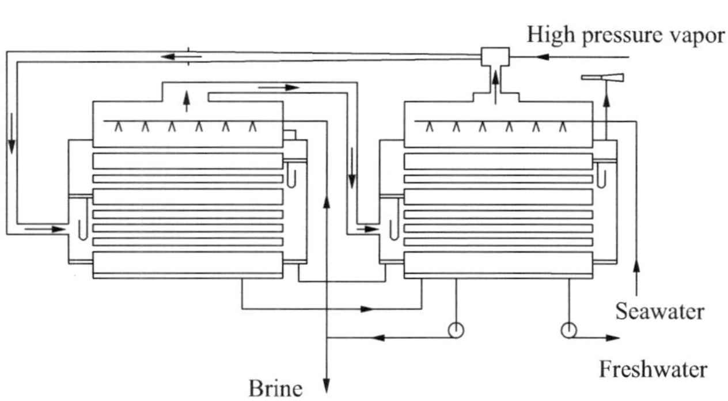

Figure 7.30
Thermal compressing double-effect evaporation system.

Table 7.4: Characteristics of two sets of thermodynamic steam distillation (TVC) system.

Main data	Sidem	Sasakura
Capacity (t/d) (kg/s)	1500 (17.36)	1500 (17.36)
Steam (kg/h) (kg/s)	9000 (2.5)	9710 (2.7)
Water-making ratio	6.94	6.436
Steam supply situation		
Pressure (10^5 Pa)	5	8
Temperature (°C)	240	Saturated
Energy consumption (kJ/kg)	389	391.2
Effective energy consumption (kJ/kg)	101.9	110.3
Specific fuel consumption (kJ/kg)	458.7	460.2

Table 7.4 shows the main features of two thermal vapor compression (TVC) systems developed by Sidem, a French company, and Sasakura, a Japanese company, respectively. The water-making ratio and energy consumption ratio per unit water desalination in two TVC systems is slightly lower than the MSF system. However, the calculation shows that under the same water ratio and energy consumption, heat transfer area required for the TVC system is lower than the MSF and the MF. In addition, the investment of the TVC system is significantly lower than the MSF system.

Fig. 7.31 is a schematic diagram of a tower HTE multifunctional device [9]. The penultimate effect of secondary steam compressed by the injector is taken as the first effect heating steam in this process. The inhalation of secondary steam only is a part of the penultimate effect of secondary steam and the rest, as the heating medium, comes into the last effect device, which is its disadvantage. According to the needs of the process, two effect devices will be selected as inlet and outlet of the jet pump. Because the loss of the HTE temperature difference is very small, using the jet pump is very favorable.

7.3.4 Advantages and Disadvantages of Compression Distillation

Vapor compression distillation has many advantages, namely high heat-to-work efficiency, small volume, no need for large-scale heat sources, and the compressor being driven by electricity or diesel engine, so it is suitable for applying in ship, island, and sea-base for seabed exploitation. However, there are some disadvantages as follows: firstly, the boiler scaling is serious, especially on the surface of the wall where a phase change happens; secondly the scale of production was limited by the compressor capacity. Thus, vapor compression distillation is more suitable for development of miniaturization, and the centrifugal compression distillation among a lot of small-scale desalination devices is a promising method.

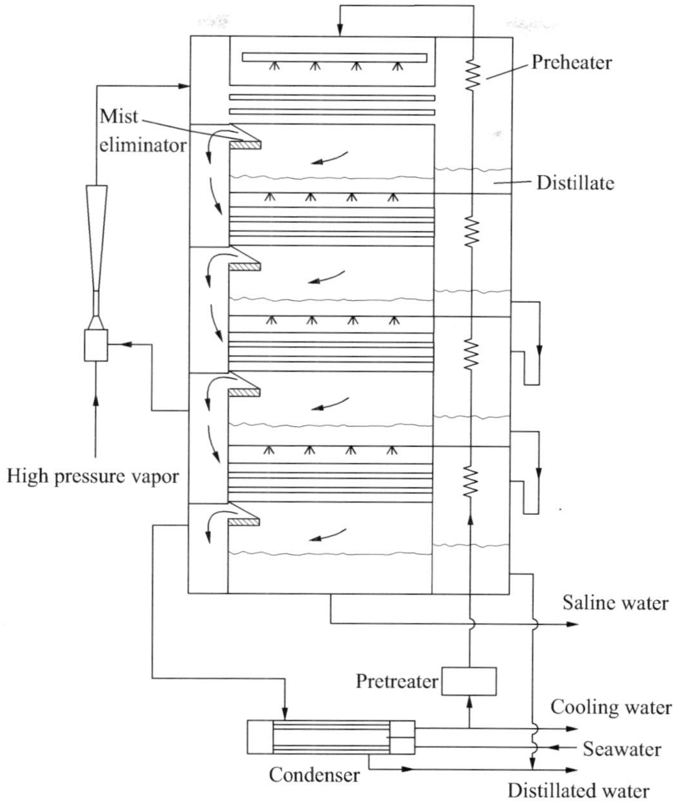

Figure 7.31
Tower horizontal tube falling film multieffect evaporator with ejector.

Because of the consumption of high-grade energy in the compression distillation process, it is not suitable for those places which have large-scale low-grade energy, such as heat energy. Therefore, the development of seawater desalination technology combined with other compression distillation technology has an important significance. Scientists have developed a variety of devices. In these devices, the VC process can run though the whole, also as a part of them, for example when VC is combined with MSF and MSB (multieffect boiling distillation, i.e., multieffect evaporation process), they give full play to their respective advantages. The features of this system are that the energy utilization is more reasonable and the overall economic benefit is significantly improved.

In order to save energy, almost all designs of the distillation process try to use VC in different forms to improve the economic performance of the system. The heat pump used in these designs can be a mechanical compressor or a jet pump. For example, in the design of VC combined with nuclear power generation, the high-pressure steam replaces a compressor and the exhaust steam can be used as the heat source of a vertical tube distillation. In the way, the heat-to-work efficiency could be further improved, and the capacity of the steam compression distillation device could be enlarged, which can create the best economic benefits.

7.3.5 Solar Compression Distillation Experiments

The energy consumption in the single VC process is mainly high-grade electrical energy and mechanical energy. With the secondary steam compressed, its pressure and temperature will be improved and the secondary once again becomes the heat steam. Because the energy consumption is mainly by mechanical work in this process, the VC system only needs a little heat, which mainly supports and keeps the system running under high temperature. The reason for this is that when the system operates in high temperature, the unit volume steam contains more water.

A common multieffect solar compression distillation process is shown in Fig. 7.32. Solar energy provides primary energy for compression distillation and makes the device operate in high temperature. In this way, it reduces the volume of steam passing though the compressor and improves compressor efficiency, thereby it reduces the pressure difference between inside and outside of the heat exchanger. When the pressure difference is small, the compressor can be replaced by a common fan machine, therefore, the cost of the whole device can be reduced.

In fact, the ideal solar compression or distillation method is to combine VC with MSF or MEB, and use solar energy as the primary energy to form a solar VC-MSF hybrid system. Solar energy can provide the heat for MSF. At a certain grade, use VC to recover the

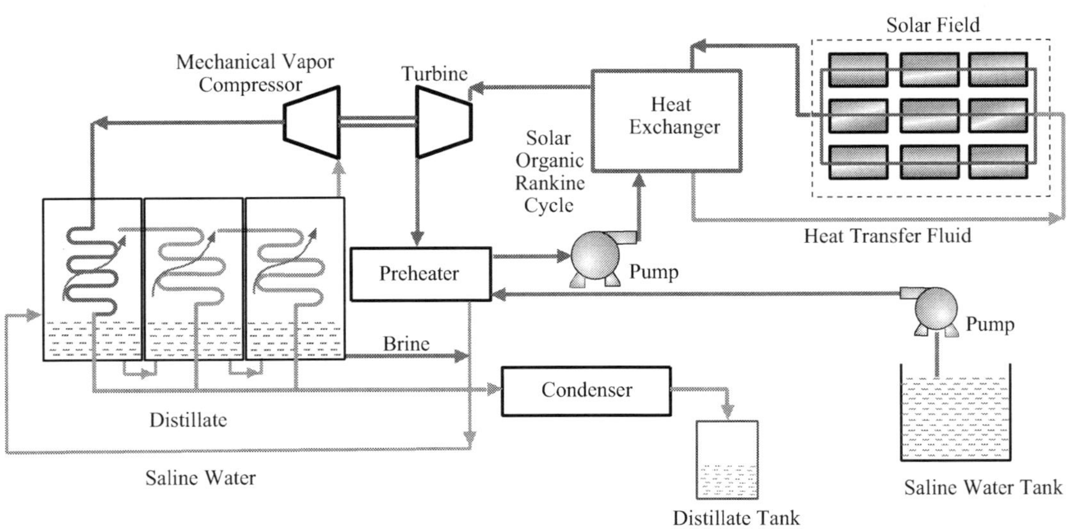

Figure 7.32
The process of multieffect solar compression distillation.

heating ability of the secondary steam and make it become the heating steam which will be reused. In this way, the heat-to-work efficiency of equipment is further improved. When the solar radiation is insufficient or at night, the system can operate continuously by using VC to supply all the energy that is needed by the system.

7.4 Desalination Using the Partial Pressure Difference Between Freshwater and Seawater

As early as the 1980s, the desalination scheme by using a refrigerator and heat pump (refrigerator-heat-pump desalination scheme: RHPDS) was described by Reali(1984). In this scheme, Reali points out that, theoretically, per kilogram freshwater production only requires 3 E mechanical energy. One E is equal to 2.5 kJ, and it is the theoretical minimum energy to separate 1 kg freshwater from seawater. The RHPDS purification technology is actually achieved by using the difference between the partial pressure of seawater and freshwater.

Both theory and experiment point out that, at the temperature range of 273−373K, the saturated vapor partial pressure of seawater is lower than saturated vapor partial pressure of freshwater for about 1.84%. In other words, if seawater and freshwater are both kept at the same temperature, freshwater will evaporate to seawater. Whereas, if the temperature of the freshwater surface is lower than the temperature of the seawater surface for a critical value, seawater will evaporate to the freshwater surface. Reali's scheme is designed to utilize the characteristic of saturated vapor partial pressure at the water surface. His basic thought is to put heat into seawater to increase the temperature of the seawater surface, and put cooling capacity into freshwater to decrease the temperature of the freshwater surface. And this status will be maintained to make sure the desalination process can be continuously run. The operation principle is shown in Fig. 7 33. Because the seawater surface is continuously evaporated, the concentration of up-level seawater is

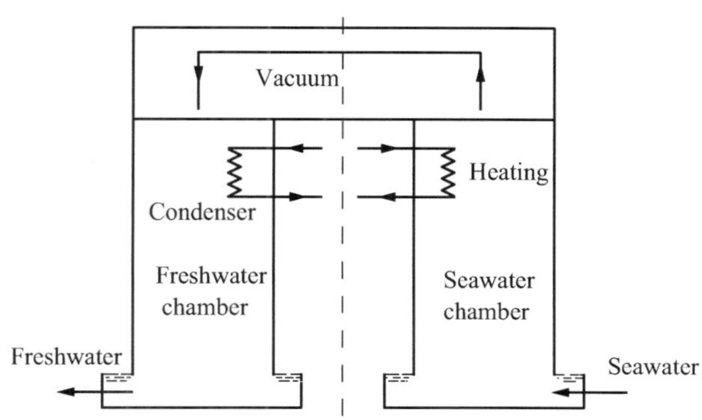

Figure 7.33
Using vapor pressure difference desalination process.

increased. This will cause the saturated vapor partial pressure of the surface to decrease, which is not conducive to further evaporation. Thus, the method shown in Fig. 7.34 can be adopted to extract high-concentration salt water.

Apparently, the evaporation process of the seawater surface shown in Fig. 7.34 decides the partial pressure of vapor within the vacuum chamber and saturated partial pressure around seawater surface. The bigger the difference above is, the more intense is the evaporation process. Many researchers have studied this process. Bemporad (1995) gave the evaporation heat transfer rate per unit seawater surface calculate function, which is shown below.

$$q_e = \frac{\alpha_m}{\rho_f}\left[f(C_s)\frac{p(T_s)}{T_s^{0.5}} - \frac{p(T_f)}{T_f^{0.5}}\right] \tag{7.33}$$

where, ρ_f is the freshwater density (kg/m^3); T_s and T_f, respectively, are the surface temperature of seawater and freshwater (K); α_m is the experimental coefficient, and the value is at the range of 10.6–10.7; p is the saturated partial pressure at the corresponding temperature; C represents concentration; $f(C)$ is the factor related to C.

And for a relatively low concentration solution, $f(C)$ can be calculated by equation below.

$$f(C) = 1 - 0.0054 \times C \tag{7.34}$$

Density of solution is the function of temperature and concentration, and it can be given by comparing to the normalized value.

$$\rho(T, C) = \rho_0\left(1 - 5 \times 10^{-4}\Delta T_0 + 8 \times 10^{-3}\Delta C\right) \tag{7.35}$$

here, ρ_0 is the density of standard solution; ΔTO and ΔCO, respectively, are the temperature difference and concentration difference at the ρ_0 state.

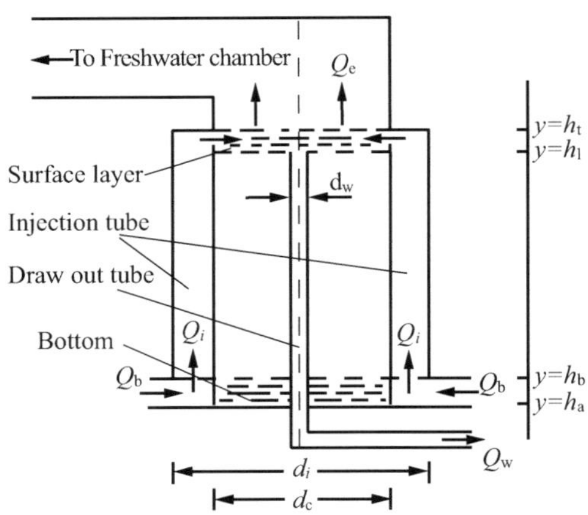

Figure 7.34
The structure of seawater chamber.

Actually, both temperature and concentration have an impact on heat capacity. However, this influence is so small that it can be neglected. The impact caused by concentration can be calculated with the following equation:

$$C_p(C) = -30.10 \times C + 4178.4 \quad (\mathrm{J/kg\ ^\circ C}) \tag{7.36}$$

Thus, the heat flow density of evaporation surface (corresponding to evaporation heat transfer rate) can be calculated as:

$$\phi_e = \rho_f h_{fg}(T_s) q_e \quad \left(\mathrm{J/m^2\ s}\right) \tag{7.37}$$

where $h_{fg}(T_s)$ is latent heat of vaporization at corresponding temperature (J/kg).

Because the temperature of vapor within a vacuum chamber is approximately equal to the temperature of seawater surface, the radiation heat transfer can be neglected. Thus, the net heat flux of vapor on evaporation surface is evaporative heat flow.

Bemporad has done the analytical calculation for the system shown in Figs. 7.33 and 7.34. The results show that the evaporation rate on seawater surface is 2.0×10^{-8} m^3/s, under a given situation that T_f is kept at 10°C; $dc = 0.25$ m; $di = 0.30$ m; $dw = 0.05$ m; potential height of freshwater and seawater is $ht = 0.1$ m; initial temperature of seawater is $T_i = 20$°C; seawater concentration is 4%; and the power of input heater is 50 W. During evaporation, the temperature of seawater is kept at 67°C. During the experiment, the required power to heat up the seawater from 20°C to 67°C is 4 W; 45 W is used to drive seawater to evaporate; and the remaining 1 W can be considered as heat loss, which means that about 98% input energy is used to evaporate seawater.

If we utilize solar power, which can be collected by solar collector, as heat input, and efficiency of solar collector is 50% under typical sunny day conditions, the total heat put into seawater is 2.8 MJ. Thus, the water production of the system per day can reach up to 1.08×10^{-3} m^3. During a day, under the given changing solar irradiance, the evaporation rate of the system changed with time can be seen in Fig. 7.35. Theoretical analysis shows that about 2.7 MJ energy (96% of total) are used to evaporate seawater. Therefore, the efficiency of this kind of desalination system is high. Estimation can also be made that, for a normal sunny day with 23 MJ/m^2 gross solar radiation intensity per day, water production per square meter can be 8.8 kg/day. It is much larger than the water production of a disc distiller.

Actually, the above system is more suitably run by a heat pump. Put the distiller of a heat pump within a freshwater sink, and put a condenser in a seawater sink. The heat released by the condenser will be used to heat up the seawater, making it evaporate. Then, the water vapor is adsorbed by the freshwater sink. The latent heat released by vapor will be captured by working fluid by evaporation. The working fluid comes back to the condenser

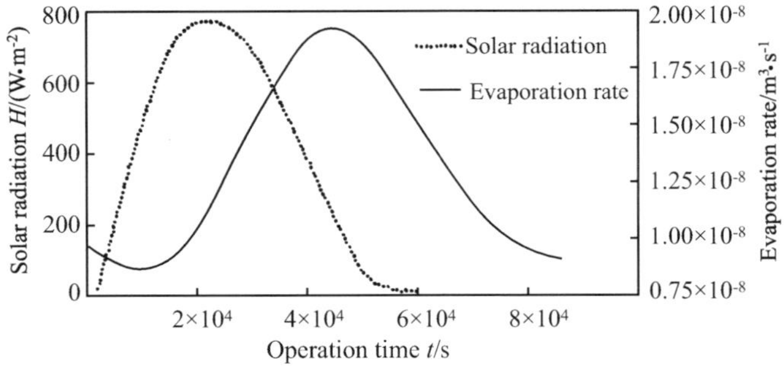

Figure 7.35
Evaporation rate and solar irradiance changes with time.

through a compressor. Thus, the energy recycle process is completed, and the energy is efficiently utilized. Fig. 7.36 represents the operating principle of heat pump desalination with horizontal tube falling film evaporation and freshwater spraying absorption. This system has a relatively large heat transfer coefficient and evaporating area because of horizontal tube falling film evaporation. And the freshwater spraying process makes the system own a large absorption area. Thus, this kind of system has relatively large water production and energy efficiency.

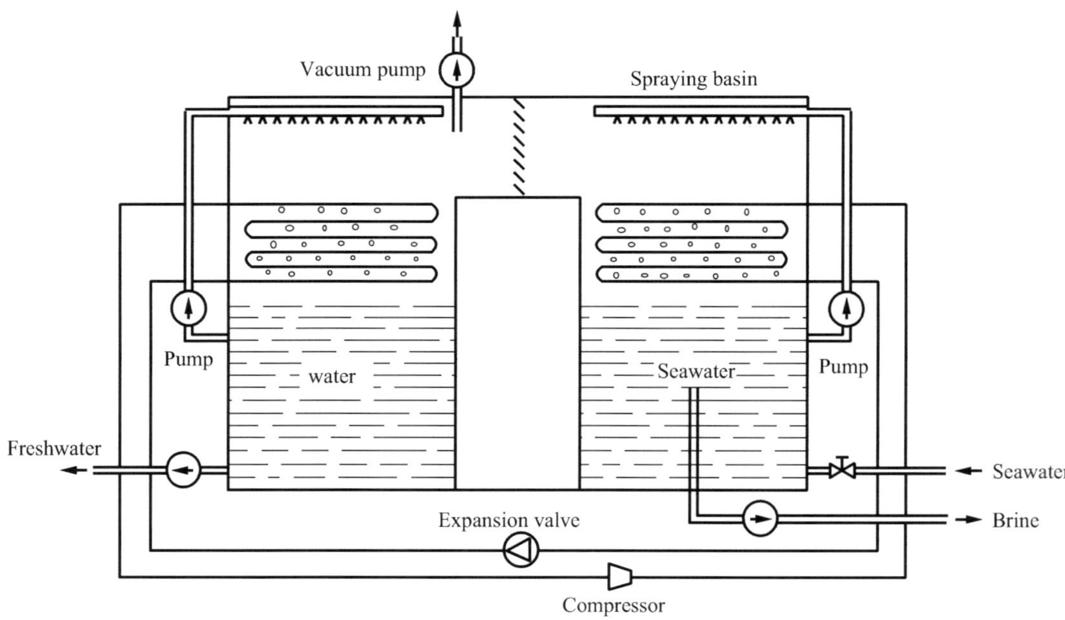

Figure 7.36
Heat pump type desalination system with horizontal tube falling film evaporation and freshwater absorbing process.

Freshwater is used to absorb vapor coming from the evaporator directly. Because there is no thermal resistance between water vapor and liquid water, the heat and mass transfer coefficient is large. This principle has received attention and been applied in traditional desalination industry, and direct-contact flash distillation system is one of the typical examples. Fig. 7.37 shows a schematic diagram of direct-contact flash distillation system. This system makes use of heated oil that comes from the heater to heat up seawater. With liquid–liquid direct contact, the heat exchange surface can become as large as possible to make sure there is enough heat transfer. Even if under the circumstance that the temperature difference is only a little, the desired heat transfer rate can still be achieved. And the water vapor generated from flash evaporation chamber can be directly condensed on water spray or film, without contacting any metal surface to make the heat transfer process. The flash evaporation and condensation processes will be repeated several times from the upper level to the lower level. Within the same chamber, because of vapor condensed, the temperature of distilled water increases. And the distilled water can flow to the next level and be processed as condensation water. With more vapor condensed, the volume of distilled water becomes larger. When the working fluid reaches the first level, both temperature and volume are increased to the maximum value. Heated distilled water

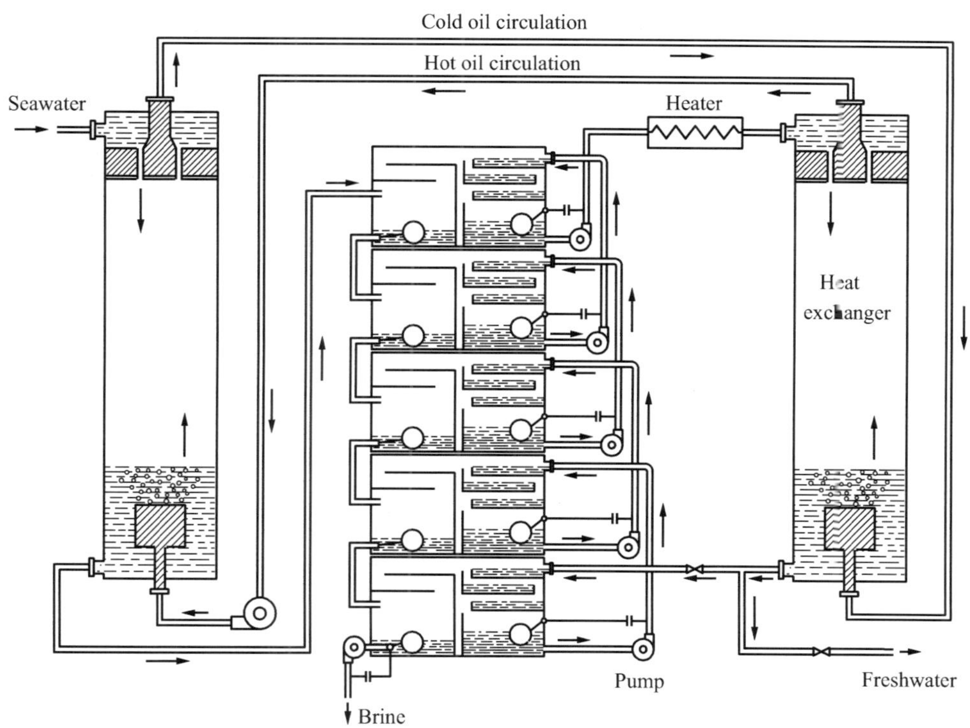

Figure 7.37
Flash distillation process by directly contacted heat.

is reheated and then pumped in to the liquid—liquid heat exchanger, giving its heat to cold oil. The temperature of the oil is thus increased, and the temperature of the water is decreased because of the direct-contact heat exchanging process. Heated oil will go to the next heat exchanger and gives its heat to seawater. The total amount of distilled water is equal to the sum of condensed vapor from each flash evaporation chamber. The required heat and heat loss of the whole process is given by a heater.

There is a small amount of oil, which can be eliminated by water vapor evaporation method. However, the oil cannot be totally removed. So the application range is limited because of water quality. And oil loss is another issue we need to considerate on. Thus, this kind of system has a problem to achieve large-scale application.

7.5 Horizontal Tube Falling Film Evaporation and Multieffect Recover Solar Desalination System

The development direction of solar desalination technology is combined with traditional solar desalination technology. Traditional technology has a relatively mature scheme, measure, materials, and management experience, whereas present technology is lower in energy utilization efficiency and water production rate of unit light area of heat collector. Especially, the water production rate of unit energy consumption of present devices is much lower than traditional industrial desalination devices. In addition, there are great difficulties in scale and industrialization. Therefore, using existing human knowledge will greatly promote the development of the solar desalination technology.

The operation principle of traditional industrial desalination technology can be divided into two types. (1) Phase transformation process. This concludes multistage flash, multiple-effect boiling, and vapor compression and so on. (2) Dialysis process. This mainly contains reverse osmosis and electrodialysis. The most significant advantage of phase transformation is the ability to reuse the latent heat of evaporation and condensation process. While latent heat preheats seawater entered into the device, part of the vapor is condensed to be freshwater. Due to the quite high cost of solar energy power generation at present, the use of solar power for the ED process for seawater desalination is not economic. Therefore, the combination of solar energy and traditional industrial seawater desalination technology is still mainly focused on the phase transformation process.

The following is a set of solar desalination system fully integrated solar energy technology and modern seawater desalination technology. Based on being fully aware of the advantage of falling film evaporation and falling film condensation heat transfer, a set of solar seawater desalination system with multieffect heat recovery and realizing the—self-balance is innovatively designed, absorbs the advantageous aspect of traditional industrial desalination technology in favor of solar energy utilization. In order to understand the performance of the system as soon as possible, an electric heating system is used to complete the steady-state simulation instead of the solar heating system.

7.5.1 Experiment Device and Operation Principle

The schematic diagram and photo of the experimental unit are shown in Fig. 7.38 [11]. Schematics of the falling film evaporator and condenser are shown in Fig. 7.39.

The experimental unit is mainly composed of a large cavity consisting of an outer shell and two round covers, a condenser, two evaporator–condensers (which are called the second-stage and first-stage evaporator–condenser, respectively) and an evaporator. They are arranged one on top of the other. The evaporation is not interflowed between each

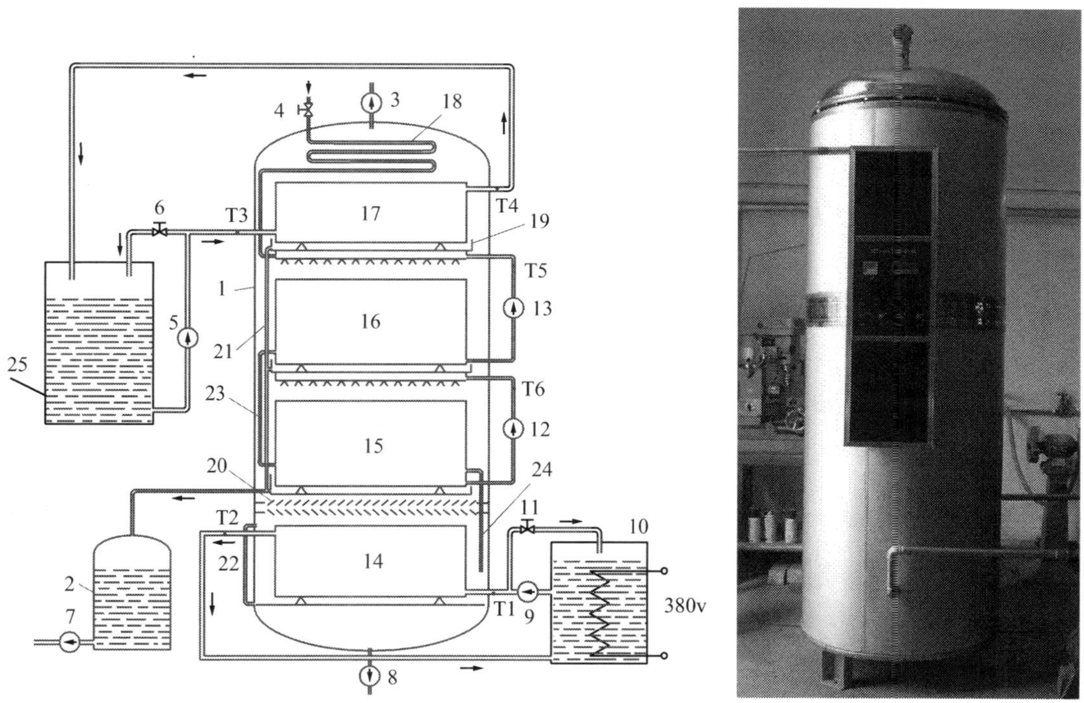

Figure 7.38

Schematic diagram and photo of the experimental unit. 1, the shell of unit; 2, freshwater tank; 3, vacuum pump; 4, valve; 5, cooling water circulation pump; 6, valve; 7, freshwater pump; 8, brine water pump; 9, heating water circulation pump; 10, heating water tank; 11, valve; 12, 13, circulation pump; 14, evaporator; 15, first-stage evaporator–condenser; 16, second-stage evaporator–condenser; 17, condenser; 18, import feed coil pipe; 19, collecting freshwater basin; 20, gas–liquid separator; 21, freshwater connection tube; 22, liquid level meter; 23, 24, brine water connection tube; 25, cooling water tank.

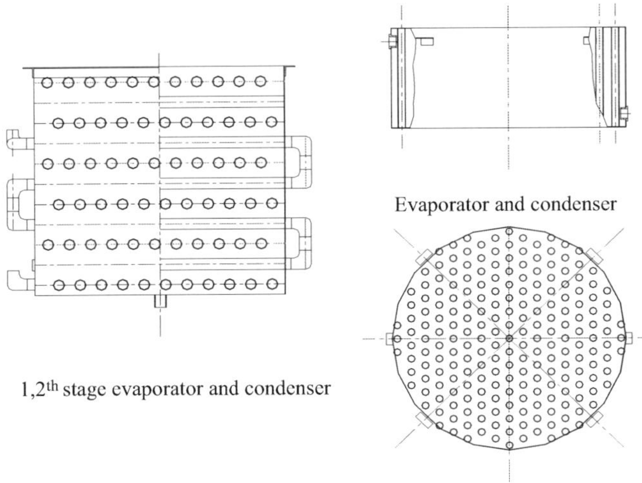

Figure 7.39
Schematics of the falling film evaporator and condenser.

other or between the two parts except for during the period of the vacuum pump operation. Under the condenser, the second evaporator—condenser and the first evaporator—condenser, a freshwater basin is equipped, respectively. The freshwater collected by every basin flows into the freshwater tank through a tube (shown in Fig. 7.38). Two sprinklers are installed over the second evaporator—condenser and the first evaporator—condenser, respectively. A circulation pump is also installed in each of the evaporator—condensers to supply brine for the sprinkler at its bottom. An evaporator—condenser is composed of 121 brass tubes with a length of 490 mm and diameter of 25 mm. These tubes are welded with two vertical surfaces, as shown in Fig. 7.39. They have an overall heat exchange area of 4.9 m^2. The condenser and the evaporator consist, respectively, of 220 and 248 tubes with a length of 300 mm and diameter of 25 mm, which are welded vertically with two covers of the cylindrical box, as shown in Fig. 7.5. Their total heat exchange areas are 4.9 and 5.5 m^2, respectively. The operation principle of the unit can be presented as follows:

1. The seawater enters the unit via valve 4 and is preheated when it flows through the spiral tube 18, with which some freshwater is produced. Finally, it mixes with the brine from pump 13 and is sprinkled to the top of the second evaporator—condenser 16. On the tubes of the second evaporator—condenser, the falling film evaporation process is carried out. The remaining brine enters the first evaporator—condenser 15 via connection tube 23 and the falling film evaporation is carried out again through the circulation pump 12. After that, the remainder of the brine enters the lowest evaporating cavity and is continuously evaporated when heated further by the evaporator 14. The strong brine is drained out by the pump 8.

2. The freshwater from the freshwater basin 19 through the connection tube 21 joining with those flowing from the second and third basins flows into the freshwater tank. The freshwater can be pumped out by the pump 7.
3. The hot water used to heat the unit comes from the solar energy system or the hot-water tank 10 and is pumped into the evaporator 14 by pump 9. After releasing the heat energy through the evaporator, the hot water reflows to hot-water tank 10 via the back-water tube.
4. The cold water used to cool the unit comes from cooling-water tank 25 and is pumped into condenser 17, where it absorbs the latent heat of vapor. After being heated, the cold water reflows to cooling water tank 25.
5. After entering the unit, the seawater will finally reach the bottom of the evaporation cavity where the brine is heated by the evaporator and the first stage of vapor is produced. The vapor rises to the space of the first evaporator–condenser 15. Then the freshwater falling film is formed on the outside surface of the evaporator–condenser and the inside surfaces of the horizontal tubes. At the same time, the vapor releases the latent heat that makes brine outside the horizontal tubes create falling film evaporation. The condensed freshwater is collected through a freshwater basin and flows to the freshwater tank 2. The substage vapor produced on the first evaporator–condenser 15 goes up to the second evaporator–condenser 16 and is condensed to form another freshwater falling film. Simultaneously, it releases the latent heat that makes the falling film of brine outside the horizontal tube of the second evaporator–condenser 16 be evaporated. Then the second substage vapor is produced. The vapor rises to the space around the condenser 17 because of the pressure difference and is condensed into falling film freshwater on the outer surface of the condenser and inner surfaces of the vertical tubes. Meanwhile the vapor releases latent heat to the cooling water flowing inside the condenser 17. Vacuum pump 3 maintains the vacuum degree of the system. The shell of the system is insulated with asbestos cloth and urethane foam with a thickness of 33 mm.

At every measuring point, a thermoelectric couple has been fixed up (shown in Fig. 7.4). The electric signal from the thermoelectric couple is sent to a microvolt meter so as to determine the temperature of the measured point. The freshwater yield is measured through a liquid level meter. The flow rates of heating water, cooling water and feedstock are measured with water meters with a relative error of 5%. The pressure inside the unit is measured through a vacuum meter with a relative error of 5%. The electrical energy consumed by pumps during the experiment is regarded as their nominal power.

7.5.2 Experimental Results and Analysis

The experiments are carried out indoors. The ambient temperature and relative humidity are 20–35°C and 40–65%, respectively. Under the steady conditions, the temperature, the freshwater yield, and other correlative parameters influencing performances of the unit have been measured. Some experimental conditions are as follows. The flow rates of the heating and cooling water are 4235 ± 100 and 3700 ± 100 kg/h, respectively. The feedstock flowrate is adjustable and varies in the range of 100–200 kg/h. The flow rate of the circulation pump is 1200 kg/h. The load of the solution outside the horizontal tubes is about 0.06 kg/m s.

7.5.2.1 Influence of Heating Water and Cooling Water Temperatures on the Freshwater Yield

At the same flowrate mentioned above, the influence of heating water temperature on the freshwater yield with different pressure is shown in Fig. 7.40. The experimental results indicate that the unit operation temperature (it is regarded as the heating water inlet temperature) is one of the important factors that influence the freshwater yield. The freshwater yield rate grows fast with the increase of heating water temperature, especially in the scope of 65–80°C. But when the temperature becomes higher, temperature becomes higher, the growth of yield rate will become slower. The possible reason is that when the temperature is getting higher and higher, the leak quantity of the vapor via the pressure balance tubes (they are used for vacuuming) increases because of the temperature difference between the stages being larger. Therefore, the diameter of the pressure balance tube should be considered carefully in the unit. If it is too big, more vapor will leak through the tube. But, if it is too small, the pressure difference between the adjacent two stages may be too big, which is not of benefit to the maintenance of the operation model of self-balance in the unit.

The influence of cooling water temperature on the freshwater yield is shown in Fig. 7.41. The results indicate that higher cooling water temperature is not helpful to increase the

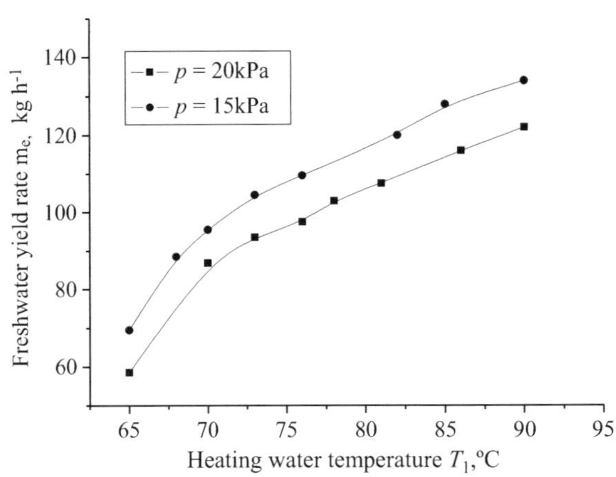

Figure 7.40
Influence of heating water temperature on the freshwater yield.

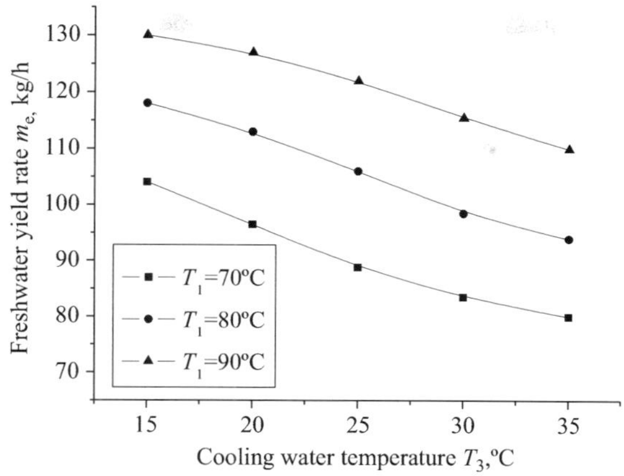

Figure 7.41
Influence of cooling water temperature on the freshwater yield.

freshwater yield, especially under low operation temperature. When the operation temperature is 70°C and if the temperature of the cooling water changes from 25 to 35°C, the yield will be reduced by 30% to 40%. Hence, the cooling water temperature should be lower.

7.5.2.2 Influence of Operation Pressure in the Unit on the Freshwater Yield

The pressure (called operation pressure) in the unit is another important factor influencing the freshwater yield. When the operation pressure decreases, the boiling point of water will decrease accordingly. That makes the brine inside the unit boil more acutely with the same heating water temperature and evaporates more water and the yield increases. The experimental results are shown in Fig. 7.42. However, it does not mean that less pressure

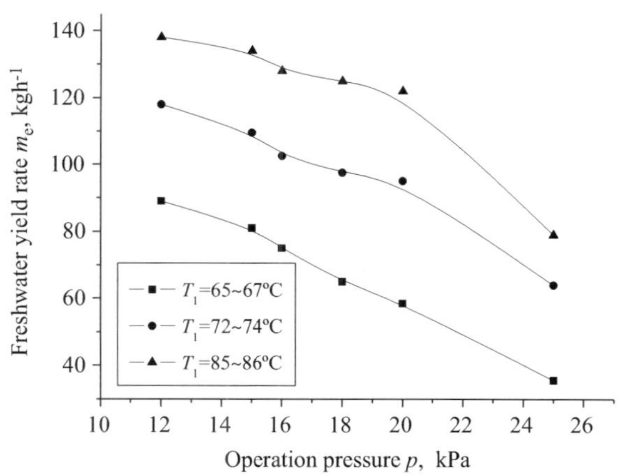

Figure 7.42
Influence of the operation pressure in the unit on the freshwater yield.

inside the unit is better. Because when the pressure inside the unit is lower, the frequency of start-up of the vacuum pump is higher, this will increase the consumption of electricity. Therefore, the operation pressure of the system should be reasonable. The experimental results show that the optimal value of the operation pressure is in the range of 14–20 kPa. The reasons are that the specific yield will obviously decrease with a pressure higher than 20 kPa and the vacuum pump will start frequently with a pressure lower than 14 kPa.

7.5.2.3 Influence of the Flow Rates of Heating Water and Cooling Water

The purpose of transmitting heating water to the unit is to provide heat energy to evaporate brine. When the temperature of heating water is unchanging, increasing the flow rate of heating water means that more heat energy is supplied to the unit. On the other hand, when the flow rate of heating water is unchanging, increasing the temperature means that more heat energy is supplied to the unit too. The cooling water is pumped into the condenser for the condensation process. Increasing the flow rates of heating or cooling water is helpful to get higher freshwater yield. However, the pump power will increase at the same time and more electric energy is needed. Therefore, under the condition of not influencing the yield, the flow rates of heating and cooling water should be as small as possible.

Provided pressure $p = 18\text{–}20$ kPa, cooling water flow rate $M_f = 3.7\ m^3/h$, cooling water temperature $T_5 = 25°C$, feedstock flow rate $m_w = 150\text{–}180$ kg/h, the variation of yield with the heating water flow rate is shown in Fig. 7.43. Provided heating water flow rate $M_s = 4.0\ m^3/h$, the change of yield with the cooling water flow rate is shown in Fig. 7.44.

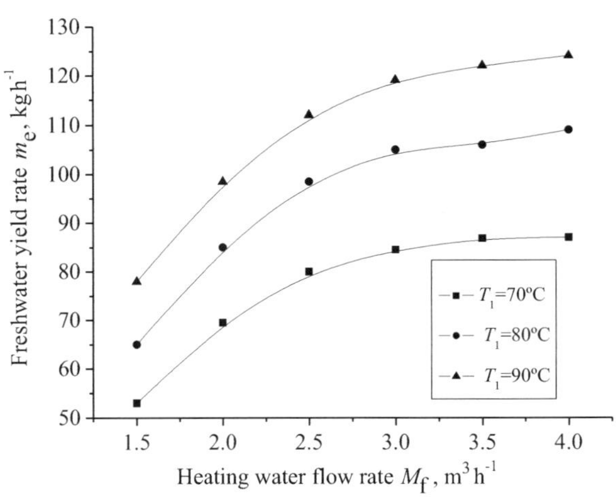

Figure 7.43
Influence of heating water flow rate on the freshwater yield.

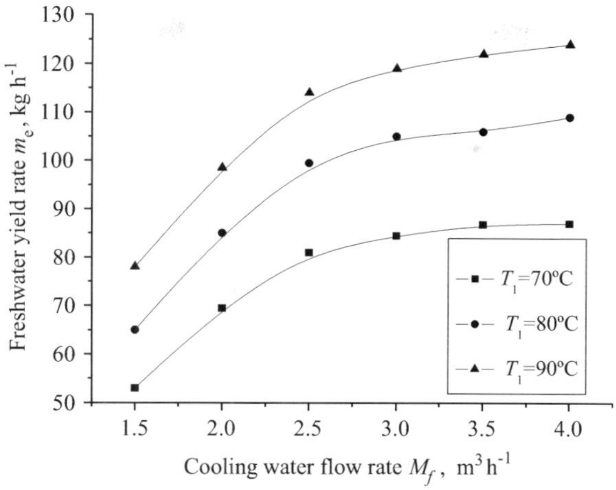

Figure 7.44
Influence of cooling water flow rate on the freshwater yield.

From Figs. 7.43 and 7.44, it can be found that when the flow rates of heating water and cooling water are less than 2.0 m^3/h, the yield is influenced considerably. There are probably two reasons for that. Firstly, when the flow rate is reduced, the input or output of heat energy transmitted through water decreases. As a result, heating or cooling driving force decreases, which causes the yield decrease. Secondly, when the flow rate is reduced, the flow states of the water in evaporator and condenser change from turbulent flow into laminar flow. Then the heat exchange rates decrease and the yield lessens. However, if the flow rates are larger than 2.0 m^3/h, the increase of the flow rates has no obvious effect on the yield. With comprehensive consideration, the flow rates of 2.0–2.5 m^3/h are appropriate and the total power of the pumps is about 200–250 W.

7.5.2.4 Influence of Operation Pressure in the Unit on the Freshwater Yield

The PR of the unit is an important indicator reflecting the unit quality. The larger the PR is, the better the unit. Based on the relationship of the input and output energy, the steady-state PR of the unit is given as follows,

$$\eta = \frac{mh_{fg}}{E} = \frac{mh_{fg}}{Q + P\Delta t} \tag{7.38}$$

Approximated $h_{fg} = 2320$ kJ/kg.

From the experimental results, it can be found that the PR of the unit is relatively large. When the unit operates in 70–75°C, its PR is about 1.8. This guarantees that the unit has a high yield. The variation of the PR with the heating water temperature under various operation pressures is shown in Fig. 7.45. The results indicate that when the other conditions remain unchanging, the higher the heating water temperature, the bigger

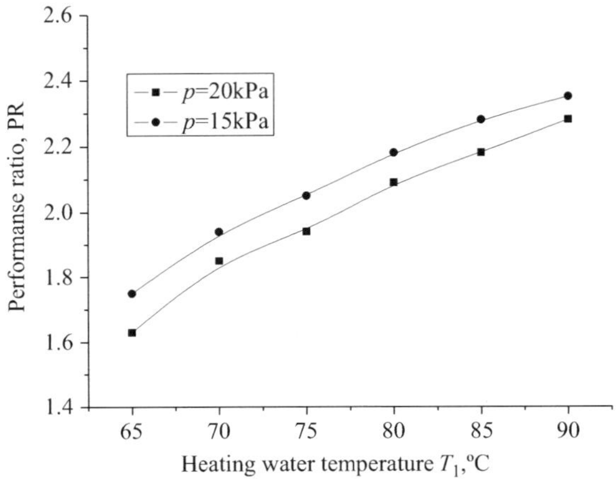

Figure 7.45
Influence of the heating water temperature of the unit on the performance ratio.

the PR. This implies that the performance of the unit is better with higher temperature of heating water. The experimental results also show that there are other factors influencing the steady PR of the unit, especially the operation pressure and feedstock flow rate. Lower pressure is in favor of the evaporation and the PR is better. The strong brine will take away some sensible heat when it is drained. If the feedstock flow rate is reduced, the PR of the unit generally rises. However, because the feedstock flow rate is limited by the crystallization of the strong brine, the flow rate cannot be too small. Usually, it is 1.0–2.0 times as much as the freshwater yield. The vapor leakage in the pressure balance tube is one of the reasons for the PR reduction. Therefore the pressure balance tube should not be too big.

7.5.2.5 The Transient-State Performance of the Unit

The transient-state performance of the unit is an important index to measure the unit stability. Reaching a steady state in a short time is very important to increase the desalination efficiency of the unit, especially for solar desalination. In order to measure the transient-state performance, when the temperature of heating water T_1 reaches 80°C, the unit is started to operate and the other parameters are kept in the range mentioned above. The temperature of every measured point inside the unit is recorded every 10 min. They are outlet temperature T_4 of cooling water, brine temperature T_5 of the second stage evaporator–condenser, and the brine temperature T_6 (shown in Fig. 7.38) of the first-stage evaporator–condenser. The temperature variations within 100 min are shown in Fig. 7.46. Under the same condition as above, the yield of the unit is recorded every 10 minutes with the heating water temperatures of 70°C, 75°C, and 80°C, respectively. The variation curve of total yield m_e with the operation time is shown in Fig. 7.47.

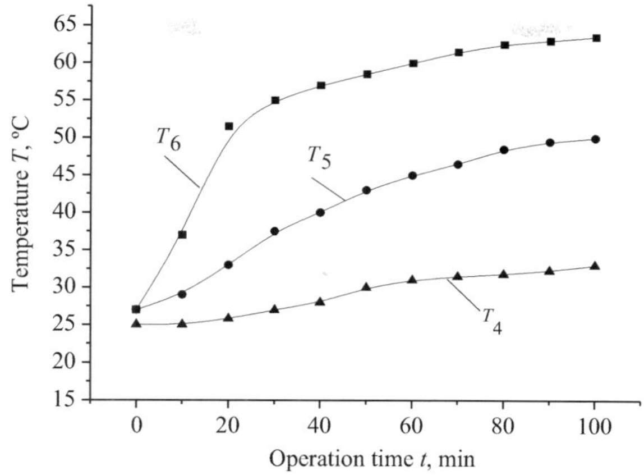

Figure 7.46
Variations of the temperature with transient-state operating time.

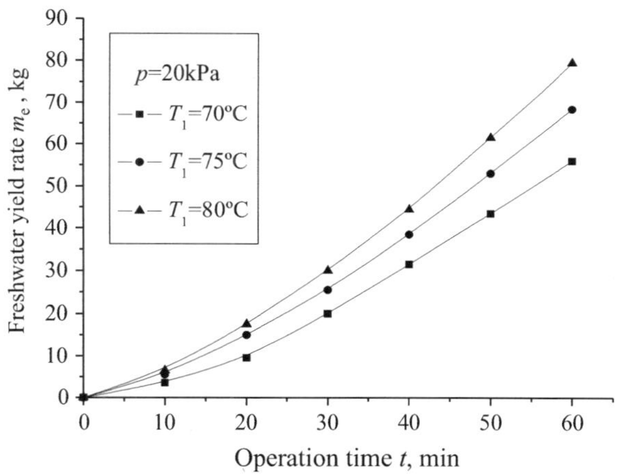

Figure 7.47
Variations of the total yield with transient-state operating time

It can be found from Fig. 7.46 that the increase in temperature is not fast because there are about 8–10 kg of brine in every stage of the unit. But during the first 20 min the temperature T_6 goes up quite fast, from 25°C to about 52°C. The brine temperatures T_5 and T_4 go up slower than T_6. However they rise placidly and still have the rising trend until 80 min.

It is shown that the brine layer is very thin and easy to evaporate after being heated owing to the application of the falling film technique. Meanwhile the heat and mass transfer processes are enhanced with the circulation pumps. Therefore the water-producing performance of the unit is better. The unit starts to output freshwater after 10 min of operation and becomes steady state about 1 h later.

7.5.3 System Operating Under Actual Weather

Using the above unit as the main equipment, a solar desalination system was built in northern China (in Qinhuangdao City). All required energy of the system comes from solar energy and wind energy. The desalination system mainly consists of eight parts: solar collection system, thermal storage tank, sea water desalination unit, cooling pool, pumping system, wind power generation system, electrical energy storage system, and automatic control system. The specific connections of various components are shown in Fig. 7.48 [12]. The location photograph of the actual operating solar desalination system is shown in Fig. 7.49.

The area of the solar collection system which consists of all-glass vacuum tubes is about 80 square meters. The heating circulated working fluid within the solar collector is freshwater. The collection system operates by way of natural and forced circulation mode. When the temperature difference of water between inlet and outlet of the water tank is less than 10°C, the natural circulation type works; when the temperature difference is more than 10°C, the forced circulation type will work. The connected pipes between the collector and the heat storage tank are 3.8 cm galvanized water pipes, height difference of inlet and outlet of the heat storage tank is 1.2 m, volume of the heat storage tank is 2 cubic meters, power of the circulating pump is 200 W, angle between the solar collector and ground is 30°. There is no reflecting plate at the bottom of the vacuum tube collector

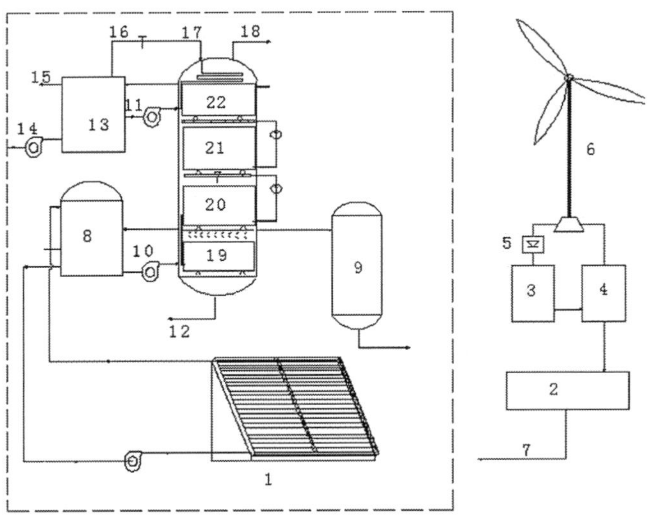

Figure 7.48
Schematic diagram of the practical desalination system. 1, solar collector; 2, control system; 3, battery; 4, inverters; 5, one-way controller; 6, wind power installation; 7, electricity; 8, hot water tank; 9, freshwater tank; 10, heating water; 11, cooling water; 12, discharging concentrated seawater; 13, seawater tank; 14, seawater inlet; 15, water overflow; 16, flowmeter; 17, seawater pipe; 18, vacuum pump; 19, evaporator; 20, first-stage evaporator–condenser; 21, second-stage evaporator–condenser; 22, condenser.

Figure 7.49
Exterior photo of the practical solar desalination system. (A) Multi-distillation tower; (B) solar desalination system.

in order to prevent sea breeze attacks, while that will lead to about a 20% decrease of heat-collecting capacity.

According to local weather information where the system built-up, the average daily hot water production capacity of this system can be estimated as shown in Table 7.5. The original temperature in the heat storage tank is assumed as 40°C, required collector water temperature as 70°C, and system collection efficiency as 45% during calculations. I is the average daily amount of solar radiation, M_f is hot water production.

The place where this desalination system is installed is rich in wind resource. Annual average wind speed is more than 7 m per second. The wind turbine power is 1 kW, inverter capacity is 1.5 kW. The power storage system consists of eight sets of 12 V × 100 Ah battery component.

Table 7.5: Average daily water production of the solar collection system in different months in a year.

	April	May	June	July	August	September	October	November
I (MJ/m^2)	16.9	18.6	18.1	19.8	16.1	15.5	12.6	9.7
M_f (m^3)	12.6	13.9	13.7	14.0	12.3	11.5	9.4	7.2

The circulating pump power of the solar collection system is 200 W. The power of the pump which inputs hot water for the desalination unit is 100 W, circulation flow rate is about 2.5 m^3/h. Cooling water pump power is 100 W, flow rate is about 2.5–3.0 m^3/h. Absolute pressure within the system is maintained from 15–22 kPa by the vacuum pump. Range of the seawater flow to the desalination unit is always about 100–200 kg/h according to the running state. The circuit of desalination device host is driven by the magnetic drive pump with a total power of 70 W. Power of each pump and the daily running time are shown in Table 7.6.

The automatic control system consists of pressure sensor, temperature sensor, water level signal acquisition device, and electromagnetic valve, etc. It can achieve fully automatic operation through reasonable control of the electromagnetic valves, pumps, and vacuum pumps. Thus it can basically be unmanned. The system will start to work when the temperature which is measured by the probe in the hot water tank is higher than 70°C. Before starting, the vacuum level, salt water level, and freshwater level of the system will be detected. If they were in the right values, the system will be automatically started. The absolute pressure of the system is set at the range of 19–22 kPa. The design capacity of the cooling water pool is 9 m^3. The pool is in an open-air setting and cooling water is used in a circulation.

7.5.3.1 Test Results in Typical Weathers

After the system has been completed, detailed testing and analysis were carried out in several typical weather conditions. Some meaningful results were gained.

Fig. 7.50 shows the variation of solar radiation with time on the inclined surface which has 30° angles different to the horizontal plane. Fig. 7.51 shows the temperature change situation of the heating and cooling water entering the system with the operation time on a day.

The results of Fig. 7.51 show that the inlet hot water temperature to the seawater desalination unit of 70°C for a long time during the day. The internal pressure was

Table 7.6: Pump power and daily electricity consumption.

No.	Device name	Device power (W)	Working hours per day (h)	Daily consumption (kWh/d)
1	Collection circulation pump	200	2.5	0.5
2	Hot water circulating pump	100	8	0.8
3	Cooling water circulation pump	100	8	0.8
4	Sprinkling pump	70	8	0.56
5	Vacuum pump	270	0.25	0.07
6	Freshwater pump	120	0.5	0.06
7	Concentrated seawater pump	120	0.5	0.06
	Total			2.9

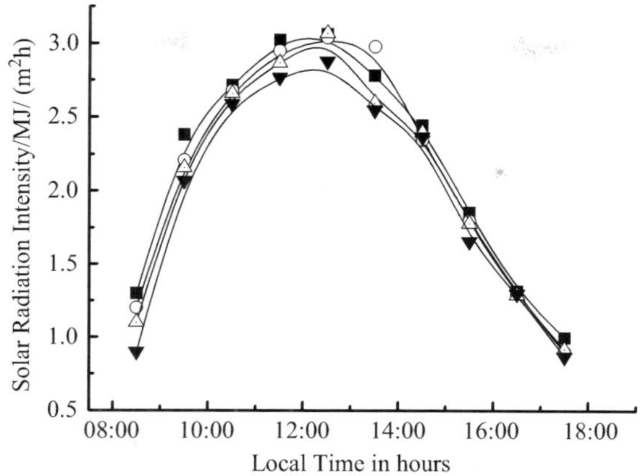

Figure 7.50

Hourly variation of the incident solar radiation intensity on the inclined area.

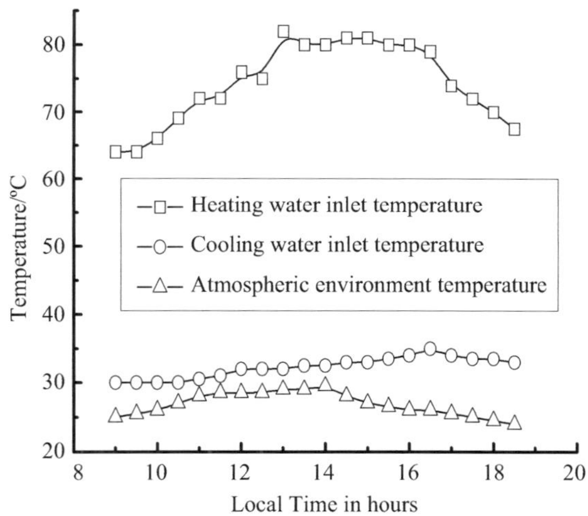

Figure 7.51

Hourly variation of the heating and cooling water temperature entering desalination unit.

adjusted relatively high, about 22 kPa. The unit failed to achieve the best efficiency of water production. And the cooling water temperature stayed at above 34°C for a long time and it increased with time. This indicated the cooling water tank design capacity might be too small and the performance of the cooling device was not very good.

The water production per hour of the device under four typical weathers is shown in Fig. 7.52. The results show that water production per hour increased rapidly with time, and reached the maximum at about 14:00–15:00, there was still a large amount of water production until 17:00. Even after the system shutdown, sometimes the freshwater output of the next morning could reach nearly 20 kg. But it is noteworthy that the sunlight

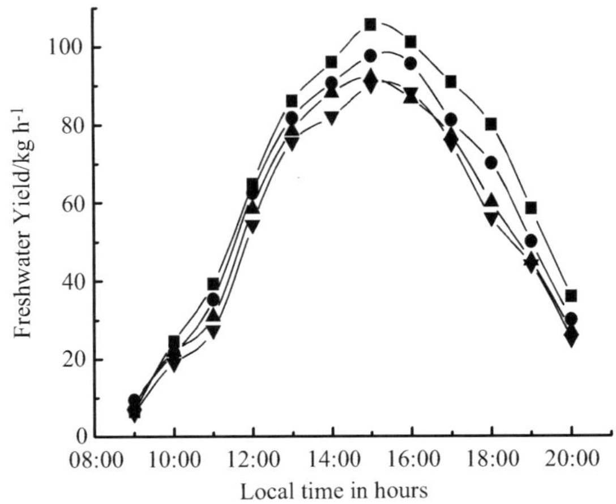

Figure 7.52
Hourly variation of the freshwater yield under typical weather conditions.

gradually becomes weaker after September or October. It is not conducive for the system to produce freshwater. Especially in the mornings when the system restarted after a few days shutdown, water production was low before 10 o'clock.

Hourly PR is one of the most important factors to estimate desalination system. The definition of hourly PR is as follows:

$$PR_h = \frac{m_i \cdot h_{fg}}{H_i + W_h} \tag{7.39}$$

where m_i is freshwater yield per hour of the system (excluding yield at night), kg/h; h_{fg} is latent heat of vaporization of water, usually 2300 kJ/kg; H_i is the solar irradiation energy per hour on the collector, kJ/h; W_h is the average electricity energy used by system per hour, kJ/h (it is equal to $W_d/24$, where W_d is daily electricity consumption of the system). According to the experimental results which are shown in Fig. 7.52, the hourly PR of the system has been calculated in Fig. 7.53 (excluding the freshwater yield at night). As shown in Fig. 7.53, the hourly PR continuously increases during the day.

Daily PR is another one of the most important factors to estimate a desalination system. Definition of daily PR is as follows:

$$PR_d = \frac{m_d \cdot h_{fg}}{H_d + W_d} \tag{7.40}$$

where m_d is freshwater yield of the system along a full day operation (including yield at night, kg); H_d is the solar radiation energy to the collector along a full day's operation, kJ/d. The total average PR values along a full day's operation of the system are 1.055, 0.972, 0.905, and 0.865 for these 4 days, respectively. The daily PR value is higher in

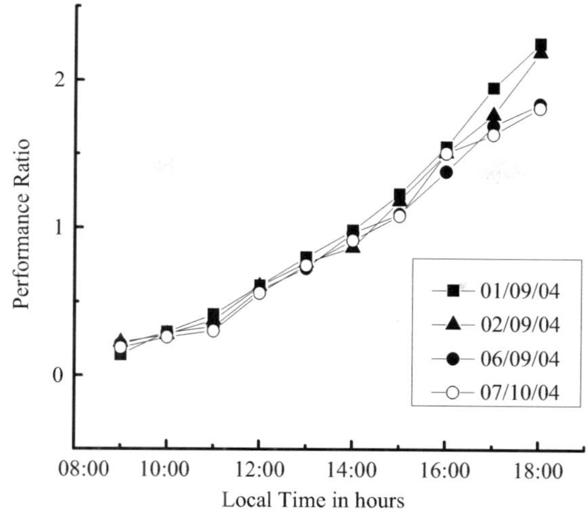

Figure 7.53
Hourly variation of the performance ratio of the system.

better weather because solar radiation intensity is stronger. Because there is no reflector at the back of the solar receivers, 20% of total radiation energy was wasted. The daily PR would be much higher if some useful improvements had been carried out to the system.

7.5.3.2 Long-Term Operational Testing of the System

Long-term monitoring had been carried out after the completion of the solar desalination system. The system was built in northern China where winter temperatures were low, generally below zero. It was not safe for working of the solar collector, so this system was shut during the winter months (December to March of the following year). When October came, the local solar radiation intensity gradually became weaker. The environment temperature gradually went lower, while the water production rate of the system also gradually decreased. After April in the following year, when the system was restarted and water production was retested because of stronger solar radiation and higher environment temperature. Fig. 7.54 shows water production distribution from April to November. Total solar energy per month per square meter and the average environment temperature per month of the solar collectors is shown in Fig. 7.55. The results of Fig. 7.54 indicate that the water productions of June, July, August, and September in a year were relatively high. The water production in July reached 25 m^3. In June and August it reached more than 22 m^3. During these months, solar irradiance was best and environment temperature was highest. However, in April, May, October, and November, the water production was slightly lower, because during these months solar irradiation was relatively weak and environment temperature was also low. The results show that it is more meaningful to set up the system at those places with good solar irradiance and high environment temperature.

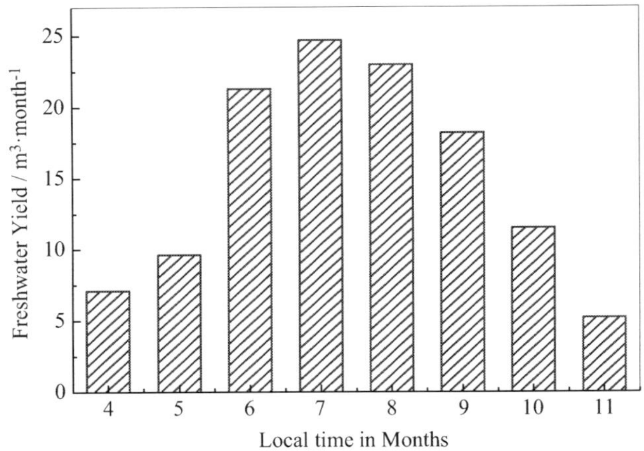

Figure 7.54
The distribution of monthly yield in 2005.

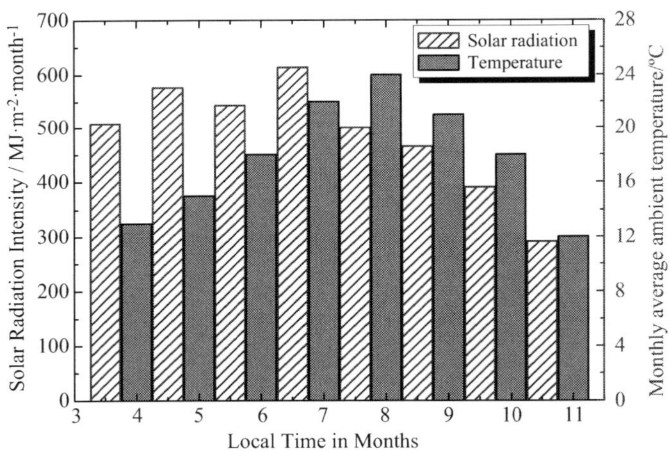

Figure 7.55
The distribution of monthly solar radiation intensity and ambient temperature in 2005.

It is noteworthy that the system was built in northern China where winter temperatures were low. The system needed to avoid the freezing problem and there were nearly 4 months unsuitable for running. If the system was built in southern China, such as Hainan Island, where winter temperatures were high, the system could run all year round. Even more, it was expected that the water production would be higher and water production costs would be lower if the system was in southern China. As a result, the system would be more conducive to be put into the market in southern China.

7.5.3.3 Other Influencing Factors

The analysis of experimental results showed that: the factors that affect system performance are various. In addition to operating pressure, hot water flow rate and temperature have a significant impact on the system performance (the details have been analyzed in Ref. [13]). There are still other factors as follows.

1. Seawater flow rate. When low-temperature seawater flows into the device, firstly it will be heat-absorbing and its temperature will increase on the condenser coil above the third-stage condenser. This will help to improve the production of freshwater. However, if the seawater flow rate into the device was too high, the operating temperature of the entire device would be lower; the water production rate of the system will be lower too. Usually it is suitable for the seawater to flow at a rate of 1.5–2.0 times freshwater production.
2. Spray amount of circulating pump. The liquid flow rate of the horizontal pipes has a direct relationship to the thickness of evaporation liquid film. So the liquid flow rate of the horizontal pipes affects flow velocity and flow pattern of saltwater outside the pipes, as a result, it affects the occurrence of the amount of steam. When the spray amount is too small, it is easy to make pipe outside generate a "dry zone"; while spray volume is too large, liquid film outside will become thick and will not be conducive to falling film evaporation. The experimental results show that the suitable liquid flow load range of falling film evaporation outside of the pipe is 0.045–0.15 kg/m s [14].
3. Sensible heat recovery of discharged seawater. The amount of discharged seawater is about 50% of the input seawater. Because the temperature of discharge brine is high, it is significant to try to recover its sensible heat. Calculation by 1 m^3 of freshwater production per day, if the recovery temperature difference is 20°C, then the daily recyclable heat can reach 83.6 MJ which is equivalent to collecting heat of collector area of 7 m^2 on sunny days (in total collector efficiency of 50%).
4. There was inadequate capacity design existing in the actual operation process of the system, for example, the designed heat storage tank of the solar collection system was too small, only 2 m^3, the seawater desalination device could not take solar heat energy instantly, resulting in the water temperature of the heat storage tank sometimes reaching nearly 90°C, at this time the collection efficiency of the solar collection system would be greatly reduced. It is estimated that an appropriate heat storage tank should be 4 m^3.

7.5.4 Conclusions

1. The system can run stably and automatically in practical fine weather. Freshwater production of per unit of solar collection area can reach up to 10 kg/m^2 d.
2. The system can work on condition that hot water temperature is in the range of 60–90°C and cooling water temperature is in the range of 20–45°C.
3. The amount of the seawater contained within the system is small, so it could be quickly heated or cooled. For this reason, the system has a good transient performance and freshwater can be produced after heating for about 20 min.

4. The requirements for feed water quality are low and this desalination system can be both used on seawater and brackish water, so it is suitable for those areas with brackish water where solar and wind energy are rich.
5. The system took closed-cycle mode and without external interference, it can produce freshwater with high purity, which is subjected to the approval of the health and epidemic prevention departments. The measured results show that the quality of water can accord with China's national drinking water standards (GB5749-85). That means the output freshwater by the system is drinkable.

7.6 Low-Temperature Multieffect Desalination System Together With Solar Pond

Solar pond has had the attention of people with advantages of large heat collection area, stable operation, and ability to store thermal energy for a long time. As the basic heat source of desalination, the solar pond has been studied by many researchers, which indicates that this method is completely feasible. The stable operating temperature of the solar pond is generally only 60–70°C, so it is suitable for a low-temperature multieffect system.

Leblanc [14] introduced a low-temperature three-effect evaporation desalination system powered by a 720 m^2 areas solar pond, capable of producing up to 2300 L of freshwater per day, wich has been designed using the model. The running structure of the whole system is shown in Fig. 7.56.

When the system is running, high-temperature water in the solar pond is first pumped into the first-effect evaporator, which releases heat by flowing in the coil immersed in the seawater and then returns to the solar pond. Seawater in the first-effect evaporator is heated to evaporate, resulting in the first-stage steam, and the steam goes through pipes up to the second-effect device, where heat coagulates it into freshwater. At the same time, the latent heat of condensation over to the seawater, where the water is again evaporated to form the second-stage steam. The second-stage steam enters the third-effect evaporator, repeating the above steps and completing the evaporation and condensation process in the third-effect device. Finally, the freshwater produced in the three-effect evaporator brings together to enter the freshwater collection tank. To improve the heat utilization rate of the system, a number of devices were designed to recover the waste heat from seawater and freshwater, and maintaining the low pressure of the system by using the vacuum pump to make the system work in a high-efficiency state.

The main specifications of a low-temperature multieffect system are summarized in Table 7.7.

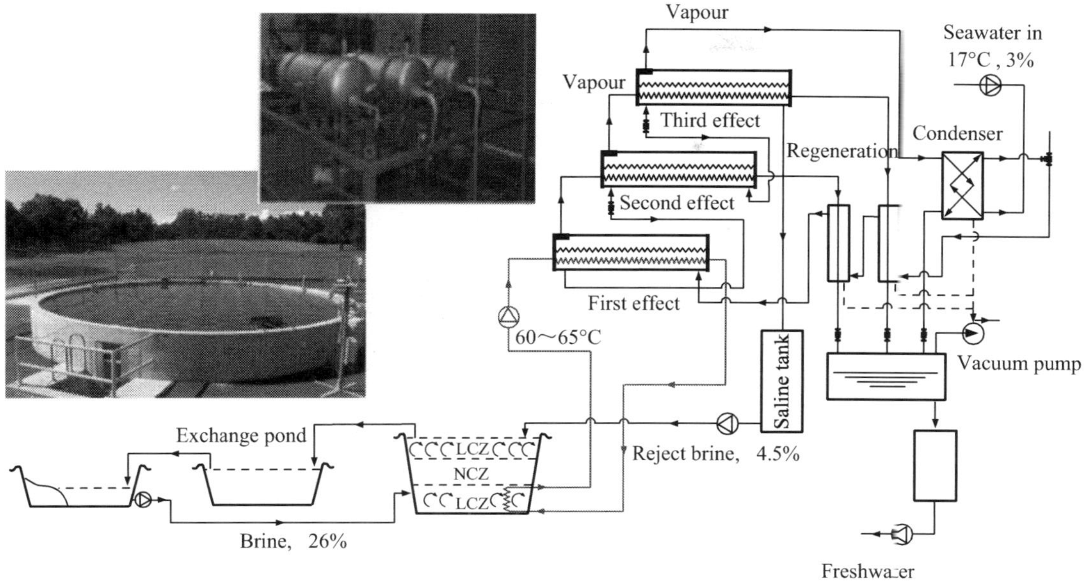

Figure 7.56
Three-effect evaporation desalination system powered by solar pond [14].

Table 7.7: Technical specifications of low-temperature multieffect device.

Project	Parameter
The number of effects	3
Evaporator shell diameter (m)	0.36
Evaporation heat transfer area (m^2)	1.0
Condenser shell diameter (m)	0.75
Intake seawater temperature (°C)	17
Feed salinity (ppm)	30,000–43,000
Length of evaporator–condenser (m)	1.0
Vacuum pressure system (liquid ring pump) (kPa)	7.0

In view of such a system, Table 7.8 shows the summary of the predicted and measured performance values, which indicate that the two are in good accordance with one another. It has a good guiding significance for other similar projects.

The thermal efficiency of the solar pond is defined as the ratio of the heat that the solar ponds supplied to the desalination device and the energy that the sun supplied to the solar pond, namely:

$$\eta_{sp} = \frac{m_{sp} c_{p_{sp}} (T_{sps} - T_{spr})}{G \times A_{sp}}$$

Table 7.8: Comparative table (predicted and experimental performance).

Performance parameters	Simulation results	Experimental results
Total intake seawater flow rate (L/min)	19.4	15.1
Heating water flow rate (L/min)	37.2	22.3
Heating fluid water supply temperature (°C)	86	86
Feed flow rate (L/min)	2.63	2.55
Feed salinity (ppm)	32,100	32,100
Distillate flow rate (L/min) (ppm)	1.49	1.32
Distillate salinity (ppm)	0	9
Specific thermal energy (kJ/kg)	925	968
Recovery ratio (%)	56.6	51.8

here, η_{sp} is the thermal efficiency of the solar pond, G is solar radiation at the surface of pond (in W/m^2), A_{sp} is the area of the solar pond (in m^2), T_{sps},T_{spr} are the temperature of the inlet and outlet of the internal heat exchanger, respectively. m_{sp} is the mass flow rate (in kg/s) and $c_{p_{sp}}$ is the specific heat of the circulating brine (in J/kg°C). Theoretical and experimental results show that the solar pond efficiency is about 15% when the heating temperature of the system is 87°C, and the efficiency rises to 20% when the heating temperature is 65°C.

For a low-temperature MED plant, the higher the heating temperature is, the higher the distillate production rate is. The distillate production rate of the system is presented as a function of the supply temperature from the solar pond in Fig. 7.57, which tells us that the heating fluid temperature and distillate production rate basically present a linear relationship.

In summary, the literature [14] introduced a low-temperature multieffect evaporation desalination system powered by a solar pond, capable of producing up to 2300 L of

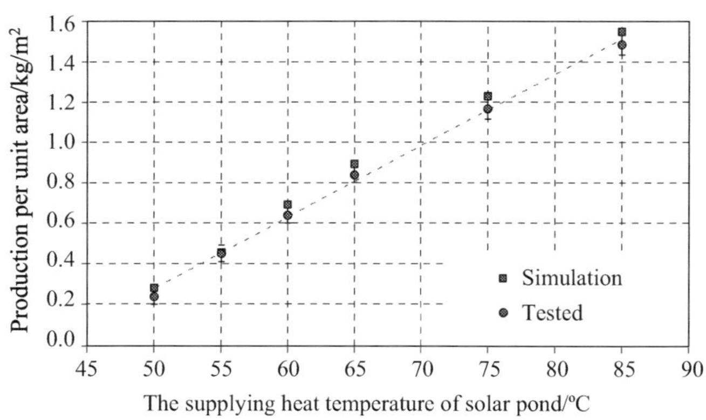

Figure 7.57
Distillate production rate at different supply temperatures.

freshwater per day. The efficiency of the system is about 14–15%. The recovery ratio is 56% and the production cost is about \$18 per kiloliter freshwater, this is relatively high. The main reason is the high cost of the solar pond. If the device is scaled, such as a commercial unit of this type with 12 effects, 10,000 m^2 area solar pond and the system life is recorded for 25 years, then the freshwater production costs immediately dropped to \$3.4–\$5.1 per kiloliter freshwater.

7.7 Reverse Osmosis Desalination System Driven by Solar Photovoltaic Panel

PV solar energy has been greatly developed in recent years and the battery prices continue to decline, while the performance has been continuously improved. Solar cells have the advantages of stable performance, simple structure, low maintenance cost, long service life, and not having moving parts, thus becoming the darling of the lack of electricity or low-electricity area. On the other hand, RO technology is developed rapidly, its share of the global seawater desalination market increased year by year and has become the mainstream of the desalination market. Therefore, the combination of solar cell and RO technology, producing freshwater from seawater, is bound to become a choice. It is said that the RO desalination system powered by PV solar energy has no technical obstacles in addition to the economic problems.

We know that RO water production is a process that consumes power. The previous section has told us that producing a kiloliter of freshwater needs 0.7 kWh of work at least. However, according to the current level of human beings, getting a kiloliter freshwater spends 5–7 times that amount of energy at least. Nevertheless, this energy is still relatively low compared to the thermal method, and the thermal method is about 30 kWh of heat energy per kiloliter freshwater. Therefore, providing work for RO technology by using solar cells is a very meaningful choice. Solar cells are suitable to use in remote areas and areas lacking electricity for its quiet, nonpolluting, and ease of system miniaturization.

For the combination of PV solar energy and RO desalination system, currently, the majority are using solar cells to supply power for high-pressure water pump of the RO system, that is to say, the two systems are completely independent. A RO desalination unit powered by PV solar energy is proposed by Eyad, as shown in Fig. 7.58 [13].

Fig. 7.58 clearly tells us that the system consists of two main subunits—the energy production subunit and the desalination subunit. The electric energy produced by the solar cell is stored partly, and the other part is converted to 220 V alternating current by inverter to drive the high-pressure pump and provide the power for the seawater desalination system. Among them, the technical features of one solar panel are shown in

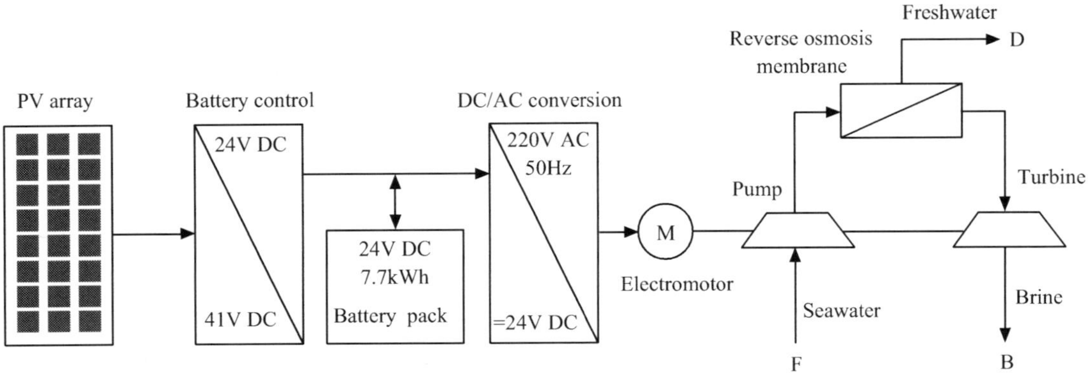

Figure 7.58
Schematic diagram of the reverse osmosis desalination unit powered by PV solar energy.

Table 7.9. The system is mainly used to simulate the desalination of brackish water with different salinity.

The power output of the energy production subunit is calculated by the following formula

$$P = \eta \cdot \eta_b \cdot A \cdot H_m = \eta_{mp} \cdot \eta_t \cdot \eta_b \cdot A \cdot H_m$$

here, P is the PV modules output power (kW), η is the overall efficiency of the cell panel, η_b is the battery efficiency, A is PV modules area (m^2), and H_m is the average daily solar irradiation (kWh/m^2 day), where

$$\eta_t = \eta_d \cdot \eta_{inv} \cdot \eta_{con}$$

here, η_d is the total efficiency of dirt, η_{inv} is the inverter efficiency, and η_{con} is the loss due to connections.

The maximum power point efficiency of the PV modules (η_{mp}) decreases with increasing module temperature. The following equation is used to take this into account:

$$\eta_{mp} = \eta_{mps} + \mu_{mp}(T_c - T_s)$$

here, η_{mps} is the module efficiency under standard operating conditions, μ_{mp} is the temperature coefficient of efficiency at the maximum power point, T_c and T_s are the actual

Table 7.9: Technical features of one solar panel of the installed PV generator.

Project	Parameters	Project	Parameters
Open circuit voltage (V_{oc})	21.7 V	Voltage at peak power (V_{mpp})	17.3 V
Short circuit current	3.27 A	Peak power (P_{max})	50 W
Current at peak power (I_{mpp})	3 A	Output voltage (V_{out})	24 V

cell temperature and the temperature at the standard operating conditions (°C), T_s generally taken as 25°C. The module efficiency at standard operating conditions is

$$\eta_{\mathrm{mps}} = \frac{I_{mps} V_{mps}}{A H_{ms}}$$

here, I_{mps} and V_{mps} are the module's current and voltage at the maximum power point under standard operating conditions, H_{ms} is the solar irradiation under standard operating conditions (kWh/m^{2}d). The temperature coefficient of efficiency at the maximum power point is approximated as follows:

$$\mu_{mp} = \eta_{mps} \frac{\mu_v}{V_{mps}}$$

μ_v is the voltage temperature coefficient (V/°C). Then, the average daily energy production equals

$$E = \eta_{mp} \cdot \eta_t \cdot \eta_b \cdot A \cdot H_m \cdot h$$

where h is the average hours of daylight in the region.

In fact, the freshwater production is not entirely determined by the output energy of the battery, but is also related to the efficiency of the desalination system, in turn, the efficiency of the desalination system is related to the concentration of brine The relationship between the energy consumption and water salinity for the desalination subunit is given by reference [13], as shown in Fig. 7.59. With brackish water of 2000 and 5000 mg/L TDS, the amount of energy required is 1.1 and 1.6 kWh/m^3, respectively.

According to the above equations and conditions, the daily amount of water production during a 1-year cycle using the same size of proposed desalination system for four

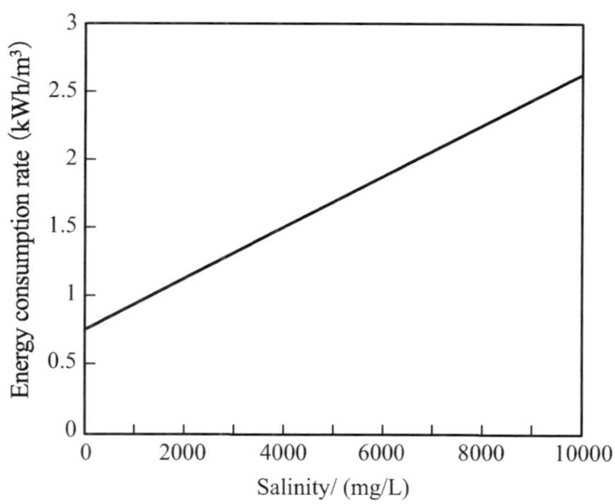

Figure 7.59
Energy consumption of the RO desalination subunit as a function of TDS.

different values of the feed water's salinity (TDS of 3000, 5000, 7000, and 10,000 mg/L) is exhibited in the literature [13]. The solar radiation data for the selected sites are shown in Fig. 7.60, and the variation of water yield with time is shown in Fig. 7.61.

Seen from the figure, the water yield of the system is extremely related to the solar radiation, which can be more than four times as much in the summer months as the winter months. The salinity of the feed has a great influence on the water yield of the system, and the yield increased significantly when the feed salinity was decreased.

In the actual RO desalination system, the pressure exerted on the feed is also an important factor affecting the performance of the system. For a solar-powered reverse osmosis

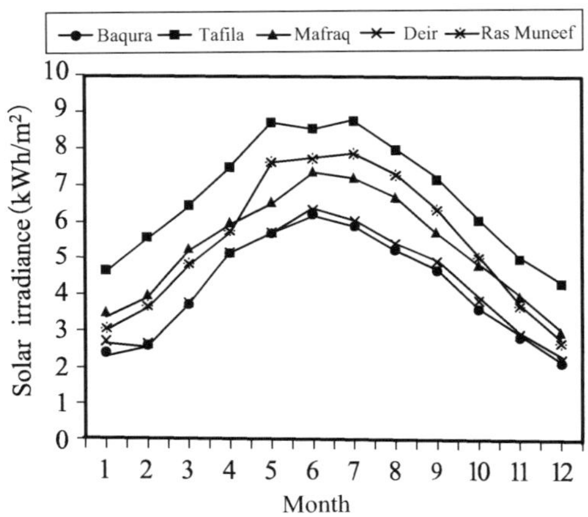

Figure 7.60
Monthly average solar radiation in Baqura, Tafila, Mafraq, Deir Alla, and Ras Muneef.

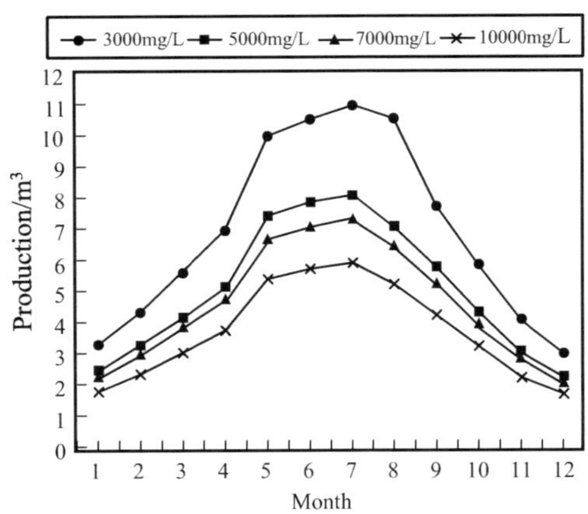

Figure 7.61
The daily water production during a 1-year cycle as a function of TDS.

desalination system, pressure controlling is the key to obtaining the maximum efficiency of the system. The feed pressure directly affects the energy consumption of the RO plant and the permeate production. It is found that for the conditions assumed in the analysis, the range of the feed pressure must be between 38 and 57.5 bars, instantaneous energy consumption will increase while improving the operating pressure, water yield will also increase. That is, operating pressure and energy consumption are inverse relation. Gandhidasan conducted a study on this process and pointed out that the relationship between the two is basically linear [15]. The effects of feed pressure on the system performance are studied and the results are shown in Figs. 7.62 and 7.63. As can be seen from the graph, the selection of operating pressure is very important for improving the efficiency of the system. Because the solar power supply is limited and unstable, it must be a reasonable choice for operating pressure to ensure the maximum efficiency of the system.

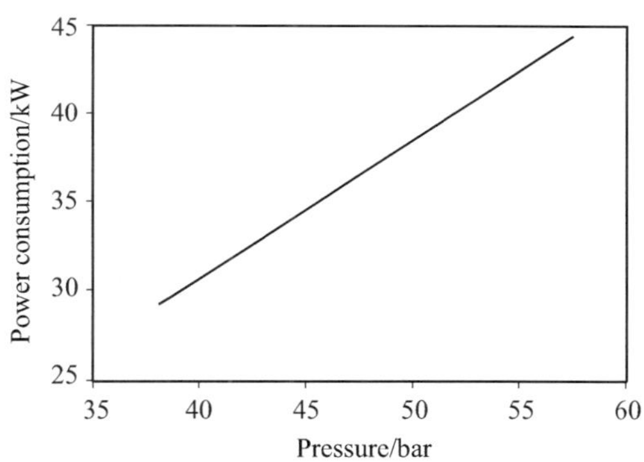

Figure 7.62
The effect of feed pressure on the power available.

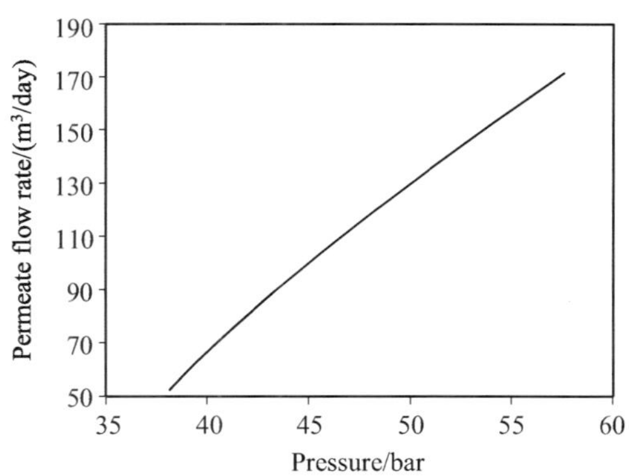

Figure 7.63
The effect of feed pressure on the permeate production rate.

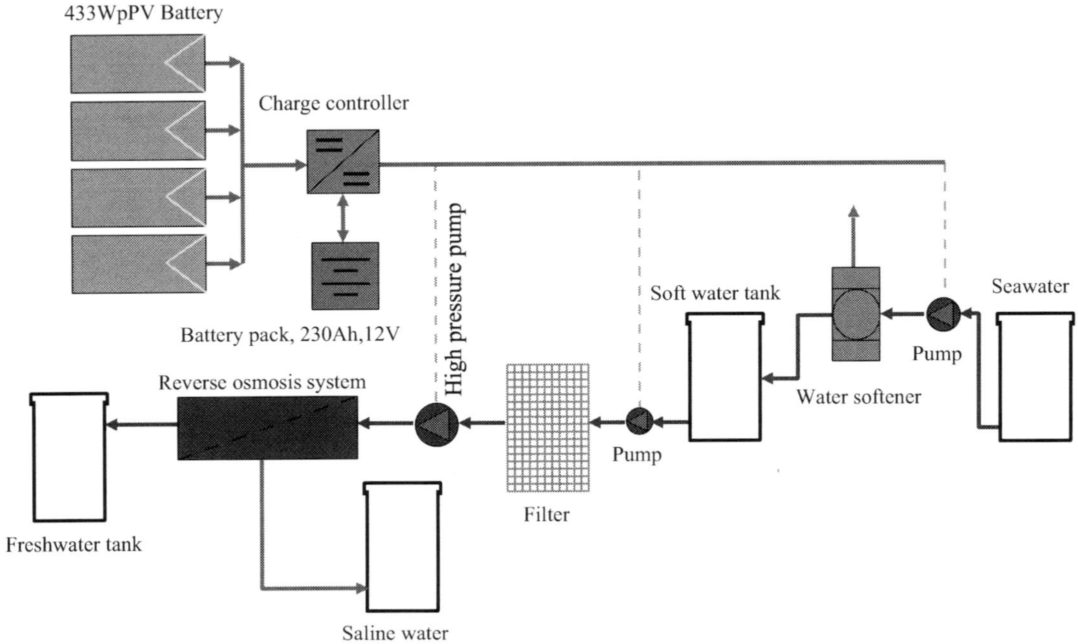

Figure 7.64
Process flow sheet of photovoltaic-powered reverse osmosis (PV-RO) brackish water desalination system.

Hazim Qiblawey introduced a typical photovoltaic-powered reverse osmosis (PV-RO) brackish water desalination system and gave the results [16]. The system is composed as Fig. 7.64.

The related parameters of solar cells in the system are as follows: 8 PV modules, 2.592 m^2 total lighting areas, 433 Wp output power (each 54 Wp). The number of membrane components is four; the electricity of all pumps of the system is provided by solar cells. The results are shown in Figs. 7.65 and 7.66.

Experimental results show that this solar RO system is completely feasible. But this type of system is still in the primary stage so far, the main function of solar cells is providing power for the system, in fact, this power can also be provided by the commercial power.

Recently, Habib Cherif proposed a kind of RO desalination unit architecture powered by a standalone PV-wind hybrid system, as shown in Fig. 7.67 [17]. From Fig. 7.67, it is not difficult to see that the desalination unit only uses the power of the PV-wind hybrid system; the PV-wind hybrid system is completely separated with the desalination system. The desalination system can continue to run by using commercial power when the PV-wind hybrid system stops running.

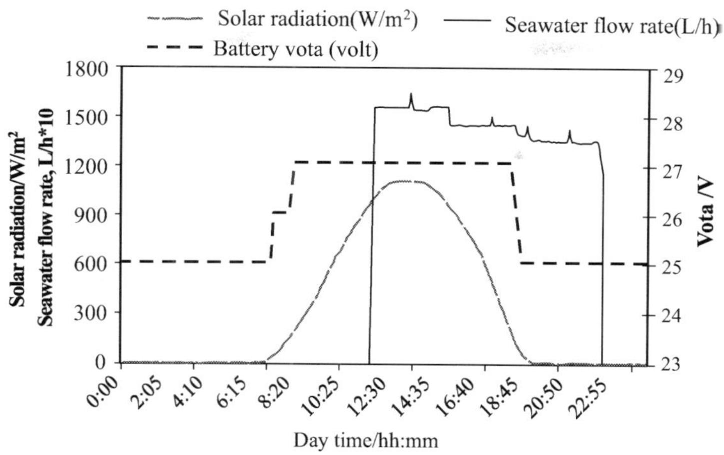

Figure 7.65
Voltage of the battery storage as a function of solar irradiation

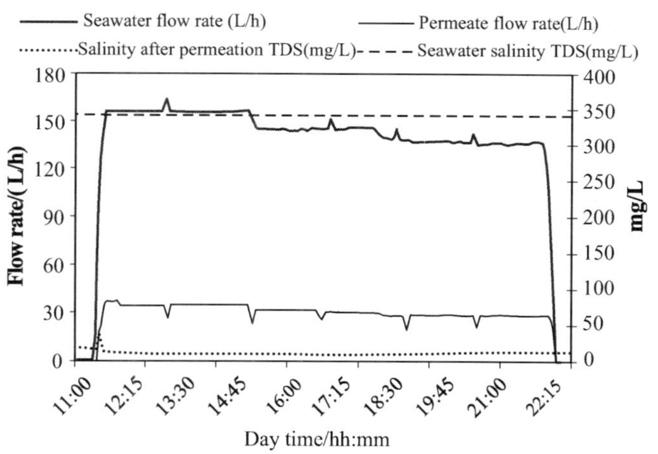

Figure 7.66
Performance of the system with brackish water with 340 ppm.

This system is of benefit for the balance of uneven distribution of solar energy in a year. Generally speaking, the sun's radiation is strong in summer and weak in winter, while the power of the wind is reversed. In winter, the wind is relatively stronger and in the summer it is weaker, so that the total seawater desalination system has a good production in a year, ensuring the supply. According to the local solar radiation data and wind power level, the annual energy supply was calculated, and the change of water yield of the year with the month was also given in the paper [17], as shown in Figs. 7.68 and 7.69, which can be used for reference.

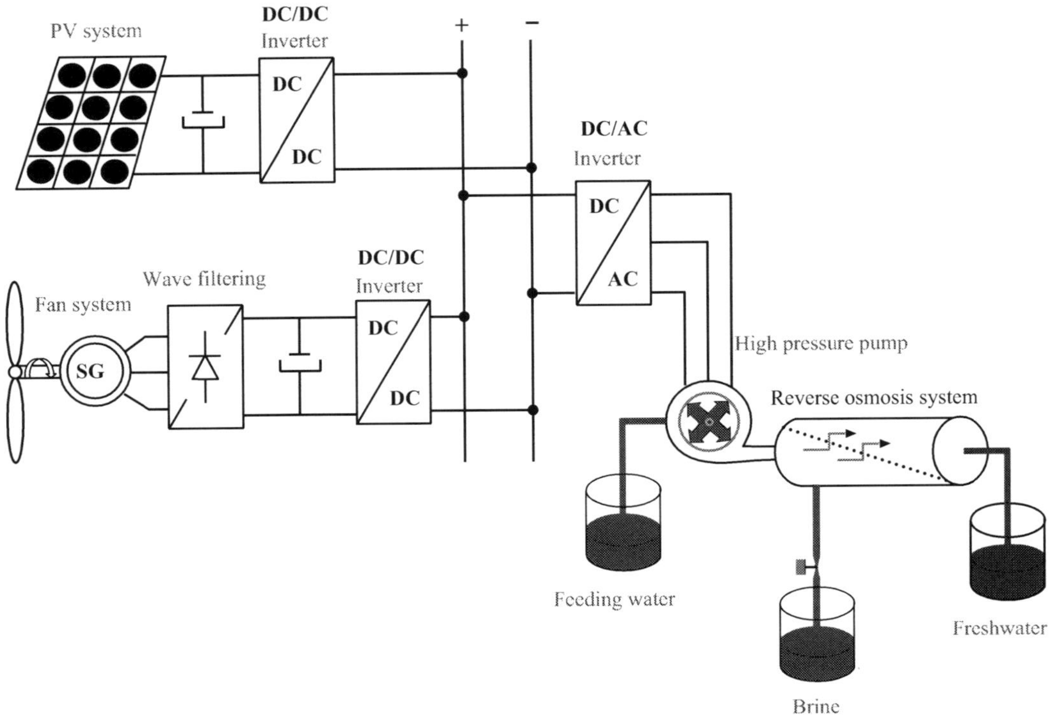

Figure 7.67
Reverse osmosis desalination unit architecture powered by a standalone PV-wind hybrid system without battery storage.

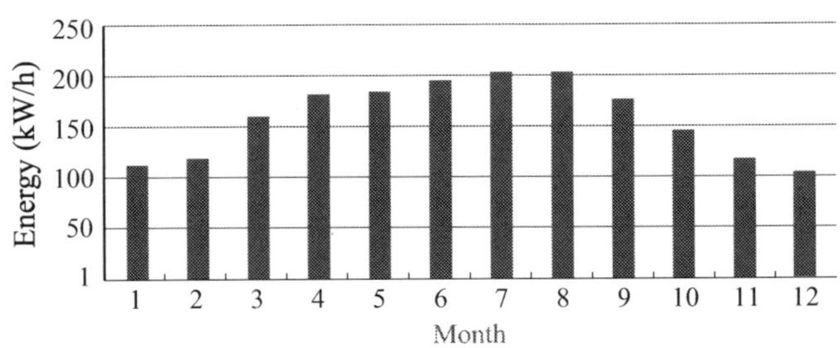

Figure 7.68
Estimated electrical energy from hybrid system.

7.8 Reverse Osmosis Desalination System Driven by Solar PV/T Device

PV thermal technology has been developed rapidly in recent years. But the efficiency of solar cells is generally less than 20%, so that more than 80% of the energy is wasted. Moreover, the excess solar energy will increase the temperature of the cell panel, which

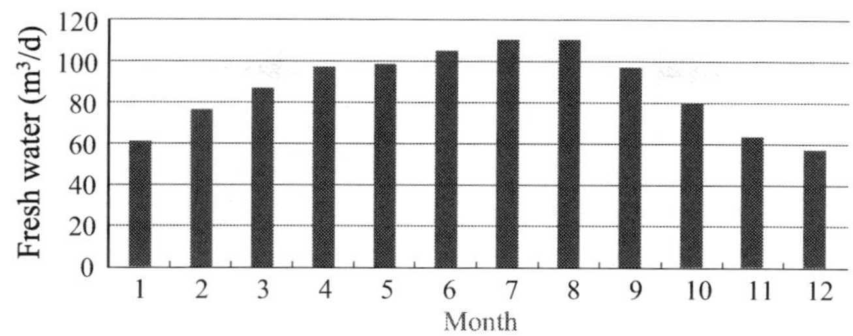

Figure 7.69
Monthly average produced potable water.

will further reduce the power generation efficiency of the cell panel. According to reports, when the temperature of the cell panel increases 1°C, the efficiency of the cell panel decreased by 0.4%. These two reasons cause people to consider the cooling problem of the cell panel and how to use the wasted energy.

Furthermore, the price of solar panels remains high. In order to save the cost and reduce the use of the cells, the concentrating PV thermal technology has been developing in recent years. Regarding concentrating PV thermal technology, supplied temperature further increases and this part of the energy can directly promote low-temperature MED systems, so as to make solar power generation and desalination systems closer together.

Gur Mittelman proposed a combined system, comprised of concentrating PV thermal collector field and multieffect evaporation desalination system. He did the theoretical analysis and simulation for this system and gave a lot of useful suggestions [18]. The coupled system is shown in Fig. 7.70.

Gur Mittelman and Dubey [19] did simulation for such a system, in which the desalination system is a low-temperature multieffect distillation system and the relevant parameters are in Table 7.10. (TBT is top brine temperature, T_{bd} is blowdown temperature, T_f is feed temperature.)

Gur Mittelman's analysis shows that the combined system is still dominant in some areas or conditions compared with the traditional RO technology. He compared photovoltaic reverse osmosis systems (PV-RO) and parabolic trough heat to a low-temperature multieffect system (PTC-MEE) with combined system, given the performance change of low concentration ratio (concentration ratio 32) system (CPVT-L) and high concentration ratio (concentration ratio 200) system (CPVT-H). A comparison of several solar desalination alternatives in a range of installed costs of \$1−6/Wp is presented in Fig. 7.71.

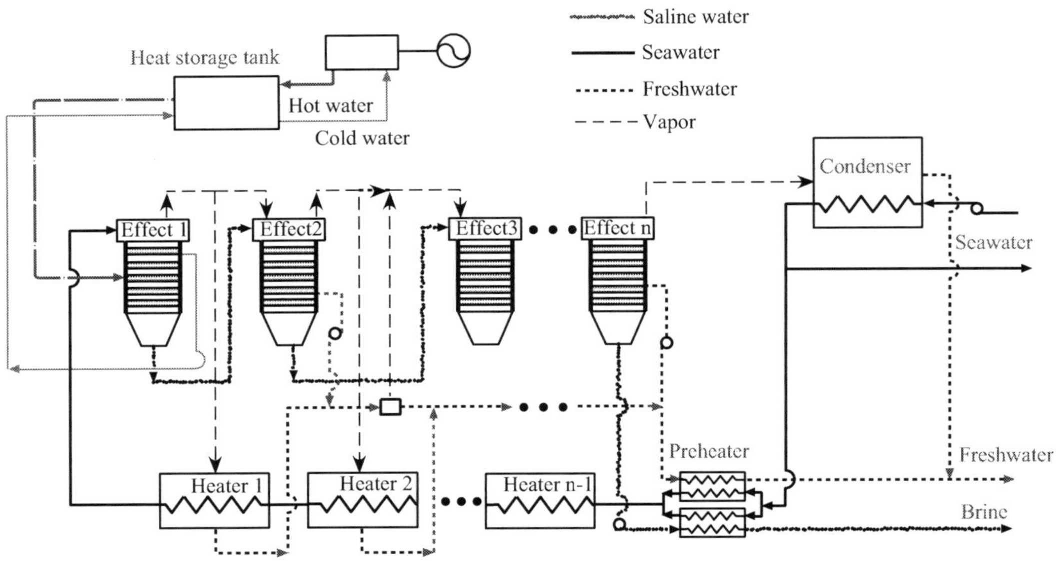

Figure 7.70
Schematic of the concentrating photovoltaic thermal-based multieffect evaporation desalination plate desalination system.

Table 7.10: Data for the multieffect evaporation desalination plant model.

Project	Parameters
Capacity	10,000 m^3/day
Number of effects	14
Seawater temperature T_{sw}	25°C
Seawater salinity of X_{SW}	42,000 ppm
Blowdown salinity X_{bd}	63,000 ppm
Heaters efficiency η	80%
TBT − T_1	5°C
$T_{bd} - T_f$	5°C

The results demonstrate that concentrating PV thermal technology desalination system is still very dominant in some intervals. When the electricity price is higher and installation cost is low, the solar system has more market potential.

In fact, the most ideal case is using electricity generated by a PV system to drive RO system. As in the discussion in the last section, photovoltaic cell powered the RO system, while the generated thermal energy can be used to heat the inlet water, so the efficiency of the RO system is improved [19]. Concentrating PV thermal technology gets higher thermal energy, meanwhile the heat can be directly available to the low-temperature multieffect

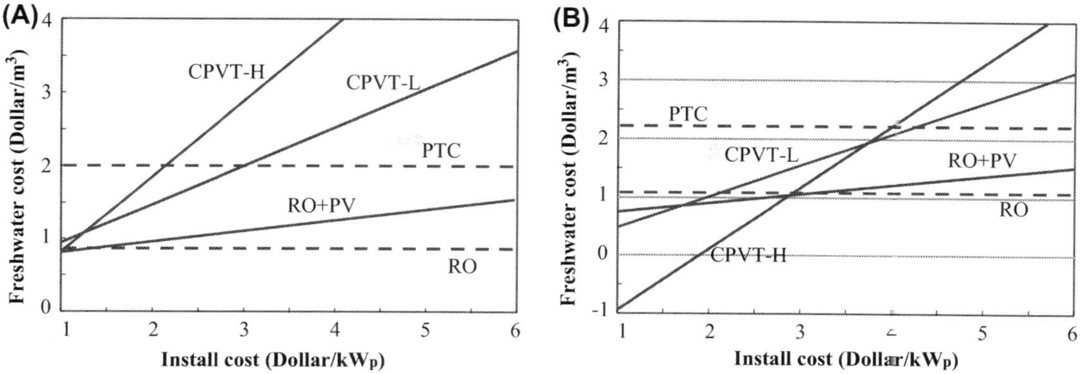

Figure 7.71
Comparison of desalinated water cost for conventional and solar technologies, as a function of PV installation cost (A) electricity price ¢ 7/kWh (B) electricity price ¢ 15/kWh.

system and the electricity can also provide power to the RO system, then getting freshwater from the two aspects, the total efficiency has been greatly increased. Researches show that the productivity of freshwater can be increased by 50% with feed water temperature from 15–45°C in the RO system. In this aspect, the scholars of our country carried out some exploration and have given some useful experience. That is heating the water in RO desalination system to around 40°C by using a solar heat collector and the heat supply tank [20]. Increasing the water temperature will increase the water productivity of the RO membrane and reduce the operating pressure of the device. This method saves energy and reduces consumption, hardly affecting the quality of the produced water, which is acceptable.

7.9 Reverse Osmosis Desalination System Driven by Solar Power Cycle System

As for the combined system discussed above, the solar energy system and the desalination system are completely independent and not closely linked. Due to the low efficiency of solar PV cells, the efficiency of the combined system is low too. So it is the biggest problem of PV solar desalination. The main reason is that the efficiency of solar cells is only about 20% and 80% of the solar energy is wasted. The electric energy generated by a solar cell panel cannot be fully used in the desalination system, it is also affected by the inverter efficiency, circuit efficiency, and other factors.

Can we use solar energy directly to work, to provide direct compression power for RO system, to reduce intermediate links and improve efficiency? Around this problem, recently, many scholars have made a new study on RO desalination directly driven by solar energy. It is considered that this scheme is feasible and has advantages in the high-temperature solar power generation system.

As early as 1978, researchers tried to combine solar collectors with RO technology. The generated heat is mainly used in an organic Rankine cycle generator to provide electricity for the RO desalination device [1]. Of course, such a system is still separated from the desalination device. P.A. Davies [21] introduced a solar organic Rankine cycle/ RO desalination system, which gets high-density energy through a Fresnel lens concentrator. He improved the mechanical structure of the system and studied the energy loss and overall performance of a RO system. The above researches still focus on the transforming of solar energy into electrical energy and then providing power for a RO system. The research of A.S. Nafey changed people's ideas. This was that the solar power system directly powered the RO system by using mechanical power instead of converting solar power to electricity. At the same time, the waste heat from the steam expansion is used to heat the seawater in the desalination device, and then the water productivity is improved [22,23]. The solar system is organically combined with the seawater desalination system.

The schematic diagram given by A.S. Nafey is shown in Fig. 7.72. The system consists of several parts: solar collector, heat recovery unit, condenser, circulating pump, reverse

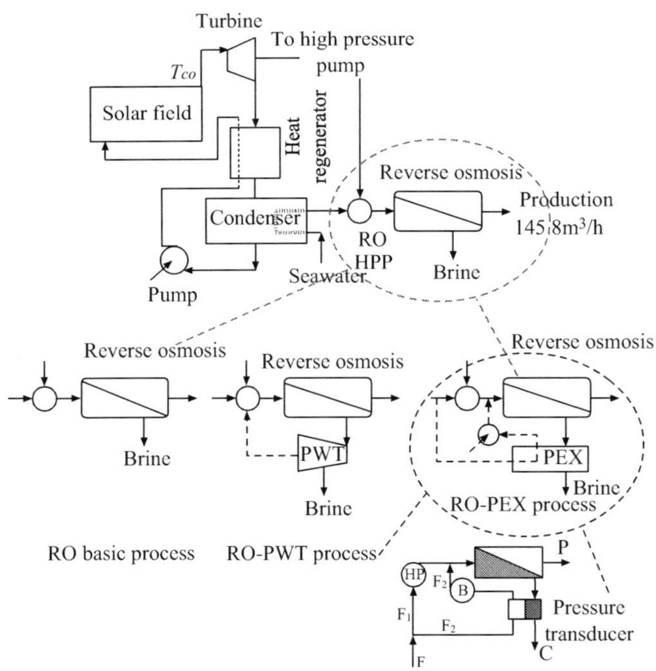

Figure 7.72
Reverse osmosis desalination system powered by solar thermal power cycle.

osmosis seawater desalination system. Solar energy drives the turbine, and then the turbine power directly drives the desalination system to get the high pressure. Operating pressure is 68 bars. In order to increase its inlet temperature, the seawater is transported to the condenser to preheat. In his model, the specifications of the desalination system are shown in Table 7.11. After analysis and calculation, he got the effect of freshwater production rate on thermo-economic product cost, as shown in Fig. 7.73.

Fig. 7.73 shows that, only when it reaches a certain size, does the system have a good economic value. When the freshwater production rate of the system is below 100 m^3/h, the production cost is high, but with the increase in productivity, the cost decreases rapidly. When the freshwater productivity is greater than 200 m^3/h, the trend of cost reduction is slow. Overall, the larger the scale is, the lower the cost of water production is. It has certain guiding significance for the development of such systems in the future.

Table 7.11: The relevant parameters in the calculation of the system.

Project	Parameter
Solar radiation (W/m^2)	850
Turbine efficiency (%)	85
Generator efficiency (%)	95
Rankine cycle pump efficiency (%)	75
Superheat temperature of solar collector (°C)	340
Condensing temperature (°C)	35
Seawater input temperature (°C)	20
Heat removal factor of heat exchanger (effectiveness)	0.8
Ambient temperature (°C)	25.4

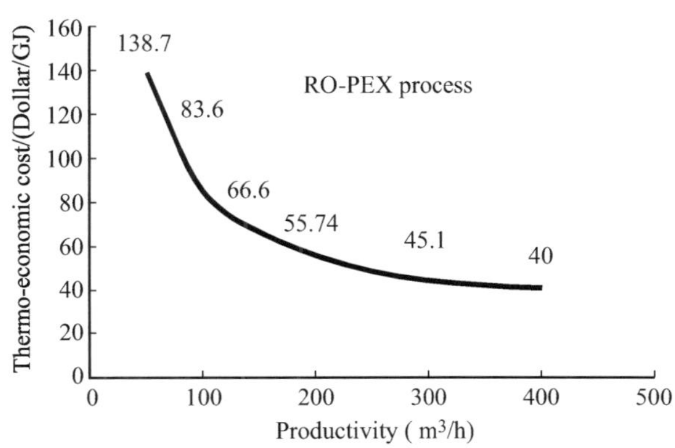

Figure 7.73
Variation of thermo-economic product cost against the variation in productivity.

7.10 Large-Scale Solar Desalination by Combination With Concentrated Solar Power: Concept and Analysis

The combination of desalination with concentrated solar power (CSP) plants (CSP + D) is considered as the most realistic opportunity for the development of large-scale desalination with solar energy. The availability of solar power and waste heat in CSP plants offers possibilities for both RO and thermal desalination systems which will produce the maximum solar application efficiency. In the case of RO the combination can be done directly by transporting the electricity. In the case of thermal desalination technologies there are different alternatives for coupling with CSP plants which can offer advantages with respect to the CSP + RO combination, such as the reduction of the cooling needs of the power cycle and the application of the waste heat of the power cycle.

For small communities, the combination of PV energy and RO desalination has already been developed into concrete applications that are starting to penetrate the market. However, the large upfront investment required for the solar power supply is a barrier when considering solar desalination for medium- or large-scale supplies of water. Furthermore, PV is hardly a dispatchable power and this is a problem for the operation of RO, especially at an industrial scale. Until large-size PV power plants can guarantee a constant power supply to the desalination system, the most promising alternative is found in the exploitation of the synergies between CSP production and desalination. The larger CSP dispatchability relies on simpler, more cost-effective and sustainable means for storing energy than PV (i.e., molten salts). However, CSP plants need large amounts of water for their own operation, which jeopardizes their potential installation in many locations where water scarcity is already a problem. The integration of water desalination technologies in CSP plants can improve the sustainability of the solar power plant, for instance by providing the water required for the cooling system. At the same time, the deployment of desalination can take advantage of the availability of solar electricity and waste heat from the CSP plant.

In the case of thermal desalination, one possibility is to reuse the low-grade energy released to the ambient in the CSP plant cooling system, which can contribute to further increasing the sustainability of the CSP + D concept, especially considering that condensation of the exhaust steam is required and its temperature is a critical issue in CSP since it affects directly the overall efficiency of the CSP plant.

Several studies related to the combination of RO and thermal desalination units with CSP plants have been published in the literature covering different areas of the world. Trieb et al. [24] were the first to present a research work about the combination of CSP and desalination, explaining the options of CSP for seawater desalination, either using electricity or steam, and proving with economic data the potential of CSP + D in Middle East and North Africa (MENA) region. Concretely, Olwig et al. [25] also made a detailed economic analysis for the combination of parabolic trough CSP (PT-CSP) plants with a low-temperature multieffect distillation (LT-MED) unit as condenser of the power cycle and with RO plants using wet cooling in the power block. Fylaktos et al. [26] compared three options: a solar power CSP plant for electricity only, electricity with RO, and electricity with LT-MED. The economic analysis was performed in terms of the levelized cost of energy, using three distinct incomes for the analysis: selling electrical energy to the grid, selling desalination seawater to the government, and selling emission allowances for avoiding CO_2 emissions.

Recently, Patricia et al. [27] proposed four CSP + D configurations and gave a detailed techno-economic analysis to them. The four different CSP + D configurations are shown in Fig. 7.74, which are:

- Configuration #1: LT-MED unit integrated into a CSP plant (Fig. 7.74(1))
- Configuration #2: LT-MED + TVC unit integrated into a CSP plant (Fig. 7.74(2))
- Configuration #3: TVC-MED unit integrated into a CSP plant (Fig. 7.74(3))
- Configuration #4: RO unit connected to a CSP plant (Fig. 7.74(4)).

In Configuration #1, the steam leaving the lower pressure turbine (LPT) is fully condensed in the first effect of the LT-MED unit, which produces freshwater through the evaporation—condensation processes that take place in each effect of the desalination plant. In this case, the steam is expanded down to a temperature of 70°C, which is higher than that of a standalone CSP plant, so only four extractions can be taken from the LPT instead of the typical five (as in Configurations #2 to #4).

In Configuration #2, part of the exhaust steam is used as the entrained vapor in a steam ejector, while the motive steam is higher-pressure steam extracted from the LPT. The motive steam compresses the entrained vapor resulting in a middle-pressure steam that is used to drive an LT-MED process. The remaining exhaust steam from the turbine is condensed through the condenser of the power cycle.

In Configuration #3, a TVC-MED is integrated into the CSP plant using high-pressure steam from LPT to feed a steam ejector coupled to an MED plant. The total exhaust steam from the turbine is condensed in the cooling system of the power cycle.

Finally, Configuration #4 corresponds to the basic combination of an RO system with a CSP plant, in which the desalination process is completely independent from the power generation.

Patricia et al. [27] established a mathematical model to compare the performance of the four configurations. Table 7.12 shows the techno-economic analysis results they obtained

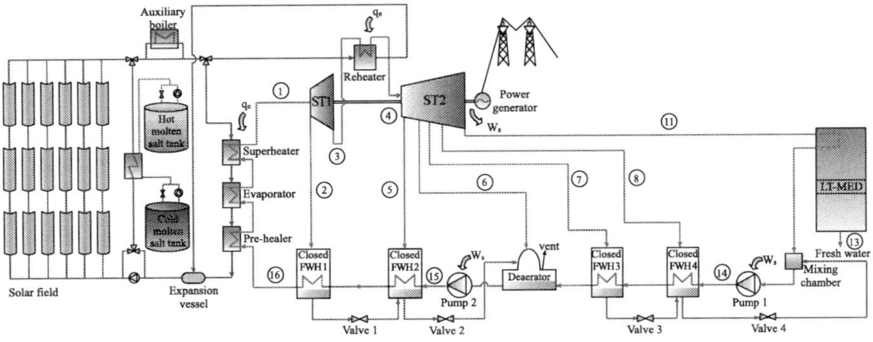

Flow diagram of Configuration #1.

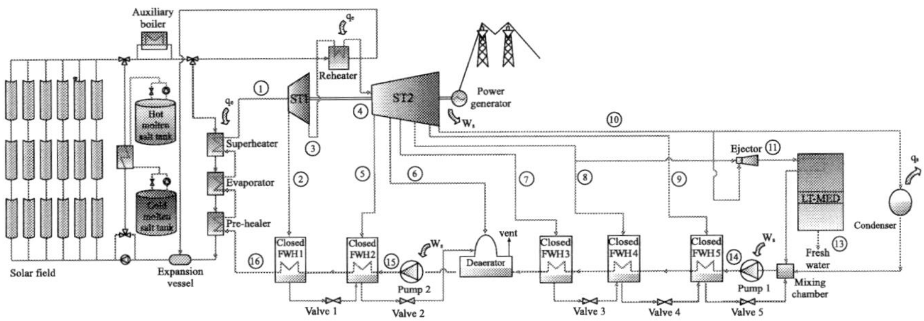

Flow diagram of Configuration #2

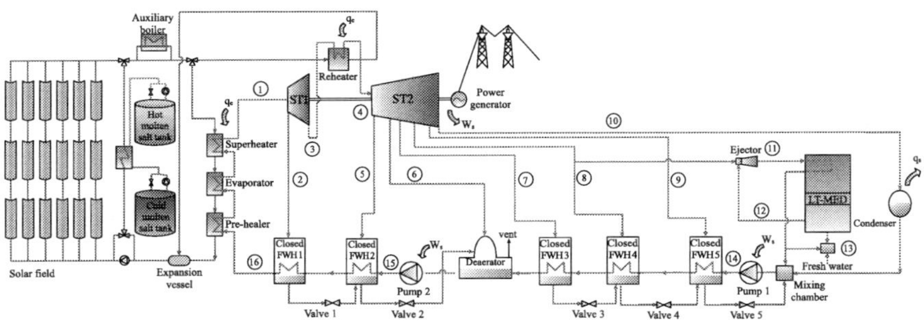

Flow diagram of Configuration #3

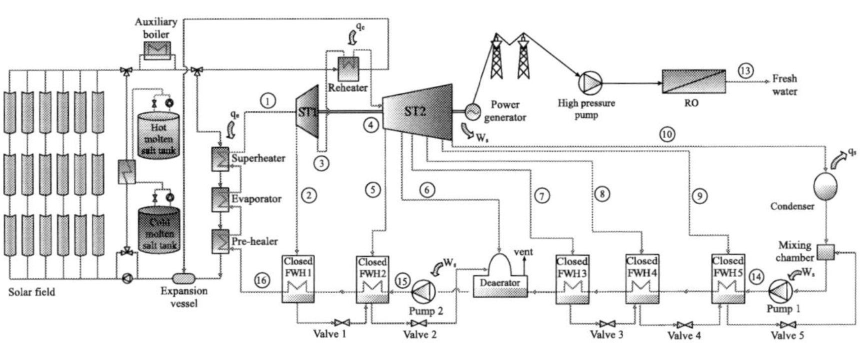

Flow diagram of Configuration #4

Figure 7.74

Four different CSP + D configurations.

Table 7.12: Results obtained from the techno-economic analysis in Abu Dhabi [8].

Options	Units	Conf#1	Conf#2	Conf#3	Conf#4
Global efficiency	%	30.41	26.42	24.46	29.65
Net output thermal capacity	MW_{th}	164	189	204	169
Gross freshwater production	m^3/day	35,950	35,985	36,012	40,448
Gross power production	MW_e	55.36	59.89	63.89	62.95
Aperture area	m^2	807,690	931,950	1,007,150	830,580
Solar fraction	%	53.96	54.09	54.13	54.11
Levelized electricity cost	c€/kWh	16.58	18.95	20.38	17.40
Levelized water cost	€/m^3	0.83	0.83	0.83	0.88

based on the ambient conditions of Abu Dhabi. These comparisons provide very good reference data for future designs.

References

[1] Ali MT, Fath HES, Armstrong PR. A comprehensive techno-economical review of indirect solar desalination. Renew Sustain Energy Rev 2011;15:4187–99.

[2] Wang Z. Solar thermal power station design (in Chinese). Chemical Industry Press; 2014.

[3] Wang J. Desalination (in Chinese). Beijing: Science press; 1978.

[4] Wang S. Desalination project (in Chinese). Beijing: Chemical Industry Press; 2003.

[5] Farwati MA. Theoretical study of multi-stage flash distillation using solar energy. Energy 1997;22(1):1–5.

[6] Mouatafa SMA, Jarrar DI, Mansy HI. Sol Energy 1985;35:333–8.

[7] Klaren DG, Pieper GA. The multi-stage-flash/fluidized bed evaporator in combination with solar energy As heating source. In: Delyannis A, Delyannis E, editors. Proceedings 6th intern. symposium fresh water from the sea. Canaria, Spain, 2; 1978. p. 31–40.

[8] Kalogirou S. Survey of solar desalination systems and system selection. Energy 1997;22(1):69–81.

[9] Zarza E, Ajona JI, Leon J, Genthner K, Gregorzewski A. In: Proceedings of the Biennial Congress of the International solar energy society, vol. 1; 1991. p. 2270–5. Part II;. Denver, CO.

[10] Huang F, Xie Y. Engineering thermodynamics principles and applications (in Chinese). Beijing: Electric Power Industry Press; 1982. p. 257–8.

[11] Zheng H, He K, Yang Y, Chen Z, Li H. Study on a multi-effects regeneration and integral-type solar desalination unit with falling film evaporation and condensation processes. Sol Energy 2006;80(9):1189–98.

[12] Chen Z, Xie G, Chen Z, Zheng H. Field test of a solar seawater desalination unit with triple-effect falling film regeneration in northern China. Sol Energy 2012;86:31–9.

[13] Hrayshat ES. Brackish water desalination by a stand alone reverse osmosis desalination unit powered by photovoltaic solar energy. Renew Energy 2008;33:1784–90.

[14] Leblanc J, Andrews J, Akbarzadeh A. Low-temperature solar–thermal multi-effect evaporation desalination systems. Int J Energy Res 2010;34:393–403.

[15] Gandhidasan P, Al-Mojel SA. Effect of feed pressure on the performance of the photovoltaic powered reverse osmosis seawater desalination system. Renew Energy 2009;34:2824–30.

[16] Qiblawey H, Banat F, Al-Nasser Q. Performance of reverse osmosis pilot plant powered by photovoltaic in Jordan. Renew Energy 2011;36:3452–60.

[17] Cherif H, Belhadj J. Large-scale time evaluation for energy estimation of stand-alone hybrid photovoltaice wind system feeding a reverse osmosis desalination unit. Energy 2011;36:6058–67.

[18] Mittelman G, Kribus A, Mouchtar O, Dayan A. Water desalination with concentrating photovoltaic/thermal (CPVT) systems. Sol Energy 2009;83:1322–34.
[19] Dubey S, Tiwari GN. Thermal modeling of a combined system of photovoltaic thermal(PV/T) solar water heater. Sol Energy 2008;82:602–12.
[20] Xianqiu Z, Mingxia D, Yong Z. Reverse osmosis water treatment employing solar energy (in Chinese) Acta Energiae Solaris Sin 2010;31(11):1419–22.
[21] Davies PA. A solar-powered reverse osmosis system for high recovery of freshwater from saline groundwater. Desalination 2011;271:72–9.
[22] Nafey AS, Sharaf MA, García-Rodríguez L. Thermo-economic analysis of a combined solar organic Rankine cycle-reverse osmosis desalination process with different energy recovery configurations. Desalination 2010;261:138–47.
[23] Sharaf MA, Nafey AS, García-Rodríguez L. Exergy and thermo-economic analyses of a combined solar organic cycle with multi effect distillation (MED) desalination process. Desalination 2011;272:135–47.
[24] Trieb F, et al. Technologies for large scale seawater desalination using concentrated solar radiation. Desalination 2009;235:33–43.
[25] Olwig R, et al. Techno-economic analysis of combined concentrating solar power and desalination plant configurations in Israel and Jordan. Desalin Water Treat 2012;41:9–25.
[26] Fylaktos N, Mitra I, Tzamtzis G, Papanicolas CM. Economic analysis of an electricity and desalinated water cogeneration plant in Cyprus. Desalin Water Treat 2015;55:2453–70.
[27] Palenzuela P, Alarcón-Padilla D-C, Zaragoza G. Large-scale solar desalination by combination with CSP: techno-economic analysis of different options for the Mediterranean Sea and the Arabian Gulf. Desalination 2015;366:130–8.

CHAPTER 8

Absorption and Adsorption Solar Desalination System

Chapter Outline

8.1 Features and Performance of Absorption and Adsorption Working Substance

Many metallic salt solutions, like LiBr, LiCl, LiI, $LiNO_3$, $CaCl_2$, $ZnCl_2$, and KNO_3, etc., are found to have strong adsorption behavior on water vapor; in other words, the corresponding saturated vapor partial pressure of the aqueous solutions of these types of substances is very low. For example, when the temperature and concentration (mass percentage) of an LiBr-H_2O solution are 25°C and 50% respectively, the partial pressure of water vapor on the surface is about only 0.8 kPa, even lower than the saturated vapor

Solar Energy Desalination Technology. http://dx.doi.org/10.1016/B978-0-12-805411-6.00008-7

partial pressure of water at 6°C. The saturated vapor pressure of water at 25°C is 3.16 kPa, which means the vapor on the water surface can still be absorbed by the LiBr solution. The lower the saturated vapor pressure for the solution is, the stronger it is to absorb water vapor. Owing to the abilities of the solutions absorbing the water vapor with a much lower temperature, they are first utilized in the areas such as air condition cooling and heat pumping and achieve great success. Currently, most of the large central air conditioners in commercial use are manufactured by those absorption-working substances. Those substances are environmentally friendly because they have many advantages, such as nontoxic, odorless, nonflammable, nonexplosive, and nonstaining for the environment, etc.

Although this kind of substance has the property that strongly absorbs water vapor, they will release water sufficiently when its temperature reaches a certain degree (this required energy in this process is mainly thermal energy). In addition the released water is pure. For instance, for an LiBr solution, the boiling point for water is 100°C and for LiBr is 1265°C at atmospheric pressure. The difference between them is more than 1000°C, so the vapor generated when this solution is boiling is almost pure water. The constituent of LiBr is little, so pure water can be obtained without further distillation. Therefore researchers applied these working substances for seawater desalination to build absorption seawater desalination system. If it is driven by the heat supply from solar energy, it is called a solar absorption desalination system. The most popular working substance used is LiBr, and other types are LiCl and $CaCl_2$ solutions.

Similar to some metallic salt solutions, some solid substances, like zeolite, activated carbon, strontium chloride, and calcium chloride, etc., have the property of strongly adsorbing water vapor, e.g., when the vapor partial pressure on surface of zeolite is 2 kPa, it can adsorb the vapor, which equals 20% of its weight. The air conditioning invented based on this feature is called adsorption air conditioning. Similar to an absorption desalination system, it was put forward that seawater can be desalinated by solid adsorption, which refers to the adsorption desalination system.

The adsorption by a solid is different from the absorption by a liquid. Adsorption can be divided into two types: one is physical adsorption (physisorption), and the other is chemical adsorption (chemisorption). Physisorption is mainly caused by the physical interactions (attraction between molecules) between adsorbate and the substrate surface. The chemical properties of both substances do not change. Because the molecules are far away, the adsorption is weak. The heat of adsorption is relatively less. Multilayer adsorption can occur, and the adsorbed gaseous substances are normally nonselective. The amount of adsorbed substances increases with the rise of adsorption surface area. Adsorption and desorption processes are fast.

In chemisorption, new types of electronic bonds will be formed by electron transfer. The chemical properties of adsorbate and adsorbent will change under certain conditions.

The features of chemisorption are strong adsorption and large heat of adsorption. The adsorption is selective and monolayer adsorption. The speed of adsorption and desorption is relatively slow.

Actually, there is no rigid boundary between physisorption and chemisorption. Under room temperature, the two kinds of adsorption can exist at the same time. Physisorption may be faster at the beginning, and it will slow down when chemisorption starts. Because desorption depends on the heating from an external heat source no matter which adsorption, both physisorption and chemisorption can be applied in seawater desalination, but the selection should be determined by the factors like the capability of adsorbing water vapor and the purity of desorption.

The most popular seawater desalination systems in research are utilizing zeolite and activated carbon as adsorbent.

8.2 Solar Absorption Desalination System

The fundamental cycle of a solar absorption desalination system is shown in Fig. 8.1. It consists of the following parts basically: generator (most of them are solar collectors at the same time), absorber, condenser, and evaporator.

The working process is that seawater comes into the evaporator through seawater inlet pipe (3) by flash evaporation and evaporates in the evaporator by absorbing heat Q_{in}.

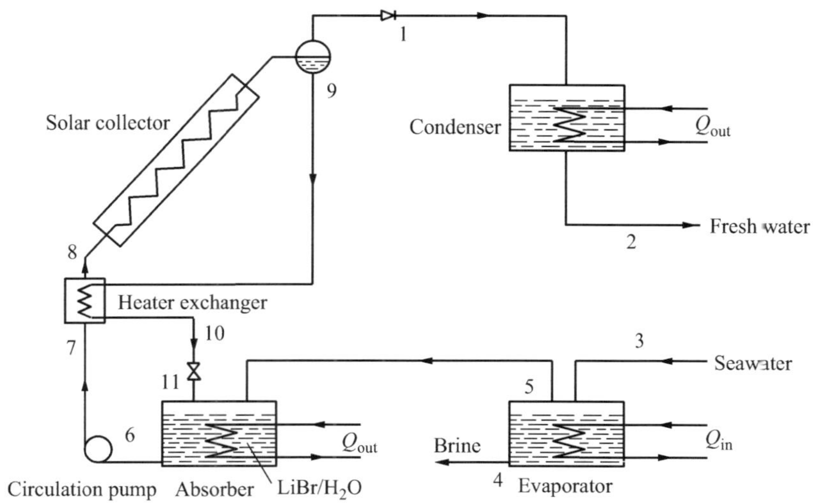

Figure 8.1
Traditional single-effect solar absorption desalination system. 1, one-way valve; 2, fresh water output; 3, seawater input; 4, brine output; 5, steam communicating pipe; 6, recycling pump; 7, heat exchanger; 8, solar collector; 9, gas–liquid separator; 10, concentrated solution pipe; 11, expansion valve.

The water vapor in the evaporator is absorbed by the salt solutions (mainly the LiBr solution currently) inside and releases the latent heat of vapor condensation Q_{out}. The latent heat is taken away by cold water. The brine left in the evaporator will be pumped out. After the solution absorbing vapor, the concentration becomes lower so that it comes into the dilute solution. Then it is pumped into a solar collector, which is a generator by pump 6, during which it goes through heat exchanger (8) to raise the temperature. In the generator, the solution temperature increases until boiling by the heating from solar energy. After the boiled solution is separated by a gas–liquid separator, gas comes into the condenser through the check valve. The strong solution returns to the absorber through the heat exchanger with releasing heat and expansion valve (10) to absorb vapor gain. The vapor flowing into the condenser will be transported to the user after condensing into fresh water. Meanwhile, it releases latent heat of condensation, which is brought away by chilled water. To improve the efficiency of this equipment, the whole system operates under negative pressure. The vacuum pressure of the system should be under 20 kPa.

Because the heat supply by solar energy has the characteristics of instability and it cannot make the system work continuously at night, the solar energy system is considered to be commonly integrated with other heat supply systems. For example, the solar absorption system can cooperate with vapor recompression to constitute a solar absorption

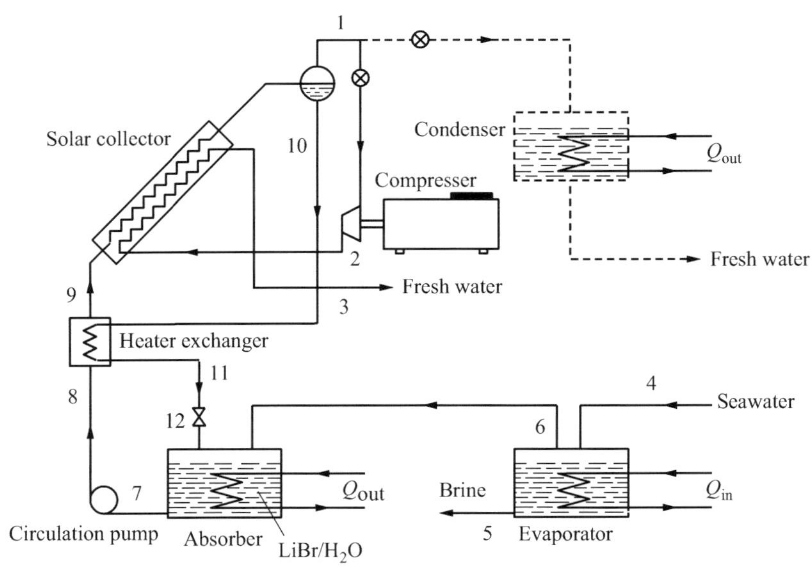

Figure 8.2
Solar absorption desalination system: 1, steam pipe and one-way valve; 2, compressor; 3, fresh water output; 4, seawater input; 5, brine output; 6, steam communicating pipe; 7, recycling pump; 8, heat exchanger; 9, dilute solution pipe; 10, the concentrated solution pipe; 11, liquid return pipe; 12, expansion valve.

desalination system with vapor compression. Fig. 8.2 demonstrates its working principle. The vapor generated from a gas–liquid separator will leave in two ways: (1) When sunlight is sufficient, vapor goes into the condenser through valve one and releases latent heat to form fresh water for users; (2) When sunlight is insufficient or at night, vapor goes into the compressor through valve two and forms high-pressure steam by compression of the compressor. It releases heat in the generator and condenses to fresh water to users through valve three. The operations of other parts are the same as shown in Fig. 8.1.

The solar absorption desalination system can be designed into a rotary structure. The whole process, like evaporation and condensation, can be finished in rotation. During the rotation of the system, the solution in the generator and absorber will form radial solution film due to centrifugal force. The seawater in the evaporator will become liquid films, which have a very high heat transfer rate because of their huge extended areas and extremely small heat capacity.

The rotary system is demonstrated in Fig. 8.3. The working substances (like water or heat oil) heated by solar energy are transported to the surface wall of the generator in the rotary system, and then the weak solutions on the wall of the generator are heated. The weak solution will evaporate after heating. The vapor will condense on the wall of the condenser, which is close to the generator, and form fresh water. Finally, it will be extracted to users from point 1. The strong solution produced by weak solution

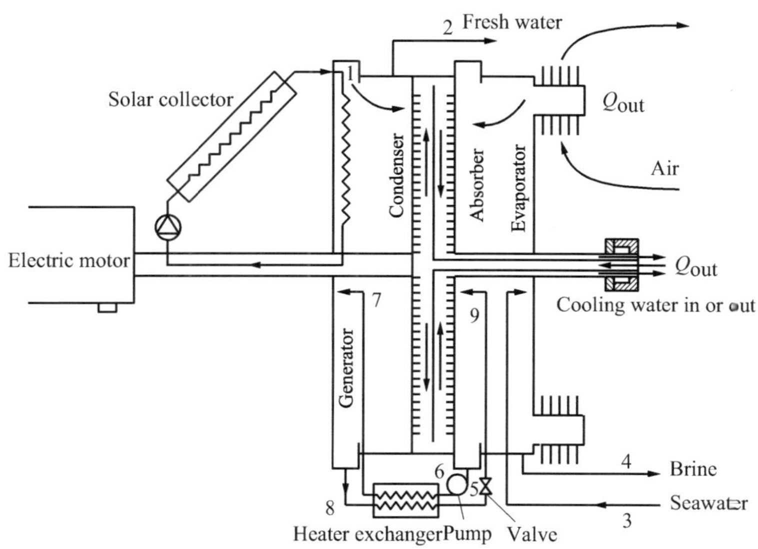

Figure 8.3
Rotation absorption solar desalination system: 1, solar collector system; 2, fresh water output pipe; 3, seawater input pipe; 4, brine output pipe; 5, expansion valve; 6, recycling pump; 7, diluted solution spraying; 8, heat exchanger; 9, concentrated solution spray.

evaporation will go into the cavity formed by the evaporator and absorber through 8 and the heat exchanger. Then it flows onto the wall of the absorber through point 9 and forms surfaces of thin films to absorb vapor from the wall of the generator. Because the distance between evaporator surface and absorber surface is short, the efficiencies of evaporation and the absorption process are very high. The latent heat released from condensation and absorption will be brought away by chilled water or cold air with the directions as shown in the diagram, by which the total seawater desalination will finish.

Riffat (1995) did the numerical analysis of the structure shown in Figs. 8.1–8.3. The flat plate solar collector was applied in his assumption (the vacuum tube collector can perform at higher efficiency). Results show that the thermal efficiency or coefficient of performance (COP) (which refers to Q_e/Q_g; Q_e is the input heat of the evaporator and Q_g is the obtained heat in the generator) can reach to 1.04. If the system efficiency of solar energy is assumed to be 50% and the collected heat per square meter under a sunny day is 23.5 MJ/day, the fresh water generated by this system for per-square meter collector can be estimated to be 5.2 kg approximately, which is the result of a single-effect system. The amount of generating fresh water will increase for a multieffect system. For the system cooperating with the compressor presented in Fig. 8.2, the COP of this system (which is Q_e/W_{comp} where W_{comp} is the power of consumption) can reach 7.9. The operating temperatures of evaporator and absorber are assumed to be 20°C and 40°C and 80°C for the generator.

The analyzed results illustrate that the solar absorption desalination system has broad application prospects. Particularly, it can be integrated with many existing systems in industry and business as the supplement processes of drying, evaporation, and boiling. For instance, it can be applied in the drying of paper, grain, medicine, milk, etc., as well as the concentration of pulp and syrup. It will contribute to many aspects of the national economy, so it is noteworthy to be studied further.

8.3 The Solar Absorption System Integrated With Industrial Seawater Desalination

8.3.1 Single-Effect Solar Absorption Desalination System

In a traditional seawater desalination area, there is no publication of the equipment that is fully commercialized or industrialized, although many researchers have been established for absorption and adsorption desalination systems. But in a cooling and air conditioning area, there are numerous relevant publications. In the area of seawater desalination, Weimberg et al. (1980) calculated and evaluated the seawater desalination processes of an LiBr absorption heat pump, multiple-effect evaporation, and vacuum freezing, etc. Results showed that the integration of absorption system and multiple-effect evaporation can improve the COP of a multieffect seawater desalination system to 18 ~ 20% [1]. It was also pointed out that the integration can deal with the miniaturization problem of the system. In 1985, Alefeld and Ziegler firstly proposed an integrated seawater desalination system, which is cooperated with an LiBr-H_2O absorption heat pump. This system consists of three stages, including seawater treatment and fresh water production [2]. In 1996, Yanniotis et al. did the theoretical simulation and experimental analysis of this kind of system [3]. Aly (1988), Fiathalah, and Aly (1991) analyzed an LiBr-H_2O heat pump forced by solar energy and intended to use that to generate high-quality steam in order to derive a multieffect evaporation (MEE) seawater desalination system [4,5]. Al-Juwayhel et al. (1997) investigated the performances of single-effect desalination systems cooperating with various heat pumps. Results demonstrated that the COP of the Absorption Vapor Compression (ABVC) system is three times higher than that of the traditional single-effect thermal evaporation and compression system [6]. The research by El-Dessouky and Ettouney (1991) also put forward that the systems integrating MEE-ABVC and MEE-Adsorption Vapor Compression (ADVC) have the COP, which is 50% higher than the COP of the MEE-TVC system (Thermal Vapor Compression), and the COP of this type of system can reach 20 [7].

Fig. 8.4 demonstrates a sketch diagram of the ABVC system integrated with solar energy for the heat supply. In this system, the solar collector provides the heat supply only, and it can be driven by traditional steam heating. In this system, the evaporator comprises a horizontal falling film tube, brine spray, foam catcher, and saltwater pool, etc. The

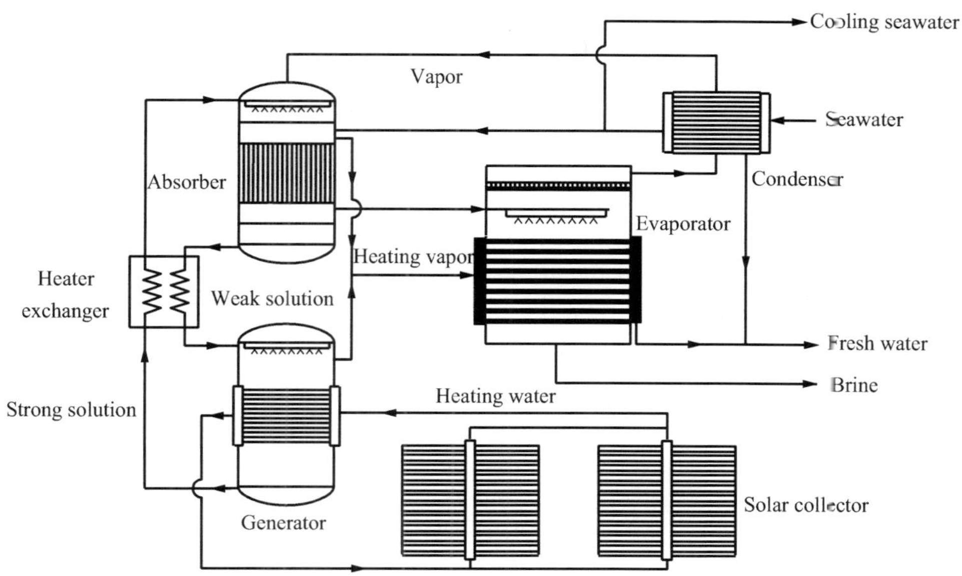

Figure 8.4
Single-effect absorption vapor compressing solar desalination system.

condenser has a similar case and horizontal tube structure, but vapor condenses on the outside surface of the tube and the vapor remained goes into the absorber. The liquid flowing in the horizontal tube in the condenser is seawater for cooling. When it flows out of the condenser, part of the seawater will be eliminated, and another part will flow into the absorber to be cooled. The absorber is the shell and tube construction as well. It utilizes the vertical tube falling film absorption. The chilled seawater brings away the latent heat released by vapor, and then the heated seawater begins spray evaporation after it goes into the evaporator.

Because properties of LiBr-H_2O solution are very stable, which has a very strong absorbing capability and a boiling point of 1265°C, compared with H_2O of 100°C, when the LiBr-H_2O solution evaporates at the temperature close to 100°C, the evaporated substance is almost H_2O. No problem of water pollution was raised. Hence almost all of the absorption seawater desalination systems apply LiBr as a working substance. In the system shown in Fig. 8.4, the LiBr-H_2O solution in the absorber absorbs the vapor from the condenser, and the concentration becomes weak. The temperature of the weak solution increases after passing through the heat exchanger. Finally, it goes into a generator and begins spray-film evaporation. The heat of the generator is supplied by the solar collector. The vapor produced in the generator is transported to the horizontal tube in the generator through pipes and has membrane heat exchange with sprayed seawater in order to force the evaporation of seawater out of the tube. The vapor in the tube will condense to fresh water for users. After the heating and evaporation in the generator, the LiBr-H_2O solution turns from weak to strong, which recovers the absorption capability of a strong solution. It will flow back to the generator through the heat exchanger and pump to repeat the absorption on vapor. In this way, this system operates in a cycle. The system produces fresh water in two components: one is inside the horizontal tube of the evaporator, and the other is outside of the tube of the condenser. If the evaporator is heated by the vapor from the boiler, some fresh water will be generated inside the horizontal tube of the generator. The heat exchanger between the generator and the absorber is mainly applied for reclaiming heat by the process during which the weak solution is cooled and flows into the strong solution in the absorber and rising the temperature of the weak solution flowing into generator. It should be noted that when the solution in the absorber absorbs vapor, it will release the amount of latent heat. Meanwhile, cold seawater is heated to a saturation state, and even part of the seawater is evaporated to vapor. The generated vapor will be transported into the horizontal tube in the evaporator to form fresh water. The foam catcher in the evaporator is utilized to prevent the seawater droplets to be taken into the condenser.

Under steady-state conditions, the analysis of energy and mass balance can be conducted. First, assuming the vapor generated in the evaporator, the absorber and generator contain no salt, which means that the brine will not splash into the vapor during evaporation. The energy releasing to surroundings from the system can be ignored; in other words, the thermal insulation of each part of the system is excellent. The physical properties of vapor are calculated at the average temperature of the calculated element.

Therefore the mass balance equation of total mass and mass of salt content (salinity) can be expressed as:

$$M_\mathrm{f} = M_\mathrm{d} + M_\mathrm{b} \tag{8.1}$$

$$M_\mathrm{b} = M_\mathrm{f}(X_\mathrm{f}/X_\mathrm{b}) \tag{8.2}$$

where M is mass flow rate; X is salt concentration; and the subscripts b, d, and f represent the states of strong brine, fresh water, and initial seawater. In Eq. (8.2), the concentration of drained brine X_b should be restricted, which is limited to be lower than 90% of the concentration of saturated $CaSO_4$ solution. Therefore the fitting equation of the relationship between X_b and T_b can be determined according to the relationships between saturation concentration and temperature of $CaSO_4$ solution:

$$X_\mathrm{b} = 0.9\left(457628.5 - 11304.11T_\mathrm{b} + 107.578T_\mathrm{b}^2 - 0.360747T_\mathrm{b}^3\right) \tag{8.3}$$

where T represents temperature, °C.

In the evaporator, the brine for falling film evaporation absorbs the latent heat of vapor coming from the evaporator and absorber and makes the vapor condensate to some fresh water. During this process, the mass of secondary steam D generated at temperature T_v is determined by the equation:

$$M_\mathrm{s} \cdot h_{\mathrm{fg,s}} = D \cdot h_{\mathrm{fg,v}} \tag{8.4}$$

where h_{fg} is latent heat of vaporization of steam and the subscripts of s and v refer to the states of hot vapor and secondary steam. In the evaporator, absorber, and generator, the boiling temperature of liquid is usually higher than the corresponding saturation temperature of vapor. Assuming the increased value at this temperature is ΔT_y, there is

$$T_\mathrm{b} = T_\mathrm{v} + \Delta T_\mathrm{y} \tag{8.5}$$

In the condenser, the condensation temperature T_c is lower than the evaporation temperature T_v. The main reason is that the boiling point will increase, which means there will be ΔT_y, but there will be pressure losses during the transportation of secondary steam, which will reduce the temperature of some secondary steam. Therefore:

$$T_\mathrm{c} = T_\mathrm{b} - \Delta T_\mathrm{y} - \Delta T \tag{8.6}$$

where ΔT represents the temperature loss that is caused by the pressure difference.

In the condenser, the total amount of imported seawater is $(M_{cw} + M_f)$, and its temperature rises from T_{cw} to T_f. Here, M_{cw} is the amount of outlet seawater after passing through the condenser, and M_f is the amount of imported seawater into the absorber after passing through the condenser. In the condenser, the energy for increasing seawater temperature comes from the released latent heat from partial condensation of vapor in the evaporator. Hence the energy balance equation can be expressed as:

$$(D - M_{ev})h_{fg,c} = (M_{cw} + M_f)C_p(T_f - T_{cw}) \tag{8.7}$$

where the subscripts of c, cw, and ev represent the states of condensed vapor, imported seawater, and the vapor flowing into the absorber. The hot vapor going into the evaporator comes from two parts: one is M_{ev} from the condenser to the absorber, and it will be released after heating by the evaporator. Another is M_{ab} generated in the absorber, which is generated by the evaporation of chilled seawater M_f. Therefore,

$$M_s = M_{ev} + M_{ab} \tag{8.8}$$

According to Fig. 8.4, the total fresh water produced by the system constitutes two parts, which are:

$$M_d = D + M_{ab} \tag{8.9}$$

The energy balance equation of the absorber is:

$$M_g h_g + M_{ev} h''_{ev} + M_f h_f = (M_g + M_{ev})h_o + M_{ab} h''_s + M_a h_a \tag{8.10}$$

where M_g is the flow rate of a strong LiBr-H_2O solution flowing into the absorber and M_a is the mass flow rate of brine leaving the absorber. The specific enthalpies of saturated vapor h''_{ev} and h''_s are obtained at the temperature in condenser T_c and the temperature of hot vapor coming into the evaporator, respectively, and h_g and h_o are the specific enthalpies of strong and weak LiBr- H_2O solution at temperatures (C_g,T_g) and (C_o,T_o), respectively. Obviously, T_g and T_o should be related to the saturation temperatures T_s and T_c of vapor. The specific enthalpies h_f and h_a of initial seawater and brine should be calculated by (T_f, X_{cw}) and (T_a, X_a). The vapor temperature in the evaporator is relevant to the temperature T_a of brine that evaporated in the evaporator clearly. In addition the increase of boiling point should be considered. Hence,

$$T_s = T_a - \Delta T_y \tag{8.11}$$

The energy balance in the generator should be the energy released by motive steam for heat supply equal to the sum of the energy increase of the LiBr-H_2O solution and the energy contained in the generated vapor, which is:

$$(M_g + M_{ev})h_o + M_m h''_m = M_g h_g + M_{ev} h''_s \tag{8.12}$$

where, M_m and h''_m represents the mass flow rate and specific enthalpy of motive steam, respectively.

The mass balance and salt balance between the imported seawater and exported brine in the absorber are provided by the following equations:

$$M_f = M_a + (M_s - M_{ev}) \tag{8.13}$$

$$X_f M_f = M_a X_a \tag{8.14}$$

Similarly, the equations of total mass balance and total salt balance of strong LiBr-H_2O solution and weak LiBr-H_2O solution are:

$$M_o = M_g + M_{ev} \tag{8.15}$$

$$M_o C_o = M_g C_g \tag{8.16}$$

where C_o and C_g are the concentrations of strong and weak LiBr- H_2O solutions.

Here the performance ratio (PR) of the system can be defined as the ratio of the amount of produced fresh water to a unit mass of motive steam, which is

$$\text{PR} = M_d/M_m \tag{8.17}$$

The equations are the energy mass balance equations of the system shown in Fig. 8.4.

The literature [7] calculated values of the seawater desalination part in this system by the provided relevant parameters. The criteria applied in the calculation are as follows:

- The designed capacity of the system is 1 kg/s.
- The temperature of compressed vapor is 80°C.
- The temperature of boiling for brine is 77°C.
- The mass fraction of a strong LiBr-H_2O solution is 0.7.
- The temperature of seawater at inlet is 25°C.
- The concentration of seawater at inlet is $X_f = 36{,}000$ ppm.
- The temperature of seawater leaving the condenser is 5°C lower than that of vapor in the condenser.
- The temperature of heated vapor imported to the generator or the temperature of hot water from the solar collector is 5°C higher than the boiling temperature of the LiBr solution.
- The temperature of seawater leaving the absorber is 5°C lower than the boiling point of the weak LiBr solution.
- The temperature difference is 3°C between the cold and hot sides of the LiBr solution heat exchanger.
- The flow rate of seawater in the condenser is 1.5 m/s.
- The concentration of produced fresh water is 0 ppm.

- The outer diameter and wall thickness of the horizontal tube in the condenser and evaporator are 0.015 and 0.005 m, respectively.
- The thermal conductivity of the horizontal tube is 0.042 kW/m °C.
- The thermal fouling resistance of the horizontal tube is 0.001 kW/(m °C).
- The flow rate of the falling film outside the horizontal tube in the evaporator is 1.5 m/s, and the thickness of the film is 0.001 m.

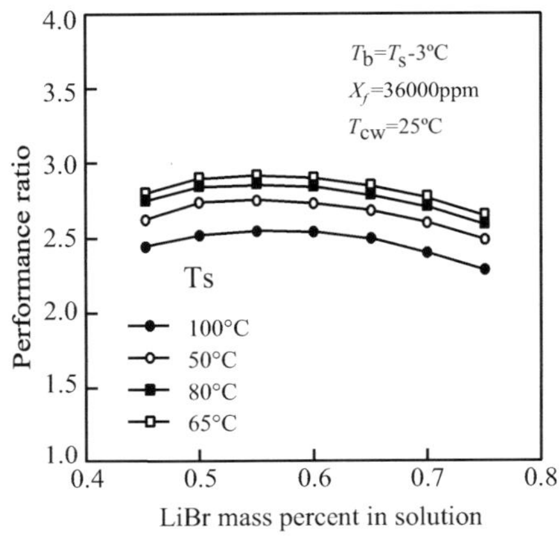

Figure 8.5
Variation of performance ratio with LiBr mass percent in a solution under a different temperature.

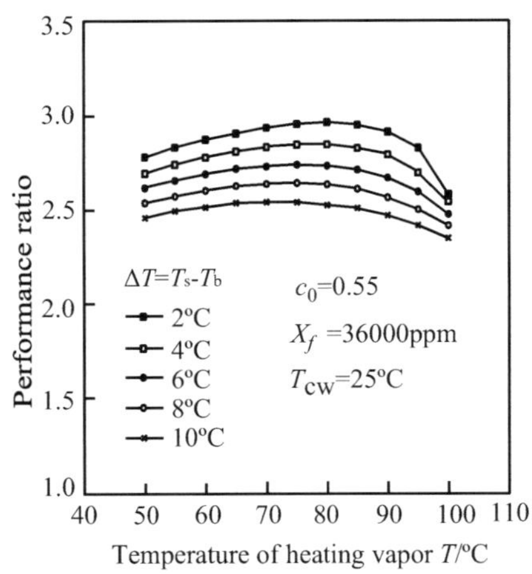

Figure 8.6
Variation of performance ratio with heating temperature in a different temperature saline difference of heating vapor and strong solution.

According to the criteria, the PR calculated for this system is shown in Figs. 8.5 and 8.6 by solving mass and energy balance equations. The performance ratio is defined as $PR = M_d/M_m$; M_d is the generating rate of fresh water, and M_m is the flow rate of heated vapor. If it is heated by the solar energy system, the performance ratio should be defined as $PR = \frac{M_d \cdot h_{fg}}{M_m(T_{in} - T_{out})c_p}$; T_{in} and T_{out} are the temperatures when heat from the solar collector comes in and leaves the generator, and c_p is the specific heat of hot water. Some of the relevant parameters used in calculation are provided in the figure as well (the detailed calculation results are shown in the literature [8]): T_b is the temperature of strong brine when it is eliminated; T_s is the temperature of heated vapor, which is imported into the condenser-evaporator; X_f is the concentration when seawater is imported; T_{cw} is the temperature when seawater is imported; and C_o is the percentage concentration of a weak LiBr-H_2O solution.

Figs. 8.5 and 8.6 present that the ABVC system has a higher production rate of fresh water, and the PR can reach higher than 2.5, that is to say, the PR of a single-effect ABVC system can achieve the PR of 2 ~ 3 effects system of a multiple-effect flashing evaporation system. It can be seen that the ABVC system has popular prospects in the future. Results also show that the temperature of hot steam and the mass fraction of a solution produced in a generator have a relatively big effect on the performance of the system. But when they reach certain values, the PR will not change significantly when they are in specific ranges. For example, if the mass fraction of a solution is between 0.45 ~ 0.7 and the temperature of hot steam is within 50°C ~ 100°C, the PR can stay around 2.5, which provides conveniences for the stable operation of the system and being heated by a solar collector. The two figures also demonstrate that the best concentration of a solution is within 50 ~ 60%, and the best temperature of secondary steam is in the range of 70 ~ 90°C.

In a word, the numerical analysis shows that the integration of absorption heat pump and single-effect vapor compression seawater desalination is possible. Theoretically, it can perform a higher fresh water production rate or higher PR which can reach 2.4 ~ 2.8. The production of it is around 50 ~ 70%; that is higher than that of a single-effect vapor compression system [9,10].

8.3.2 Multiple-Effect Solar Absorption Seawater Desalination System [10]

In order to further enhance the fresh water production rate of a solar absorption vapor compression (SABVC) desalination system or reuse the thermal energy imported in the system, the SABVC system can also be designed as a multiple-effect structure, or the SABVC and MEE system can be combined to form a multiple-effect absorption vapor

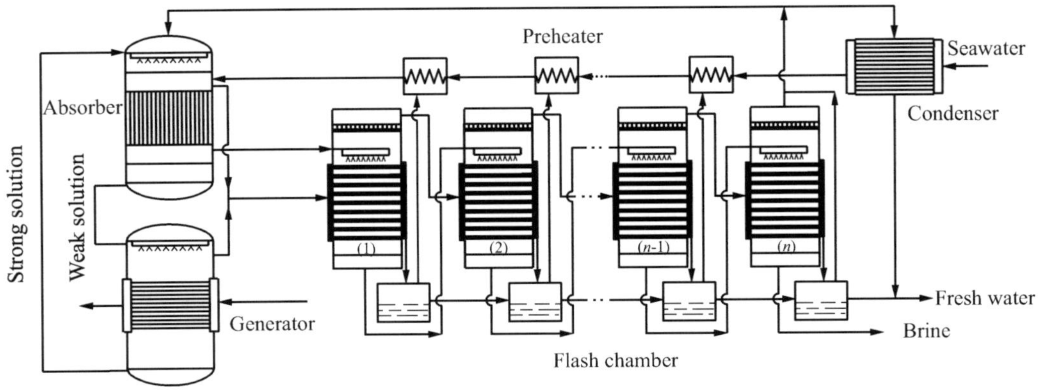

Figure 8.7
Multieffect absorption vapor compression solar desalination [10].

compression system. Fig. 8.7 provides the diagram of this novel system. This structure adds a seawater preheater and a fresh water collector compared with a single-effect structure, and other components of it are similar to the single-effect structure (only the single evaporator is changed into multiple evaporators). The vapor released by the first evaporator will go into the next evaporator and so on, rather than the condenser. Until the last evaporator, the vapor released from the last evaporator will go into the condenser in order to preheat the imported seawater and reclaim the latent heat of vapor. The part of the vapor emitted from the last evaporator will be transported to the absorber. Absorption is an exothermic process, which requires cooling by imported seawater. It causes the seawater to heat up and generate some vapor. Then the weak LiBr-H_2O solution is transported into the generator in which the hot vapor sent by the solar system is heated to vaporize in order to achieve the concentration of the solution. The strong solution will be pumped into the absorber to be absorbed again. The vapor produced in this process will be transported to the first evaporator with the vapor generated in the absorber together, by which the operation cycle is completed. This structure can also be applied in steam driven by industrial waste heat.

Theoretical calculation results illustrate that the thermal performance of the facility can be significantly improved after utilizing a multiple-effect structure. It has a higher PR when compared with the simple multiple-effect evaporation [10]. Under a steady state, only for the seawater desalination system, the comparisons of PRs of the MEE-ABVC system and the MEE system under different operating conditions and different effective numbers are shown in Fig. 8.8, where n refers to effective number. The results in Fig. 8.8 present that the PR of the system with quadruple-effect increases by 100% compared to the single-effect ABVC system, especially when the temperature of secondary steam (the steam exported from the generator) is in higher ranges. When it applies 12-effect, its PR can

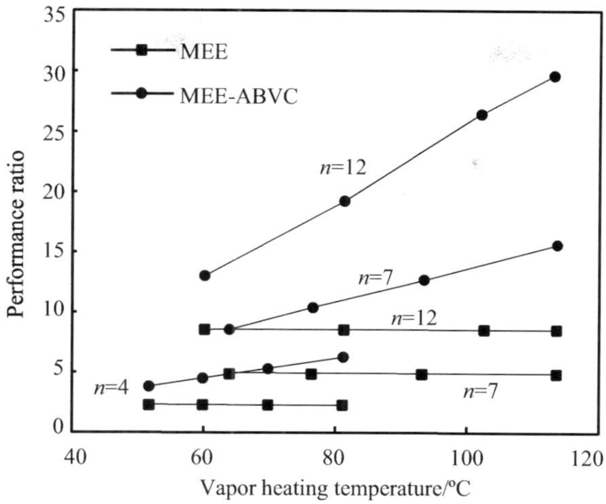

Figure 8.8
Performance ratio changes with the vapor heating temperature of the MEE-ABVC system.

reach up to 27 if the temperature of secondary steam is 115°C (here it should be high-pressure steam heating). It is worth mentioning that when the operating temperature rises, the PR of the MEE-ABVC system will rise with the temperature increase. However, the PR of a simple MEE system will not increase and will have different degrees of reduction. The reason of it is that in the MEE-ABVC system, the amount of latent heat of vapor absorbed will decrease with the temperature rise. The results imply the further investigation and analysis of the system integrating MEE and ABVC has realistic significance.

8.4 Practical Test and Evaluation of a Multiple-Effect Solar Absorption Seawater Desalination System

Based on the principles of falling film evaporation and condensation, the literature [11,12] designed a solar absorption seawater desalination system with quadruple-effect regeneration. The solar collector system is replaced by an electric water heating tank. An analog experiment was taken for this system. During the experiment, the performance under transient state and steady state of the system was tested, and the fresh water production rate and PR under different operating temperatures and pressures were given. Results show that because horizontal and vertical tube falling film evaporation and falling film condensation are utilized in this system, most of the latent heat of steam and the sensible heat of partial brine are reused many times. In addition, the evaporation on the last effect is enhanced because the absorbing working substance does the active absorption on the vapor of the last effect so that it reclaims the enthalpy of vapor. Hence the system has a relatively high PR. When the temperature of the supplied hot water is 80°C and the internal pressure of the system is 15 kPa, the PR of the quadruple-effect system can reach up to 3.0, which presents good advantages of the absorption system.

8.4.1 Structure of Experimental Facilities

The introduction before pointed out that falling film evaporation and falling film condensation have a relatively high heat transfer coefficient. The heat transfer surface can be saved a lot under the same thermal loads so that they are widely applied in the seawater desalination area. This facility mainly utilizes the following measures to enhance heat and mass transfer processes: falling film evaporation outside horizontal tube, falling film condensation inside horizontal tube, and falling film evaporation and absorption.

This facility mainly consists of the following parts: absorber, generator, evaporator, heat exchanger, and storage tank (for strong and weak brine). The internal structure and dimensions of them are introduced as follows:

1. The generator and absorber have similar structures. The absorber has one more pipe inlet for importing steam, as shown in Fig. 8.9. Both of them have the total same size. They have a shell-and-tube vertical heat exchanger whose inner diameter is 400 mm and height is 1200 mm. The diameter and length of vertical tubes applied in the interior are 28/32 and 750 mm, respectively. The total number of vertical tubes is 43. Hence the area of heat transfer surface through the pipe wall is 2.84/3.24 m^2. In order to enhance heat transfer and make the falling film solution wet pipe wall more sufficient, a

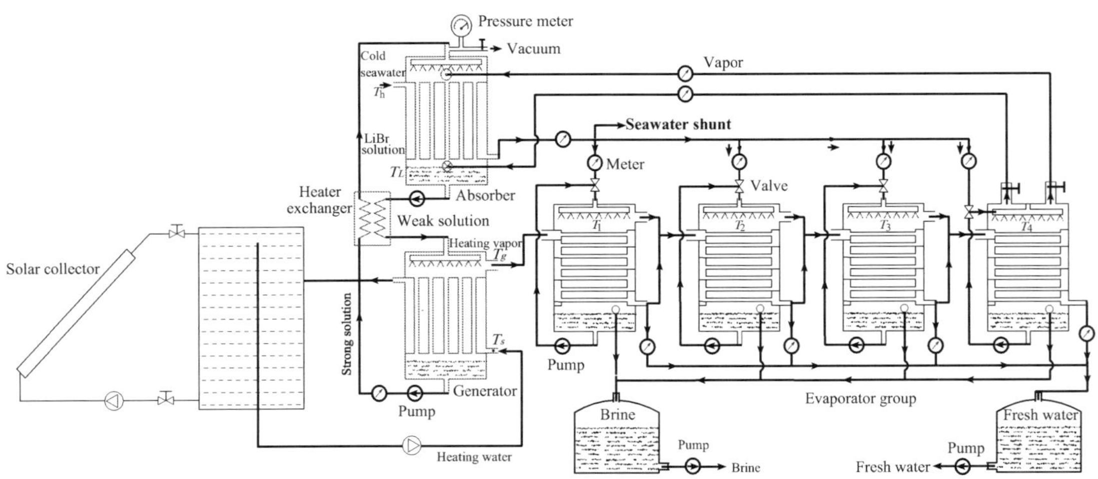

Figure 8.9
Schematic diagram of the structure and operation principle of the device.

solenoid made by iron wire with a length of 2 m and a diameter of Φ2.5 mm is inserted in each pipe. The solenoid is close to the pipe wall due to the effect of expansionary force. It provides the downstream effect to make the falling film solution wet internal wall of the tube more uniform. Meanwhile, to guarantee each tube has solution flowing down, siphon cotton is positioned on the top of the upper orifice. The solution will flow down in each tube as long as there is solution in the upper end of the heat exchanger. Both the generator and absorber have circular spray decks on top. The diameter of the spray deck is 360 mm, and the height of it is 30 mm. The diameter of the spray hole is 2 mm, the distance between holes is 2 mm × 2 mm, and the distance between spray hole and evaporator tube is 250 mm.

2. There are four evaporators in total, which can be joined in assembly or connected separately. Hence the system can change from single-effect, double-effect, and triple-effect to quadruple-effect. The external height of the evaporator is 0.8 m and both its width and length are 0.5 m. There are 56 metal tubes with 0.45 m as length and 28/32 mm as diameter traversed inside the evaporator. The tubes in the adjacent two layers are staggered, by which each brine droplet can reach the external walls of the evaporation tubes to achieve falling film evaporation. The areas of internal and external heat transfer surfaces of tubes are 2.22/2.53 m^2, respectively.
   Each generator has a rectangular spray deck, the base dimension of which is 0.46 × 0.46 m. The height of it is 0.03 m. The diameter of the spray hole is 2 mm, the distance between holes is 2 mm × 2 mm, and the distance between spray hole and evaporator is 0.2 m.
3. The heat exchanger consists of heat pipes in order to be the tube heat exchanger. The hot solution flows in the inner tube. The total area of the heat transfer surface is 0.138 m^2.
4. All of the tubes for solution flow are the polyethylene aluminum composite pipes with Φ15 mm. The pipelines for water (including brine and fresh water) utilize a polyethylene aluminum composite pipe with Φ20 mm. The diameter for the pipelines used for the steam flow is Φ60 mm. The pipelines in other components are the same.

8.4.2 Working Principles of Experimental Facilities

The working principles of experimental facilities can be expressed in several stages as follows:

1. Cold seawater flows into the absorber through the upper inlet. Then it flows out from the bottom outlet after passing through the cavity formed by the external walls of the vertical tubes and case. Finally, it flows into the quadruple-effect evaporator by flowmeter as the seawater waiting for evaporation to desalination. The rest of it will leave the facility.

2. The seawater flowing into the evaporator will be forced to circulate into the spray deck by a circulator pump. It will then be sprayed onto the external wall of horizontal tubes for falling film evaporation. The strong brine without evaporation will flow to the storage tank of strong brine, due to the water level difference, and be discharged out of the facility by the pump.
3. The LiBr-H_2O solution in the absorber absorbs the vapor from the fourth evaporator, and its concentration becomes weak. The temperature of the weak solution rises by passing through the heat exchanger. Then the solution flows into the generator for spray falling film heating evaporator. The thermal energy required by the generator is supplied by the solar simulator. To improve the absorption efficiency, there is forced falling film absorption in the absorber.
4. The vapor produced in the generator is transported into the horizontal tube of the first evaporator. Meanwhile, its latent heat is released to increase the temperature of the brine outside the surface of the horizontal tube, which has falling film evaporation to force the evaporation process. The vapor in the tube will condense to form fresh water and flow into a fresh water storage tank through the flowmeter. Finally, it will be transported out of the facility by pump. After the heating evaporation by generator, the LiBr-H_2O solution changes from weak to strong so that the absorption capability is recovered. After passing through the heat exchanger, it returns to the absorber to absorb vapor again, by which the system can achieve cyclic operation.
5. The secondary steam generated in falling film evaporation outside the horizontal tube of the first generator flows into the horizontal tube of the next evaporator through the vapor communication pipe. It also releases latent heat to the brine with the falling film evaporator outside the tube to increase the brine temperature so that it forces the brine to evaporate again. The vapor in the tube will condense to fresh water and flow into the fresh water storage tank by flowmeter. In the same manner, thermal energy will be transferred, and fresh water will be produced in the following levels. In addition, the latent heat released by vapor will be reclaimed by each next level in order to reuse the thermal energy supplied into the system. The fresh water generated in each level will be collected in the fresh water storage tank to become the total amount of fresh water production of the facility.
6. The heat exchanger between the generator and absorber is mainly used for reheating the weak solution flowing into the generator and cooling the strong solution flowing into the absorber at the same time to reclaim part of the heat. It provides a heat-regenerative effect.
7. When the vapor produced by the last evaporator flows into the absorber, it releases latent heat to the chilled water flowing inside the absorber to increase its temperature. Thus it increases the temperature of the seawater imported into the evaporator and reclaims some heat. The vacuum degree is maintained by vacuum pump. The case of the

facility is insulated by asbestos cloth with 30 mm thickness and a special insulation board.

8. The working process of the solar collector is as follows. The solar collector generates thermal energy by absorbing solar radiation and heats the water in the collector. When the temperature of aqueous media is higher than a designed value, bypass valve two is opened and bypass valve one is closed. The hot water in the collector flows to the upper part of the hot water storage tank, and the water flowing at the bottom is heated by the collector. When the temperature of aqueous media is lower than a designed value, bypass valve one is opened and bypass valve two is closed at the same time. The aqueous media is heated in circulated heating by the collector. The launch of the water pump of hot water circulation is controlled by the controller. When the solar radiation is higher than the designed value, the water pump for hot water circulation is launched after comparing the temperature of aqueous media and the temperature of the surroundings in order to ensure solar radiation can increase the temperature of aqueous media in the collector, or the water pump is not launched. The hot water storage tank connects the seawater desalination system and the solar collector. After the hot water in high temperature is pumped to the seawater desalination system for heat exchange, the hot water in low temperature will flow back to the hot water storage tank. The hot water storage tank is sealed. The system is designed to be a confined system. The highest protection temperature is 115°C. The system will release pressure through the relief valve when the temperature exceeds this temperature. Then the capacity of the hot water storage tank can guarantee the continuous operation of the seawater desalination system at night or when the solar radiation is not sufficient.

8.4.3 Results and Analysis of the Experiment

The experiment of the seawater desalination part is undertaken indoors. An electric water heater is used to initiate the hot water produced by the solar collector. The surrounding temperature is within 20−25°C, and relative humidity is within 40−85%. The seawater for the experiment is made by mixing tap water with salt. In the experiment, the operating parameters are provided as follows: the flow rate of heated water is 820 ± 50 kg/h; the flow rate of chilled water is 600 ± 30 kg/h; and the flow rate of seawater flowing into the facility is adjustable, which is around 40−80 kg/h. The circulation flow rate of circulator pumps in each level is set to be 240−280 kg/h, and the average operating temperature is 10 kPa; the temperature at chilled water inlet is 18 ± 0.5°C; and the operating temperature is defined as the temperature of hot water flowing into the facility.

8.4.3.1 Transient Water Production Performance of the Device

When the temperature of hot water that is initiated as solar hot water reaches 75°C (sometimes it is 85°C), the hot water is imported into the facility. In an hour and a half,

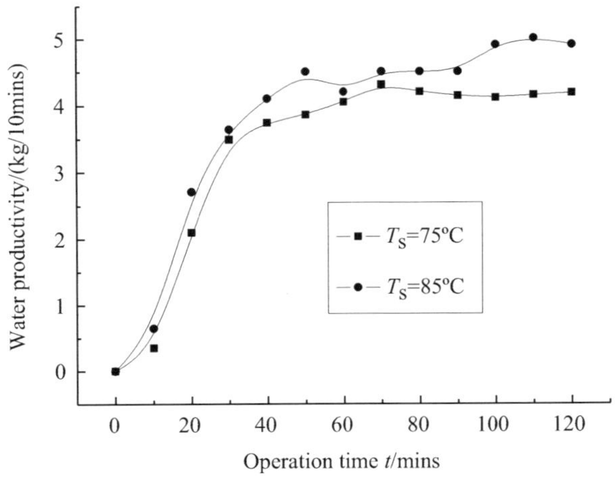

Figure 8.10
Variation of total water yield rate with time.

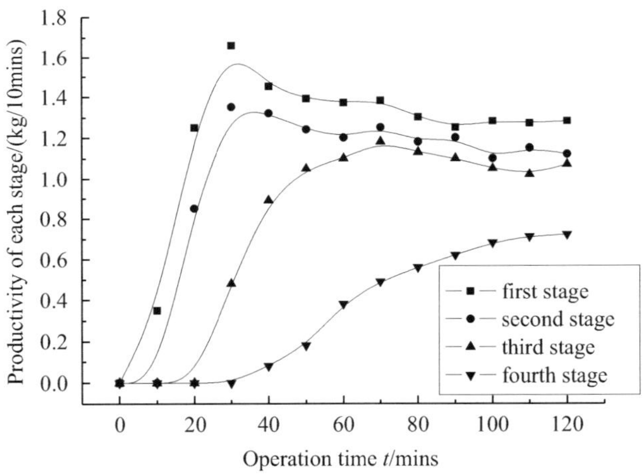

Figure 8.11
Variation of total water yield rate in each stage with time.

the changing processes of the total amount of generated water and the amount of generated water in each level in each 10 min are shown in Figs. 8.10 and 8.11.

Results in Fig. 8.10 demonstrate that the total amount of water production in each 10 min will achieve a stable value when the facility operates for around 40 min and stays constant, although at this time the water production rates at the third and fourth generator do not reach the maximum value. Results in Fig. 8.11 illustrate that the water production rate of the first and second evaporator reaches the maximum value when the facility operates for

40 min. There will be slight decreases later and then it stays stable. The water production of the third and fourth evaporator will achieve stability approximately after 60 min.

The reason for the decreases after the water generation rate reaches the maximum value of the first and second evaporator could be explained as follows: at the beginning of operation, the evaporators after are cold, which is better for the evaporation of the previous evaporator so that the production rate increases rapidly; however, when it operates for a while, the temperatures of the evaporators after the rise and the thermal resistances increase. It is not good for the seawater evaporation outside the horizontal tube of the previous evaporator. The condensation of vapor in tubes is affected, and the production rate of fresh water is decreased.

8.4.3.2 Transient Performance of the Device With Temperature

When the temperature of analog solar hot water reaches 75°C, import the hot water to the facility and start timing. The temperature change with time of each typical test point on the facility is shown in Fig. 8.12, where T_g is the vapor temperature in generator and T_1, T_2, T_3, and T_4 are the temperatures of vapor in the first, second, third, and fourth evaporator. Fig. 8.12 shows that the system achieves steady state after operating 60 ~ 80 mins.

8.4.3.3 Performance of Device Under Steady State

The transient performance is an important index to evaluate the stability and reliability of the facility. Under the steady state, the more the fresh water production is, the better the system performance. Thus it is significant to test the steady-state performance of the system.

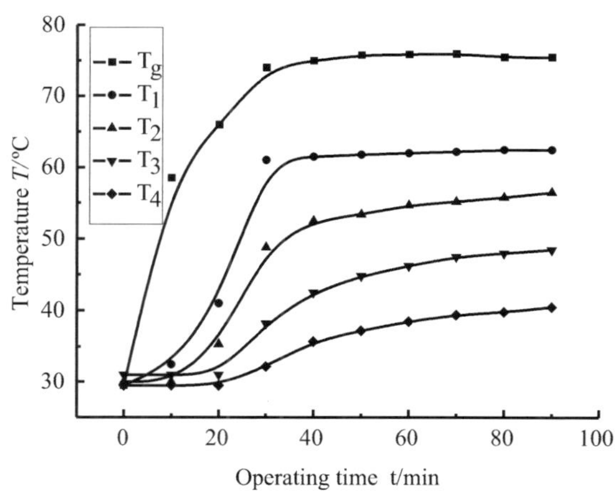

Figure 8.12
Variation of temperature with time in each measuring point.

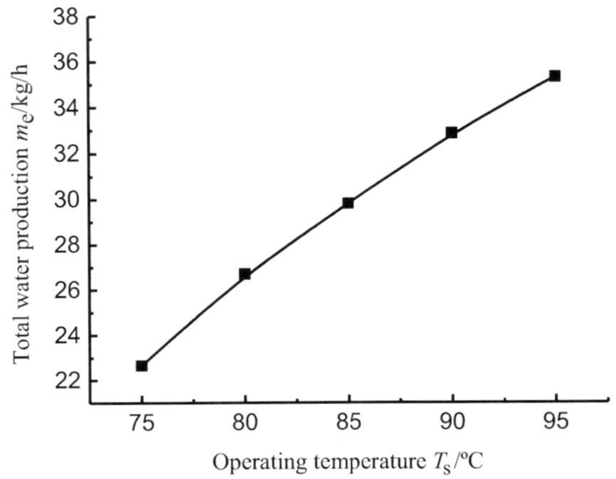

Figure 8.13
Effect of operating temperature on total water yield.

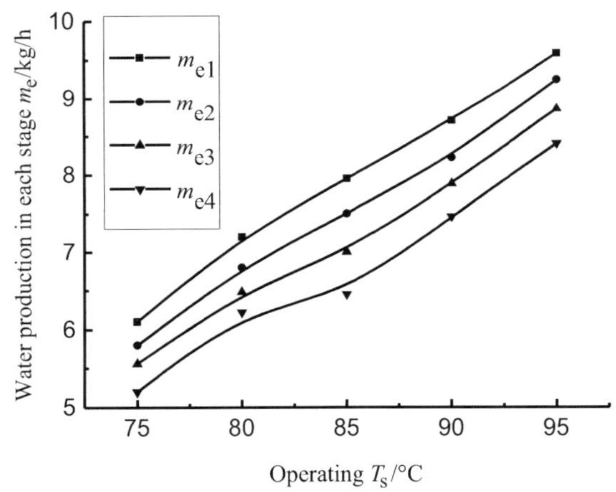

Figure 8.14
Effect of operating temperature on water yield in each stage.

When the provided pressure of the facility is about 10 kPa and the flow rate of hot water is 820 ± 50 kg/h, the effects of hot water temperature on total fresh water production are illustrated in Fig. 8.13. The amount of fresh water production of evaporators in each level is shown in Fig. 8.14. Both of the figures demonstrate that when the operating temperature rises, the quantities of fresh water production in the whole facility and the evaporator in each level increase. In addition, they have linear relationships to temperature increases.

Under the provided conditions, when the facility operates under the steady state, the temperature distributions of typical temperature test points on the facility are shown in

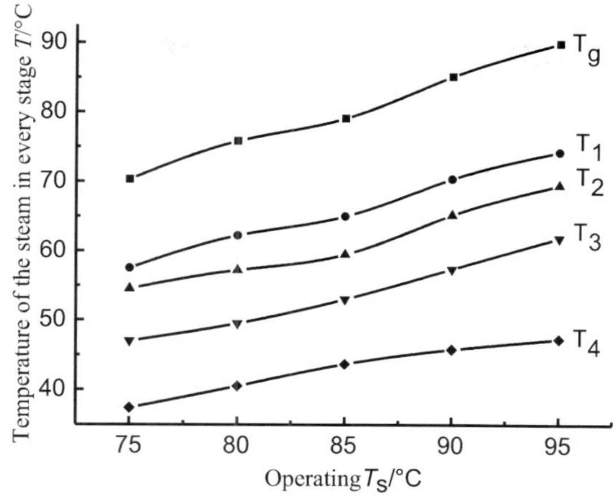

Figure 8.15
Effect of heating temperature on steam temperature in each evaporator.

Fig. 8.15, which demonstrates that the temperature of the vapor in the generator is the highest and the temperature of the vapor in the fourth evaporator is lowest, but both of them rises with the increases of hot water temperature. The temperature difference between T_g and T_1 is the biggest, and then is the difference between T_3 and T_4. The reason for the relatively big temperature difference between T_g and T_1 may be the smaller steam pipe connecting the generator and the first evaporator. The diameter of this pipe should be increased. The reason for the large temperature difference between T_3 and T_4 may be the active absorption of the absorber on the vapor generated in the fourth evaporator, which means that the absorption seawater desalination system can make the last evaporator operate under a lower temperature and reclaim the latent heat of the vapor in the last evaporator in order to enhance the thermal utilization efficiency of the facility. Actually, the lower operating temperature of the last level of the facility enhances the evaporation and condensation in the evaporators in the previous levels, and then the efficiency of thermal mass transfer of the whole facility can be improved.

8.4.3.4 Performance Ratio for the Facility Under Steady State

According to the relationships between input and output energy, the PR of this facility under the steady state can be expressed by the following equation:

$$\eta = \frac{m_e h_{fg}}{E} = \frac{m_e h_{fg}}{Q + P\Delta t} \tag{8.18}$$

where m_e is the total amount of fresh water production of this system, h_{fg} is the latent heat of water vaporization, and the approximation of it is $h_{fg} = 2320$ kJ/kg. E is the total

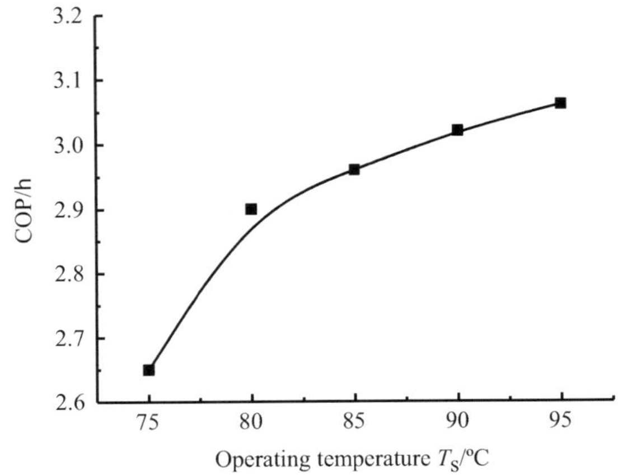

Figure 8.16
Effect of operating temperature on coefficient of performance (COP).

energy input to the system, Q is the total input thermal energy, P is the power of the pump, and Δt is operating time.

Experimental results show that the facility has a relatively high PR generally. When it operates at the temperature around 80–90°C, the PR can achieve 3.0 approximately, which guarantees the relatively high fresh water production rate. Fig. 8.16 provides the PR change with the operating temperature (the temperature of imported hot water). It shows that when the other conditions are the same, the higher operating temperature causes higher COP. Thus this system has greater advantages when it operates under high temperature.

The experiment results also illustrate that there are many factors to affect the stable performance of the system, especially the operating pressure, flow rate of supplied seawater, etc. The reduction of pressure is better for evaporation so that it helps to improve the COP of the system. Part of the sensible heat will be taken away by brine discharge. When the flow rate of supplied seawater decreases, the COP of the system rises generally, but the flow rate of the supplied seawater is controlled by the crystallization of strong brine. The flow rate cannot be too slow, which should be 1.5–2.0 times the production rate of fresh water.

8.4.4 Simulation Calculation During the Heating of Solar Collector

Based on the hourly data of direct and diffuse sunlight, the solar collection efficiency can be calculated. Establish the model of environmental temperature change and the collection and dissipation of the solar collector storage tank. The capital cost of facility manufacture and electricity price are based on the market price. The analysis is undertaken in accordance with experiment data. The following assumptions will be applied in analog calculation [12]:

1. The hourly solar radiation distribution is uniform in one hour.
2. The size of the hot water storage tank changes with the increase of hot water capacity on the basis of a certain relationship. The insulation material and thickness of the storage tank do not change.
3. The high temperature hot water in the solar collector and the low temperature cold water in the desalination system are fully mixed after they flow into the hot water storage tank.
4. Environmental temperature is the daily average temperature.
5. The heat release of connection pipes is not considered.

The relation curve of the experiment between operating temperature T_s (°C) and water production volume is given before. The following equations can be obtained by curve fitting:

$$M_e = 49.05 - 252.38 \bigg/ \left(1 + \exp\left(\frac{T_s - 13.96}{24.40}\right)\right) \tag{8.19}$$

$$\mathrm{PR} = \begin{cases} -23.9 + 0.639T_s - 0.0038T_s^2 & T_s \leq 80 \\ 0.515 + 0.0458T_s - 0.0002T_s^2 & 80 \leq T_s \leq 115 \end{cases} \tag{8.20}$$

The collected thermal energy Q_c (kJ/h) by solar collector per unit time can be estimated by the equation:

$$Q_c = K_1 K_2 A_c \times I_t \times \eta_c \tag{8.21}$$

where A_c, I_t, and η_c are the total area of the solar collector (m^2), the total solar radiation vertically projected onto a unit area of the solar collector in a unit time (kJ/h), and solar collection efficiency; K_1 and K_2 are the correction factors of vertical incident angle and horizontal incident angle for the solar collector. I_t projection can be calculated by the hourly direct and diffuse solar radiation, and η_c is calculated by the equation:

$$\eta_c = a + b\frac{T_{ic} - T_a}{I_t} + c\frac{(T_{ic} - T_a)^2}{I_t} \tag{8.22}$$

where a, b, and c are multinomial coefficients that are related to manufacturer and specification. T_{ic} (°C) is the temperature of the aqueous medium at the inlet of the solar collector. T_a (°C) is ambient temperature. The values of coefficients here are: $a = 0.803$, $b = 2.01$, and $c = 0.0034$.

The heat is transported to the hot water storage tank by the aqueous medium after being collected by the solar collector. When the temperature of hot water reaches the designed temperature, it will be pumped to the desalination system to generate fresh water.

According to the earlier assumption, the heat balance equation of the hot water storage tank changing with time t can be expressed as:

$$(m_c c_p)(T_{is} - T_s) + (M_s c_p)(T_o - T_s) - Q_L = \frac{\rho V_t c_p}{3600}\frac{dT_s}{dt} \tag{8.23}$$

where m_c and M_s (kg/h) are the flow rate of circulated hot water in the solar collector and desalination system. V_t (m^3) is the volume of the hot water storage tank, Q_L (kJ/h) is the amount of released heat of the hot water storage tank in a unit time, ρ is water density, which is 1000 kg/m^3, c_P is specific heat capacity with the value of 4.18 kJ/kg °C, and t (s) represents time.

For the solar collector and seawater desalination system, the following equations can be obtained without the consideration of heat losses on connection pipe:

$$(m_c c_p)(T_{is} - T_s) = Q_c \tag{8.24}$$

$$[(M_s c_P)(T_o - T_s) + P_s]\text{PR} = M_e h_{fg} \tag{8.25}$$

where P_s (kW) is the mechanical power consumed during the operation of the seawater desalination system with the value of 0.75 kW, h_{fg} is the latent heat of water vaporization, which is $h_{fg} = 2320$ kJ/kg approximately, T_{is} is the temperature of hot water flowing into the hot water storage tank, and T_{st} is the temperature of hot water supplied to the seawater desalination system. The water production rate of the desalination system M_e and PR can be calculated by Eqs. (8.19) and (8.20). In accordance with the earlier assumptions and equations, the changing of the amount of annual fresh water production and the cost of fresh water with the area of the solar collector can be simulated by a computer. They are shown in Figs. 8.17 and 8.18, respectively. When the solar collector and the facility operates with the best matching parameters, which means the area of the solar collector is

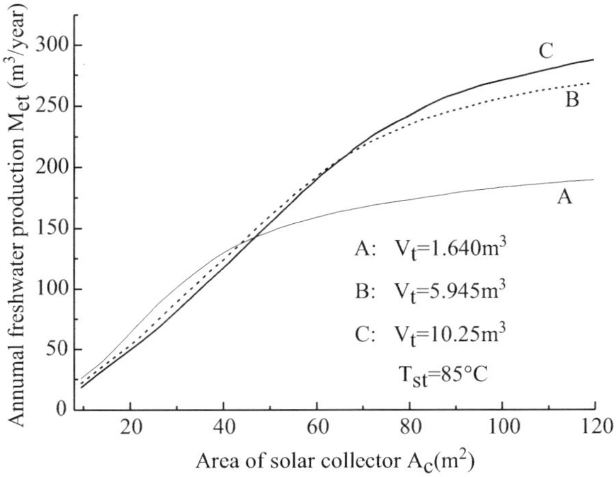

Figure 8.17
Variation of annual output of fresh water with the area of solar collector [12].

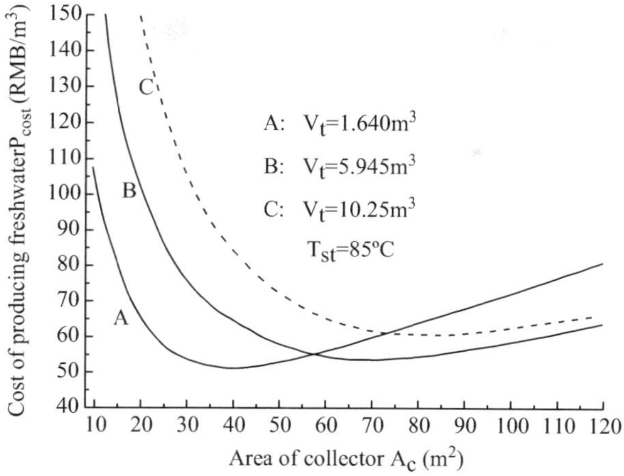

Figure 8.18
Variation of the cost of producing fresh water with the area of solar collector [12].

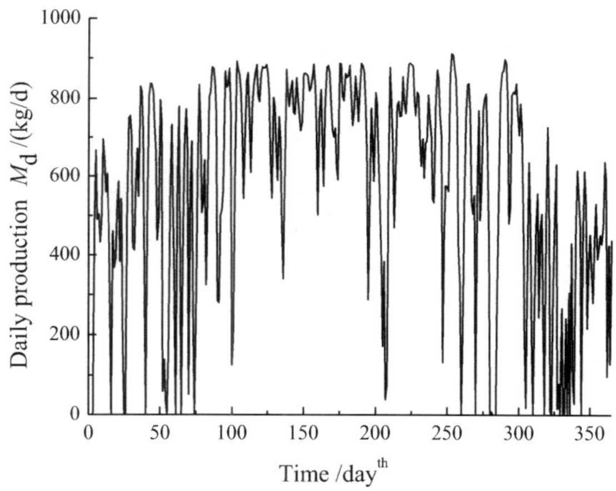

Figure 8.19
Daily fresh water production in a year [12].

64.41 m^2, the volume of the hot water storage tank is 4920 L, the working temperature is 77°C, and the pause temperature is 76°C, the total amount of annual fresh water production can be calculated as 211.45 m^3. According to the analog calculation, the change of daily and monthly fresh water production capacity with time is shown in Figs. 8.19 and 8.20. It can be seen that the biggest daily and monthly water production volumes are 913.2 L and 23.5 m^3.

8.5 Adsorption Solar Seawater Desalination System

The adsorption and desorption bed for vapor can be made with the advantages of the properties that can strongly adsorb water vapor of some solid substances like zeolite and

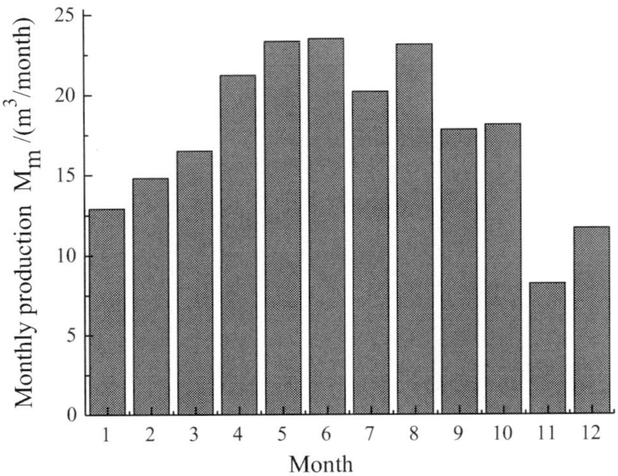

Figure 8.20
Monthly fresh water production in a year [12]

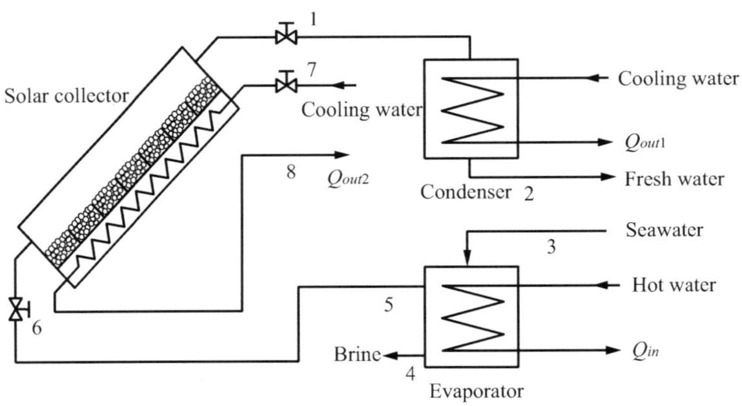

Figure 8.21
Interval adsorption solar desalination system.

activated carbon, etc. With the drive on adsorption and desorption by the solar collector, the solar adsorption seawater desalination system can be established. Fig. 8.21 provides a diagram of a batch-operated single-effect solar adsorption desalination system.

In a facility, the adsorption bed containing the adsorbent can receive sunlight directly. Therefore the solar collector is not only a generator (corresponding to desorption), but also an adsorber (corresponding to vapor adsorption). The two processes are conducted separately. At night, heat the evaporator by hot water; in other words, import Q_{in}.

Meanwhile, import seawater to the evaporator through point 3. The seawater will be heated to evaporate at this time. Vapor will flow into the adsorption bed in the solar collector through point 5 and valve 6. The adsorption bed adsorbs the vapor from the generator and releases latent heat. The latent heat will be taken away by the cooling system under the bed, which means the Q_{out2} is output. The strong brine formed in the evaporator will be discharged from the facility by point 4. Finally, the adsorption bed will achieve saturation by adsorbing vapor completely.

At daytime, the solar radiation enhances gradually. The temperature in the adsorption bed increases with the thermal collection by the solar collector. When the partial pressure of vapor on the surface of the adsorbent is higher than the saturation partial pressure of vapor in the condenser, open valve 1 and the vapor will go into the condenser to form fresh water through valve 1. The latent heat released during the condensation of vapor will be taken away by chilled water, which means the Q_{out1} will be released. The total cyclic process will be finished by now.

Because the adsorbents are solid, they are difficult to move in the facility. Hence the adsorption system cannot operate continuously like the absorption system, but several adsorption beds can be made to achieve conducting adsorption and desorption alternatively in order to make the system work continuously. Fig. 8.22 demonstrates a system like this.

This desalination system separates the solar collector and adsorber and consists of a hot water storage tank for working at night and at the time without efficient solar radiation.

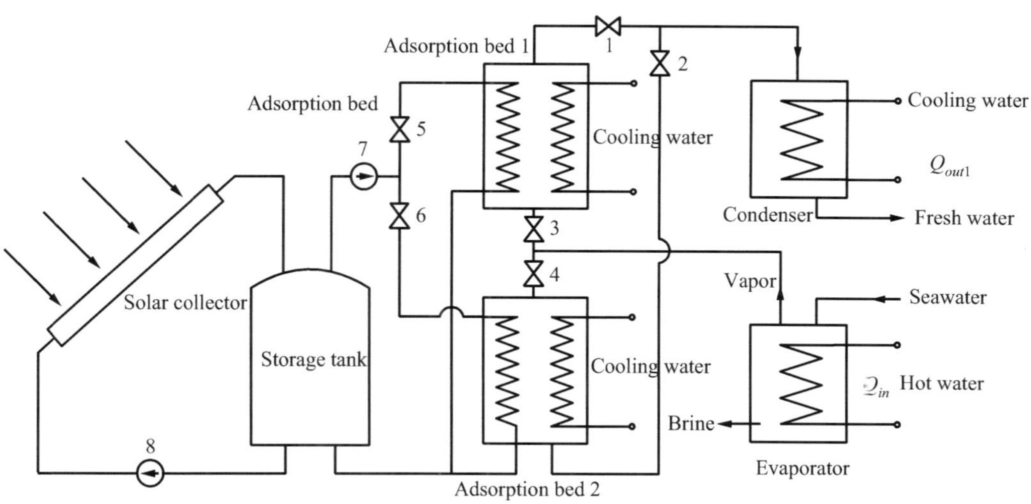

Figure 8.22
Double adsorption bed solar desalination system.

The two adsorbers work as generator and absorber alternatively to make the system operate continuously. The operating process is as follows:

1. Seawater flows into the evaporator. After being heated by coils, vapor is generated. The concentrated seawater left will be discharged from the facility through the outlet shown in Fig. 8.22.
2. The vapor produced in the generator goes into the upper adsorption bed through a steam pipe and solenoid valve 3 and is adsorbed there. Solenoid valve 4 is closed at this time.
3. When the upper adsorption bed is saturated, pump 7 transports the hot water in the hot water storage tank into the heating coil in the upper adsorption bed to heat the adsorption bed. At the moment, valve 5 is opened and valve 6 is closed.
4. When the saturation partial pressure of vapor in the upper adsorber is higher than the partial pressure of vapor in the condenser, valve 1 is opened and valve 2 is closed. Vapor flows into the condenser to condense, and the condensed fresh water will be transported to users through an outlet. The released latent heat will be taken away by chilled water.
5. Close valve 5 and stop heating when the upper adsorption bed desorbs completely. Meanwhile, import chilled water into the cooling coil in the adsorption bed to cool down the upper adsorption bed. The partial pressure of vapor in the upper adsorption bed decreases gradually. Until it is lower than the partial pressure of the vapor in the condenser, open valve 2 and 4, and close valve 1 and 4. By this time, the vapor from the evaporator goes into the upper adsorption bed. The whole process will repeat as the earlier description.
6. During the adsorption in the upper adsorption bed, desorption is conducted in the lower adsorption bed. The two beds have totally opposite processes and do not affect each

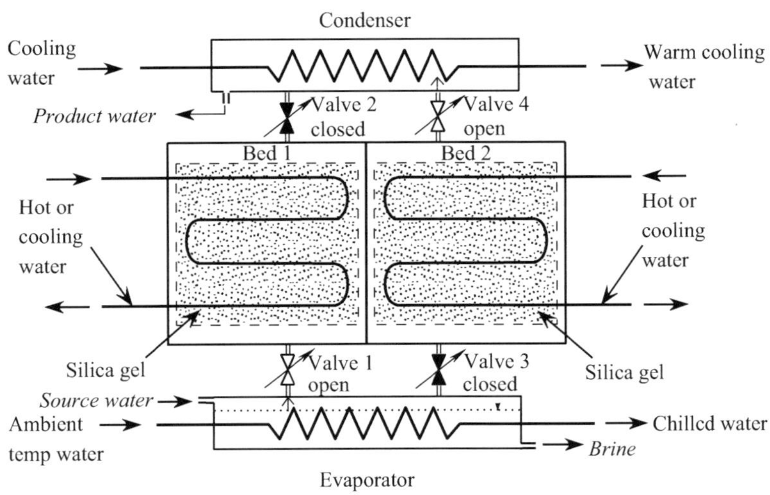

Figure 8.23
Schematic of the two-bed adsorption desalination system.

other. When they are heated and cooled respectively, open and close valve 5 and 6 in time.

The adsorption and desorption that happen in the system consisting of two adsorption beds can be simplified into the physical model, as shown in Fig. 8.23 [13]. It contains three main parts: evaporator, adsorption bed, and condenser. In the diagram, each bed constitutes a thermodynamic cycle in an adsorption period and desorption period. The processes in the beds at the left and right are opposite. When the system starts operating, seawater is transported to the evaporator. In order to enhance evaporation, thermal energy, like the heat from the solar collector, can be added into the evaporator. At the moment, valve 1 is open. Vapor is transported to bed 1 to be adsorbed. The heat released during adsorption is taken away by the chilled water in the heat exchanger in bed 1. When the progress of adsorption is finished, close valve 1 and valve 2. Import high-temperature hot water into the coils in bed 1. When desorption is finished, the previous process will be repeated in bed 2. In order to save energy as much as possible, a heat exchanger can be added between bed 1 and bed 2 to work as a preheater because the process from high to low temperature in each bed contains a lot of heat, which can be used to preheat each other to save the cost on thermal energy. In a bed, there are close relationships among the vapor pressure p, working substance temperature T, and the moisture content X of a unit working substance. In the whole circulation, their relationships are shown in Fig. 8.24 [13].

The working process in Fig. 8.24 can be expressed as follows:

Process 1−2. Assuming the starting point of system is 1, the working substance (for example, the system utilizes silica gel) is at the highest temperature T_1, i.e., 90°C, and the

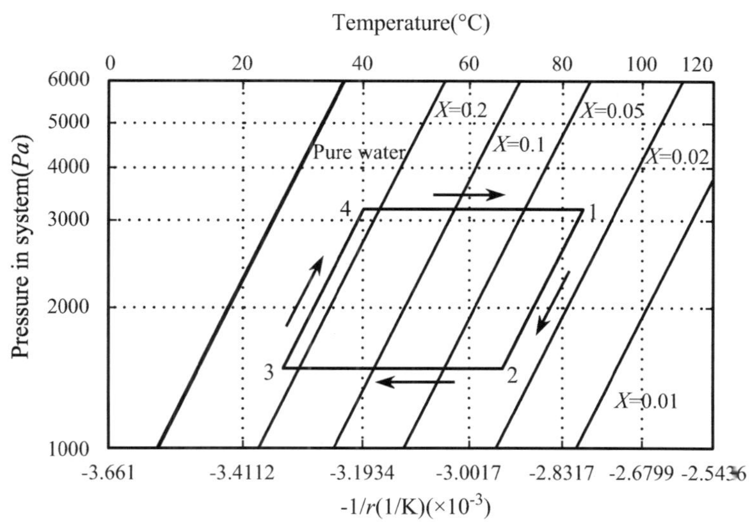

Figure 8.24
P-T-X diagram of an adsorption bed operation under a steady-state condition.

corresponding vapor pressure is p_1. During this process, the moisture content of working substance is constant, p_2 is the pressure of saturated vapor in the evaporator, and the temperature of the working substance at point 2 is around 60–70°C.

Process 2–3. At point 2, the valve connecting the adsorption bed and the evaporator is opened. Chilled water continues circulating in the bed to decrease temperature. At the moment, the vapor pressure in the bed reduces. The vapor in the evaporator goes into the bed and then it is adsorbed. Meanwhile, the seawater is vaporized to maintain the vapor pressure to be stable. During this process, the pressures of the evaporator and adsorption bed are almost constant, $p_2 = p_3$. The temperature in the bed decreases continuously until it drops toT_3. At point 3, the moisture content of the working substance reaches to the maximum X_3.

Process 3–4. At point 3, valve 1 will be closed, and the circulation of chilled water in the bed is stopped. Hot water will be transported into the bed to heat the working substance. The vapor pressure in the bed increases at the same time, but the moisture constant stays stable $X_3 = X_4$ until the temperature is increased to the condition when the partial pressure of vapor equals the partial pressure of vapor in the condenser. Then valve 2, connecting the adsorption bed and the condenser, is opened to condense vapor.

Process 3–4. When the process reaches point 4, valve 2 is opened. The vapor is desorbed by the working substance and condensed by the condenser to liquid water. By this time, hot water still circulates in the bed and chilled water circulates in the condenser. During this process, the vapor pressure in the bed stays constant approximately, and the moisture content decreases. The temperature increases until achieving T_1. The moisture content is at its minimum value X_1. The whole cyclic process is finished by now.

Without the consideration of the heat capacity of the facility, and assuming the temperature differences of heat transfer for all of the heat exchangers in this system are 5°C, as well as the working substance is silica gel, the literature [13] provides the changes of water production rate, unit energy consumption for producing water, and the temperature of chilled water for the system are shown in Figs. 8.25 and 8.26. Results in Fig. 8.25 demonstrate that the temperatures of heating and condensation are significant for the unit energy consumption of water production. For instance, when the chilled water is at 30°C and the heating temperature is lower than 80°C, the unit energy consumption increases rapidly. But when the chilled water is at 20°C, the change of unit energy consumption with heating temperature is not obvious. Generally speaking, for a provided heating temperature, the reduction of chilled water benefits the improvement of water production, and the change is noticeable. The higher the heating temperature is, the more the water production gets. It also implies that the system has advantages at higher temperatures. The results in Fig. 8.26 also verify it. When the temperature of chilled water reaches a certain value, the fresh water cannot be generated. Therefore the chilled water temperature must be controlled.

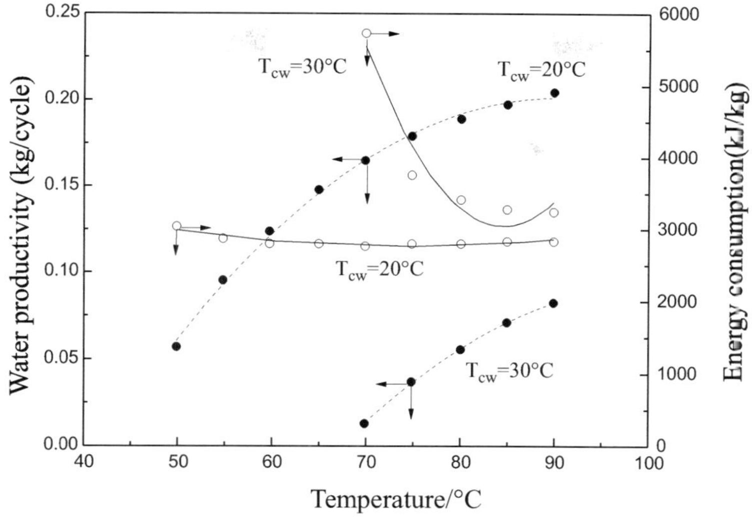

Figure 8.25

Effect of inlet temperature of hot water on water yield and energy consumption when the cooling water temperature is 20°C and 30°C, respectively.

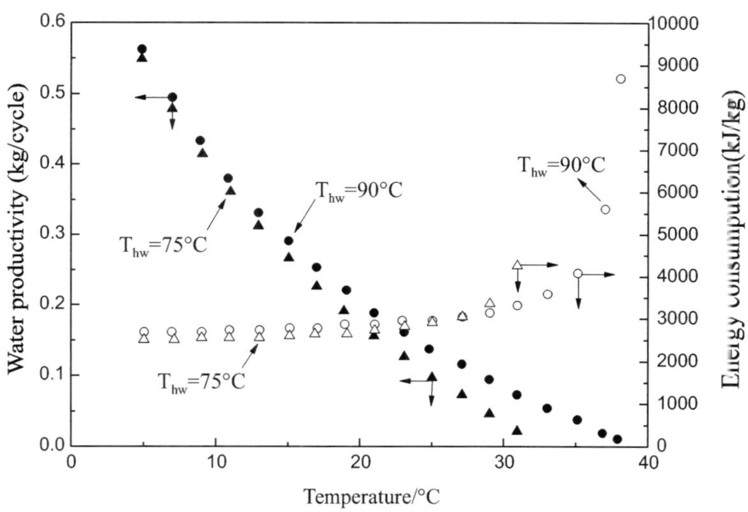

Figure 8.26

The effect of the condenser cooling water temperature entering the adsorption bed and the condenser on water productivity *r* and energy consumption when two different heat source temperatures are 75°C and 90°C, respectively.

The literature resources about solar adsorption Seawater desalination system are still few. The main reason is that the heating temperature required in adsorption and desorption is high, and the efficiency of the solar collector decreases when it operates under high temperature. The efficiency of the whole system is not high. But in theory, if an appropriate working substance is selected, it is possible to significantly reduce the operating temperature of desorption so that the adsorption desalination system can have better performance. Therefore the cruxes of this technology are on designing the adsorber, decreasing operating temperature, enhancing heat and mass transfer, and enhancing energy recovery and reuse.

8.6 Solar Absorption System Combined With Industrial Seawater Desalination Technology

8.6.1 Single-Effect Solar Adsorption Desalination System

Adsorption and desorption heat pumps are environmentally friendly facilities. In addition, these kind of systems can be forced by renewable energy or industrial waste heat. Under appropriate conditions, they perform fairly high thermal efficiency. They also have the advantages of a simple structure without moving components, long lifetimes, no need for shockproof and noise cancellation, etc. Therefore the research about adsorption and desorption heat pumps have been brought to the attention of scientists all over the world in recent years. It is expected that the traditional compression system will be replaced by them. Adsorption and desorption heat pumps have been utilized in air conditioning and ice-making areas. In seawater desalination aspect, they have been developed toward practical application.

Traditional adsorption seawater desalination is considered to be integrated with a vapor compression system. It is called an adsorption vapor compression system and is simplified as ADVC. In an air-conditioning area, adsorption refrigeration requires the heat source with the higher temperature, which is higher than 100°C generally. There are few facilities driven by solar energy. They are mainly driven by industrial heat waste with higher temperatures, for example, exhaust gas waste heat from an internal combustion engine. In the seawater desalination area, although the performance is better when the temperature of heat resource is higher, the performance is fairly good at 70 ~ 90°C. Thus solar energy is an ideal way to drive the system. In addition, solar collection technology has had great developments, like the evacuated tube solar collector, evacuated tube heat pipe solar collector, and evacuated tube collector integrated with parabolic trough, etc. The developments of a large amount of high-temperature solar collection technologies provide conveniences for the solar adsorption seawater desalination system.

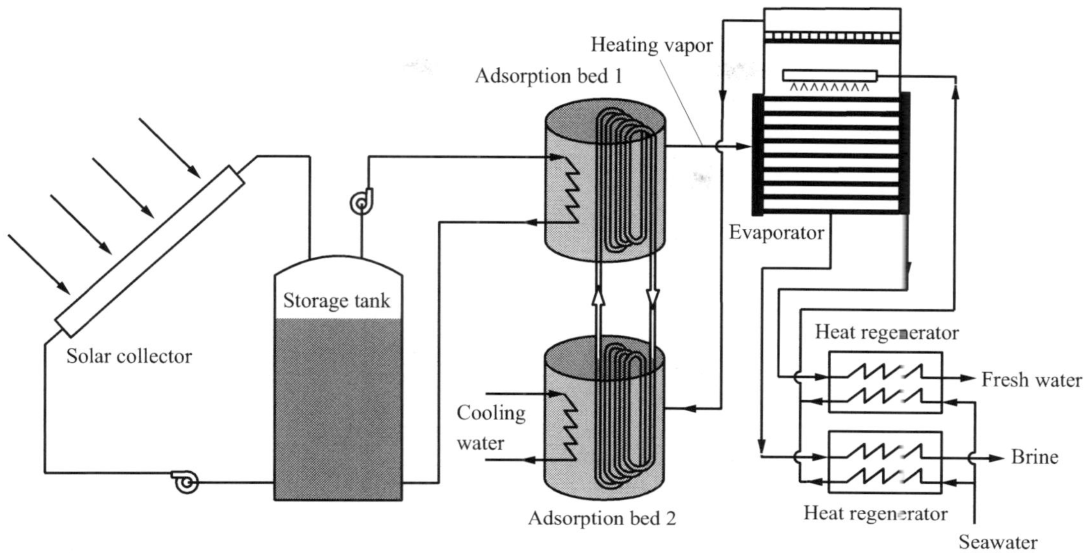

Figure 8.27
Double adsorption bed single-effect solar desalination [10].

Fig. 8.27 provides a diagram of the adsorption desalination system driven by solar energy. Because it cooperates with common seawater desalination technology (vapor compression), it is called SADVC (solar adsorption vapor compression) system. The system includes an evaporator (also works as condenser), two adsorption beds, two seawater preheaters, and a heat exchanger to transfer heat between the adsorption bed and desorption bed. In this system, the evaporator and condenser are combined as one component to replace the traditional separated evaporator and condenser in the traditional absorption heat pump. The seawater preheated is designed in disk construction to transfer heat among the imported seawater, condensed vapor, and concentrated seawater. The adsorber works as a partial condenser, which means it can absorb and exhaust the remaining heat. The operating process of the system is described as follows:

At the beginning, assume the adsorption bed 1 is cold and saturated for absorbing water. At the moment, the mass of bed 1 is the sum of adsorbent mass M_Z and adsorbed water mass M_d at temperature T_a. Assume the second adsorption bed is dry and hot, the temperature of which is T_c. The cold bed temperature T_a must be lower than the temperature of the saturation temperature for adsorbed water. The temperature of hot bed T_c equals the temperature of vapor generally flowing to the evaporator. When the circulating fluid starts to flow between the two beds, the process begins. The first bed is heated and the second bed is cooled at the same time. During this process, the vapor generated in bed 1 (secondary steam) is transported to and condensed in the horizontal tube in the evaporator. The heat is exchanged to the seawater with falling film condensation outside the tube. Seawater is heated and evaporated to produce vapor. Vapor is transported to the second bed to be absorbed. When the first bed releases all the vapors it absorbed, the adsorption of the second bed is going to reach a saturated state. Then the heating and cooling are reversed for the two beds, and the previous process is repeated in order to make the system operate continuously. In the horizontal tube in the evaporator, most of the vapors are condensed to fresh water. The fresh water and remained vapor are transported to the seawater preheated and then further cooled to condensation. They are finally transported to the users. The heat will be transferred to the seawater imported to the facility. The seawater outside of the horizontal tube in the evaporator is concentrated after evaporation. The concentrated seawater is transported to another seawater preheater through a pipe and transfers most of the sensible heat to the imported seawater. At last, it will be discharged from the facility to finish the whole circulation.

For the system shown in Fig. 8.27, the literature [14,15] provided energy and mass balance analysis under the following consumptions:

1. System is under a steady state. Two adsorbers and desorbers operate in alternation and conduct opposite processes.
2. The pressure in the adsorber is uniform. The vapor pressure and temperature of adsorbent can be calculated by the energy balance equation.
3. The temperatures of working substances in the adsorber and desorber are uniform.
4. There are no energy losses to the surroundings.
5. The density, heat transfer coefficient, and flow rate of the fluid are constant.
6. The amount of adsorbed vapor in the second bed equals the amount of generated vapor in the first bed.
7. The adsorption and desorption rates are equal and constant.
8. The heat transfer rate between the two beds is constant.

Therefore the energy balance equations of the system are as follows:

The total material balance and salt balance is:

$$M_{\mathrm{f}} = M_{\mathrm{d}} + M_{\mathrm{b}} \tag{8.26}$$

$$M_{\mathrm{b}} = M_{\mathrm{f}}(X_{\mathrm{f}}/X_{\mathrm{b}}) \tag{8.27}$$

Energy balance of the preheater is:

$$M_{\mathrm{f}}C_{\mathrm{p}}(T_{\mathrm{f}} - T_{\mathrm{cw}}) = M_{\mathrm{d}}[H(T_{\mathrm{d}}) - H(T_{\mathrm{O}})] + M_{\mathrm{b}}C_{\mathrm{p}}(T_{\mathrm{b}} - T_{\mathrm{O}}) \tag{8.28}$$

Energy balance of the evaporator:

$$M_{\mathrm{f}}C_{\mathrm{p}}(T_{\mathrm{b}} - T_{\mathrm{f}}) + M_{\mathrm{d}}h_{\mathrm{fg,v}} = M_{\mathrm{d}}h_{\mathrm{fg,d}} + M_{\mathrm{d}}C_{\mathrm{pv}}(T_{\mathrm{s}} - T_{\mathrm{d}}) \tag{8.29}$$

Relationship between boiling temperature of brine in the facility and steam temperature:

$$T_b = T_v + \Delta T_y \tag{8.30}$$

During the cooling process of the second bed, temperature changes from T_{e2} to T_a. The energy balance equation of this process is:

$$M_{cw}C_p(T_{cwo} - T_{cwi}) = M_z C_{pz}(T_{e2} - T_a) \tag{8.31}$$

The heat transferred from the second bed to the first bed is:

$$Q_{21} = M_d h_{fg,v} + M_z C_{pz}(T_c - T_{e2}) + M_d[H(T_v) - H(T_a)] \tag{8.32}$$

The energy required for heating the first bed is:

$$Q_{21} + M_m h_{fg,m} = M_d h_{fg,s} + M_z C_{pz}(T_c - T_a) + M_d[H(T_c) - H(T_a)] \tag{8.33}$$

Combine Eqs. (8.32) and (8.33):

$$M_m h_{fg,m} = M_d(h_{fg,d} - h_{fg,v}) + M_z C_{pz}(T_{e2} - T_a) + M_d[H(T_c) - H(T_v)] \tag{8.34}$$

Combine Eqs. (8.32) and (8.34):

$$M_m h_{fg,m} = M_d(h_{fg,d} - h_{fg,v}) + M_{cw}C_p(T_{cwo} - T_{cwi}) + M_d C_p(T_c - T_v) \tag{8.35}$$

Define the heat transfer efficiency of circulating fluid in the heat exchanger:

$$\eta = \frac{M_z C_{pz}(T_{e2} - T_a)}{M_{cw}C_p(T_{e2} - T_{cwi})} \tag{8.36}$$

The energy balance of circulating fluid in the heat exchanger is:

$$M_z C_{pz}(T_{e2} - T_a) = M_{cw}C_p(T_{cwo} - T_{cwi}) \tag{8.37}$$

According to Eqs. (8.36) and (8.37):

$$T_{e2} = [T_{cwo} - T_{cwi}(1 - \eta)]/\eta \tag{8.38}$$

With the consideration of the relationship between the inlet—outlet temperature of chilled water and the temperature of the seawater imported to the facility, there is:

$$T_{cwi} = T_{cw} \tag{8.39}$$

$$T_{cwo} = T_{e2} - \Delta T \tag{8.40}$$

The relationship between pressure and temperature of the adsorber and descrber is:

$$\ln(p) = a + \frac{b}{T} \tag{8.41}$$

where $a = 20.49 - 60.4\alpha + 787\alpha^2 - 2.14 \times 10^3\alpha^3$.

In the adsorber, the mass balance of the water vapor is:

$$M_z = M_m/(\alpha_g - \alpha_a) \tag{8.42}$$

The physical significances of symbols used in the earlier equations are similar to those in the previous section. P and T are the balance pressure and temperature in the adsorber and desorber in units 100 P_a (mbar) and K. The valve of P equals the pressure of vapor supplied to the evaporator. In the adsorber, temperature T is equal to T_c. α refers to adsorbed vapor by per kg zeolite. T_{cw} is the temperature of imported seawater in °C, C_{pz} is the specific heat of zeolite in kJ/kg K, ΔT is the temperature difference of the heat exchanger in the adsorber in °C, T_{e1} is the temperature of the bed when desorption starts, and T_{e2} is the temperature when adsorption starts in °C. The symbols with subscripts m, s, and z represent the states of the main heating steam supplied to the facility and heating steam produced by the desorber and zeolite.

According to the analysis and combining the following criteria, the literature [7] provides numerical calculations for the system shown in Fig. 8.11. The relevant criteria are as follows:

- The concentration of concentrated seawater discharged from the system is 70,000 ppm (concentration ratio $\approx$2).
- The concentration of seawater supplied to the facility is 42,000 ppm.
- The temperature of seawater supplied to the facility is 25°C.
- The water production rate of the system is 1 kg/s.
- The temperature of strong brine at the bottom of the evaporator is $T_b = 65$°C.
- The specific heat at the atmospheric pressure of vapor is 1.884 kJ/kg °C.
- The adsorption working substance is zeolite, whose specific heat is 0.9 kJ/kg °C.
- The efficiency of heat exchanger between adsorption beds is 0.9.

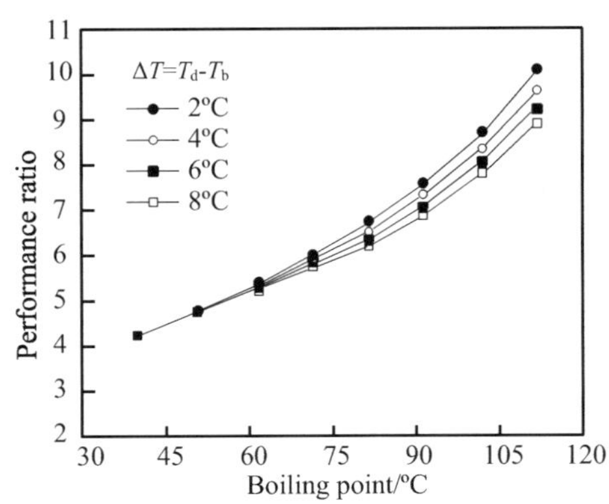

Figure 8.28
Variation of performance ratio with boiling point in different temperature difference.

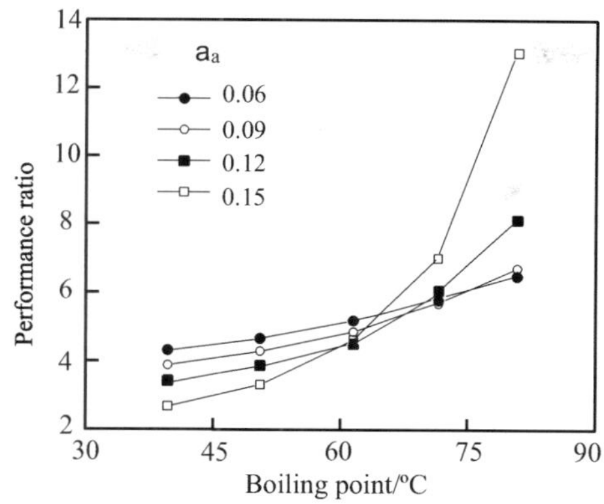

Figure 8.29
Variation of performance ratio with boiling point in different adsorption ratio.

- The adsorption rate of adsorption working substance is $\alpha_a = 0.14$ kg(water)/kg(zeolite).
- The temperature difference between the cold and hot sides of the heat exchanger between adsorption beds is 5°C.
- The temperature of seawater after passing through a preheater is $T_f = (T_b - 2)$°C.
- The temperature of adsorber is $T_c = (T_s - 5)$°C, and T_s is the temperature of vapor when coming into the evaporator.
- The temperature of the main heating steam flowing into the adsorption bed is $T_m = (T_c - 5)$°C.

Based on the criteria, the relationship between the PR of the ADVC seawater desalination system and the temperature of secondary steam going into the evaporator is shown in Figs. 8.28 and 8.29 by solving the mass and energy balance equations of each component. Other relevant criteria are provided in the figures as well, where T_d is the temperature when fresh water is discharged and α_a is the adsorption rate of adsorption working substance. The calculations are discussed in the literature [10] in detail.

Results in Figs. 8.28 and 8.29 illustrate that the ADVC system has a higher water production rate. When the temperature of vapor imported to the evaporator is higher than 60°C, the performance can reach to 4.0 or higher. When the operating temperature is much higher, the advantages of ADVC are more obvious. For example when the boiling temperature of seawater in the evaporator is around 100°C, the PR of a single-effect ADVC system can achieve 8.0, which is three times more than a single-effect ABVC system. Therefore many researchers think that the ADVC system is one of the most efficient single-effect vapor compression seawater desalination systems.

This kind of system utilizes many traditional components structurally, like the evaporator and seawater preheater. There are proven techniques for these facilities.

Thus the ADVC system is easy to be integrated with the traditional seawater desalination technology, which is one of the reasons that it has broad prospects. The key technologies of it are in investigating high-efficiency adsorption and desorption beds and making them work alternatively when the temperature and pressure are satisfied. In addition, the required boiling temperature of seawater is not high for this facility; the PR is 5 ~ 6 when the boiling temperature is at 75°C. Although the temperature in the desorber is 10°C higher than that, it will be about 85°C, which is relatively easy to achieve by the evacuated tube solar collector. Hence the ADVC system can use solar energy to operate, which is an environmentally friendly facility.

8.6.2 Multiple-Effect Solar Adsorption Desalination System

The theoretical calculation shows that when the boiling temperature T_b in the evaporator reaches to around 75°C, the PR of a single-effect ADVC system can achieve higher than 5.0. Although this value is fairly high, it is still a waste to discharge the concentrated seawater at 75°C and condensed fresh water in a horizontal tube out of the facility. For reusing energy and improving system performance, the ADVC system and MEE system can be integrated as a multiple-effect system as a MEE-ABVC system.

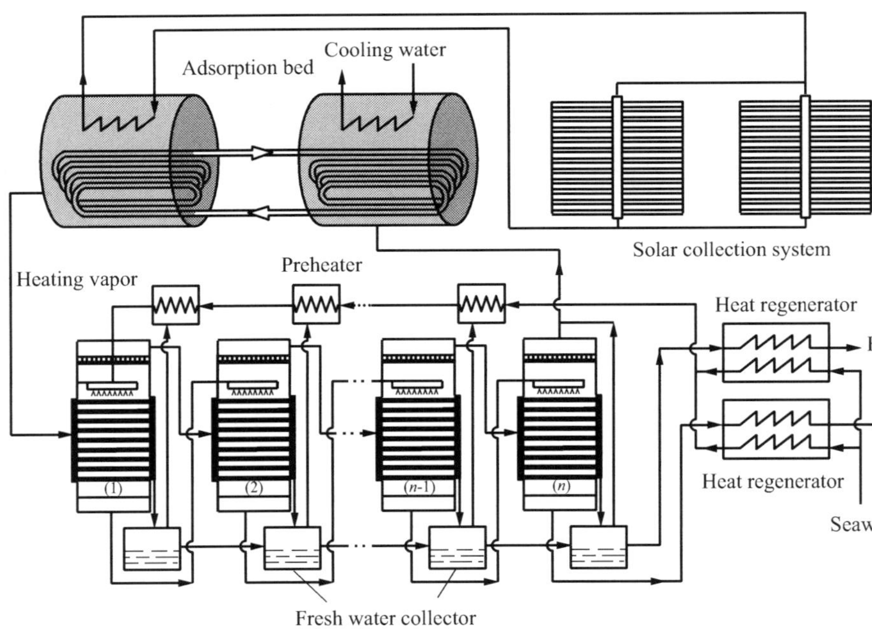

Figure 8.30
Adsorption multieffect vapor compression solar desalination system [10].

Obviously, this system can be driven by the solar collector. Fig. 8.30 illustrates an operation diagram of the MEE-ADVC system [10].

Similar to the single-effect ADVC system, the MEE-ADVC system includes the main parts of the adsorption bed, desorption bed, evaporator, heat exchanger, and seawater preheater, etc., but the number of seawater preheaters and fresh water collectors is increased, as shown in Fig. 8.30. The vapor from the desorption bed goes into the first evaporator and heats the seawater with falling film evaporation outside the heat pipe. The generated secondary steam is transported to the second evaporator to heat the seawater with falling film evaporation. Then the generated vapor here is transported to the third evaporator to continue this process until the last evaporator. Finally, the secondary steam generated in the last evaporator will be transported to the adsorption bed to be adsorbed to finish the whole circulation of vapor.

The theoretical calculation implies that the MEE-ADVC system has excellent PR. Under the same levels, its PR is one time higher than that of the single-effect MEE system, which is the most significant advantage of the MEE-ADVC system. Under the operating conditions with different temperatures and levels, the comparison between the MEE-ADVC and MEE system is shown in Fig. 8.31 [10], in which n refers to the number of effects. Results in the figure present that for the four-effect systems, the PR of the MEE-ADVC system is one time higher than that of the MEE system; the PR of the seven-effect MEE-ADVC system equals the PR of the 12-effect MEE system. This provides the foundation for saving materials of the MEE-ADVC system to simplify structure. On the other hand, with the increase of operating temperature, for

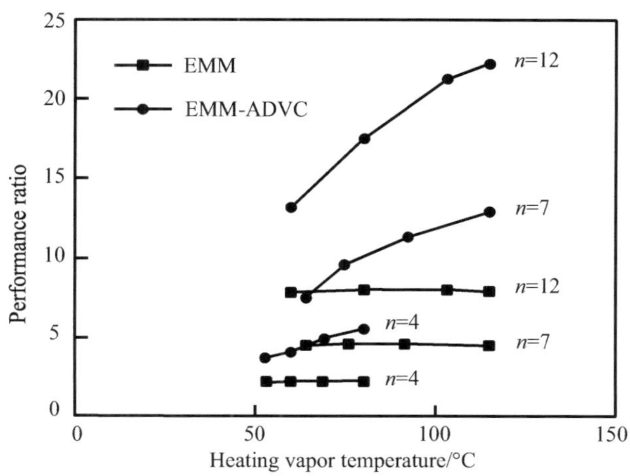

Figure 8.31
Variation of performance ratio with heating vapor temperature.

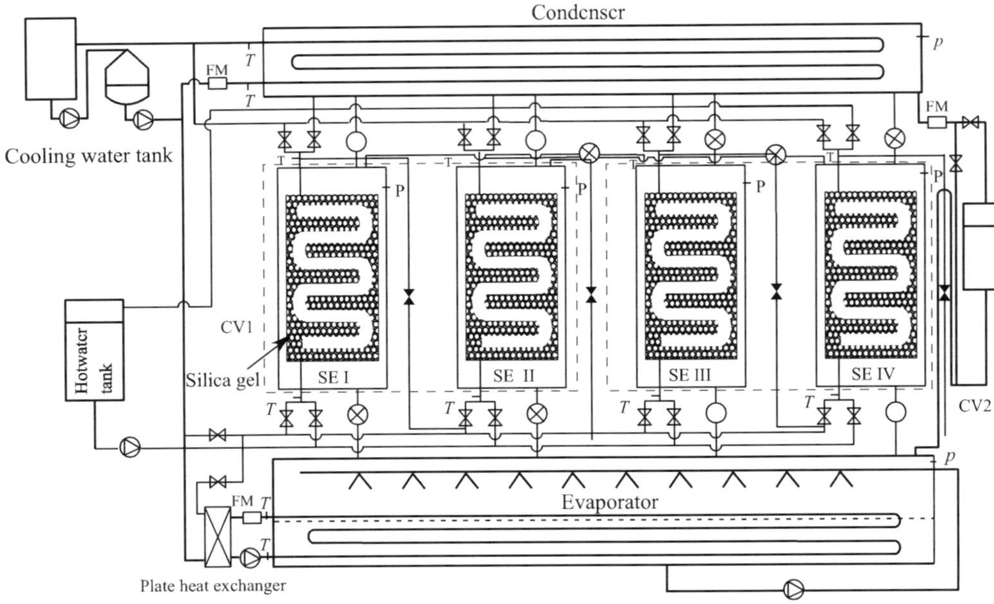

Figure 8.32
Diagram of combination of adsorption cooling and desalination system with four-bed modes.

Figure 8.33
Pictorial view of the pilot four-bed adsorption cooling-cum desalination plant.

example, improving the temperature from 80°C to 115°C, the PR of MEE changes a little, but the PR of MEE-ADVC system improves significantly. It implies that the MEE-ADVC system has better performance at high temperatures, and it is more suitable to working under high temperatures. The reason for it is that under higher temperatures, the latent heat of the adsorbing vapor is smaller.

Table 8.1: Relevant parameters of the adsorption seawater desalination process [19].

Term	Parameter	Temperature (°C)	Flow rate (kg/s)
Mass of silica gel of each bed (kg)	36		
Area of flat solar collector (m^2)	215		
Circulation time (min)	40		
Closing time (s)	20		
Temperature of hot water inlet (desorber)		65–85	0.8
Temperature of chilled water (adsorber)		29.5	0.8
Temperature of chilled water (condenser)		30	2
Temperature of cold water inlet in evaporator		30	1.9

In order to verify the performance of the absorption seawater desalination system, Kim Choon Ng et al. designed a compound cooling and desalination system with four adsorbers. The main structure is illustrated in Fig. 8.32 [13]. Fig. 8.33 is the photo of the experiment facility. The research is undertaken for the different operating conditions when the silica gel is applied as the adsorption substance [16–18]. The operating results of the system are simulated when it utilizes low-temperature heat sources, including solar energy to supply heat.

Under the operating conditions where solar energy supplies thermal energy, the relevant parameters for analog calculation and experiment are shown in Table 8.1. In experiment and calculation, only two beds of the system are in use. The water production rate and

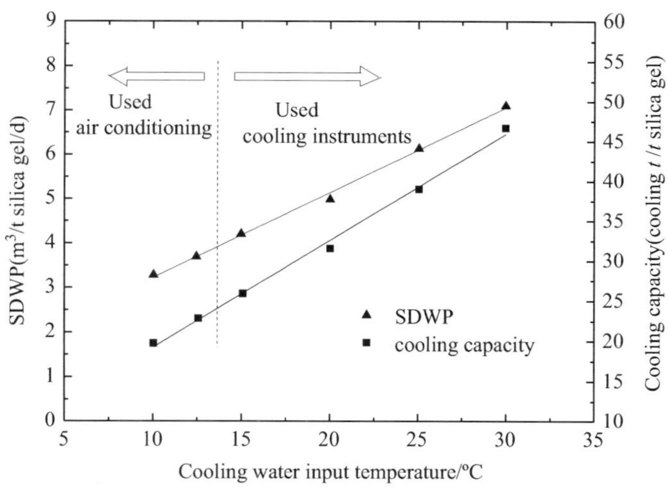

Figure 8.34
Effects of cooling water temperature on the cooling capacity and water yielc of the device (heating temperature at 85 °C) [19].

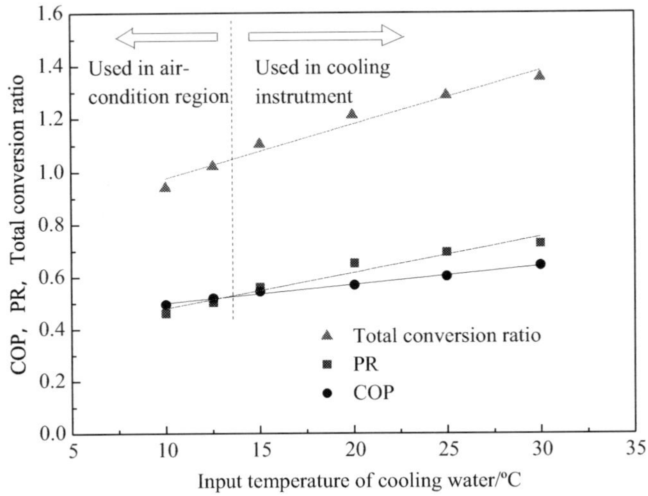

Figure 8.35
Effect of cooling water temperature on the performance of the device (heating temperature 85 °C) [19].

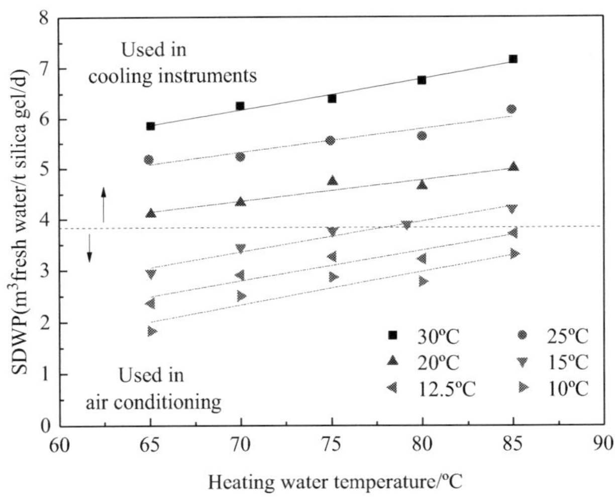

Figure 8.36
Effect of heating source temperature on the water yield of two adsorption beds at different cooling water temperatures [19].

cooling capacity per unit working substance are mainly tested and analyzed. The effects by cooling water temperature, evaporation temperature, and heating temperature on the system are investigated. At the different temperatures of chilled water (the water temperature at evaporator outlet), the water production rate per unit working substance is demonstrated in Fig. 8.34, in which the SDWP refers to the amount of fresh water

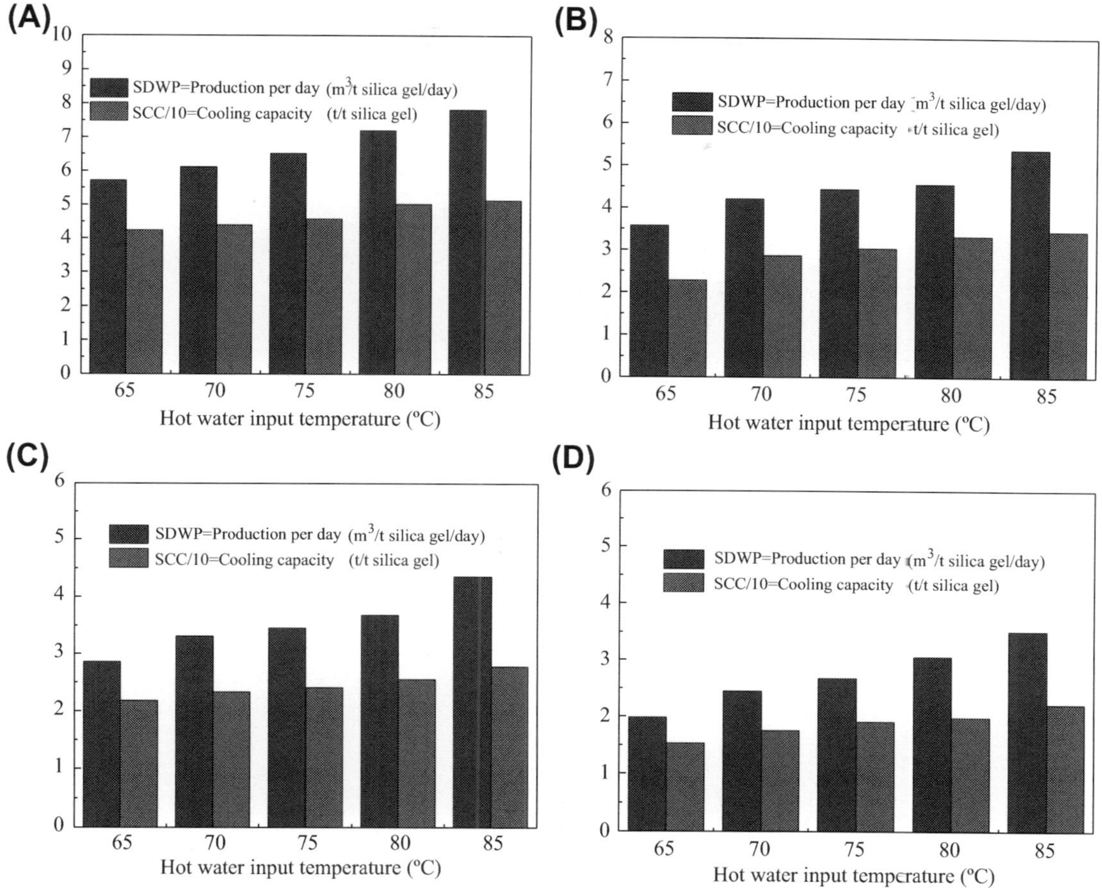

Figure 8.37

Effect of heating water temperature on the cooling capacity and water yield at different cooling water temperatures (four adsorption beds at the same time) [19]: (A) cooling water temperature is 30°C, (B) cooling water temperature is 20°C, (C) cooling water temperature is 15°C, and (D) cooling water temperature is 10°C.

production for a ton of silica gel per day. The calculated system COP, PR, and total conversion rate are shown in Fig. 8.35. The effects of different heating temperature on water yield are illustrated in Fig. 8.36.

If the waste heat is used to drive the system and the four beds are working together, the water production rate and cooling capacity of a unit working substance under different heating and cooling conditions can be tested and calculated as shown in Fig. 8.37.

Under practical operating conditions, waste heat is utilized to replace solar energy, and the performance of actual water production of the system is tested. During the experiment, the

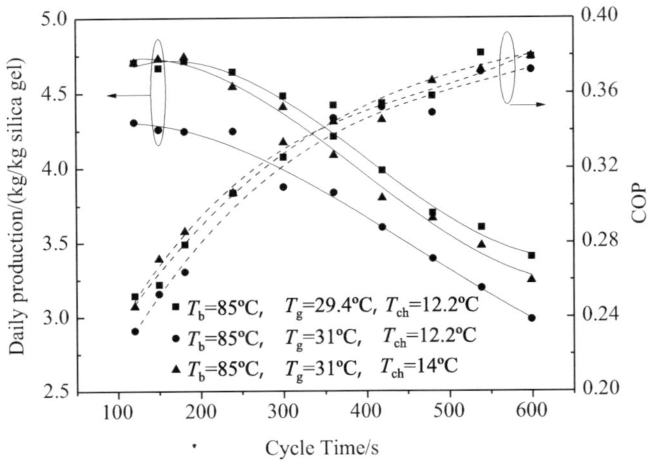

Figure 8.38
Effect of cycle time period on the water production and coefficient of performance (COP) [18].

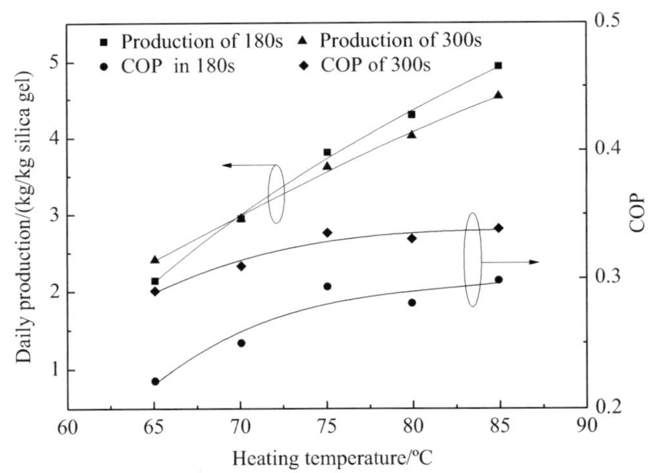

Figure 8.39
Effect of heating water temperature on water production and coefficient of performance (COP) [18].

temperatures at hot water inlet, cold water inlet, and chilled water inlet are 85°C, 29.4°C, and 12.2°C, respectively. Other operating parameters or conditions are similar to the operating conditions analyzed earlier. The tested water production rate and the variation of COP with time are illustrated in Fig. 8.38. COP is defined as the ratio of cooling energy taken away by chilled water in the evaporator to the thermal energy supplied by hot water, which is:

$$COP = \frac{\int dQ_{\mathrm{evap}}dt}{\int dQ_{\mathrm{hot}}dt} \tag{8.43}$$

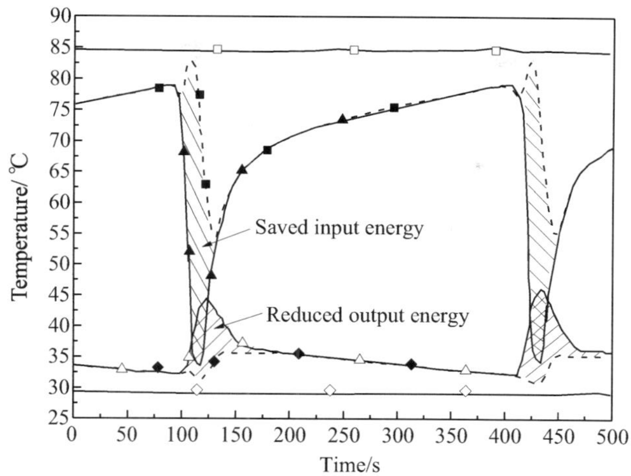

Figure 8.40
The relationship between the temperatures of a component and times when the system operates at different stages [18].

where $dQ_{evap} = m_{chilled}C_p(T_{chilled\text{-}in} - T_{chilled\text{-}out})$ is the cooling energy taken away by the evaporator (or the thermal energy supplied to the evaporator), and $dQ_{hot} = m_{hot}C_p(T_{hot\text{-}in} - T_{hot\text{-}out})$ is the thermal energy supplied to the system. Results in Fig. 8.38 demonstrate that with the increase of circulation time, water production decreases but COP increases. The lower the temperature of chilled water is, the better the performance, which is consistent with the previous discussion. The relationship between water production rate and heat supply temperature is shown in Fig. 8.39. The higher heat supply temperature results in better performance, as well as higher COP. The system has advantages at high temperatures.

But if it is driven by solar energy, the efficiency of the solar system will decrease when it operates at high temperatures. Thus the comprehensive consideration must be taken to select an appropriate heating temperature. Fig. 8.40 provides the temperature changes with time of each component in different operating stages and at different times of the system. It clearly shows the temperature tendencies of different components in the system, which is periodic variations.

In conclusion, the integration of the adsorption seawater desalination system and the solar system should be further investigated. The cooperation between the high-temperature heating system and the multiple-effect evaporator should be explored to improve the efficiency of the integrated system. This kind of system has a fairly high PR, but it utilizes low-quality thermal energy. It is an environmentally friendly and completely green seawater desalination system.

References

[1] Weinberg J, Ophir A, Fisher U. Coupling of multi-effect distillation with vacuum freezing to reduce energy cost in sea water desalination. In: Proceedings of the 7th int. symposium on fresh water from the sea, vol. I; 1980. p. 283–92.

[2] Alefeld G, Ziegler F. Advanced heat pump and air-conditioning cycles for the working pair H_2O/LiBr: industrial applications. ASHRAE Tech. Data Bull. Desalination 1997;114:253–75.

[3] Yanniotis S, Pilavachi PA. Mathematical modeling and experimental validation of an absorption-driven multiple-effect evaporator. Chem Eng Technol 1996;19:448–55.

[4] Aly SE. Vapour compression distillation using waste heat absorption systems. Desalination 1988;68:57–68.

[5] Fathalah K, Aly SM. Theoretical study of a solar powered absorption/MED combined system. Energy Convers 1991;31:529–44.

[6] Al-Juwayhel F, El-Dessouky HT, Ettouney HM. Analysis of single-effect evaporator desalination systems driven by vapor compression heat pumps. Desalination 1997;114:253–75.

[7] El-Dessouky HT, Ettouney HM. Simulation of combined multiple effect evaporation-vapor compression desalination processes. In: 1st IDA int. desalination conference in Egypt, Cairo, Egypt; September 1997.

[8] Hamed OA, Zamamiri AM, Aly S, Lior N. Thermal performance and exergy analysis of thermal vapor compression desalination system. Energ Convers Manage 1996;37:379–87.

[9] Kuehn TH, Ramsey JW, Threlkeld JL. Thermal environmental engineering. 3rd ed. New York: Prentice Hall; 1998.

[10] El-Dessouky HT, Ettouney HM. Fundamentals of salt water desalination. New York: Elsevier Science B. V.; 2002. p. 111–49.

[11] Zhengliang L, Hongfei Z, et al. Experimental study on an absorption type solar desalination system with falling film evaporation and multi-effects regeneration. Acta Energiae Solaris Sin 2007;28(11):1189–93.

[12] Ziqiana C, Hongfeia Z, Chaochen M, et al. Experiment and optimal parameters of a solar heating system study on an absorption solar desalination unit. Desalination Water Treat 2009;1:128–38.

[13] Wua JW, Biggsb MJ, Hu EJ. Thermodynamic analysis of an adsorption-based desalination cycle. Chem Eng Res Des 2010;88:1541–7. Applied Thermal Engineering 25(2005) 2780–2789.

[14] El-Dessouky HT, Alatiqi I, Bingulac S, Ettouney HM. Steady-state analysis of the multiple effect evaporation desalination process. Chem Eng Tech 1998;21:15–29.

[15] El-Dessouky HT, Ettouney HM. Single effect thermal vapor compression desalination process: thermal analysis. Heat Transf Eng 1999;20:52–68.

[16] Ng KC, Thu K, Chakraborty A, Saha BB, Chun WG. Solar-assisted dual-effect adsorption cycle for the production of cooling effect and potable water. Int J Low-Carbon Technol 2009;4:61–7.

[17] Ng KC, Thu K, Saha BB, Chakraborty A. Study on a waste heat-driven adsorption cooling cum desalination cycle. Int J Refrig 2012;35:685–93.

[18] Wang X, Ng KC. Experimental investigation of an adsorption desalination plant using low-temperature waste heat. Appl Therm Eng 2005;25(17–18):2780–9.

[19] Thu K, et al. Performance investigation of a waste heat-driven 3-bed 2-evaporator adsorption cycle for cooling and desalination. Int J Heat Mass Transf 2016;101:1111–22.

CHAPTER 9

Solar Concentrating Directly to Drive Desalination Technologies

Chapter Outline

9.1 Introduction

9.1.1 Challenges in Solar Desalination Technology

People have spared no effort to utilize solar energy for seawater desalination in the past 100 years, but solar desalination technologies have not been widely used until now. The reason is that it is not economical when compared with traditional sweater desalination system. In terms of capital cost, the cost of producing water by the seawater reverse osmosis (SWRO) system is about ¥5−8 (CNY)/ton; the cost of producing water by the low-temperature multiple-effect seawater desalination, which commonly incorporates with an electric power station, is around ¥8−12/ton; the cost of producing water by a solar seawater desalination system is higher than ¥20/ton, in view of the lifetime of 20 years and 12 L/day per square meter of the thermal collector, and the cost may be higher in practice. With respect to scale, the water production by a relatively large solar desalination system in the world is dozens of tons per day only, which is much lower than that by the SWRO system, generating tens of thousands of tons of water per day. It is also much

Solar Energy Desalination Technology. http://dx.doi.org/10.1016/B978-0-12-805411-6.00009-9

lower than the water production by the low-temperature multiple-effect desalination system, which produces thousands to tens of thousands of tons per day. Hence these problems affect the development of the solar seawater desalination system. What restricts the economy of it and its large scale? These are all thoughtful questions.

Nowadays, it tends to integrate modern industrial seawater desalination technologies with solar energy technologies. Many researchers work in this area. However, the former studies still cannot solve the economic and scale problems of the solar desalination system. Moreover, the efficiency is low when the solar collector is in the high-temperature range (≥100°C), while the efficiency of the desalination system is low in the low-temperature range, which presents the problem of "structural" mismatch. It can be explained as follows: although the solar thermal technology becomes increasingly mature, such as a vacuum tube collector, trough parabolic collector, and medium-temperature large solar pond, etc., the temperature of the collector can reach higher than 100°C. However, the collection efficiency will decrease significantly, or the cost of facility is too high. For instance, the large trough parabolic solar collector can obtain the thermal energy when the temperature is higher than 200°C, but the cost of tracking and manufacture are very high and the long pipes for heat transportation are required, which leads to big heat losses. In addition, it requires conduction oil as the heat-transfer working fluid, which is not suitable to be used to heat seawater directly, because scaling is easy to occur on the surface of the heat exchanger. If use the vapor generated by heating fresh water to drive traditional seawater desalination facility. Extra links of heat exchange that is adds up to extra investment cost.

Incorporating a modern industrial seawater desalination system with solar technologies is good for improving the total efficiency of the solar desalination system by taking advantage of advanced experiences and technologies of current seawater desalination technology. In this aspect, there are many practical applications already. For example, the seawater desalination system with the area of 680 m^2 in Mexico has the desalination capacity of 12 m^3/day, and its operating temperature of the solar collector is within 35°C ~ 110°C [1]. The seawater desalination project in France has the solar collector area of 670 m^2 and water production amount of 40 m^3/day [2]. The United Arab Emirates built the biggest multistage flash distillation seawater desalination factory, where the output of a single machine achieves 45,400–57,000 m^3/day [3]. Kuwait has a solar multistage distillation system, which can produce 10 tons of fresh water per day [4]. All of these focus on increasing the efficiency of the solar energy system, reusing thermal energy, and rising the heat exchange coefficient, etc., which makes contributions to improving the total efficiency and economy for the solar seawater desalination system.

Although there are numerous studies about solar desalination systems, which have been put forward all over the world, the critical problems still exist; for example, the high building cost of investing in a solar desalination system causes the bad economy, which is difficult to be accepted by customers. In view of the current solar desalination project, the ratio of the cost of solar collectors and desalination is around 1:1. A solar desalination system producing 10 tons of water per day requires solar collectors with an area of 800 m^2 (all-glass evacuated tube solar collectors) that cost $110,000 USD (in China), and the cost of seawater desalination facilities is also around $110,000 USD. In winter the temperature of the evacuated tube collector decreases, so that the annual water production is about 2800 tons. If the lifetime of the facilities is 20 years, the price of generating water is higher than $4/ton. With taking operating costs and bank interest into account, the cost of producing water will increase, which will be much higher than the cost, ¥5–8/ton, of that in the reverse osmosis system.

By analyzing the working principles and operating processes of past solar desalination systems, they have the disadvantage of not only the mismatch of temperature for the solar collector and desalination facility, but also the following points:

1. There is severe separation between the components in the system. The separation of numerous components like the collector, accumulator, pipe, heat exchanger, and seawater desalination equipment results in high manufacture and administrative costs, which are difficult to reduce.
2. The large-scale solar desalination system requires big areas for the solar collector. However, transporting collected energy to desalination equipment needs a lengthy pipe, which causes huge heat losses. The energy is transferred in various processes. First, the luminous energy is converted to thermal energy. Thermal energy is stored in a heat-conducting medium, and then the medium transports thermal energy to the power accumulator, where the thermal energy is transferred to another medium. Thermal energy is taken by this new medium to the desalination apparatus and then transferred to seawater to promote evaporation on seawater. The common processes are shown in Fig. 9.1, in which the solar desalination is divided into at least three sections: the solar collection part, the energy storage part, and the desalination part. Thermal resistance exists in every process and provides efficiencies. The product reduces the efficiency of the whole system significantly. It is the main factor that prevents decreasing the cost of the solar desalination system.
3. The way of desalination determines the high cost. Current solar desalination tends to incorporate with modern industrial seawater desalination technology, which has mature experience and design philosophy. It is expected to combine solar technology with low-temperature multiple-effect technology, because the low-temperature multiple-effect technology has the advantages of high water production ration and high thermal efficiency. However, the problems of thermal resistance and cost for the collector, accumulator, heat exchanger, and pipe, etc. still exist, which determines the cost of water production to be high. In addition, low-temperature multiple-effect technology requires vacuum technology that causes high costs on building and maintenance, which are difficult to be reduced.

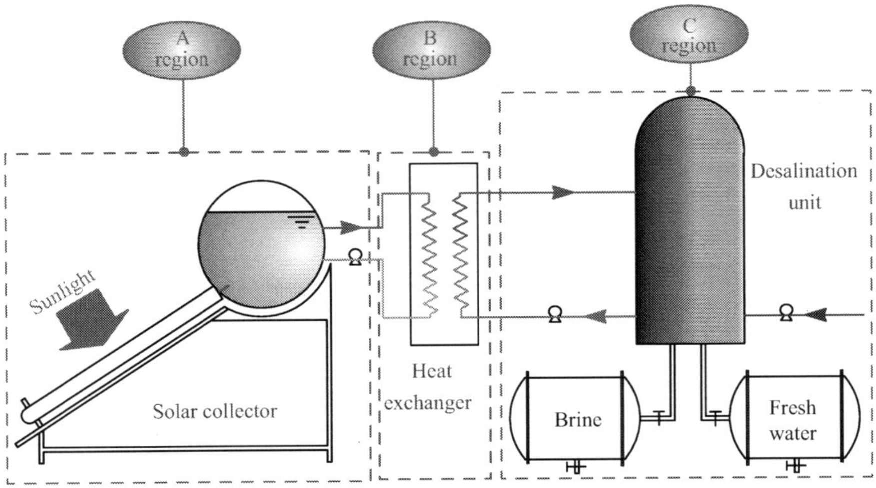

Figure 9.1
Working process of a common solar desalination plant.

4. Most traditional industrial seawater desalination systems utilize the distillation process with negative pressure and large space, which is good for enhancing vaporization. However, the materials for making facilities must have bearing capacity, corrosion resistance, and salt spray resistance, such as titanium alloy. Therefore the manufacture cost of the facility becomes very expensive to increase the investment of the solar energy system.

A series of studies [5–8] focusing on improving the efficiency of a single system were conducted, for example, establishing high-efficiency solar concentration and receiving system, designing high-efficiency seawater desalination system, etc. All of these works promote the reduction of solar desalination system cost and improvement of system efficiency. Nevertheless, they do not solve the large-scale and low-cost problems of the solar desalination system. There are not many innovative thoughts in improving system efficiency, reducing manufacture cost of the building system, putting forward the novel desalination method and achieving large-scale by reducing the resistance of heat and mass transfer for the whole system. Most of the approaches introduced later are basic ideas.

To conclude, the separated components of the traditional solar desalination system lead to increasing cost and heat and mass transfer resistance between the components results in decreasing system efficiency. The seawater desalination methods are so traditional that the costs of material and manufacture rise. The cost problem can still not be solved by integrating solar technology with traditional desalination technology, because it has already been determined by concentration and desalination methods.

The key to achieve the low cost and large scale of the solar desalination system is to overcome the weaknesses discussed before and incorporate solar concentration technology and traditional seawater desalination technology efficiently.

9.1.2 A New Way to Solve These Problems

Considering the problems discussed before, concentrating light to heat seawater desalination directly is a noteworthy technology, the features of which are to transmit light to seawater directly to make the seawater evaporate, and then it reuses the latent heat from vapor condensation. Therefore the concentrator, collector, accumulator, pipework, and desalination apparatus are combined to one simplified system so that the energy transfer resistances are reduced, as well as the nonmetallic material can be used as much as possible to achieve the objectives of resisting corrosion and low cost. It is clearly described in Fig. 9.2. The most important improvement of it is to utilize heating even sunlight to promote seawater evaporation by transporting light to desalination equipment directly, which reduces heat-transfer processes. When the evaporation is done, it should be managed to incorporate with traditional industrial seawater desalination technologies to reuse the latent heat from vapor condensation to improve the efficiency of the whole system.

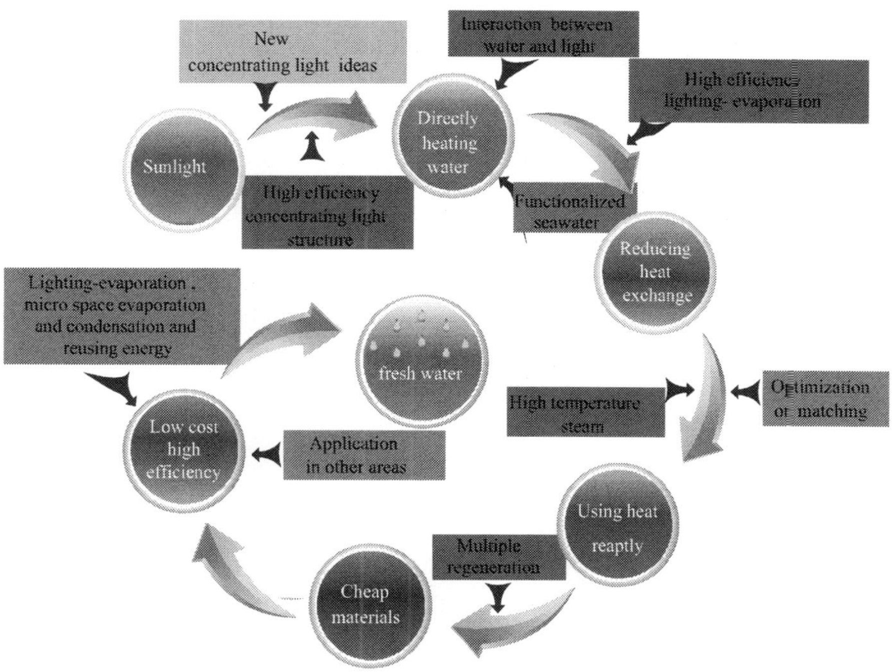

Figure 9.2
An innovative high efficiency solar desalination system.

Actually, this kind of solar desalination system is not a new idea. It has existed in nature for a long time. For instance, when sunlight strikes the sea, water vapor will generate and form rain. This is a typical case of seawater evaporating due to direct irradiation. The most classic evaporation systems using sunlight or heating directly are the tray solar distillers, but they do not achieve several times spotlight so that the concentration ratio of them is only one. The operating temperatures are relatively low as well. The efficiency of tray solar distiller is 35% approximately. However, they have several advantages, such as stable operations, easy-to-obtain materials, and simple structures so that some people still use it today. Thus improving concentration ratio, rising operating temperature, and reusing latent heat are the future directions of these kinds of systems. To raise the operating temperature of the distiller, Joshua [9] proposed a tray solar distiller with concentrator, as shown in Fig. 9.3, and simulated and tested the performance of the prototypes for it. It was found that for a reflector, its width is 2.5 times larger than the still, so the output per unit area per day roughly triples with only about 10% increase in cost and moderate maintenance (weekly tilts), indicating that compound parabolic concentrators have a significant economic advantage in producing solar distilled water.

In the solar concentrator, many large or ultralarge concentrators were invented, such as solar parabolic dish concentrator, parabolic trough solar concentrator, Fresnel reflector solar concentrator, and concentrated solar tower, etc. The costs of them are decreasing rapidly as well. Utilizing these solar concentration technologies for a direct light-driven solar desalination system is a great opportunity, and there are some preliminary explorations that have been taken for it. Some new plans have been put forward, which is good for the improvement of solar desalination.

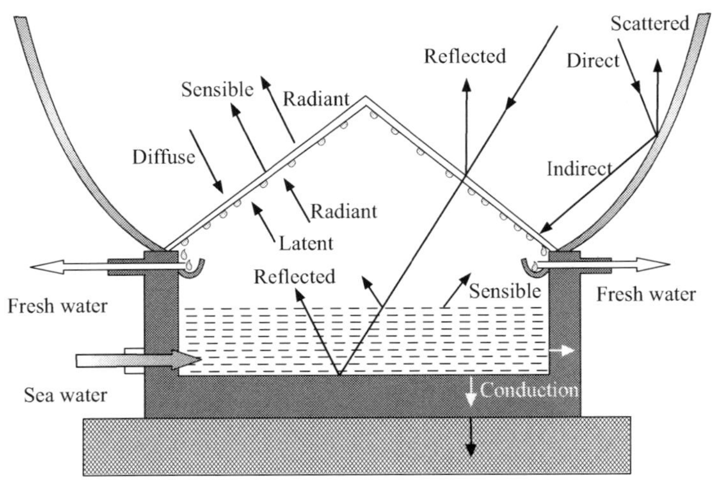

Figure 9.3
A standard single-effect basin still with a compound parabolic concentrator.

9.2 Concentrated Light-Driven Solar Desalination System

As discussed earlier, the most traditional solar distiller is a concentrated light-driven solar desalination system, which utilizes concentrated sunlight to heat seawater directly for desalination, but the concentration ratio of it is only one, so the operating temperature is low. The mass transfer efficiency is not high because the heat and mass transfer processes in the distiller depend on natural convection. Some distillers use some measures to strengthen heat and mass transfer. However, some factors, like seawater in a tray, have big thermal inertia and the latent heat from vapor condensation is not used efficiently, because the efficiency of the distiller cannot be improved and the water production per unit area is low, which restricts its popularization. On the other hand, it has many outstanding advantages, such as simple structure, easy-to-obtain materials, easy-maintenance, and stable operations. The concentrated light-driven solar desalination systems are attempting to enhance these advantages to weaken its disadvantages, especially by increasing its concentration ratio to rising its operating temperature.

The systems introduced in the following parts are typical trials of a concentrated light-driven solar desalination system. Many of them have preliminary experiment results or even only have initial ideas, but all of them are worthy of investigation in further research.

9.2.1 Solar Desalination Systems Directly Driven by Dish Concentrator

Utilizing a solar parabolic dish concentrator is a common approach to obtain super high temperature by solar energy in commercial and academic areas. This kind of concentrator has the advantages of high concentration ratio and high flux density distribution.

Bechir Chaouchi et al. [10] designed and built a small solar desalination unit equipped with a parabolic concentrator, as shown in Fig. 9.4. The experimental device was covered with rectangular stainless steel sheet segments with a thickness of 1 mm. The absorber is mounted at its focus, which is shaped like a cylindrical vase, with a receiving surface of 0.013 m^2 and a geometric concentration of 195. This absorber is completely insulated except the part lit by the solar rays reflected by the parabolic surface.

The results show that the maximum efficiency corresponds to the maximum solar lightning obtained toward 14:00. At that hour, the boiler was nearly in a horizontal position, which maximizes the offered heat-transfer surface. The experimental results were given in Fig. 9.5 and compared with those calculated theoretically. According to Fig 9.5, the production rate increases at the beginning of the day to reach their maximum toward 14-hand decreases thereafter. This evolution is closely linked to solar lightening, which is

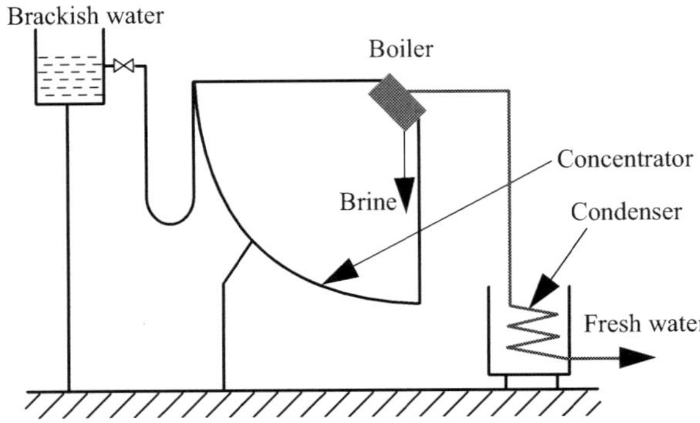

Figure 9.4
Experimental device with a parabolic concentrator.

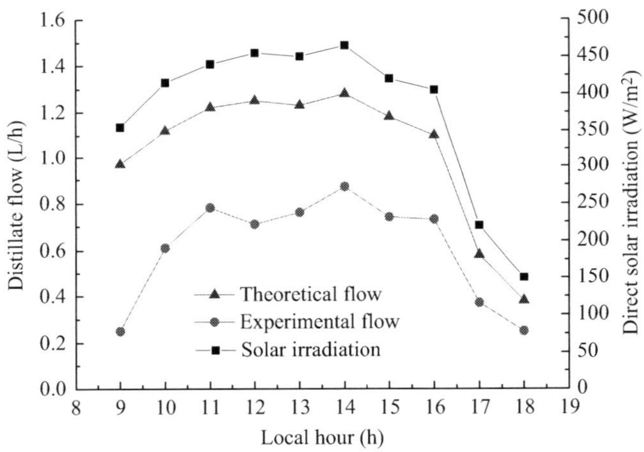

Figure 9.5
Experimental and theoretical distillate flow rates and solar irradiation versus local time.

responsible for this production and therefore has a similar rate. This figure shows a difference between theoretical and experimental flows. This deviation can be explained by the fact that in the morning, only a small part of the absorbing surface is covered with water because of the strong tilt in addition to the geometry imperfection and the sun's manual follow up.

It is easy to be found from Fig. 9.5 that the experimental results are not so good. However, it is a new attempt. In the case of the desalination process, to achieve higher yield, the concentrator is coupled with the solar still by means of increasing water temperature in the basin. The natural circulation mode or forced circulation mode is used to supply water or

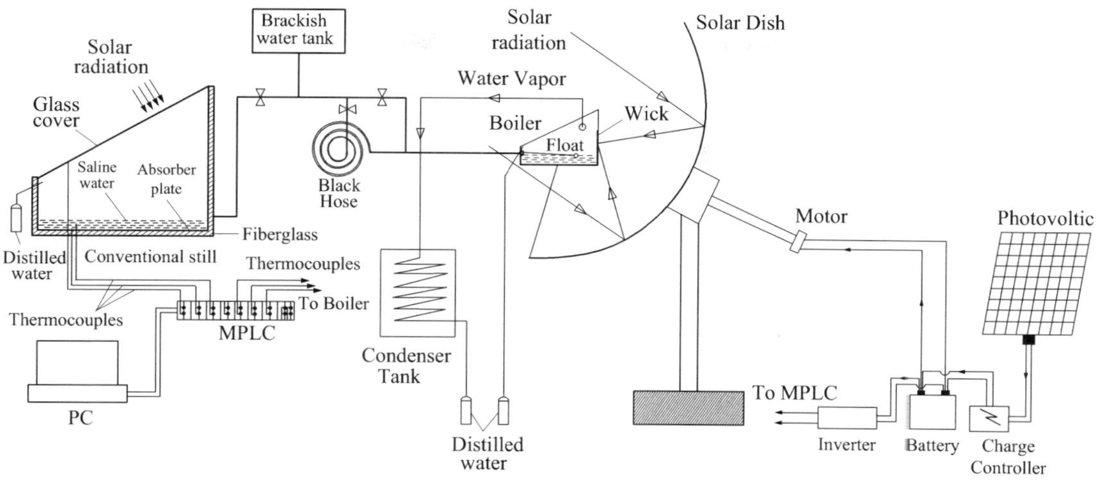

Figure 9.6

Schematic diagram of the solar dish concentrator and conventional solar still [12].

oil to the trough receiver pipe. This kind of system was reported by Singh [11]. Omara (2013) also designed a solar dish concentrator (SDC) coupled with a conventional solar still (CSS), as shown in Fig. 9.6. The developed solar thermal desalination system consists of a brackish water tank, CSS (single slope basin solar still), SDC, boiler, heat exchanger (condenser), control unit (two-axes tracking system), PV system, simple solar collector, and modular programmable logic control. Here, a mini single-slope air-tight solar still is designed and installed at the focus of the dish concentrator, which is used as a boiler. The developed desalination system is evaluated and compared with the CSS.

The dish-shaped concentrator reflects and concentrates incoming direct normal insolation to a receiver (boiler) located at the focal point of the dish. Inside the boiler, the brackish water is vaporized by the intense heat created by the concentrated sunlight on the bottom and the backside of the boiler, on which a distributed capillary wick is covered. The generated water vapor flows up and condenses in the coupling heat exchanger, where it gives out the evaporation latent heat to the working fluid (cold water). Also the condensate vapor on the boiler glass cover is collected out.

The results indicated that the daily average of distillate water was 6.7 L/m^2/day for the SDC with preheating of brackish water, while the distillate productivity was 3.0 L/m^2/day for CSS. In Omara's study, the daily average efficiency of SDC and CSS was 68% and 34%, respectively. The increase in distillate production for SDC is about 244% and 347% higher than that of CSS without and with preheating, respectively.

Using the dish concentrator directly to drive the solar desalination has several shortcomings: (1) the solar parabolic dish concentrator cannot be too big with the

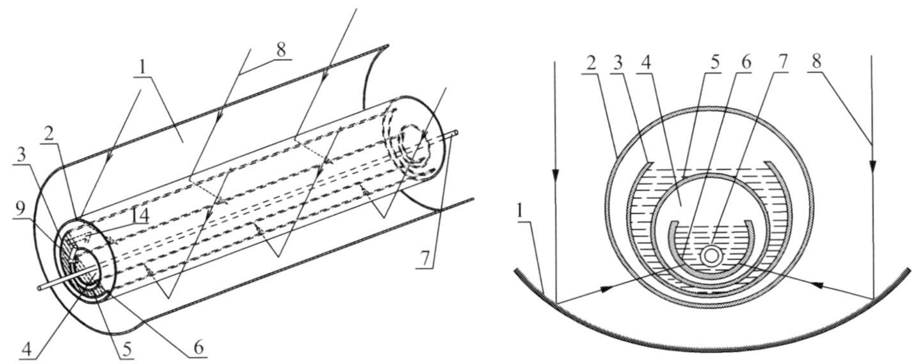

Figure 9.7
Concentrating multieffect tube solar distiller: 1-solar concentrator; 2-tubular case; 3-water trough; 4-evaporation-condensation chamber; 5-sleeving; 6-sunlight collector; 7-heating pipe; 8-sunlight; 9-seawater inlet pipe; 14-communication pipe.

restriction of dish area. It is not applied in the disc solar distiller usually. (2) It requires a two-axis tracking system to follow the sun. If the collector is too large, it will affect the system performance. Thus the authors of this book proposed a multieffect tube solar distiller with trough parabolic concentrator, as shown in Fig. 9.7. They did preliminary experiments. The good results imply that it is worthy to be explored further.

9.2.2 The Solar Desalination System Directly Driven by Linear Fresnel Lens Concentrator

For the linear solar collector, the most common types are trough parabolic reflector and linear Fresnel reflector (LFR). The working principles of trough parabolic reflector are shown in Fig. 2.55 and of LFR are illustrated in Fig. 2.62. A feature of both is that the focal lines of them are above reflectors. There will be many difficulties if absorbers are installed, especially the evaporators for seawater desalination, on focal lines. A study [13] proposed an idea of utilizing a curved Fresnel lens to concentrate solar energy, which creates the conditions for installing the evaporator under systems, especially on the ground. In addition, the cylindrical Fresnel lenses solar collector put forward by the literature [13] has the advantage of that its absorber need not move, which is lacking by traditional trough parabolic concentrator. Considering this advantage, we proposed several novels solar desalination systems, which incorporate the cylindrical Fresnel lens solar collector with the seawater desalination system, as demonstrated in Fig. 9.8. All of them transmit sunlight onto seawater directly to drive evaporation. The desalination system illustrated in Fig. 9.8 is based on the humidification and dehumidification of air, which consists of a curved Fresnel lens concentrator, secondary concentrator, transparent glass cover, humidifier, condenser, plate for salt water, air passages, circulating pump, and control valve. The humidifier contains a sprinkler, black porous material, and plate for salt

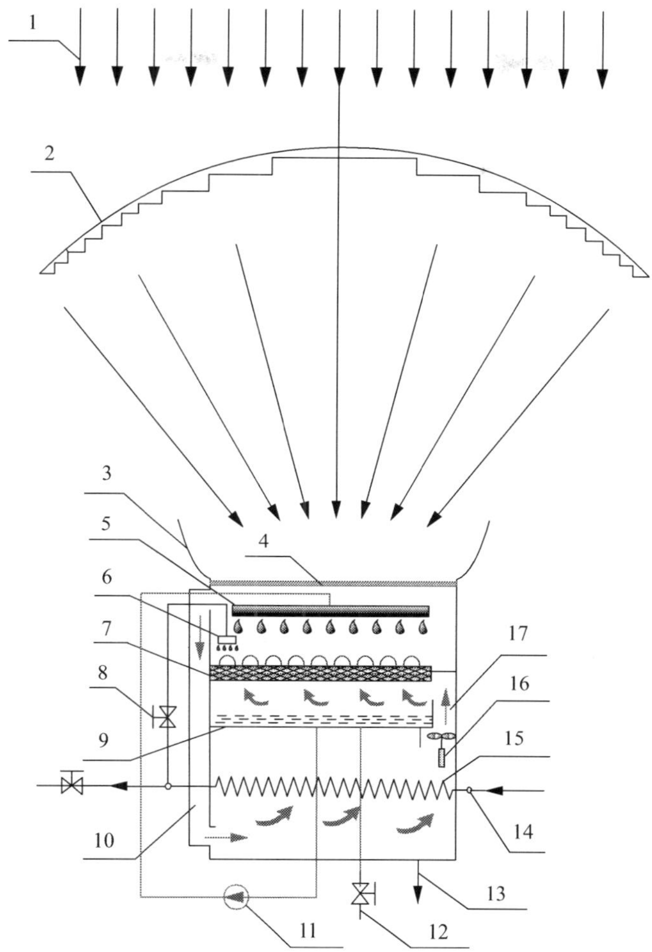

Figure 9.8
A unit of liner concentrated light-driven heating humidification–dehumidification desalination system. 1-incident sunlight; 2-Fresnel concentrator; 3-secondary concentrator; 4-glass cover; 5-sprinkler; 6-water replenisher; 7-humidifier; 8-control valve; 9-saltwater plate; 10-hot air duct; 11-circulating pump; 12-outlet for discharging strong brine; 13-outlet of fresh water; 14-inlet of seawater; 15-condenser; 16-fan; 17-cold air duct.

water. The humidifier and condenser are separated into two spaces to form independent units.

The working principles of this system can be described as follows: the incident sunlight is concentrated onto the glass cover through the Fresnel lens concentrator and secondary concentrator, and it goes onto the surface of the humidifier made by porous ceramics. Seawater is sprayed onto the humidifier surface from the sprinkler, and then it evaporates due to irradiation. The evaporated vapor lets heat and moisture exchange with air to form saturated moist air, which goes into the condensation chamber to form fresh water with the fan driving. The brine in the saltwater plate is pumped to the sprinkler by the circulating pump for spraying to be reused. The super concentrated brine will be discharged from the pipework under the plate.

The system presented in Fig. 9.8 can be connected in a series to comprise the multiple-effect seawater desalination system shown in Fig. 9.9. With connecting several sunlight-driven heating humidification–dehumidification desalination system units in series to form a multiple-effect solar humidification–dehumidification desalination system, this system can achieve the processes of multiple-level humidification and dehumidification. The Fresnel lens concentrator collects sunlight to the transparent glass cover (4) to heating humidifier (7) directly. The saltwater plates (9) in different effect levels are connected by a pipe to guarantee that the amount of water they have are the same. Saltwater plate (9) collects dropped seawater from humidifier (7). The collected seawater is sprayed on humidifiers through circulating pump (11) at the bottom and sprinkler (5), by which the isothermal heating of seawater is finished. The vapor formed in the humidifier and the saturated moist air formed by heat and moisture exchange between vapor and air circulate among humidification–dehumidification units in different effect with the drive of inner fan (16) in the first and last effects. Then multiple humidification and dehumidification processes of air are finished. The process can be explained as follows: the fan (16) in the first effect blows the air in high temperature and humidity in humidifier (7) into the condensation chamber of the second effect. Then the air exchanges heat with feed seawater inside the condenser to generate some fresh water and preheats feed seawater at

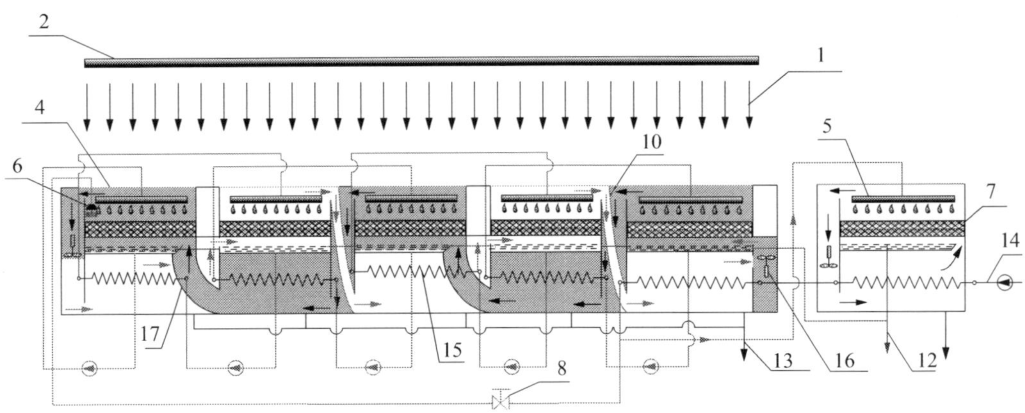

Figure 9.9
Liner concentrated light-driven multiple-effect humidification–dehumidification desalination system: 1-incident sunlight; 2-Fresnel concentrator; 4-glass cover; 5-sprinkler; 6-water replenisher; 7-humidifier; 8-control valve; 10-hot air duct; 12-outlet for discharging strong brine; 13-outlet of fresh water; 14-inlet of seawater; 15-condenser; 16-fan; 17-cold air duct.

the same time. The cooled moist air goes into the humidifier (7) of the third effect and comprises high-temperature and high-humidity air again by mixing the high-temperature vapor. Then it is transported to the dehumidification chamber of the forth effect to condensate. The air after condensation flows to the fifth effect humidifier. The air will flow level by level until the last one. Finally, the fan of the last effect drives the air again to pass through the humidification and dehumidification chambers of previous levels one by one to complete an enclosed air circulation. In the dehumidification chamber of each level, the high-temperature and high-humidity air exchanges heat with feed seawater; fresh water is generated when the temperature is lower than the dew point and preheats feed seawater at the same time. It achieves a multiple-effect isothermal heating process and improves utilization efficiency of solar energy. In a word, in this system, the structures of the dehumidification and humidification unit in each effect are totally the same. Air is heated continuously in each level in the enclosed space, and then it is dehumidified. It continues circulating with the effects of the fan.

To further improve the energy efficiency of the system, the level for reclaimed energy can be set. Thus feed seawater flows into the humidification–dehumidification unit of reclaimed energy. This unit is not within the concentrated area of concentrator (2). It is only applied to reclaim the energy of some preheated seawater from the condenser of the first effect.

Here, the receiver of sunlight is the evaporator of the desalination system. Combining several systems in one significantly decreases the thermal resistance and reduces the components of the system. It can be seen that the linear solar concentrator can be utilized for concentrating light for direct heating. It has these features: each effect has isothermal evaporation, operating in high temperature and multiple back-heating and the amount of imported and discharged seawater is very small for all effects except the last one, which guarantees the sensible heat losses taken away by strong brine are the minimum.

9.2.3 The Solar Desalination System Directly Driven by Tower Concentrator

The concentrated solar tower is applied in a large-scale solar power station. This kind of concentration technology becomes more mature. However, some places in the world have more serious problems with water scarcity than with electricity scarcity. Utilizing a concentrated solar tower for seawater desalination has been brought to the attention of researchers.

Abdelkader [14] studied a solar desalination system with multieffect humidification-dehumidification cycles, which was coupled with a central solar receiver, as shown in Fig. 9.10. In this system, saline water is warmed through the solar central receiver and then it entered into a desalination chamber. The air circulation in the humidification chamber is provided by natural convection.

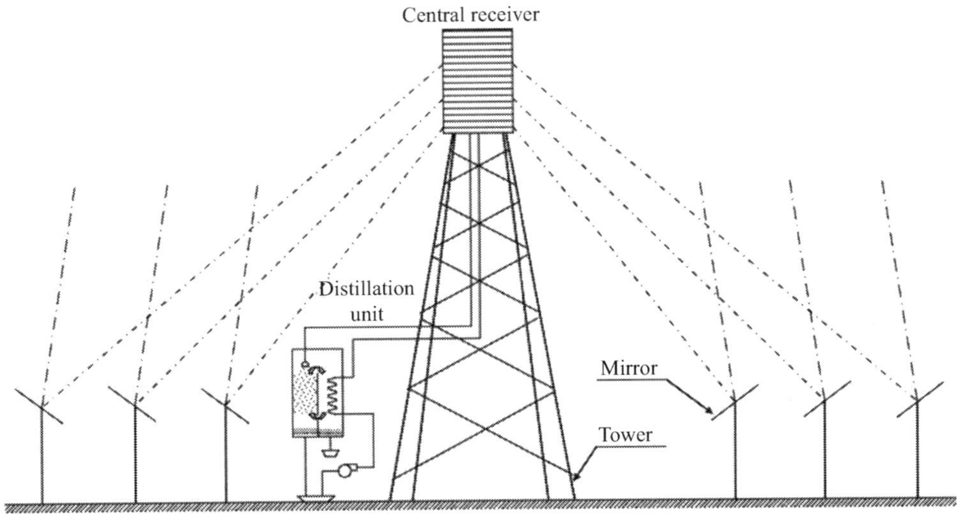

Figure 9.10
Schematic diagram of the test rig. http://www.tandfonline.com/doi/full/10.1080/19443994.2013.808394 [14].

Although the system put forward by Abdelkader is a small experimental system and the advantages shown from results are limited, some researchers introduced the ideas of establishing a high-concentration light-driven evaporation solar desalination system, which is mainly applied on a heliostat concentrated solar tower. This system uses heliostat to reflect sunlight. The solar tower collects light, and the strong light heats the functionalized seawater (adding some ceramic particles can make seawater have a better ability of absorbing light) to generate high-temperature vapor. Vapor is transported to a traditional seawater desalination system for desalination, and salt is produced by strong brine generated from the system. Based on the theoretical calculation, this method has a higher economic value than the simple power generation.

Fig. 9.11 demonstrates the diagram of a compound seawater desalination system that combines a concentrated solar tower with a traditional low-temperature multiple-effect seawater desalination system. The working principle of it can be described as follows: the heliostat (3) reflects sunlight onto the concentrated solar tower (8); sunlight strikes the seawater evaporator (12) with arrayed transparent pipes through a glass case (11). This evaporator contains a black absorber and black ceramic particles, whose density is the same as water to absorb sunlight. Seawater is sprayed in the evaporating pipe from the top and flows out from the bottom of the pipe. The structure and working principles of the evaporating pipe are presented in Fig. 9.12. The functionalized seawater is heated to produce high-temperature vapor with the effects of high-density solar radiation. The desalination of vapor happens when vapor passes through a foam-catching net (9). The desalted vapor flows into multiple-effect distillers for vapor condensation. The condensed

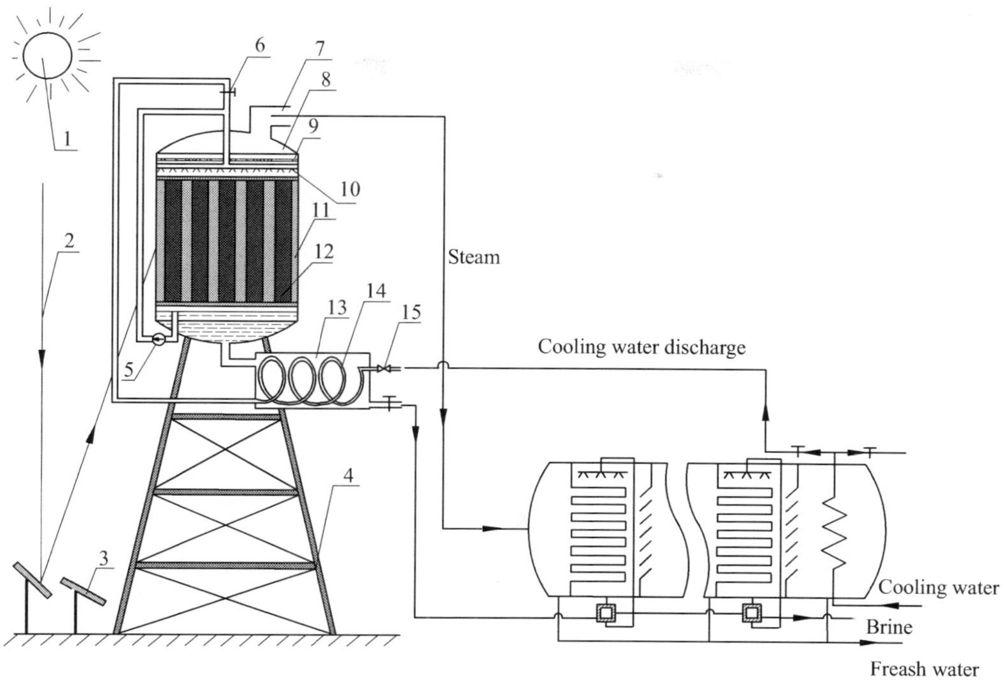

Figure 9.11
Heliostat concentrated solar tower desalination system: 1-sun; 2-solar radiation; 3-heliostat; 4-tower support; 5-pump; 6-valve; 7-vapor outlet; 8-concentrated evaporation tower; 9-foam catching net; 10-seawater sprinkler; 11-transparent glass cover; 12-semitransparent seawater evaporation pipe; 13-hot water return tank; 14-seawater heat exchanging pipe; 15-check valve.

vapor forms fresh water and flows to a fresh water storage tank through the output pipe for users. The seawater that does not evaporate comes together at the bottom of the solar tower (8). Some seawater is pumped back to the tower by pump (5) to repeat evaporation again, and other seawater flows to the heat exchange tank (13) from the bottom of the tower to preheat low-temperature seawater by its thermal energy. The seawater in the pipe (14) for seawater heat exchange is sprayed into the solar tower from the valve (6). The check valve (15) is used to prevent the return of seawater due to pressure.

Compared with traditional solar seawater desalination technology, the concentrated solar tower desalination system has the following advantages:

1. Using heliostat to concentrate sunlight without long pipework reduces the investment and heat losses of the pipe.
2. Utilizing ceramic particles or other methods to functionalize seawater to let seawater absorb sunlight and evaporate directly omits the heat exchanger and reduces heat-transfer processes, so that the concentration efficiency of this system is improved.

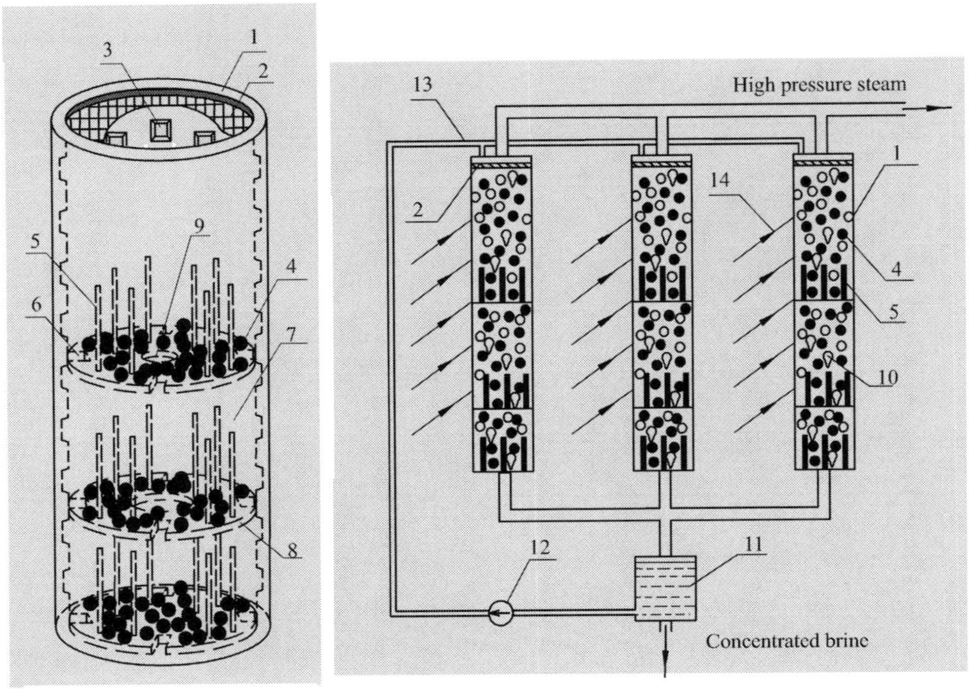

Figure 9.12
Ceramic particles and vertical tubes compound light-driven seawater heater: 1-double-layered transparent evacuated glass tube; 2-spill-proof net; 3-particle-blocking net; 4-dark ceramic particles with the same density as water; 5-tubular fins; 6-diversion trench; 7-light-driven heating evaporation chamber; 8-split baffle; 9-central circular hole; 10-seawater bubbles; 11-strong brine tank; 12-pump 13-circulating seawater; 14-high-density solar radiation.

3. Functionalized particles move in water and collide with each other so that the incrustations on the distiller surface are ground away. The problem of incrustations in high-temperature heating seawater is completely solved.
4. The heat exchange section on the concentrated solar tower, in other words, the facilities contacting seawater, can be made of nonmetallic materials, e.g., the heat pipe that absorbs sunlight to make seawater evaporation can be made of quartz glass. The corrosion resistance of the whole system is enhanced.
5. The traditional seawater desalination system can be used as the seawater treatment facilities in the end. The traditional system can be used for large scale, has mature technologies and high efficiency.

In addition, the concentrated solar tower can be integrated with a traditional multiple-stage flash evaporation system to achieve 24/7 operation by adding an auxiliary heating system.

In a general view of the components for the concentrated solar tower desalination system, the heliostat concentration system and seawater desalination system in the end are relatively mature. Only the part of directly generating vapor from seawater in the tower needs to be further investigated. Some people proposed an idea to apply black ceramic particles or other substances to functionalize seawater in the tower to make it have a better ability to absorb light and use sunlight to heat seawater to let it evaporate. It is demonstrated in Fig. 9.12 in detail, which presents a vertical tubular seawater desalination vapor generator consisting of numerous transparent tubes. Seawater is in transparent tubes. There are black ceramic particles, whose density is almost the same as in seawater. They absorb the strong light incidents into transparent tubes to convert it to thermal energy. Thermal energy is transferred to seawater to make it evaporate. Then vapor is transported to the next stage seawater desalination system. The brine repeats, circulating until its concentration reaches a certain level.

This design achieves the direct heat transfer between light and water, which reduces the heat-transfer processes and does not require the heat exchanger. It provides a new direction for solar desalination. The whole system can be made of nonmetallic materials so that the corrosion and costs are decreased. Owing to the continuous movement and collision of ceramic particles, the incrustations on particles and internal pipe wall are removed, which guarantees the transparency of tubes and optical absorption of particles.

Based on the advantages discussed earlier, it is significant to put forward a concentrated light-driven heating solar seawater desalination system, which can be large-scale. Considering the economy of it, the concentrated solar tower desalination system can be used as an example for discussion. First, compare it with a concentrated solar tower electricity generation system. For a 5 MW concentrated solar tower power generation system, generating 1 MW electricity requires 8 MW thermal energy with temperature higher than 400°C (this is an average level). Actually, the heat loss of solar tower power generation is much more than the solar tower seawater desalination system, because for the desalination system, the required vapor temperature in the tower is within 150°C ~ 210°C. If a 5 MW power station provides 40 MW thermal energy, this tower generates 60 tons vapor per hour (ignoring the sensible heat of seawater; the input seawater is preheated by strong brine and output fresh water, and the sensible heat of vaporization is 2400 kJ/kg). Importing the entire vapor into 10-effect multiple-stage flash evaporation or multiple-effect distillation traditional seawater desalination systems can produce around 480 tons of fresh water (adding the water condensed from initial vapor) with the system performance ratio of seven. If the working hours per day are 8 h, the fresh water generated per day reaches 3840 tons, which can be regarded as large-scale.

Second, the economy of a concentrated solar tower desalination system is obvious. For a calculation based on the residential electricity price in Beijing, China, which is around $0.10/kWh, the income per day of power generation under good weather is $4000 (solar power system operates 8 h per day). The fresh water produced per day is 3840 tons.

The water price in Beijing is \$1/ton, so the income from fresh water is \$3840/day. It can be seen that the economies of utilizing a concentrated solar tower to generate electricity and desalinate seawater are almost the same. However, the costs of establishing and maintaining a solar desalination system are much lower than that of a solar power generation system. In addition, for a solar desalination system, the required stability is not high, and heat dissipation is not severe; a solar tower desalination system can utilize strong brine to produce salt and conduct other business. Thus it can be implied that a concentrated solar tower desalination system has better economy. Actually, the manufacture costs of a tower desalination system are much lower than that of a tower power generation system, so its economy must be better than a power generation system.

Therefore it is significant to investigate the large scale and low cost of high-temperature solar desalination based on the experienced traditional industrial desalination system and concentrated solar tower technology. Another key to achieve it is the functionalization of seawater, which makes seawater absorb sunlight directly. The researches of the seawater functionalization process are required.

An air humidification–dehumidification seawater desalination system has many advantages, such as no need for vacuum, seawater can be controlled within a small range, and the requirements of seawater pretreatment are not high. For a small-scale system, it can be designed to use a concentrated solar tower incorporated with a light-driven heating humidification–dehumidification desalination system. To integrate two systems very well, some researchers proposed a concentrated heating multiple-effect heat recovery matrix air humidification–dehumidification solar desalination system, as shown in Fig. 9.13. It can reuse the sensible heat from vapor condensation in a humidification–dehumidification solar desalination system, and then the cost of a unit volume of fresh water produced by the solar desalination system will reduce. The efficiency of utilizing solar energy in a seawater desalination system is improved as well. It is suitable to be applied in outlying land and islands, especially in the places where the solar radiation is strong.

The solar desalination system, as shown in Fig. 9.13, contains a dehumidifier, ventilated liquid collector, light absorber, wind tunnel, humidification chamber, and supporters, etc. The feed seawater flows into the facility from the right-hand side at the bottom. Following the direction of seawater flow, from bottom to top and right to left, the humidification and dehumidification chambers in the facility are named stage 1, stage 2, stage 3, and stage 4.

Its working principles can be described as follows: several humidification and dehumidification chambers are connected in a series. Feed seawater flows in the dehumidification chamber 18-1 of stage 1 through inlet pipe 14 and then flows into the dehumidification chamber 18-2 of stage 2. The seawater from the stage 2 dehumidifier is split into two parts, and the main part returns to the humidification chamber 17-1 of

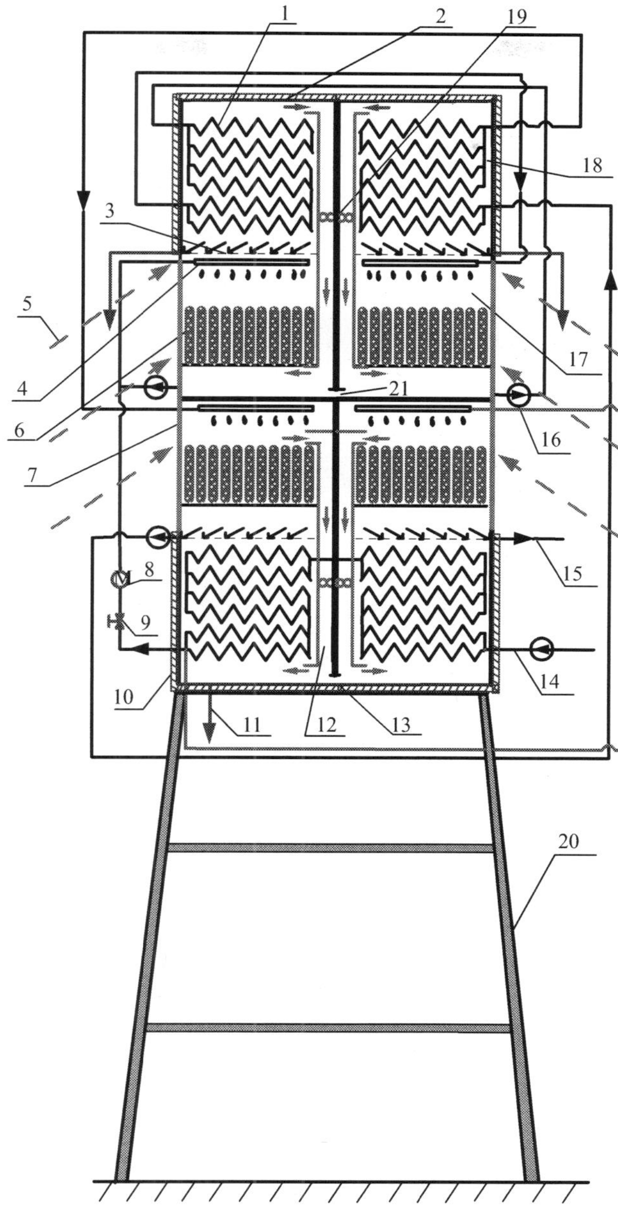

Figure 9.13
Concentrated solar tower multiple-effect air humidification–dehumidification system: 1-dehumidifier; 2-case; 3-ventilated liquid collector; 4-sprinkler; 5-incident sunlight; 6-sunlight absorber; 7-transparent window; 8-flowmeter; 9-control valve; 10-insulation materials; 11-fresh water outlet; 12-wind tunnel; 13-fresh water flow path; 14-feed seawater; 15-discharge strong brine; 16-circulating pump; 17-humidification chamber; 18-dehumidification chamber; 19-fan; 20-supporter.

stage 1 and is sprayed in the chamber by the sprinkler. Another part joins the brine at the bottom in stage 4 through control valve (9) and flowmeter (8), and then the converged fluid is sprayed to the stage 4 humidification chamber and evaporates with the radiation of sunlight. The brine at the bottom of stage 2 humidification chamber 17-2 flows to the stage 3 dehumidifier (1-3) for condensation through the pipework and serially connected pump, and then it returns to the stage 2 humidification chamber 17-2 to be sprayed. Similarly, the brine at the bottom of stage 3 humidification chamber flows to the stage 4 dehumidifier (1-4) for condensation through the pipework and serially connected pump, and then it returns to the stage 3 humidification chamber 17-3 to be sprayed. It should be noticed that in this facility, the cases of humidification chambers are transparent. Sunlight can occur into them through transparent window (7) to heat seawater or brine. Thus this facility can concentrate the high-intensity sunlight to heat seawater for desalination. It is worthy of further exploration.

9.3 Floating Solar Desalination Film

From the age of the sail, people have invented various types of small floating seawater desalination facilities, which are used for self-rescue. For different reasons, many ships sank all over the world every year. During the period when film technology was still developing, using solar energy directly to desalinate seawater was the only way for self-rescue.

Researchers in different countries designed many different seawater desalination facilities. The most common one is utilizing a disc distiller floating on the sea to obtain fresh water. Fig. 9.14 demonstrates transparent equipment like jellyfish tying on a lifeboat. It is considered to be used for collecting fresh water by solar energy. It is simple in principle,

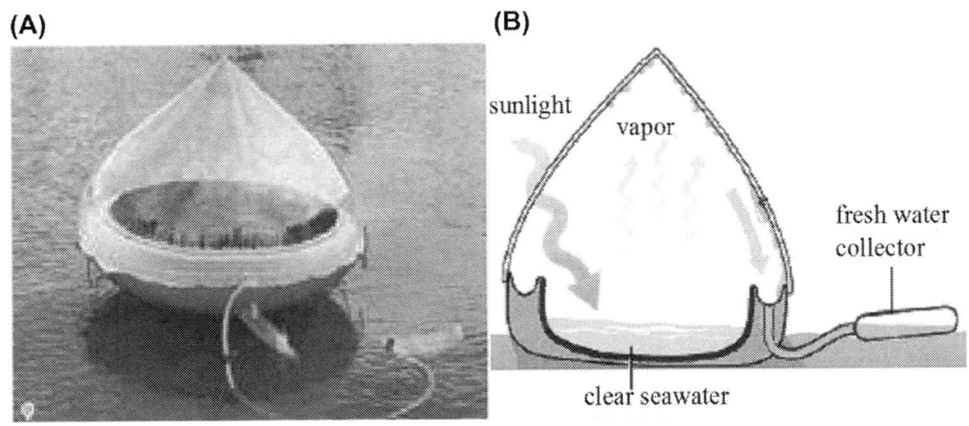

Figure 9.14
A common emergency fresh water collector on the sea (A) (B).

Figure 9.15
Watercone fresh water collector.

an airbag is arranged at the bottom and floating in the sea. The hot seawater caused by sunlight coming through thin film. As long as the cover has a suitable inclination like cone, condensed water will flow down the cover by gravity, gather in the water trough, and flow though the weep hole out of the still, then fresh water product can be obtained. Another fresh water collector called the watercone is illustrated in Fig. 9.15, which utilizes sunlight to separate fresh water from salt water. It has a simple structure and works as long as sunlight exists. Around 1.6 L of drinking water can be obtained by collecting water vapor under high temperatures after injecting seawater.

There are still many similar small solar desalination facilities applied in sea survival. Each one has special advantages, but the efficiency of using energy is low, which is not suitable to compose a large-scale desalination facility on the sea.

The main problem faced by researchers is to utilize solar energy to produce fresh water on a large scale without occupying land area, especially to produce fresh water on the sea. On the other hand, many coastal countries have large intertidal zones, which cannot be cultivated due to the lack of fresh water. Hence developing agriculture in intertidal zones or in the sea is expected to develop directions for many countries. For example, the area of costal intertidal zones in China is more than 2,000,000 ha, and the area of estuarine intertidal zones increases 20,000−30,000 ha per year, which provides excellent conditions and great potential for developing seawater agriculture. Fields and gardens on the sea are the expectation from people for seawater agriculture.

Actually, researchers did many trials of seawater agriculture. The floating farms on ships on the sea are a typical case. The ship is designed for planting vegetables and crops, feeding aquatic products, and providing sufficient green food for the ships and people living in the sea city. To provide the required fresh water for planting and feeding, an advanced fresh water generation system is required on the ship. In addition, this system should be driven by renewable energy such as wind energy or solar energy instead of traditional energy. The seawater after desalination should meet the standard of drinking water to be used for production and drinking.

It can be seen that fresh water is the key problem in seawater agriculture. If a traditional seawater desalination system is utilized, large amounts of traditional energy are required. However, if a solar desalination system is installed, it will contend for solar energy with crops. Using a traditional seawater desalination system is also limited with the consideration of the high cost and low efficiency of a traditional solar desalination system.

To solve this problem, the concept and design of floating solar desalination films were proposed. This kind of film concentrates sunlight by the tiny holes on the film and evaporates seawater directly by solar radiation. The greatest advantage of it is to incorporate the solar collector and seawater desalination facility together. It has a simple structure and is easy to maintain. In addition, it can be made of transparent nonmetallic materials for which the cost is low.

The working principles of floating solar desalination film can be explained as follows:

First, it is a film floating on the sea. The thickness of it can be designed based on requirements. There are tiny structures that concentrate sunlight. The concentrated sunlight heats the seawater directly and makes it evaporate. This film can be stitched on a large scale to achieve large-scale seawater desalination, as shown in Fig. 9.16.

The processes of sunlight concentration inside the solar desalination film vary. They can be refraction or reflection. The two typical concentrated light-driven evaporation processes are presented in Fig. 9.17. The concentration and desalination processes in Fig. 9.17A: film contains tiny concentration lens (1), middle partition (3), tiny concentration lens (4), supporting cylindrical chamber (2), and some other components. The top of (2) seals the upper part (10) of the film, and the bottom of (2) seals the lower part (5) of the film. The seawater inlet pipe (6) is positioned in the lower part (5) of the film. The seawater inlet pipe (6) is filled with a bundle of capillary tubes. The function of the bundle of

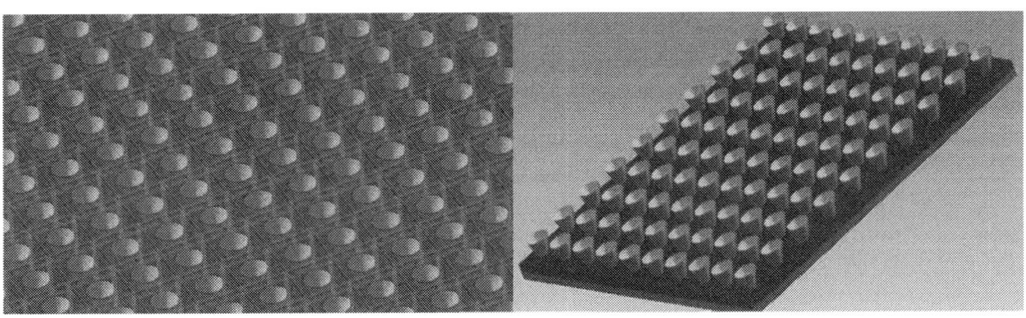

Figure 9.16
Solar desalination film and its internal structure.

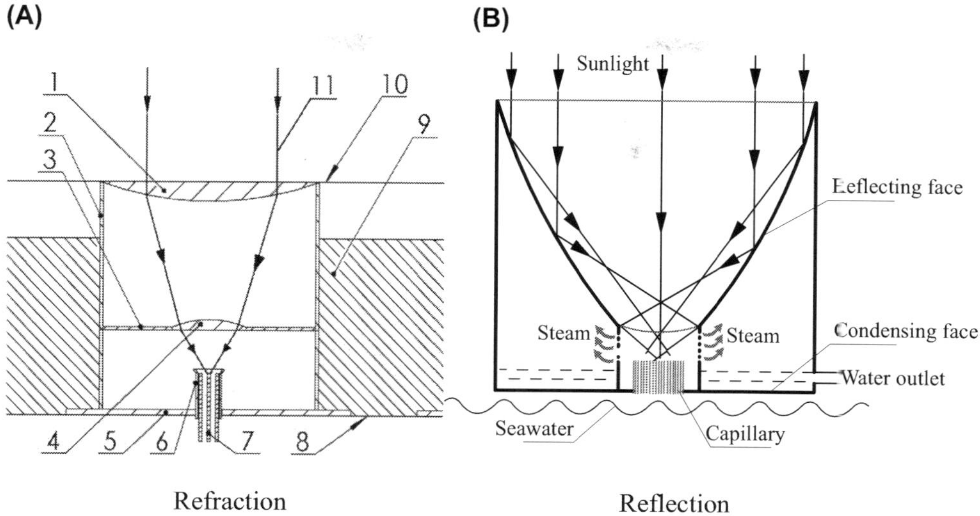

Figure 9.17

The process of concentration and evaporation in microstructure: (A) refraction and (B) reflection. 1-tiny concentration lens; 2-supporting cylindrical chamber: 3-middle partition; 4-tiny concentration lens; 5-lower part of film; 6-seawater inlet pipe; 7-evaporating tubes; 8-lower part of film; 9-porous materials; 10-upper part of film; 11-incident light.

evaporating tubes (7) is to absorb seawater up to the top of (6) to evaporate and to prevent a large amount of seawater flowing in. When seawater is heated to evaporate at the top of (6), the concentration and density of it increases. The strong brine then flows back to the sea due to gravity, which can prevent salt crystals precipitating at the top of (6). There are water storage spaces surrounding (2). For preventing the shaking of generated fresh water, hydrophilic porous materials (9) can be applied in those spaces. It should be noticed that because the chamber has a certain pressure, seawater cannot flow into the chamber when the film vibrates or waves on the sea; in other words, it has an automatic shake-prevention function, as water cannot flow into an overturned bottle on the sea.

During the operation of the film, it is located on the sea with sunshine. Sunlight is concentrated onto the tiny concentration lens (4) through tiny concentration lens (1) and then it is concentrated onto the top of pipe bundle (7) by (4). Evaporation pipe bundle (7) extracts seawater to the top, and then the seawater is heated by the concentrated light spot from concentrated lens (4) to generate water vapor. Vapor condenses in the surroundings, especially condensing on the inner surface of the bottom of film (5), which contacts seawater directly due to the low temperature there. The convex surface of tiny concentration lens (4) is upward-facing to avoid vapor returning evaporation pipe bundle (7) after condensation on convex surface. Then condensed fresh water is absorbed by capillary fibers and transported to hydrophilic porous materials (9) for storage. When the amount of fresh water reaches a certain level, users can collect fresh water from (9) directly.

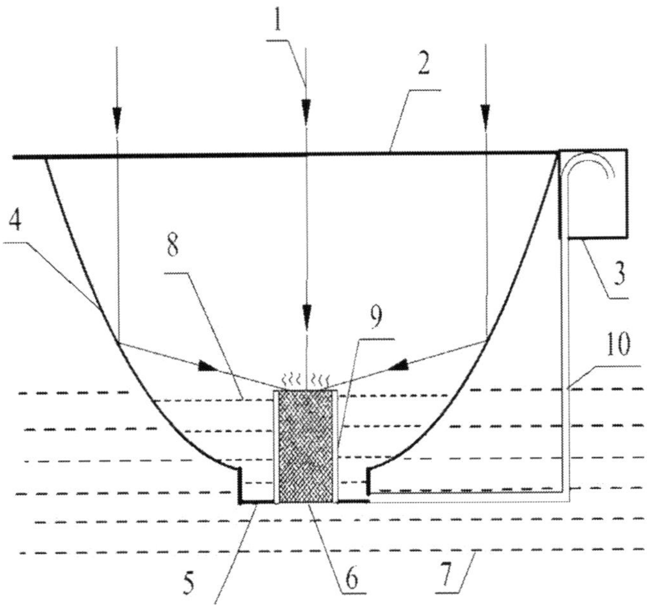

Figure 9.18
Light-driven evaporation pulsed self-collecting microsolar desalination unit. 1-incident light; 2-transparent cover; 3-collecting sink; 4-concentration-condensation surface; 5-water collecting sink; 6-hydrophilic porous materials; 7-seawater; 8-fresh water; 9-seawater inlet tube; 10-discharge tube.

Fig. 9.17B illustrates a reflective concentration process, which has similar desalination principles. The sunlight comes into the facility from the top surface and is concentrated to the light outlet at the bottom of the concentrator after reflections on curved reflective surfaces. Finally, the concentrated light is received by the top of evaporation tubes to evaporate a large amount of water vapor. A transparent cover is added at the light outlet of the concentrator to prevent vapor going into concentration spaces. The generated vapor flows to the condensation chamber from air holes. Then vapor condenses in the condensation chamber, especially on the base of the chamber because the base contacts seawater and has the lowest temperature. Finally, fresh water is exported to users by a fresh water outlet.

Seawater desalination film has many different structures. Some of them are fancy. Fig. 9.18 presents a concentrated light-driven evaporation pulsed self-collecting solar desalination unit. The working principles are as follows: incident light (1) goes through a transparent cover (2) and is reflected by the inner surface of a concentration-condensation surface (4). It is concentrated on the top of a transparent seawater inlet tube (9). Because the (9) is filled with hydrophilic porous materials (6), seawater (7) is extracted to the top of seawater inlet tube (9) and heated by the concentrated sunlight to evaporate. Vapor condensates to fresh water at the bottom of the inner surface of concentration-condensation surface (4)

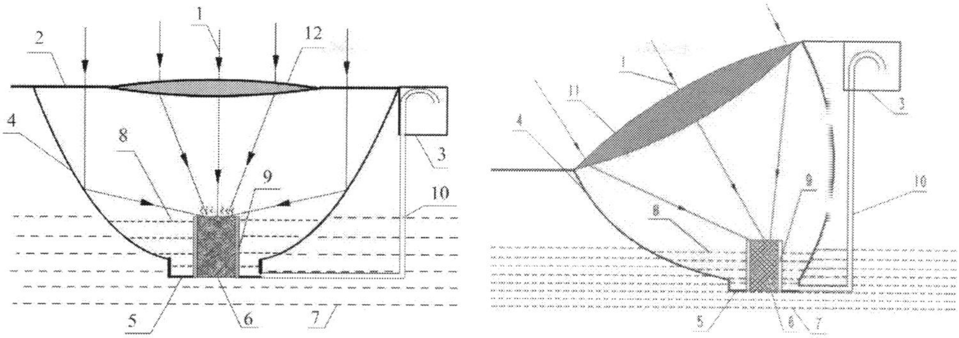

Figure 9.19
Seawater desalination unit with various structure shapes. 1-incident light; 2-transparent cover; 3-collecting sink; 4-concentration-condensation surface; 5-water collecting sink; 6-hydrophilic porous materials; 7-seawater; 9-seawater inlet tube; 10-discharge tube; 11,12-condensing lens

whose temperature is relatively low. The part (5) is the water-collecting sink. As the lower temperature part of the concentration-condensation chamber base is covered with fresh water, the heat-transfer areas between the vapor and condensation wall reduce, which leads to the condensation rate of vapor decreasing. The gas temperature and pressure in the chamber increase. When the gas pressure reaches a certain level, the pressure in the chamber can transport the condensed water in the chamber to the fresh water-collecting sink (3) through pulsed water discharge tube (10), and then the fresh water is transported to users through collecting sink (3). When most fresh water is pumped out of the evaporation-condensation chamber, the liquid level decreases, condensation area increases, pressures in the chamber reduce, vapor condensation rate rises, and fresh water in the chamber increases again. Finally, it becomes a "condensation-pumped out-condensation" automatically pulsed water discharge cycle.

Similarly, the structure shown in Fig. 9.19 can be considered better for collecting solar energy or saving money. For the northern latitudes, the sunlight beaming from the south is more than that from the north. If the tiny concentration holes are designed in slant, they can receive more sunlight. In addition, to reduce the manufacture cost, the chamber can be designed in a long trough shape, which is easy to manufacture. The tiny concentration method can be designed as the integration of refraction and reflection. In short, there are many ways to design the tiny concentration holes, but the key to achieve its low cost and high efficiency is to enhance the heat and mass transfer inside.

Based on the working principles mentioned earlier, these kinds of solar desalination units can comprise a large-scale desalination film by various arrangements, as demonstrated in Fig. 9.20. Although the fresh water produced by solar energy is not much, there are numerous units in a film so that the amount of cumulative fresh water is considerable.

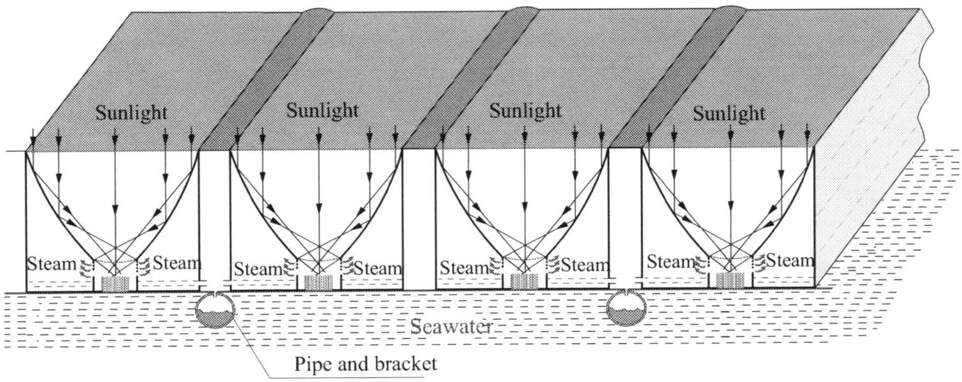

Figure 9.20
Schematic diagram of solar desalination membrane structure.

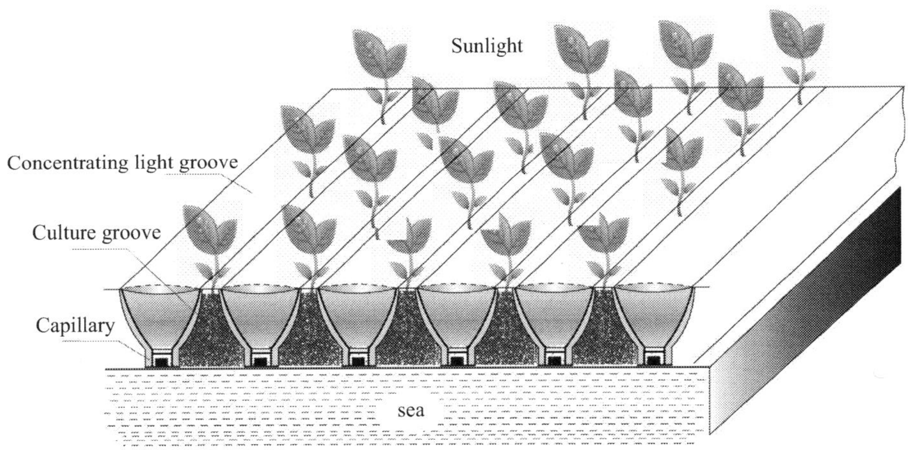

Figure 9.21
Schematic diagram of marine garden planting film.

This film can be made of transparent organic materials. Users cannot only collect fresh water to drink directly but can utilize the produced seawater for seawater agriculture, which means to plant crops or vegetables on the film directly to achieve seawater agriculture, as illustrated in Fig. 9.21. If this ideal comes true, it will be a revolution of seawater agriculture.

It is worthy to point out that although this solar desalination film can float on the sea, it may be struck by a surge. Hence it is suitable to position the film in the limited seawater pond with tide-protection facilities. The ideal condition is that there are some supporters and the sea level is controlled.

Here, the highest water production efficiency of this solar desalination film can be predicted. Assume the solar irradiation is 1 kW/m^2 and the efficiency of the tiny concentration system is 65%. The collected thermal energy is 2340 J/m^2. For ignoring the sensible heat required when seawater is heated, the thermal energy can produce 1 kg vapor approximately. If all vapors condensate, there will be 1 kg fresh water, which means 1 kg of fresh water can be produced by per square meter film per hour. If the film works 8 h per day, 8 kg of fresh water can be produced by per square meter film per day, which is a considerable number. In addition, if part of the thermal energy from vapor condensation can be reclaimed, it will have a better performance. The manufacture cost is around \$40/m^2, which implies its feasibility in economy. If it is applied in seawater agriculture, part of the film will be shaded by plants, but the film can also meet the requirements of planting plants on the sea with the reasonable calculation or water storage methods (plant shade area is less at the beginning).

9.4 The Enhanced Methods of Concentrating Direct Heating Seawater for Evaporation

The purpose of proposing the concentrated light-driven seawater evaporation process is to simplify the complicated solar desalination system and utilize nonmetallic materials for structure to reduce the cost of the solar desalination system and improve its efficiency. However, seawater is transparent and does not react with solar radiation directly. The transmission and absorption of strong light in seawater must be considered to solve the problem of utilizing concentrated light to heat seawater. To enhance the strong light to be absorbed by seawater, there must be other absorbers in seawater, which may be the black particles, black surfaces, or dyed seawater. They can convert luminous energy to thermal energy. The current methods of enhancing the light absorption of seawater are summarized in Fig. 9.22, which include adding a solid surface (like columns or panels) with the light absorption ability in seawater, adding black solid particles with similar density with seawater, and adding pigment in seawater, etc. They will be discussed in the following section.

9.4.1 General Rules of Sunlight Transmission in Seawater

When sunlight irradiates onto the sea, it will lead to reflection and transmission. Reflection is easy to estimate because it is a one-off event. But the transmission into seawater attenuates gradually. The solar irradiance is different at different depths in the sea.

The sensible heat loss created by reflection is determined by the incident angle of sunlight. On a flat water surface, the reflection loss factor ρ_b of direct solar radiation on the seawater surface can be expressed as follows:

$$\rho_b = 0.5\left[\frac{\sin^2(i-r)}{\sin^2(i+r)} + \frac{\mathrm{tg}^2(i-r)}{\mathrm{tg}^2(i+r)}\right] \tag{9.1}$$

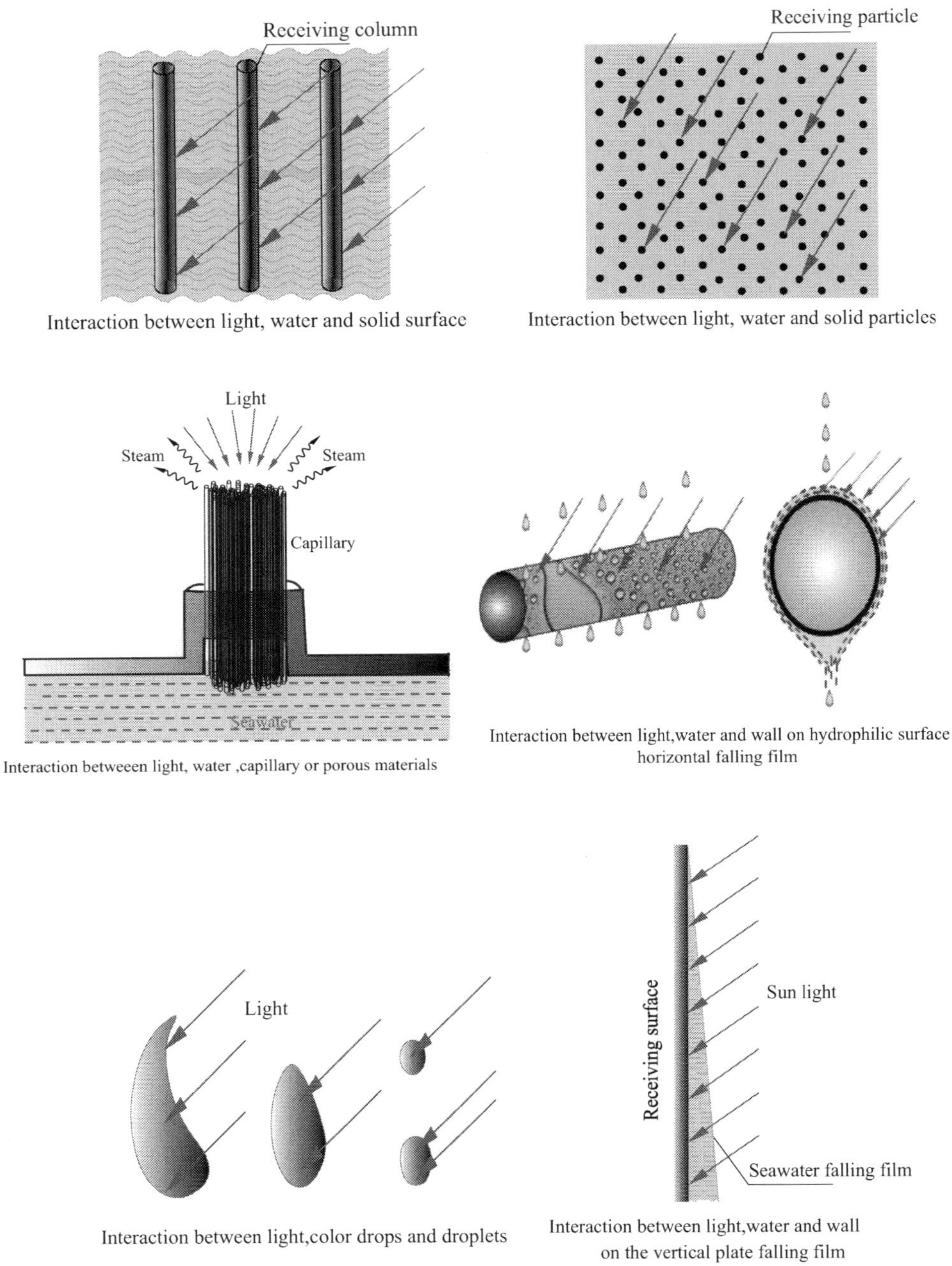

Figure 9.22
Interaction between light beam, water, and solid surface.

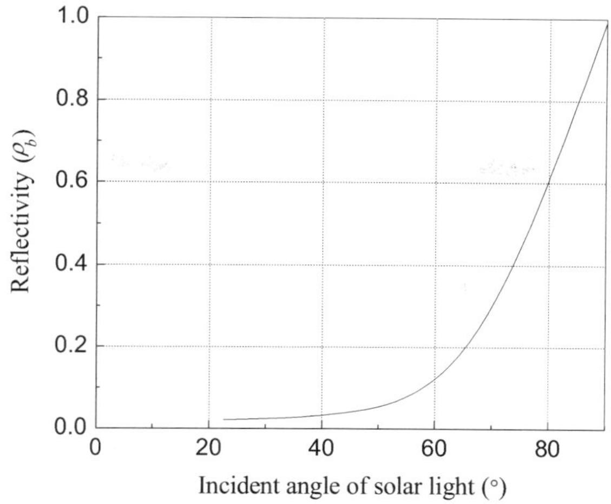

Figure 9.23
Variation of reflectivity with incident angle of solar light.

where i, r are the angles of incidence and refraction and the refractive index of seawater is $\sin i/\sin r = 1.34$.

According to Eq. (9.1), ρ_b will not be too big if the incident angle of sunlight is not big (e.g., when $i = 60°$, $\rho_b = 0.06$). However, if the incident angle increases further, the value of ρ_b will rise rapidly. The effects of incident angle on reflection loss are shown as Fig. 9.23 approximately, which can be used as a reference (it is drawn based on Eq. (9.1)).

For the diffused sunlight, Weinberger [15] provided the average reflection loss factor is about 0.07, which is obtained with the assumption that the diffused solar radiations from the sky dome in different directions are uniform.

According to calculation results, the reflection losses on the water surface are not large. However, these values are only the approximate basis. The actual conditions are more complicated. For example, the calculations are based on the flat water surface. When there are waves on the water surface, the reflection losses will increase significantly. It is still difficult to quantitatively estimate the effects on reflection losses of the solar pond surface by waves.

When sunlight comes into seawater, the absorption of light is affected by various factors. It is related to not only the wavelength of light, but also the transparency of water. According to the data offered by Defant [16], the solar radiation with a short wavelength can transmit to the very deep places in water, but seawater is opaque for the near-infrared radiation. It seems that these features of seawater prevent more solar radiation reaching the seawater, but they also protect the rapid losses of collected thermal energy under the sea.

Table 9.1: Relationship between received solar radiation under water and distance of transmission [16].

	Distance of transmission in water x (m)				
Wavelength (μ_m)	0	0.01	0.1	1.0	10
0.2–0.6	23.7	23.7	23.6	22.9	17.2
0.6–0.9	36	35.3	30.5	12.9	0.9
0.9–1.2	17.9	12.3	0.8	–	–
More than 1.2	22.4	1.7	–	–	–

Table 9.2: Attenuation coefficients of rays in different wavelength intervals in water.

Wavelength interval (μ_m)	n	η_n	μ_n (m)
0.2–0.6	1	0.237	0.032
0.6–0.75	2	0.193	0.45
0.75–0.9	3	0.167	3.0
0.9–1.2	4	0.179	35

As approximate calculations, Rabl and Nieslen [17] provided the equation presenting the relationship between the solar irradiation I underwater and the distance of transmission x (m):

$$I = \sum_{n=1}^{4} \eta_n \exp(-\mu_n x) \tag{9.2}$$

where η_n, μ_n are the constants corresponding to certain wavelength intervals. The values of them are shown in Table 9.2.

When sunlight comes into water, the attenuation rate of sunlight in clean seawater is provided by a simpler equation based on the literature [18]:

$$\frac{I}{I_0} = \left[0.36 - 0.08 \ln\left(\frac{x}{cosr}\right)\right] \tag{9.3}$$

where I_0 is the solar irradiation when sunlight occurs in seawater, r is the angle of refraction when sunlight occurs in the sea. When the transmission distance x is longer than 0.01 m, the results obtained by Eqs. (9.2) and (9.3) agree to the values provided in Table 9.1 [19].

Therefore the solar irradiation underwater at depth of x (m) is

$$I(x) = (1 - \overline{\rho}) \cdot I \tag{9.4}$$

where I is calculated by Eq. (9.2) or (9.3) and $\overline{\rho}$ is the weighted average of $\overline{\rho_b}$ and $\overline{\rho_d}$.

9.4.2 Adding Black Particles as Absorbers in Water

Zehui Chang [20,21] investigated the features of black particles, which are utilized as absorbers in seawater, and obtained considerable results. The diameters of particles are millimeter-sized. The selected ceramic particles are easily obstructed by a mesh sieve to prevent them flowing out with brine or flowing to the next-stage facilities with vapor. In this area, Safwat Nafey [22] and Yao Zeng [23] did similar research. They explored the problem that adding black particles increase the evaporation of surface water and proved that adding dyes or black charcoal can improve the fresh water production by 25–30% [24], which implies the feasibility of increasing fresh water production by functionalizing seawater.

After adding black solid particles in seawater, the suspended solid particles can be considered as individual absorbers. Particles convert luminous energy to thermal energy as long as they receive light. The interactions between light and particles under strong light are shown in Fig. 9.24. The strong light not only heats water but makes particles move in water when they are heated, which is called thermophoresis. Thus a characteristic of this method is that the seawater is heated uniformly rather than only the water surface is heated.

Assuming the particles are uniformly distributed in water, particles are regular sphere, and the absorption of light by water is ignored, the light-blocked area on particles to the water along the light transmission direction in a unit distance can be expressed as:

$$dS_d = S_r \times dx \times R \times \pi \times \left(\frac{d}{2}\right)^2 \tag{9.5}$$

where S_r is the area of incident light (m^2); S_d is the area that particles receive light (m^2); R is the number of particles in a unit volume of water (ind./m^3); d is the diameter of particle (m); dx is the differential element of length along the light transmission direction (m); and R can be calculated as follows,

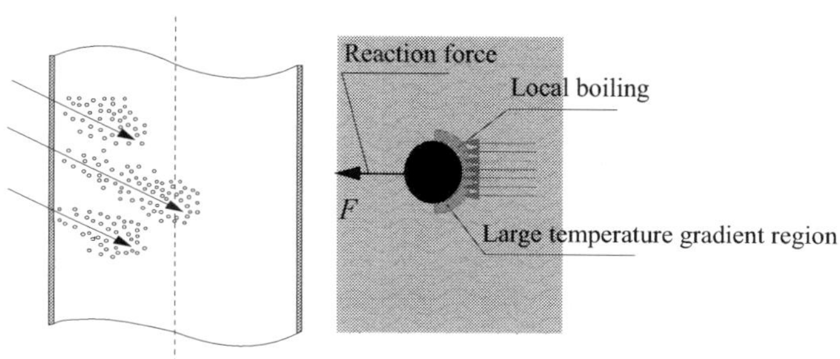

Figure 9.24
Interaction process of light particles in seawater.

$$R = \frac{m}{\rho} \times \frac{1}{V} \times \frac{1}{V_l} \tag{9.6}$$

where m is the mass of the particles added in water (kg); ρ is the density of particles (kg/m^3); V is the volume of tested water (m^3); and V_l is the volume of a single particle (m^3).

The light area after light transmitting underwater in distance of dx can be described as:

$$S_{t1} = S_r - dS_d = S_r - S_r \times dx \times R \times \pi \times \left(\frac{d}{2}\right)^2 \tag{9.7}$$

where S_{t1} is the area after light transmits in water in the first dx (m^2).

The light area after light transmitting in water in "n" propagation length is

$$S_m = S_{t(n-1)} - S_{t(n-1)} \times dx \times R \times \pi \times \left(\frac{d}{2}\right)^2 \tag{9.8}$$

Therefore the final transmittance of water can be expressed by:

$$\tau = \frac{S_m}{S_r} \times 100\% = \frac{S_{t(n-1)} - S_{t(n-1)} \times dx \times R \times \pi \times \left(\frac{d}{2}\right)^2}{S_r} \times 100\% \tag{9.9}$$

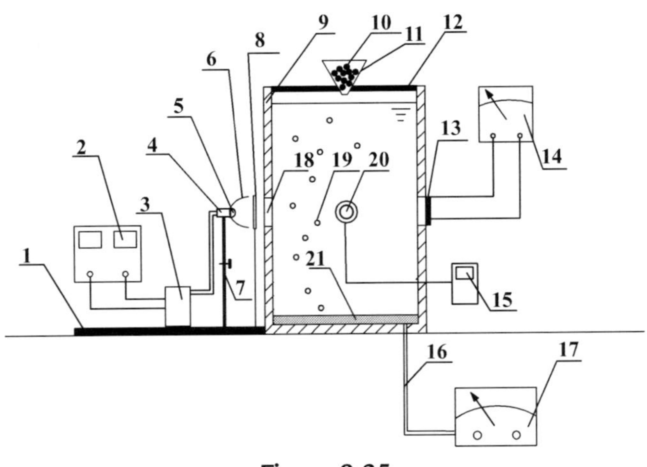

Figure 9.25
Experiment system diagram of black particles blocking light in water: 1-guideway; 2-dc stabilized power supply; 3-electronic ballast; 4-heat-dissipation ring; 5-light source; 6-parabolic reflector; 7-tripod; 8-lens; 9-experiment trough; 10-functional particles; 11-dropping funnel; 12-light protection panel; 13-cilicon photocell; 14-microcurrent detector; 15-photometer; 16-gas-guide tube; 17-air compressor; 18-light guide hole; 19-air bubble; 20-sensor of photometer; 21-microporous ceramic plate.

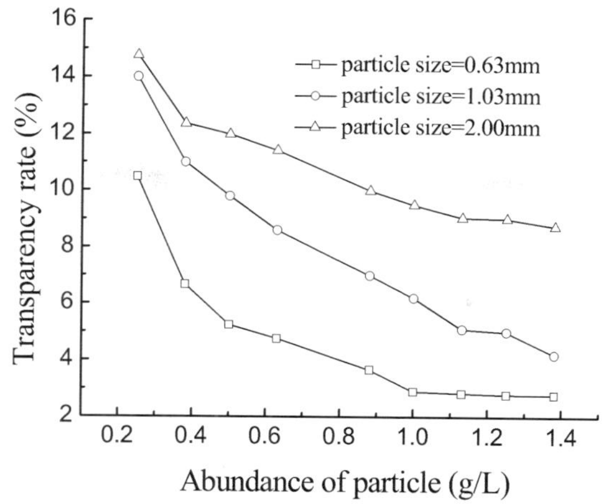

Figure 9.26
Effects on transmittance by abundance of different particles [21].

To compare with theoretical results, Zehui Chang [20,21] established an experiment system, as presented in Fig. 9.25. The capabilities to block light in seawater of black particles are tested. The relationships between the densities of black particles and the efficiencies of blocking lights are obtained, as shown in Fig. 9.26.

In the experiment, the particles are uniformly distributed in water by the disturbance from tiny bubbles. The distance of light transmission in water is 20 cm, the area of incident light is 11.2 cm^2, and the density of the particle is 890 kg/m^3 (when it is immersed in water, its density is 1000 kg/m^3approximately, which is close to the density of water).

According to Fig. 9.26, the transmittances of water with three different particles have similar variation tendencies to the change of abundance. The transmittances decrease with the increase of particle abundances. When the particle abundances are smaller, the transmittances reduce faster, and when the particle abundances are larger, they reduce slower. The transmittance reduction of the water containing functionalized particles with a diameter of 0.63 mm is the biggest with the increase of abundances, which means that this type of water has the best extinction properties.

The transmittance of lighting transmitting 10 cm in water can be calculated by substituting experiment parameters into Eq. 4.23. The comparisons between theoretical and experimental results are demonstrated in Fig. 9.27A and B. It can be seen that the transmittances of water decrease with the rising of particle abundances for both theoretical and experimental results. The two results agree with each other when the diameter of the test particle is 0.63 mm. Experimental results are larger than theoretical results only when particle abundances are bigger than 3.0 g/L. The main reason is that when particle abundances increase, particles block each other so that the effective light-blocked areas

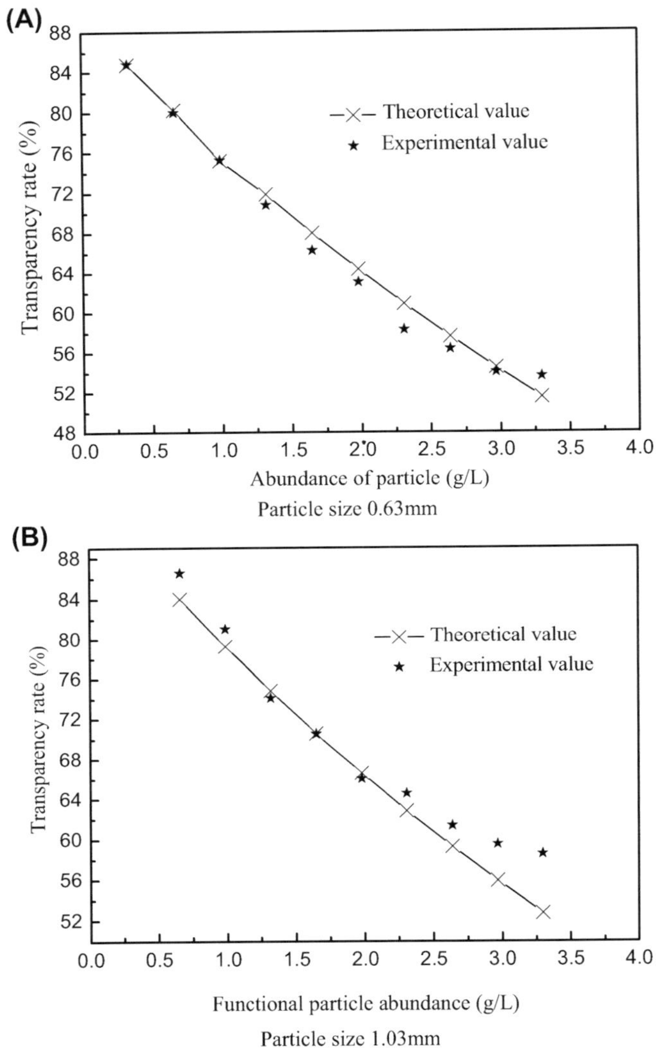

Figure 9.27

Comparison between theoretical and experimental results of transmittances in functionalized water [20]: (A) particle size 0.63 mm and (B) particle size 1.03 mm.

reduce. Thus the experimental transmittances are larger than theoretical values. This phenomenon appears more obviously in the functionalized water with the particles whose diameter is 1.03 mm.

9.4.3 Absorption Enhancement by Colored Water

Much research shows that the distance of light transmitting under water is mainly affected by the turbidity of water. The clearer the water is, the longer light transmits. Wang and

Seyed-Yagoobi [25] obtained the fitting equation of the transmittance of solar radiation with the consideration of water turbidity (unit: NTU), as presented in the following equation:

$$\tau(\varphi, x) = [0.58 - 0.076\ \ln(100x)] \cdot \left[1 - 0.1975x(\varphi - 0.3) + 0.0144x(\varphi - 0.3)^2\right] \quad (9.10)$$

where $0.3 \leq \varphi \leq 4.5$ NTU, τ is the transmittance of seawater, which is calculated when using seawater turbidity of 0.3 NTU as reference, and φ is the turbidity of seawater.

It can be seen that when increasing the turbidity of seawater artificially, the light absorption of seawater will increase. Therefore adding dyes into seawater can improve the seawater absorption. Although this method is not suitable for a large-scale seawater desalination system, especially for the flowing sea that leads to environmental pollution, it can still be considered for the desalination system with standing seawater, such as a distiller.

9.4.4 Adding the Black Columns or Plates in Water as the Receivers

For a large-scale concentrated light solar desalination system, establishing black adsorption surfaces may be the only possible way to improve the absorption efficiency of seawater to solar radiation. For example, several black columns can be set in a solar tower desalination system to increase the ability of the water to absorb light.

When light passes through water and reaches the black absorbing surface, it will be absorbed and converted to thermal energy. Then thermal energy is transferred to water by convection to increase the water temperature, as shown in Fig. 9.28.

To solve this problem at the macrolevel, the arrangement of vertical columns should be considered. On one hand, water should absorb heat uniformly; on the other hand, light should be prevented to escape out of the water. At the microlevel, three processes need to be explored, namely, the energy transformation of strong light on surface, the temperature

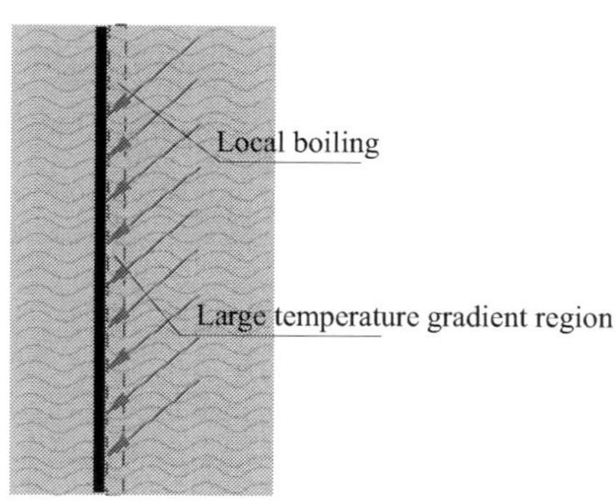

Figure 9.28
Interaction process of the light and wall in seawater.

rising even boiling processes of the seawater near wall surfaces, and the heat-transfer processes of a large temperature gradient.

Obviously, utilizing falling film evaporation is an ideal choice to reduce the thermal inertia of the seawater system, like using falling film evaporation outside round tubes or on the external surfaces of vertical panels. During falling film evaporation, because the mass of seawater is relatively low, the temperature rises rapidly. It is easy to obtain a high temperature to speed up the evaporation of seawater. However, it should be noticed that because the water in falling film evaporation is thin and has smaller thermal inertia, under this condition, strong light may cause local boiling, seawater splashing, and even dry zones, which should be avoided.

Because capillary tubes or porous materials connect to seawater, it is another possible method to select capillary tubes or porous materials as the absorber of seawater. Both of them are appropriate to be utilized in small-space light-driven evaporations, such as the solar desalination film referred to before.

In a word, the real seawater is transparent. Although the attenuation of light in water has an exponential relationship, the attenuation is slow. Even though the water absorbs 60% sunlight, the required water thickness is larger than 1 m approximately. To enhance the light absorption of seawater, it is necessary to add absorbers in seawater, which can significantly improve the absorption on sunlight of seawater.

References

[1] Von Medeazza GLM. "Direct" and socially-induced environmental impacts of desalination. Desalination 2005;185(5):57.
[2] Spiegler KS, Laird ADK. Principles of desalination. 2nd ed. New York: Academic Press; 1980.
[3] Dreizin Y, Tenne A, Hoffman D. Integrating large scale seawater desalination plants within Israel's water supply system. Desalination 2008;220(3):132.
[4] Kalogirou S. Survey of solar desalination systems and system selection. Energy 1997;22(1):69–81.
[5] Ansari O, Asbik M, Bah A, et al. Desalination of the brackish water using a passive solar still with a heat energy storage system. Desalination 2013;324:10–20.
[6] Ghaffour N, Reddy VK, Abu-Arabi M. Technology development and application of solar energy in desalination: MEDRC contribution. Renew Sustain Energy Rev 2011;15(9):4410–5.
[7] Baig H, Antar MA, Zubair SM. Performance evaluation of a once-through multi-stage flash distillation system: impact of brine heater fouling. Energy Convers Manag 2011;52:1414–25.
[8] Ahsan A, Imteaz M, Rahman A, et al. Design, fabrication and performance analysis of an improved solar still. Desalination 2012;292:105–12.
[9] Pearce JM, Denkenberger DC. Numerical simulation of the direct application of compound parabolic concentrators to a single effect basin solar still. Proceedings of the solar cookers International Conference in Granada, Spain July 2006;217:12–6.
[10] Chaouchi B, Zrelli A, Gabsi S. Desalination of brackish water by means of a parabolic solar concentrator. Desalination 2007;217:118–26.
[11] Singh SK, Bhatnagar VP, Tiwari GN. Design parameters for concentrator assisted solar distillation system. Energy Convers Manag 1996;37(2):247–52.

[12] Omara ZM, Eltawil MA. Hybrid of solar dish concentrator, new boiler and simple solar collector for brackish water desalination. Desalination 2013;326:62–8.
[13] Hongfei Z, Xinglong MA. Performance test of a solar concentrating system with cylindrical Fresnel lens. In: 14th International Conference on sustainable energy technologies – SET; 2015.
[14] Abdelkader M. Investigation of multi-effect humidification-dehumidification solar desalination system coupled with solar central receiver. In: The 2nd International Conference on water resources & arid environment; 2006.
[15] Weinberger H. Sol Energy 1964;8:45. "Development of the Gel Pond Technology".
[16] Defant A. Physical oceanography, vol. 1. New York: Pregamon Press; 1961. p. 53.
[17] Rabl A, Nieslen CE. Sol Energy 1973;14:433. "The Use of Waste Heat in a Sloar Still".
[18] Bryant HC, Colbeck I. A solar pond for London. Sol Energy 1977;19(3):321–2.
[19] Wang YF, Akbarzadeh A. A study on the transient behaviour of solar ponds. Energy December 1982;7(12):1005–17.
[20] Chang Z, Zheng H, Hou J, Jia Y, He F. Adding black particles in brackish water reduces water transmission properties of solar desalination system. In: Transaction of the Chinese Society of Agricultural Engineering, vol. 24; 2013. p. 204–10.
[21] Chang Z-H, Zheng H-F, Hou J, HE F-B. Extinction performance research on the functioned water in the concentrating light and direct-heating solar desalination system. J Eng Thermophys 2014;05:839–43.
[22] Safwat Nafey A, Abdelkader M, Abdelmotalip A, Mabrouk AA. Enhancement of solar still productivity using floating perforated black plate. Energy Convers Manag 2002;43:937–46.
[23] Zeng Y, Yao J, Horri BA, Wang K, Wu Y, Li D, et al. Solar evaporation enhancement using floating light-absorbing magnetic particles. Energy Environ Sci 2011;4:4074–8.
[24] Rajvanshi AK. Effects of various dyes on solar distillation. Sol Energy 1981;27:51–65.
[25] Wang J, Seyed-Yagoobi J. Effects of water turbidity and salt concentration levels on penetration of solar radiation under water. Sol Energy 1994;52:429–38.

CHAPTER 10

The Benefit Evaluation and Material Selecting

Chapter Outline

10.1 The Economic Feasibility of Solar Desalination System

Although the solar desalination technology is still at an early stage, its invaluable potential has stimulated the research by many scholars. Multiple commercialized or semicommercialized solar desalination systems have been set up around the world to supply a large amount of potable water, providing valuable experience for the further development of solar desalination technology. Statistical data in Fig. 10.1 give the proportion of different solar desalination systems in the whole industry, which show the photovoltaic-reverse osmosis (PV-RO) system has the largest share, accounting for 40% of the total volume. People prefer PV-RO because of its stable system and convenient maintenance, regardless of the high cost of the technology. The next are the solar membrane distillation (MD), multieffect desalination (MED), and multistage flash (MSF) with a total share of 36%, which becomes a nonnegligible mainstream and important direction in the future.

Solar Energy Desalination Technology. http://dx.doi.org/10.1016/B978-0-12-805411-6.00010-5

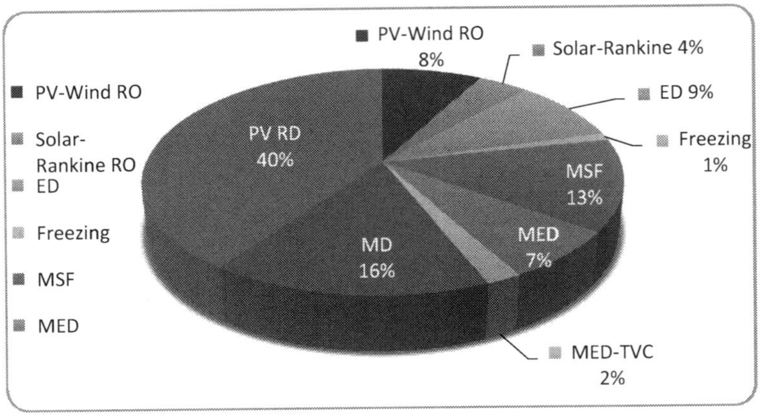

Figure 10.1
Share of various solar desalination [1].

The economic feasibility is very important for the development of solar desalination technology. At present, most of the solar desalination technology still cannot compare to the conventional one. The main problems include high initial investment, overcomplicated technology, and poor stability, etc. It can be said that the economic feasibility is the bottleneck restricting the development of solar desalination technology. A technically feasible device cannot be developed without economic feasibility. With the improvement of technology and the use of advanced materials, solar desalination costs have been greatly reduced, and some can almost be paralleled with that of the conventional desalination device, especially in the Middle East Gulf countries and regions with abundant sunlight and long coastlines but with a lack of drinking water.

The study by Jimmy Leblanc et al. shows that the combination of MED and the solar collector is one of the most economical desalination methods. In his paper Leblanc found by calculation that the fresh water production cost varies significantly among desalination systems with different scales: the larger the scale, the lower the cost. When the system produces 1000 m^3 fresh water per day, the production cost will be around \$1.2/m^3; when the production scale is close to 10,000 m^3/day, the cost of producing 1000 m^3 fresh water will be decreased to \$0.92; and if the system is further extended to the scale that the drinking water for the whole Saudi Arabian country can be produced, the production cost will be lowered to \$0.69. At the same time, he also calculated the water production cost of the PV-RO technology, which differs widely in various treatment concentration and scales of installed capacity, but all the fresh water costs per m^3 are above \$10 [2].

Al-Hallaj and Selman JR et al. have been studying the fresh water cost limit of the solar desalination technology. In their book, they figure out the cost limits of fresh water obtained from the solar desalination method, which are shown in Table 10.1 [3].

Table 10.1: The cost of the solar desalination method and the comparison with the cost of conventional reverse osmosis [3].

System type	Conventional reverse osmosis		Solar-MED		Solar-thermal membrane binding assay	
Water production (m^3/day)	20,000	200,000	20,000	200,000	20,000	20,000
Initial investment (mil·$)	20.0	160.0	48.0	380.0	32.0	250.0
Water production costs ($/(d·$m^3$))	1000	800	2400	1900	1600	1250
Unit cost of water production ($/$m^3$)	0.77	0.66	0.89	0.71	0.79	0.65

As shown in Table 10.1, under technically feasible conditions, there is no big difference between the water production costs by solar desalination and conventional Reverse Osmosis (RO) method, if the desalination system is large enough. Therefore solar desalination is a very promising technology.

Suri RK and Al-Marafie AMR et al. have done the calculation of the costs needed to produce 1 m^3 fresh water by different solar desalination methods, including three aspects, i.e., water production by using traditional energy, the combination of solar and traditional energy, and purely by solar energy. The result suggests that the utilization ratio is 75% by solar energy and 90% by traditional energy, which is shown in Table 10.2.

RO water production driven by traditional energy has an obvious price advantage over solar energy for now, as indicated in Table 10.2. When using solar energy for desalination,

Table 10.2: The water production cost comparison between using traditional energy and solar energy [4].

Parameter	Conventional desalination plant	Conventional desalination plant combined with solar energy	Solar desalination plant
	MSF		
Annual water (m^3)	328	274	274
Unit production cost of water ($/$m^3$)	1.75	1.79	2.84
	RO		
Annual water (m^3)	328	274	274
Unit production cost of water ($/$m^3$)	1.30	5.70	12.05

cost of the RO technology is several times higher than that of MSF. So the PV-RO method shows diseconomy by solar desalination.

Based on experiments, Riffel and Carvalho [5] and Gandhidasan and Al-Mojel [6] believe that though the PV-RO technology has a good developing prospect, it is only applicable to some small-scale desalination plants due to the limitation of cost. By comparison, the Thermal Desalination-Solar Energy (TD-SE) technology possesses more spreading value because it can be directly combined with the conventional thermal desalination method.

In the paper, A. Lamei et al. suggested the calculation formula of desalination cost under the condition of 250–50,000 m^3/day water production:

$$C_p = 1.6 - 3.42 \times 10^{-5} Q + 9.76p + 0.14f_e \tag{10.1}$$

where C_p, is the cost of the product water ($\$/m^3$); Q, is the water yield of the desalination system (m^3/day); p, is the price of energy ($/kW h); and f_e, is (kW h/m^3).

Since the solar desalination technology is still at its starting stage, cost of the product water is higher than other desalination technologies using traditional energy. But with the gradual awareness of the clean and sustainable solar desalination method and the increasingly mature technology, cost of the product water by this method will be getting close to or even lower than that of the current conventional desalination methods.

From another viewpoint, solar desalination technology has not only economic returns, but also social benefits such as saving the fossil energy resources and reducing the emission of carbon dioxide to protect the environment.

10.2 The Economic Evaluation Method of Solar Desalination System

Solar energy is a plentiful renewable resource. The sheer volume of the solar radiation that reaches the Earth's surface is sufficient enough to meet the foreseeable demand for energy from human beings. However, economic feasibility, as well as technical feasibility, is important to achieve the large-scale development and utilization of solar energy. Similar to other energies such as petrol, natural gas, coal, and nuclear power, the development and utilization cannot be done without investment, which includes two main types: one is initial investment, and the other is operation and maintenance cost. Generally speaking, the operation cost of the solar device is relatively low, while the initial investment is obviously high. The reason for the low operation cost is that the solar system requires less fuel and electricity cost, and the high initial investment is caused by the discreteness and intermittency of solar energy. The energy flow density of solar energy is relatively low, and the radiation reaching land surface is less than 1 kW/m^2. Solar devices need a large lighting area and connecting pipes, while conventional energy may only require less equipment to meet the same demand. In order to provide energy under the condition without or lack of sunlight, it is necessary to install energy storage equipment or an energy supply system in some cases. In addition, a water pump, fan, heat exchanger, and control equipment are always required in the solar system, and the purchase expense, freight, and installation cost of all this equipment are regarded as an initial investment. The operation costs also include the labor and material fee for operation and maintenance, besides the fuel and electricity cost to run the equipment complex.

There is no unified theory about how to evaluate the economic performance of the solar desalination system by now. The first reason is that the solar desalination system is generally very complex, especially for the device combined with the existing conventional system, which is usually composed of many parts; some researchers even hold the view that the solar desalination system is investment-intensive. Secondly, with few researchers and no convincing and broad evaluation theories, the evaluation method varies with different researchers. Some pay more attention to economic performance, while others consider social benefits at the same time. Therefore no unified economic evaluation theory has been established so far.

Purely from an economic perspective, the solar desalination system can be treated as an investment-return system, with the construction being the investment and the fresh water being the product. Then it is comparatively easy to economically evaluate such a system regardless of other problems. However, a prediction model of the system's production capacity has to be built in the first place, which is closely related to the local climate conditions, manpower, and material costs. So the economic performance of the solar desalination system is not only about the system structure (which decides the initial investment), but also related to the point of use and user, which is crucial to the yield of product water. So the same system would have different economic returns for different locations and users. To sum up, the economic evaluation of the solar desalination system should include the factors of location, technology, economic parameters, solar energy drive (i.e., partial or complete), and also the scale of the system.

10.2.1 Timeliness of Funds

There are many ways to evaluate the economic effects of the solar desalination system, but the differences between investment and operation costs are not defined in most of the existing evaluation methods. In fact, the value of capital changes with time. The interest payment needs to be done year-by-year if the money is from a bank loan. And if interest can be obtained by bank deposit, it can also be gained from other investments. As seen from the economic point of view, if money is to be invested in solar energy utilization devices, only when the economic returns are no less than the interests mentioned earlier, can the investment be considered reasonable. To take this factor into account, it is necessary to convert the capital investment and return at different time into the value at the same period.

The method of discounting to present value is used below [7].

¥ F is the present value, $P = F(1+i)^n$ is the equivalent value of F after n years.where i is the annual interest rate expressed as decimal.

In other words, when ¥ P is converted to present value

$$F = \frac{P}{(1+i)^n} \tag{10.2}$$

Suppose investment cost of a solar utilization device is C, and P_n is the accumulative return within the service life converted into present value, then the net present value of investment return is equal to ($P_n - C$). Obviously, this investment is favorable only if ($P_n - C$) is positive. To compare the solar energy device with the conventional energy device, take C as an increment of the capital invested in the former, then P_n will be the present value of the accumulative saving of the operation cost. Generally, residual discount value of the equipment could continue to operate and should be accounted into P_n.

If the annual saving of the operation cost (e.g., fuel cost) is ¥ Y when a solar energy device has entered service for n years, then the present discount value of the accumulative saving is:

$$\frac{Y}{1+i} + \frac{Y}{(1+i)^2} + \cdots + \frac{Y}{(1+i)^{n-1}} + \frac{Y}{(1+i)^n} \tag{10.3}$$

Therefore the total value of accumulative saving in n years will be:

$$P_n = Y\left[\frac{(1+i)^n - 1}{i(1+i)^n}\right] = Y\left[\frac{1-(1+i)^{-n}}{i}\right] \tag{10.4}$$

It should be noted that the inflation rate and fuel price increase rate are not considered in the formula above. If the fuel price goes up year-by-year, then the annual saving of operation cost will not be constant. Suppose the annual saving of operation cost is ¥ Y based on the current fuel price, and the fuel price will increase by an annual interest rate of ii (ii is still expressed in a decimal). Then the accumulative saving would be:

$$Y(1+i_i),\ Y(1+i_i)^2, \cdots,\ Y(1+i_i)^{n-1},\ Y(1+i_i)^n \tag{10.5}$$

Obviously, if the rate of inflation is taken into consideration, a similar formula can also be deduced, and then the present discount value of the accumulative saving of fuel cost would be:

$$P_n = \frac{Y(1+i_i)}{(1+i)} + \frac{Y(1+i_i)^2}{(1+i)^2} + \cdots + \frac{Y(1+i_i)^{n-1}}{(1+i)^{n-1}} + \frac{Y(1+i_i)^n}{(1+i)^n} \tag{10.6}$$

If both interest rate and inflation rate or fuel price increase rate are considered, then an effective interest rate parameter can be defined as:

$$i_e = \frac{1+i}{1+i_i} - 1 = i - i_i \tag{10.7}$$

Thus after taking all the factors into consideration, the total saving value would be:

$$P_n = Y\left[\frac{(1+i_e)^n - 1}{i_e(1+i_e)^n}\right] = Y\left[\frac{1-(1+i_e)^{-n}}{i_e}\right] \tag{10.8}$$

10.2.2 Economic Analysis of an Installed Solar Desalination System

As mentioned in the previous section, the economic performance of the solar desalination system is not only related to the system material and structure, but also the installation location and user, so economic assessment of the system is usually pretty complicated. However, for an installed system, the environment is fixed, and its water production efficiency is also decided by the structure and material, so it can be analyzed as a static system. The analysis method is similar to other industrial systems.

Total investment of the solar desalination system in the year of installation generally includes the following parts: (1) investment of the solar heat-collecting system f_1; (2) investment of the main engine (including the control system) f_2; (3) investment of the required electric assist system f_3; (4) investment of construction f_4; and (5) investment of total maintenance fee during service period f_5.

In fact, the investments mentioned earlier are connected. For example, if the investment of the solar heat-collecting system is increased, the investment on the main engine of desalination system f_2 may be reduced accordingly, so f_1 and f_2 are interrelated. Moreover, f_3 and f_1 are also interlinked. For instance, if the electric assist system is powered by wind energy, though the electricity consumption of the solar desalination system is not too much, the wind power generation system still keeps running during the winter and nighttime and provides some spare electric power to consumers, by which part of the investment can be recovered. In winter, the solar collector might not be able to reach a temperature of 70°C to generate the desalination system, but partial investment recovery can also be achieved by heat water supply (the desalination system should be closed if the temperature is too low in winter).

Therefore a well-designed solar desalination system is supposed to be a multipurpose utilization system, and environmental benefits should also be included in the system returns. In order to analyze in detail the economic performance of the desalination system, suppose C is the total investment of the current year, P is the total return of the system during n year of service period, i is the annual interest rate, D is the price of the product water, and e is the annual increment of fuel price or water price, then the total investment C would be:

$$C = f_1 + f_2 + f_3 + f_4 + f_5 \tag{10.9}$$

If $P - C > 0$, then the system has its investment value; otherwise, the investment value is zero.

Preliminary economic analysis of the case study of a solar desalination system is as follows [8].

1. Solar heat-collecting system
   Assume investment of the solar heat-collecting system at the beginning of the construction period is f_1. Besides the return of the heat supply to the desalination system P_{11}, part of the other returns P_{12} can be obtained by providing overplus energy, and the annual maintenance cost is f_{51}. In the jth year of operation, convert the economic index (i.e., investment or return) into the present year of investment, then the total investment of the solar collector during the service period is:

$$F_1 = f_1 + \sum_{j=1}^{n} \frac{f_{51}}{(1+i)^j} \tag{10.10}$$

   Total return is:

$$P_1 = \sum_{j=1}^{n} (P_{11} + P_{12})(1+e)^j \tag{10.11}$$

2. Desalination main engine system
   Suppose the initial investment in the construction period is f_2, annual maintenance cost is f_{52}, annual water production return is P_{21}, and return of other by-products is P_{22}, then the total investment and return during the service period are:

$$F_2 = f_2 + \sum_{j=1}^{n} \frac{f_{52}}{(1+i)^j} \tag{10.12}$$

$$P_2 = \sum_{j=1}^{n} (P_{21} + P_{22})(1+e)^j \tag{10.13}$$

3. Electric assist system
   Suppose investment on the electric assist system in the present year of construction is f_3, and annual maintenance cost is f_{53}. The system can supply electricity to the consumer directly as well as to the desalination system, so its returns can be divided into two parts. P_{31} represents the return of the desalination system and P_{32} represents the return of the electricity supplier to the consumer. Thus the total investment of the system during the service period is:

$$F_3 = f_3 + \sum_{j=1}^{n} \frac{f_{53}}{(1+i)^j} \tag{10.14}$$

   Total return is:

$$P_3 = \sum_{j=1}^{n} (P_{31} + P_{32})(1+e)^j \tag{10.15}$$

4. Construction investment
   Assume investment in the current year of construction is f_4, and annual maintenance cost is f_{54}, so total investment during the service period is:

$$F_4 = f_4 + \sum_{j=1}^{n} \frac{f_{54}}{(1+i)^j} \tag{10.16}$$

Suppose the final residual value is equal to the disposal cost, i.e., the value and disposal cost can be neglected, then total investment F of the whole system during the life cycle is:

$$F = F_1 + F_2 + F_3 + F_4 = f_1 + f_2 + f_3 + f_4 + \sum_{j=1}^{n} \frac{f_{51} + f_{52} + f_{53} + f_{54}}{(1+i)^j} \tag{10.17}$$

And total lifetime return of the system would be:

$$P = P_1 + P_2 + P_3 = \sum_{j=1}^{n} (P_{11} + P_{12} + P_{21} + P_{22} + P_{31} + P_{32})(1+e)^j \tag{10.18}$$

Let A denote the area of the solar heat-collecting system and p denotes the investment price of the heat collector per unit area, and then the total investment of the heat-collecting system would be:

$$f_1 = A \cdot p \tag{10.19}$$

Furthermore, let η_s be the annual average efficiency of the collector, when the total volume of the solar radiation per square meter is H and total areas of the solar collector is A, then total binding energy of the solar heat-collecting system would be:

$$E = \eta_s \cdot H \cdot A \tag{10.20}$$

If the coal price of the present year is r_1, then the current return of the solar heat-collecting system would be:

$$C_{10} = E \cdot r_1 = \eta_s \cdot H \cdot A \cdot r_1 \tag{10.21}$$

Then let β_1 be the proportionality constant of the system's heat supply to the consumer, then

$$C_{12} = C_{10} \cdot \beta_1 \tag{10.22}$$

This is the return of the system's heat supply to the consumer.

By using the same analytical method, return of the electric assist system supplying electricity to the consumer can be obtained:

$$C_{32} = \eta_p \cdot H \cdot r_3 \cdot \beta_3 \cdot A_e \tag{10.23}$$

where η_p is the generating efficiency of the electric system, A_e is the lighting area of the solar panel, H is the solar radiation value of the current year (assuming that photovoltaic cells are used in the electric assist system), r_3 is the electricity price, and β_3 is the ratio of the power system directly supplying heat to consumers. If the wind power system is taken as auxiliary electric, then

$$C_{32} = H' \cdot r_3 \cdot \beta_3 \tag{10.24}$$

where H' is the total power generation amount of the present year.

Main sources of the return of by-product from the main engine system include the salt generated from desalinated seawater or other chemical materials. Suppose r_2 is the price of the chemical materials, and m_c is the total by-product volume, thus:

$$C_{22} = m_c \cdot r_2 \tag{10.25}$$

The main product of the solar desalination system is pure water, and the pure water value is the primary benefit of the desalination system. Let r_4 be the pure water price of the current year: if annual water production is me, then total return of the fresh water is:

$$C_{21} = m_e \cdot r_4 \tag{10.26}$$

All the analysis leads to the total return of the system during its service period, and it can be converted into the capital quantity of the present construction year:

$$C = \sum_{j=1}^{n} (A\eta_5 H r_1 \beta_1 + m_e \cdot r_4 + m_c \cdot r_2 + \eta_p A_e H \cdot r_3 \cdot \beta_3)(1+e)^j \tag{10.27}$$

If the wind power generation system is used:

$$C = \sum_{j=1}^{n} (\eta_s \cdot H \cdot A \cdot r_1 \beta_1 + m_e r_4 + m_c r_2 + H' r_3 \beta_3)(1+e)^j \tag{10.28}$$

10.2.3 Case Calculation of Two Actual Systems

Case 1: An integrated system with a solar collector and desalination main engine

Take an actual solar desalination system as example to calculate its investment and return, and investigate the water production cost of the system. Parameter values in the calculation are as follows [8]:

$$\eta_s = 45\%;\ H \approx 5400\ \text{MJ/year} \cdot \text{m}^2;\ A = 80\ \text{m}^2;\ r_1 = 0.0717\ \text{RMB/MJ}(1\ \text{ton of coal} = \text{RMB}\ 1500);$$

$\beta_1 = 0.2$, $m_e = 300$ ton/year, r_4 is the desired value, $m_c = 0$, i.e., by-products are not considered in the system.

Auxiliary power of the desalination system is provided by wind generator, based on the local wind situation, actual total generated power per year $H' = 1861.5$ kW h, and electricity price $r_3 = 1$ RMB/kW h,$\beta_s = 0.35$. Assuming that the bank interest rate is 5% (low loan interest for new energy), escalating coefficient of energy price $e = 5\%$, and service period of the system is 15 years, then total return within the system life cycle can be obtained accordingly:

$$C = \sum_{j=1}^{n} (0.45 \times 5400 \times 80 \times 0.0717 \times 0.2 + 300 \times r_4 + 1861.5 \times 1 \times 0.35)(1+0.05)^j$$

Table 10.3: Total initial investment of the system.

Item	f_1	f_2	f_3	f_4	f_5	Subtotal F
Investment amount (RMB)	51,000	50,000	21,000	5000	15,000	132,000

Furthermore, with a known construction investment amount of the system, investment of separate parts of the system and total maintenance cost (RMB1,000 per year by estimation) can be calculated as shown in Table 10.3:

In order to calculate water production cost of the system, suppose the total investment is equal to the total return within the service period, then:

$$127{,}000 + \sum_{j=1}^{15} \frac{15{,}000}{15 \times (1+0.05)^j} = \sum_{j=1}^{15} (3437.925 + 300 \times r_4) \times 1.05^j$$

By solving the above formula, $r_4 = 8.75$ RMB/ton.

This means water production cost of the solar desalination system is approximately 8.75 RMB/ton, which is much lower than the pure water price, but still higher than the tap water price of 4 RMB/ton.

As indicated in the calculation, there are several ways to improve the solar desalination system and make it more economical: (1) to use the system in the areas with higher solar radiation; (2) to improve the efficiency of the solar heat-collecting system; and (3) to prolong the service life by reducing the part prices of the system.

Case 2: A multistage basin type solar still with multistage condensation enhancement surface

Take the example of an actually made multistage basin type solar still with a multistage condensation enhancement surface; the economic analysis of the investment on each component of the system is as follows:

1. Solar vacuum tube heat-collecting system
   Suppose the initial investment of the solar heat-collecting system is f_1, annual return of heat supply to the desalination system is C_{11}, and maintenance cost per year is f_{51}, and then convert these indexes into the value of the present year of investment, so total investment of the solar heat-collecting system within the service period is:

$$F_1 = f_1 + \sum_{j=1}^{n} \frac{f_{51}}{(1+i)^j} \tag{10.29}$$

   Total return is:

$$C_1 = \sum_{j=1}^{n} C_{11} \cdot (1+e)^j \tag{10.30}$$

   where j is bank interest and e is the price inflation rate of energy or fresh water.
2. Components of the evaporator

Suppose the investment of the construction period is f_2, the annual maintenance fee is f_{52}, and the annual return of the product fresh water is C_{21}, then its total investment and return during the service period would be:

$$F_2 = f_2 + \sum_{j=1}^{n} \frac{f_{52}}{(1+i)^j} \tag{10.31}$$

$$C_2 = \sum_{j=1}^{n} (C_{21})(1+e)^j \tag{10.32}$$

As for the whole solar still, the area of the solar collector in the system is 4.0 m^2, the lighting area of the evaporator surface is 1.2 m^2, and the evaporation area inside the system is 1.4 m^2. The solar energy utilization factor of the whole device, i.e., performance coefficient average of the solar still PR = 1.4 (the device is three-stage, and efficiency of the heat collector has been considered). Price of investment in the solar collector per unit area is $p = 800$, and total investment of the heating system:

$$f_1 = A_1 \cdot p = 4\,p = 3200 \text{ RMB}$$

The solar collector has no maintenance cost, i.e., $f_{51} = 0$.

Total investment of the evaporator is RMB4, 000 (contact arts of this component with seawater is made of a copper titanium alloy sheet). Suppose the annual maintenance cost of this component is RMB100 (used for cleaning and daily management), i.e., $f_{52} = 100$:

$$C = C_2 = \sum_{j=1}^{n} (C_{21})(1+e)^j \tag{10.33}$$

Total investment within the service life is:

$$F = F_1 + F_2 = f_1 + f_2 + \sum_{j=1}^{n} \frac{f_{52}}{(1+i)^j} = 7200 + \sum_{j=1}^{n} \frac{100}{(1+0.05)^j} \text{ (RMB)}$$

Value setting of each parameter in the calculation is as follows:

$H \approx 5400$ MJ/(year × m^2); Then total fresh water production per year is (the system has two parts of the area receiving sunlight):

$$m_e = (A_1 + A_2)H \times PR/h_{fg} = 5.2 \times 5400 \times 10^3 \times 1.4/2300 = 17,092 \text{ (kg)}$$

Suppose the selling price of fresh water is ϕ RMB/ton, the fresh water price increase coefficient $e = 5\%$, and the service period of the device is 20 years, then total income of the product water in 20 years would be:

$$C = \sum_{j=1}^{20} m_e\, \phi(1+e)^j = \sum_{j=1}^{20} 17.092\, \phi(1+0.05)^j \tag{10.34}$$

Let the total investment be equal to the total return within the service period, then:

$$\sum_{j=1}^{20} 17.092\, \phi(1+0.05)^j = 7200 + \sum_{j=1}^{20} \frac{100}{(1+0.05)^j}$$

Solve the above formula, $\phi = 14.23$ RMB/ton.
The water production cost of this solar desalination system is 14.23 RMB/ton, which is still high compared with the tap water cost of 4 RMB/ton. However, the device can automatically produce pure water to provide drinking water to the islanders, basically without any labor management during the whole operation period. From this perspective, this system possesses a very good application prospect.

10.2.4 Economic Analysis of the Solar Desalination System Under Site Selection State

Once the construction of the solar desalination system is finished, it is relatively easy to do the economic analysis. But for the systems still in the site selection stage before construction, the analysis would be much more difficult. Related calculation models have been established by some researchers, and in most of the models, the factors of the environment and human resources are also considered.

Different solar desalination systems have different life circles, which may be relatively long for large-scale systems and shorter for small systems. Many researchers believe the system average life is 20 years at present. Annual cost of the system in service depends on the following factors: annual interest and amortization rate, expressed as IA (in the percentage of investment amount); the annual cost of transporting seawater into the evaporating system (C_s); and annual maintenance and repair fee by average, *MR* (in the percentage of investment amount). Taking into account all these factors, the reference document [4] gives the formula for calculating water production price, i.e., the average cost of distilled water (RMB/m^3) can be expressed as:

$$C = \frac{F(R_a + M_r + T_i + Q_c + MC_s)}{A(Y_d + Y_c)} \tag{10.35}$$

where A is the distillation area (m^2); c is the labor price during the operation period (RMB/man-hour); C_s is the cost needed to transport seawater into the distillation system (RMB/m^3); I is the total investment amount (RMB); Q is the annual labor quantity (man-hour); Y_c is the annual quantity of collected rainwater per unit area (in regards to the rainwater-collecting system) (m^3/m^2); Y_d is the annual output of distilled water per unit area (m^3/m^2); and TI is the average annual tax and insurance cost, percentage of the investment amount (%). IA (Annual Interest and Amortization Rate) is a function of average annual interest rate r (%), expressed as:

$$IA = r\left[1 + \frac{1}{(1 + r/100)^n - 1}\right] \tag{10.36}$$

where n is the payback period (in years).

As for the solar multieffect distillation system driven by the solar collector or solar cell, if the seawater concentration is between 5000 and 35,000 ppm, the reference document [9] gives a prediction formula of distilled water cost:

$$C = \frac{N(R_a + 1)}{868(1 + N)}\left[\frac{1.246(2 + N)C_h}{h(\Delta T - aN)} + 1\right] + \frac{0.199(R_a + 1)C_c}{H_s(1 + N)} + \frac{0.266(R_a + 1)}{1 + N} + \frac{7.54}{1 + N} \tag{10.37}$$

where a is the boiling point increased number caused by brine concentration (k); C_c is the cost of the solar collector (RMB/m^2); C_h is the unit heat transfer cost of the evaporator (RMB/m^2); h is the overall heat transfer coefficient of the evaporator (kW/$m^2 \cdot$K); H_s is the daily solar irradiation (MJ/m^2d); N is the number of effects; and ΔT is the temperature difference of adjacent effects (K).

As discussed earlier, it is complicated to calculate the economic cost of the solar still, as calculation formulas vary with different researchers from different perspectives. Taking a wide view of the various formulas for different stills, the calculated water production cost is generally between 0.52 and 2.99 US dollars/m^3, depending on which type of device and analysis method is used. Nevertheless, it is worthwhile to note that large water production does not necessarily mean that the still is more economical. That is to say, both efficiency and economy must be met for any commercially feasible solar desalination system, and the optimum water yield should balance the cost increase of investment and maintenance.

By making an economic analysis of the solar desalination system, a more reasonable decision can be made, the solar device can be more economical, and better service will be provided to the customer. However, the solar systems with more economic performance are faced with four challenges: climate, solar energy technology, cost, and scale. On top of that, how to match the distillation system with the solar system is also challenging, because the large-scale solar collector area requires a large seawater desalination plant, which makes the match more difficult.

10.2.5 System Scale Influence on the Fresh Water Cost

Scale has an importance influence on the economic performance for many systems, as well as the solar desalination system. However, since the solar system can be combined with any other forms of desalination systems, the structures can be very different, so it is difficult to describe its scale economy with only one form.

There are a number of factors, e.g., inside or outside the system and man-made factors etc., affecting the solar desalination cost. For some large-scale systems, rather than going up, the efficiency will drop even lower than that of the small systems, and so efficiency should also be considered for systems in scale. Due to the advancement of solar desalination technology, generally the water production cost is ever decreasing, especially for the technology at a large scale. Table 10.1 shows how the water production costs of some systems decrease with the increasing scales, which is worth learning.

In recent years, people have never stopped striving for the large-scale system. The previous small-scale solar desalination systems have always been denounced and have received less attention from the government. The match and link between the solar system and the industrialized seawater desalination system, especially the increase of efficiency and temperature of the solar heat-collecting system, has created favorable conditions for the combination of solar energy and desalination systems and has brought great hope for solar desalination at a large scale; for example, systems like the solar tower-type desalination system and the solar electron-water binocular system have promising prospects.

10.2.6 Multiple-Factor Analysis of the Fresh Water Cost

Doubtlessly, the factors with an impact on the fresh water cost of the solar desalination system are plenty. Even for a built solar desalination system, good effects cannot be gained without the proper usage. Therefore a full understanding of the influencing factors of solar desalination cost is very important. The solar desalination system is comprised of two main parts: the solar heat-collecting system and the seawater desalination system. The solar heating system and the desalination device are each regarded as a subset to be analyzed on the factors affecting their performance, while the weather conditions and environmental factors are regarded as another subset included in the solar system.

The solar radiation is unstable, and the energy collected by the solar system per unit time is changing, so the type of solar collector, collector area, and capacity of the heat storage tank may directly affect the fresh water production efficiency and cost of the desalination device. Therefore to investigate the matching relation between the desalination device and the parameters of the solar collector and to calculate the optimum parameter values are crucial to the optimization of the seawater desalination device. Generally speaking, when temperature of the heat energy produced by the solar collector increases, the efficiency of the collected solar energy will decrease accordingly, while the situation for the seawater desalination part is exactly reversed, i.e., the higher the operational temperature, the better the effect. But considering the thermal conversion efficiency, the temperature at 120−140°C would be better in general. Thus in order to get the best desalination effect and reduce the fresh water production cost, it is necessary to match the solar system with the seawater desalination system.

To analyze from the principal contradiction, the main factors which have an impact on the cost of the solar heating system include the type of solar collector, the collector area, and the capacity of the heat storage tank, etc., so other basic parameters can be assumed as fixed. The type of collector determines the ultimately achievable operation temperature and the relationship between efficiency and temperature; the collector area decides the total amount of the obtained heat or the operation temperature; and the capacity of the heat storage tank determines the operation temperature and the operation duration able to be extended. Generally, the producer of the solar collector will give the efficiency curve, so combined with the environmental factors and the solar radiation data, the total amount of the energy supply and temperature will be calculated, which can finally be connected with the desalination device by the heating temperature T_{si} and regeneration temperature T_{so}, when only the two indexes reflecting the performance of the solar heating system are considered: the solar collector area A_c and the capacity of the heat storage tank V_t. As for the seawater desalination device, the water production curve while running steadily will also be provided upon delivery out of the factory. Its overall performance is reflected by the curves of $\dot{m}_{et}$-T_{si} and PR-T_s developed from the steady-state test. Therefore by investigating the effect on the fresh water cost P_m from the factors of solar collector area A_c, capacity of heat storage tank V_t, desalination device operation starting temperature T_{st}, and suspending temperature T_{br}, parameters of the solar heating system can be optimized, and an understanding of the solar desalination system can also be deepened A simple calculation is derived as follows:

The thermal energy $\dot{Q}_c$(kJ/h) obtained per unit time by the solar collector is calculated by the following formula:

$$\dot{Q}_c = 3.6\, K_1 K_2 A_c \times G_n \times \eta_c \tag{10.38}$$

where K_1 and K_2 are the correction factor of radial and lateral correction angles, respectively, which can be found in general solar energy-related books; A_c is the collector area (m^2); G_n is the solar irradiance or radiation intensity ($W \times m^{-2}$); and η_c is the solar collector efficiency (can be obtained from producer).

The hot water heated by the solar collector is stored in the heat storage tank; when hot water temperature increases gradually and reaches the set value, the hot water will be pumped into the desalination system to produce fresh water. The heat balance equation showing how the heat storage tank changes with time t is:

$$\left(\dot{m}_c c_p\right)\left(T_{co} - T_{ci}\right) - \left(\dot{M}_f c_p\right)\left(T_{si} - T_{so}\right) - 3600\ \dot{Q}_{Lt} = 3600\ \rho V_t c_p \frac{dT_{si}}{dt} \quad (10.39)$$

Regarding the combination of the solar collector and desalination system, without considering heat radiation of the connection pipeline, the following equations can be obtained:

$$\dot{Q}_c = \left(\dot{m}_c c_p\right)\left(T_{co} - T_{ci}\right) \quad (10.40)$$

$$\left(\dot{M}_f c_p\left(T_{si} - T_{so}\right) + 3.6P_o\right)PR = \dot{m}_{et}\ h_{fg} \quad (10.41)$$

where $\dot{m}_{et}$ and PR are taken from the factory experimental result.

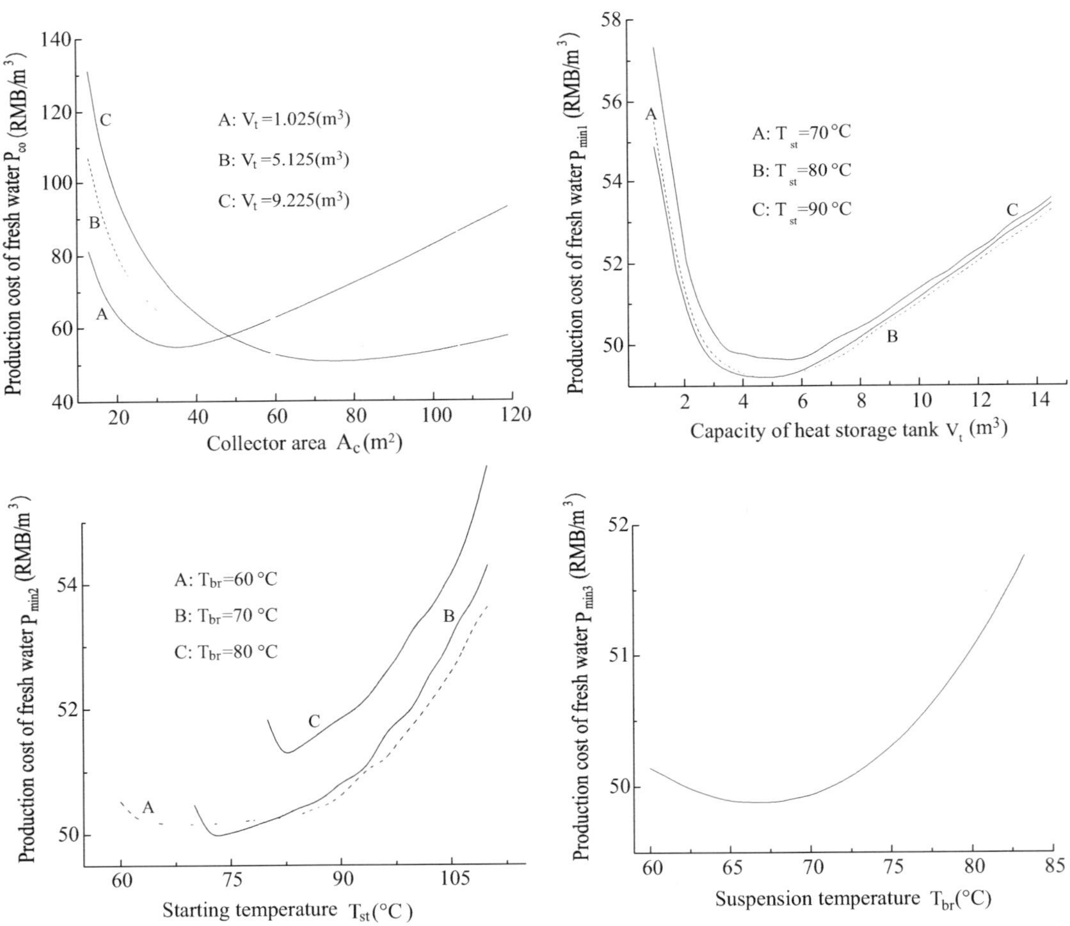

Figure 10.2
Some effect of the operation and design parameter to the water production cost.

Suppose F_s is the total cost of the solar desalination device, y is the device service period, F_e is the annual power consumption fee, and M_{et} is the annual production of fresh water, then the annual average cost P_m(RMB/m^3) for producing 1 m^3 fresh water would be:

$$P_m = (F_s/y + F_e)/M_{et} \tag{10.42}$$

Assuming as follows: the electricity price is 0.5 RMB/kW, equipment lift-span is 15 years, price of the solar collector is 1315 RMB/m^2, and the heat pipe evacuated tubular collector is selected as the solar collector. The analysis in the reference document [4] presents some curves of a small-scale solar low-temperature multieffect system, as shown in Fig. 10.2.

The results in Fig. 10.2 indicate that there might be several factors affecting the minimum value of the fresh water cost, and the minimum cost curve can also be obtained by the combination of these factors, but the analysis may be more complicated [10].

10.3 The Material Selecting of Solar Desalination System

The cost of solar desalination equipment is largely decided by the cost of its materials. There are two ways that the material performance will affect the equipment cost: service life and the water productivity. Due to the particularity of the work environment, seawater corrosion resistance of the material directly determines the service life of the desalination equipment. On the other hand, for the thermal desalination equipment, heat transfer performance of the interior metal has a large effect on the water productivity. The metals with good heat transfer performance can improve water productivity and in some way reduces the fresh water cost. Therefore material selection is very important to seawater desalination equipment.

10.3.1 Corrosion of Metals Caused by Seawater

Seawater is a kind of neutral electrolyte solution with various salts and a certain amount of oxygen, so most metals (except Mg and its alloys, whose negative electrode potential values are high) have electrochemical characteristics while corroding in seawater. With the exception of Mg and its alloys, all other metals exhibit oxygen depolarization corrosion upon exposure to seawater. Taking brass as an example, the oxidation in the anode area is as follows:

$$Cu \rightarrow Cu^{2+} + 2e \tag{10.43}$$

Reduction reaction in the cathode area is:

$$\frac{1}{2}O_2 + H_2O + 2e \rightarrow 2OH^- \tag{10.44}$$

Electrons are created in the anode area and consumed in the cathode area, which lead to electric corrosion called microcell corrosion, and most metal corrosion in the seawater are such corrosion. Another accelerated metal corrosion with low electrode potential and caused by couple action is called couple corrosion or macrocell corrosion.

The seawater physicochemical properties, which are largely related to the metal corrosion speed, include salinity, conductance, temperature, chlorinity, pH value, dissolved oxygen, flow velocity, and marine organisms, etc.

10.3.2 The Influence Factors of Seawater Corroding Metallic Materials

1. Salinity

There are a lot of salts mainly in sodium chloride dissolved in the seawater, and generally seawater is seen as the solution with 3% or 3.5% NaCl. The salt content of seawater is usually expressed in salinity or chlorinity. Salinity refers to the total grams of solid saline materials dissolved in 1000 g seawater, and chlorinity refers to the grams of chloride ion in 1000 g seawater; these two parameters are usually expressed in percentage or thousandth. The conversion relation between the seawater salinity ($S\%_{oo}$) and chlorinity ($Cl\%_{oo}$) is as follows:

$$S\%_{oo} = 1.80655\ Cl\%_{oo} \tag{10.45}$$

The seawater salinity will directly affect its electrical conductivity and oxygen content. The higher the salinity, the bigger the conductivity, but the oxygen content will decrease. It is clear from the electrochemical reaction equation that when the conductivity goes up, reaction rates of the cathode and anode will speed up, as does the corrosion, but at the

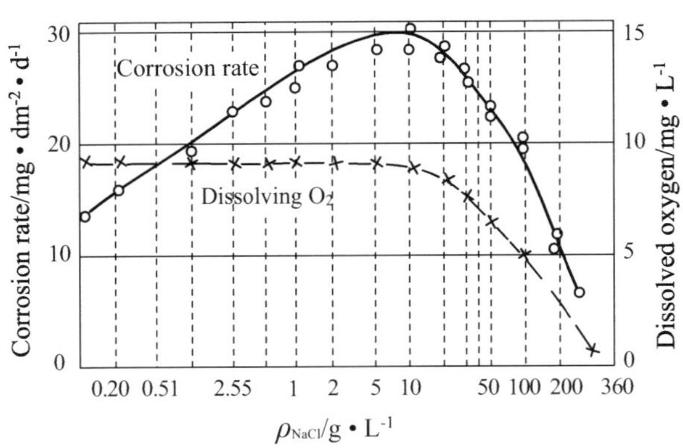

Figure 10.3
The relation between corrosion speed of steel and the salt content ρ_{NaCl} [11].

same time, the oxygen content will decrease and the electrochemical reaction and material corrosion rate will slow down. So there exists a salt content where the corrosion speed of metal is the fastest. The relation between corrosion speed of steel and the salt content that is reported in the literature is shown in Fig. 10.3 for reference.

2. Seawater electrical conductivity

Seawater electrical conductivity is mainly determined by the salinity and temperature, and it will get higher with the increase of salinity and temperature; also, the microcell and macrocell corrosion speeds of metal will increase accordingly in seawater.

3. Seawater Temperature

Seawater temperature directly decides the electrochemical reaction speed. The temperature increase will speed up the cathode and anode reaction process. When the seawater temperature rises, diffusion speed of the oxygen in seawater will get faster and electrical conductivity will increase. All of these will accelerate the corrosion of metals. On the other hand, the increase of temperature will lower the oxygen content in the seawater and generate the protective calcareous scale, which will slow down the corrosion. The effect of seawater temperature on the corrosion of metal is relatively complicated. The reference document [12] investigates the temperature effect on the corrosion speed of iron in 3% saline solution. With the rise in temperature, the amount of iron corrosion will increase and reach a maximum at 80°C, above which the speed of corrosion will decrease due to the reduction of dissolved oxygen. However, if the dissolved oxygen cannot be released in a closed system, the corrosion speed will increase as the temperature rises, which is particularly worth noting. Table 10.4 lists the corrosion data of several conventional metals in cold and hot seawater.

As indicated in this figure, the corrosion speeds of almost all the metals will increase with the temperature. It's to be noted that when the temperature gets higher, the corrosion

Table 10.4: Corrosion data comparison in cold and hot seawater [12].

Material	Corrosion speed (10^{-2} mm/a)	
	Cold seawater (27°C)	Hot seawater (86°C)
Monel 400	0.76	1.02
304 Stainless steel	7.62	0.25
Carbon steel	73.66	55.88
70/30Cu-Ni	3.81	3.56
90/10Cu-Ni	7.11	2.79
Aluminum brass	4.83	6.35
Navy brass	3.05	4.83

speeds of some metals are astonishing, which are several or tens of times as those in a normal temperature. The aluminum brass or the admiralty brass often used at present has good resistance to seawater corrosion and thermal conductivity, which is worth mentioning.

4. Seawater pH value

The bigger the seawater pH value is, the slower it corrodes the steel materials. Seawater pH value affects the corrosion speed of desalination equipment by influencing formation of the calcareous scale: the higher the pH value is, the easier the calcareous scale forms, and the less seawater corrosion of the equipment will be. But on the other hand, the scale formed on the equipment will affect its heat transfer and have a big effect on the fresh water productivity, which is unfavorable for the entire system. In fact, the seawater pH value fluctuates within a narrow range, and it will not have a significant effect on the corrosion of equipment.

5. Dissolved oxygen in seawater

As seen from the basic mechanism of seawater corrosion, the oxygen concentration in seawater is an importance factor affecting its corrosiveness. The higher the concentration is, the stronger the corrosion on equipment will be. Moreover, the oxygen concentration of seawater is highly correlated with its temperature. Table 10.5 shows the solubility of oxygen with different seawater salinities under different temperatures.

As indicated in the figure, on the premise of no change of seawater salinity, the solubility of oxygen gets smaller with the increase of temperature. When the seawater temperature rises from 0°C to 30°C, the solubility of oxygen is almost halved, and it also decreases with the increase of seawater salinity under the same temperature. The effect of dissolved oxygen on the seawater corrosiveness is mainly reflected in its temperature and salinity, and the effect of oxygen content on the corrosiveness varies with the kinds of metals. As for those metals with no passivation in seawater, such as carbon steel, low-alloy steel and cast iron, the increase of oxygen content will speed up their corrosion; while for those metals which resist seawater corrosion by the passive film on surface, like aluminum and stainless steel, the increase of oxygen content will accelerate the formation and

Table 10.5: Oxygen solubility in seawater under normal pressure [13], mL/L.

	Salt concentration (%)					
Temperature (°C)	0.0	1.0	2.0	3.0	3.5	4.0
0	10.30	9.65	9.00	8.36	8.04	7.72
10	8.02	7.56	7.09	6.63	6.41	6.18
20	6.57	6.22	5.88	5.52	5.35	5.17
30	5.57	5.27	4.95	4.65	4.50	4.34

Table 10.6: Some common thermal conductivity of metallic materials [14].

	Thermal conductivity, λ (W/m·°C)				Thermal conductivity, λ (W/m·°C)		
Name	**300K**	**400K**	**600K**	**Name**	**300K**	**400K**	**600K**
Pure aluminum	237	240	231	Low-carbon steel	63.9	58.7	48.8
Aluminum alloy	168–177	174–186	185–1186	Silico-manganese steel	41	42.2	39.7
Pure copper	401	393	379	Stainless steel	13.4–15.1	15.2–17.3	18.3–20.0
Bronze	52	52	59	Zinc	116	111	103
Brass	110	137	149	Silver	429	425	412
Pure iron	80.2	69.5	54.7	Titanium 16	11.03	11.41	13.06
				Titanium 17	7.82	8.54	10.84

maintenance of the passive films on the metal surface, which inhibits the corrosion from seawater.

10.3.3 Heat Conductivity Influence on the Material Selection

As for the thermal desalination equipment, heat exchange between the fresh water vapor and seawater will directly influence its water production performance. Take the low-temperature multieffect desalination equipment as an example. Heat exchange processes inside the equipment, and the vapor generated from seawater evaporation condenses inside the heat exchange tube, then transfer the heat to the next-stage seawater (which will spray outside the heat exchange tube) at the same time. The vapor condenses in the condenser tube on the one hand; on the other hand, the seawater evaporates outside the tube, and it proceeds in turn like this. Thus materials selected for the heat exchanger of low-temperature multieffect equipment should be resistant to seawater corrosion and possess good thermal conductivity.

Performances of some metals often used at present and with high thermal conductivity can be found in Table 10.6. Certainly, with the development of science, especially the alloy technology, the high strength heat conductors available in the field of seawater desalination are far more than these, and even some engineering plastics have already entered scientists' field of vision.

As for the solar thermal desalination equipment, especially the solar still, in order to prevent heat loss, it is necessary to choose proper heat insulation materials as well as high-quality metals. There are many alternative insulation materials, but their coefficients of thermal conductivity are usually considered as constants in many references. Actually, the coefficient of thermal conductivity is not a constant, but it changes with temperature and relates to humidity for some materials. For the commonly used heat insulation materials

Table 10.7: The thermal conductivity of some heat insulation materials [14].

Name	Thermal conductivity at 300K, λ (W/m·°C)	Name	Thermal conductivity at 300K, λ (W/m·°C)
Air	0.26	Bakelite	1.4
Water	0.62	Magnesia brick	3.8
Glass	1.4	Clay	1.9
Polyurethane foam	0.033	Cotton	0.06
Asphalt	0.062	Diatomite	0.25
Foam glass	0.058	Refractory bricks	1.0
Glass fiber	0.043	Loose vermiculite	0.068
Asbestos cement board	0.58	Plywood	0.12
Perlite	0.053	Ordinary brick	0.72
Mineral wool	0.046	Cement mortar	0.072
Ultra-fine glass wool	0.033	Sawdust board	0.087
Hardwood	0.16	Magnesium oxide	0.051
Softwood	0.12	Polystyrene	0.029

like pearlite, mineral wool, and foam glass, their coefficients of heat conductivity increase quickly with increasing temperature. Because all of these are porous insulation materials, when the temperature rises, radiation and convection heat transfer will increase according and enlarge the effective thermal conductivity of the material. For example, taking mineral wool as an insulation material, when its bulk density v is between 180 and 250 kg/m^3, its thermal conductivity changes with temperature as follows:

$$\lambda = 0.047 + 0.00058\, t[\mathrm{W/(m \cdot {}^\circ C)}] \tag{10.46}$$

where t is the Celsius temperature. Obviously, it will bring considerable error while calculating the rate of heat transfer of such a kind of material, if the changing relation of the thermal conductivity and temperature is not considered.

If the variation scope of the temperature is not too large, the relation between the thermal conductivity λ and temperature T is generally regarded as linear, i.e.:

$$\lambda = \lambda_0(1 - \beta T) \tag{10.47}$$

where λ_0 is some thermal conductivity value at normal temperatures; β is the temperature coefficient.

If the changing relation between thermal conductivity and temperature is not linear, but its functional relation $\lambda(T)$ is known, then the heat transfer can be calculated by taking the average value of thermal conductivity within a certain temperature range. The calculation formula for the average thermal conductivity is as follows:

$$\lambda_m = \frac{1}{T_1 - T_2} \int_{T_2}^{T_1} \lambda(T)dT \tag{10.48}$$

For ease of reference, Table 10.7 shows how the thermal conductivity changes with the temperature for some commonly used materials.

10.3.4 Other Performance of Metals' Influence on Material Selection

As for some desalination devices working under negative pressure, material strength and density should also be considered when selecting the shell materials. There are some other structures requiring high-strength-weight ratio to meet their particular function requirements. Considering the equipment installation and transportation, the materials with low density should be selected. Usually a seawater desalination plant needs not only one type of material, and galvanic corrosion may also occur. First of all, it is better to select materials of the same type or with close galvanic series to avoid such a problem. Secondly, when selecting the welding material, in order to prevent galvanic corrosion between the welding seam and the parent material, the welding seam metal should show cathodic, and the potential difference between the two is supposed to be as little as possible. When the two metal materials with big potential difference connect, corresponding protection measures should be taken to prevent galvanic corrosion.

Therefore material selection for the seawater desalination device should fully consider its service life and corresponding costs, as well as the work environment and requirements to be met.

Different desalination devices have different work environments, so the requirements of materials are not the same. The work environment and operation principle should be first understood when selecting the device materials; besides, the service life and cost should also be comprehensively considered. For example, the stacked tray desalination device operates at normal pressure, while the low-temperature multieffect or multistage flash desalination method requires a vacuum to improve water productivity. Moreover, their working principles are different, and so the materials needed vary with the different methods of seawater desalination.

For the normal basin type still, if it is fixed, nonmental material can be selected as the basin to save cost and resist corrosion. But for the multistage stacked tray desalination device, especially for the multistage basin type still, which is frequently mobile, corrosion resistance must be considered, so metal materials with good heat conductivity need to be used as the basin. Nevertheless, most of these devices are under all static condition and normal pressure, so their work environments are relatively favorable compared to other seawater desalination devices. It can be seen by analyzing the working principle of the multistage stacked tray desalination device that, except the continuous contact between the basins of all stages and seawater, other components of the main part keep clear of the seawater. So service life of the basin determines the life of the device. Therefore basins of the multistage stacked tray desalination device should be made of titanium alloys with good seawater corrosion resistance, while other parts can select the galvanized sheet and stainless steel with cheaper prices.

However, the low-temperature multieffect solar desalination systems are in a different situation, so differential measures should be taken for their material selection.

The evaporator is the main part of the low-temperature multieffect seawater desalination device, including the shell and heat exchange tube, etc. The heat exchange tube should be able to not only resist seawater corrosion, but also have good heat conductivity. As indicated in Tables 10.4 and 10.6, the heat exchange tube should be made of aluminum brass for its corrosion resistance and heat conductivity. In fact, aluminum brass or admiralty brass is used as a heat transfer material in all the low-temperature multieffect seawater desalination plants at present. Considering the strength and service life needed, more expensive special stainless steels and/or titanium are also used.

Fortunately, with the advancement of technology, some new and low-cost materials are being developed to replace the conventional heat transfer materials. For example, the aluminum alloy tube developed by the Israeli company IDE has been applied in the low-temperature multieffect seawater desalination device for 20 years, although some gaps exist between its service life and that of the aluminum brass heat transfer tube. It's only around one-third of the brass tube price and one-fourth of the weight, and the hydrophilicity of its horizontal tube falling film is better. This material is used in some seawater desalination plants in China.

Since the low-temperature multieffect desalination device generally operates under negative pressure, the shell material should have good load performance, as well as some seawater corrosion resistance function. The shell of the evaporator is usually made of the "304 stainless steel," which has good corrosion resistance and load performance. However, the cost of the stainless steel is relatively high. Some companies have been trying to use a special concrete to replace stainless steel as the shell material of the evaporator, which can largely increase the equipment service life to greatly lower the water production cost. There are some devices with anticorrosion coating inside the carbon steel material, which enables easy machining and standardized assembling, so many companies would choose such devices.

Another important factor causing the high fresh water cost is the high device manufacture cost, in which metals constitute a large proportion. With the development of the material industry, a great deal of new materials with a low price and high quality will be applied in the desalination device, and the manufacture cost will be extensively decreased.

References

[1] Ali MT, Fath HES, Armstrong PR. A comprehensive techno-economical review of indirect solar desalination. Renew Sustain Energy Rev 2011;15:4187–99.
[2] Leblanc J, Andrews J, Akbarzadeh A. Low-temperature solar–thermal multi-effect evaporation desalination systems. Int J Energy Res 2010;34:393–403.
[3] Al-Hallaj S, Parekh S, Farid MM, Selman JR. Solar desalination with humidification – dehumidification cycle: review of economics. Desalination 2006;195(1–3):169–86.
[4] Goosen MFA, Sablani SS, Shayya WH. Thermodynamic and economic considerations in solar desalination. Desalination 2000;129:63–89.
[5] Riffel DB, Carvalho PCM. Small-scale photovoltaic-powered reverse osmosis plant without batteries: design and simulation. Desalination 2009;247:378–89.
[6] Gandhidasan P, Al-Mojel SA. Effect of feed pressure on the performance of the photovoltaic powered reverse osmosis seawater desalination system. Renew Energy 2009;34:2824–30.
[7] Xinshi G, et al. Solar engineering—principles and applications. Beijing: Academic Journal Publishing Press; 1988.
[8] Zhili C. Research on the desalination technology and system based on the solar heat method[d] [Chongqing: A Thesis Submitted to Chongqing University in Partial Fulfillment of the Requirement for the Degree of Doctor of Engineering]. 2009.
[9] Singh AK, Tiwari GN. Int J Ambient Energy 1995;16(2):89.
[10] Ziqian C. The theoretical research on solar desalination technology[d] [Beijing: A Thesis Submitted to Beijing Institute of Technology in Partial Fulfillment of the Requirement for the Degree of Doctor of Engineering]. 2007.
[11] Rizhang Z, et al. Corrosion science of metal. Metallurgical Industry Press; 1989.
[12] Baoming W. Metal corrosion theory and application. Chemical Industry Press; 1984.
[13] Schumacher M, Li Dachao, et al. Seawater corrosion handbook. National Defence Industry Press; 1985.
[14] Incropera FP, et al. Fundamentals of heat and mass transfer. John Wiley & Sons; 2006.

Appendix

Appendix Table 1A: Thermodynamic properties of saturated water and unsaturated vapor (arranged by temperature).

Temperature	Pressure	Specific volume		Enthalpy		Gasification latent heat	Entropy	
		Liquid	Steam	Liquid	Steam		Liquid	Steam
t (°C)	p (MPa)	v' (m³/kg)	v'' (m³/kg)	h' (kJ/kg)	h'' (kJ/kg)	γ (kJ/kg)	s' [kJ/(kg K)]	s'' [kJ/(kg K)]
0.00	0.0006112	0.00100022	206.154	−0.05	2500.51	2500.6	−0.0002	9.1544
0.01	0.0006117	0.00100021	206.012	0.00	2500.53	2500.5	0.0000	9.1541
1	0.0006571	0.00100018	192.464	4.18	2502.35	2498.2	0.0153	9.1278
2	0.0007059	0.00100013	179.787	8.39	2504.19	2495.8	0.0306	9.1014
4	0.0008135	0.00100008	157.151	16.82	2507.87	2491.1	0.0611	9.0493
5	0.0008725	0.00100008	147.048	21.02	2509.71	2488.7	0.0763	9.0236
6	0.0009352	0.00100010	137.670	25.22	2511.55	2486.3	0.0913	8.9982
8	0.0010728	0.00100019	120.868	33.62	2515.23	2481.6	0.1213	8.9480
10	0.0012279	0.00100034	106.341	42.00	2518.90	2476.9	0.1510	8.8988
12	0.0014025	0.00100054	93.756	50.38	2522.57	2472.2	0.1805	8.8504
14	0.0015985	0.00100080	82.828	58.76	2526.24	2467.5	0.2098	8.8029
15	0.0017053	0.00100094	77.910	62.95	2528.07	2465.1	0.2243	8.7794
16	0.0018183	0.00100110	73.320	67.13	2529.90	2462.8	0.2388	8.7562
18	0.0020640	0.00100145	65.029	75.50	2533.55	2458.1	0.2677	8.7103
20	0.0023358	0.00100185	57.786	83.86	2537.20	2453.3	0.2963	8.6652
22	0.0026444	0.00100229	51.455	92.23	2540.84	2448.6	0.3247	8.6210
24	0.0029846	0.00100276	45.884	100.59	2544.47	2443.9	0.3530	8.5774
25	0.0031687	0.00100302	43.362	104.77	2546.29	2441.5	0.3670	8.5560
26	0.0033625	0.00100328	40.997	108.95	2548.10	2439.2	0.3810	8.5347
28	0.0037814	0.00100383	36.694	117.32	2551.73	2434.4	0.4089	8.4927
30	0.0042451	0.00100442	32.899	125.68	2555.35	2429.7	0.4366	8.4514
35	0.0056263	0.00100605	25.222	146.59	2564.38	2417.8	0.5050	8.3511
40	0.0073811	0.00100789	19.529	167.50	2573.36	2405.9	0.5723	8.2551
45	0.0095897	0.00100993	15.2636	188.42	2582.30	2393.9	0.6386	8.1630
50	0.0123446	0.00101216	12.0365	209.33	2591.19	2381.9	0.7038	8.0745
55	0.015752	0.00101455	9.5723	230.24	2600.02	2369.8	0.7680	7.9896
60	0.019933	0.00101713	7.6740	251.15	2608.79	2357.6	0.8312	7.9080
65	0.025024	0.00101986	6.1992	272.08	2617.48	2345.4	0.8935	7.8295
70	0.031187	0.00102276	5.0443	293.01	2626.10	2333.1	0.9550	7.7540
75	0.038565	0.00102582	4.1330	313.96	2634.63	2320.7	1.0156	7.6812
80	0.047376	0.00102903	3.4086	334.93	2643.06	2308.1	1.0753	7.6112
85	0.057818	0.00103240	2.8288	355.92	2651.40	2295.5	1.1343	7.5436
90	0.070121	0.00103593	2.3616	376.94	2659.63	2282.7	1.1926	7.4783
95	0.084533	0.00103961	1.9827	397.98	2667.73	2269.7	1.2501	7.4154
100	0.101325	0.00104344	1.6736	419.06	2675.71	2256.6	1.3069	7.3545
110	0.143243	0.00105156	1.2106	461.33	2691.26	2229.9	1.4186	7.2386
120	0.198483	0.00106031	0.89219	503.76	2706.18	2202.4	1.5277	7.1297
130	0.270018	0.00106968	0.66873	546.38	2720.39	2174.0	1.5346	7.0272
140	0.361190	0.00107972	0.50900	589.21	2733.81	2144.6	1.7393	6.9302
150	0.47571	0.00109046	0.39286	632.28	2746.35	2114.1	1.3420	6.8381

Continued

Appendix Table 1A: Thermodynamic properties of saturated water and unsaturated vapor (arranged by temperature).—cont'd

Temperature	Pressure	Specific volume		Enthalpy		Gasification	Entropy	
		Liquid	Steam	Liquid	Steam	latent heat	Liquid	Steam
t (°C)	p (MPa)	v′ (m³/kg)	v″ (m³/kg)	h′ (kJ/kg)	h″ (kJ/kg)	γ (kJ/kg)	s′ [kJ/(kg K)]	s″ [kJ/(kg K)]
160	0.61766	0.00110193	0.30709	657.62	2757.92	2082.3	1.9429	6.7502
170	0.79147	0.00111420	0.24283	719.25	2768.42	2049.2	2.0420	6.6661
180	1.00193	0.00112732	0.19403	763.22	2777.74	2014.5	2.1396	6.5852
190	1.25417	0.00114136	0.15650	807.56	2785.80	1978.2	2.2358	6.5071
200	1.55366	0.00115641	0.12732	852.34	2792.47	1940.1	2.3307	6.4312
210	1.90617	0.00117258	0.10438	897.62	2797.65	1900.0	2.4245	6.3571
220	2.31783	0.00119000	0.086157	943.46	2801.20	1857.7	2.5175	6.2846
230	2.79505	0.00120882	0.071553	989.95	2803.00	1813.0	2.6096	6.2130
240	3.34459	0.00122922	0.059743	1037.2	2802.88	1765.7	2.7013	6.1422
250	3.97351	0.00125145	0.050112	1085.3	2800.66	1715.4	2.7926	6.0716
260	4.68923	0.00127579	0.042195	1134.3	2796.14	1661.8	2.8837	6.0007
270	5.49956	0.00130262	0.035637	1184.5	2789.05	1604.5	2.9751	5.9292
280	6.41273	0.00133242	0.030165	1236.0	2779.08	1543.1	3.0668	5.8564
290	7.43746	0.00136582	0.025565	1289.1	2765.81	1476.7	3.1594	5.7817
300	8.58308	0.00140369	0.021669	1344.0	2748.71	1404.7	3.2533	5.7042
310	9.8597	0.00144728	0.018343	1401.2	2727.01	1325.9	3.3490	5.6226
320	11.278	0.00149844	0.015479	1461.2	2699.72	1238.5	3.4475	5.5356
330	12.851	0.00156008	0.012987	1524.9	2665.30	1140.4	3.5500	5.4408
340	14.593	0.00163728	0.010790	1593.7	2621.32	1027.6	3.6586	5.3345
350	16.521	0.00174008	0.008812	1670.3	2563.39	893.0	3.7773	5.2104
360	18.657	0.00189423	0.006958	1761.1	2481.68	720.6	3.9155	5.0536
370	21.033	0.00221480	0.004982	1891.7	2338.79	447.1	4.1125	4.8076
372	21.542	0.00236530	0.004451	1936.1	2282.99	346.9	4.1796	4.7173
373.99	22.064	0.003106	0.003106	2085.9	2085.87	0.0	4.4092	4.4092

This table is taken from Yan Jialu, Yu Xiaofu. Thermodynamic properties table of water and vapor. Beijing: High Education Press; 1995.

Appendix Table 1B: Thermodynamic properties of saturated water and unsaturated vapor (arranged by pressure).

Pressure	Temperature	Specific volume		Enthalpy		Gasification	Entropy	
		Liquid	Steam	Liquid	Steam	latent heat	Liquid	Steam
p (MPa)	t (°C)	v′ (m³/kg)	v″ (m³/kg)	h′ (kJ/kg)	h″ (kJ/kg)	γ (kJ/kg)	s′ [kJ/(kg K)]	s″ [kJ/(kg K)]
0.001	6.9491	0.0010001	129.185	29.21	2513.29	2484.1	0.1056	8.9735
0.002	17.5403	0.0010014	67.008	73.58	2532.71	2459.1	0.2611	8.7220
0.003	24.1142	0.0010028	45.666	101.07	2544.68	2443.6	0.3546	8.5758
0.004	28.9533	0.0010041	34.796	121.30	2553.45	2432.2	0.4221	8.4725
0.005	32.8793	0.0010053	28.191	137.72	2560.55	2422.8	0.4761	8.3930
0.006	36.1663	0.0010065	23.738	151.47	2566.48	2415.0	0.5208	8.3283
0.007	38.9967	0.0010075	20.528	163.31	2571.56	2408.3	0.5589	8.2737
0.008	41.5075	0.0010085	18.102	173.81	2576.06	2402.3	0.5924	8.2266
0.009	43.7901	0.0010094	16.204	183.36	2580.15	2396.8	0.6226	8.1854
0.010	45.7988	0.0010103	14.673	191.76	2583.72	2392.0	0.6490	8.1481
0.015	53.9705	0.0010140	10.022	225.93	2598.21	2372.3	0.7548	8.0065
0.020	60.0650	0.0010172	7.6497	251.43	2608.90	2357.5	0.8320	7.9068

Appendix Table 1B: Thermodynamic properties of saturated water and unsaturated vapor (arranged by pressure).—cont'd

Pressure	Temperature	Specific volume		Enthalpy		Gasification	Entropy	
		Liquid	Steam	Liquid	Steam	latent heat	Liquid	Steam
p (MPa)	*t* (°C)	*v*′ (m³/kg)	*v*″ (m³/kg)	*h*′ (kJ/kg)	*h*″ (kJ/kg)	γ (kJ/kg)	*s*′ [kJ/(kg K)]	*s*″ [kJ/(kg K)]
0.025	64.9726	0.0010198	6.2047	271.96	2617.43	2345.5	0.8932	7.8298
0.030	69.1041	0.0010222	5.2296	289.26	2624.56	2335.3	0.9440	7.7671
0.040	75.8720	0.0010264	3.9939	317.61	2636.10	2318.5	1.0260	7.6688
0.050	81.3388	0.0010299	3.2409	340.55	2645.31	2304.8	1.0912	7.5928
0.060	85.9496	0.0010331	2.7324	359.91	2652.97	2293.1	1.1454	7.5310
0.070	89.9556	0.0010359	2.3654	376.75	2659.55	2282.8	1.1921	7.4789
0.080	93.5107	0.0010385	2.0876	391.71	2665.33	2273.6	1.2330	7.4339
0.090	96.7121	0.0010409	1.8698	405.20	2670.48	2265.3	1.2695	7.3943
0.100	99.634	0.0010432	1.6943	417.52	2675.14	2257.6	1.3023	7.3589
0.120	104.810	0.0010473	1.4287	439.37	2683.26	2243.9	1.3609	7.2978
0.140	109.318	0.0010510	1.2368	458.44	2690.22	2231.8	1.4110	7.2462
0.150	111.378	0.0010527	1.15953	467.17	2693.35	2226.2	1.4338	7.2232
0.160	113.326	0.0010544	1.09159	475.42	2696.29	2220.9	1.4552	7.2016
0.180	116.941	0.0010576	0.97767	490.76	2701.69	2210.9	1.4946	7.1623
0.200	120.240	0.0010605	0.88585	504.78	2706.53	2201.7	1.5303	7.1272
0.250	127.444	0.0010672	0.71879	535.47	2716.83	2181.4	1.6075	7.0528
0.300	133.556	0.0010732	0.60587	561.58	2725.26	2163.7	1.6727	6.9921
0.350	138.891	0.0010786	0.52427	584.45	2732.37	2147.9	1.7278	6.9407
0.400	143.642	0.0010835	0.46246	604.87	2738.49	2133.6	1.7769	6.8961
0.450	147.939	0.0010882	0.41396	623.38	2743.85	2120.5	1.8210	6.8567
0.500	151.867	0.0010925	0.37486	640.35	2748.59	2108.2	1.8610	6.8214
0.600	158.863	0.0011006	0.31563	670.67	2756.66	2086.0	1.9315	6.7600
0.700	164.983	0.0011079	0.27281	697.32	2763.29	2066.0	1.9925	6.7079
0.800	170.444	0.0011148	0.24037	721.20	2768.86	2047.7	2.0464	6.6625
0.900	175.389	0.0011212	0.21491	742.90	2773.59	2030.7	2.0948	6.6222
1.00	179.916	0.0011272	0.19438	762.84	2777.67	2014.8	2.1388	6.5859
1.10	184.100	0.0011330	0.17747	781.35	2781.21	1999.9	2.1792	6.5529
1.20	187.995	0.0011385	0.16328	798.64	2784.29	1985.7	2.2166	6.5225
1.30	191.644	0.0011438	0.15120	814.89	2786.99	1972.1	2.2515	6.4944
1.40	195.078	0.0011489	0.14079	830.24	2789.37	1959.1	2.2841	6.4683
1.50	198.327	0.0011538	0.13172	844.82	2791.46	1946.6	2.3149	6.4437
1.60	210.410	0.0011586	0.12375	858.69	2793.29	1934.6	2.3440	6.4206
1.70	204.346	0.0011633	0.11668	871.96	2794.91	1923.0	2.3716	6.3988
1.80	207.151	0.0011679	0.11037	884.67	2796.33	1911.7	2.3979	6.3781
1.90	209.838	0.0011723	0.104707	896.88	2797.58	1900.7	2.4230	6.3583
2.00	212.417	0.0011767	0.099588	908.64	2798.66	1890.0	2.4471	6.3395
2.50	223.990	0.0011973	0.079949	961.93	2802.14	1840.2	2.5543	6.2559
3.00	233.893	0.0012166	0.066662	1008.2	2803.19	1794.9	2.6454	6.1854
3.50	242.597	0.0012348	0.057054	1049.6	2802.51	1752.9	2.7250	6.1238
4.00	250.394	0.0012524	0.049771	1087.2	2800.53	1713.4	2.7962	6.0688
4.50	257.477	0.0012694	0.044052	1121.8	2797.51	1675.7	2.8607	6.0187
5.00	263.980	0.0012862	0.039439	1154.2	2793.64	1639.5	2.9201	5.9724
6.00	275.625	0.0013190	0.032440	1213.3	2783.82	1570.5	3.0266	5.8885
7.00	285.869	0.0013515	0.027371	1266.9	2771.72	1504.8	3.1210	5.8129
8.00	295.048	0.0013843	0.023520	1316.5	2757.70	1441.2	3.2066	5.7430

Continued

Appendix Table 1B: Thermodynamic properties of saturated water and unsaturated vapor (arranged by pressure).—cont'd

Pressure	Temperature	Specific volume: Liquid	Specific volume: Steam	Enthalpy: Liquid	Enthalpy: Steam	Gasification latent heat	Entropy: Liquid	Entropy: Steam
p (MPa)	t (°C)	v' (m³/kg)	v'' (m³/kg)	h' (kJ/kg)	h'' (kJ/kg)	γ (kJ/kg)	s' [kJ/(kg K)]	s'' [kJ/(kg K)]
9.00	303.385	0.0014177	0.020485	1363.1	2741.92	1378.9	3.2854	5.6771
10.0	311.037	0.0014522	0.018026	1407.2	2724.46	1317.2	3.3591	5.6139
12.0	324.715	0.0015260	0.014263	1490.7	2684.50	1193.8	3.4952	5.4920
14.0	336.707	0.0016097	0.011486	1570.4	2637.07	1066.7	3.6220	5.3711
16.0	347.396	0.0017099	0.009311	1649.4	2580.21	930.8	3.7451	5.2450
18.0	357.034	0.0018402	0.007503	1732.0	2509.45	777.4	3.8715	5.1051
20.0	365.789	0.0020379	0.005870	1827.2	2413.05	585.9	4.0153	4.9322
22.0	373.752	0.0027040	0.003684	2013.0	2084.02	71.0	4.2969	4.4066
22.064	373.99	0.003106	0.003106	2085.9	2085.87	0.0	4.4092	4.4092

This table is taken from Yan Jialu, Yu Xiaofu. Thermodynamic properties table of water and vapor. Beijing: High Education Press; 1995.

Appendix Table 2: Thermodynamic properties of unsaturated water and superheated vapor.

p	0.001 MPa			0.005 MPa			0.01 MPa		
Saturation parameters	t_s = 6.949°C, v' = 0.0010001, v'' = 129.185, h' = 29.21, h'' = 2513.3, s' = 0.1056, s'' = 8.9735			t_s = 32.879°C, v' = 0.0010053, v'' = 28.191, h' = 137.72, h'' = 2560.6, s' = 0.4761, s'' = 8.3930			t_s = 45.799°C, v' = 0.0010103, v'' = 14.673, h' = 191.76, h'' = 2583.7, s' = 0.6490, s'' = 8.1481		
t (°C)	v (m³/kg)	h (kJ/kg)	s [kJ/(kg K)]	v (m³/kg)	h (kJ/kg)	s [kJ/(kg K)]	v (m³/kg)	h (kJ/kg)	s [kJ/(kg K)]
0	0.0010002	−0.05	−0.0002	0.0010002	−0.05	−0.0002	0.0010002	−0.04	−0.0002
10	130.598	2519.0	8.9938	0.0010003	42.01	0.1510	0.0010003	42.01	0.1510
20	135.226	2537.7	9.0588	0.0010018	83.87	0.2963	0.0010018	83.87	0.2963
40	144.475	2575.2	9.1832	28.854	2574.0	8.4366	0.0010079	167.51	0.5723
50	149.096	2593.9	9.2412	29.783	2592.9	8.4961	14.869	2591.8	8.1732
60	153.717	2612.7	9.2984	30.712	2611.8	8.5537	15.336	2610.8	8.2313
80	162.956	2650.3	9.4080	32.566	2649.7	8.6639	16.268	2648.9	8.3422
100	172.192	2688.0	9.5120	34.418	2687.5	8.7682	17.196	2686.9	8.4471
120	181.426	2725.9	9.6109	36.269	2725.5	8.8674	18.124	2725.1	8.5466
140	190.660	2764.0	9.7054	38.118	2763.7	8.9620	19.050	2763.3	8.6414
150	195.227	2783.1	9.7511	39.042	2782.8	9.0078	19.513	2782.5	8.6873
160	199.893	2802.3	9.7959	39.967	2802.0	9.0526	19.976	2801.7	8.7322
180	209.126	2840.7	9.8827	41.815	2840.5	9.1396	20.901	2840.2	8.8192
200	218.358	2879.4	9.9662	43.662	2879.2	9.2232	21.826	2879.0	8.9029
250	241.437	2977.1	10.1625	48.281	2977.0	9.4195	24.136	2976.8	9.0994
300	264.515	3076.2	10.3434	52.898	3076.1	9.6005	26.448	3078.0	9.2805
350	287.592	3176.8	10.5117	57.514	3176.7	9.7688	28.755	3176.6	9.4488
400	310.669	3278.9	10.6692	62.131	3278.8	9.9264	31.063	3278.7	9.6064
450	333.746	3382.4	10.8176	66.747	3382.4	10.0747	33.372	3382.3	9.7548
500	356.823	3487.5	10.9581	71.362	3487.5	10.2153	35.680	3487.4	9.8953
600	402.976	3703.4	11.2206	80.594	3703.4	10.4778	40.296	3703.4	10.1579

p	0.050 MPa			0.10 MPa			0.20 MPa		
Saturation parameters	t_s = 81.339°C v' = 0.0010299, v'' = 3.2409 h' = 340.55, h'' = 2645.3 s' = 1.0912, s'' = 7.5928			t_s = 99.634°C v' = 0.0010431, v'' = 1.6943 h' = 417.52, h'' = 2675.1 s' = 1.3028, s'' = 7.3589			t_s = 120.240°C v' = 0.0010605, v'' = 0.88590 h' = 504.78, h'' = 2706.5 s' = 1.5303, s'' = 7.1272		
t (°C)	v (m³/kg)	h (kJ/kg)	s [kJ/(kg K)]	v (m³/kg)	h (kJ/kg)	s [kJ/(kg K)]	v (m³/kg)	h (kJ/kg)	s [kJ/(kg K)]
0	0.0010002	0.00	−0.0002	0.0010002	0.05	−0.0002	0.0010001	0.15	−0.0002
10	0.0010003	42.05	0.1510	0.0010003	42.10	0.1510	0.0010002	42.20	0.1510
20	0.0010018	83.91	0.2963	0.0010018	83.96	0.2963	0.0010013	84.05	0.2963
40	0.0010079	167.54	0.5723	0.0010078	167.59	0.5723	0.0010078	167.67	0.5722
50	0.0010121	209.36	0.7037	0.0010121	209.40	0.7037	0.001012	209.49	0.7037
60	0.0010171	251.18	0.8312	0.0010171	251.22	0.8312	0.0010170	251.31	0.8311
80	0.0010290	334.93	1.0753	0.0010290	334.97	1.0753	0.0010290	335.05	1.0752
100	3.4188	2682.1	7.6941	1.6961	2675.9	7.3609	0.0010434	419.14	1.3068
120	3.6078	2721.2	7.7962	1.7931	2716.3	7.4665	0.0010603	503.76	1.5277
140	3.7958	2760.2	7.8928	1.8889	2756.2	7.5654	0.93511	2748.0	7.2300
150	3.8895	2779.6	7.9393	1.9364	2776.0	7.6128	0.95968	2768.6	7.2793
160	3.9830	2799.1	7.9848	1.9838	2795.8	7.6590	0.98407	2789.0	7.3271
180	4.1697	2838.1	8.0727	2.0783	2835.3	7.7482	1.03241	2829.6	7.4187
200	4.3560	2877.1	8.1571	2.1723	2874.8	7.8334	1.08030	2870.0	7.5058
250	4.8205	2975.5	8.3547	2.4061	2973.8	8.0324	1.19878	2970.4	7.7076
300	5.2840	3075.0	8.5364	2.6388	3073.8	8.2148	1.31617	3071.2	7.8917
350	5.7469	3175.9	8.7051	2.8709	3174.9	8.3840	1.43294	3172.9	8.0618
400	6.2094	3278.1	8.8629	3.1027	3277.3	8.5422	1.54932	3275.8	8.2205
450	6.6717	3381.8	9.0115	3.3342	3381.2	8.6909	1.66546	3379.9	8.3697
500	7.1338	3487.0	9.1521	3.5656	3486.5	8.8317	1.78142	3485.4	8.5108
600	8.0577	3703.1	9.4148	4.0279	3702.7	9.0946	2.01301	3701.9	8.7740

p	0.50 MPa			0.80 MPa			1.0 MPa		
Saturation parameters	t_s = 151.867°C v' = 0.0010925, v'' = 0.37490 h' = 640.55, h'' = 2748.6 s' = 1.8610, s'' = 6.8214			t_s = 170.444°C v' = 0.0011148, v'' = 0.24040 h' = 721.20, h'' = 2768.9 s' = 2.0464, s'' = 6.6625			t_s = 179.916°C v' = 0.0011272, v'' = 0.19440 h' = 762.84, h'' = 2777.7 s' = 2.1388, s'' = 6.5859		
t (°C)	v (m³/kg)	h (kJ/kg)	s [kJ/(kg K)]	v (m³/kg)	h (kJ/kg)	s [kJ/(kg K)]	v (m³/kg)	h (kJ/kg)	s [kJ/(kg K)]
0	0.0010000	0.46	−0.0001	0.0009998	0.77	−0.0001	0.0009997	0.97	−0.0001
10	0.0010001	42.49	0.1510	0.0010000	42.78	0.1510	0.0009999	42.98	0.1509
20	0.0010016	84.33	0.2962	0.0010015	84.61	0.2961	0.0010014	84.80	0.2961
40	0.0010077	167.94	0.5721	0.0010075	168.21	0.5720	0.0010074	168.38	0.5719
50	0.0010119	209.75	0.7035	0.0010118	210.01	0.7034	0.0010117	210.18	0.7033
60	0.0010169	251.56	0.8310	0.0010168	251.81	0.8308	0.0010167	251.98	0.8307
80	0.0010288	335.29	1.0750	0.0010287	335.53	1.0748	0.0010286	335.69	1.0747
100	0.0010432	419.36	1.3066	0.0010431	419.59	1.3064	0.0010430	419.74	1.3062
120	0.0010601	503.97	1.5275	0.0010600	504.18	1.5272	0.0010599	504.32	1.5270
140	0.0010796	589.30	1.7392	0.0010794	589.49	1.7389	0.0010793	589.62	1.7386
150	0.0010904	632.30	1.8420	0.0010902	632.48	1.8417	0.0010901	632.61	1.8414
160	0.38358	2767.2	6.8647	0.0011018	675.72	1.9427	0.0011017	675.84	1.9424
180	0.40450	2811.7	6.9651	0.24711	2792.0	6.7142	0.19443	2777.9	6.5864
200	0.42487	2854.9	7.0585	0.26074	2838.7	6.8151	0.20590	2827.3	6.6931

Continued

Appendix Table 2: Thermodynamic properties of unsaturated water and superheated vapor.—cont'd

p	0.50 MPa			0.80 MPa			1.0 MPa		
Saturation parameters	$t_s = 151.867°C$, $v' = 0.0010925$, $v'' = 0.37490$, $h' = 640.55$, $h'' = 2748.6$, $s' = 1.8610$, $s'' = 6.8214$			$t_s = 170.444°C$, $v' = 0.0011148$, $v'' = 0.24040$, $h' = 721.20$, $h'' = 2768.9$, $s' = 2.0464$, $s'' = 6.6625$			$t_s = 179.916°C$, $v' = 0.0011272$, $v'' = 0.19440$, $h' = 762.84$, $h'' = 2777.7$, $s' = 2.1388$, $s'' = 6.5859$		
t (°C)	v (m³/kg)	h (kJ/kg)	s [kJ/(kg K)]	v (m³/kg)	h (kJ/kg)	s [kJ/(kg K)]	v (m³/kg)	h (kJ/kg)	s [kJ/(kg K)]
250	0.47432	2960.0	7.2697	0.29310	2949.2	7.0371	0.23264	2941.8	6.9233
300	0.52255	3063.6	7.4588	0.32410	3055.7	7.2316	0.25793	3050.4	7.1216
350	0.57012	3167.0	7.6319	0.35439	3161.0	7.4078	0.28247	3157.0	7.2999
400	0.61729	3271.1	7.7924	0.38426	3266.3	7.5703	0.30658	3263.1	7.4638
450	0.66420	3376.0	7.9428	0.41388	3372.1	7.7219	0.33043	3369.6	7.6163
500	0.71094	3482.2	8.0848	0.44331	3479.0	7.8648	0.35410	3476.8	7.7597
600	0.80408	3699.6	8.3491	0.50184	3697.2	8.1302	0.40109	3695.7	8.0259

p	2.0 MPa			3.0 MPa			4.0 MPa		
Saturation parameters	$t_s = 212.417°C$, $v' = 0.0011767$, $v'' = 0.099600$, $h' = 908.64$, $h'' = 2798.7$, $s' = 2.447$, $s'' = 6.3395$			$t_s = 233.893°C$, $v' = 0.0012166$, $v'' = 0.066700$, $h' = 1008.2$, $h'' = 2803.2$, $s' = 2.6454$, $s'' = 6.1854$			$t_s = 250.394°C$, $v' = 0.0012524$, $v'' = 0.049800$, $h' = 1087.2$, $h'' = 2800.5$, $s' = 2.7962$, $s'' = 6.0688$		
t (°C)	v (m³/kg)	h (kJ/kg)	s [kJ/(kg K)]	v (m³/kg)	h (kJ/kg)	s [kJ/(kg K)]	v (m³/kg)	h (kJ/kg)	s [kJ/(kg K)]
0	0.0009992	1.99	0.0000	0.0009987	3.01	0.0000	0.0009982	4.03	0.0001
10	0.0009994	43.95	0.1508	0.0009989	44.92	0.1507	0.0009984	45.89	0.1507
20	0.0010009	85.74	0.2959	0.0010005	86.68	0.2957	0.0010000	87.62	0.2955
40	0.0010070	169.27	0.5715	0.0010066	170.15	0.5711	0.0010061	171.04	0.5708
50	0.0010113	211.04	0.7028	0.0010108	211.90	0.7024	0.0010104	212.77	0.7019
60	0.0010162	252.82	0.8302	0.0010158	253.66	0.8296	0.0010153	254.50	0.8291
80	0.0010281	336.48	1.0740	0.0010276	337.28	1.0734	0.0010272	338.07	1.0727
100	0.0010425	420.49	1.3054	0.0010420	421.24	1.3047	0.0010415	421.99	1.3039
120	0.0010593	505.03	1.5261	0.0010587	505.73	1.5252	0.0010582	506.44	1.5243
140	0.0010787	590.27	1.7376	0.0010781	590.92	1.7366	0.0010774	591.58	1.7355
150	0.0010894	633.22	1.8403	0.0010888	633.84	1.8392	0.0010881	634.46	1.8381
160	0.0011009	676.43	1.9412	0.0011002	677.01	1.9400	0.0010995	677.60	1.9389
180	0.0011265	763.72	2.1382	0.0011256	764.23	2.1369	0.0011248	764.74	2.1355
200	0.0011560	852.52	2.3300	0.0011549	852.93	2.3284	0.0011539	853.31	2.3268
250	0.111412	2901.5	6.5436	0.070564	2854.7	6.2855	0.0012514	1085.3	2.7925
300	0.125449	3022.6	6.7648	0.081126	2992.4	6.5371	0.058821	2959.5	6.3595
350	0.138564	3136.2	6.9550	0.090520	3114.4	6.7414	0.066436	3091.5	6.5805
400	0.151190	3246.8	7.1258	0.099352	3230.1	6.9199	0.073401	3212.7	6.7677
450	0.163523	3356.4	7.2828	0.107864	3343.0	7.0817	0.080016	3329.2	6.9347
500	0.175666	3465.9	7.4293	0.116174	3454.9	7.2314	0.086417	3443.6	7.0877
600	0.199598	3687.8	7.6991	0.132427	3679.9	7.5051	0.098836	3671.9	7.3653

p	5.0 MPa			6.0 MPa			7.0 MPa		
Saturation parameters	t_s = 263.980°C, v' = 0.0012861, v'' = 0.039400, h' = 1154.2, h'' = 2793.6, s' = 2.9200, s'' = 5.9724			t_s = 275.625°C, v' = 0.0013190, v'' = 0.032400, h' = 1213.3, h'' = 2783.8, s' = 3.0266, s'' = 5.8885			t_s = 285.869°C, v' = 0.0013515, v'' = 0.027400, h' = 1266.9, h'' = 2771.7, s' = 3.1210, s'' = 5.8129		
t (°C)	v (m^3/kg)	h (kJ/kg)	s [kJ/(kg K)]	v (m^3/kg)	h (kJ/kg)	s [kJ/(kg K)]	v (m^3/kg)	h (kJ/kg)	s [kJ/(kg K)]
0	0.0009977	5.04	0.0002	0.0009972	6.05	0.0002	0.0009967	7.07	0.0003
10	0.0009979	46.87	0.1506	0.0009975	47.83	0.1505	0.0009970	48.80	0.1504
20	0.0009996	88.55	0.2952	0.0009991	89.49	0.2950	0.0009985	90.42	0.2948
40	0.0010057	171.92	0.5704	0.0010052	172.81	0.5700	0.0010043	173.69	0.5696
50	0.0010099	213.63	0.7015	0.0010095	214.49	0.7010	0.001009[illegible]	215.35	0.7005
60	0.0010149	255.34	0.8286	0.0010144	256.18	0.8280	0.0010140	257.01	0.8275
80	0.0010267	338.87	1.0721	0.0010262	339.67	1.0714	0.0010258	340.46	1.0708
100	0.0010410	422.75	1.3031	0.0010404	423.50	1.3023	0.0010399	424.25	1.3016
120	0.0010576	507.14	1.5234	0.0010571	507.85	1.5225	0.0010565	508.55	1.5216
140	0.0010768	592.23	1.7345	0.0010762	592.88	1.7335	0.0010756	593.54	1.7325
150	0.0010874	635.09	1.8370	0.0010868	635.71	1.8359	0.0010867	636.34	1.8348
160	0.0010988	678.19	1.9377	0.0010981	678.78	1.9365	0.0010974	679.37	1.9353
180	0.0011240	765.25	2.1342	0.0011231	765.76	2.1328	0.0011223	766.28	2.1315
200	0.0011529	853.75	2.3253	0.0011519	854.17	2.3237	0.0011510	854.59	2.3222
250	0.0012496	1085.2	2.7901	0.0012478	1085.2	2.7877	0.0012460	1085.2	2.7853
300	0.045301	2923.3	6.2064	0.036148	2883.1	6.0656	0.029457	2837.5	5.9291
350	0.051932	3067.4	6.4477	0.042213	3041.9	6.3317	0.035225	3014.8	6.2265
400	0.057804	3194.9	6.6448	0.047382	3176.4	6.5395	0.039917	3157.3	6.4465
450	0.063291	3315.2	6.8170	0.052128	3300.9	6.7179	0.044143	3286.2	6.6314
500	0.068552	3432.2	6.9735	0.056632	3420.6	6.8781	0.048110	3408.9	6.7954
600	0.078675	3663.9	7.2553	0.065228	3655.7	7.1640	0.055617	3647.5	7.0857

p	8.0 MPa			9.0 MPa			10.0 MPa		
Saturation parameters	t_s = 295.048°C, v' = 0.0013843, v'' = 0.023520, h' = 1316.5, h'' = 2757.7, s' = 3.2066, s'' = 5.7430			t_s = 303.385°C, v' = 0.0014177, v'' = 0.020500, h' = 1363.1, h'' = 2741.9, s' = 3.2854, s'' = 5.6771			t_s = 311.037°C, v' = 0.0014522, v'' = 0.018000, h' = 1407.2, h'' = 2724.5, s' = 3.3591, s'' = 5.6139		
t (°C)	v (m^3/kg)	h (kJ/kg)	s [kJ/(kg K)]	v (m^3/kg)	h (kJ/kg)	s [kJ/(kg K)]	v (m^3/kg)	h (kJ/kg)	s [kJ/(kg K)]
0	0.0009962	8.08	0.0003	0.0009957	9.08	0.0004	0.0009952	10.09	0.0004
10	0.0009965	49.77	0.1502	0.0009961	50.74	0.1501	0.0009956	51.70	0.1500
20	0.0009982	91.36	0.2946	0.0009977	92.29	0.2944	0.0009973	93.22	0.2942
40	0.0010044	174.57	0.5692	0.0010039	175.46	0.5688	0.0010035	176.34	0.5684
50	0.0010086	216.21	0.7001	0.0010082	217.07	0.6996	0.0010078	217.93	0.6992
60	0.0010136	257.85	0.8270	0.0010131	258.69	0.8265	0.0010127	259.53	0.8259
80	0.0010253	341.26	1.0701	0.0010248	342.06	1.0695	0.0010244	342.85	1.0688
100	0.0010395	425.01	1.3008	0.0010390	425.76	1.3000	0.0010385	426.51	1.2993
120	0.0010560	509.26	1.5207	0.0010554	509.97	1.5199	0.0010549	510.68	1.5190
140	0.0010750	594.19	1.7314	0.0010744	594.85	1.7304	0.0010738	595.50	1.7294
150	0.0010855	636.96	1.8337	0.0010848	637.59	1.8327	0.0010842	638.22	1.8316
160	0.0010967	679.97	1.9342	0.0010960	680.56	1.9330	0.0010953	681.16	1.9319
180	0.0011215	766.80	2.1302	0.0011207	767.32	2.1288	0.0011199	767.84	2.1275

Continued

Appendix Table 2: Thermodynamic properties of unsaturated water and superheated vapor.—cont'd

p	8.0 MPa			9.0 MPa			10.0 MPa		
Saturation parameters	t_s = 295.048°C v' = 0.0013843, v'' = 0.023520 h' = 1316.5, h'' = 2757.7 s' = 3.2066, s'' = 5.7430			t_s = 303.385°C v' = 0.0014177, v'' = 0.020500 h' = 1363.1, h'' = 2741.9 s' = 3.2854, s'' = 5.6771			t_s = 311.037°C v' = 0.0014522, v'' = 0.018000 h' = 1407.2, h'' = 2724.5 s' = 3.3591, s'' = 5.6139		
t (°C)	*v* (m^3/kg)	*h* (kJ/kg)	*s* [kJ/(kg K)]	*v* (m^3/kg)	*h* (kJ/kg)	*s* [kJ/(kg K)]	*v* (m^3/kg)	*h* (kJ/kg)	*s* [kJ/(kg K)]
200	0.0011500	855.02	2.3207	0.0010490	855.44	2.3191	0.0011481	855.88	2.3176
250	0.0012443	1085.2	2.7829	0.0012425	1085.3	2.7806	0.0012408	1085.3	2.7783
300	0.024255	2784.5	5.7899	0.0014018	1343.5	3.2514	0.0013975	1342.3	3.2469
350	0.029940	2986.1	6.1282	0.025786	2955.3	6.0342	0.022415	2922.1	5.9423
400	0.034302	3137.5	6.3622	0.029921	3117.1	6.2842	0.026402	3095.8	6.2109
450	0.038145	3271.3	6.5540	0.033474	3256.0	6.4835	0.029735	3240.5	6.4184
500	0.041712	3397.0	6.7221	0.036733	3385.0	6.6560	0.032750	3372.8	6.5954
600	0.048403	3639.2	7.0168	0.042789	3630.8	6.9552	0.038297	3622.5	6.8992

p	15.0 MPa			20.0 MPa			30.0 MPa		
Saturation parameters	t_s = 342.196°C v' = 0.0016571, v'' = 0.010300 h' = 1609.8, h'' = 2610.0 s' = 3.6836, s'' = 5.3091			t_s = 365.789°C v' = 0.0020379, v'' = 0.0058702 h' = 1827.2, h'' = 2413.1 s' = 4.0153, s'' = 4.9322					
t (°C)	*v* (m^3/kg)	*h* (kJ/kg)	*s* [kJ/(kg K)]	*v* (m^3/kg)	*h* (kJ/kg)	*s* [kJ/(kg K)]	*v* (m^3/kg)	*h* (kJ/kg)	*s* [kJ/(kg K)]
0	0.0009928	15.10	0.0006	0.0009904	20.08	0.0006	0.0009857	29.92	0.0005
10	0.0009933	56.51	0.1494	0.0009911	61.29	0.1488	0.0009866	70.77	0.1474
20	0.0009951	97.87	0.2930	0.0009929	102.50	0.2919	0.0009887	111.71	0.2895
40	0.0010014	180.74	0.5665	0.0009992	185.13	0.5645	0.0009951	193.87	0.5606
50	0.0010056	222.22	0.6969	0.0010035	226.50	0.6946	0.0009993	235.05	0.6900
60	0.0010105	263.72	0.8233	0.0010084	267.90	0.8207	0.0010042	276.25	0.8156
80	0.0010221	346.84	1.0656	0.0010199	350.82	1.0624	0.0010155	358.78	1.0562
100	0.0010360	430.29	1.2955	0.0010336	434.06	1.2917	0.0010290	441.64	1.2844
120	0.0010522	514.23	1.5146	0.0010496	517.79	1.5103	0.0010445	524.95	1.5019
140	0.0010708	598.80	1.7244	0.0010679	602.12	1.7195	0.0010622	608.82	1.7100
150	0.0010810	641.37	1.8262	0.0010779	644.56	1.8210	0.0010719	651.00	1.8108
160	0.0010919	684.16	1.9262	0.0010886	687.20	1.9206	0.0010822	693.36	1.9098
180	0.0011159	770.49	2.1210	0.0011121	773.29	2.1147	0.0011048	778.72	2.1024
200	0.0011434	858.08	2.3102	0.0011389	860.36	2.3029	0.0011303	865.12	2.2890
250	0.0012327	1085.6	2.7671	0.0012251	1086.2	2.7564	0.0012110	1087.9	2.7364
300	0.0013777	1337.3	3.2260	0.0013605	1333.4	3.2072	0.0013317	1327.9	3.1742
350	0.011469	2691.2	5.4403	0.0016645	1645.3	3.7275	0.0015522	1608.0	3.6420
400	0.015652	2974.6	5.8798	0.0099458	2816.8	5.5520	0.0027929	2150.6	4.4721
450	0.018449	3156.5	6.1408	0.0127013	3060.7	5.9025	0.0067363	2822.1	5.4433
500	0.020797	3309.0	6.3449	0.0147681	3239.3	6.1415	0.0086761	3083.3	5.7934
600	0.024882	3580.7	6.6757	0.0181655	3536.3	6.5035	0.0114310	3442.9	6.2321

This table is taken from Yan Jialu, Yu Xiaofu. Thermodynamic properties table of water and vapor. Beijing: High Education Press; 1995.

Appendix Table 3: Density, specific heat capacity and thermal conductivity of common metallic materials.

Material name	20°C			Thermal conductivity, k [W/(m K)]									
				Temperature (°C)									
	Density, ρ (kg/m^3)	Specific heat capacity, c [J/(kg K)]	Thermal conductivity, k [W/(m K)]	−100	0	100	200	300	400	600	800	1000	1200
Pure aluminum	2710	902	236	243	236	240	238	234	228	215			
Duralumin (96Al-4Cu, Micro Mg)	2790	881	169	124	160	188	188	193					
Aluminum Alloy (92Al-8Mg)	2610	904	107	86	102	123	148						
Aluminum Alloy (87Al-13Si)	2660	871	162	139	158	173	176	180					
Beryllium	1850	1758	219	382	218	170	145	129	118				
Pure copper	8930	386	398	421	401	393	389	384	379	366	352		
Aluminum bronze (90Cu −10Al)	8360	420	56		49	57	66						
Bronze (89Cu −11Sn)	8800	343	24.8		24	28.4	33.2						
Brass (70Cu −30Zn)	8440	377	109	90	106	131	143	145	148				
Copper Alloy (60Cu −40Ni)	8920	410	22.2	19	22.2	23.4							
Gold	19,300	127	315	331	318	313	310	305	300	287			
Pure iron	7870	455	81.1	96.7	83.5	72.1	63.5	56.5	50.3	39.4	29.6	29.4	31.6
Eminem iron	7860	455	73.2	82.9	74.7	67.5	61.0	54.8	49.9	38.6	29.3	29.3	31.1
Gray cast iron ($\omega_c \approx$ 3%)	7570	470	39.2		28.5	32.4	35.8	37.2	36.6	20.8	19.2		
Carbon steel ($\omega_c \approx$ 0.5%)	7840	465	49.8		50.5	47.5	44.8	42.0	39.4	34.0	29.0		
Carbon steel ($\omega_c \approx$ 1.0%)	7790	470	43.2		43.0	42.8	42.2	41.5	40.6	36.7	32.2		
Carbon steel ($\omega_c \approx$ 1.5%)	7750	470	36.7		36.8	36.6	36.2	35.7	34.7	31.7	27.8		
Chrome steel ($\omega_{cr} \approx$ 5%)	7830	460	36.1		36.3	35.2	34.7	33.5	31.4	28.0	27.2	27.2	27.2
Chrome steel ($\omega_{cr} \approx$ 13%)	7740	460	26.8		26.5	27.0	27.0	27.0	27.6	28.4	29.0	29.0	
Chrome steel ($\omega_{cr} \approx$ 17%)	7710	460	22		22	22.2	22.6	22.6	23.3	24.0	24.8	25.5	
Chrome steel($\omega_{cr} \approx$ 26%)	7650	460	22.6		22.6	23.8	25.5	27.2	28.5	31.8	35.1	38	
Ch-Ni steel (18-20Cr/8-12Ni)	7820	460	15.2	12.2	14.7	16.6	18.0	19.4	20.8	23.5	26.3		
Ch-Ni steel (17-19Cr/9-13Ni)	7830	460	14.7	11.8	14.3	16.1	17.5	18.8	20.2	22.8	25.5	28.2	30.9

Continued

Appendix Table 3: Density, specific heat capacity and thermal conductivity of common metallic materials.—cont'd

	20°C			Thermal conductivity, k [W/(m K)]									
				Temperature (°C)									
Material name	Density, ρ (kg/m^3)	Specific heat capacity, c [J/(kg K)]	Thermal conductivity, k [W/(m K)]	−100	0	100	200	300	400	600	800	1000	1200
Nickel steel ($\omega_{Na} \approx 1\%$)	7900	460	45.5	40.8	45.2	46.8	46.1	44.1	41.2	35.7			
Nickel steel ($\omega_{Na} \approx 3.5\%$)	7910	460	36.5	30.7	36.0	38.8	39.7	39.2	37.8				
Nickel steel ($\omega_{Na} \approx 25\%$)	8030	460	13.0	—									
Nickel steel ($\omega_{Na} \approx 35\%$)	8110	460	13.8	10.9	13.4	15.4	17.1	18.6	20.1	23.1			
Nickel steel ($\omega_{Na} \approx 44\%$)	8190	460	15.8		15.7	16.1	16.5	16.9	17.1	17.8	18.4		
Nickel steel ($\omega_{Na} \approx 50\%$)	8260	460	19.6	17.3	19.4	20.5	21.0	21.1	21.3	22.5			
Mn steel ($\omega_{Mn} \approx 12\sim13\%$ $\omega_{Na}3\%$)	7800	487	13.6			14.8	16.0	17.1	18.3				
Mn steel ($\omega_{Mn} \approx 0.4\%$)	7860	440	51.2			51.0	50.0	47.0	43.5	35.5	27		
Tungsten steel ($\omega_W \approx 5\sim6\%$)	8070	436	18.7		18.4	19.7	21.0	22.3	23.6	24.9	26.3		
Pb	11,340	128	35.3	37.2	35.5	34.3	32.8	31.5					
Mg	1730	1020	156	160	157	154	152	150					
Mo	9590	255	138	146	139	135	131	127	123	116	109	103	93.7
Nickel	8900	444	91.4	144	94	82.8	74.2	67.3	64.6	69.0	73.3	77.6	81.9
Platinum	21,450	133	71.4	73.3	71.5	71.6	72.0	72.8	73.6	76.6	80.0	84.2	88.9
Silver	10,500	234	427	431	428	422	415	407	399	384			
Tin	7310	228	67	75	68.2	63.2	60.9						
Titanium	4500	520	22	23.3	22.4	20.7	19.9	19.5	19.4	19.9			
Uranium	19,070	116	27.4	24.3	27	29.1	31.1	33.4	35.7	40.6	45.6		
Zinc	7140	388	121	123	122	117	112						
Zirconium	6570	276	22.9	26.5	23.2	21.8	21.2	20.9	21.4	22.3	24.5	26.4	28.0
Wolfram	19,350	134	179	204	182	166	153	142	134	125	119	114	110

This table is taken from Yang Shiming, Tao Wenquan. Heat transfer. Beijing: High Education Press; 1998.

Appendix Table 4: Common density and thermal conductivity of insulation and building materials.

Material name	Temperature, t (°C)	Density, ρ (kg/m^3)	Thermal conductivity, k [W/(m K)]
Expanded perlite bulk material	25	60–300	0.021–0.062
Asphalt perlite	31	233–282	0.069–0.076
Phosphate expanded perlite products	20	200–250	0.044–0.052
Water glass expanded perlite product	20	200–300	0.056–0.065
Wool products	20	80–150	0.035–0.038
Exfoliated vermiculite	20	100–130	0.051–0.07
Asphalt vermiculite tube	20	350–400	0.081–0.10
Asbestos powder	22	744–1400	0.099–0.19
Asbestos tiles	21	384	0.099
Asbestos rope		590–730	0.10–0.21
Asbestos fiber		35–230	0.055–0.077
Asbestos	30	770–1045	0.10–0.14
Magnesium carbonate asbestos dust		240–490	0.077–0.086
Diatomite asbestos dust		280–380	0.085–0.11
Ash brick	27	458–589	0.12–0.22
Slag wool	30	207	0.058
Glass wire	35	120–492	0.058–0.07
Glass wool felt	28	18.4–38.3	0.043
Soft wood	20	105–437	0.044–0.079
Wood fiber fiberboard	25	245	0.048
Straw pulp plate	20	325–365	0.068–0.084
Stick board	25	108–147	0.056–0.11
Sugarcane plate	20	282	0.067–0.072
Kwai core panel	20	95.5	0.05
Corn stalk board	22	25.2	0.065
Cotton	20	117	0.049
Wire	20	57.7	0.036
Sawdust	20	179	0.083
Rigid foam	30	29.5–56.3	0.041–0.048
Flexible foam	30	41–162	0.043–0.056
Aluminum spacer layer (5 layer)	21		0.042
Brick (to create a state)	25	1860	0.87
Red brick	35	1560	0.49
Pine (vertical wood grain)	15	496	0.15
Pine (parallel to grain)	21	527	0.35
Cement	30	1900	0.30
Concrete slabs	35	1930	0.79
Acid-resistant concrete slab	30	2250	1.5∼1.6

Continued

Appendix Table 4: Common density and thermal conductivity of insulation and building materials.—cont'd

Material name	Temperature, t (°C)	Density, ρ (kg/m^3)	Thermal conductivity, k [W/(m K)]
Yellow sand	30	1580–1700	0.28~0.34
Soil	20		0.83
Ceramic tile	37	2090	1.1
Glass	45	2500	0.65~0.71
Polystyrene	30	24.7–37.8	0.04~0.043
Granite		2643	1.73~3.98
Marble		2499–2707	2.7
Mica		290	0.58
Scale	65		1.31~3.14
Ice	0	913	2.22
Clay	27	1460	1.3

This table is taken from: Fu Qinsheng, He Yalin, Basis and application of thermodynamics and heat transfer, 2006, Beijing, Chemical Industry Press.

Appendix Table 5: Relationship between thermal conductivity and temperature of several refractory insulation.

Material name	Maximum allowable temperature materials (°C)	Density, ρ (kg/m^3)	Thermal conductivity, k [W/(m K)]
Ultrafine glass wool felt, tube	400	18–20	$0.033 + 0.000230t$ ①
Slag wool	550–600	350	$0.0674 + 0.000215t$
Cement vermiculite products	800	420–450	$0.103 + 0.000198t$
Cement perlite products	600	300–400	$0.0651 + 0.000105t$
Fly ash foam brick	300	500	$0.099 + 0.0002t$
Plate glass wool cloth sewn	600	100	$0.0314 + 0.000198t$
A Grade diatomite products	900	500	$0.0395 + 0.00019t$
B Grade diatomite products	900	550	$0.0477 + 0.0002t$
Expanded perlite	1000	55	$0.0424 + 0.000137t$
Porous calcium silicate products	650	≤250	$0.041 + 0.0002t$
Refractory brick	1350–1450	1800–2040	$(0.7\sim0.84) + 0.00058t$
Lightweight refractory bricks	1250–1300	800–1300	$(0.29\sim0.41) + 0.00026t$
Super lightweight refractory bricks (1)	1150–1300	540–610	$0.093 + 0.00016t$
Super lightweight refractory bricks (2)	1100	270–330	$0.058 + 0.00017t$
Silica brick	1700	1900–1950	$0.93 + 0.0007t$
Magnesia brick	1600–1700	2300–2600	$2.1 + 0.00019t$
Chrome brick	1600–1700	2600–2800	$4.7 + 0.00017t$

① t represents the average temperature of the material.

This table is taken from: Fu Qinsheng, He Yalin, Basis and application of thermodynamics and heat transfer, 2006, Beijing, Chemical Industry Press.

Appendix Table 6: Thermo-physical properties of dry air ($p = 1.01325 \times 10^5$ Pa).

t (°C)	P (kg/m³)	c_p [kJ/(kg°C)]	$\lambda \times 10^2$ [W/(m°C)]	$\alpha \times 10^6$ (m² s)	$\eta \times 10^6$ [W/(m s)]	$v \times 10^6$ (m² s)	Pr
−50	1.584	1.013	2.04	12.7	14.6	9.23	0.728
−40	1.515	1.013	2.12	13.8	15.2	10.04	0.728
−30	1.453	1.013	2.20	14.9	15.7	10.80	0.723
−20	1.395	1.009	2.28	16.2	16.2	11.61	0.716
−10	1.342	1.009	2.36	17.4	16.7	12.43	0.712
0	1.293	1.005	2.44	18.8	17.2	13.28	0.707
10	1.247	1.005	2.51	20.0	17.6	14.16	0.705
20	1.205	1.005	2.59	21.4	18.1	15.06	0.703
30	1.165	1.005	2.67	22.9	18.6	16.00	0.701
40	1.128	1.005	2.76	24.3	19.1	16.96	0.699
50	1.093	1.005	2.83	25.7	19.6	17.95	0.698
60	1.060	1.005	2.90	27.2	20.1	18.97	0.696
70	1.029	1.009	2.96	28.6	20.6	20.02	0.694
80	1.000	1.009	3.05	30.2	21.1	21.09	0.692
90	0.972	1.009	3.13	31.9	21.5	22.10	0.690
100	0.946	1.009	3.21	33.6	21.9	23.13	0.688
120	0.898	1.009	3.34	36.8	22.8	25.45	0.686
140	0.854	1.013	3.49	40.3	23.7	27.80	0.684
160	0.815	1.017	3.64	43.9	24.5	30.09	0.682
180	0.779	1.022	3.78	47.5	25.3	32.49	0.681
200	0.746	1.026	3.93	51.4	26.0	34.85	0.680
250	0.674	1.038	4.27	61.0	27.4	40.61	0.677
300	0.615	1.047	4.60	71.6	29.7	48.33	0.674
350	0.566	1.059	4.91	81.9	31.4	55.46	0.676
400	0.524	1.068	5.21	93.1	33.0	63.09	0.678
500	0.456	1.093	5.74	115.3	36.2	79.38	0.687
600	0.404	1.114	6.22	138.3	39.1	96.89	0.699
700	0.362	1.135	6.71	163.4	41.8	115.4	0.706
800	0.329	1.156	7.18	188.8	44.3	134.8	0.713
900	0.301	1.172	7.63	216.2	46.7	155.1	0.717
1000	0.277	1.185	8.07	245.9	49.0	177.1	0.719
1100	0.257	1.197	8.50	276.2	51.2	199.3	0.722
1200	0.239	1.210	9.15	316.5	53.5	233.7	0.724

This table is taken from:Yang Shiming, Tao Wenquan, Heat transfer, 1998, Beijing, High Education Press.

Appendix Table 7: Thermal physical properties of saturated water.

t (°C)	$p \times 10^{-5}$ (pa)	ρ (kg/m^3)	h' (kJ/kg)	c [kJ/(kg K)]	$\lambda \times 10^2$ [W/(m K)]	$\alpha \times 10^8$ (m^2 s)	$\eta \times 10^6$ [W/(m s)]	$v \times 10^6$ (m^2 s)	$\alpha_v \times 10^4$ (K^{-1})	$\gamma \times 10^4$ (N/m)	Pr
0	0.00611	999.9	0	4.212	55.1	13.1	1788	1.789	−0.81	756.4	13.67
10	0.01227	999.7	42.04	4.191	57.4	13.7	1306	1.306	+0.87	741.6	9.52
20	0.02338	998.2	83.91	4.183	59.9	14.3	1004	1.006	2.09	726.9	7.02
30	0.04241	995.7	125.7	4.174	61.8	14.9	801.5	0.805	3.05	712.2	5.42
40	0.07375	992.2	167.5	4.174	63.5	15.3	653.3	0.659	3.86	696.5	4.31
50	0.12335	988.1	209.3	4.174	64.8	15.7	549.4	0.556	4.57	676.9	3.54
60	0.19920	983.1	251.1	4.179	65.9	16.0	469.9	0.478	5.22	662.2	2.99
70	0.3116	977.8	293.0	4.187	66.8	16.3	406.1	0.415	5.83	643.5	2.55
80	0.4736	971.8	355.0	4.195	67.4	16.6	355.1	0.365	6.40	625.9	2.21
90	0.7011	965.3	377.0	4.208	68.0	16.8	314.9	0.326	6.96	607.2	1.95
100	1.013	958.4	419.1	4.220	68.3	16.9	282.5	0.295	7.50	588.6	1.75
110	1.43	951.0	461.4	4.233	68.5	17.0	259.0	0.272	8.04	569.0	1.60
120	1.98	943.1	503.7	4.250	68.6	17.1	237.4	0.252	8.58	548.4	1.47
130	2.70	934.8	546.4	4.266	68.6	17.2	217.8	0.233	9.12	528.8	1.36
140	3.61	926.1	589.1	4.287	68.5	17.2	201.1	0.217	9.68	507.2	1.26
150	4.76	917.0	632.2	4.313	68.4	17.3	186.4	0.203	10.26	486.6	1.17
160	6.18	907.0	675.4	4.346	68.3	17.3	173.6	0.191	10.87	466.0	1.10
170	7.92	897.3	719.3	4.380	67.9	17.3	162.8	0.181	11.52	443.4	1.05
180	10.03	886.9	763.3	4.417	67.4	17.2	153.0	0.173	12.21	422.8	1.00
190	12.55	876.0	807.8	4.459	67.0	17.1	144.2	0.165	12.96	400.2	0.96
200	15.55	863.0	852.8	4.505	66.3	17.0	136.4	0.158	13.77	376.7	0.93
210	19.08	852.3	897.7	4.555	65.5	16.9	130.5	0.153	14.67	354.1	0.91
220	23.20	840.3	943.7	4.614	64.5	16.6	124.6	0.148	15.67	331.6	0.89
230	27.98	827.3	990.2	4.681	63.7	16.4	119.7	0.145	16.80	310.0	0.88
240	33.48	813.6	1037.5	4.756	62.8	16.2	114.8	0.141	18.08	285.5	0.87
250	39.78	799.0	1085.7	4.844	61.8	15.9	109.9	0.137	19.55	261.9	0.86
260	46.94	784.0	1135.7	4.949	60.5	15.6	105.9	0.135	21.27	237.4	0.87
270	55.05	767.9	1185.7	5.070	59.0	15.1	102.0	0.133	23.31	214.8	0.88
280	64.19	750.7	1236.8	5.230	57.4	14.6	98.1	0.131	25.79	191.3	0.90
290	74.45	732.3	1290.0	5.485	55.8	13.9	94.2	0.129	28.84	168.7	0.93
300	85.92	712.5	1344.9	5.736	54.0	13.2	91.2	0.128	32.73	144.2	0.97
310	98.70	691.1	1402.2	6.071	52.3	12.5	88.3	0.128	37.85	120.7	1.03
320	112.90	667.1	1462.1	6.574	50.6	11.5	85.3	0.128	44.91	98.10	1.11
330	128.65	640.2	1526.2	7.244	48.4	10.4	81.4	0.127	55.31	76.71	1.22
340	146.08	610.1	1594.8	8.165	45.7	9.17	77.5	0.127	72.10	56.70	1.39
350	165.37	574.4	1671.4	9.504	43.0	7.88	72.6	0.126	103.7	38.16	1.60
360	186.74	528.0	1761.5	13.984	39.5	5.36	66.7	0.126	182.9	20.21	2.35
370	210.53	450.5	1892.5	40.321	33.7	1.86	56.9	0.126	676.7	4.709	6.79

This table is taken from: Yang Shiming, Tao Wenquan, Heat transfer, 1998, Beijing, High Education Press.

Appendix Table 8: Thermophysical properties of dry saturated water steam.

t (°C)	$p \times 10^{-5}$ (pa)	ρ'' (kg/m^3)	h'' (kJ/kg)	r (kJ/kg)	c [kJ/(kg K)]	$\lambda \times 10^2$ [W/(m K)]	$\alpha \times 10^8$ (m^2 s)	$\eta \times 10^6$ [W/(m s)]	$\nu \times 10^6$ (m^2 s)	Pr
0	0.00611	0.004847	2501.6	2501.6	1.8543	1.83	7313.0	8.022	1655.01	0.815
10	0.01227	0.009396	2520.0	2477.7	1.8594	1.88	3881.3	8.424	896.54	0.831
20	0.02338	0.01729	2538.0	2454.3	1.8661	1.94	2167.2	8.840	509.9	0.847
30	0.04241	0.03037	2556.5	2430.9	1.8744	2.00	1265.1	9.218	303.53	0.863
40	0.07375	0.05116	2574.5	2407.0	1.8853	2.06	768.45	9.620	188.04	0.883
50	0.12335	0.08302	2592.0	2382.7	1.8987	2.12	483.59	10.022	120.72	0.896
60	0.19920	0.1302	2609.6	2358.4	1.9155	2.19	315.55	10.424	80.07	0.913
70	0.3116	0.1982	2626.8	2334.1	1.9364	2.25	210.57	10.817	54.57	0.930
80	0.4736	0.2933	2643.5	2309.0	1.9615	2.33	145.53	11.219	38.25	0.947
90	0.7011	0.4235	2660.3	2283.1	1.9921	2.40	102.22	11.621	27.44	0.966
100	1.0130	0.5977	2676.2	2257.1	2.0281	2.48	73.57	12.023	20.12	0.984
110	1.4327	0.8265	2691.3	2229.0	2.0704	2.56	53.83	12.425	15.03	1.00
120	1.9854	1.122	2705.9	2202.3	2.1198	2.65	40.15	12.798	11.41	1.02
130	2.7013	1.497	2719.7	2173.1	2.1763	2.76	30.46	13.170	8.80	1.04
140	3.614	1.967	2733.1	2144.1	2.2408	2.85	23.28	13.543	6.89	1.06
150	4.760	2.548	2745.3	2113.1	2.3145	2.97	18.10	13.896	5.45	1.08
160	6.181	3.260	2756.6	2081.3	2.3974	3.08	14.20	14.249	4.37	1.11
170	7.920	4.123	2767.1	2047.8	2.4911	3.21	11.25	14.612	3.54	1.13
180	10.027	5.160	2776.3	2013.0	2.5958	3.36	9.03	14.965	2.90	1.15
190	12.551	6.397	2784.2	1976.6	2.7126	3.51	7.29	15.298	2.39	1.18
200	15.549	7.864	2790.9	1938.5	2.8428	3.68	5.92	15.651	1.99	1.21
210	19.077	9.593	2796.4	1898.3	2.9877	3.87	4.86	15.995	1.67	1.24
220	23.198	11.62	2799.7	1856.4	3.1497	4.07	4.00	16.338	1.41	1.26
230	27.976	14.00	2801.8	1811.6	3.3310	4.30	3.32	16.701	1.19	1.29
240	33.478	16.76	2802.2	1764.7	3.5366	4.54	2.76	17.073	1.02	1.33
250	39.776	19.99	2800.6	1714.4	3.7723	4.84	2.31	17.446	0.873	1.36
260	46.943	23.73	2796.4	1661.3	4.0470	5.18	1.94	17.848	0.752	1.40
270	55.058	28.10	2789.7	1604.8	4.3735	5.55	1.63	18.280	0.651	1.44
280	64.202	33.19	2780.5	1543.7	4.7675	6.00	1.37	18.750	0.565	1.49
290	74.461	39.16	2767.5	1477.5	5.2528	6.55	1.15	19.270	0.492	1.54
300	85.927	46.19	2751.1	1405.9	5.8632	7.22	0.96	19.839	0.430	1.61
310	98.700	54.54	2730.2	1327.6	6.6503	8.06	0.80	20.691	0.380	1.71
320	112.89	64.60	2703.8	1241.0	7.7217	8.65	0.62	21.691	0.336	1.94
330	128.63	76.99	2670.3	1143.8	9.3613	9.61	0.48	23.093	0.300	2.24
340	146.05	92.76	2626.0	1030.8	12.2108	10.70	0.34	24.692	0.266	2.82
350	165.35	113.6	2567.8	895.6	17.1504	11.90	0.22	26.594	0.234	3.83
360	186.75	144.1	2485.3	721.4	25.1162	13.70	0.14	29.193	0.203	5.34
370	210.54	201.1	2342.9	452.6	76.9157	16.60	0.04	33.989	0.169	15.7
374.15	221.20	315.5	2107.2	0.0	∞	23.79	0.0	44.992	0.143	∞

This table is taken from:Yang Shiming, Tao Wenquan, Heat transfer, 1998, Beijing, High Education Press.

Appendix Table 9: Thermophysical properties of several gases ($p = 1.01325 \times 10^5$ Pa).

Gas name	t (°C)	ρ (kg/m³)	c [kJ/(kg K)]	$\lambda \times 10^2$ [W/(m K)]	$a \times 10^2$ (m²/h)	$\eta \times 10^6$ [kg/(m s)]	$v \times 10^2$ (m² s)	Pr
Hydrogen (H_2)	−50	0.1064	13.82	14.07	34.4	7.355	69.1	0.72
	0	0.0869	14.19	16.75	48.6	8.414	96.8	0.72
	50	0.0734	14.40	19.19	65.3	9.385	128	0.71
	100	0.0636	14.49	21.40	84.0	10.277	162	0.69
	150	0.0560	14.49	23.61	105	11.121	199	0.68
	200	0.0502	14.53	25.70	128	11.915	237	0.66
	250	0.0453	14.53	27.56	152	12.651	279	0.66
	300	0.0415	14.57	29.54	178	13.631	321	0.65
Nitrogen (N_2)	−50	1.485	1.043	2.000	4.65	14.122	9.5	0.74
	0	1.211	1.043	2.407	6.87	16.671	13.8	0.72
	50	1.023	1.043	2.791	9.42	18.927	18.5	0.71
	100	0.887	1.043	3.128	12.2	21.084	23.8	0.70
	150	0.782	1.047	3.477	15.3	23.046	29.5	0.69
	200	0.699	1.055	3.815	18.6	24.811	35.5	0.69
	250	0.631	1.059	4.129	22.1	26.674	42.3	0.69
	300	0.577	1.072	4.419	25.7	28.241	49.1	0.69
Carbon dioxide (CO_2)	−50	2.373	0.766	1.105	2.2	11.28	4.8	0.78
	0	1.912	0.829	1.454	3.3	13.83	7.2	0.78
	50	1.616	0.875	1.830	4.7	16.18	10.0	0.77
	100	1.400	0.921	2.221	6.2	18.34	13.1	0.76
	150	1.235	0.959	2.628	8.0	20.40	16.5	0.74
	200	1.103	0.996	3.059	10.1	22.36	20.3	0.72
	250	0.996	1.030	3.512	12.3	24.22	24.3	0.71
	300	0.911	1.063	3.989	14.8	25.99	28.5	0.59
Oxygen (O_2)	−100	2.192	0.917	1.465	2.7	12.94	5.9	0.80
	−50	1.694	0.917	1.884	4.4	16.18	9.6	0.79
	0	1.382	0.917	2.291	6.5	19.12	13.9	0.77
	50	1.168	0.925	2.687	8.9	21.97	18.8	0.76
	100	1.012	0.934	3.035	11.6	24.61	24.3	0.76

Appendix Table 10: Common materials surface emissivity.

Material categories and surface conditions	Temperature t (°C)	Emissivity ε
Aluminum: polishing, purity 98%	200–600	0.04–0.06
Industrial board	100	0.09
Crude board	40	0.07
Heavily oxidized	100–550	0.20–0.33
Foil, shiny	100–300	0.06–0.07
Brass: highly polished	250	0.03
Polishing	40	0.07
Lackluster board	40–250	0.22
Oxidation	40–250	0.46–0.56

Appendix Table 10: Common materials surface emissivity.—cont'd

Material categories and surface conditions	Temperature t (°C)	Emissivity ε
Chromium: polishing pad	40–550	0.08–0.27
Copper: highly polished copper	100	0.02
Polishing	40	0.04
Mild polishing	40	0.12
Lackluster	40	0.15
Black oxide	40	0.76
Gold: highly polished gold	100–600	0.02–0.035
Steel: low-carbon steel, polished	150–500	0.14–0.32
Steel, polished	40–250	0.07–0.10
Steel, rolling	40	0.66
Plate, rough, heavily oxidized	40	0.80
Cast iron, treated epidermis	40	0.70–0.80
Cast iron, new working plane	40	0.44
Cast iron, oxide	40–250	0.57–0.66
Cast iron, polished	200	0.21
Wrought iron, smooth	40	0.35
Wrought iron, dark oxide	20–360	0.94
Stainless steel, polished	40	0.07–0.17
Stainless steel, repeated heating and cooling	230–930	0.50–0.70
Asbestos: asbestos plate	40	0.96
Asbestos cement	40	0.96
Asbestos watt	40	0.97
Brick: crude brick	40	0.93
Refractory brick	1000	0.75
Light dust	40	0.95
Clay: sintering	100	0.91
Concrete: roughened surface	40	0.94
Glass: flat glass	40	0.94
Quartz glass (thickness 2 mm)	250–550	0.96–0.66
Borosilicate glass	250–550	0.94–0.75
Plaster	40	0.80–0.90
Snow	−3	0.82
Ice: smooth surface	0	0.97
Water: more than 0.1 mm thick	40	0.96
Mica	40	0.75
Paint: all kinds of paint	40	0.92–0.96
White paint	40	0.80–0.95
Bright black paint	40	0.9
Paper: white	40	0.95
Rough roofing tar paper mat	40	0.90
Porcelain: glazed	40	0.93
Rubber: hard	40	0.94
Snow	−12 to −6	0.82
Boiler slag	0–1000	0.97–0.70
Plastering walls	20	0.94
Various wood	40	0.80–0.92

This table is taken from:Yang Shiming, Tao Wenquan, Heat transfer, 1998, Beijing, High Education Press.

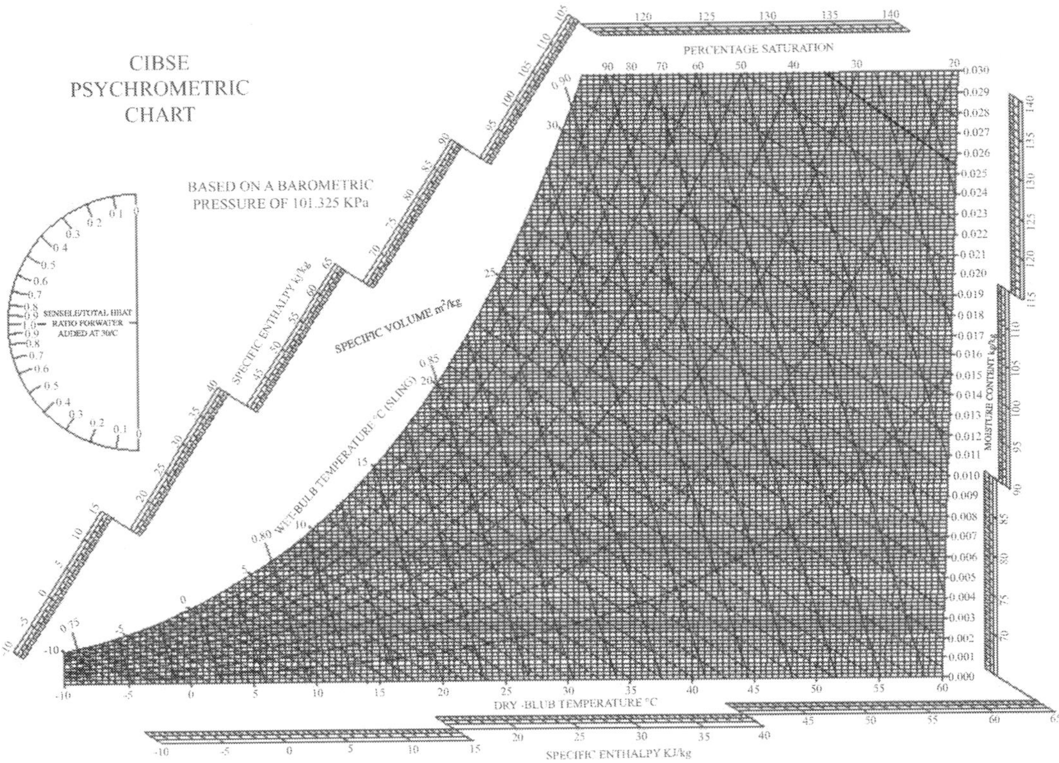

Appendix Figure 1
Moist air psychrometric chart ($p = 0.1$ MPa).

Index

'*Note*: Page numbers followed by "f" indicate figures and "t" indicate tables.'

N

Kolibri 360 GmbH
Ferdinand-Jühlke-Straße 7
99095 Erfurt, Deutschland
produktsicherheit@kolibri360.de